Geophysics and Environmental Physics

The series *Geophysics and Environmental Physics* (GEP), reports new developments in geophysics research and teaching with a high quality and the explicit aim to summarize and communicate current knowledge in an accessible way. Books published in this series are conceived as bridging material between graduate textbooks and the forefront of research.

The research fields covered by the series are:

- Atmospheric Physics, Environmental Physics, Biogeophysics and Physics of Climate;
- Earth Magnetism and Rock Physics, Seismology and Physics of Volcanoes;
- Geodynamics and Tectonics;
- Physical Oceanography and Hydrology;
- Surface Processes;
- Geodesy;
- Instrumentation and Data Systems;
- Nonlinear Processes in Geophysics;
- Planetary and Solar System Physics.

Proposal should be sent to a member of the Editorial Board, or directly to the Responsible Editor at Springer:

Dr Lisa Scalone - lisa.scalone@springernature.com

Ulrich Platt · Jochen Stutz

Differential Optical Absorption Spectroscopy

Principles and Applications

Second Edition

Springer

Ulrich Platt
Institute for Environmental Physics
Heidelberg University
Heidelberg, Germany

Jochen Stutz
Department of Atmospheric and Oceanic Sciences
University of California Los Angeles
Los Angeles, CA, USA

ISSN 2948-2186 ISSN 2948-2194 (electronic)
Geophysics and Environmental Physics
ISBN 978-3-662-68637-9 ISBN 978-3-662-68639-3 (eBook)
https://doi.org/10.1007/978-3-662-68639-3

This Springer imprint is published by the registered company Springer-Verlag GmbH, DE, part of Springer Nature.
The registered company address is: Heidelberger Platz 3, 14197 Berlin, Germany

Preface to the Second Edition

Since the publication of the first edition of this book in 2008, the use of DOAS has become much more widespread and many applications that were just emerging when the first edition was written, such as multi-axis DOAS, cavity-enhanced DOAS, imaging DOAS, and satellite DOAS, are now well established. New developments on the software side, as well as in hardware, have further advanced DOAS, making new applications possible. Consequently, we felt that it was time to update and expand our original book to bring these new developments to the reader.

When reading the first edition we found that several chapters were outdated, missing many of the exciting developments in the past 15 years. The sheer volume of scientific literature on DOAS made this an exhaustive task. In some areas, such as MAX-DOAS and satellite applications of DOAS, the literature has grown so enormously that only examples can be given, and we apologize to all the colleagues who have contributed to the progress of DOAS and whose valuable contributions could not be included in this book.

Because the motivation to study atmospheric composition and chemistry with DOAS and the underlying physical principles have not substantially changed, Chaps. 1–5 have relatively few changes. The introduction into the DOAS principle in Chap. 6 has been updated with new figures and some extended explanations. DOAS technology has undergone more substantial changes, such as the use of new imaging spectrographs and Fabry–Perot interferometers, and the emergence of more sophisticated Cavity Enhanced DOAS instruments. Chapter 7 has, therefore, been expanded to include these new developments. Several new numerical tools to analyze atmospheric absorption spectra have been added to Chap. 8. These include the determination of instrument function from Fraunhofer spectra, the additive polynomial, methods for choosing the optimal spectral interval for a particular application (Vogel–Sihler Plot), the linearization of spectral misalignment correction, the Tilt Effect, and Pukite Terms. We also include the application of Principal Component Analysis (PCA) for DOAS evaluations. The growing use of MAX-DOAS and satellite DOAS motivated us to considerably revise Chap. 9. Our particular focus was on a more detailed explanation of airmass factor calculations with modern radiative transfer models and the initialization of these models.

Methodologies for the retrieval of vertical column densities from satellites as well as concentration profiles from MAX-DOAS are now described in more detail. A discussion on the 'Vibrational Ring Effect' was also included. The most substantial updates in this second edition were made to Chaps. 10 (Sample Application of 'Active' DOAS with Artificial Light Sources) and 11 (Sample Application of 'Passive' DOAS). Both sections now include applications of DOAS that go beyond pure research, including the growing use of active DOAS to monitor industrial emissions at the fenceline and passive DOAS from different platforms as a tool to quantify anthropogenic and volcanic emissions. The satellite section in Chap. 11 has been considerably expanded to demonstrate the important role these observations play in monitoring and studying our atmosphere on a global scale. Finally, we included a section on Solar Induced Fluorescence, which uses an approach that is largely based on the same principles as DOAS to observe plant photosynthesis from local to global scales. Many of the developments in the past 15 years were mentioned in Chap. 12 (DOAS: Yesterday, Today, and Tomorrow) of the first edition of this book. We have attempted to look further into the future of DOAS with a revised Chap. 12.

The second edition of this book would not have been possible without the support of our families. We would also like to thank Maja Rüth for the help with the figures and we are most thankful to Yasmine Stutz and Christina Schmidt for proofreading the manuscript.

Heidelberg, Germany
Los Angeles, USA

Ulrich Platt
Jochen Stutz

Preface to the First Edition

Light is the essential source of information about the surrounding world for most of us. We see objects of different brightness and colors, or, in scientific terms, we distinguish objects by the way they reflect light of different wavelengths with varying efficiency. It was exciting for both of us to learn that scientific instruments can expand our senses and provide information from this very same radiation that goes far beyond the simple recognition of objects. The ability to "see" the composition of the air surrounding us and above our heads, to us and others, is one of the most fascinating aspects of the application of modern technology to study the atmosphere. This fascination drove many of the applications of the Differential Optical Absorption Spectroscopy (DOAS) method presented in this book and continues to be the motivation for many current developments.

DOAS is an elegant and powerful analytical method to study the atmosphere and is based on the relatively simple principles of classical absorption spectroscopy. It has therefore always been quite surprising to us that this method has not found wider use in atmospheric research and air-quality monitoring.

We have introduced many university students and researchers to DOAS, often wishing that we could improve our efforts with a comprehensive text that describes both the theoretical basis and the practical applications of the method. Through our collaborations with air-quality monitoring agencies, we also realized that few of our partners had a deeper understanding of the requirements and the benefits of this method, explaining why DOAS was not considered as an air-quality monitoring technique by these agencies.

Many of the principles of DOAS and examples of scientific applications have been published in the literature over the past three decades, while some of the basics such as hardware and software design can be found only in unpublished documents or have been communicated orally in the few research groups specializing in this method.

Over the years it has become clear to us that the lack of a comprehensive text was an obstacle to the advancement of DOAS. This book emanated from the desire to provide such a text for students, researchers at universities and in industry, as well as air-monitoring agencies wishing to employ this method.

The development and first application of DOAS dates back more than 30 years. We would like to thank the large number of colleagues who have made contributions to the improvement of the method and the development of new uses over the years. In particular, we want to acknowledge Dieter Perner, with whom Ulrich Platt first introduced the idea of DOAS, and Dieter Ehhalt, who supported this new development. Fruitful discussions with John Noxon, who had similar ideas at the time, lead to improvements in this method. John Burrows, who so stubbornly and resourcefully pursued the idea of DOAS in space and trusted us that it would work, deserves much of the credit for the development of satellite DOAS instruments. We are also in debt to Jim Pitts and Paul Crutzen, who immediately saw the potential of the technique. This book would have been impossible to write without the help of all the students and researchers who have worked in our groups. Among them we would like to single out Gerd Hönninger, who played a crucial role in the development of the MAX-DOAS technique, and who tragically passed away during a research visit to Alaska. We would also like to thank Yasmine Stutz, who helped us in the preparation of this book.

Finally, and most of all, we would like to thank our families for their continuing support and great patience with us.

Heidelberg, Germany — Ulrich Platt
Los Angeles, USA — Jochen Stutz

Contents

1 Introduction

The atmosphere is a thin layer of gases separating the earth's surface from space. It protects the surface from radiation, provides a blanket that elevates the surface temperature through the natural greenhouse effect, is responsible for the redistribution of water and heat, provides us with oxygen, and removes harmful gaseous compounds. These properties are essential to the earth's ability to support life. For example, it is believed that only after the ozone layer formation 0.5–1 billion years ago was harmful solar UV radiation sufficiently reduced to allow life to leave the oceans and further develop on land. While the atmosphere has undergone natural variations during the past 4.5 billion years, the arrival and expansion of human societies over the recent millennia have led to changes in the chemistry and physics of the atmosphere. Industrialisation, which began approximately 200 years ago, combined with widespread land-use change have had a marked impact on almost all aspects of the earth's environment, including the composition of the atmosphere, for example by increasing levels of trace gases, such as carbon dioxide, oxides of nitrogen and tropospheric ozone. Human influence on the Earth system has become so dominant that we consider it a new age, the "Anthropocene".

Scientific interest in the chemistry and physics of the atmosphere has expanded. While this interest was initially driven by scientific curiosity, over the past few decades research has been increasingly motivated by the urgency to understand the response of the atmosphere to anthropogenic influence.

As the atmosphere is the medium in which most plants, animals, and humans live, the 'health' of the atmosphere is of great societal concern. Numerous examples have shown the impact that emission and transformation of gases have on the atmosphere. In particular, urban air quality has been a concern for many years. Beginning with observation of increased numbers of deaths in industrialised areas in 19th century Europe, air pollution was first recognised as a serious threat after the London 'killer-smog' in December 1952. A new kind of air pollution was simultaneously

U. Platt and J. Stutz, *Differential Optical Absorption Spectroscopy*, Geophysics and Environmental Physics, https://doi.org/10.1007/978-3-662-68639-3_1

recognised in Los Angeles, which lead to serious plant damage, and lung and eye irritation in humans. Although recognised for half a century, urban and regional air quality problems remain among the most challenging issues of the 21st century. After seven decades of air pollution research, political efforts to improve air quality, and an increasing public awareness, more than 100 million people in the United States still live in or near areas where ozone and particulate matter (PM) levels exceed national air quality standards (Zhang et al. 2019, https://www.lung.org/research/sota/key-findings), with racial/ethnic minorities and lower-income groups experiencing the highest exposure levels (Jbaily et al. 2022; Tessum et al. 2021). It is also estimated that 97% of Europe's urban population is exposed to elevated levels of PM and nitrogen dioxide (European Environmental Ageny 2023, www.eea.europa.eu/en/topics/in-depth/air-pollution). Air pollution is also an increasing problem in developing countries that strive to reach western living standards through rapid industrialisation (Fuller et al. 2022). Secondary pollutants, such as ozone, particulate matter, and most likely nitrogen dioxide, are known to have detrimental effects on human health (Manisalidis et al. 2020; Fuller et al. 2022). The World Health Organization, WHO, estimates that 6.7 million premature deaths per year are due to ambient and household air pollution (https://www.who.int/health-topics/air-pollution, 2023).

Another well-known example of human impact on the composition of the atmosphere is stratospheric ozone depletion events over Antarctica. While scientists had anticipated a thinning of the ozone layer due to the release of man-made chemicals summarised under the term chlorofluorocarbons (CFCs), discovery of the extent and severity of this 'ozone hole' in 1985 was a surprise. The urgent need to protect the ozone layer, which was supported by a sound scientific assessment of the depletion mechanism, led to a number of international treaties to curb the release of CFCs. Although the ozone hole will probably not 'close' until the year 2050, the ban of CFCs is the first successful international effort to abate a severe change in the atmospheric composition on a global scale.

The most publicly discussed topic of atmospheric change in the 21st century is the anthropogenic impact on the climate system of our earth. In this context, climate change refers not only to temperature but to all aspects of the atmospheric system, including weather and rain patterns, the fate of global ice sheets, the frequency of severe weather events, etc. The fact that the earth's climate is altered by a change in atmospheric composition is well known (Arrhenius 1896). The analyses of various paleorecords, such as ice cores, have shown that atmospheric composition and climate have been correlated over the past 100,000 years. Major volcanic eruptions also perturbed the climate on shorter time scales. For example, weather pattern changes and a global temperature decrease of up to 1.2 °C were observed for 5 years after the eruption of Krakatau in 1883. These natural variations in the earths climate have been ongoing throughout the entire history of our atmosphere.

It is clear that human activities have changed the climate of our planet in an unprecedented way during the past two centuries. Observations show a 1.1 °C increase in global surface temperature over the past 150 years, a shrinking of Arctic sea ice volume, a retreat of glaciers, and many other examples that all support this conclusion (Calvin et al. 2023). The primary instigators of these changes are trace gases that trap infrared radiation. These trace gases are released from human activities or formed as a consequence of these emissions. Emission and formation of small particles suspended in air also plays an important role. Although uncertainties about details of the processes influencing our climate system remain, it is clear that the possible consequences of these changes are severe, and that strategies to mediate these changes are needed.

Efforts to understand the fundamental physical and chemical processes that control the atmosphere are thus not only motivated by scientific curiosity, but also by the desire to assess and mitigate anthropogenic impact. Abatement strategies for air pollution, climate change, and other environmental problems must be based on a sound scientific understanding, as well as an accurate quantification of the changes and the efficiency of remediation efforts.

Measurement techniques are an important source of knowledge about atmospheric chemistry and physics. Scientific interest in the quantitative determination of atmospheric trace constituents dates back more than 150 years, when Christian Schönbein measured the concentration of atmospheric ozone, which at that time was considered beneficial for human health. Since then, a wide variety of analytical methods to study trace gas concentrations have been developed. Spectroscopic techniques (initially introduced by Kirchhoff and Bunsen in Heidelberg around 1859) have been on the forefront of scientific research and have made major contributions to many aspects of atmospheric research. Discoveries of the UV absorption of atmospheric ozone by Cornu and Hartley around 1880, the ozone layer by Dobson (1925) and Götz (1926) in the 1920s, and methane and carbon monoxide in the earth's atmosphere by Migeotte (1949) were some of the early milestones.

Since then spectroscopic methods have played an important role in the study of the chemical and physical processes dominating the atmospheric composition. Among other discoveries, they elucidated the central role of free radicals by making the first unambiguous identification of the hydroxyl radical (OH) in the nightglow (Bates and Nicolet 1950), and the first detection of tropospheric OH (Perner et al. 1976). The discovery and study of the ozone hole was also driven by spectroscopic techniques. The first report of the ozone hole was based on the long-term spectroscopic monitoring of ozone in Antarctica (Farman et al. 1985). The identification and quantification of various halogen radicals such as ClO, OClO, and BrO shed light on the catalytic processes responsible for the destruction of ozone.

Among the many spectroscopic techniques, differential optical absorption spectroscopy (DOAS) has proved to be one of the most powerful methods to measure a wide variety of trace gases. It is particularly useful for the observation of highly reactive trace gases, and allowed the first detection of radicals such as OH, NO_3, OClO, BrO, IO, ClO, etc. both in the troposphere and the stratosphere. The calibration-free absolute measurements, and the unequivocal identification of these and other trace gases, make this method a unique tool for the investigation of atmospheric chemistry and physics. DOAS applications now include smog chamber and laboratory studies, monitoring emissions from volcanoes and industrial facilities, atmospheric chemistry process studies, and satellite measurements of global trace gas levels.

In particular satellite measurements have become an indispensable tool to validate models, to study atmospheric processes on regional to global scales, and to monitor the composition of Earth's atmosphere. In addition to trace gas measurements, DOAS-techniques have been expanded to quantify chlorophyll fluorescence, a measure for the Gross Primary Production (GPP) of the biosphere, determine aerosol optical depth and extinction profiles, and provide high-resolution imaging of trace gas distributions.

This book provides an introduction to the DOAS technique and an overview of the current state of this measurement method. It was written for readers who want to become familiar with the DOAS technique, and therefore provides the necessary background to appreciate and understand the breadth of DOAS applications. The physical and technical principles that form the basis of DOAS are conveyed for readers with a background in the natural sciences. The book is intended to present the details of this method for future and established DOAS users. It provides enough information to make educated use of DOAS instruments, and aid in the construction of new instruments and the improvement of the analytical methods. The book is divided into twelve chapters. A short review of the chemistry of the troposphere and the stratosphere is given in Chap. 2, introducing the trace gases and concepts needed to understand the various applications of DOAS. The third chapter will briefly convey the fundamentals of atomic and molecular spectroscopy that form the basis of all spectroscopic methods. Chapter 4 describes the propagation of radiation in the open atmosphere. A review of various analytical methods to measure atmospheric trace constituents is presented in Chap. 5. Chapter 6 gives a general introduction to DOAS. The technical aspects in the construction of DOAS instruments, the mathematical basics of the analytical techniques, and radiative transfer methods are presented in Chaps. 7 through 9, respectively. Chapters 10 and 11 present a number of examples for the application of DOAS in the troposphere and the stratosphere. A look into the future of DOAS is provided at the end of the book in Chap. 12.

2 Atmospheric Chemistry

Air quality, the health of the ozone layer, and the earth's climate are closely tied to the composition of the atmosphere and the chemical transformation of natural and anthropogenic trace gases. Although the purpose of this book is not to cover the full breadth of atmospheric chemistry, this chapter provides a short introduction on this topic to motivate and clarify the applications of the differential optical absorption spectroscopy (DOAS) technique. A much more detailed treatment of atmospheric chemistry and physics can be found in specialised textbooks, e.g. Finlayson-Pitts and Pitts (2000); Holloway and Wayne (2010); Akimoto (2016); Seinfeld and Pandis (2016). This chapter is mostly restricted to chemical reactions occurring between gas molecules, although a few surface reactions are also presented. It should be noted that much of our knowledge of atmospheric chemistry comes from studies employing sophisticated instrumentation for the detection of atmospheric trace gases, including the DOAS technique.

Much of the motivation to study atmospheric chemistry and the composition of the atmosphere is based on various man-made environmental problems that have emerged over the past two centuries. Human activities have upset the natural balance of the atmosphere by influencing the trace gas and aerosol composition on local, regional, and global scales. The most significant atmospheric environmental problems include:

- 'London Type' smog was first recognised as an environmental problem in the 19th century (Brimblecombe and Heymann 1998). The emission of soot particles and the formation of aerosols consisting of sulfuric acid, which were produced from oxidation of sulfur dioxide, SO_2, emitted from combustion sources, had a severe impact on human health. During the so-called 'London Killer Smog' in 1952, approximately 3000 excess deaths were counted in London during the first three days of the event, and an additional 9000 excess death have been estimated for

U. Platt and J. Stutz, *Differential Optical Absorption Spectroscopy*, Geophysics and Environmental Physics, https://doi.org/10.1007/978-3-662-68639-3_2

the following two months (Bell and Davis 2001). It should be noted that the term 'Smog' was coined to describe the interaction of Smoke and Fog (Smog) that formed the strong haze observed in London during winter time.

- Crop damage and human discomfort were the first indication of high levels of oxidants during sunny and hot days in Los Angeles (Haagen-Smit 1952; Haagen-Smit and Fox 1954; Finlayson-Pitts and Pitts 2000). The formation of this 'Los Angeles Type' smog was found to be primarily due to photochemical formation of large amounts of ozone, carbonyl compounds, and organic aerosol from car exhaust and industrial emissions of nitrogen oxides, carbon monoxide, and volatile organic compounds (VOCs) (Haagen-Smit 1952; Finlayson-Pitts and Pitts 2000). Despite many decades of research and mitigation activities, Los Angeles type smog remains one of the most common air pollution problems in urban areas today.
- Forest decline and lake acidification were signs of another man-made environmental problem. Increased emissions of SO_2 and nitrogen oxides, NO_X, followed by their gas and aqueous phase oxidation to sulfuric and nitric acid, causes 'Acid rain' which leads to the deposition of these acids to various ecosystems (Charlson and Rhode 1982). While acid rain has successfully been reduced in Europe and the United States, it remains a problem in many developing countries.
- The health of the atmospheric ozone layer has always been a concern of atmospheric scientists. In 1971, Johnston (1971) predicted the loss of stratospheric ozone due to emission of NO_X by a planned fleet of supersonic passenger aircraft. Molina and Rowland (1974) warned of a gradual loss of stratospheric ozone due to catalysed ozone destruction processes by halogen species transported to the stratosphere in the form of extremely stable halogen-containing compounds, so called chlorofluorocarbons (CFCs), used as coolants or spray can propellants. Their predictions became reality in 1985, when Farman et al. (1985) discovered the stratospheric 'ozone hole' over Antarctica. This recurring phenomenon reduces the thickness of the stratospheric ozone layer over Antarctica to less than one third of its normal level every Antarctic Spring (e.g. Farman et al. 1985; Solomon 1999). Since then steps have been taken to stop the emissions of CFCs (Montreal Protocol and amendments) and there is evidence for a beginning decline of CFC mixing ratios in the atmosphere. Despite some setbacks, signs of recovery of stratospheric ozone are evident (see e.g. Fang et al. 2019; Braesicke et al. 2018). However, it is estimated that it will take another three decades before the ozone hole will close.
- One of the most serious environmental problems is the impact of human activities on the climate of our planet. The potential of global warming caused by IR-active 'greenhouse' gases such as CO_2, CH_4, N_2O, O_3, CFCs (e.g. $CFCl_3$ and CF_2Cl_2), and the direct and indirect effects of chemically generated aerosol, have been known for some time (e.g. Arrhenius 1896; Rhode et al. 1997; IPCC 2002; IPCC 2023; Calvin et al. 2023). Human activities, and thus the emission of the various greenhouse gases, over the past century have reached levels where an impact on global surface temperatures, the global water cycle, ocean levels, etc. has become evident (IPCC 2023; Calvin et al. 2023). The global increase of tropospheric

ozone, which likely started in the middle of the 19th century, also contributes to this phenomenon. It is thought to be caused by catalytic ozone production due to widespread emission of oxides of nitrogen and hydrocarbons (HC) (Volz and Kley 1988; Staehelin et al. 1994).

Many of these phenomena are closely related to the atmospheric chemistry presented in this chapter. We begin this chapter by introducing the structure of the atmosphere and its composition (Sect. 2.1), which have an important influence on atmospheric chemistry. The composition of the atmosphere is influenced by emissions of natural and man-made trace species (Sect. 2.2), which undergo chemical transformation forming secondary trace gases and particles. In the troposphere this chemistry is dominated by ozone and other oxidants, i.e. hydroxyl/peroxy radicals, nitrogen oxides, and volatile organic carbons (Sects. 2.3–2.7). Sulfur chemistry (Sect. 2.8) also plays an important role as a source of acidity in rain, as well as the formation of particles . The chemistry of reactive halogen species (RHS) in the troposphere influences ozone levels on regional and global scales (Sect. 2.9). Many of the radical species introduced in Sects. 2.4–2.9 contribute to the chemical degradation of emitted trace gases and are thus responsible for the the 'self-cleaning' power of the atmosphere (Sect 2.9). The last section of this chapter (Sect. 2.10) is dedicated to the chemistry of stratospheric ozone, the global ozone layer, which protects the earth's surface from UV harmful radiation, and the processes that lead to the formation of the Antarctic ozone hole.

2.1 Atmospheric Structure and Composition

For practical purposes, air can be viewed as an ideal gas. The relationship between pressure p, absolute temperature T, and volume V for a given number of moles n (1 mol is equal to $\mathrm{N_A} = 6.023 \times 10^{23}$ molecules of air) is given by the ideal gas law:

$$pV = nRT \tag{2.1}$$

where $R = 8.315\,\mathrm{J\,mole^{-1}K^{-1}}$ denotes the universal gas constant.

At standard sea-level pressure ($p_0 = 1.013 \times 10^5\,\mathrm{Nm^{-2}}$) and temperature ($T_0 = 273.15$ K or 0.0° Celsius), each cubic centimetre ($\mathrm{cm^3}$) of air contains 2.69×10^{19} molecules. At higher altitudes, z, the atmospheric pressure, $p(z)$ drops exponentially from its value at sea level:

$$p\left(z\right) = p_0 e^{-\frac{Mgz}{RT}} = p_0 e^{-\frac{z}{z_S}} \tag{2.2}$$

where $M = 0.02897\,\mathrm{kg\,mole^{-1}}$ is the mean molar mass of air, $\mathrm{g} = 9.81\mathrm{m/s^2}$ is the acceleration of gravity, and $z_S = \frac{RT}{Mg} \approx 7 \pm 1$ km denotes the 'scale height' of the atmosphere.

The vertical structure of the atmosphere can be divided into different regions according to several criteria:

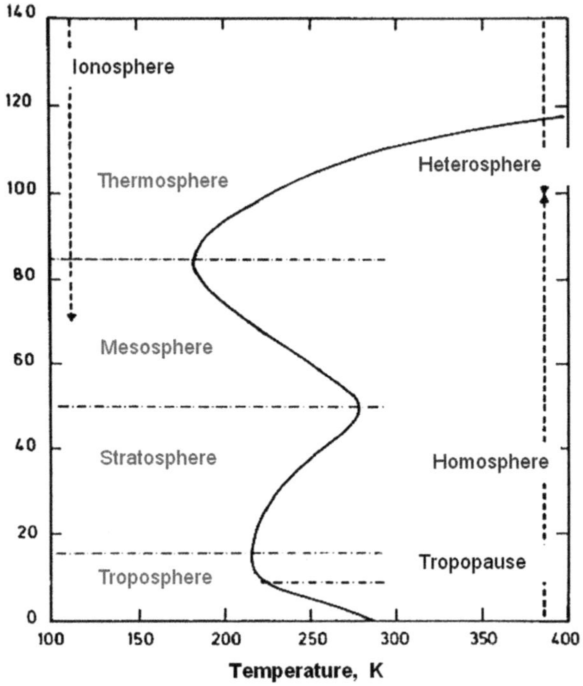

Fig. 2.1 The layers of the atmosphere, divided according to temperature, mixing mechanism, and degree of ionisation. Reprinted with permission from Brasseur and Solomon (1986), ©Springer Science+Business Media B.V. 2005. All rights reserved

- The thermal structure
- The dominating mixing process
- The ionisation state
- The (trace gas) chemical composition.

While there is a monotonic drop in pressure with altitude, the vertical temperature profile shows distinct variations, with several maxima and minima leading to a division of the atmosphere into a number of distinct compartments (see Fig. 2.1):

The *troposphere* is the lowest layer of the atmosphere, and is relatively well mixed. It extends from the surface to about 8 km altitude at the poles, and from the surface to about 18 km at the equator. The troposphere can be further subdivided into:

- The atmospheric boundary layer (BL), which roughly covers the lowest 1–2 km of the atmosphere. Air movement and mixing in the BL are influenced by friction on the earth's surface.
- The free troposphere, which extends from the top of the BL to the tropopause.

The *stratosphere* is a poorly (but still turbulent) mixed layer extending above the troposphere to about 50 km altitude. Common subdivisions of the stratosphere include:

Table 2.1 The main constituents of the (unpolluted), dry atmosphere

Gas	Chemical formula	Mixing ratio by volume	Mixing ratio vol. %
Nitrogen	N_2	0.7808	78.08
Oxygen	O_2	0.2095	20.95
Argon	Ar	$9.3 \cdot 10^{-3}$	0.93
Carbon dioxide	CO_2	$0.42 \cdot 10^{-3}$	0.042
Neon	Nc	$18 \cdot 10^{-6}$	0.0018
Helium	He	$5.2 \cdot 10^{-6}$	0.00052
Methane	CH_4	$1.7 \cdot 10^{-6}$	0.00017
Krypton	Kr	$1.1 \cdot 10^{-6}$	0.00011
Xenon	Xe	$0.9 \cdot 10^{-6}$	0.00009
Hydrogen	H_2	$0.5 \cdot 10^{-6}$	0.00005
Dinitrogen oxide	N_2O	$0.3 \cdot 10^{-6}$	0.00003

- The tropical tropopause layer
- The lower and middle stratosphere
- The upper stratosphere

The *upper atmosphere* extending above the stratopause is subdivided into the mesosphere, thermosphere, and exosphere.

Despite the vertical temperature and pressure variation up to about 100 km altitude, the composition of the atmosphere is fairly constant with respect to its main components, hence its designation *homosphere*. This is due to the dominance of turbulent mixing over diffusive mixing up to this altitude. Higher up, the exponentially decreasing pressure, and thus exponentially increasing mean free path length of air molecules, causes molecular diffusion to dominate over turbulence. The three most abundant species, which constitute more than 99.9% of dry air (see Table 2.1), are molecular nitrogen and oxygen, as well as the noble gas argon. In addition, the atmosphere contains water vapour between 10^{-4}% and a few % (by volume). Other species with comparable abundances are carbon dioxide, and all remaining nobel gases. With the exception of water vapour, methane and carbon dioxide, the species listed in Table 2.1 have atmospheric residence times in excess of 1000 years, and are therefore referred to as 'permanent' constituents. In addition to the above species, the atmosphere contains a multitude of other species at much lower abundance. They are called 'trace gases' and include methane (CH_4, $\approx 0.00019\%$), molecular hydrogen (H_2, $\approx 0.00005\%$), and nitrous oxide (N_2O, $\approx 0.00003\%$) among others.

2.1.1 Trace Species in the Atmosphere

In addition to the main gaseous constituents (Table 2.1), the atmosphere contains a large number of trace gases. Only a few trace species (e.g. CH_4) exceed a mixing ratio of one molecule in a million air molecules. However, a very large number of species (e.g. ozone, carbonyl sulfide, difluoro–dichloro–methane, methyl chloride, and methyl bromide) are present at mixing ratios around or below one molecule in a billion air molecules. Atmospheric chemistry is predominately concerned with the fate of these trace species, which, despite their low concentration, have a very noticeable impact on our atmosphere.

A very important class of non-gaseous components in the atmosphere is small particles of liquid or solid matter dispersed and suspended in air: the *aerosol*. The literal translation of the word aerosol is 'solution in air'. Since a solution includes both a solvent and a solute, one might insist that the term aerosol should only be used to denote the entire system, including both the suspended particles and the carrier gas. However, today the term aerosol is used in the literature synonymously with 'suspended particles', excluding the carrier gas.

An important property of a solution is that it is a stable system. Applying a similar criterion to aerosol particles means that only those particles which remain suspended in air for a sufficiently long time to be transported by the wind over reasonably long distances can be part of an aerosol. The atmospheric residence time of large aerosol particles is limited by their settling velocity, which is the cause of sedimentation. The settling velocity varies approximately with the square of the particle radius, and therefore the upper limit for the size of aerosol particles is relatively sharp. Particles with radii much larger than about 10 μm have settling velocities too large to behave like true aerosol particles, and are thus classified as 'coarse dust particles'. (A spherical particle of 10 μm diameter has a settling velocity of about 2.42 cm s^{-1}, or 87 m h^{-1}, and thus will be rapidly removed from the atmosphere by sedimentation).

Defining the lower limit for the existence range of aerosol particles is more difficult. Clearly, aerosol particles must exceed the size of individual molecules, which have radii in the range of fractions of a nanometre (e.g. the gas kinetic radius of N_2 or O_2 is ≈ 0.2 nm). Nucleation theory teaches us that clusters composed of several condensable molecules (e.g. a mixture of sulfuric acid and water molecules) can only be stable when they exceed a certain critical size. Before reaching that critical size, the clusters tend to evaporate again. It appears attractive to use that critical size as a lower limit for the definition range of aerosol particles in the atmosphere. However, the critical size is not uniquely defined, but depends on atmospheric conditions, in particular on temperature, relative humidity, and the oxidation rate of SO_2, which yields the condensable sulfuric acid molecules. Very often the critical size of these so-called 'secondary aerosols' is in the range of 1 nm.

2.1.2 Quantification of Gas Abundances

The amount of trace gases present in the atmosphere can be quantified by using two different descriptions. First, we define the concentration of a trace gas as the amount of trace gas per volume of air (at a given temperature):

$$c = \frac{\text{Amount of trace gas}}{\text{Volume of air}} \tag{2.3}$$

where 'amount' refers to either mass (c_m), number of molecules (c_n), or number of moles (c_M). Examples for units of concentration are micrograms per m^3 or molecules per cm^3. The latter is also known as 'number density' of a gas. The partial pressure of a species is also a measure of its concentration.

Second, we can define the *mixing ratio* of a trace gas as the ratio of the amount of a trace gas to the amount of air, including the trace gas:

$$x = \frac{\text{Amount of trace gas}}{\text{Amount of air + Trace gas}} \approx \frac{\text{Amount of trace gas}}{\text{Amount of air}} \tag{2.4}$$

At typical atmospheric trace gas mixing ratios of $<10^{-6}$, the distinction between the 'amount of air' and 'amount of air + trace gas' is so small that it can be neglected for practical purposes. It is necessary to distinguish whether the 'amount' is in volume, number of moles, number of molecules, or mass.

An example of a mixing ratio is parts per million (ppm) by volume:

$$x_V = \frac{\text{Unit volume of trace gas}}{10^6 \text{ unit volumes of (air + trace gas)}} \text{ppm}$$

For smaller trace gas mixing ratios, x_V is given as parts per billion (ppb) and parts per trillion (ppt), which are analogously defined as:

$$x_V = \frac{\text{Unit volume of trace gas}}{10^9 \text{ unit volumes of (air + trace gas)}} \text{ppb}$$

$$x_V = \frac{\text{Unit volume of trace gas}}{10^{12} \text{ unit volumes of (air + trace gas)}} \text{ppt}$$

While mixing ratios could also be given by mass, this is rarely done in atmospheric chemistry. Nevertheless, in the literature the terms ppmv, ppbv, and pptv are sometimes used to stress the fact that volume mixing ratios are understood. Similar to volume mixing ratios are molar mixing ratios:

$$x_M = \frac{\text{Moles of trace gas}}{\text{Mole of (air + trace gas)}} \tag{2.5}$$

It should be noted that x_M is the new SI unit for mixing ratios. Since air under ambient conditions can be regarded, in good approximation, as an ideal gas, for practical purposes a mixing ratio x_M specified in moles per mole equals the mixing ratio x_V. Thus, the terms micromole per mole, nanomole per mole, and picomole per mole are essentially equivalent to ppm, ppb, and ppt by volume, respectively.

A common problem in atmospheric chemistry is the conversion of units of the amounts of trace gases. For example, for trace gas i we obtain the following conversion between the number density c_n (in molecules per cm^3) and mass concentration c_m (in grams per cm^3):

$$(c_m)_i = \frac{c_n \cdot M_i}{N_A} \tag{2.6}$$

Table 2.2 The different units for the abundance of atmospheric trace gases ($T = 293.15$K, $p =$ 101325 Pa)

Trace gas	Molecular mass g/mole	Mixing ratio x_V ppb	Number density c_n molec. cm^{-3}	Concentration c_m $\mu g\, m^{-3}$
O_3	48.00	1.000	$2.503 \cdot 10^{10}$	1.995
		0.501	$1.254 \cdot 10^{10}$	1.000
SO_2	64.06	1.000	$2.503 \cdot 10^{10}$	2.662
		0.376	$0.941 \cdot 10^{10}$	1.000
NO	30.01	1.000	$2.503 \cdot 10^{10}$	1.251
		0.799	$2.000 \cdot 10^{10}$	1.000
NO_2	46.01	1.000	$2.503 \cdot 10^{10}$	1.912
		0.532	$1.33 \cdot 10^{10}$	1.000
CH_4	16.04	1.000	$2.503 \cdot 10^{10}$	0.667
		1.500	$3.755 \cdot 10^{10}$	1.000
CH_2O	30.03	1.000	$2.503 \cdot 10^{10}$	1.248
		0.801	$2.005 \cdot 10^{10}$	1.000
CO	28.01	1.000	$2.503 \cdot 10^{10}$	1.164
		0.859	$2.150 \cdot 10^{10}$	1.000

where M_i denotes the molecular mass of the species i in g per mole and N_A is Avogadro's number with $N_A = 6.0221420 \times 10^{23}$ molecules $mole^{-1}$. Thus, the conversion between number density and mass concentration is different for species with different molecular (atomic) mass.

The conversion of number density c_n (in molecules per cm^3) into the corresponding (volume) mixing ratio is given by:

$$x_V = c_n \frac{V_0}{N_A} \quad \text{or} \quad c_n = x_V \cdot \frac{N_A}{V_0} \tag{2.7}$$

where V_0 denotes the molar volume in cm^3 for the pressure p and temperature T at which the number density c_n was measured. For standard conditions ($p_0 =$ 101325 Pa = 1 Atmosphere, $T = 273.15$K) the molar volume is $V_0 = 22414.00\, cm^3$ $mole^{-1}$. For arbitrary temperature and pressure conditions, we can use:

$$x_V = c_n \frac{1}{N_A} \frac{T}{p} \frac{p_0}{T_0} V_0 = c_n \frac{1}{N_A} \cdot \frac{RT}{p} = c_m \frac{1}{M_i} \cdot \frac{RT}{p} \tag{2.8}$$

or

$$c_m = x_V \cdot M_i \cdot \frac{p}{RT} \tag{2.9}$$

Table 2.2 gives a few examples for the conversion of the different units in which the abundance of trace gases is customarily expressed.

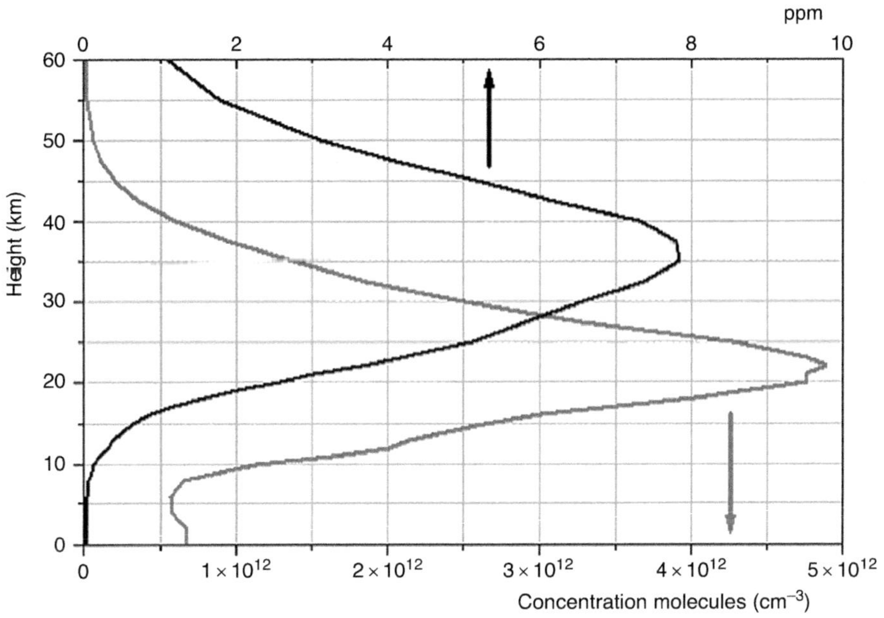

Fig. 2.2 Vertical profile of ozone in troposphere and stratosphere (US Standard Atmosphere), both in mixing ratio (in ppm or μmole per mole), peaking at about 36 km and concentration (molecules per cm^3), peaking at about 22 km

It should be noted that mixing ratios, x_V and x_M, are independent of temperature and pressure. Consequently, they are conserved during vertical transport in the atmosphere. In contrast, the concentration c depends on both parameters and changes when air is transported. However, trace gas concentrations are relevant for the calculation of chemical reaction rates and radiative properties (such as UV absorption) of the atmosphere. Spectroscopic measurement techniques (including DOAS) also give results in number density or concentration, not mixing ratios.

Thus both types of units are useful and must be converted into each other as needed. Atmospheric trace gas profiles might look quite different when viewed in terms of mixing ratios or concentrations, as illustrated by the example of the atmospheric ozone profile in Fig. 2.2. The concentration (in molecules per cm^3) has a noticeable, essentially constant level in the troposphere, and a maximum at about 22 km in the lower stratosphere. In contrast, tropospheric mixing ratios (in ppm or μmole per mole) are very low and peak at about 36 km.

Another category for specifying trace gas abundances is the column density S, which is defined as the integral of the number density along a certain path in the atmosphere:

$$S = S_{\mathrm{n}} = \int_{\mathrm{Path}} c_{\mathrm{n}}(s)\, ds \tag{2.10}$$

The unit of S_n, and column densities in general, is molecules per cm^2. In analogy to the mass mixing ratio c_m, the mass column density S_m can be defined as:

$$S_m = \int_{\text{Path}} c_m(s)\, ds \tag{2.11}$$

Typical units for mass column densities are μg/cm^2. The conversion between S_n and S_m is analogous to the conversion between c_n and c_m.

A frequently used path is the total (vertical) atmospheric column density V, where the concentration is integrated vertically from the surface to infinity:

$$V = \int_0^\infty c_n(z)\, dz \tag{2.12}$$

2.1.3 Chemical Kinetics and Lifetime of Trace Gases in the Atmosphere

The atmosphere is not a static system with respect to its components. Rather, gases are released to the atmosphere or formed by chemical transformations, and removed again by chemical degradation or deposition on the ground. A useful quantity in this context is the steady-state (average) lifetime, or residence time, of a species in the atmosphere. Chemical transformations in the gas-phase occur in four different reaction pathways (see e.g. Seinfeld and Pandis 2016). The first is the first-order or unimolecular reaction of a compound, here called A, into its reaction products, here called B and C:

$$A \rightarrow B + C \tag{2.13}$$

The change in the concentrations of A, B and C, follow a simple differential equation, where t denotes time and compound symbols in square brackets denote concentration of the compound (e.g. [A]):

$$\frac{d[A]}{dt} = -\frac{d[B]}{dt} = -\frac{d[C]}{dt} = -k_1 \times [A] \tag{2.14}$$

The rate constant, k_1, has units of s^{-1}.

In second-order or bimolecular reactions two reactants, A and B, are transformed into one or more products, for example C and D:

$$A + B \rightarrow C + D \tag{2.15}$$

The temporal change of the concentrations of the reactants and products again follows a differential equation:

$$\frac{d[A]}{dt} = \frac{d[B]}{dt} = -\frac{d[C]}{dt} = -\frac{d[D]}{dt} = -k_2 \times [A] \times [B] \tag{2.16}$$

In this case the unit of the rate constant k_2 is $cm^3 molec.^{-1} s^{-1}$. Both k_1 and k_2 are usually temperature dependent.

Third-order or termolecular reactions involve a third molecule, M, that participates in the reaction, but is not consumed:

$$A + B + M \rightarrow C + D + M \quad \text{or} \quad A + B \stackrel{M}{\rightarrow} C + D \tag{2.17}$$

In the atmosphere M is an air molecule with a concentration far in excess of the reactants. Consequently, its concentration can be assumed to be constant. It should be noted here that termolecular reactions are not three-body reactions, but rather proceed through the formation of an excited complex AB^* that is energetically stabilised by the consequent collision with M. The temporal evolution of the reactants in termolecular reaction is described by the same differential equation as for bimolecular reactions (Eq. 2.16), where the dependence on [M] is folded into the temperature and pressure dependent rate constant k_3.

The rate constant $k_1(T)$, $k_2(T)$ and $k_3(T, p)$ for atmospheric reactions are typically determined in laboratory experiment and can be found in the scientific literature. Various compilations of rate constants relevant to atmospheric chemistry are available at: NASA Panel for Data Evaluation Burkholder et al. (2016) at https://jpldataeval.jpl.nasa.gov/ and IUPAC Subcommittee on Gas Kinetic Data Evaluation for Atmospheric Chemistry Atkinson et al. (2004, 2006) at https://iupac-dev.ipsl.jussieu.fr/. The last type of reaction is somewhat unique to atmospheric chemistry. These photochemical reactions occur when a photon of sufficient energy, or low enough wavelength, interacts with a molecule, leading to the excitation or dissociation:

$$A + h\nu(\lambda < \lambda_{lim}) \rightarrow B + C \tag{2.18}$$

It is common to include the term "$h\nu$" in this equation to clarify that this reaction is photolytic. In addition, the wavelength range at which the reaction can occur is also often added. The rate of the reaction is given by:

$$\frac{d[A]}{dt} = -\frac{d[B]}{dt} = -\frac{d[C]}{dt} = -J_A \times [A] \tag{2.19}$$

where J_A is the molecule specific photolysis frequency. The photolysis frequency depends on the actinic flux, $L(\lambda)$, the absorption cross-section of A, $\sigma(\lambda)$, and the quantum yield of the reaction $\Phi(\lambda)$:

$$J_A = \int L(\lambda) \times \sigma(\lambda) \times \Phi(\lambda) \times d\lambda \tag{2.20}$$

The actinic flux depends on many environmental parameters, such as solar position, presence of clouds, aerosol extinction, albedo, etc., and is either measured or calculated using atmospheric radiative transfer models (Seinfeld and Pandis 2016). To fully describe atmospheric chemistry, a large set of chemical reactions has to be considered. This can be achieved by simultaneously solving the differential equations associated with each reaction, using numerical models. The reader is referred to literature to learn more about these atmospheric chemistry models (see for example Jacobson 2005; Brasseur and Jacob 2017).

The concept of a (pseudo) steady-state has been developed to provide quantitative understanding of chemical sub-systems in the atmosphere without solving differential equation systems. In cases where the concentration of atmospheric species are not (or only slowly) changing with time, one can approximate the time differential $d[A]/dt = 0$, thus changing the differential equations (Eqs. 2.14, 2.16, 2.19) to simple equations. The pseudo steady-state approach often allows derivation of simple equations that provide insight into the chemical mechanisms in the atmosphere. The results of some of such calculation will be shown in the remainder of this chapter.

If the concentration of a species has reached a steady-state value, and therefore its rate of production P (for instance $d[C]/dt$ in Eq. 2.16) equals its rate of destruction D, its lifetime can be defined as the ratio:

$$\tau = \frac{c}{D} = \frac{c}{P} \tag{2.21}$$

Assuming the production P to be constant and the destruction D to be proportional to c, i.e. $D = c/\tau$ with the constant of proportionality $1/\tau$, we obtain:

$$c\left(t\right) = P \cdot \tau \cdot \left(1 - e^{-\frac{t}{\tau}}\right) \xrightarrow[\tau \to \infty]{} P \cdot \tau \tag{2.22}$$

The constant τ becomes clear by calculating $c(\tau) = 1 - 1/e \approx 63.2\%$ of the final value $c(\infty) = c(t \to \infty) = P \cdot \tau$. In other words, the lifetime τ is the time constant with which the trace gas concentration approaches its steady state.

Figure 2.3 gives approximate lifetimes of atmospheric trace constituents. The variation of the concentration of a gas in space and time also depends on its lifetime. For instance, the long-lived components such as N_2, O_2, or noble gases, with lifetimes of thousands of years, show very little variation, whereas the concentrations of short-lived species may vary considerably with time and location. The quantitative relationship of residence time and variation of the mixing ratio of a gas was first established by Junge (1963).

2.2 Direct Emission of Trace Gases to the Atmosphere

As described above (Sect. 2.1.3), most gaseous (and particulate) components of our atmosphere are in a balance between continuous addition to the atmosphere and removal from it. To understand the levels of trace gases in the atmosphere and their chemistry, it is important to study their sources or, in the case of gases which are formed chemically, the sources of their precursors. This is an important issue with respect to air pollution since human activities have added new trace gas sources over the past centuries. These sources have to be compared with natural emission sources in order to assess their impact on the composition of the atmosphere. While this section only gives a very brief overview of sources of trace gases that are of

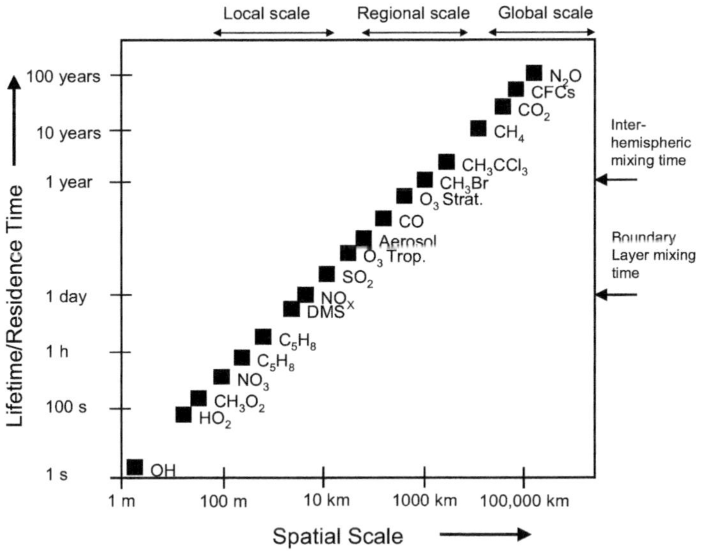

Fig. 2.3 Average lifetimes (residence times) of gases in the atmosphere range from seconds (and below) to millennia. Accordingly, transport can occur on scales reaching from a few metres to the global scale

interest in this book, the accurate determination of emissions is a crucial aspect of atmospheric chemistry.

Emission sources are typically subdivided into natural and anthropogenic processes. The following list gives examples of the types of sources found in the atmosphere:

Natural emissions include:

- Emission of gases from soil (e.g. methane, oxides of nitrogen NO and N_2O)
- Emission of gases by vegetation (primarily volatile organic compounds (VOC))
- Emission by biomass burning (VOCs, CO, NO_X, etc.)
- Marine emissions (Dimethyl sulfide (DMS), iodine, sea salt aerosol, etc.)
- Volcanic emissions (CO_2, SO_2, HCl, BrO, etc.)
- Formation in thunderstorms (NO)

Anthropogenic emissions result from human activities, and include:

- Emission during combustion processes (e.g. traffic, power stations, and industry). The emitted gases include oxides of nitrogen (NO and NO_2), carbon monoxide, and VOCs
- Emission of VOCs by industrial processes (e.g. manufacturing, refineries, coatings, etc.)

- Agricultural gas emissions from fertilised soil and e.g. rice growing: oxides of nitrogen (NO and N_2O) and methane
- Emission by biomass burning: VOCs, CO, and NO_X
- Emission of gases, by vegetation: primarily VOCs

We will briefly discuss the source and sink strengths, as well as average mixing ratios and lifetimes (residence times), of a series of important atmospheric trace gases, which have their dominant sources at the surface. The discussion is subdivided into nitrogen, sulfur, and carbon-containing species.

2.2.1 Nitrogen Species

Fixed nitrogen species play an important role in the chemistry and the climate of the atmosphere (Sect. 2.5). Here we focus on the three most important nitrogen species that are emitted into the atmosphere—nitrous oxide (N_2O), nitrogen oxides (NO and NO_2), and ammonia (NH_3).

2.2.1.1 N_2O

Nitrous oxide is an important greenhouse gas and also the dominant source of reactive nitrogen in the stratosphere. N_2O is naturally emitted by bacteria in the soil and the ocean during the nitrogen fixation process (Table 2.3). Human activities have lead to an increase of these sources, predominantly due to the intensification and expansion of agriculture. As a consequence, N_2O mixing ratios have increased from $\sim$275 ppb before the year 1800 to over 330 ppb today (Tian et al. 2020).

2.2.1.2 NO_X ($NO + NO_2$)

The significance of nitrogen oxides will be discussed in detail in Sect. 2.5. NO and NO_2 are produced in a large number of natural and anthropogenic processes. Particularly important are those processes where air is heated to high temperatures, such as in combustion processes (in internal combustion engines), forest fires, or lightning strikes. Natural sources of nitrogen oxides produce about 13–31 million tons of nitrogen fixed as NO_X per year globally (Table 2.3). The largest natural contributions come from brush and forest fires, lightning storms, and soil emission. Smaller contributions come from diffusion from the stratosphere and oxidation of ammonia (Ehhalt and Drummond 1982; Lee et al. 1997).

Anthropogenic NO_X emissions can be grouped into three categories that essentially all originate from combustion processes. The largest source is the stationary combustion of fossil fuel (power stations, industry, and home heating). A further important source is emission by automobiles. This emission additionally occurs in densely populated areas at very low emission height. Finally, part of the NO_X emission from forest fires and from soil is due to anthropogenic influence (intentional biomass burning and artificial fertilisation).

Table 2.3 Tropospheric sources, sinks, and mixing ratios of nitrogen species [Source strength in million tons of fixed nitrogen (Tg N) per year (from Lee et al. 1997, Ehhalt 1999)]

Species	Mixing ratio, ppb	Major sources	Average source strength Tg/year	Major Sinks	Sink strength Tg/year	Atm. life time
Nitrous oxide, N_2O	330	Soil emission Ocean emission Anthropogenic	10 6 9	Loss to stratosphere	25	110a
Nitrogen oxides, NO, NO_2	0.03–5	Fossil fuel combustion Biomass burning Soil emission/ microbial production Thunderstorms Aircraft emissions	21 8 10 5 0.8	Oxidation by OH and O_3 to HNO_3	44	2d
Ammonia, NH_3	0.1 marine 5 continent	Domestic animals Emission from vegetation Ocean emission Fertiliser use	26 6 9 8	Dry deposition Conversion to NH_4+ aerosol	17 36	5d

h = hour, d = day, a = year

The contributions of the various NO_X sources are summarised in Table 2.3. On a global scale, the contributions of natural and anthropogenic NO_X sources are about equal in magnitude. The spatial distribution of strong NO_X sources, as derived from satellite-based DOAS measurements (Chap. 11) (Leue et al. 2001; Georgoulias et al. 2019; Beirle et al. 2019a.) Natural NO_X sources are much more equally distributed over the surface of the earth than the anthropogenic sources, which are concentrated on a very small fraction of the earth's surface. For instance, in Germany the natural contribution to the total NO_X emission was small (NO_X emission 2000: 0.9 million tons nitrogen, as N (Schmölling 1983; Fricke and Beilke 1993)).

2.2.1.3 NH_3

Ammonia plays a crucial role in the atmosphere as it is the only significant atmospheric base that neutralises acids such as sulfuric and nitric acids. The ammonium ion, which is formed upon the uptake of ammonia on particles, is an important part of the aerosol. Ammonia is emitted predominately by livestock wastes and fertilised soils. Emissions from the ocean and vegetation also play a role (Table 2.3). Modern cars with catalytic converters also emit ammonia. In fact NH_3 can be the dominant fixed-nitrogen species emitted by cars (see e.g. Bishop and Stedman 2015).

2.2.2 Sulfur Species

The increase of sulfur emissions since the onset of industrialisation has led to numerous environmental problems, such as London-type smog and acid rain. Natural sulfur is emitted through biological processes in the soil and the ocean in its reduced form as carbonyl sulfide (COS), hydrogen sulfide (H_2S), and DMS (CH_3SCH_3). Volcano eruptions can also contribute to the natural sulfur emissions, in the form of sulfur dioxide, SO_2 (Table 2.4). The emissions of DMS dominate the global sulfur budget and are therefore particularly important (Sect. 2.7).

Anthropogenic emissions of SO_2 originate primarily from fossil fuel combustion. Over the past 200 years, sulfur emissions have sharply increased. As a response to the growing acid rain problem, the emission of sulfur has been greatly reduced in industrialised countries since 1975 (Fig. 2.4). This is due to stack gas desulfurisation measures, and also to the change of the economies in the eastern European countries. On a global scale, however, the decrease in sulfur emission is slow, due to continued coal combustion in Asian countries (Fig. 2.4).

The sulfur species that are most important in the atmosphere, their typical concentrations, atmospheric residence times, and their degradation mechanisms are summarised in Table 2.4.

Table 2.4 Tropospheric sources, sinks, and mixing ratios of sulfur species [Source strength in million tons of sulfur (Tg S) per year (from Ehhalt 1999)]

Species	Mixing ratio, ppb	Major sources	Source strength Tg/year	Major sinks	Sink strength Tg/year	Atm. life time
Carbonyl sulfide, COS	0.5	Soil emission	0.3	Uptake by vegetation	0.4	7 a
		Ocean emission	0.3	Flux to stratosphere	0.15	
		CS_2 + DMS ox.	0.5			
Hydrogen sulfide, H_2S	0.005–0.09	Soil emission Vegetation Volcanoes	0.5 1.2 1	Reaction with OH	2.7	3 d
Dimethyl sulfide, DMS, CH_3SCH_3	0.005–0.1	Ocean emission Soil emission	68 2	Reaction with OH, NO_3, BrO	70	2 d
Sulfur dioxide, SO_2	0.02–0.09 (marine), 0.1–5 cont.	Fossil fuel burning	160	Dry deposition, reaction with OH, liquid phase	100	4 d
		Volcanoes	14	Oxidation to $SO_4^=$	140	
		Sulfide oxidation	70	Wet deposition		

h = hour, d = day, a = year

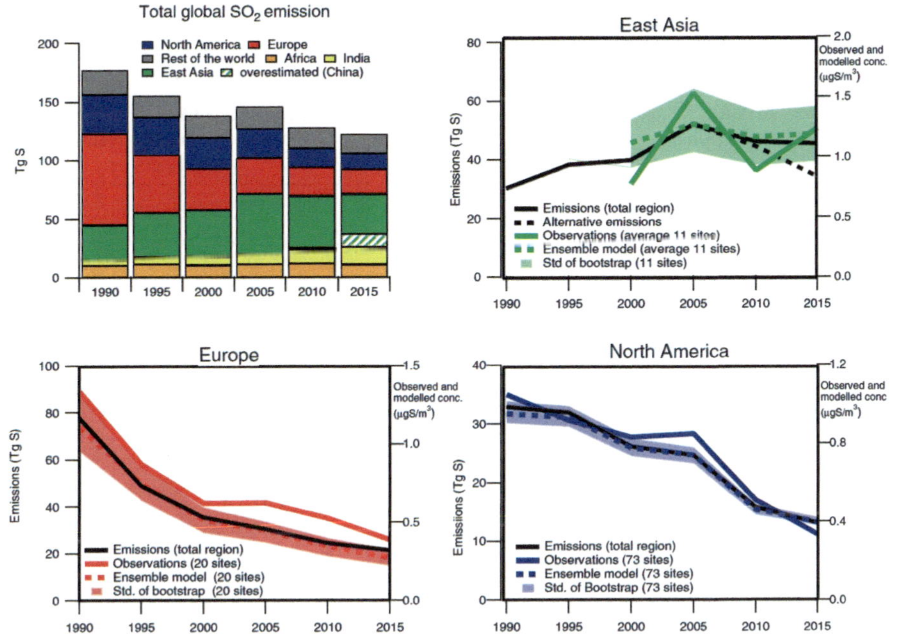

Fig. 2.4 Regional trends of global sulfur emissions (in Tera-grams (Tg, or million tons) of sulfur per year) during the past two decades. Reprinted under CC BY 4.0 license from Aas et al. (2019). ©The Author(s)

2.2.3 Carbon-Containing Species

A very large number of carbon-containing trace species are present in the atmosphere. The most important carbon-containing species is carbon dioxide (CO_2). While CO_2 plays a crucial role as a greenhouse gas, it undergoes little chemistry in the atmosphere, and its sources will not be a topic of this section. The second most abundant carbon species is methane (CH_4), which, as will be discussed in Sect. 2.4, is chemically degraded in the atmosphere, and directly participates in the formation of tropospheric ozone. Methane is also a greenhouse gas. It is difficult to distinguish natural and anthropogenic sources of methane as the emissions predominately stem from animals, including farm livestock, and wetlands, which includes rice paddies (Table 2.5). Fossil fuel production, storage, and transport also contributes to today's methane emissions. The expansion of agriculture and industry over the past two centuries has led to an increase of methane sources and a rise of methane mixing ratios from ~750 ppb to over 1900 ppb today (see e.g. Nisbet et al. 2023).

Carbon monoxide (CO) is a product of incomplete combustion of natural material, for example in forest fires, and fossil fuels. In addition, CO is a product of the chemical degradation of methane and volatile organic carbon (VOC) species in the atmosphere. Because CO is toxic and participates in the formation of photochemical smog, various

Table 2.5 Important organic trace gases, approximate mixing ratio, emission to the atmosphere, removal from the atmosphere, and average lifetime (residence time) τ in the atmosphere [see (2.22)]

Species	Mixing ratio, ppb	Major sources	Source strength Tg/year	Major sinks	Sink strength Tg/year	Atm. life time
Methane, CH_4	1900	Rice fields	75	Reaction with OH	490	8 a
		Domestic animals	100	Export to strato-sphere	40	
		Swamps/marshes	200	Soil uptake	30	
		Biomass burning	40			
		Fossil fuel	100			
Carbon monox-ide, CO	200 (NH) 50 (SH)	Anthropogenic emission	440	Reaction with OH	2400	0.2 a
		Biomass burning	770	Flux to strato-sphere	100	
		CH_4 oxidation	860	Soil uptake	400	
		Ox. of natural VOC	610			
Isoprene, C_5H_8	0.6–2.5	Emission from deciduous trees	570	Reaction with OH, ozonolysis	570	0.2 d
Terpenes, $C_{10}H_{16}$	0.03–2	Emission from coniferous trees	140	Reaction with OH, ozonolysis	140	0.4 d
Total VOC		Fossil fuel	65	Reaction with OH	850	
		Biomass burning	35			
		Foliar emission	744			
		Oceanic emission	6			

The table is based on Ehhalt (1999), but has been extended
h = hour, d = day, a = year

measures have been taken to reduce its emissions from combustion sources. The most prominent example is the catalytic converter of automobiles, which was primarily designed to reduce CO emissions.

Hydrocarbons constitute a large class of carbon-containing species in the atmosphere. Often the terms non-methane hydrocarbons (NMHC) or volatile organic carbons (VOCs) are used to exclude CH_4 from the hydrocarbon class. The number of different species is very large and we will consider only the most important members of this group here. On a global scale, the emission by vegetation of isoprene, C_5H_8, and to a lesser extent, terpenes, $C_{10}H_{16}$, dominate the emissions of VOC.

These species are highly reactive in the atmosphere and participate in the formation of ozone and particles. Globally, natural emissions of other species are of lesser importance. Anthropogenic (and natural) emissions of VOC are an essential part of Los Angeles-type smog. A complex mixture of different trace gases is emitted by fossil fuel production and consumption, both through incomplete combustion and the loss of VOC vapours before combustion, industrial processes, refining of oil, solvents, etc. The emitted mixture, which contains alkanes (ethane, propane, etc.), alkenes (ethane, propene, etc.), alkynes and aromatics (benzene, toluene, etc.), and a variety of oxidised VOC (formaldehyde, acetaldehyde, etc.), depends on the emissions signatures of the different sources, as well as the distribution of these sources. For details on urban VOC emissions, we refer the reader to atmospheric chemistry textbooks, such as Finalyson-Pitts and Pitts (2000) or Seinfeld and Pandis (2016).

2.3 Ozone in the Troposphere

Ozone is a key compound in the chemistry of the atmosphere. In the troposphere it is a component of smog, poisonous to humans, animals and plants, as well as a precursor to cleansing agents (such as the OH radical, see Sect. 2.4.1). Tropospheric ozone is also an important greenhouse gas.

Ozone is formed by two distinctly different mechanisms in the troposphere and stratosphere. In the stratosphere O_2 molecules are split by short-wave UV radiation into O-atoms, which combine with O_2 to form O_3. This process is the core of the 'Chapman Cycle' (Chapman 1930; see Sect. 2.10.1). Until the late 1960s, it was believed that tropospheric ozone originated from the stratosphere. Today we know that large amounts of O_3 are formed and destroyed in the troposphere. Influx of O_3 from the stratosphere is only a minor contribution to the tropospheric ozone budget.

Model calculations (IPCC 2023) put the cross tropopause flux of O_3 at 277–535 Mt a^{-1} , while they derive ozone formation rates in the troposphere at 4510–4937 Mt a^{-1}. The formation is largely balanced by photochemical destruction in the troposphere amounting to 3948–4442 Mt a^{-1}. Another, relatively small contribution to the O_3 loss is deposition to the ground, modelled at 846–996 Mt a^{-1}. An overview of tropospheric ozone levels, trends, and budgets is found in several assessment reports, e.g. Fleming et al. (2018); Gaudel et al. (2018); Tarasick et al. (2019); IPCC (2023).

2.3.1 Mechanism of Tropospheric Ozone Formation

In the early 1950s, it became clear that, under certain conditions in the atmosphere near the ground, high concentrations of ozone are formed (e.g. Haagen-Smit 1952). In fact, it could be shown in 'smog-chamber' experiments that large amounts of ozone are formed when mixtures of NO_X and VOC are exposed to solar UV radiation. While the phenomenon of ozone formation as a function of VOC and NO_X in illuminated mixtures was empirically found in the 1960s, the exact mechanism only became clear much later (Weinstock 1969; Crutzen 1970; Levy et al. 1971). Ozone formation in the troposphere is initiated by the production of $O(^3P)$ from NO_2 photolysis (indicated by the term hν, see Chap. 3).

Under clear sky conditions at noontime, the average lifetime of the NO_2 molecule is on the order of 2 min [$j_{NO2} = j_2 \approx 8 \times 10^{-3} s^{-1}$; e.g. Junkermann et al. (1989)]:

$$NO_2 + h\nu \rightarrow NO + O(^3P) \tag{2.1}$$

This reaction is followed by the rapid recombination of $O(^3P)$ with O_2:

$$O(^3P) + O_2 + M \rightarrow O_3 + M \tag{2.2}$$

Where M denotes any atmospheric molecule as described above. At high pressure (and thus high M and O_2 concentrations) in the troposphere, other reactions of $O(^3P)$, in particular its reaction with O_3, are negligible. Therefore, for each photolysed NO_2 molecule, an ozone molecule is formed. Reactions Eqs. 2.1 and 2.2 are essentially the only source of ozone in the troposphere. However, ozone is often rapidly oxidised by NO back to NO_2:

$$O_3 + NO \rightarrow NO_2 + O_2 \tag{2.3}$$

Reactions Eqs. 2.1–2.3 lead to a 'photo-stationary' state between O_3, NO, and NO_2. The relation between the three species can be expressed by the *Leighton Relationship* (Leighton 1961):

$$\frac{[NO]}{[NO_2]} = \frac{j_{2.1}}{k_{2.3} \cdot [O_3]} \tag{2.23}$$

where $j_{2.1}$ denotes the photolysis frequency of NO_2 and $k_{2.3}$ denotes the rate constant for the reaction of ozone with NO. For typical ozone mixing ratios of 30 ppb, the $[NO]/[NO_2]$ ratio during daytime near the ground is on the order of unity. The reaction cycle formed by Eqs. 2.1–2.3 does not lead to a net formation of ozone. However, any reaction that converts NO into NO_2 without consuming an O_3 molecule interferes with this cycle and leads to net ozone production. Thus the key factor in tropospheric O_3 formation is the chemical conversion of NO to NO_2.

In the troposphere, the conversion of NO to NO_2 without O_3 occurs through a combination of the reaction cycles of HO_X ($OH + HO_2$), peroxy radicals, and NO_X (Fig. 2.5). In these cycles, OH radicals are converted to HO_2 or RO_2 radicals, through their reaction with CO or HC. HO_2 and RO_2, on the other hand, react with NO to reform OH, thus closing the HO_X/RO_X cycle. This reaction also converts NO to NO_2 (see also Sect. 2.5.1), which is then photolysed back to NO (2.1). The oxygen atom formed in the NO_2 photolysis then reacts with O_2 to form ozone (2.2). The process, shown in Fig. 2.5, therefore acts like a machine that, in the presence of NO_X and sunlight, converts the 'fuel' CO and HC into CO_2, water, and ozone. Because, HO_X and NO_X are recycled, this catalytic ozone formation can be quite efficient (Crutzen 1970). HO_X radicals are always present in the sunlit atmosphere (Sect. 2.4); they are, for example, formed through the photolysis of ozone in the presence of water vapour. The cycles are only interrupted if either a NO_X or a HO_X is removed from the

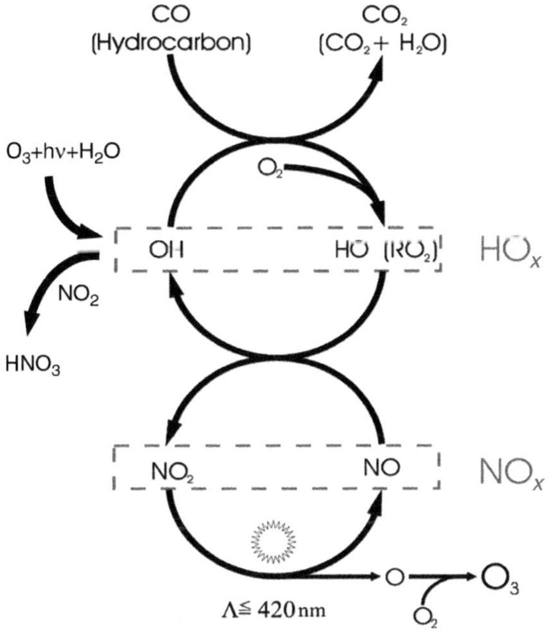

Fig. 2.5 Ozone formation in the troposphere is catalysed by hydrogen radicals ($OH + HO_2 = HO_X$), peroxy radicals, and NO_X

respective cycles, for example by the reaction of OH with NO_2, or the self-reactions of HO_2 and RO_2.

In background air, fuel for ozone formation is always present in the form of methane (mixing ratio of ~1.9 ppm, see e.g. Nisbet et al. 2023) and CO, which is formed as a degradation product of CH_4 (Sect. 2.4.1). However, in clean air, the NO_X level might be very low, and thus insufficient to act as catalyst. This effect is outlined in more detail in Sect. 2.5.1.

2.3.2 Ozone Formation in Urban Centres and Downwind

In many urban centres, car exhaust dominates as a source of pollution. Emissions from home heating and industry can also contribute considerably. These sources emit large amounts of NO_X (mostly in the form of NO) and VOC, the main ingredients of ozone formation. As a consequence, ozone formation rates can be orders of magnitude higher than in slightly polluted air outside urban centres. On the other hand, in urban centres there are also effective ozone sinks:

- Reaction of O_3 with freshly emitted NO
- Reaction of O_3 with olefins.

Overall, the ozone lifetime can be quite short (on the order of a few hours) in urban areas. Therefore, there are usually large diurnal variations of the O_3 level in these regions. In fact, in urban centres the ozone level near the ground frequently reaches zero a few hours after sunset and remains at zero for the rest of the night. After sunrise, enhanced vertical mixing replenishes some O_3 from higher layers, thus initiating radical chemistry again (see Fig. 2.8). In addition, photolysis of formaldehyde, nitrous acid, and nityl chloride (see Sects. 2.4.1, 2.5.1, and 2.8.1) can considerably speed up radical processes, and thus growth of the O_3 mixing ratio in the morning.

In a series of experiments where the concentrations of NO_X and VOC were independently varied, the ozone formation after a fixed time was found to vary with the initial levels of both groups of species in a characteristic way: At a given initial level of NO_X, the O_3 increases linearly with the initially present level of VOC. In other experiments with higher initial VOC levels, the amount of O_3 levels off. Increasing initial VOC even further does not lead to higher O_3 production. This regime is called 'NO_X limited'. Conversely, at a given initial level of VOC, the O_3 increases linearly with the initially present level of NO_X. At higher initial NO_X levels, the amount of O_3 levels off. Increasing initial NO_X even further in this 'VOC limited' or 'NO_X saturated' regime does not lead to higher O_3 production. In fact, due to removal of OH by reaction with NO_2 (Fig. 2.5), O_3 formation will be reduced. Figure 2.6 illustrates this relationship. The lines of constant ozone formation as a function of the VOC and NO_X composition of the initial mixture (or VOC and NO_X

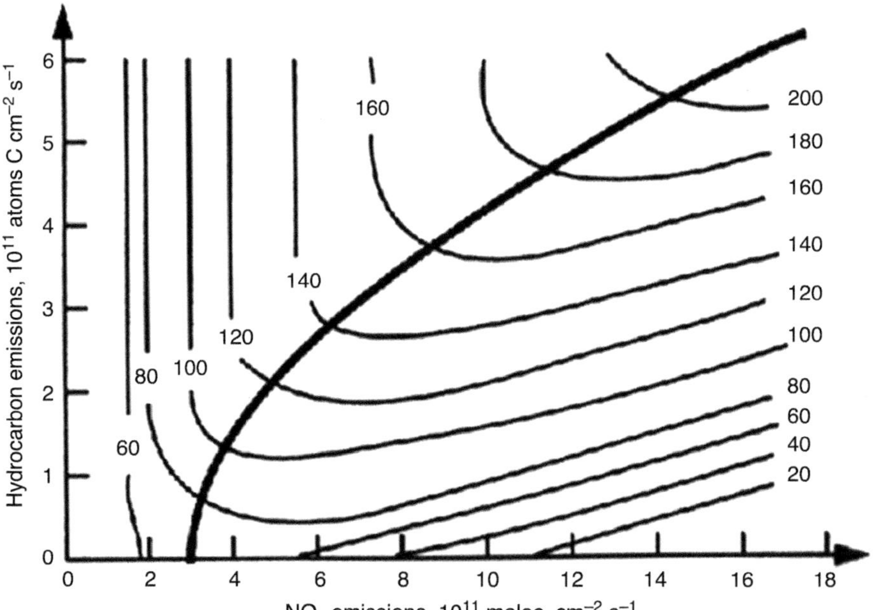

Fig. 2.6 Ozone concentrations simulated by a regional photochemical model as a function of NO_X and hydrocarbon (VOC) emissions. The thick line separates the NO_X limited regime (top left) from the VOC limited regime (bottom right). Reprinted with permission from Sillman et al. 1990, ©American Geophysical Union. All rights reserved

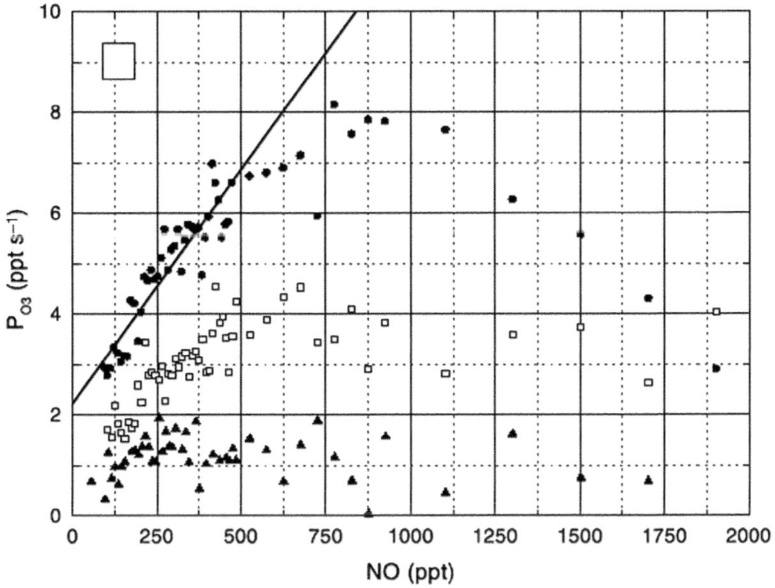

Fig. 2.7 Ozone production rates (P_{O3}) calculated from simultaneous observations of NO, NO_2, O_3, OH, HO_2, H_2CO, actinic flux, and temperature during the 1999 'Southern Oxidant Study' (June 15 to July 15) at Cornelia Fort Airpark, Nashville, Tennessee. Averaged P_{O3} is plotted as a function of the NO mixing ratio. The data were placed into three P_{HOx} bins: high ($0.5 < P_{HOx} < 0.7$ ppt s^{-1}, circles), moderate ($0.2 < P_{HOx} < 0.3$ ppt s^{-1}, squares), and low ($0.03 < P_{HOx} < 0.07$ ppt s^{-1}, triangles), and then averaged as a function of NO. All three P_{HOx} regimes demonstrate the expected generic dependence on NO. P_{O3} increases linearly with NO for low NO (<600 ppt NO), and then P_{O3} becomes independent of NO for high NO (>600 ppt NO). The crossover point between NO_X-limited and NO_X-saturated O_3 production occurs at different levels of NO in the three P_{HOx} regimes. Reprinted with permission from Thornton et al. 2002, ©American Geophysical Union. All rights reserved

emission) are called ozone isopleths. Measurements in the open atmosphere also illustrate this dependence of the ozone production on the NO_X level (an example is shown in Fig. 2.7).

When considering which 'mixture' of HC/VOC and NO_X is most effective in ozone production (at a given level of solar UV), NO_X and VOC must be present in a certain ratio $R_{X0} = [NO_X]/[VOC]$, indicated by the ridge line in Fig. 2.6. Often this maximum is not reached in urban centres, where the primary emissions occur. Because the atmospheric lifetimes of VOC and NO_X are different, R_{X0} for an airmass changes with time. In addition, it takes several hours for high ozone levels to be formed. Consequently, the highest O_3 levels are often found downwind of urban centres. This is illustrated in Fig. 2.8 which shows the diurnal variation of ozone levels in different types of air masses at a rural site and in a city.

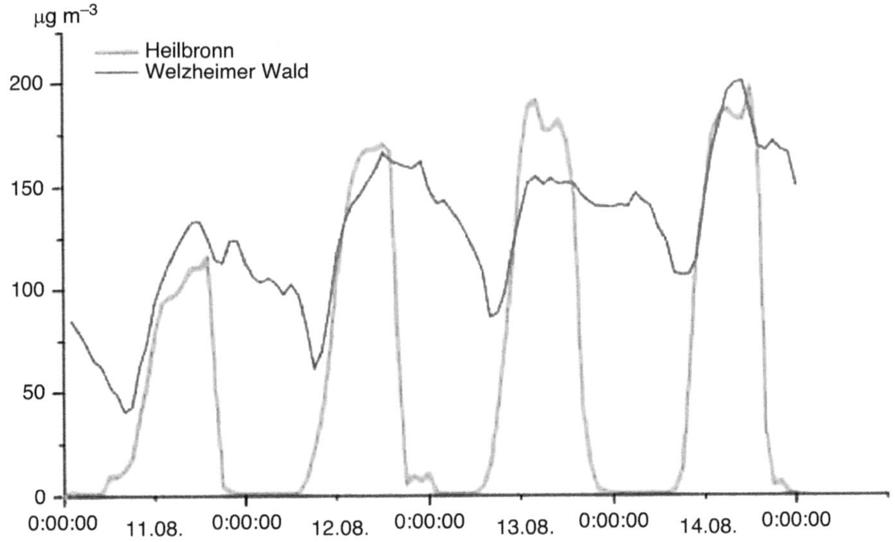

Fig. 2.8 Diurnal variation of ozone levels in different types of air masses during August 11–14, 2000: Forest (Weltzheimer Wald, Germany) and city of Heilbronn (southern Germany). *Source* Landesanstalt für Umweltschutz Baden-Württemberg and UMEG

2.4 Radical Processes in the Atmosphere

Free radicals are the driving force for most chemical processes in the atmosphere. Since the pioneering work of Weinstock (1969) and Levy (1971), photochemically generated HO_X radicals (hydrogen radicals, $OH + HO_2$) have been recognised to play a key role in tropospheric chemistry. In particular, hydrogen radicals:

- initiate the degradation and removal of most oxidisable trace gases emitted into the atmosphere
- give rise to the formation of strongly oxidising agents (mostly in the troposphere), such as ozone or hydrogen peroxide
- catalytically destroy stratospheric ozone (see Sect. 2.10)
- are difficult to remove once they are generated, since radical-molecule reactions tend to regenerate radicals.

Today we have an enormous amount of direct and indirect evidence of the presence of HO_X radicals (see, for example, Ehhalt et al. 1991; Wennberg et al. 1998; Platt et al. 2002 ; Rohrer and Berresheim 2006; Kanaya et al. 2012; Griffith et al. 2016), and the importance of HO_X for atmospheric chemistry is well established. Nevertheless, the possible role of other radicals, beginning with the (historical) idea of the impact of oxygen atoms $O(^3P)$ or excited oxygen molecules $O_2(^1\Delta)$, has been the topic of past and current investigations. The nitrate radical, NO_3, (see Sect. 2.5.2 and e.g. Ng et al. 2017), halogen atoms, and halogen oxide radicals, BrO, IO, and ClO (Sect. 2.8), can make a considerable contribution to the oxidising capacity of the troposphere. For instance, reaction with NO_3 or BrO can be an important sink of DMS in marine environments. In addition, nighttime reactions of nitrate radicals with organic species and NO_X play an important role for the removal of these species. NO_3 chemistry can also be a source of peroxy radicals (such as HO_2 or CH_3O_2), and even OH radicals (Sect. 2.5.2). Chlorine atoms have also been recognised as important oxidants in marine and urban areas, where they are formed through the photolysis of $ClNO_2$ that is produced from the heterogeneous reaction of N_2O_5 with aerosol halides (Osthoff et al. 2008; Thornton et al. 2010; Young et al. 2013, 2014). Table 2.6 shows an overview of the most important radical species in the troposphere and their significance for atmospheric chemistry. The details of the chemistry of NO_3 and reactive halogen species will be discussed in the following sections. Here, we will concentrate on the tropospheric chemistry of hydroxyl radicals.

Table 2.6 Free radical cycles pertinent to tropospheric chemistry, and key processes influenced or driven by reaction of those radicals

	Species	Significance
HO_X cycle	OH	Degradation of most volatile organic compounds (VOC)
		Key intermediate in O_3 formation
		$NO_X \Rightarrow NO_Y$ conversion
	HO_2	Intermediate in O_3 formation
		Intermediate in H_2O_2 formation
	RO_2	Intermediate in ROOR' formation
		Aldehyde precursor
		PAN precursor
		Intermediate in O_3 formation
NO_3 cycle	NO_3	Degradation of certain VOC (olefins, aromatics, DMS, etc.)
		$NO_X \Rightarrow NO_Y$ conversion (via N_2O_5 or DMS-reactions)
		RO_2 precursor (nighttime radical formation)
XO_X cycle	XO (X = Cl, Br, I)	Catalytic O_3 destruction (cause of 'Polar Trop. Ozone Hole')
		Degradation of DMS (BrO)
		Change of the NO_2/NO (Leighton) ratio
	X	Degradation of (most) VOC (Cl)
		Initiates O_3 formation
		RO_2 precursor
		Initiates particle formation (IO_X)

2.4.1 Sources of Hydrogen Radicals (OH and HO_2)

Hydroxyl radicals are probably the most important free radicals in the atmosphere. The degradation of most (but not all) oxidisable trace gases (such as HC or CO) is initiated by OH reaction (cf. Platt 1999). OH radicals are therefore sometimes called the 'cleansing agent' of the atmosphere. Figure 2.9 shows a simplified outline of the HO_X ($OH + HO_2$) and RO_2 cycle.

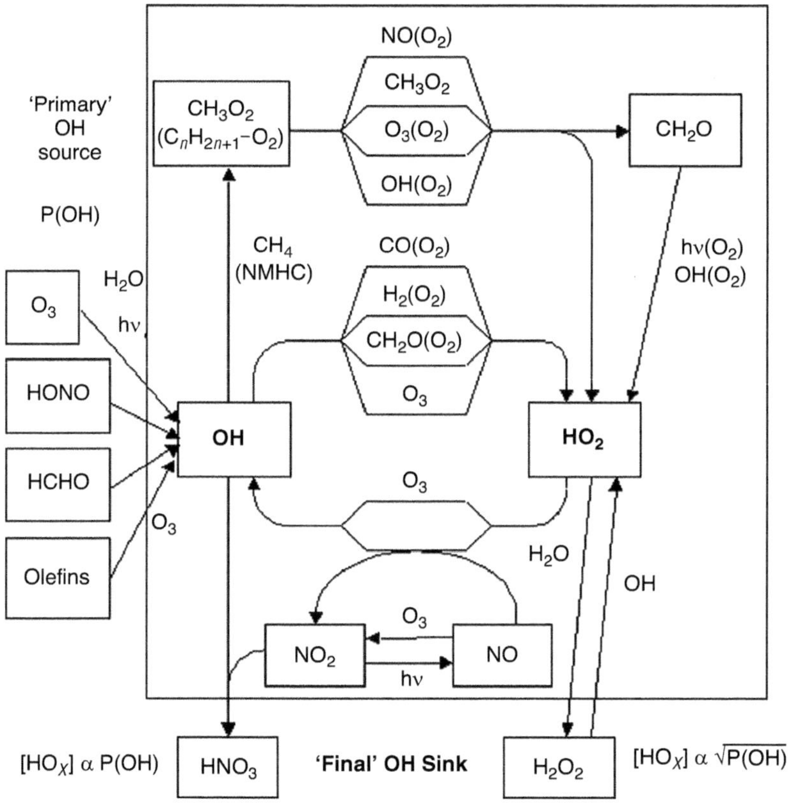

Fig. 2.9 A simplified outline of the $HO_X(OH + HO_2)$ and RO_2 cycles in the troposphere. The very reactive OH attacks most oxidiseable species (e.g. CO, hydrocarbons). Typical OH lifetimes are therefore below 1 s. Consequently, noontime concentrations only reach 10^6–10^7 cm^{-3} (0.04–0.4 ppt). These reactions usually lead to the formation of HO_2 or RO_2, which, in turn, can be reconverted to OH by reaction with NO or O_3, thus preserving HO_X. Alternatively HO_X is destroyed by self-reaction of HO_2, producing H_2O_2. The only other 'final' HO_X sink of importance is the $OH + NO_2$ recombination. In a very simplified picture, one can imagine the HO_X reservoir to be fed by Eq. R2.4 followed by Eq. R2.6 (rate of OH production = P_{OH}) and drained by either HO_2 self-reaction (if NO_X is low) or $OH + NO_2$ recombination (at high NO_X). All other reactions (inside the box) only interconvert OH and HO_2 (partly via RO_2). Note that $[HO_X] \propto \sqrt{P_{OH}}$ at low NO_X and $[HO_X] \propto P_{OH}$ at high NO_X (i.e. $NO_X > 1$ ppb)

Several primary OH production mechanisms are known:

(a) Globally, the most important process forming OH is initiated by UV photolysis of ozone to form electronically excited oxygen atoms:

$$O_3 + h\nu(\lambda < 320\,\text{nm}) \rightarrow O(^1D) + O_2(^1\Delta) \quad \text{(R2.4a)}$$

Recent research shows that photolysis of vibrationally excited ozone can considerably enhance the rate of reaction (R2.4a). In addition, some contribution comes from the channel:

$$O_3 + h\nu(\lambda < 420\,\text{nm}) \rightarrow O(^1D) + O_2(^3\Sigma) \quad \text{(R2.4b)}$$

(see e.g. Talukdar et al. 1998). The excited oxygen atoms $O(^1D)$ from (R2.4) can be deactivated by collision with N_2, O_2, or H_2O (i.e. any molecule 'M'):

$$O(^1D) + M \rightarrow O(^3P) + M \quad \text{(R2.5)}$$

A certain fraction X of the excited oxygen atoms, however, will react according to:

$$O(^1D) + H_2O \rightarrow OH + OH \quad \text{(R2.6)}$$

Assuming $k_{2.5a}$ and $k_{2.5b}$ denote the rate constants of the reaction of $O(^1D)$ with N_2 or O_2, respectively, and $k_{2.6}$ the rate constant of reaction (R2.6), we obtain for X:

$$X = \frac{k_{2.6} \cdot [H_2O]}{k_{2.5a} \cdot [N_2] + k_{2.5b} \cdot [O_2] + k_{2.6} \cdot [H_2O]} \quad (2.24)$$

In the lower troposphere, X is typically in the range of 0.05–0.1. At very high relative humidity, up to 0.2 can be reached. The production rate of OH radicals, $P(OH)$, is accordingly calculated from the O_3-concentration and the photolysis frequency [see (2.26) below] for the above process (R2.4a):

$$\frac{d}{dt}\left[O(^1D)\right] = [O_3] \cdot J_1 \quad (2.25)$$

$$P(OH) = \frac{d}{dt}[OH] = 2 \cdot X \cdot [O_3] \cdot J_1 \quad (2.26)$$

(b) Another important source of OH is the reaction of HO_2 radicals with NO:

$$HO_2 + NO \rightarrow OH + NO_2 \quad \text{(R2.7)}$$

An analogous reaction of HO_2 with ozone (R2.18) also yields OH. HO_2 radicals are largely formed by OH reactions. The consequences of this 'radical recycling' will be analysed below. In addition, the photolysis of aldehydes forms HO_2. This process is particularly significant in urban areas, where the oxidation of VOC and direct emissions lead to high aldehyde levels. A number of non-photochemical mechanisms also produce HO_2 radicals. Among these, reactions of NO_3 radicals with certain organic species, such as olefins, phenols, or DMS, are particularly important (e.g. Platt et al. 1990, 2002; Geyer et al. 2003a; Geyer and Stutz 2004, see Sect. 2.5.2).

(c) A further OH source is the photolysis of HONO (nitrous acid, see Sect. 2.5.3, reaction R2.31), which is of particular importance in polluted air. HONO is formed by heterogeneous processes, i.e. chemical reactions on surfaces, such as building walls or aerosol particles (see, e.g. Platt 1986; Alicke et al. 2002, 2003; Zhou et al. 2001, 2002a; Kleffmann 2007; Tuite et al. 2021), as well as biologically in soil (Su et al. 2011; Oswald et al. 2013, Sect. 2.5).

(d) The ozonolysis of olefins also produces OH radicals (Atkinson et al. 1992; Atkinson and Aschmann 1993; Paulson et al. 1997, 1999):

$$\text{Olefin} + O_3 \rightarrow OH + \text{products} \tag{R2.8}$$

An ozone molecule adds across the C = C double bond to form a primary ozonide, which decomposes to form vibrationally excited carbonyl oxide and carbonyl products, see e.g. (Finlayson-Pitts and Pitts 2000; Paulson et al. 1999). The OH production yields of these reactions vary from 7 to 100%, depending on the structure and size of the alkene (Paulson et al. 1997, 1999). Note that this source (though usually weak when compared with others) is independent of light, and thus also provides radicals at night.

(e) Finally, photolysis of hydrogen peroxide produces OH radicals:

$$H_2O_2 + h\nu \rightarrow OH + OH \tag{R2.9}$$

In contrast to processes (b) and (c), the latter source would only be relevant in air masses with very low NO_X pollution, since only under these conditions can significant amounts of H_2O_2 be formed.

2.4.2 Temporal Variation of the HO_X Source Strength

The photolysis frequency $J_{2.4}$, and thus the strength of the main OH source [P(OH)], varies with the time of day, cloudiness, and seasons. Consequently, the OH source strength is highly variable in the atmosphere. In addition, other environmental variables, such as the water vapour concentration, influence the formation of OH radicals. Figure 2.10, for example, shows the diurnal variation of OH concentrations at a mid-latitude, rural, site. The OH levels follow the diurnal behaviour of the O_3 photolysis rate closely. The seasonal dependence of the OH source strength at mid-latitudes is shown in Fig. 2.11.

In polluted air, the diurnal variation of the ozone concentration (with a maximum in the early afternoon) has an additional influence on OH. Since the rate of R2.4 is proportional to the ozone concentration, P(OH) should also be proportional to $[O_3]$.

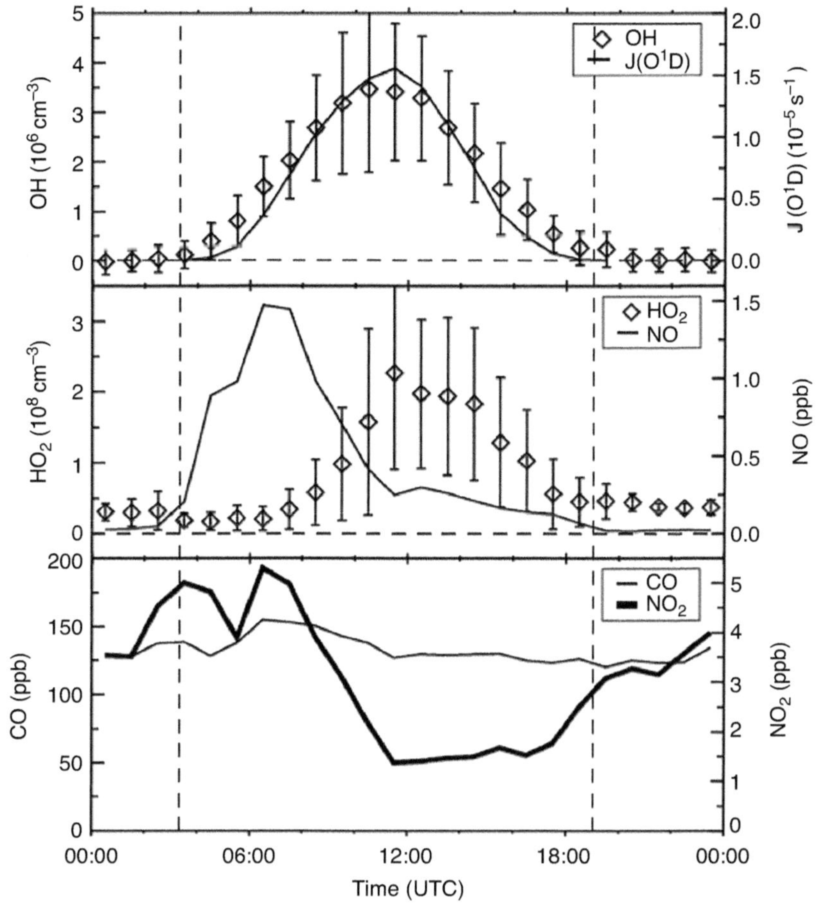

Fig. 2.10 Average diurnal OH, HO_2, and $J(O^1D)$ levels as observed during the BERLIOZ field experiment near Berlin in summer 1998. The OH concentration follows the solar radiation closely, while the levels of HO_2 are also influenced by the concentrations of NO. Reprinted with permission from Holland et al. 2003, ©American Geophysical Union. All rights reserved

However, a higher O_3 level will shift the NO/NO_2 stationary state towards higher NO_2 (see Sect. 2.5.1), and thus will reduce the OH concentration.

2.4.3 Chemistry of Hydrogen Radicals (OH and HO_2)

Hydroxyl radicals react with most oxidisable gases in the atmosphere, for instance with carbon monoxide:

$$OH + CO \rightarrow CO_2 + H \qquad (R2.10)$$

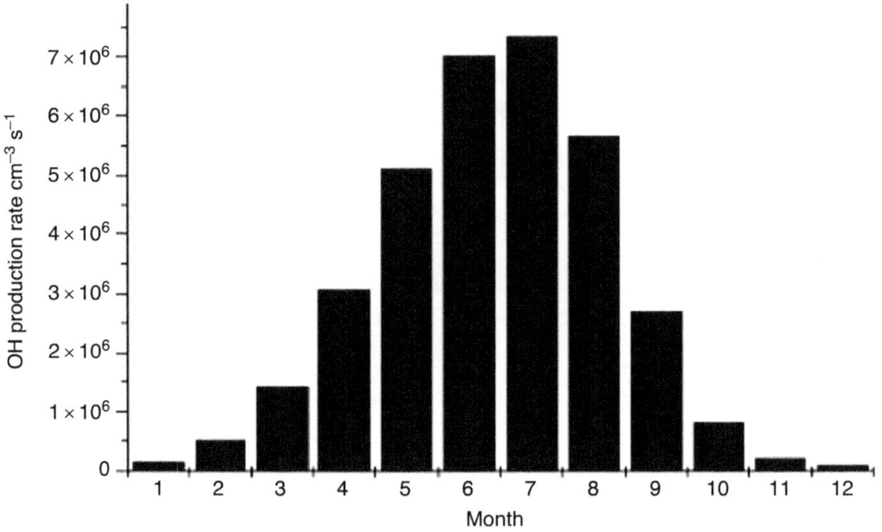

Fig. 2.11 The seasonal variation of the average diurnal OH source strength under cloud-free conditions in mid-latitudes (Germany)

In this reaction, both CO_2 and hydrogen atoms are produced. In the atmosphere, the hydrogen atoms immediately form hydroperoxyl radicals, HO_2:

$$H + O_2 \xrightarrow{M} HO_2 \tag{R2.11}$$

As products of the degradation of VOCs, peroxy radicals are also produced. An example is methane degradation:

$$OH + CH_4 \rightarrow CH_3 + H_2O \tag{R2.12}$$

$$CH_3 + O_2 \xrightarrow{M} CH_3O_2 \tag{R2.13}$$

Further important reactions of OH are the oxidation of molecular hydrogen and formaldehyde (and higher aldehydes):

$$OH + H_2 \rightarrow H_2O + H \tag{R2.14}$$

$$OH + CH_2O \rightarrow CHO + H_2O \tag{R2.15}$$

$$CHO + O_2 \rightarrow HO_2 + CO \tag{R2.16a}$$

In these reactions, the primary product is also HO_2 (R2.14 followed by R2.11, or R2.15 followed by R2.16a, respectively)

Thus, in all of the above reactions, and in fact most OH reactions, free radicals (here HO_2 radicals, which are easily converted to OH) are regenerated. An important exception to this rule is the reaction of OH with oxides of nitrogen:

$$OH + NO_2 \xrightarrow{M} HNO_3 \tag{R2.17}$$

The nitric acid produced in this reaction will (in the troposphere) be removed from the gas phase by various processes, such as dry deposition to the ground or rainout. Figure 2.9 shows the simplified reaction scheme of the hydrogen radicals HO_X ($OH + HO_2$), as outlined above.

The balance between OH sources and sinks determines the levels of OH and HO_2 and their respective lifetimes. Figure 2.12 shows the OH and HO_2 budget as a function of the NO_X concentration. The OH concentration shows a non-monotonous behaviour as a function of the NO_X level: At very low NO_X (below e.g. 0.1 ppb) essentially each OH radical reacting with CO (or HC) is converted to HO_2 (or an organic peroxy radical, RO_2). These radicals are, in turn, lost by radical–radical interaction. Thus, the OH concentration remains relatively low. At medium NO_X levels (about 1–2 ppb) most HO_2 and RO_2 radicals react with NO to re-form OH. Therefore, OH levels are much higher than in the case of low NO_X. At very high NO_X levels (10 ppb or higher) most of the OH radicals react with NO_2 to form HNO_3, and therefore OH levels are low again.

As a consequence, OH levels show a maximum at around 1 ppb of NO_X, while HO_2 levels drop monotonously with increasing NO_X. This behaviour is illustrated in Fig. 2.13.

Although its concentration (in clean air) can be up to two orders of magnitude larger than that of OH (see Fig. 2.13), the reactivity of the HO_2 radical is much lower than that of the OH radical. However, the HO_2 radical is an important OH reservoir, and it is also the precursor for atmospheric H_2O_2. In addition, there may be a direct role of HO_2 in liquid-phase chemistry, for instance in cloud droplets (Chameides and Davis 1982; Chameides 1984; Lelieveld and Crutzen 1990, 1991; Sommariva et al. 2006; Tan et al. 2020; Kohno et al. 2022).

Only four HO_2 reactions play a role in the troposphere. Their significance depends on the levels of NO_x:

(1)

$$HO_2 + NO \rightarrow OH + NO_2 \tag{R2.7}$$

(2)

$$HO_2 + O_3 \rightarrow OH + 2O_2 \tag{R2.18}$$

(3)

$$HO_2 + HO_2 \rightarrow H_2O_2 + O_2 \tag{R2.19}$$

(4)

$$HO_2 + OH \rightarrow H_2O + O_2 \tag{R2.20}$$

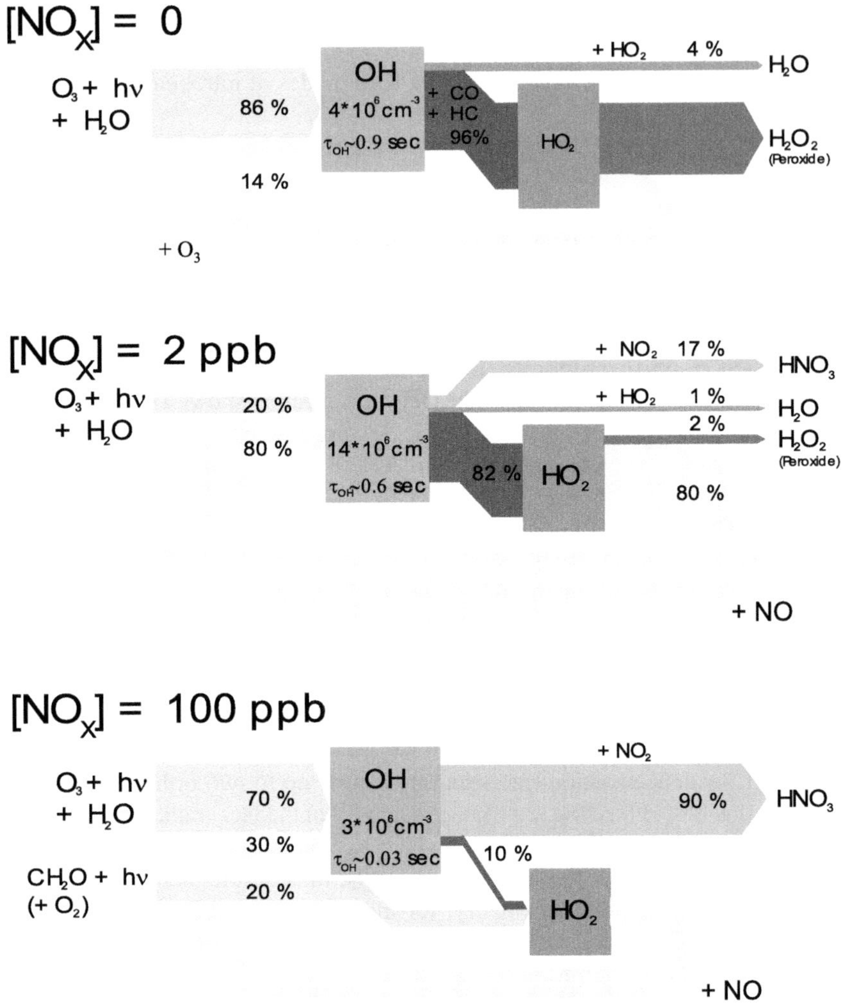

Fig. 2.12 The relative strength of OH sources and sinks, and the resulting OH concentration as a function of the NO_X ($NO + NO_2$) level. Here, three cases are shown: (**a**) At very low NO_X, essentially each OH radical reacting with CO (or hydrocarbons) is converted to HO_2 (or an organic peroxy radical, RO_2). These radicals are, in turn, lost by radical–radical interaction, and thus the OH concentration remains relatively low. (**b**) At medium NO_X levels (about 1 ppb), most HO_2 and RO_2 radicals react with NO to reform OH. Therefore OH levels are much higher than in case (**a**). (**c**) At very high NO_X levels, most of the OH radicals react with NO_2 to form HNO_3, and therefore OH levels are low again

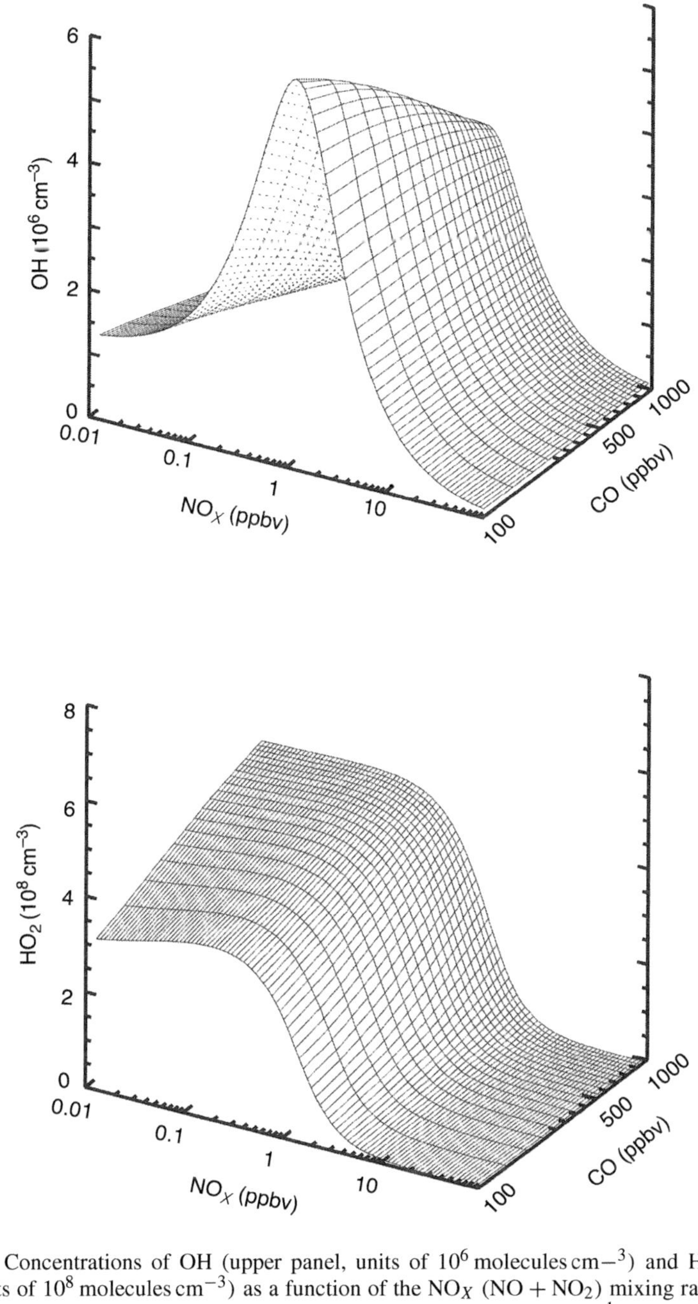

Fig. 2.13 Concentrations of OH (upper panel, units of 10^6 molecules cm$-^3$) and HO_2 (lower panel, units of 10^8 molecules cm^{-3}) as a function of the NO_X ($NO + NO_2$) mixing ratio for typical conditions (37.6 ppb ozone, 88.7 ppb CO, 0.92 ppb CH_2O, $J(O_3 \rightarrow O^1D) = 3 \times 10^{-5}\,s^{-1}$, $J(NO_2) = 9.1 \times 10^{-3}\,s^{-1}$). The HO_2 concentration, and therefore H_2O_2 production rate, are highest in unpolluted air at NO_X levels below a few 100 ppt. Reprinted with permission from Ehhalt 1999, ©Springer. All rights reserved

The first two reactions reduce HO_2 radicals to OH. The third and forth reactions can be regarded as 'final sinks' for the HO_2 radical (and thus for HO_X). While hydrogen peroxide can be photolysed to yield two OH radicals (R2.9), this is a slow process. Thus rainout and washout of H_2O_2 is much more likely, since it is highly water soluble. A further sink of H_2O_2 is reaction with OH:

$$H_2O_2 + OH \rightarrow H_2O + HO_2 \tag{R2.21}$$

Hydrogen peroxide is an important oxidant in the liquid phase, e.g. in raindrops, or cloud or haze droplets. Reactions of H_2O_2 in cloud or haze droplets contribute significantly to the oxidation of SO_2 (S(IV)) to sulfate (S(VI)). Hydrogen peroxide is essentially only formed via reaction R2.19 (i.e. HO_2 recombination). A small contribution is assumed to come from the reaction of olefins with ozone (Becker et al. 1990). Assuming reaction R2.19 is the dominant source, high rates of H_2O_2 production (and thus usually also high levels) are to be expected under conditions of high HO_2. Thus, the highest H_2O_2 levels are to be expected in air masses with high humidity and ozone, and low NO.

2.5 Oxides of Nitrogen in the Atmosphere

Oxides of nitrogen are key species in atmospheric chemistry. They regulate many trace gas cycles and influence the degradation of most pollutants in clean air, as well as in polluted regions.

NO_X concentration has a strong influence on the atmospheric level of hydroxyl radicals (Sect. 2.4.1), which, in turn, are responsible for the oxidation processes of most trace gases. In addition, NO_X is a catalyst for tropospheric ozone production (Sect. 2.3). NO_X (or acids formed from them) can also react with VOC degradation products to form organic nitrates or nitrites (e.g. peroxy acetyl nitrate (PAN) or methyl nitrite), as well as nitrosamines (Platt et al. 1980b). These species can be much more detrimental to human health than the primary oxides of nitrogen. Finally, the thermodynamically most stable degradation product of all atmospheric oxides of nitrogen, nitric acid, is (besides sulfuric acid) the main acidic component in 'acid rain'.

An overview of the most important oxidised nitrogen species in the atmosphere is given in Fig. 2.14. The main reaction pathways between the various species are indicated by arrows. Oxides of nitrogen are primarily emitted in the form of NO (plus some NO_2) and N_2O. While N_2O is a very inert species and therefore plays no role for the chemical processes in the troposphere, NO reacts rapidly with natural ozone to form NO_2. On the other hand, NO_2 will be destroyed (photolysed) by sunlight during the day, and therefore a stationary state between these two NO_X species will be established. NO_2 then further reacts with OH radicals or NO_3, ultimately forming nitric acid (or aerosol nitrate).

The 'classic' role of oxides of nitrogen in atmospheric chemistry (Sect. 2.5.1) has been known since the 1980s. Since then, the importance of this class of species has been demonstrated by various observations:

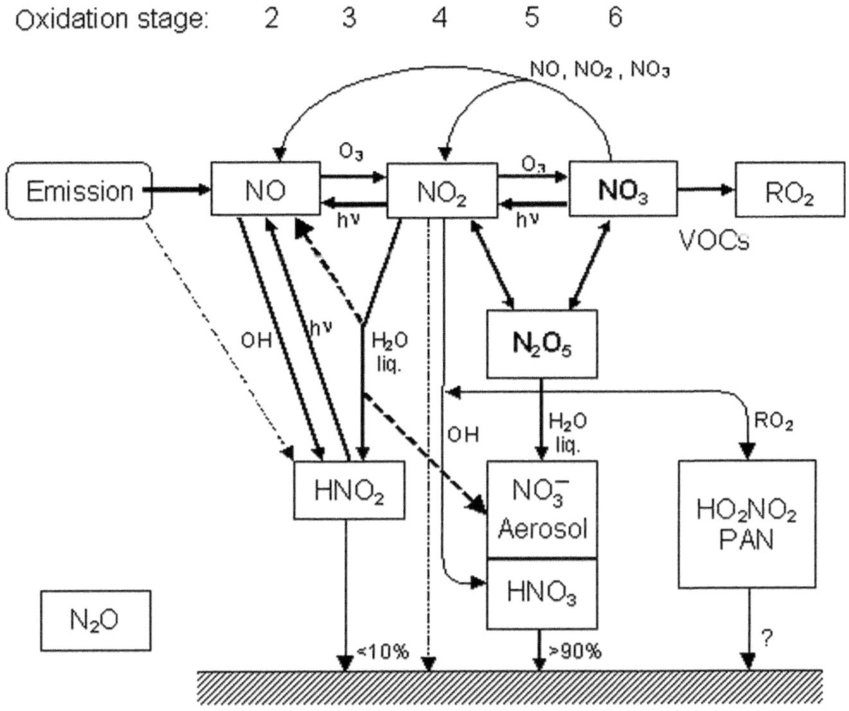

Fig. 2.14 Simplified overview of the NO_X reaction scheme in the atmosphere. Arrows indicate main reaction pathways

- The confirmation of an additional NO_X degradation pathway via heterogeneous and homogeneous hydrolysis of N_2O_5 at night (Heintz et al. 1996; Mentel et al. 1996; Wahner et al. 1998a; Geyer et al. 2001c; Brown et al. 2003; Stutz et al. 2004b; Brown and Stutz 2012; Ng et al. 2017)
- The non-photochemical production of peroxy radicals (HO_2 and RO_2) via NO_3–VOC reactions as a further consequence of elevated NO_X levels in the atmosphere (Platt et al. 1990; LeBras et al. 1993; Mihelcic et al. 1993; Geyer et al. 1999; Geyer and Stutz 2004; Brown and Stutz 2012; Ng et al. 2017)
- The establishment of HONO photolysis as another significant OH source in polluted air (Perner and Platt 1979; Harris et al. 1982; Platt 1986; Harrison et al. 1996; Alicke et al. 2002, 2003; Ren et al. 2006, 2013; Griffith et al. 2016).
- The observations of elevated concentrations of $ClNO_2$ showed that heterogeneous reactions of N_2O_5 releasing this chlorine atom precursor are active throughout the atmosphere. The chlorine atoms released from $ClNO_2$ photolysis have a noticeable impact on radical chemistry and ozone formation (Osthoff et al. 2008; Thornton et al. 2010; Young et al. 2013, 2014).

A detailed knowledge of the cycle of oxidised nitrogen species has great practical importance, as NO_X chemistry participates in the formation of ozone and particles. It thus impacts human health directly or through these secondary pollutants.

2.5.1 Classical Chemistry of Oxides of Nitrogen in the Atmosphere

In this section, we briefly summarise the central aspects of the classical chemistry of oxides of nitrogen in the atmosphere (see also Fig. 2.14 and Table 2.3). We focus here primarily on NO and NO_2. The chemistry of the NO_3 radical and of nitrous acid will be discussed separately in Sects. 2.5.2 and 2.5.3.

The transformation pathways between the atmospheric nitrogen reservoir are closely coupled to the reactions of ozone (as discussed in Sect. 2.3) and to the cycles of hydrogen radicals (OH and HO_2). An important question in this context is: Are NO_X reactions sources or sinks for ozone? The simple system described by reactions (2.1)–(2.3) (Sect. 2.3) neither destroys nor produces ozone. However, in the atmosphere there are additional reactions oxidising NO to NO_2 without consuming ozone. Thus, the NO_X level ultimately determines whether there is net ozone formation or destruction. In the unpolluted (background) atmosphere, the processes oxidising NO to NO_2 without consuming O_3 (and thus producing O_3 via reactions (2.1) and (2.2)) are reactions of NO with peroxy radicals (HO_2, CH_3O_2, and higher RO_2). In the simple case of HO_2, we have:

$$NO + HO_2 \rightarrow NO_2 + OH \tag{R2.7}$$

The HO_2 radicals are produced, for example, by reaction of OH radicals with CO ((R2.10), followed by (R2.11)). At the same time, the reactions (R2.7), (R2.10), and (R2.11) facilitate another stationary state in the atmospheric concentration ratio of the hydrogen radicals OH and HO_2, (HO_X). Since ozone also reacts with HO_2 radicals (but not with other peroxy radicals), the HO_X system also destroys ozone:

$$O_3 + HO_2 \rightarrow OH + 2O_2 \tag{R2.18}$$

Whether the above-described interplay of the NO_X and HO_X systems leads to net ozone formation (as shown in Fig. 2.7) or O_3 destruction depends on the relative rates of the reactions (R2.7) and (R2.18), and thus on the NO concentration. At a NO concentration around 1/3000 of the O_3 concentration (i.e. $[NO] \approx 0.01$ ppb at 30 ppb O_3), the rates of both reaction pathways are equal. In other words, ozone formation equals ozone destruction. At higher NO_X (and thus NO) levels net O_3 formation occurs, while at lower NO levels O_3 is photochemically destroyed. In regions influenced by industrial activities, where NO levels can reach many ppb, O_3 formation by far exceeds destruction. This can also be seen from the observation that ozone levels in the (more industrialised) northern hemisphere exceed those in the southern hemisphere by about 50%. Figure 2.7 shows model calculated and observed ozone formation rates as a function of the NO level.

Oxides of nitrogen influence the concentrations of OH and HO_2 radicals in the atmosphere by several mechanisms. The most important removal process for OH radicals is their recombination with NO_2. This reaction (R2.6) leads to final removal of OH from the radical chain. Consequently, the OH concentration is reduced by high levels of NO_X (and thus NO_2). On the other hand, the reaction of NO with HO_2 radicals reduces the stationary state $[HO_2]/[OH]$ ratio (reactions (R2.7), (R2.10), and (R2.11)), and consequently leads to higher OH radical concentrations. This relationship is illustrated in Figs. 2.12 and 2.13.

The crucial step initiating the eventual removal of nitrogen oxides from the atmosphere is their conversion to nitric acid (HNO_3) or aerosol nitrate, which are subsequently removed from the atmosphere by dry and wet deposition (see Table 2.3). The direct (dry) deposition of nitrogen oxides to the ground is of minor importance (<10%). Nitric acid and aerosol nitrate are formed in reactions of NO_2:

- Directly by reaction with OH followed by partitioning of HNO_3 between the gas and particle phase (R2.17).
- Indirectly by reaction of NO_2 with O_3, forming NO_3 (R2.22), which can either abstract a hydrogen atom from VOCs or further react with NO_2 to form N_2O_5, (R2.26) the anhydride of nitric acid, which in turn is taken up on particles to form aerosol nitrate (reaction R2.27)

2.5.2 Tropospheric Chemistry of Nitrate Radicals, NO_3

At night, when levels of OH radicals are low, other oxidants play an important role in the troposphere (Brown and Stutz 2012; Ng et al. 2017). Among those, the NO_3 radical plays an important role for the budget of NO_X and for the concentrations of certain organic compounds. Nitrate radicals are formed by reaction of NO_2 with ozone (Fig. 2.14)

$$NO_2 + O_3 \rightarrow NO_3 + O_2 \qquad \text{(R2.22)}$$

which constitutes the only relevant source of NO_3 in the troposphere. The rate of NO_3 production, P_{NO3}, can be calculated from the concentration of NO_2 and O_3:

$$\frac{d\,[NO_3]}{dt} = [NO_2] \cdot [O_3] \cdot k_{2.22} = P_{NO3} \qquad (2.27)$$

During the daytime, NO_3 radicals are effectively destroyed by photolysis:

$$NO_3 + h\nu \rightarrow NO + O_2 \qquad \text{(J2.23a)}$$

$$\rightarrow NO_2 + O \qquad \text{(J2.23b)}$$

(photolytic lifetime is about 5 s), and rapid reaction with NO ($k_{2.24} = 2.6 \times 10^{-11}$ $cm^3 molec.^{-1} s^{-1}$ at 293K):

$$NO_3 + NO \rightarrow NO_2 + NO_2 \qquad (R2.24)$$

This leads to low NO_3 levels during daytime (Geyer et al. 2003b).
At night, when photolysis is unimportant and NO levels are usually low, the NO_3 loss rate is much smaller and thus its concentration is larger. Loss of NO_3 due to gas-phase reaction with nitrogen dioxide or the NO_3 self-reaction is slow. Similarly, the unimolecular decomposition of NO_3 is probably of minor importance in the atmosphere (Johnston et al. 1986; Russel et al. 1986). Gas-phase reaction of NO_3 with water vapour is endothermic.

Primary nighttime destruction mechanisms are thus reaction with organic species (Table 2.7) or heterogeneous loss (i.e. reaction at the ground or at the surface of aerosol or cloud particles) of either NO_3 or N_2O_5.

2.5.2.1 NO_3 + VOC Reactions

NO_3 reacts with a variety of organic compounds in the atmosphere, such as biogenic VOC, including isoprene and terpenes, and DMS.
NO_3 reactions with alkanes and aldehydes proceeds via H abstraction (Atkinson 1990):

$$NO_3 + RH \rightarrow R + HNO_3 \qquad (R2.25)$$

The alkyl radicals produced in reaction R2.25 are transformed to peroxy radicals (RO_2) by reaction with O_2.

Reactions with alkenes (including isoprene and terpenes) proceed via NO_3 addition to a C = C double bond. Further reaction steps are O_2 addition, followed by either reaction with other peroxy radicals or NO_2. Reasonably stable end products are organic nitrates and peroxy radicals (Atkinson 1990). A consequence of these reactions is the degradation of floral scents, which—in turn—could reduce the attraction of nighttime pollinating insects to flowers (Chan et al. 2024).

Reactions of NO_3 with aromatic compounds are also important. While the reaction with benzene is quite slow (rate constant on the order of 10^{-17} $cm^3 molec.^{-1} s^{-1}$), hydroxy-substituted benzenes react very fast, with reaction rate constants on the order of 10^{-12} cm^3 $molec.^{-1}$ s^{-1} (furans and phenol) to 10^{-11} cm^3 $molec.^{-1}$ s^{-1} (cresols). Products were found to include nitro-phenolic species (e.g. Wang and Li 2021).

Finally, NO_3 radicals react rapidly with DMS (reaction rate constant 10^{-12} cm^3 $molec.^{-1} s^{-1}$). This can be significant for the budgets of both DMS and NO_3, in particular in the marine atmosphere. The primary step of this reaction is H abstraction, leading to the formation of nitric acid and peroxy radicals (Butkovskaya and LeBras 1994; Platt and LeBras 1997). The exact reaction mechanism is, however, still subject to investigations.

Table 2.7 Reaction of OH, NO_3, O_3, and Cl with VOC (at 25°C) (from Atkinson, 1994)

Species	Reaction rate constant 10^{15} cm^3 $molec.^{-1}$ s^{-1}			
	With OH	With NO_3	With Ozone	With Cl
Alkanes				
Methane	6.14	<0.001	<0.00001	100
Ethane	254	0.0014	<0.00001	59000
Propane	1120	0.017	<0.00001	137000
n-Butane	2190	0.046	<0.00001	205000
n-Pentane	4000	0.04	<0.00001	280000
Alkenes				
Ethene	8520	0.21	0.0017	99000
Propene	26300	9.4	0.013	230000
1-Butene	31400	12	0.011	220000
2-Methyl-2-Butene	68900	9300	0.423	
2,3-Dimethyl-2-Butene	110000	57000	1.16	
Alkynes				
Ethyne	5100	0.0082	<0.00001	
Aromatics				
Benzene	1230	<0.03	<0.00001	
Toluene	5960	0.068	<0.00001	
Phenol	26300	3780		
o-Cresol	42000	13700	0.00026	
m-Cresol	64000	9740	0.00019	
p-Cresol	47000	10700	0.00047	
Benzaldehyde	12900	2.6	No data	
Isoprene	101000	678	0.013	460000
Terpenes				
α—Pinene	53700	6160	0.0866	480000
β—Pinene	78900	2510	0.015	
3-Carene	88000	9100	≈0.037	
Limonene	171000	12200	≈0.2	
Sabinene	117000	10000	≈0.086	
α—Terpinene	363000	180000	8.47	
Myrcene	215000	11000	0.47	
Aldehydes				
Formaldehyde	9370	0.6	<0.00001	73000
Acetaldehyde	15800	2.7	<0.00001	79000

2.5.2.2 Heterogeneous Loss of NO_3 and N_2O_5

In the atmosphere, NO_3 and NO_2 are in a chemical equilibrium with N_2O_5:

$$NO_2 + NO_3 \overset{M}{\longleftrightarrow} N_2O_5 \tag{R2.26}$$

The equilibrium concentration of N_2O_5 is given by:

$$[N_2O_5] = K \cdot [NO_2] \cdot [NO_3] \tag{2.28}$$

Under 'moderately polluted' conditions (NO_2 mixing ratio of about 1 ppb) and temperatures around 293K, NO_3 and N_2O_5 concentrations are roughly equal (e.g. Wängberg et al. 1997). While homogeneous reactions of N_2O_5 with water vapour and other atmospheric trace gases are believed to be extremely slow (Mentel et al. 1996), heterogeneous reaction with water contained in aerosol or cloud particles will convert N_2O_5 to nitric acid:

$$N_2O_5 + H_2O \xrightarrow{het} HNO_3 \tag{R2.27}$$

Laboratory studies have determined the reactive uptake coefficient, γ, for NO_3 in the range of $1.5 \times 10^{-4} - 0.5$, depending on the composition of liquid/solid, while γ for N_2O_5 is in the range of $1.5 \times 10^{-4} - 0.05$ (Brown and Stutz 2012). Field observations of $\gamma(N_2O_5)$ are in the range of 0.001 on mixed organic and sulfate aerosol to 0.02 on sulfate rich aerosol (Brown and Stutz 2012).

2.5.2.3 Atmospheric NO_3 Lifetime

From measured concentrations of O_3, NO_2, and NO_3, the atmospheric lifetime of NO_3 (τ_{NO3}, see Sect. 2.1.3), limited by the combination of any first-order loss process, can be calculated (Heintz et al. 1996; Geyer et al. 2001a,b,c, 2002; Brown et al. 2003; Geyer and Stutz 2004; Brown and Stutz 2012; Sobanski et al. 2016):

$$NO_3 + Z \rightarrow \text{products} \tag{R2.28}$$

Assuming pseudo stationary state conditions with respect to NO_3, τ_{NO3} becomes

$$\tau_{NO3} = \frac{[NO_3]}{P_{NO3}} = \frac{[NO_3]}{[NO_2] \cdot [O_3] \cdot k_{22}} \tag{2.29}$$

Since $k_{2.22}$ is known from laboratory measurements, and the concentrations of O_3, NO_2, and NO_3 can be simultaneously measured (for instance, by the DOAS technique), τ_{NO3} can readily be calculated for various atmospheric conditions. It must be emphasised that the steady-state assumption is an approximation, which is quite good for higher atmospheric temperatures, but becomes less accurate when it is cold.

The NO_3 lifetime can also be limited by any irreversible loss of N_2O_5 (for instance $k_{2.27} > 0$), because both species are in a rapidly established thermodynamic equilibrium (R2.28, the time constant to reach equilibrium, is on the order of 1 min at ambient temperatures). Under conditions where both NO_3 and N_2O_5 are lost, the observed NO_3 lifetime τ_{NO3} is given by:

$$\tau_{NO3} = \frac{1}{\frac{1}{\tau'_{NO3}} + \frac{[NO_2]}{K} \cdot \frac{1}{\tau_{N2O5}}} \tag{2.30}$$

In the case of negligible loss of $N_2O_5 (\tau_{N2O5} \to \infty)$, $\tau'_{NO3} = \tau_{NO3}$. For conditions where direct loss of NO_3 molecules can be neglected ($\tau'_{NO3} \to \infty$), (2.30) reduces to:

$$\tau_{NO3} = \frac{K}{[NO_2]} \cdot \tau_{N2O5} \tag{2.31}$$

The simple stationary state assumption predicts an observed NO_3 lifetime τ_{NO3} inversely proportional to the NO_2 concentration in cases where only N_2O_5 is lost. Since NO_3 formation is proportional to $[NO_2]$ (R2.22), $[NO_3]$ should become independent of the NO_2 concentration under such conditions.

Formation of Hydrogen Radicals by NO_3 + VOC Reactions

An interesting consequence of the nighttime reaction of NO_3 with organic species is a considerable contribution to the HO_X budget (Platt et al. 1990; Mihelcic et al. 1993; Carslaw et al. 1997; Geyer et al. 2001a, b; Geyer and Stutz 2004).

In circumstances where NO levels are very low, peroxy radicals undergo reactions with other RO_2 radicals, O_3, or NO_3:

$$RO_2 + NO_3 \to RO + NO_2 + O_2 \tag{R2.29}$$

In urban areas, where ubiquitous NO emissions and vertical transport can lead to an active mixing of NO_3, N_2O_5, and NO, the likely fate of RO_2 radicals is reaction with NO (Geyer and Stutz 2004):

$$RO_2 + NO \to RO + NO_2 \tag{R2.30}$$

This gives rise to a radical chemistry which is similar to that during the day. However, the region in which NO_3 and NO can coexist is often shallow and it moves vertically throughout the night depending on the NO emission strength and vertical stability (Geyer and Stutz 2004).

2.5.3 Nitrous Acid, HONO in the Atmosphere

Nitrous acid, HONO is an important oxidised nitrogen compound, the source of which is still somewhat enigmatic. The photolysis of HONO is an important source for OH radicals in the polluted urban atmosphere. In addition, HONO is toxic (Beckett et al. 1995) and its chemistry leads to the formation of carcinogenic nitrosamines, (Shapley 1976; Famy and Famy 1976; Pitts 1983). HONO formation at chamber walls is a candidate for an intrinsic 'chamber dependent' radical formation, which is observed in most smog chamber experiments (e.g. Gleason and Dunker 1989; Rohrer et al. 2004). The photolysis of HONO leads to a formation of an OH radical and NO (e.g. Stockwell and Calvert 1978; Stutz et al. 2000):

$$HONO + h\nu \rightarrow OH + NO \quad (300\,nm < \lambda < 405\,nm) \qquad (R2.31)$$

The photolysis of HONO begins early in the morning, when HONO concentrations are typically highest due to accumulation in the absence of photolysis over night. The combination of elevated HONO concentrations at sunrise and fast photolysis results in a peak in the production of OH which surpasses other HO_X sources, such as O_3 and HCHO photolysis (Alicke et al. 2002, 2003). In many environments HONO photolysis also plays an important role as a primary OH source throughout the day (Kleffmann 2007; Ren et al. 2006; Griffith et al. 2016).

Nitrous acid was first identified and quantified in the atmosphere by DOAS (Perner and Platt 1979; Platt and Perner 1980). It has since been found in polluted regions (e.g. Perner and Platt 1979; Platt et al. 1980b; Kessler et al. 1981; Harris et al. 1982; Alicke et al. 2002, 2003; Zhou et al. 2002a; Aumont et al. 2003; Ren et al. 2003; Kleffmann et al. 2005; Wong et al. 2011, 2012; Pinto et al. 2014; Wu et al. 2014; Ryan et al. 2018; Theys et al. 2020; Wang et al. 2020; Tang et al. 2020; Jordan and Osthoff 2020), in background air (e.g. Tang et al. 2019; Wang et al. 2019b, 2020; Andersen et al. 2023), and in the polar BL (Li 1994; Zhou et al. 2001). Typical HONO mixing ratios are in the range of 0 – 15 ppb (Calvert et al. 1994; Lammel and Cape 1996; Alicke 2000; or Akimoto and Tanimoto 2021). Generally, nitrous acid concentrations scale with the NO_2 concentration, and therefore with the degree of pollution (Nakashima et al. 2017; Jordan and Osthoff 2020). The diurnal variation of the mixing ratio of HONO is dominated by its photolysis, which leads to lower levels during the day. During the night concentrations increase, often reaching a constant value in the early morning. The amount of HONO formed is often correlated with the stability of the nocturnal boundary layer, as inferred for example from radon data (Febo et al. 1996; Wong et al. 2011, 2012; Tuite et al. 2021).

Several studies have discussed the importance of HONO photolysis as an OH source (e.g. Platt et al. 1980b; Harrison et al. 1996; Stutz et al. 2002b; Alicke et al. 2002, 2003). Alicke et al. (2002) showed OH production rates of up to 3×10^7 molec. $cm^{-3}\,s^{-1}$ in Milan, Italy. Many other observations confirm these results (e.g. Harris et al. 1982; Sjödin 1988; Ryan et al. 2018; Garcia-Nieto et al. 2018; Griffith et al. 2016). Model studies (e.g. Harris et al. 1982) show that HONO photolyis can lead to an increase in the maximum daytime ozone concentration of up to 55%, if 4% of the initial NO_X is present as HONO (Los Angeles case, initial NO_x: 0.24 ppm), compared with a calculation without nitrous acid. Jenkin et al. (1988) showed HONO photolysis can account for a fivefold increase in OH at 6:00 AM, a 14% increase in OH present at the daily maximum (noon), and a 16% increase in net photochemical ozone production. In addition, the formation of ozone begins earlier in the day due to the HONO photolysis in the morning.

HONO is formed through various chemical mechanisms in the atmosphere. The only important gas-phase reaction forming nitrous acid is reaction 2.32 (Stuhl and Niki 1972; Pagsberg et al. 1997; Zabarnick 1993; Nguyen et al. 1998):

$$OH + NO(+M) \rightarrow HONO(+M) \tag{R2.32}$$

Gas-phase nitrous acid producing pathways, such as the reaction of NO_2 with HO_2, have been reported to be of minor importance (e.g. Howard 1977; Tyndall et al. 1995). The bulk of urban HONO is suspected to be formed from either of the mechanisms summarised in reaction (R2.33) or (R2.34), involving only NO_X and water vapour (Perner and Platt 1979; Kessler and Platt 1984; Platt 1986).

$$NO_2 + NO + H_2O \Leftrightarrow 2\,HONO \tag{R2.33}$$

$$2NO_2 + H_2O \Leftrightarrow HONO + HNO_3 \tag{R2.34}$$

These reactions (and their reverse reactions) appear to proceed heterogeneously on surfaces (Kessler 1984; Kessler et al. 1981; Sakamaki et al. 1983; Lammel and Perner 1988; Notholt et al. 1991, 1992; Junkermann and Ibusuki 1992; Ammann et al. 1998; Calvert et al. 1994; Longfellow et al. 1998; Goodman et al. 1999; Kalberer et al. 1999; Veitel 2002; Veitel et al. 2002). A number of laboratory studies have shown reaction (R2.33) to be insignificant (Sakamaki et al. 1983; Pitts et al. 1984b; Jenkin et al. 1988; Svensson et al. 1987; Kleffmann et al. 1998). Several field observations, where the presence of high ozone at night excluded NO, or low NO was documented, confirm this result (e.g. Kessler and Platt 1984; Harrison and Kitto 1994).

The exact mechanism of the heterogeneous formation of HONO summarised in reaction (R2.34) is unknown. Several studies have shown that it is first order in NO_2 and water (Svensson et al. 1987; Jenkin et al. 1988; Kleffmann et al. 1998; Finlayson-Pitts et al. 2003). Neither the reaction rate constants nor the nature of the surface is known, which makes calculation of the OH production by models difficult. In addition, reactions on 'urban surfaces' like organic aerosols or soot aerosols have been suggested (e.g. Gutzwiller et al. 2002; Bröske et al. 2003). In fact, reactions of NO_2 on asphalt or roof tile surfaces appear to be sufficiently rapid to explain observations (Trick 2004; Wong et al. 2011; Tuite et al. 2021).

Several photolytic mechanisms have been proposed, including the photolytic reduction of surface-adsorbed HNO_3/NO_3^- to HONO, the photolysis of aerosol nitrate (Ye et al. 2016; Andersen et al. 2023), the photoenhanced conversion of NO_2 on urban surfaces, HONO formation in soils, etc. (Tsai et al. 2018; Akimoto and Tanimoto 2021; Tuite et al. 2021). The most important gas-phase chemical removal process for HONO, besides photoloysis, is reaction R2.35. A few percent of HONO is expected to be destroyed by OH radicals:

$$OH + HONO \rightarrow H_2O + NO_2 \tag{R2.35}$$

Some studies reported determinations of the strength of direct emission of nitrous acid from combustion processes (Kessler and Platt 1984; Pitts et al. 1984a; Kirchstetter et al. 1996; Ackermann 2000; Winer and Biermann 1994). Up to 1% of the emitted NO_X was found as nitrous acid, making this source important in heavily polluted areas with high amounts of traffic. Kirchstetter et al. (1996) report a $HONO/NO_X$ emission ratio of 0.35% for a north-American car fleet, while Ackermann (2000) and Kurtenbach et al. (2002) found 0.65% of NO_X emitted as HONO in a traffic tunnel in Germany.

2.6 Tropospheric Chemistry of VOCs

Organic compounds are an extremely large class of chemicals, which encompasses known, as well as yet unknown, species that play a significant role in the atmosphere (Goldstein and Galbally 2007; Park et al. 2013). As discussed in Sect. 2.3, organic compounds provide the fuel for ozone formation. In addition, chemical reactions with organic carbons can lead to the formation of particles, so called secondary organic aerosol (Seinfeld and Pandis 2016). They also impact radical budgets (Sect. 2.4) and thus influence radical chemistry.

The chemistry of VOCs is a complex topic which, by far, exceeds the scope of this book. Here, we briefly discuss typical levels of organics in the atmosphere, and in particular in urban areas, and their initial reactions with various radical species. VOCs can be subdivided into a number of different classes encompassing hydrocarbons (HC) and oxidised species (Table 2.7):

- alkanes (C_nH_{2n+2})
- alkenes or olefins (C_nH_{2n})
- alkynes (C_nH_{2n-2})
- aromatics, which contain one or more benzene rings (C_6H_6)
- HC containing more than one double bond, in particular isoprene (C_5H_8) and terpenes
- oxidised species: aldehydes, ketones, alcohols, and organic acids

Members of each class often undergo similar chemistry. We will therefore discuss their initial reactions steps in a general way.

Alkanes are molecules containing carbon and hydrogen molecules, where the carbon molecules are held together by single molecular bonds. Consequently, the degradation of alkanes by OH proceeds by abstraction of one hydrogen atom analogous to the reaction of methane (R2.36 and R2.37):

$$OH + RH \rightarrow R + H_2O \tag{R2.36}$$

where R denotes an alkyl radical (C_nH_{2n+1}), e.g. C_2H_5. The rate constants for some alkane–OH reactions are given in Table 2.7.

Similar to CH_3 (see Sect. 2.4.1), alkyl radicals immediately react with molecular oxygen to form organic peroxy radicals:

$$R + O_2 \xrightarrow{M} RO_2 \tag{R2.37}$$

The fate of peroxy radicals in the atmosphere are self-reaction (forming peroxides), reaction with O_3, and, mostly in polluted areas, reaction with NO:

$$RO_2 + NO \rightarrow RO + NO_2 \tag{R2.38}$$

Alkoxy radicals, RO, react with O_2 to form aldehydes and HO_2 radicals:

$$RO + O_2 \rightarrow R'CHO + HO_2 \tag{R2.39}$$

where R′ denotes an alkyl radical with one less C atom ($C_{n-1}H_{2n-1}$).

Initial reactions of alkanes with NO_3 radicals and Cl atoms also proceed by hydrogen abstraction, followed by reactions Eqs. R2.37–R2.39 (Table 2.7). Ozone only reacts very slowly with alkanes and its reactions are unimportant in the atmosphere (Table 2.7).

Alkenes are HC which contain one or more carbon double bonds. While hydrogen abstraction can still occur, the predominant first step in alkene + OH reactions is the addition of the OH radical to the double bond, followed by addition of O_2 to the resulting peroxy radical:

$$OH + RCH_2CH_2 \xrightarrow{M} RCH_2CH_2OH \tag{R2.40}$$

$$RCH_2CH_2OH + O_2 \xrightarrow{M} O_2RCH_2CH_2OH \tag{R2.41}$$

Alkenes with aliphatic chains can also (though less likely) react by hydrogen abstraction from the aliphatic chain. The rate constants for some of the more abundant alkenes with OH radicals are listed in Table 2.7. Similar to the alkane reaction chain, the $O_2RCH_2CH_2OH$ radical can react with NO forming a β-hydroxyalkoxy radical:

$$O_2RCH_2CH_2OH + NO \xrightarrow{M} ORCH_2CH_2OH \tag{R2.42}$$

The $ORCH_2CH_2OH$ radical then further decays into hydroxyl or alkoxy radicals and aldehydes.

Reactions of alkenes with NO_3 radicals proceed primarily through the addition of NO_3 to the double bond. The radical formed in the NO_3 addition reacts further with O_2 to form a β-nitratoalkyl peroxy radical, which can then react with NO to form nitrates, hydroxyl radicals, and other products. Alkenes can also react with O_3. Although these reactions are slower than those of OH, they can be important at night, when OH levels are low. Their significance also stems from the fact that

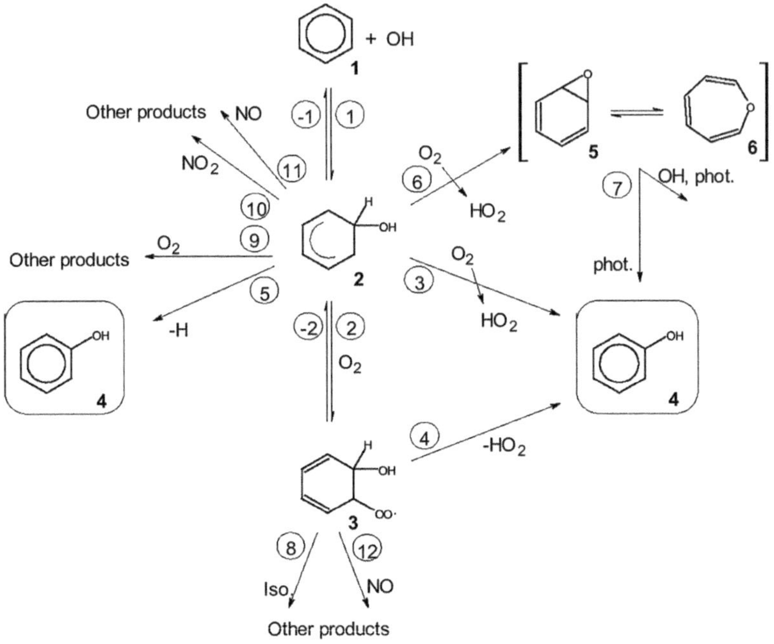

Fig. 2.15 The initial steps in the OH initiated oxidation of benzene (1). Currently proposed loss processes of the aromatic-OH adduct (2) and the peroxy radical (3) (which are in rapidly established equilibrium) are shown that lead in part to the formation of phenol (4). Intermediate species are indicated by bold numbers. Similar schemes for the phenol forming pathways can also be drawn for the alkyl-substituted aromatics. Reprinted from R. Volkamer, PhD Thesis, University of Heidelberg, 2001

OH radicals are formed from the reaction of ozone and larger alkenes (Paulson et al. 1999). Cl atom reactions also proceed through Cl addition to the double bond. Details of the reaction chains for individual alkenes can be found in Finlayson-Pitts and Pitts (2000).

Alkynes contain carbon triple bonds. Their reactions proceed in similar ways as those of alkenes. An important product of these reactions is glyoxal (CHOCHO).

For monocyclic *aromatics* the primary reaction pathway is also OH addition. The resulting radical can either lose a hydrogen atom (through HO_2 formation) and form a phenol, or the ring can open leading to rapid degradation of the products (see Fig. 2.15) (Volkamer et al. 2002). In the latter case, glyoxal was found to be an important product (Volkamer et al. 2001).

Isoprene and terpenes are olefins with more than one double bond; they are very reactive towards OH, NO_3, O_3, and Cl, with rate constants of reactions with OH radicals approaching the collision limit. Because these molecules have a complex chemical structure, their reactions mechanisms are quite complex, and beyond the scope of this book. The reader is referred to a review by Atkinson and Arey (1998). A compilation of kinetic and photochemical data is found at Mellouki et al. (2021).

Aldehydes, such as formaldehyde, glyoxal, and acetaldehyde, react with OH, NO_3, and Cl by abstraction of the weakly bonded aldehydic hydrogen (see for example Eq. 2.15). The radical formed in this reaction then reacts with O_2, similar to the reactions described above. Photolysis is also an important sink of aldehydes. Ketones react by abstraction of the H atom from the alkyl chains, similar to the reactions of alkanes.

2.7 Tropospheric Chemistry of Sulfur Species

Sulfur compounds play an important role in the atmosphere, in clean air as well as in polluted areas (e.g. Rotstayn and Lohmann 2002). Sulfur species with low vapour pressure (such as sulfuric acid or methane sulfonic acid) have a large influence on particle formation in the atmosphere. The legendary London Smog largely consisted of sulfuric acid aerosol. The most important atmospheric sulfur species, their typical mixing ratios, residence times, and degradation pathways are summarised in Table 2.8. In this section we will briefly discuss the chemistry of these species and the impact of sulfur chemistry on the climate.

2.7.1 Sulfur Dioxide SO_2

Sulfur dioxide is oxidised in the gas phase and, owing to its relatively high solubility in water, by reactions in atmospheric liquid water, such as cloud, fog, or aerosol (see Fig. 2.16).

The most important gas-phase reaction is oxidation, which is initiated by OH addition:

$$SO_2 + OH \xrightarrow{M} HSO_3 \tag{R2.43}$$

$$HSO_3 + O_2 \rightarrow SO_3 + HO_2 \tag{R2.44}$$

$$SO_3 + H_2O \xrightarrow{M} H_2SO_4 \tag{R2.45}$$

In this reaction sequence, discovered by Stockwell and Calvert (1983), the number of HO_X radicals is conserved, as in many other OH-initiated oxidation processes. Overall, OH oxidation of SO_2 facilitates about one third of the total SO_2 degradation.

The oxidation of SO_2 in the liquid phase also plays an important role (see Sect. 2.4.1). Gaseous SO_2 is in an equilibrium with aqueous $SO_3^{=}$:

$$SO_{2(g)} + H_2O_{(liq)} \leftrightarrow SO_2 \cdot H_2O_{(liq)} \tag{R2.46}$$

$$SO_2 \cdot H_2O_{(liq)} \leftrightarrow HSO^-_{3\,(liq)} + H^+_{(liq)} \tag{R2.47}$$

$$HSO_3^- \leftrightarrow SO_3^{=} + H^+ \tag{R2.48}$$

Table 2.8 Sulfur compounds in the atmosphere

Species	Chemical formula	Typ. mixing ratio, ppb	Atmospheric life-time	Degraded by reaction with
Sulfur dioxide	SO_2	0.01–100	1–9 d	OH, liquid-phase oxidation
Hydrogen sulfide	H_2S	0–300	~3 days	OH
Methyl mercaptan	CH_3SH		0.2 d	OH
Dimethyl sulfide (DMS)	CH_3SCH_3	0.01–1	0.3–3 d	OH, NO_3, BrO
Dimethyl disulfide	CH_3SSCH_3		~1day	Like DMS
Dimethyl sulfoxide (DMSO)	CH_3SOCH_3	0–0.02	Days	OH, deposition
Methansulfonic acid (MSA)	CH_3SO_3H	0–0.06		OH, gas–particle conv.
Carbon Disulfide	CS_2	0.025 ± 0.01	12 d	OH
Carbonyl sulfide	COS	0.51 ± 0.07	≈6 years	OH, O
Sulfurous acid	H_2SO_3			Liquid phase
Sulfuric acid	H_2SO_4	10^{-4}–10^{-3}	0.01 day	Gas to particle conversion
Sulfates, e.g. Ammonium sulfate	$(NH_4)SO_4$ Aerosol		2–8 d	Deposition
Sulfur hexafluoride	SF_6	≈ 0.01	>1000 years	Ion reactions
Short lived intermediates				
Sulfur trioxide	SO_3	10^{-5}	10^{-6} s	H_2O
SH-radicals	SH		Seconds	O_3, O_2
SO-radicals	SO		<3 ms	O_2

HSO_3^- and $SO_3^=$ can be further oxidised to H_2SO_4 or sulfate, for example by hydrogen peroxide. This oxidation mechanism can be summarised as:

$$HSO_3^- \rightarrow \cdots \rightarrow SO_4^= \tag{R2.49}$$

The sulfate formed is eventually transported to the ground together with the cloud or fog droplets, thus contributing to 'acid rain' or 'acid haze'.

Finally, SO_2 can also react directly with the ground. This process is termed dry deposition (in contrast to processes where precipitation events play a role). About 30–40% of the SO_2 is removed from the atmosphere by dry deposition.

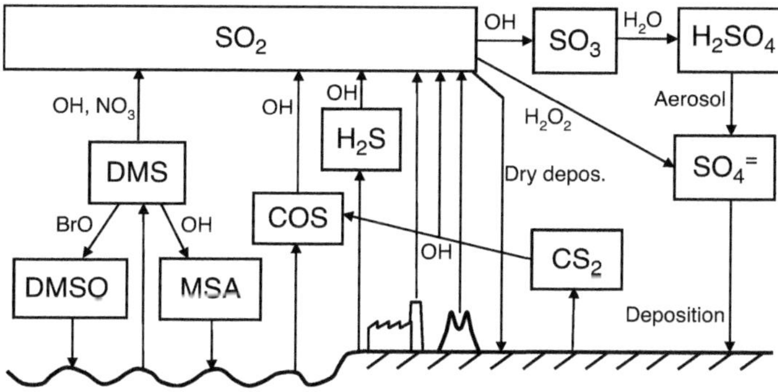

Fig. 2.16 Overview of the cycles of sulfur species in the atmosphere. The surface ocean (lower left corner) is relevant as source of atmospheric sulfur compounds

2.7.2 Reduced Sulfur Species: DMS, COS, CS_2, H_2S

Most sulfur compounds, with the exception of SO_2, are emitted in a reduced form and are rapidly oxidised in the atmosphere. These natural sulfur emissions are in the form of the following species:

Hydrogen Sulfide (H_2S) is present at mixing ratios of up to 300 ppb in the atmosphere. It is a by-product of the degradation of proteins and smells like rotten eggs. The oxidation of H_2S is initiated by reaction with OH:

$$H_2S + OH \rightarrow H_2O + SH \qquad \text{(R2.50)}$$

The reaction pathways of the SH radical are currently unclear. While an analogy to the OH radical should exist, (cf. sulfur $\leftrightarrow$ oxygen), analogous reactions of SH with methane and CO are endothermic (Becker et al. 1975b). The reaction of SH with molecular oxygen is assumed to be very slow. However, SH reacts with ozone, forming SO radicals. The SO radicals finally react with O_2 or ozone to form SO_2.

Carbonyl Sulfide, COS, is mostly produced by the photolysis of sulfur-containing organic species (e.g. of amino acids) in the surface layer of the ocean (e.g. Ulshöfer and Andreae 1990). Once released to the water, the majority of COS is degraded by hydrolysis to H_2S (timescale of about 10–12 h). The remaining COS is emitted to the atmosphere. There is also some emission from soils and marshes (Ulshöfer 1995). In addition to the direct emission, degradation of DMS and CS_2 by OH radicals is a source of COS (Chin and Davis 1993):

$$CS_2 + OH \rightarrow SCSOH \qquad \text{(R2.51)}$$

$$SCSOH + O_2 \rightarrow COS + SO_2 + H \qquad \text{(R2.52)}$$

Carbonyl sulfide is removed from the atmosphere by plant uptake and reaction with OH. The lifetime of COS in the troposphere is on the order of 2–3 years, and thus only small deviations from its average mixing ratio of 500 ppt are found (Ulshöfer 1995).

Vertical transport of COS plays an important role as a sulfur source in the stratosphere (much like N_2O for the transport of oxidised nitrogen, CFCs for chlorine, and halons for Bromine, see Sect. 2.10.2). The degradation of COS in the stratosphere partially sustains the stratospheric sulfate—aerosol layer [named after its discoverer Christian Junge, 'Junge-layer' (Junge et al. 1961)].

Carbon Disulfide (CS_2) is naturally released via volcanoes, the ocean, marshes and forests, and industrial activities (Watts 2000). Atmospheric mixing ratios are on the order of 15–35 ppt. Recently, CS_2 was found by DOAS in the city of Shanghai at levels up to 1.2 ppb (Yu et al. 2004). CS_2 is degraded in the atmosphere by OH oxidation (R2.51).

Dimethyl Sulfide (CH_3SCH_3) is produced by biological processes in the ocean (Andreae et al. 1985). Besides sporadic volcanic eruptions, oceanic DMS emissions are the largest natural source of sulfur in the atmosphere. Due to the important role of sulfur in the formation of aerosol particles and cloud condensation nuclei (CCN) DMS has received considerable attention (see also Sect. 2.7.3).

The degradation mechanisms of this species are not fully elucidated to date. Reactions of free radicals are probably largely responsible for its degradation. A role of OH in the degradation is likely. The first step in OH initiated degradation of DMS is OH abstraction from one of the methyl groups, or OH addition to the sulfur atom (Yin et al., 1990a, b).

$$OH + CH_3SCH_3 \rightarrow H_2O + CH_3SCH_2 \quad \text{(R2.53)}$$

$$OH + CH_3SCH_3 + M \rightarrow CH_3S(OH)CH_3 + M \quad \text{(R2.54)}$$

Intermediate products in this reaction chain are dimethyl sulfoxide (DMSO), CH_3SOCH_3, and $CH_3SO_2CH_3$. Stable end products are sulfuric acid and methane sulfonic acid (CH_3SO_3H). The involvement of NO_3 radicals has also been suggested (Winer et al. 1984; Platt and Le Bras 1997), where the H abstraction channel appears to be predominant (Butkovskaya and Le Bras 1994):

$$NO_3 + CH_3SCH_3 \rightarrow HNO_3 + CH_3SCH_2 \quad \text{(R2.55)}$$

In addition, there are several reports of a possible role of halogen oxide radicals, in particular BrO (Toumi 1994). The product of the BrO–DMS reaction is DMSO:

$$BrO + CH_3SCH_3 \rightarrow Br + CH_3SOCH_2 \quad \text{(R2.56)}$$

While sulfuric acid and methane sulfuric acid form particles, DMSO does not. Thus, the fraction of DMS degraded by BrO may determine the efficiency of particle formation in marine areas, as discussed by von Glasow et al. (2004).

2.7.3 Influence of Sulfur Species on the Climate

The low vapour pressure sulfur compounds, H_2SO_4 and methane sulfonic acid, will quickly condense after their formation in the gas phase (e.g. by reaction R2.45) forming small particles, which can act as cloud condensation nuclei (CCN) and thus impact cloud formation. Therefore, biological emission of sulfur species could affect cloud formation and thus our climate. This theory was proposed in 1987 by Charlson, Lovelock, Andreae and Warren (referred to as the CLAW hypothesis after the authors, initials) (Charlson et al. 1987; Bates et al. 1987). The authors postulate the following mechanism: Due to microbiological activities in the ocean, volatile sulfur species (essentially DMS) are released. DMS is oxidised by photochemical processes in the atmosphere to sulfuric acid or methane sulfonic acid. Due to their low vapour pressure, these species condense to form small particles (mean radius ~0.07 μm), which can act as CCN. In large areas of the worlds oceans, the supply of CCN is so small that cloud formation can be limited by CCN availability and not by the water vapour supply (about 200 CCN cm^{-3} are required, corresponding to a DMS mixing ratio of less than 100 ppt). The extent of cloud cover now feeds back on the insolation and thus the ocean surface temperature. If the biological DMS formation is positively correlated with the ocean surface temperature this process could form a biological negative feedback loop which effectively stabilises the surface temperature of earth [see also Gaia Hypothesis (Lovelock 1979)]. Since its publication, the CLAW hypothesis has been under intense debate. For instance, Schwartz (1988) argued that the emission of gaseous sulfur species in the northern hemisphere (but not in the southern hemisphere) largely increased during the last 100 years due to anthropogenic activities without major change in cloud albedo or temperature (Schwartz 1988). Furthermore, Green and Hatton (2014) come to the conclusion that the CLAW-hypothesis is an oversimplification of the complex processes associated with biological sulfur release and its effects on climate.

However, sulfate aerosol is known to influence the climate by direct reflection of sunlight after major volcanic eruptions, which deposit large amounts (megatons) of SO_2 in the stratosphere. These events were proven to cool the climate in a measurable and sometimes severe way.

2.8 Chemistry of Halogen Radicals in the Troposphere

Motivated in part by the discovery of BrO and IO in the troposphere by DOAS in the 1990s (Hausmann et al. 1994; Alicke et al. 1999), interest in the chemistry of reactive halogen species has been increasing (see e.g. Von Glasow et al. 2007; Saiz-Lopez and von Glasow 2012; Simpson et al. 2015). Reactive halogen chemistry has now been observed in many regions of the troposphere (see Simpson et al. 2015 for a review).

Reactive chlorine compounds can be found in many different environments. It has been estimated that the extra-tropical Cl-atom concentration is on the order of $2.6-18 \times 10^3$ (Platt et al. 2004; Wingenter et al. 1996). Higher concentrations of $1-10 \times 10^4\,cm^{-3}$ have been found in the arctic using the indirect hydrocarbon clock technique (Jobson et al. 1994; Ramacher et al. 1997, 1999; Röckmann et al. 1999). Direct observations of ClO are sparse, but up to 15 ppt have been reported over the Great Salt Lake (Stutz et al. 2002b) and in volcanic plumes (Lee et al.

2005a; Bobrowski 2005). In volcanic plumes indirect evidence for the presence ClO was derived from OClO observations (Gliß et al. 2015). Two different Cl-atoms precursors have been observed at high levels in the troposphere. Cl_2 is present in the Artic at levels up to 400 ppt (Liao et al. 2014), in marine environments at mixing ratios of up to 35–150 (Spicer et al. 1998; Lawler et al. 2011; Liao et al. 2014), and at similar levels in polluted atmospheres downwind of power plants (Riedel et al. 2013). $ClNO_2$ is formed when high NO_X is mixed with aerosol chloride, for example in coastal cities, and has been observed at mixing ratios of more than 1 ppb in the morning (Osthoff et al. 2008; Thornton et al. 2010; Young et al. 2014). Both Cl_2 and $ClNO_2$ photolyse rapidly to Cl-atoms in the troposphere and thus drive active chlorine radical chemistry (Young et al. 2013).

Reactive bromine species can be present during polar sunrise, with BrO mixing ratios of 0–45 ppt (Jobson et al. 1994; Ramacher et al. 1997, 1999; Hausmann and Platt 1994; Tuckermann et al. 1997; Hegels et al. 1998; Martínez et al. 1999; Hönninger and Platt 2002; Frieß et al. 2004a; Hönninger et al. 2004c; Liao et al. 2012; Wang and Pratt 2017). HOBr has also been observed in the Arctic with mixing ratios up to 10–35 ppt (Liao et al. 2012). Even higher BrO mixing ratios, up to 176 ppt, can be present above salt lakes, such as the Dead Sea, the Great Salt Lake, and the Salar de Uyuni (Hebestreit et al. 1999; Matveev et al. 2001; Stutz et al. 2002b; Wagner et al. 2001b; Hönninger et al. 2004b; Holla et al. 2015). The highest reactive bromine levels have been observed in volcanic plumes, where BrO reached up to 1000 ppt (Bobrowski et al. 2003; Bobrowski 2005). Bromine levels in the marine boundary layer and the free troposphere remain uncertain, although various studiesd have reported up to $\approx$ 2ppt of BrO (Leser et al. 2002; Saiz-Lopez 2004a; Frieß et al. 1999; van Roozendael et al. 2002; Pundt et al. 2002; Read et al. 2008).

Reactive iodine species have been observed at salt lakes (Zingler and Platt 2005) and in coastal areas with IO mixing ratios typically in the 1–10 ppt range (Alicke et al. 1999; Mahajan et al. 2009; Tuite et al. 2018). Maxima of up to 35 ppt and up to 302 ppt of I_2 have been observed close to kelp (Huang et al. 2010; Whalley et al. 2007). Mixing ratios over the ocean are typically in the range of 0–3 ppt (Saiz-Lopez et al. 2005; Read et al. 2008; Mahajan et al. 2012, 2019; Prados-Roman et al. 2015). The presence of reactive iodine species has also been confirmed in polar regions through the detection of IO (e.g. Wittrock et al. 2000a; Frieß et al. 2001; Saiz-Lopez et al. 2007; Mahajan et al. 2021), I_2 (Raso et al. 2017), and its link to new particle formation (O'Dowd et al. 2002; Baccarini et al. 2020).

The presence of these RHS in the troposphere has many consequences. Among those, the most dramatic is ozone destruction during polar sunrise which can lead to complete loss of BL ozone in the Arctic and Antarctica. This 'Polar Tropospheric Ozone Hole' was the first hint for tropospheric halogen chemistry (Oltmans and Komhyr 1986; Bottenheim et al. 1986, 1990; Barrie et al. 1988; Hausmann et al. 1994; Platt and Lehrer 1997; Barrie and Platt 1997; Platt and Hönninger 2003). In addition, other disturbances of tropospheric chemistry can occur, as detailed in Sect. 2.8.3.

2.8.1 Tropospheric Sources of Inorganic Halogen Species

The sources of RHS (X, X_2, XY, XO, HOX, $XONO_2$, HX, where X = Cl, Br, I) in the troposphere are the degradation of organic halogen compounds and the volatilisation of halogen ions (X^-) from sea salt aerosol, surface salt deposits, and the ocean surface.

Fully halogenated compounds (such as CF_2Cl_2 or CF_2ClBr), which are the main source of RHS in the stratosphere, are photolytically stable in the troposphere. RHS can therefore only be released from less stable precursors (Cauer 1939), such as partially halogenated organic compounds like CH_3Br, or polyhalogenated species, such as $CHBr_3$, CH_2Br_2, CH_2I_2, CH_2BrI (Cicerone 1981; Schall and Heumann 1993; Khalil et al. 1993; Schauffler et al. 1998; Carpenter et al. 1999, 2001), or even I_2 and HOI (Saiz-Lopez et al. 2004a; Peters 2004; Peters et al. 2005; Carpenter et al. 2013). While some of these species, such as CH_3Br, are emitted by up to 50% from anthropogenic sources, most of them, and in particular polyhalogenated species, are only emitted from biological sources, predominately in the ocean or in coastal areas.

The atmospheric lifetime of halogenated organics varies widely. The lifetime of methyl bromide, CH_3Br, is on the order of 1 year, bromoform, $CHBr_3$, has a lifetime of several days, while diiodo methane CH_2I_2 is photodegraded in minutes (Wayne et al. 1995; Yvon and Butler 1996; Davis et al. 1996; Carpenter et al. 1999). It is possible that even less stable halocarbon species are emitted, which, probably due to their instability, have escaped detection (Carpenter et al. 1999).

The release of inorganic halogen species from sea salt or salt solutions (e.g. sea salt aerosol, sea salt deposits, or the ocean), appears to proceed by three main pathways (Fig. 2.17):

(1) Strong acids can release HCl (but not HBr) from sea-salt halides. Under certain conditions (see below), HX can be heterogeneously converted to RHS.
(2) Oxidising agents may convert Br^- or Cl^- to gaseous Br_2 or BrCl. In particular, HOX (i.e. HOBr and HOCl) is such an oxidant. Direct photochemical (Oum et al. 1998a) or non-photochemical oxidation of halides by O_3 may also occur (Oum et al. 1998b; Hirokawa et al. 1998), though probably quite slowly. The release of I_2 and HOI following the deposition of ozone on the ocean surface has recently been demonstrated by Carpenter et al. (2013); MacDonald et al. (2014); Sherwen et al. (2016).
(3) Oxidised nitrogen species, in particular N_2O_5 and NO_3, and perhaps even NO_2, can react with sea-salt bromide or chloride to release HBr, HCl, or photolabile species, in particular $ClNO_2$ and $BrNO_2$ (Finlayson-Pitts and Johnson 1988; Finlayson-Pitts et al. 1990; Behnke et al. 1993, 1997; Rudich et al. 1996; Schweitzer et al. 1999; Gershenzon et al. 1999; Osthoff et al. 2008; Thornton et al. 2010).

Because these mechanisms are closely linked to the atmospheric cycling of RHS, we present further information in the next section.

2.8.2 Tropospheric Cycles of Inorganic Halogen Species

In this section, we review the reaction cycles of inorganic halogen species in the troposphere (Figs. 2.17 and 2.18), (Wayne et al. 1995; Platt and Janssen 1996a; Platt and Lehrer 1997; Lary et al. 1996; Platt and Hönninger 2003; von Glasow Crutzen 2003).

Following release, inorganic and organic halogen species are photolysed to form halogen atoms which then predominately react with ozone:

$$X + O_3 \rightarrow XO + O_2 \qquad (R2.57)$$

Typical lifetimes of halogen atoms (X) due to R2.57 at tropospheric background O_3 levels are around 0.1 s for Cl, and on the order of 1 s for Br and I. Halogen atoms are regenerated in a series of reactions including photolysis of XO, which is of importance for X = I, Br, and, to a minor extent, Cl:

$$XO + h\nu \rightarrow X + O \qquad (R2.58)$$

where the photolysis frequencies $J_{2.58} \approx 3 \times 10^{-5}\,s^{-1}$, $4 \times 10^{-2}s^{-1}$, $0.2s^{-1}$ for X = Cl, Br, and I, respectively. The reaction of XO with NO also recycles halogen atoms:

$$XO + NO \rightarrow X + NO_2 \qquad (R2.59)$$

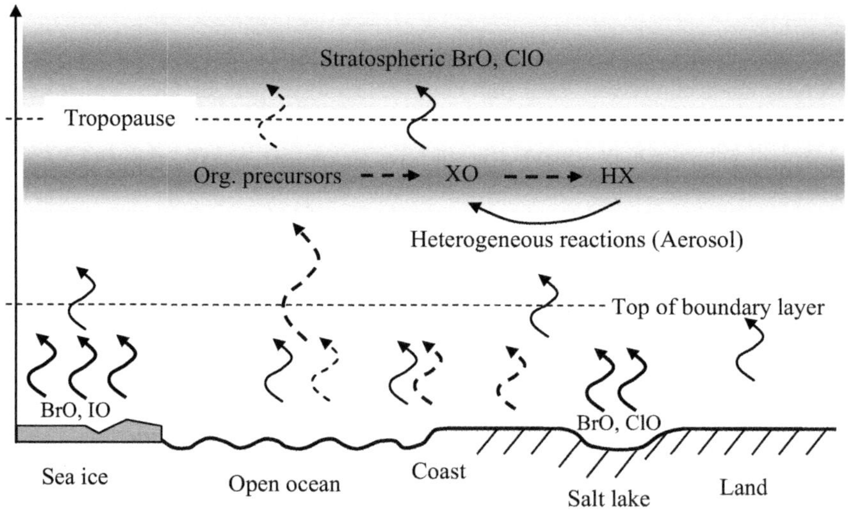

Fig. 2.17 The occurrence of reactive halogen species (XO, X = Cl, Br, I) in the atmosphere: BrO is found above polar sea ice during springtime, IO (and BrO) are present at most coastal areas. It is suspected that there is a layer of BrO in the free troposphere (see von Glasow et al. 2004; Schmidt et al. 2016). *Drawn arrows*: emission of RHS, *dashed arrows*: emission of organohalogens

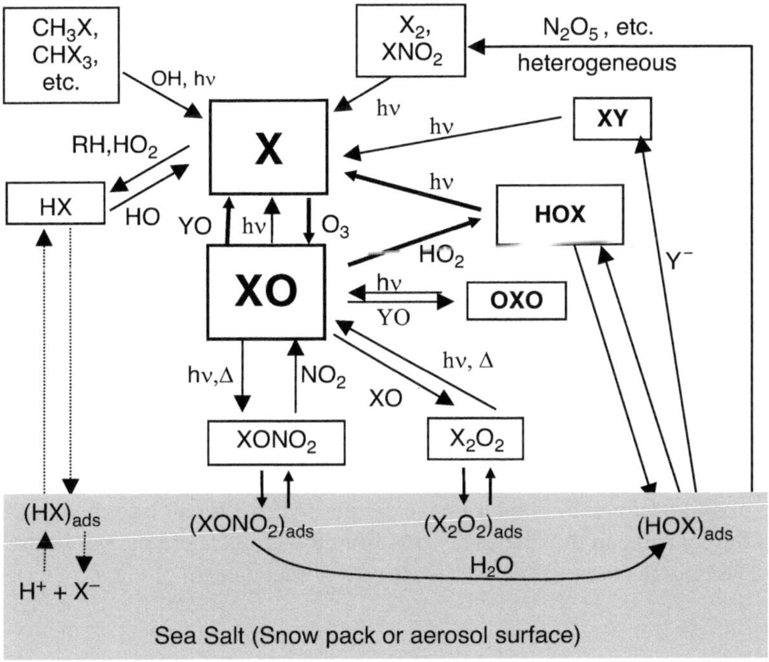

Fig. 2.18 Reaction cycles of reactive halogen species in the troposphere. X = Cl, Br, or I

This reaction leads to a shift of the NO/NO_2 photostationary state in the atmosphere, by providing a shortcut in the normal $NO/NO_2/O_3$ cycle (2.1/2.2/2.3). This has consequences for the photochemistry in the troposphere and, in particular, ozone formation (Stutz et al. 1999).

The self-reactions of XO (or reaction with another halogen oxide YO) also play a role:

$$XO + YO \rightarrow X + Y + O_2 \tag{R2.60a}$$

$$XO + YO \rightarrow XY + O_2 \tag{R2.60b}$$

$$XO + YO \rightarrow OXO + O \tag{R2.60c}$$

$$XO + YO \xrightarrow{M} X_2O_2 \tag{R2.60d}$$

Reactions R2.57, R2.60a, R2.60b, R2.60d, and the photolysis of the halogen molecules formed in R2.60 constitute a catalytic cycle which destroys ozone in the troposphere. The efficiency of this cycle is determined by the amount of RHS present and the rates of R2.60a, R2.60b, R2.60d. This type of halogen-catalysed ozone destruction has been identified as the prime cause of polar BL ozone destruction (Oltmanns and Komhyr 1986; Barrie et al. 1988; Hausmann et al. 1994; Barrie and Platt 1997).

Halogen atoms can also react with saturated (Cl) or unsaturated HC (Cl and Br) to form hydrogen halides, e.g.:

$$RH + Cl \rightarrow R + HCl \qquad (R2.61)$$

Here, R denotes an organic radical. An alternative is the reaction of halogen atoms with formaldehyde (or higher aldehydes) or HO_2, also leading to the conversion of halogen atoms to hydrogen halides:

$$X + HO_2 \rightarrow HX + O_2 \qquad (R2.62)$$

$$X + HCHO \rightarrow HX + CHO \qquad (R2.63)$$

The former reaction occurs for X = Cl, Br, I, the latter only for X = Cl, Br. In particular, Cl is a very strong oxidant and a Cl concentration of 10^4 cm^{-3} can considerably contribute to the oxidation capacity of the troposphere.

The main loss of RHS is their conversion to hydrogen halides which are highly water soluble, and can thus be easily lost from the atmosphere by wet or dry deposition. The only relevant gas-phase 'reactivation' mechanism of HX is reaction with OH:

$$HX + OH \rightarrow X + H_2O \qquad (R2.64)$$

Only at high OH concentration and in the absence of deposition processes, such as in the free troposphere, does this reaction play a role. HX can also be recycled through the aerosol phase, as we will discuss below.

Other important inorganic halogen species are HOX and $XONO_2$. HOX is formed via the reaction of peroxy radicals with XO (Canosa-Mas et al. 1999):

$$XO + HO_2 \rightarrow HOX + O_2 \qquad (R2.65)$$

Formation of HOX is followed by its photolysis:

$$HOX + h\nu \rightarrow X + OH \qquad (R2.66)$$

Reactions R2.57, R2.65, and R2.66 form another catalytic cycle that destroys ozone during the day. This cycle dominates at low RHS levels, such as those found in the marine BL. The catalytic ozone destruction cycle through reactions R2.60a, R2.60b, R2.60d is important at higher RHS levels, such as those found in polar regions. Reactions R2.65 and R2.66 also lead to a shift in the HO_2/OH ratio in the troposphere, and thus directly influence the oxidising capacity of the troposphere (Bloss et al. 2005).

Another important reaction cycle involving HOBr is the liberation of gaseous bromine species (and to a lesser extent chlorine species) from sea-salt halides (Fan and Jacob 1992; Tang and McConnel 1996; Vogt et al. 1996):

$$HOBr + (Br^-)_{Surface} + H^+ \rightarrow Br_2 + H_2O \qquad (R2.67)$$

The reaction appears to occur at appreciable rates only at pH < 6.5 (Fickert et al. 1999). The required H^+ can be supplied by strong acids, such as H_2SO_4 and HNO_3 from anthropogenic or natural sources. Reaction R2.67, together with R2.65, R2.57, and the photolysis of Br_2, form a cycle where Br^-, for example, in sea ice, blowing snow, or the aerosol, is volatilised. Because the uptake of one HOBr molecule leads to the formation of two Br atoms, which, ignoring any Br loss, can form two HOBr molecules (R2.65), this cycle is autocatalytic. The explosion-like behaviour of this cycle has lead to the term 'Bromine Explosion', which is believed to be responsible for the high reactive bromine levels found in the Arctic (Platt and Lehrer 1995; Platt and Janssen 1996a; Wennberg 1999).

The reaction of halogen oxides with NO_2 forms halogen nitrate, $XONO_2$:

$$XO + NO_2 \xrightarrow{M} XONO_2 \tag{R2.68}$$

Bromine nitrate is assumed to be quite stable against thermal decay, but is readily photolysed and may be converted to HOX by heterogeneous hydrolysis (van Glasow et al. 2002):

$$XONO_2 + H_2O \xrightarrow{M} HNO_3 + HOX \tag{R2.69}$$

or to Br_2 or BrCl by heterogeneous reaction with HY:

$$XONO_2 + HY \xrightarrow{\text{Surface}} HNO_3 + XY \tag{R2.70}$$

Overall, $XONO_2$ is probably of minor importance at the low NO_X levels typically found in the free troposphere, but it can play an important role in polluted air, for instance at polluted coastlines. There, $XONO_2$ can be an important source of HOX, and $BrONO_2$ might contribute to bromine explosion events.

Under conditions of elevated NO_X, halogen release also occurs via the reaction:

$$N_2O_5(g) + NaX(s) \rightarrow NaNO_3(s) + XNO_2(g) \tag{R2.71}$$

(e.g. Finlayson-Pitts et al. 1989). This pathway was identified as a major source of $ClNO_2$ by Osthoff et al. (2008). It has since been confirmed to occur at many other locations (for example Thornton et al. 2010; Young et al. 2014). The photolysis of $ClNO_2$ is an important source of highly reactive Cl-atoms in the atmosphere (Young et al. 2013; Simpson et al. 2017).

The XNO_2 formed in the reaction R2.71 may also further react with sea salt (Schweitzer et al. 1999):

$$XNO_2(g) + NaX(s) \rightarrow NaNO_2(s) + X_2(g) \tag{R2.72}$$

This reaction sequence would constitute a dark source of halogen molecules. The direct reaction of NO_3 on a salt surface can also release reactive halogens, for example in the form of halogen atoms:

$$NO_3(g) + NaX(s) \rightarrow NaNO_3(s) + X(s) \tag{R2.73}$$

(Seisel et al. 2010; Gershenzon et al. 1999). The uptake coefficient of NO_3 by aqueous solutions of NaBr and NaCl was found by Rudich et al. (1996) to be near 0.01 for sea water, while Seisel et al. (2010) found up to 0.05 for dry NaCl.

2.8.3 Potential Impact of Inorganic Halogen Species on Tropospheric Chemistry

The role of reactive halogens in the troposphere has been a controversial issue and remains a poorly understood aspect of atmospheric chemistry. However, it is becoming clear that the RHS are present in many parts of the troposphere. At the levels observed in the atmosphere, RHS can have a noticeable effect on several aspects of tropospheric chemistry and the climate, see e.g. Saiz-Lopez et al. 2023. These include:

(1) RHS, and in particular reactive bromine and iodine can readily destroy tropospheric ozone through catalytic cycles (Solomon et al. 1994a; Platt and Janssen 1996a; Davis et al. 1996; Sander et al. 2003; von Glasow et al. 2004; Schmidt et al. 2016). This has consequences for atmospheric chemistry and, considering that ozone is a greenhouse gas, also the global climate (e.g. Roscoe et al. 2001). For instance in a GEOS-chem model study Schmidt et al. (2016) find a 14% decrease in tropospheric ozone due to multiphase Cl- and Br-radical chemistry.
(2) Reactions R2.65, and R2.66 will lead to the conversion of HO_2 to OH, and thus reduce the HO_2/OH ratio, with consequences for the atmospheric oxidation capacity and thus indirectly for the lifetime of methane, another important greenhouse gas. For instance, in a model study Schmidt et al. (2016) find a 11% decrease in the tropospheric OH level and a 16% increase in CH_4 lifetime. The effect has, however, proven to be difficult to quantify (Stone et al. 2018).
(3) The presence of highly reactive chlorine atoms, and to a lesser extent bromine atoms, has a direct impact on the oxidation capacity of the troposphere (Platt et al. 2004; Young et al. 2014, 2013; Wang et al. 2019b).
(4) The reaction of BrO with DMS might be important in the unpolluted remote marine BL, where the only other sinks for DMS are the reaction with OH and NO_3 radicals (Toumi 1994; von Glasow and Crutzen 2003, 2004; Breider et al. 2010; Chen et al. 2018). For instance, Chen et al. (2018) find a 12% contribution of BrO to DMS oxidation in a GEOS-Chem global modelling study. The presence of BrO may thus have an impact on cloud formation, as described in Sect. 2.7.3 above.
(5) Deposition of mercury was found to be enhanced by the presence of reactive bromine species (e.g. Schmidt et al. 2016), in particular in polar regions (e.g. Barrie and Platt 1997; Schroeder et al. 1998). This process appears to be one of the most important pathways for toxic mercury compounds to enter the arctic (Lindberg et al. 2002), and possibly also other ecosystems.
(6) Iodine species are involved in particle formation in the marine BL (Hoffmann et al. 2001; O'Dowd et al. 2002; Jimenez et al. 2003; Burkholder et al. 2004; Allan

et al. 2015; Baccarini et al. 2020; He et al. 2021) and also influence the ozone chemistry (Saiz-Lopez et al. 2014).

At present, more measurements are needed to ascertain the distribution and levels of RHS in the troposphere and to assess the significance of RHS on local, regional, and global atmospheric chemistry and our climate.

2.9 Oxidation Capacity of the Atmosphere

The term oxidation capacity, sometimes also called oxidative power, refers to the capability of the atmosphere (or rather a part thereof) to oxidise (or otherwise degrade) trace species emitted into it. This ability is crucial for the removal of trace species, such as the greenhouse gas methane, and is thus often also referred to as the 'self-cleaning' capacity of the atmosphere. Although there is no general definition, the oxidation capacity is frequently associated with the abundance of OH (see e.g. Alexander and Mickley 2015). However, as explained above, many other oxidants (including O_2 and O_3), as well as free radicals other than OH, can contribute to the oxidation capacity of the atmosphere. While the oxidation capacity is quite a popular concept (Li et al. 2021; Jung et al. 2023), it is nevertheless difficult to define.

One difficulty in defining the term oxidation capacity lies in the fact that a given agent (e.g. OH radicals) might act quite differently on various pollutants. Perhaps the best definition for the oxidation capacity C_i of a radical R_i (e.g. $R_i = NO_3$, OH, O_3, Cl) with respect to the degradation (oxidation) of species X_j (e.g. volatile organic species, CO, and NO_X), by R_i, can be described as:

$$C_i = \sum_j \left[X_j\right] \cdot [R_i] \cdot k_{ij} \tag{2.32}$$

where $[R_i]$ denotes the concentration of radical species i with the reaction rate constant k_{ij} towards species X_j (Geyer et al. 2001; Platt et al. 2002). 24h average values of the oxidation capacity are often calculated.

Table 2.9 summarises typical and maximum tropospheric concentrations of the radicals discussed, and shows minimum and typical fractional lifetimes of key VOCs against degradation by the various free radicals.

As an example, the relative contribution of the atmospheric oxidants OH, NO_3, and ozone to the 24 h average of the degradation of organic species in the atmosphere, as observed during the BERLIOZ 1998 campaign at Papstthum near Berlin, Germany, is shown in Fig. 2.19. When all VOCs (including CH_4 and CO) are considered, OH reactions constitute about three quarters of the oxidation capacity. For NMHC, NO_3 contributes almost one third of the oxidation capacity (Geyer et al. 2001).

Open questions include the role of heterogeneous reactions in radical cycles. While the contribution of heterogeneous reactions to HO_X chemistry is unclear to date, it is likely that heterogeneous reactions play an important role in the atmospheric

Table 2.9 Estimated lifetime (hours) for selected VOC

Species	OH (av.)	OH (peak)	NO_3 (av.)	NO_3 (peak)	ClO (av.)	ClO (peak)	Cl (av.)	Cl (peak)	BrO (av.)	Br (av.)
Conc. molec. cm^{-3}	6×10^5	1×10^7	1.5×10^8	1×10^{10}	10^6	10^8	10^3	10^5	10^8	5×10^6
DMS	110	6.3	1.7	0.03	3×10^4	280	870	8.7	10^c	26
n-Butane	190	11	2.8×10^4	430	–		1400	14		
C_2H_4	56	3.4	1×10^4	160			2800	28		
Isoprene	4.6	0.3	1.9	0.03	–		<1100	<11		0.9
Toluene	77	4.6	3.1×10^4	460	–					
CH_4	7×10^4	4×10^3	$>5 \times 10^4$	>700	$>7 \times 10^7$	$>7 \times 10^5$	3×10^6	3×10^4		

cycles of most other radicals. Another interesting question is whether the effect of several oxidising species is additive. In fact, it is possible that the presence of halogen oxides may reduce the OH concentration due to an increase in the photostationary state NO_2/NO ratio. In addition, it is well possible that there are other, yet unrecognised, mechanisms involving the above or other free radicals.

2.10 Stratospheric Ozone Layer

The role of atmospheric ozone as a filter for the suns harmful UV radiation has lead to an early interest in ozone chemistry in the atmosphere. Spectroscopic measurements by Gordon Miller Bourne Dobson in 1925 showed an ozone column density larger than expected from assuming the O_3 mixing ratio found near the surface to be constant throughout the atmosphere (e.g. Dobson and Harrison 1926; Dobson 1968). Subsequent spectroscopic measurements using the so-called 'Umkehr' effect by Paul Götz proved the theory of an ozone layer and determined the altitude of its maximum at 25 km (Götz et al. 1934). This stratospheric ozone layer contains 90% of the atmosphere's ozone and is thus responsible for the majority of the absorption of solar UV radiation.

2.10.1 Stratospheric Ozone Formation: The Chapman Cycle

In the late 1920s, Sidney Chapman proposed a reaction scheme which explained the observed vertical profile of ozone, with relatively low mixing ratios in the troposphere and a maximum around 25 km altitude. This reaction scheme is now commonly called the 'Chapman Mechanism' (Chapman 1930). The initial process is the photolysis of

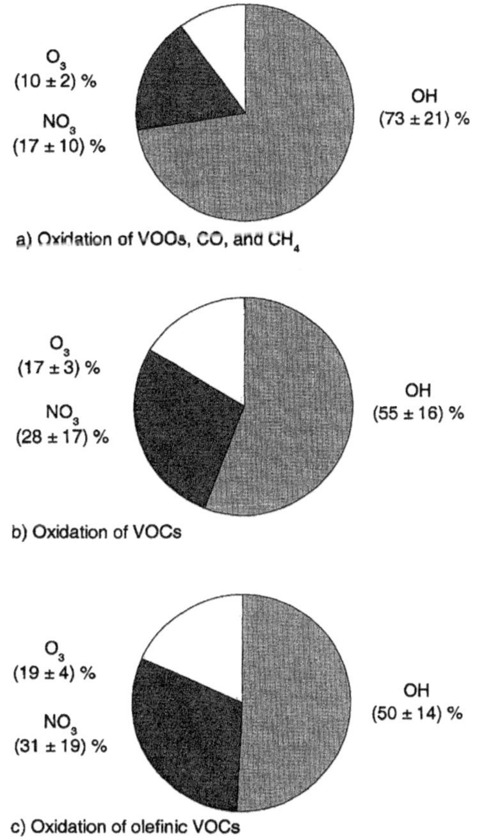

Fig. 2.19 The relative contribution of the atmospheric oxidants OH, NO_3, and ozone to the degradation of organic species in the atmosphere on a 24h basis. Case study during the BERLIOZ 1998 campaign at Papstthum near Berlin, Germany. (**a**) VOCs including CH_4 and CO, (**b**) non-methane VOCs, (**c**) alkenes only. Reprinted with permission from Geyer et al. 2001b, ©American Geophysical Union. All rights reserved

oxygen molecules to form two oxygen atoms in their ground state (indicated by the spectroscopic notation (^{3}P), see Chap. 3). In the stratosphere, sufficiently energetic UV light (i.e. light with wavelengths below 242 nm) is available to photolyse oxygen molecules:

$$O_2 + h\nu(\lambda < 242\,\text{nm}) \rightarrow O(^3P) + O(^3P) \tag{R2.74}$$

The oxygen atoms can react in three ways: (1) Recombine with an oxygen molecule to form ozone. Since two particles (O and O_2) combine to make one (O_3), collision with a third body (M, likely N_2 or O_2,) is required to facilitate simultaneous conservation of energy and momentum. The reaction is therefore pressure dependent (see Sect. 2.1.3):

$$O(^3P) + O_2 + M \rightarrow O_3 + M \tag{R2.75}$$

Alternatively, (2) the oxygen atom can react with an existing ozone molecule:

$$O(^3P) + O_3 \rightarrow O_2 + O_2 \tag{R2.76}$$

Finally, (3) the recombination of two oxygen atoms to form molecular oxygen is possible but largely unimportant in the stratosphere:

$$O(^3P) + O(^3P) + M \rightarrow O_2 + M \quad \text{(R2.77)}$$

In addition to the 'primary' production of O atoms by the photolysis of O_2, the photolysis of O_3 also provides 'secondary' O atoms. In fact, photolysis of ozone molecules occurs at a much higher rate than that of oxygen molecules:

$$O_3 + h\nu(\lambda < 612\,\text{nm}) \rightarrow O(^3P) + O_2 \quad \text{(R2.78)}$$

In summary, these reactions lead to a steady-state O_3 level in the stratosphere, in which the O atom production via reactions Eqs. R2.74 and R2.78 is in balance with their destruction via recombination with O_2 and reaction with O_3. This set of reactions explains the formation of a layer of ozone with a maximum concentration in the lower stratosphere. In the lower stratosphere and troposphere, the rate of O_2 photolysis, and thus the ozone formation rate, becomes extremely low (despite the much higher O_2 concentration there). However, O_3 destruction still occurs via O_3 photolysis, which takes place at much longer wavelengths, and the reaction of $O + O_3$. This explains why the ozone concentration should increase with height (in fact the Chapman mechanism predicts zero O_3 formation in the troposphere). On the other hand, in the upper part of the stratosphere the recombination of $O + O_2$ (reaction Eq. 2.2) becomes slower, since the concentration of air molecules necessary as a 'third body' (M) in the recombination of $O + O_2$ (reaction Eq. 2.2) reduces proportionally with the atmospheric pressure. Thus, despite increasing levels of UV radiation, the O_3 concentration (and also the mixing ratio) will eventually decrease with altitude. Figure 2.20 depicts the ozone profile predicted by the Chapman cycle.

2.10.2 Stratospheric Ozone Chemistry: Extension of the Chapman Cycle

The Chapman mechanism gave a satisfactory explanation for the occurrence of an ozone layer. However, detailed investigation during the 1960s of the elementary reactions and photolysis processes involved revealed that quantitatively the mechanism overestimates the O_3 levels by about a factor of two, see Fig. 2.20. It subsequently became clear that there are many other trace gas cycles affecting stratospheric O_3 levels. In particular, a group of reactions were found to catalyse the elementary reaction of $O + O_3$ (reaction R2.76 above). These reaction sequences follow the general scheme:

$$O_3 + Z \rightarrow ZO + O_2 \quad \text{(R2.79)}$$

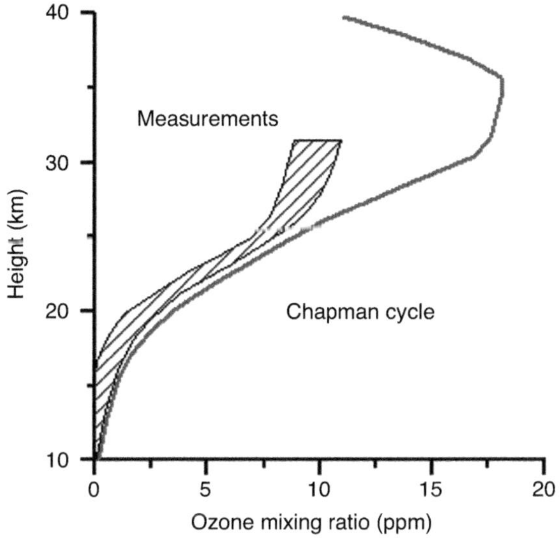

Fig. 2.20 The stratospheric ozone profile according to the Chapman Mechanism, compared to observations (figure after Röth 1994, redrawn)

$$ZO + O(^3P) \rightarrow Z + O_2 \tag{R2.80}$$

With the net result,

$$O(^3P) + O_3 \xrightarrow{Z} O_2 + O_2 \tag{R2.81}$$

where Z (and ZO) denotes a species acting as catalyst Z for reaction R2.77. In the stratosphere, the following species are active (see Table 2.10): Cl (ClO), Br (BrO), NO (NO_2), or OH (HO_2). Inclusion of these reactions brings observations and model calculations in very good agreement. Figure 2.21 illustrates the contribution to the ozone destruction of each of the catalysts at different altitudes.

All species listed in Table 2.10 play an important role in stratospheric chemistry. The *oxides of nitrogen* (NO, NO_2, see Sect. 2.5) are formed in the stratosphere by reaction of excited oxygen atoms $O(^1D)$ with N_2O (which can be regarded as a 'transport species' for oxidised nitrogen into the stratosphere):

$$\begin{aligned} N_2O + O(^1D) &\rightarrow NO + NO \\ &\rightarrow N_2 + O_2 \end{aligned} \tag{R2.82}$$

The excited oxygen atoms $O(^1D)$ are formed in the UV photolysis of O_3 by (R2.4). As in the troposphere, a photo-stationary state between NO, NO_2, and O_3 is established and conversion of NO_2 to N_2O_5 will take place. Under 'normal' stratospheric conditions, loss of N_2O_5 occurs due to photolysis, leading to an increase of NO_X during the course of a day (only during polar winter do heterogeneous reactions of N_2O_5, e.g. R2.99 become important).

Table 2.10 'Catalysts' destroying stratospheric ozone

Catalyst Z	ZO	Stratospheric source
NO	NO_2	Degradation of N_2O from the troposphere
OH	HO_2	$O^1D + H_2O$
Cl	ClO	Degradation of (mostly man-made) organo chlorides
Br	BrO	Degradation of CH_3Br, and other organo bromine species

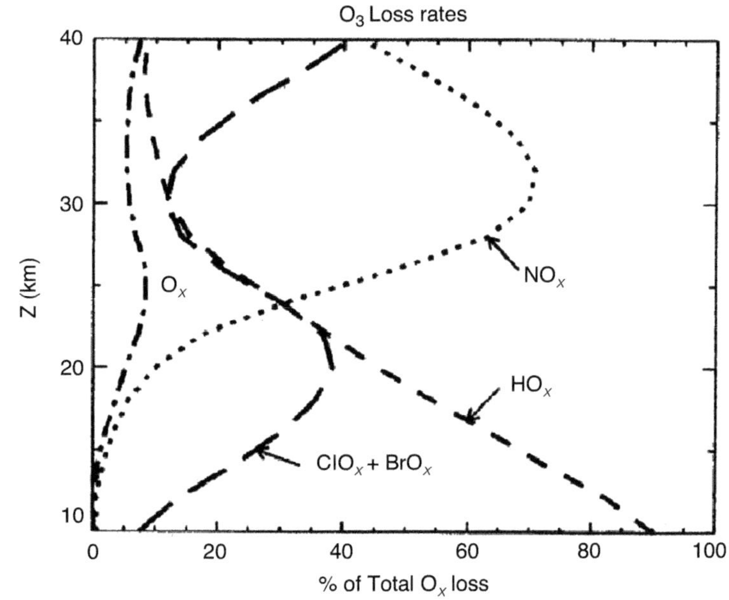

Fig. 2.21 The contribution of the ClO_X, BrO_X, NO_X, and HO_X, catalysed destruction of stratospheric ozone to the original Chapman Mechanism as a function of altitude. Reprinted with permission from from Portmann et al. 1999,

Halogen atoms are released by the degradation of long-lived halocarbon species (e.g. CH_3Cl, $CFCl_3$, CH_3Br, etc.), which are emitted naturally or anthropogenically in the troposphere and then transported into the stratosphere. Halogen atoms (denoted as X = F, Cl, Br, I), which are released in the degradation processes, react quickly with O_3 to form halogen oxides (see Sect. 2.8). In addition, halogen atoms can react with O_2 by association reactions to form halogen peroxide radicals. These radicals are unstable and dissociate by the reverse of their formation reaction, so that under atmospheric conditions the XO_2 species exists in equilibrium with X and O_2.

$$X + O_2(+M) \rightarrow XO_2(+M) \quad (R2.83)$$

The stability of XO_2 decreases in the series F > Cl > Br > I. The equilibrium for atmospheric conditions lies in favour of the atom, and XO_2 reactions do not play a significant role in stratospheric chemistry.

A rapid reaction occurs between XO and NO:

$$XO + NO \rightarrow X + NO_2 \quad (R2.84)$$

NO_2 is rapidly photolysed to produce atomic oxygen, which reforms ozone, and thus the overall effect of this reaction coupled with the $X + O_3$ reaction on ozone is neutral. On the other hand, this reaction will lead to a higher stationary state X/XO ratio, and thus (at least in the case of chlorine) enhance the rate of $X \rightarrow HX$ conversion by reaction of halogen atoms with HC (i.e. CH_4). Consequently, the presence of nitrogen oxides tends to increase the HX/XO_X ratio, and thus reduce the ozone depletion by halogen radicals.

The chlorine oxide radical undergoes self-reaction to form a dimer, Cl_2O_2 (Cox and Derwent, 1979). The ClO dimer plays a major part in the catalytic destruction of ozone occurring in the polar lower stratosphere during wintertime, which in Antarctica leads to the formation of the ozone hole (see Sect. 2.10.3 below). The relevant cycle is:

$$ClO + ClO + M \rightarrow Cl_2O_2 + M \quad (R2.85)$$

$$Cl_2O_2 + h\nu \rightarrow Cl + ClO_2 \quad (R2.86)$$

$$ClO_2 + M \rightarrow Cl + O_2 \quad (R2.87)$$

$$Cl + O_3 \rightarrow ClO + O_2 \quad (R2.88)$$

The ClO dimer is relatively unstable, and the cycle is only important at low temperatures ($T \approx 200$K) when the reverse decomposition of Cl_2O_2:

$$Cl_2O_2 + M \rightarrow ClO + ClO + M \quad (R2.89)$$

is slow compared with its photolysis.

Symmetric chlorine dioxide is formed in the stratosphere from the coupled reaction of BrO with ClO:

$$BrO + ClO \rightarrow OClO + Br \quad (R2.90a)$$

The discovery of atmospheric OClO using ground-based DOAS spectroscopy at the McMurdo Sound base in Antarctica (Solomon et al. 1987b), was the first positive indication that partitioning of atmospheric chlorine between its active and inactive forms was greatly perturbed in the Antarctic ozone hole. Chlorine dioxide is rapidly photolysed by visible light to yield O atoms

$$OClO + h\nu \rightarrow O + ClO \quad (R2.91)$$

and thus is not a catalytic agent for O_3 destruction. A consequence of its rapid photolysis are low OClO concentrations during daytime.

Observations of the evolution of stratospheric OClO during sunset allow deduction of information on the BrO concentration in that BrO is an essential 'catalyst' for its formation. On the other hand, BrO is lost by BrCl production via reaction:

$$BrO + ClO \rightarrow BrCl + O_2 \quad \text{(R2.90b)}$$

while the third channel of reaction Eq. R2.90 does not change the XO reservoir, but leads to the destruction of ozone:

$$BrO + ClO \rightarrow Cl + Br + O_2 \quad \text{(R2.90c)}$$

Since $k_{2.90a}/(k_{2.90a} + k_{2.90b} + k_{2.90c}) \approx 0.08$, the late night OClO concentration cannot become larger than $1/0.08 \approx 12$ times the daytime BrO concentration (e.g. Wahner and Schiller 1992).

The behaviour of halogen oxides in the stratosphere is strongly influenced by the chemistry of temporary reservoir species containing chlorine and bromine. Temporary reservoirs are molecules which are formed and broken down at a time scale on the order of 1 day, contain halogens, and are not active in O_3 destruction. For example, chlorine nitrate is an important reservoir for ClO_x. It is formed by the reaction of ClO with NO_2 (R2.68), and active chlorine is released by photolysis; which occurs mainly by the reaction:

$$ClONO_2 + h\nu \rightarrow Cl + NO_3 \quad \text{(R2.92a)}$$

or

$$ClONO_2 + h\nu \rightarrow ClO + NO_2 \quad \text{(R2.92b)}$$

The other important temporary reservoir for chlorine is HOCl, formed by reaction of ClO with HO_2:

$$ClO + HO_2 \rightarrow HOCl + O_2 \quad \text{(R2.93)}$$

Chlorine is released from HOCl by photolysis. The major channel of this reaction is:

$$HOCl + h\nu \rightarrow HO + Cl \quad \text{(R2.94)}$$

In an analogous way ClO may react with CH_3O_2, producing methylhypochlorite CH_3OCl, or other products (Helleis et al. 1993, 1994):

$$ClO + CH_3O_2 \rightarrow CH_3OCl + O_2 \quad \text{(R2.95a)}$$

$$\rightarrow Cl + \text{products} \quad \text{(R2.95b)}$$

Cycles involving photolytic production of Cl from $ClONO_2$ and HOCl can lead to depletion of ozone in the lower stratosphere, and these may be significant under some circumstances (see for instance Toumi and Bekki 1993). Similarly, reaction Eq. R2.95a, followed by the photolysis of CH_3OCl or reactions of type R2.95b, could cause ozone loss (Crutzen et al. 1992).

In the polar stratosphere and in circumstances of elevated concentrations of volcanic aerosol, heterogeneous reactions of $ClONO_2$ and HOCl on polar stratospheric clouds (PSCs; particles containing nitric acid trihydrate and/or water ice) lead to gross perturbation of the distribution of ClO_x between inactive reservoirs and active ozone depleting forms. The reactions involved are:

$$ClONO_2 + H_2O(s) \overset{\text{surface}}{\rightarrow} HOCl + HNO_3(s) \quad \text{(R2.96)}$$

$$ClONO_2 + HCl(s) \rightarrow Cl_2 + HNO_3(s) \quad \text{(R2.97)}$$

$$HOCl + HCl(s) \rightarrow H_2O(s) + Cl_2 \quad \text{(R2.98)}$$

$$N_2O_5 + HCl(s) \rightarrow ClNO_2 + HNO_3(s) \quad \text{(R2.99)}$$

Here, (s) refers to the chemical species adsorbed or absorbed in the solid or liquid phase.

As bromine is less tightly bound than chlorine, only a small fraction of the bromine released from bromocarbons is sequestered in the form of HBr and $BrONO_2$, rendering this atom very effective for ozone loss (e.g. Lary 1995). In particular, the combined Br–Cl catalytic cycles are very efficient in depleting ozone, and can therefore cause equal or even larger ozone destruction than chlorine alone in the lower stratosphere. Although there are significant human sources of bromine, the contemporary abundance of total stratospheric Br is only about 0.5% of that of Cl (e.g. Schauffler et al. 1998; Wamsley et al. 1998). The organic precursor species of inorganic bromine, $\approx$ 50% being CH_3Br, halons, $CHBr_3$, etc. are overall slowly decreasing. When the precursor species reach the lower stratosphere, they are photolysed and release inorganic bromine (Br_Y = BrO, HBr, HOBr, $BrONO_2$, BrONO, Br, and Br_2). Due to the lower binding energy of HBr compared with HCl, and to the more rapid photolysis of Br_2, BrCl, HOBr, $BrONO_2$ compared to their Cl analogues, BrO is the most abundant bromine species during the daytime and constitutes between 50% and 70% of total Br_Y. In mid-latitude summer with high stratospheric NO_2 levels, BrO is less abundant ($\approx$50% of the total Br_Y), while BrO can amount up to 70% of Br_Y at high latitudes in a denoxified stratosphere. This is because the Br_Y reservoir species $BrONO_2$ is less abundant in a stratosphere with low NO_X loading.

The possible participation of iodine chemistry in stratospheric processes is still under debate. Owing to the much shorter atmospheric lifetimes of iodine-containing compounds (e.g. CH_3I emitted from the oceans) compared with their Cl- and Br-containing analogues, it has generally been assumed that iodine does not reach the stratosphere in significant quantities, and the major focus of studies of iodine photochemistry has been the troposphere, as discussed in detail earlier. However, recent observations by airborne DOAS have reported low levels of IO in the lower stratosphere (Volkamer et al. 2015; Saiz-Lopez et al. 2015) that could be responsible for additional ozone destruction. Iodine mediated ozone destruction is expected to be very efficient, because potential iodine reservoir species (HI, HOI, or $IONO_2$) are less stable than their Cl- and Br-containing counterparts (Solomon et al. 1994b). In addition, coupling of the chemistry of IO with ClO and BrO is more efficient than the XO self-reaction cycles.

2.10.3 Stratospheric Ozone Hole

In 1985, Farman et al. (1985) observed a decrease of the total ozone column over Antarctica after polar sunrise. Satellite observations revealed that this ozone depletion occurred over the entire Antarctic continent. It soon became clear that, during Antarctic winter, special conditions prevail which lead to the release of reactive chlorine from its 'reservoir compounds' (HCl and $ClNO_3$) by heterogeneous reactions, i.e. chemical processes at the surface of particles in the stratosphere (see reactions Eqs. R2.96–R2.99) (e.g. Solomon 1999). This massive activation, which brings the fraction of reactive chlorine from around 1% to near 50% of total chlorine, leads to dramatic ozone losses in the lower stratosphere. A few examples are illustrated in Fig. 2.22, which shows normal ozone profiles recorded in August (Antarctic winter) and disturbed profiles in October (Antarctic spring). There is nearly complete ozone loss in the altitude range from about 14–24 km.

A number of factors contribute to the formation of the ozone hole: The primary cause is the increase in stratospheric chlorine and bromine levels due to the release

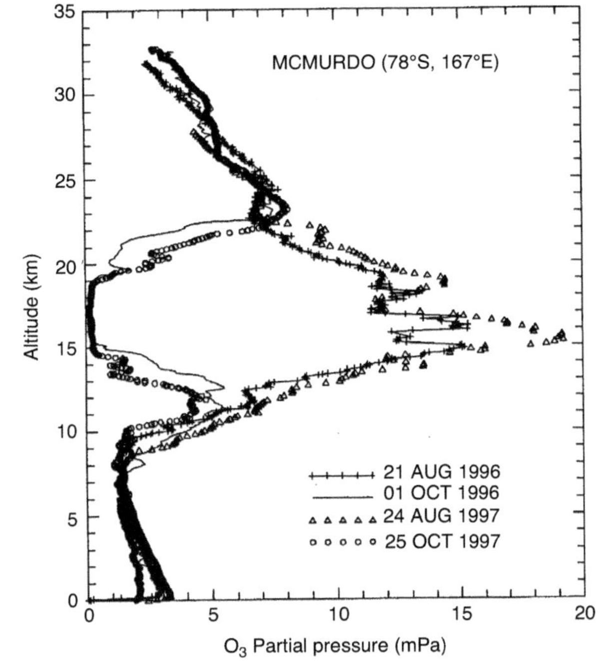

Fig. 2.22 Ozone profiles in August (Antarctic winter) 1996 and October (Antarctic spring) of 1997. Reprinted with permission from Nardi et al. 1999, ©American Geophysical Union. All rights reserved

of man-made CFCs, such as CF_3Cl and CF_2Cl_2. These compounds, which have no sinks in the troposphere, have been accumulating since their introduction in the 1930s. Because of their chemical stability, CFCs were widely used as propellants, in refrigerators, as solvents, and in other industrial applications. Most man-made halocarbons transported to the stratosphere contain more than one halogen atom, and the release of halogen atoms occurs in stages, with intermediate formation of the carbonyl halides (e.g. CFClO). For all halogens, the sink is provided by transport of the most stable (and therefore most abundant) forms (e.g. HX) to the troposphere.

A polar vortex forms in the winter polar stratosphere, which inhibits gas-exchange with lower latitudes. As a consequence, colder polar air cannot mix with warmer lower latitude air, leading to very cold stratospheric temperatures. In addition, ozone from lower latitudes cannot replenish ozone destroyed by halogen catalysed cycles. The polar vortex is very stable over Antarctica, and can exist from June to November. The northern polar vortex is less stable and is typically not stable for more than a few weeks. The more unstable northern vortex is the main reason why the ozone depletion over the north pole has not reached the same extent as that over the south pole.

Very cold stratospheric temperatures develop over Antarctica in winter due to radiative cooling and lack of solar heating. In addition, the lack of air exchange with mid-latitudes also contributes to the low temperature. Once these temperatures fall below a certain value ($\sim$ 195K), PSCs form. The clouds are made out of nitric acid tri-hydrate and water. PSCs provide a surface on which the chlorine reservoir species, such as HCl and $ClONO_2$ can be converted to Cl_2 (R2.97 and R2.98). In addition, PSC formation removes nitrogen oxides from the gas phase, further increasing the amount of reactive halogens. These heterogeneous processes proceed for several months, as long as PSCs are present.

The rapid photolysis of Cl_2 releases reactive chlorine, in the form of Cl atoms, at polar sunrise. Cl reacts quickly with O_3 forming ClO, which then participates in the catalytic ozone destruction cycles described earlier. ClO is the most important halogen oxide in the stratosphere: its concentration ranges from 10–100 ppt in the undisturbed lower stratosphere to 1–2 ppb under conditions of disturbed stratospheric chemistry, i.e. the ozone hole, where more than 50% of the available chlorine can be present as ClO. Under the conditions of very high ClO levels and low temperatures, the ClO self-reaction to form a dimer, Cl_2O_2 (reaction (R2.85) above (Cox and Derwent 1979; Molina and Molina 1987) becomes the dominant catalytic destruction cycle of ozone, leading ultimately to the formation of the ozone hole. It should be noted that cycles involving BrO also contribute to the ozone destruction, although to a lesser extend, since the partitioning between active and reservoir species favours the active forms already in the undisturbed stratosphere. The break-up of the polar vortex in December marks the disappearance of the ozone hole, as warmer and ozone-rich air masses are transported over Antarctica.

The formation of the ozone hole is a recurring phenomenon, which starts every Antarctic winter.

2.10.4 Recovery of Stratospheric Ozone

Since the signing of the Montreal Protocol in 1987 and its amendments in the following years the emissions of CFCs and hydrochlorofluorocarbons (HCFCs) has been reduced significantly and their atmospheric abundance is dropping. The chlorine loading of the troposphere (Montzka et al. 1996) and stratosphere (Rinsland et al. 2003) peaked before 2000, and has since begun to decline significantly (GAW-Report-278 2023). Figure 2.23 shows a time series of the atmospheric chlorine loading as a consequence of the Montreal Protocol and London-, Copenhagen-, and Montreal amendments. The total amount of chlorine present as transport, reservoir, and active species is expressed as Equivalent Effective Stratospheric Chlorine, EESC, from 1950 to 2022 (measured data), and projections to 2100. The impact of the various international treaties to eliminate or reduce CFC and HCFC emissions on EESC is clearly illustrated. However, the chlorine loading in the stratosphere is only slowly decreasing and will not return to natural levels for many decades.

The ozone recovery has been slow, and is difficult to detect due to short observational records and atmospheric variability. It is also partly concealed by atmospheric

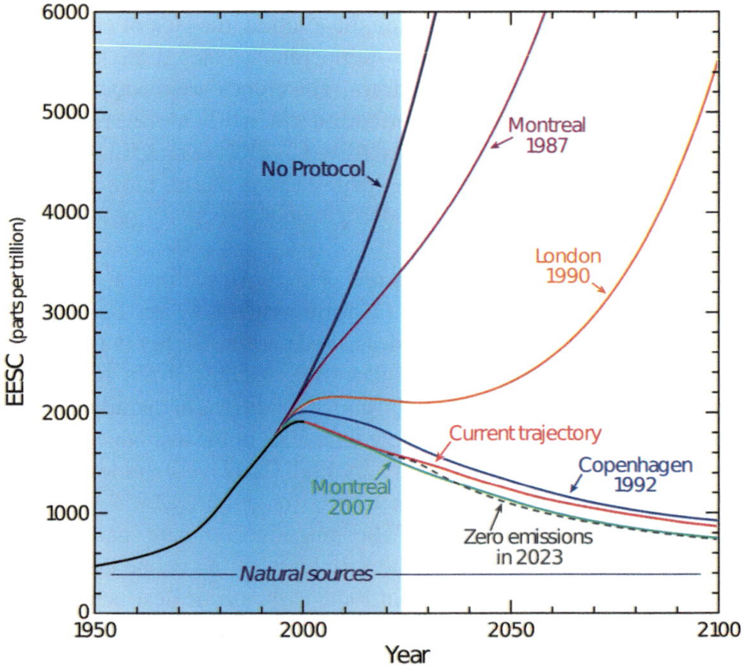

Fig. 2.23 Measured time series of the atmospheric chlorine loading (expressed as Equivalent Effective Stratospheric Chlorine, EESC) from 1950 to 2022, with projections to 2100. The expected effect of the Montreal 1987 treaty and amendmends in 1990, 1992, and 2007 for the limitation of CFCs. Reprinted from Salawitch et al. (Lead Author) (2023). Image provided by NOAA Chemical Sciences Laboratory, Boulder, Colorado, USA. (https://csl.noaa.gov/)

changes caused e.g. by continuously rising levels of greenhouse gases (see e.g. Chipperfield et al. 2017). In addition, there were setbacks due to 'pirate production' of CFCs (see e.g. Fang et al. 2019) and the increasing production of 'very short lived halogen species' (a somewhat misleading denomination for species with a lifetime shorter than 6 months), in particular CH_2Cl_2 and $CHCl_3$ (see e.g. Lauther et al. 2022). These difficulties are illustrated by the occurrence of marked ozone holes in the years 2020–2022 (Kessenich et al. 2023). Nevertheless, signs of recovery of stratospheric ozone are now visible (see e.g. Fang et al. 2019; Braesicke et al. 2018; Bodeker and Kremser 2021). It is estimated that it will take many decades before the ozone hole will close again (GAW-Report-278 2023).

Interaction of Molecules with Radiation

3

Atoms and molecules can exist in many states with respect to their electron configuration, angular momentum, parity, and energy. Transitions between these states are frequently associated with absorption or emission of electromagnetic radiation. Historically, most of the insight into these different states of molecules and atoms has been gained by studying the interaction of atoms and molecules with radiation. However, interaction of radiation with matter can also be used to determine the presence and abundance of molecules or atoms in a sample or the atmosphere.

3.1 Electromagnetic Radiation and Matter

We know electromagnetic radiation in many forms, which are solely distinguished by their wavelength or frequency (see Fig. 3.1). The wavelength λ is connected to the frequency ν by:

$$c = \nu \cdot \lambda$$

where $c = 2.99792458 \times 10^8\,\mathrm{m/s} \approx 3 \times 10^8\,\mathrm{m/s}$ denotes the speed of light.

Radiation of very short wavelength ($\lambda < 30\,\mathrm{nm}$) is called γ-radiation or, at somewhat longer wavelengths, X-radiation (X-rays). At longer wavelengths, we speak of the ultraviolet and infrared ranges framing the visible spectral range extending from about 400–780 nm. Radiation of longer wavelengths is known as sub-mm wave ($\lambda < 1\,\mathrm{mm}$), microwave, and radio wave radiation.

It is a well-known result of quantum mechanics that some aspects of electromagnetic radiation have to be described as behaviour of waves, while others can only be understood when describing radiation as a shower of particles, the photons. Depending on the energy E of a photon, which is given by $E = h\nu = hc/\lambda$, (with $h \approx 6.626176 \times 10^{-34}$ Js denoting Planck's constant) there are different forms of inter-

U. Platt and J. Stutz, *Differential Optical Absorption Spectroscopy*, Geophysics and Environmental Physics, https://doi.org/10.1007/978-3-662-68639-3_3

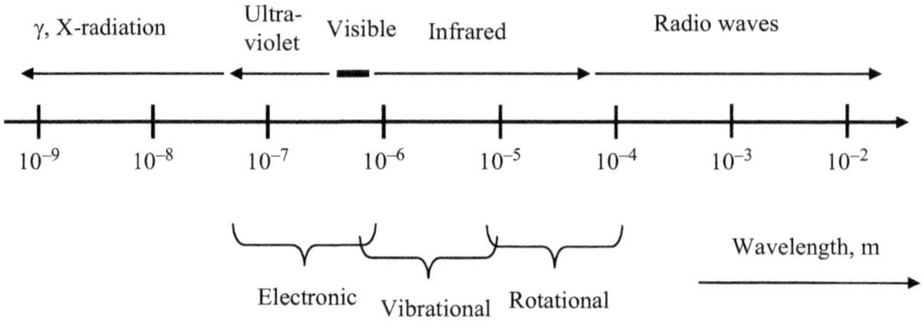

Fig. 3.1 The electromagnetic spectrum—radio waves to X-rays. The types of transition in molecules or atoms (electronic, vibrational, rotational) induced by radiation of different wavelength ranges are indicated

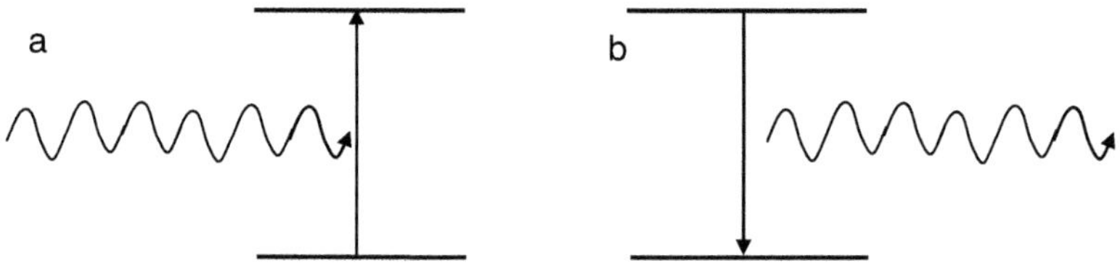

Fig. 3.2 Interaction of radiation with matter: Upon absorption of a photon with appropriate energy, the atom or molecule enters an excited state (**a**). Return to the ground state can occur by collisions with other molecules or atoms, eventually converting the photon energy into heat, or by emitting the photon (**b**)

action of radiation with matter: X-rays interact largely by ionising atoms, while UV- and visible radiation will lead to reconfiguration of the outer electron shell of atoms or molecules. This process is frequently called electronic excitation. Radiation with less energy will excite vibrational or rotational states in molecular gases, as indicated in Figs. 3.1 and 3.2 and detailed in Sect. 3.3.1. Excited atoms or molecules can return to the ground state as a result of collisions with other molecules, eventually converting the energy of the absorbed photon into heat, or by re-emitting the photon (see Fig. 3.2b).

3.2 Energy Levels and Transitions in Atoms

Atoms and molecules are quantum mechanical multi-particle systems (for details see e.g. Eisberg and Resnick 1985); their individual states are discrete and described by a set of quantum numbers. The states relevant to chemistry and optical spectroscopy are determined by the configuration of the outer (loosely bound) electrons.

In atoms, the energy levels E_n are determined by the quantum numbers n, l, m, s, m_s of a single outer electron, where l and m enumerate components of the orbital angular momentum of the electron, while s and m_s characterise its intrinsic angular momentum or spin. For small atoms, E is in good approximation given by the "Rydberg formula":

$$E_n = -\frac{R_{\mathrm{Ry}}}{n^2} \tag{3.1}$$

where R_{Ry} denotes the Rydberg constant for the particular atom. Note that, by convention, binding energies are negative; they can be thought of as the energy that was released when free electrons were attached to the nucleus (or nuclei in the case of a molecule). The energy of a photon exchanged (i.e. absorbed or emitted) by an atom (or molecule) then corresponds to the difference of the energy of two particular states. Applying Planck relationship to the difference between the two states, ΔE_{el}, and frequency ν (and thus wavelength) of a photon we obtain:

$$\Delta E_{\mathrm{el}} = E_{n_1} - E_{n_2} = R_{\mathrm{Ry}} \left(\frac{1}{n_1^2} - \frac{1}{n_2^2} \right) = h\nu_{1,2} \tag{3.2}$$

where n_1, n_2 denote the lower, upper state of the atom and $\nu_{1,2}$ the frequency of the photon, respectively. The quantity ΔE_{el} signifies the fact that we deal with an "electronic" transition, i.e. one where the configuration of the electron shell changes.

Frequently, the state of an atom (or molecule) is determined by several outer electrons. The relevant quantum numbers are the sum of the individual orbital angular momenta $L = \Sigma l_{\mathrm{i}}$, and the sum of the electron spins $S = \Sigma s_{\mathrm{i}}$. Also, frequently the relevant quantum number is one (j, or J) that characterises the sum of orbital angular momentum and electron spin. States with the same main quantum number n and different quantum numbers L or S (or J) have slightly different energies, the so called "fine structure" of the transitions.

3.3 Energy Levels and Transitions in Molecules

Besides electronic states of different energy, molecules have two excitation schemes not found in atoms: first, the rotation of the entire molecule and thus angular momentum of the entire molecule. The rotation of the molecule should not to be confused with the orbital angular momentum due to the electrons (which can be additionally present in molecules just as in atoms and spin of the constituent elementary particles of the molecules). Second, the vibration of the atoms within the molecule relative to each other.

The typical transition energies are:

- Electronic transitions on the order of 1 eV. Corresponding wavelengths for transitions between an electronically excited state and the ground state correspond to the visible- or near UV spectral ranges.
- Vibrational transitions on the order of 0.1 eV. Corresponding wavelengths for transitions between a vibrationally excited state and the ground state are in the infra red (IR) spectral range.

- Rotational transitions on the order of 10^{-3}–10^{-2} eV. Corresponding wavelengths for transitions between rotationally excited state and the ground state are in the sub-mm or microwave range.

The electronic states of molecules are designated by the following (traditional) designations:

$$[\text{State}]^{2S+1}[\text{AM}]_{\Lambda+\Sigma}$$

where:

- [State] is a code for the particular electronic state of the molecule, with [State] $= X$ denoting the lowest energy (ground) state
- S denotes total electron spin just as in the case of atoms;
- [AM] is the code for the angular momentum. This is either the projection of the total orbital momentum on the molecule's axis or a combination of the total orbital momentum with the rotation of the molecule, depending on the coupling case. [AM] $= 1, 2, 3, \ldots$ is coded as Σ, Π, $\Delta, \ldots$.
- $\Lambda + \Sigma = |M_L| + M_S$ denotes the projection of the total angular momentum of the electrons on the axis connecting the nuclei of the molecule.

For instance, the ground state of the NO molecule with the angular momentum 2 and $S = 1/2$ is designated as:

$$X^2\Pi$$

3.3.1 Rotational Energy Levels and Transitions

The angular momenta (total orbital angular momentum, total electron spin, "nuclear" rotation) in a molecule can couple in different ways, which are denoted as "Hund's Coupling Cases" to form a total angular momentum **J** and associated quantum number J.

Rotational energy levels in a molecule are given by:

$$E_j = B \cdot J(J+1) \tag{3.3}$$

where $B = \frac{\hbar^2}{2\Theta}$ denotes the rotational constant of the particular molecule and rotation mode (axis) with the moment of inertia Θ with respect to this axis. In first approximation, Θ is assumed to be independent of J (rigid rotor model). However, molecules are not rigid, i.e. the atoms within a molecule can move compared to each other (Sect. 3.3.2), therefore Θ will increase somewhat at higher values of J compared to its value at low rotational levels. As a consequence $B \propto 1/\Theta$ will decrease.

Transitions are subject to specific selection rules: The difference in the angular momentum quantum number of initial and final states is given as $\Delta J = \pm 1$, since the photon exchanged with the atom/molecule has a spin (intrinsic angular momentum) of unity. The transitions are denoted as "P-branch" ($\Delta J = -1$), "Q-branch" ($\Delta J =$

0, which only can occur if electronic or vibrational transitions take place at the same time), and "R-branch" ($\Delta J = +1$). Consequently, the photon energy $h\nu = \Delta E$ of allowed transitions is given by the energy difference of two consecutive states:

$$\Delta E_j = E_{j+1} - E_j = B \cdot [(J+1) \cdot (J+2) - J \cdot (J+1)] = 2B(J+1) \propto J \tag{3.4}$$

In other words, the energy of absorption or emission lines varies proportionally to the rotational quantum number of the ground state; the spectra consist of lines with equidistant spacing (the energy difference of two neighbouring lines being just $2B$). The energy differences between consecutive states are on the order of $10^{-3} - 10^{-2}$ eV, the corresponding wavelengths of photons exchanged in such "purely rotational" transitions in the sub-mm- or microwave ranges. These energies are on the order of the thermal kinetic energy of molecules at room temperature, thus most molecules are rotationally excited under ambient conditions (see below).

3.3.2 Vibrational Energy Levels

Vibration of molecules can, in first approximation, be treated as harmonic oscillation, where the vibrational energy levels are given by:

$$E_\nu = \left(\nu + \frac{1}{2}\right) \cdot \hbar\omega_0 \tag{3.5}$$

where $\nu = 0, 1, 2, \ldots$ denotes the vibrational quantum number (vibration level) and $1/2\hbar\omega_0$ is the zero-point energy of the molecular oscillator. Thus energies of different vibrational states vary in proportion to the vibrational quantum number ν. In addition to vibrational excitation, a molecule can also be rotationally excited (in fact is very likely to be at ambient temperatures). Thus each vibrational transition splits into a series of rotational lines (see Fig. 3.3). Energies of vibrational transitions are on the order of 0.1 eV, with the corresponding wavelengths in the IR spectral range.

3.3.3 Electronic Energy Levels

Analogous to atoms, molecules can exist in several, distinct configurations of the electronic shell; each configuration has its own set of vibrational and rotational levels, as shown in Fig. 3.3. The situation is further complicated by the fact that the moment of inertia Θ, and thus the rotational constant B, can be different in the ground and excited electronic states (usually denoted as B'' and B', respectively). This is largely due to different inter-atomic distances in the molecule in its different electronic states. As a consequence, the rotational line spacing is no longer equidistant. Rather, the lines tend to group towards a certain wavelength either below ("blue shaded") or above ("red shaded") the wavelength given by electronic and vibrational states alone. The line positions are given by the "Fortrat Parabolas" determined by the different rotational constants (B'' and B').

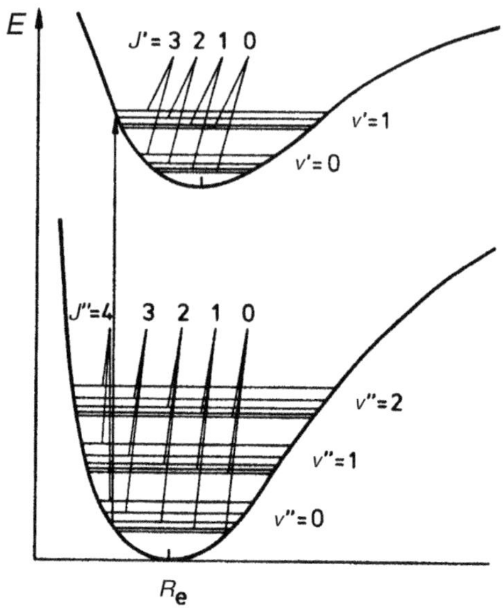

Fig. 3.3 Sketch of ro-vibrational energy levels of two electronic states of a (diatomic) molecule. Electronic energies are given as a function of the distance between the nuclei of the atoms within the molecule. They are minimal at a certain distance R_e, which is usually different in the different electronic states. The equidistant, horizontal groups of lines denote the lowest rotational levels of the vibrational states (ground state: $\nu'' = 0, 1, 2$, excited state: $\nu' = 0, 1$). Note that the energy separation of the rotational states is exaggerated in comparison to the vibrational energies

3.4 Population of States

The relative population $n(E)$ of excited states of an atom or molecule with the energies E_1, E_2 above the ground state (with $E = E_1$) is given by the Boltzmann distribution:

$$\frac{n(E_2)}{n(E_1)} = \frac{G_2}{G_1} \cdot e^{-\frac{E_2 - E_1}{kT}} \tag{3.6}$$

Here T denotes the absolute temperature and k the Boltzmann constant ($k = 1.38 \times 10^{-23}$ J/K), while G_1, G_2 are the statistical weights or degeneration factors (i.e. the number of different states with the same energy E) of the respective states.

In the case of an atom, $G_n = 2n^2$, while vibrational excitation in molecules is not degenerated (i.e. $G_\upsilon = 1$). Rotational states in molecules are degenerated by $G_J = 2J + 1$. Thus the population of rotational states is given by:

$$\frac{n(E_2)}{n(E_1)} = \frac{2J_2 + 1}{2J_1 + 1} \cdot e^{-\frac{E_2 - E_1}{kT}} = \frac{2J_2 + 1}{2J_1 + 1} \cdot e^{-\frac{B(J_2(J_2+1) - J_1(J_1+1))}{kT}} \tag{3.7}$$

With $E(J) = B \cdot J(J + 1)$, where B is the rotation constant of the particular molecule and rotation axis (see above).

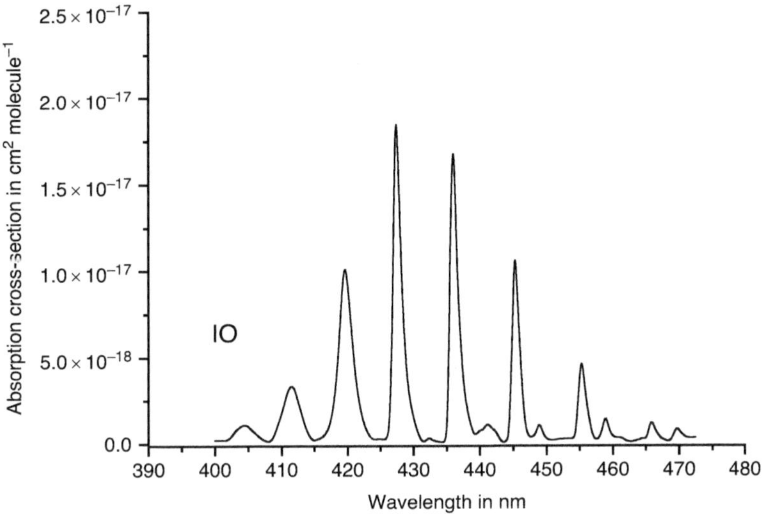

Fig. 3.4 Example of the (ro-) vibrational structure of the electronic spectrum of the iodine oxide (IO) molecule. Shown is the absorption cross-section as a function of wavelength

The kinetic energy of a molecule of around 0.03 eV at room temperature is comparable to or larger than the lowest rotational energy level, but smaller than the lowest vibrational energy level of most molecules. Therefore, several rotational states are usually populated, but only a small fraction of the molecules is vibrationally excited.

3.5 Molecular Spectra

As outlined above, the UV–visible spectra of molecules exhibit vibrational and rotational structures. Often, the rotational structure consists of many lines that are spaced closer than their line-width. This leads to a quasi-continuous absorption within a vibrational "band". Figure 3.4 shows the cross-section spectrum of the IO molecule as an example. A striking feature of the spectrum is the different strengths of its vibrational bands. This can be explained by the Franck–Condon principle stating that the electronic transition from the ground state to the excited state upon absorption of a photon by the molecule is much faster than the time needed for one vibration. Since the distance between the nuclei is usually different in the ground state vs. electronically excited state, the molecule will most likely find itself in a certain, excited vibrational state. Transitions to higher and lower vibrational states are less likely, thus having a smaller absorption cross-section, as shown in Fig. 3.4.

3.6 Broadening Mechanisms and Line Width of Absorption Lines

We first consider an excited atom or molecule with the energy E_0 above the ground state (with $E = 0$). According to classical electrodynamics, the energy of the excited electrical dipole will decrease exponentially with time t due to emission of electromagnetic radiation:

$$E(t) = E_0 \mathrm{e}^{-\delta \mathrm{t}} \tag{3.8}$$

where δ denotes some damping constant, and $\tau = 1/\delta$ is the time in which the energy is decayed to 1/e (with e = 2.7182. denoting the base of the natural logarithm) of its initial value. Accordingly, the corresponding amplitude of the electric field $\widehat{E}(t)$ of the emitted radiation will exponentially decrease with time t:

$$\widehat{E}(t) = \widehat{E}_0 \mathrm{e}^{-\delta \mathrm{t}} \cdot \cos(\omega_0 t + \varphi) \tag{3.9}$$

Here ω_0 denotes the frequency of the oscillation, which is related to the energy E_0 of the excited state, $\omega_0 = 2\pi E_0/h$, while φ is the phase of the oscillation. Since the amplitude $E(t)$ is damped, the radiation cannot be truly monochromatic. This is quantified by calculating the Fourier analysis of $E(t)$:

$$E(t) = \frac{1}{\sqrt{2\pi}} \int_{-\infty}^{\infty} a(\omega) \mathrm{e}^{i\omega t} d\omega \tag{3.10}$$

with the spectral function $a(\omega)$:

$$a(\omega) = \frac{E_0}{2} \left[\frac{\mathrm{e}^{ij}}{i(\omega - \omega_0) - \frac{\delta}{2}} - \frac{\mathrm{e}^{-ij}}{i(\omega + \omega_0) + \frac{\delta}{2}} \right] \tag{3.11}$$

Since $\delta \ll \omega_0$ at frequencies ω near the resonance frequency ω_0, we can neglect the second term in the square bracket of (3.11). We obtain the radiation intensity $I(\omega)$ as the square of $a(\omega)$ (and normalising the integral over I to unity):

$$I(\omega) d\omega = I_0 \frac{\frac{\delta^2}{4}}{(\omega - \omega_0)^2 - \frac{\delta^2}{4}} d\omega \tag{3.12}$$

This equation is known as the Lorentz distribution; its maximum is at $\omega = \omega_0$ while $I(\omega) = 0.5 \cdot I(\omega_0)$ at $\omega = \omega_0 \pm \delta/2$, i.e. δ denotes the full width at half maximum (FWHM) of the distribution.

3.6.1 The Natural Line Width

The classical image of an oscillating dipole described above cannot directly be applied to an excited atom or molecule; however, the correspondence principle suggests that, at least at large quantum numbers, (3.12) should be a good approximation. In practice it turns out that (3.12) can be used to describe the profile of an emission line of an isolated atom when using an empirical δ. In addition, there might be more than one level which an excited atom or molecule can decay to, thus $\delta_{\mathrm{L}} = \Sigma\delta_{\mathrm{i}}$ has to be used instead of δ. The individual δ_{i} describe the transition frequency to the states (including the ground state) to which the excited state under consideration can decay into. Thus, we obtain the description of the natural line shape of an isolated atom or molecule, i.e. the intensity distribution centred at ω_0:

$$I_{\mathrm{L}}(\omega)d\omega = I_0 \cdot \frac{\frac{\delta_{\mathrm{L}}^2}{4}}{(\omega-\omega_0)^2 + \frac{\delta_{\mathrm{L}}^2}{4}} d\omega \tag{3.13}$$

with the natural line width δ_L (FWHM), which essentially depends on the lifetime of the excited state of the molecule, but *not* directly on the energy difference of the transition and thus the frequency of the emitted (absorbed) radiation. For allowed transitions, the natural lifetime is on the order of 10^{-8} s, corresponding to $\delta_{\mathrm{L}} = 10^8$ Hz. For radiation of $\lambda = 400$ nm wavelength ($\omega = 2\pi c/\lambda \approx 4.7 \times 10^{15}$ Hz), this would amount to $\Delta\omega/\omega \approx 2.1 \times 10^{-8}$ or $\Delta\lambda \approx 8.5 \times 10^{-6}$ nm (≈ 0.01 pm).

3.6.2 Pressure Broadening (Collisional Broadening)

Collisions of molecules (among themselves or with other types of molecules in the gas) will reduce the lifetime of the excited state below the value given by the radiation transition alone, see (3.13), which determines the natural line width. The shape of a pressure broadened line is therefore Lorentzian as given in (3.12), with $\delta = \delta_{\mathrm{P}}$ as the width of the pressure broadened line. In principle, the collisional damping constant δ_{p} is given by the product of the gas kinetic collision frequency, z_{AB}, and the deactivation probability, p_{AB}, per collision (here A refers to the species under consideration, B to the gas providing the pressure). Since p_{AB} can be different for each molecular species, the amount of pressure broadening depends not only on the molecular species under consideration but also on the surrounding gas. Of particular interest are the self broadening (the species itself is the pressure gas, $p_{\mathrm{AB}} = p_{\mathrm{AA}}$) and air broadening (species A is occurring in traces in air, which is species B). In the latter case, the deactivation is almost exclusively occurring in collisions between air molecules and species A molecules. The collision frequency is directly proportional to the product of gas density (and thus to pressure $p\times T^{-1}$) and to the average molecular speed (and thus to $T^{1/2}$). In summary, the pressure-broadened line width δ_{P} is therefore given by:

$$\delta_P(p,T) = \delta_P(p_0,T_0) \cdot \frac{p}{p_0} \cdot \sqrt{\frac{T_0}{T}} = \delta_0 \cdot \frac{p}{p_0} \cdot \sqrt{\frac{T_0}{T}} \tag{3.14}$$

Here $\delta_P(p_0, T_0) = \delta_0$ denotes the pressure broadening at some reference pressure and temperature. Typical values at one atmosphere for pressure broadening of small molecules in the near UV are $\Delta\lambda \approx 1$ pm.

3.6.3 Doppler Broadening

In reality, the atoms or molecules of a gas are not at rest but permanently moving. This Brownian motion has two main consequences:

1. The energy E can be not only dissipated by radiation, but also by collisions with other molecules present in the gas. This effect manifests itself in an increased $\delta_P > \delta_L$, as discussed above. Since the effect can be described in terms of a damped oscillation, the line shape is also given by (3.13).
2. The Doppler effect will change the frequency ω_0 of the emitted radiation according to (in first approximation):

$$\omega = \omega_0 \cdot \left(1 + \frac{v_x}{c}\right) \tag{3.15}$$

where v_x denotes the velocity component of the emitting molecule with respect to the observer and c the speed of light. The distribution of an individual component of the velocity $N(v_x)dv_x$ (denoting the number N of molecules that have the x-component of the velocity in the range of $v_x \ldots v_x + dv_x$) is Gaussian (this should not be confused with the distribution of the absolute value of the velocity, which is given by the Maxwell–Boltzmann distribution).

$$N(v_x)\,dv_x = \text{const} \cdot e^{-\frac{mv_x^2}{2RT}}\,dv_x \tag{3.16}$$

where m denotes the atomic or molecular weight (kg/mol) and R the universal gas constant. The intensity distribution centred at ω_0 follows as:

$$I_D(\omega)d\omega = I_0 \cdot e^{-\frac{mc^2}{2RT} \cdot \frac{(\omega_0-\omega)^2}{\omega_0^2}} \cdot d\omega = I_0 \cdot e^{-\frac{\delta_D}{2} \cdot (\omega_0-\omega)^2} \cdot d\omega \tag{3.17}$$

with the FWHM of a purely Doppler broadened line:

$$\delta_D = \omega \frac{2\sqrt{2R\ln 2}}{c} \cdot \sqrt{\frac{T}{m}} = \omega \cdot K_D \cdot \sqrt{\frac{T}{m}} \tag{3.18}$$

The constant $K_D = 2\sqrt{2R\ln 2}/c$ in the above equation has the value of $2.26 \times 10^{-8}\ (\text{kg/mol})^{1/2}/\text{K}^{1/2}$. For radiation with $\lambda = 400$ nm, at $T = 300$ K and a molecular species with the molecular weight of air (0.03 kg/mol), we obtain $\delta_D/\omega_0 = \Delta\omega/\omega = 2.26 \times 10^{-8}$, or a Doppler width of $\Delta\lambda = 9 \times 10^{-4}$ nm (about 1 pm).

3.6.4 Realistic Broadening in the UV- and Visible Spectral Ranges

Of the three broadening mechanisms discussed earlier, the natural line width, i.e. natural lifetime of the excited state, leads to very small broadening (typ. 0.01 pm), while the effect of pressure broadening and Doppler broadening are roughly equal (typ. 1 pm) in the UV–visible spectral region. Note that the two types of line shapes discussed above (Lorentzian and Gaussian) have quite different properties:

(1) The intensity (or absorption cross-section) of a Lorentzian line, e.g. (3.13), only decays with the square of the deviation from the centre frequency (proportional to $1/(\omega - \omega_0)^2$), and thus a large fraction of the total emission (or total absorption) is in the wings of the line. In comparison, the Gaussian profile, e.g. (3.17), decays exponentially (proportional to $\exp[-(\omega - \omega_0)^2]$), and consequently there is very little emission (or absorption) in the wings of the line.
(2) In the case of Lorentzian (i.e. natural- or pressure) broadening, the half-width (3.13) is independent of the frequency ω (or wavelength λ), while Gaussian (i.e. Doppler) broadening (3.18) is proportional to ω (or $1/\lambda$). Therefore, Gaussian broadening is, under atmospheric conditions, usually negligible at low frequencies, i.e. in the far IR or microwave range, becomes noticeable in the near IR and dominates in the short-wavelength UV.

As can be seen from the examples above, the natural line width is usually negligible compared to Doppler and/or pressure broadening. In the visible and UV spectral range, Doppler (Gaussian) broadening is comparable to pressure (Lorentzian) broadening. The resulting line shape can be obtained by convoluting the Gaussian and Lorentzian line shapes, which is known as Voigt shape (neglecting the natural line width):

$$I_V(\omega, \delta_D, \delta_P) = \int_{-\infty}^{\infty} I_D(\omega', \delta_D) \cdot I_P(\omega - \omega', \delta_P)\, d\omega' \tag{3.19}$$

Since the convolution can only be performed by numerical calculation, approximations of the Voigt shape have been developed for describing line shapes where both broadening mechanisms contribute. Figure 3.5 shows a comparison of the three line shapes.

Since absorption is the inverse process to emission, the formulas derived above also describe the line shapes of atomic- or molecular absorption lines, and thus the absorption cross-section $\sigma(\omega)$ [or $\sigma(\lambda)$] will show the same wavelength dependence as $I(\omega)$ in (3.19).

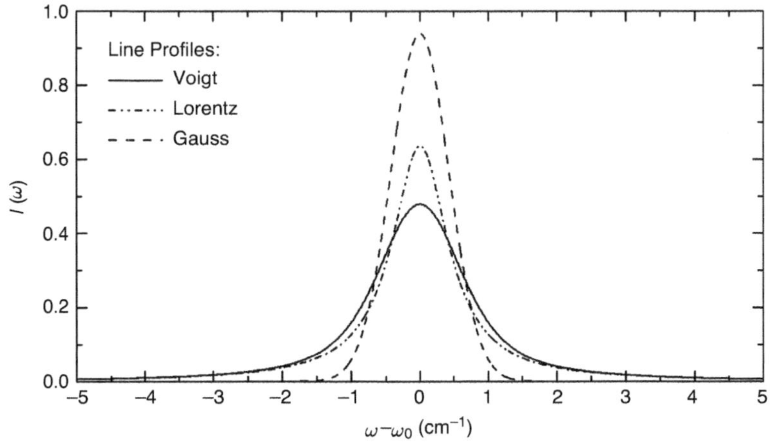

Fig. 3.5 Different line shapes: Lorentz (due to collision broadening), Doppler (or Gaussian, due to thermal motion of molecules), and Voigt shape resulting from simultaneous Doppler and collision broadening

3.7 Spectroscopic Techniques for Chemical Analysis

The interaction of radiation with matter provides a powerful tool for a wide variety of investigations, including the structure of atoms and molecules and the chemical composition of complex mixtures like the atmosphere. In principle, we distinguish between two different experimental approaches of spectroscopy: Fluorescent techniques and absorption spectroscopy.

3.7.1 The Fluorescence Techniques

The emission of radiation by excited atoms or molecules is called fluorescence. If bound states are involved in the excited state and ground state, the radiation has a characteristic set of wavelengths with the photon energy corresponding to the energy difference of the participating states (see Sects. 3.2 and 3.3).

The energy to change the molecule from the ground state to its excited state (i.e. the energy then radiated by the molecule) can be supplied by various mechanisms including thermal excitation (at ambient temperatures in the mid-infrared), electron bombardment, chemical reactions, or absorption of radiation. The latter mechanism, (light)-induced fluorescence, is employed in several measurement instruments for atmospheric trace gases (e.g. OH). A general outline of the principle is given in Fig. 3.6. In the simplest case (Fig. 3.6a), the number of fluorescence photons emitted by the molecules (or atoms) in a sample volume illuminated by the excitation radiation is proportional to the number of exciting photons multiplied by the number of

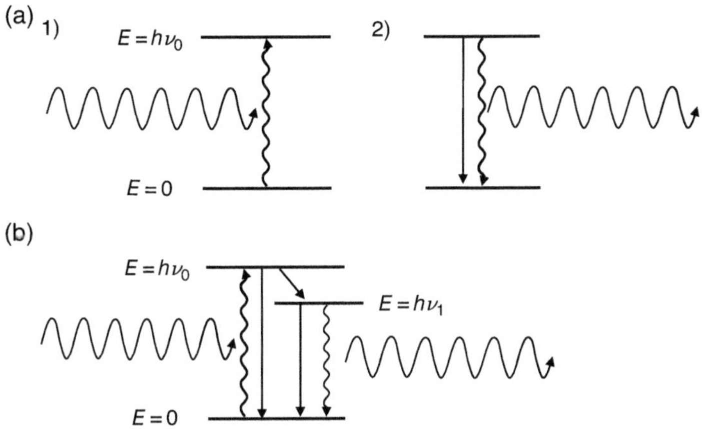

Fig. 3.6 The principle of trace gas detection by induced fluorescence spectroscopy. Scheme (**a**) uses two energy levels within the molecule, the energy difference of which is just $\Delta E = h\nu_0$. After excitation (1) by a photon with energy $E = h\nu_0$, the excited state can lose its energy (be quenched, 2) either by collision with other molecules (*straight line*) or radiating another photon with the energy $E = h\nu_0$ (*undulated line*). Scheme (**b**) uses three energy levels within the molecule. The energy difference between the lowest and the highest level is $\Delta E = h\nu_0$. After excitation by a photon with energy $E = h\nu_0$, the excited state can lose its energy completely or partially by collision with other molecules. The observed photon has a lower energy (longer wavelength) $E = h\nu_1$

trace gas molecules with matching transitions. In practice, geometrical factors, the absorption cross-section, and quenching also must be taken into account.

3.7.2 Absorption Spectroscopy

Absorption spectroscopy makes use of the absorption of electromagnetic radiation by matter (Fig. 3.7). Quantitatively, the absorption of radiation is expressed by Lambert–Beer's law (or Bouguer–Lambert law, see Sect. 6.1):

$$I(\lambda) = I_0(\lambda)\exp[-L \cdot \sigma(\lambda) \cdot c] \tag{3.20}$$

where $I_0(\lambda)$ denotes the initial intensity emitted by some suitable source of radiation, while $I(\lambda)$ is the radiation intensity after passing through a layer of thickness L, where the species to be measured is present at the concentration (number density) c. The quantity $\sigma(\lambda)$ denotes the absorption cross-section at the wavelength λ. The

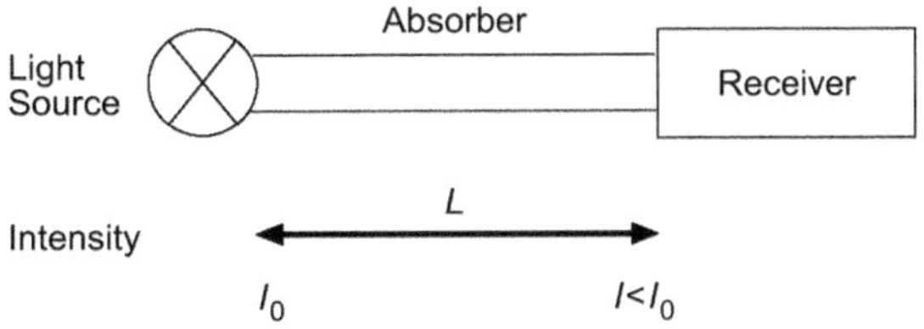

Fig. 3.7 The basic principle of absorption spectroscopic trace gas detection

absorption cross-section is a characteristic property of any species. The absorption cross-section $\sigma(\lambda)$ can be measured in the laboratory, while the determination of the light-path length L is in many cases trivial. Once those quantities are known, the trace gas concentration c can be calculated from the measured ratio $I_0(\lambda)/I(\lambda)$:

$$c = \frac{\log\left(\frac{I_0(\lambda)}{I(\lambda)}\right)}{\sigma(\lambda) \cdot L} = \frac{D(\lambda)}{\sigma(\lambda) \cdot L} \tag{3.21}$$

The expression

$$D(\lambda) = \log\left(\frac{I_0(\lambda)}{I(\lambda)}\right) \tag{3.22}$$

is called the 'optical density' of a layer of a given species. Note that in the literature, the decadic as well as the natural logarithm are used in the definition of the optical density. In this book we always use the natural logarithm.

Radiation Transport in the Atmosphere

4

Absorption based measurements of trace gases in the open atmosphere have considerable advantages, such as avoidance of wall losses and potential for remote sensing. In order to understand the implication of light-path arrangements, an in-depth knowledge of the transmission of radiation in the atmosphere is necessary. The basic issues are extinction of radiation in the atmosphere and possible generation of scattered light. This chapter describes the basic quantities pertinent to radiation transport and the fundamental laws governing radiation transport in absorbing and scattering media, i.e. the atmosphere.

4.1 Basic Quantities Related to Radiation Transport

A light source will emit a certain amount of energy W in the form of radiation.

1. The radiant flux $\Phi(\lambda)$ (sometimes called radiant power) is defined as radiation energy per unit time (regardless of the direction of its emission):

$$\Phi(\lambda) = \frac{\text{Radiated energy}}{\text{Time interval}} = \frac{dE(\lambda)}{dt} \quad \text{unit: } \frac{\text{Ws}}{\text{s}} = \text{W} \tag{4.1}$$

   The radiant flux is measured in joules/second or Ws/s = watt.

2. The irradiance $B(\lambda)$ is defined as the radiant flux $\Phi(\lambda)$ received by an 'illuminated' area A_e:

$$B(\lambda) = \frac{\Phi(\lambda)}{A_e} \quad \text{unit: } \frac{\text{W}}{\text{m}^2} \tag{4.2}$$

U. Platt and J. Stutz, *Differential Optical Absorption Spectroscopy*, Geophysics and Environmental Physics, https://doi.org/10.1007/978-3-662-68639-3_4

3. The radiation intensity $I_R(\lambda)$ (Ω = solid angle):

$$I_R(\lambda) = \frac{\Phi(\lambda)}{\Omega} \quad \text{unit: } \frac{\text{W}}{\text{sr}} \tag{4.3}$$

   In the following we will omit the subscript R, and use I for the radiant intensity or simply 'intensity'.

4. The radiance L_R or brightness (A_s = radiating area) of a light source:

$$L_R(\lambda) = \frac{\Phi(\lambda)}{\Omega \cdot A_s} \quad \text{unit: } \frac{\text{W}}{\text{sr} \times \text{m}^2} \tag{4.4}$$

All areas (A_e and A_s) are assumed to be oriented perpendicular to the direction of propagation of the radiation. In addition, radiation quantities are frequently given as wavelength or frequency-dependent quantities, i.e. the specific radiance $L_R(\lambda)\,d\lambda$.

4.2 Interaction Processes of Radiation in the Atmosphere

There are a multitude of processes in which radiation interacts with the atmosphere:

- Absorption, i.e. radiation is removed from the radiation field and converted into some other form of energy, such as heat or chemical energy (e.g. upon photolysis of a molecule).
- Elastic scattering changes the direction of propagation of an individual photon, but not its energy. Scattering can be due to air molecules (Rayleigh scattering) or aerosol particles (Mie scattering) present in air.
- Inelastic scattering: As in elastic scattering, the direction of a photon is changed, but inelastic scattering also changes its energy. Inelastic scattering by molecules is called Raman scattering; here, the energy of the scattered photon can be reduced by the amount of energy transferred to the scattering molecule (Stokes scattering). Similarly, energy can be transferred from the (thermally excited) molecules to the photon (anti-Stokes scattering).
- Thermal emission from air molecules and aerosol particles: The emission at any given wavelength cannot exceed the Planck function (or emission from a black body) for the temperature of the atmosphere; thus, noticeable thermal emission only takes place at infrared wavelengths longer than several micrometres. Due to Kirchhoff's law, absorbing gases (such as CO_2, H_2O and O_3, but not the main components of air N_2, O_2 and Ar) can emit radiation.
- Aerosol fluorescence: Excitation of molecules within aerosol particles by radiation can result in (broad band) fluorescence. This process will not be discussed further here.

4.2.1 Absorption and Extinction Processes

Radiation is absorbed by molecules in the atmosphere (such as ozone, oxygen, nitrogen dioxide, or water vapour) or aerosol (such as soot). The absorption of solar energy and thermal emissions from the earth surface in the atmosphere are key processes in the climate system of earth. Absorption of solar UV radiation at wavelengths below about 300 nm by the stratospheric ozone layer (see Chap. 2) is an important prerequisite for life on land.

The quantitative description of atmospheric absorption follows Lambert-Beer's law (Sect. 3.7.2 and 6.2), where the specific absorption of a gaseous or particulate compound is represented by the absorption cross-section $\sigma(\lambda)$. As explained in the next sections, scattering of radiation in a narrow beam (see Fig. 4.3) can also be described using the Lambert-Beer equation by introducing a scattering cross-section. Because absorption and scattering often occur in the same air mass, it is common to refer to the sum of both as extinction, and the sum of absorption and scattering cross-section as the extinction cross-section.

4.2.2 Rayleigh Scattering

Elastic scattering (i.e. scattering without change of the photon energy) by air molecules is called Rayleigh scattering. In narrow beams, scattering will weaken the intensity along the beam in the same fashion as molecular absorption. This extinction can thus be described by Lambert-Beer's law and a scattering (or for a narrow beam extinction) cross-section.

The Rayleigh scattering cross-section $\sigma_R(\lambda)$ in cm^2 is given by (Rayleigh 1899):

$$\sigma_R(\lambda) = \frac{24\pi^3}{\lambda^4 N_{air}^2} \cdot \frac{\left(n_0(\lambda)^2 - 1\right)^2}{\left(n_0(\lambda)^2 + 2\right)^2} \cdot F_K(\lambda) \approx \frac{8\pi^3}{3\lambda^4 N_{air}^2} \cdot \left(n_0(\lambda)^2 - 1\right)^2 \cdot F_K(\lambda) \tag{4.5}$$

where λ denotes the wavelength in cm, $n_0(\lambda)$ is the wavelength-dependent index of refraction of air, N_{air} is the number density of air (e.g. 2.4×10^{19} molec. cm^{-3} at 20 °C, 1 atm), and $F_K(\lambda) \approx 1.061$ is a correction for anisotropy (polarisability of air molecules).

Note that $n_0(\lambda)^2 - 1 \approx 2(n_0(\lambda) - 1) \propto N_{air}$, since $n_0 \approx 1$ [in fact $n_0(550\,\text{nm}) = 1.000293$], and $n_0 - 1 \propto N_{air}$; thus, $\sigma_R(\lambda)$ is essentially independent of N_{air}.
For simple estimates, the Rayleigh scattering cross-section can be written as:

$$\sigma_R(\lambda) = \sigma_{R0} \cdot \lambda^{-4}, \quad \sigma_{R0} \approx 4.4 \times 10^{-16}\,\text{cm}^2\,\text{nm}^4 \text{ for air} \tag{4.6}$$

On the basis of (4.6), a better expression of the Rayleigh scattering cross-section (in cm^2) is given by Nicolet (1984), (note that here λ has to be given in μm):

$$\sigma_R(\lambda) \approx \frac{\sigma_{R0}}{\lambda^{4+x}}, \quad \sigma_{R0} = 4.02 \cdot 10^{-28} \tag{4.7}$$

with $x = 0.04$ for $\lambda > 0.55\,\mu\text{m}$
$x = 0.389\lambda + 0.09426/\lambda - 0.3328$ for $0.2\,\mu\text{m} < \lambda < 0.55\,\mu\text{m}$.
An even more accurate treatment is found in Penndorf (1957).

The Rayleigh extinction coefficient $\varepsilon_R(\lambda)$ is then given by:

$$\varepsilon_R(\lambda) = \sigma_R(\lambda) \cdot N_{\text{air}} \tag{4.8}$$

The angular distribution of the scattered intensity is described by the scattering phase function, which is normalised to 1 when integrated over all scattering angles. Scattering has a cylindrical symmetry along the light propagation axis and thus only one angle is needed to define the phase function. By convention the forward scattering direction, i.e. in direction of the light propagation, is assigned a scattering angle of zero, while the backwards scattering angle is 180°. The Rayleigh phase function (see Fig. 4.1) is given by:

$$P(\cos\vartheta) = \frac{3}{4}\left(1 + \cos^2\vartheta\right) \tag{4.9}$$

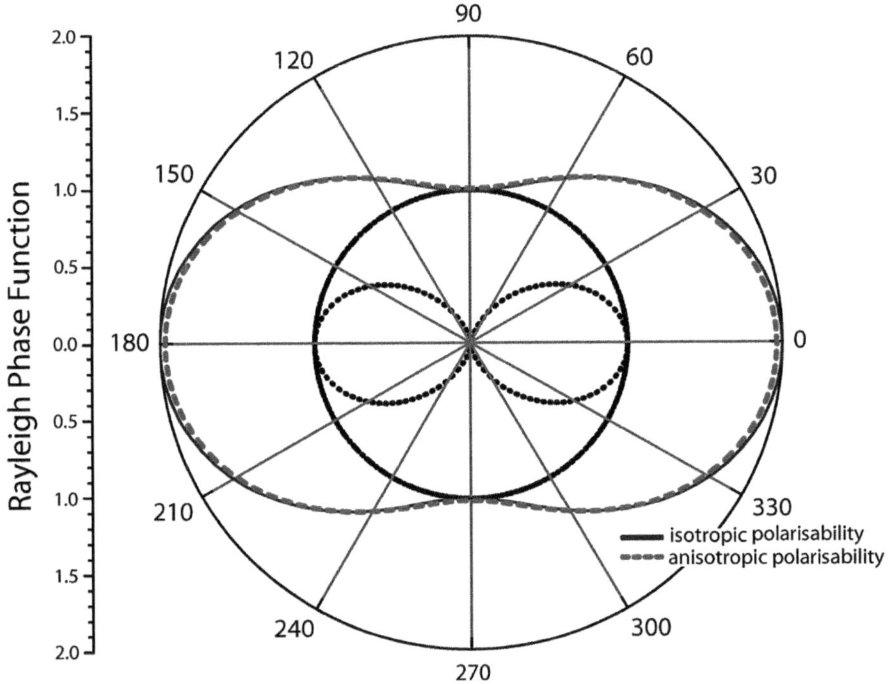

Fig. 4.1 Polar diagram of the Rayleigh scattering phase function $\Phi(\vartheta)$ for non-polarised incident light. The contribution of light polarised parallel to the scattering plane shows the $\sin^2\vartheta'$ dependence of a Hertz dipole (*dotted line*), with $\vartheta' = \pi/2 - \vartheta$ being the angle between dipole axis and the direction of the incident radiation, while the contribution of light polarised perpendicular to the scattering plane is independent of ϑ (*solid line*). Reprinted from Filsinger, Diploma Thesis, University of Heidelberg, 2004

Table 4.1 Comparison of the total cross-section for the different scattering types for 770 nm, 273 K

Scattering type	Cross-section (cm^2)	Ratio (%)
Rayleigh scattering	1.156×10^{-27}	100
O_2 RRS	7.10×10^{-29}	6.1
N_2 RRS	2.94×10^{-29}	2.5
Air RRS	3.82×10^{-29}	3.3
VRS	$\approx 10^{-30}$	0.1

Rayleigh scattered radiation is polarised. The first term in the right side of Eq. 4.9 describes the polarisation perpendicular to the reference plane defined by a light source, scattering molecule, and observer. It has no angular dependence, as shown with the solid circular line in Fig. 4.1. In contrast, the parallel polarisation direction has a $\cos^2\theta$ dependence (second term in Eq. 4.9), that has minimum at $\theta = 90°$ and maxima in forward and backward direction (dashed line in Fig. 4.1). The sum of the two polarisation direction has a distinct peanut shaped phase function (Fig. 4.1).

It should be noted that for a more precise description of the phase function the anisotropy of the polarisability has to be taken into account. P (cos ϑ) thus becomes (Penndorf 1957):

$$P\,(\cos\vartheta) = 0.7629 \cdot \left(0.9324 + \cos^2\vartheta\right) \tag{4.10}$$

4.2.3 Raman Scattering

While Rayleigh (and Mie) scattering can be regarded as 'elastic' scattering processes, where no energy is transferred between the scattering particle and the photon, 'inelastic' scattering occurs if the scattering particle (i.e. molecule) changes its state of excitation during the scattering process. A part of the photons energy is then passed from the photon to the molecule (Stokes lines, S-branch) or vice versa (Anti-Stokes, O-branch). The term rotational Raman scattering (RRS) is used if only the rotational excitation is affected ($\Delta v = 0$). If the vibrational state also changes, the term (rotational-) vibrational Raman scattering (VRS) is used ($\Delta v = \pm 1$). Only discrete amounts of energy given by the difference between the discrete excitation states of a molecule can be absorbed/emitted. For air (oxygen and nitrogen), RRS frequency shifts of up to $\pm 200\,\text{cm}^{-1}$ occur; for VRS a vibrational shift of $\pm 2331\,\text{cm}^{-1}$ for nitrogen and $\pm 1555\,\text{cm}^{-1}$ for oxygen must be added. The VRS is one order of magnitude weaker than the RRS, see Table 4.1.

In this section, we give a quantitative description of RRS and VRS by O_2 and N_2 (Bussemer 1993; Burrows et al. 1996; Haug 1996; Sioris and Evans 1999). The power density $I_{v,\,J\to v',\,J'}$ given in W/m^2 scattered into the full solid angle 4π involving a transition ($v, J \to v', J'$) is given by (Schrötter and Klöckner 1979):

$$I_{v,\,J\to v',\,J'} = I_0 \cdot \sigma_{v,\,J\to v',\,J'} \cdot L \cdot N \cdot g_J \cdot (2J+1) \cdot \frac{1}{Z} \cdot e^{-E(v,\,J)/kT} \tag{4.11}$$

where I_0 is the incident power density, N is the number of molecules in the scattering volume, L is the length of the latter, and g_J is the statistical weight factor of the initial rotational state due to the nuclear spin. J and v are the rotational and vibrational quantum numbers, respectively. The factor $(2J+1)$ accounts for the degeneracy due to the magnetic quantum numbers while $\exp(-E(v, J)/kT)$ accounts for the population of the initial state of the molecule at temperature T. The state sum Z is given by the product of the rotational state sum Z_{rot} and the vibrational state sum Z_{vib}. The absolute cross-section in (4.11) is given by $\sigma_{v,\,J\to v',\,J'}$ which can be obtained by integration of the differential cross-section $d\sigma_{v,\,J\to v',\,J'}/d\Omega$ over the entire solid angle Ω (note that the term differential refers to the solid angle). The energy of the molecule is characterised by the vibrational (v) and rotational (J) quantum numbers (see Sect. 4.2.3) and is given by

$$E(v, J) = E_{vib}(v) + E_{rot}(J) = hc\tilde{v}\left(v + \frac{1}{2}\right) + hcBJ(J+1) \tag{4.12}$$

assuming no coupling between rotation and vibration. B denotes the rotational constant and $\tilde{\nu} = 1/\lambda = \nu/c$ is the wave number in 1/cm of the ground vibration. Allowed transitions are $\Delta J = 0, \pm 2$, resulting in the Q-, O- and S-branches, and $\Delta v = 0, \pm 1$ for vibrational transitions. Due to the temperatures in the earth's atmosphere, only the ground vibrational state is occupied significantly; thus, leading to only Stokes transitions of the vibrational states. The differential cross-section for an incident light beam with the wave number $\tilde{\nu}$ can be written as:

$$\frac{d\sigma_{v,J\to v',J'}}{d\Omega} = 8\pi^4 \cdot (\tilde{\nu}_{in} + \tilde{\nu}_{v,J\to v',J'})^4 \cdot \left\{ \sum_{ij} \overline{[\alpha_{ij}]^2}_{v,J\to v',J'} \cdot f_{ij}(\Theta) \right\} \tag{4.13}$$

where $\overline{[\alpha_{ij}]^2}_{v,J\to v',J'}$ are the spatial averages of the squares of the polarisability tensor, as given by Long (1977), see Table 4.2. The invariants of this tensor are the average polarisability a, responsible for the isotropic part of the scattered light, and the anisotropy γ for pure rotational transitions, responsible for the anisotropic part, and a' and γ' for vibrational and rotational transitions, respectively (see Young 1981). The Placzek–Teller coefficients $b_{J\to J'}$ of Table 4.2 can be found in Penney et al. (1974) as:

$$\begin{aligned} b_{J\to J} &= \frac{J(J+1)}{(2J-1)(2J+3)} \\ b_{J\to J+2} &= \frac{3(J+1)(J+2)}{2(2J+1)(2J+3)} \\ b_{J\to J-2} &= \frac{3J(J-1)}{2(2J+1)(2J-1)} \end{aligned} \tag{4.14}$$

The geometric factors $f_{ij}(\Theta)$ in (4.13) are determined by the relative orientations of the incident light polarisation, the polarisability tensor and the direction of observation, given by the scattering angle Θ. For incident unpolarised light, they can be

Table 4.2 Squares of the polarisability tensor (from Long 1977)

Transition	$[\alpha_{ii}]^2{}_{v,J\to v',J'}$	$[\alpha_{ij}]^2{}_{v,J\to v',J'}$
Cabannes line $\Delta v = 0,\ \Delta J = 0$	$a^2 + \frac{4}{45} b_{J,J} \cdot \gamma^2$	$\frac{1}{15} b_{J,J} \cdot \gamma^2$
O-, S-branch $\Delta v = 0,\ \Delta J = \pm 2$	$\frac{4}{45} b_{J,J\pm 2} \cdot \gamma^2$	$\frac{1}{15} b_{J,J\pm 2} \cdot \gamma^2$
Q-branch $\Delta v = 1,\ \Delta J = 0$	$(a'^2 + \frac{4}{45} b_{J,J} \cdot \gamma'^2) \cdot \frac{h}{8c\pi^2 \nu}$	$(\frac{1}{15} b_{J,J} \cdot \gamma'^2) \cdot \frac{h}{8c\pi^2 \tilde{\nu}}$
O-, S-branch $\Delta v = 1,\ \Delta J = \pm 2$	$(\frac{4}{45} b_{J,J\pm 2} \cdot \gamma'^2) \cdot \frac{h}{8c\pi^2 \tilde{\nu}}$	$(\frac{1}{15} b_{J,J\pm 2} \cdot \gamma'^2) \cdot \frac{h}{8c\pi^2 \tilde{\nu}}$

expressed by the normalised phase functions $P^{iso}(\Theta)$ and $P^{aniso}(\Theta)$ according to a, a' and γ, γ', respectively:

$$
\begin{aligned}
p^{iso}(\Theta) &= \frac{3}{4}(1 + \cos^2 \Theta) \\
p^{aniso}(\Theta) &= \frac{3}{40}(13 + \cos^2 \Theta)
\end{aligned}
\tag{4.15}
$$

The total cross-sections of the Q-branch of the vibrational transition can be calculated (Schrötter and Klöckner 1979) as:

$$
\begin{aligned}
\sigma^{\mathrm{Q,total}}(T) &= \sum_{v,J} \sigma_{v,J\to v',J} \cdot \frac{1}{Z} \cdot g_J \cdot (2J+1) \cdot e^{-E(v,J)/kT} \\
&= \frac{128\pi^5 (\tilde{\nu}_{in} + \tilde{\nu})^4}{9(1 - e^{hc\tilde{\nu}/kT})} \left[3a'^2 + \frac{2}{3}\gamma'^2 \cdot S(T)_{\Delta J=0} \right] \cdot \frac{h}{8c\pi^2 \tilde{\nu}}
\end{aligned}
\tag{4.16}
$$

Using the weighting factors $S(T)$:

$$
S(T)_{\Delta J} = \sum_J \frac{1}{Z_{rot}} \cdot g_J \cdot (2J+1) \cdot b_{J,J+\Delta J} \cdot e^{-E_{rot}(J)/kT}
\tag{4.17}
$$

which depend slightly on the temperature; the numerical values are given in Table 4.3. The observed wavelength shifts in scattered light spectra depend on the involved transitions and are presented in Table 4.4. The exact values for a' and γ' of O_2 and N_2 are not found in the literature, but the total differential cross-sections of the Q-branch for the O_2 and N_2 have been measured by several authors (for references, see Schrötter and Klöckner 1979). Thus, either a' or γ' can be eliminated from (4.16) yielding (within a relative error of 5%):

(i) $a' = 0 \Rightarrow \gamma'^2{}_{N2} = 6.79 \times 10^{-5}\ \mathrm{cm^4\,kg^{-1}}$

$$\gamma'^2{}_{O2} = 4.44 \times 10^{-5}\ \mathrm{cm^4\,kg^{-1}}$$

Table 4.3 Numerical values of the weighting factors $S(T)$, as used in the calculation of the total cross-section (4.17) of the Q-branch

Temperature	$S(T)_{\Delta J=0}$		$S(T)_{\Delta J=2}$		$S(T)_{\Delta J=-2}$		$\Sigma_{\Delta J=0,\pm 2} S(T)_{\Delta J}$	
	N_2	O_2	N_2	O_2	N_2	O_2	N_2	O_2
234 K	0.2561	0.2617	0.4450	0.4311	0.2989	0.3068	1.0000	0.9997
300 K	0.2552	0.2592	0.4369	0.4246	0.3078	0.3147	0.9999	0.9985
Relative deviation (%)	0.4	1.0	1.8	1.5	−2.9	−2.5	0.01	−0.12

Table 4.4 Observed wave number shifts in scattered light spectra as a function of the transition involved

Vibrational transitions	Rotational transitions		Wave number shifts
$\Delta v = 0$	Q-branch	$\Delta J = 0$	$\tilde{\nu} = 0$
	S-branch	$\Delta J = +2$	$\tilde{\nu} = 4B(J + 3/2)$
	O-branch	$\Delta J = -2$	$\tilde{\nu} = +4B(J - 1/2)$
$\Delta v = 1$	Q-branch	$\Delta J = 0$	$\tilde{\nu}_{N2} = -2331\ \mathrm{cm}^{-1}$
	S-branch	$\Delta J = +2$	$\tilde{\nu}_{O2} = -1555\ \mathrm{cm}^{-1}$
	O-branch	$\Delta J = -2$	$\tilde{\nu} = \tilde{\nu}_{N2/O2} - 4B(J + 3/2)$
			$\tilde{\nu} = \tilde{\nu}_{N2/O2} + 4B(J - 1/2)$

(ii) $\gamma' = 0 \Rightarrow a'^2{}_{N2} = 2.7 \times 10^{-6}\ \mathrm{cm}^4\ \mathrm{kg}^{-1}$

$$a'^2{}_{O2} = 1.8 \times 10^{-6}\ \mathrm{cm}^4\ \mathrm{kg}^{-1}$$

In scenario (i), the isotropic part of the vibrational band is omitted. Therefore, only the anisotropic part is present. Since for linear molecules the Q-branch contains a quarter of the entire scattered light of the vibrational band (Weber and Anderson 1973), the entire scattered radiance of the vibrational band can be calculated from the radiance of the Q-branch (see Table 4.5). In scenario (ii), the anisotropic part is set to zero, and thus there are no O- and S-branches in this vibrational band.

Table 4.5 shows the relative differential cross-sections of the elastic component (Cabannes line), the rotational Raman and vibrational Raman parts, setting the contribution of the elastic component to 100%. The values were derived for a scattering angle of 0° (or 180°) and depend slightly on the incident wavelength (neglecting the weak wavelength dependence of the averaged polarisability a and the anisotropy γ).

Table 4.5 The relative differential cross-sections of the elastic component (Cabannes line)

Scenario	Relative differential cross-section $d\sigma/d\Omega$		
	Cabannes ($\Delta v = 0$, $\Delta J = 0$): RRS ($\Delta v = 0$, $\Delta J = \pm 2$): VRS ($\Delta v = 1$, $\Delta J = 0, \pm 2$)		
	N_2	O_2	Air ($0.8 \cdot N_2$–$0.2 \cdot O_2$)
$a' = 0, \gamma' \neq 0$	100:1.98:0.27	100:5.31:0.32	100:1.58:0.22–100:1.01:0.06
$a\prime \neq 0, \gamma' = 0$	100:1.94:0.07	100:5.31:0.08	100:1.58:0.06–100:1.01:0.02

4.2.4 Polarisation Properties of Vibrational Raman Scattered Light and Line Filling in

The polarisation properties of the isotropic and anisotropic components of Raman scattered light are described by:

$$\begin{aligned} \frac{I_{\text{parallel}}}{I_{\text{perp}}}(\text{anisotropic}) &= \frac{6+\cos^2\Theta}{7} \\ \frac{I_{\text{parallel}}}{I_{\text{perp}}}(\text{isotropic}) &= \cos^2\Theta \end{aligned} \tag{4.18}$$

where the terms parallel and perpendicular refer to the plane defined by the sun, the scattering point and the observer. Thus, essentially only the isotropic part leads to enhanced polarisation of scattered light, especially at large scattering angles. Since the Q-branch of the vibrational band of the Raman scattered light consists of an isotropic fraction, this will lead to an enhanced degree of polarisation in the centre of a Fraunhofer line. But due to the small cross-section of VRS, this enhancement is very small (e.g. ≅ 0.3% for the Ca-I-line at 422.7 nm and a resolution of 0.01 nm) and occurs only at large scattering angles. Thus, the observed high degrees of polarisation at small solar zenith angles (SZA) cannot be attributed to VRS (Clarke and Basurah 1989).

Due to the relatively small cross-section of vibrational–rotational Raman scattering, the additional filling in of Fraunhofer or terrestrial absorption lines is about 10% of the filling in due to RRS (see Table 4.5).

4.2.5 Scattering and Absorption of Radiation by Particles (Mie Scattering)

Radiation may be absorbed and scattered by particles. These processes are described by the absorption and scattering coefficients $\varepsilon_a(r, \lambda)$ and $\varepsilon_S(r, \lambda)$, respectively:

$$dI = -I(\lambda) \cdot \varepsilon_a(r, \lambda)\, ds \tag{4.19}$$

$$dI_S = -I(\lambda) \cdot \varepsilon_S(r, \lambda)\, ds \tag{4.20}$$

$I(\lambda)$ and $dI(\lambda)$ are the radiation flux and its change after passing through a layer of the aerosol of thickness *ds*, respectively. We define the 'single scattering albedo' of the aerosol as:

$$A_S = \frac{\varepsilon_S(r, \lambda)}{\varepsilon_a(r, \lambda) + \varepsilon_S(r, \lambda)} \tag{4.21}$$

and the extinction coefficient as:

$$\varepsilon_M(r, \lambda) = \varepsilon_a(r, \lambda) + \varepsilon_S(r, \lambda) \tag{4.22}$$

Mie scattering was first described in a comprehensive way by Gustav Mie and is defined as the interaction of light with (particulate) matter of dimensions comparable to the wavelength of the incident radiation (Mie 1908). It can be regarded as the radiation resulting from a large number of coherently excited elementary emitters (i.e. molecules) in a particle. Since the linear dimension of the particle is comparable to the wavelength of the radiation, interference effects occur. The most noticeable difference compared to Rayleigh scattering is the typically much weaker wavelength dependence (typically proportional to $\lambda^{-1.3}$) and a strong dominance of the forward direction in the scattered light. The calculation of the Mie scattering cross-section can be very complicated (involving summing over slowly converging series), even for spherical particles, but even more so for particles of arbitrary shape. Generally, the Mie phase function, P(cos ν), depends on the ratio between radius r and wavelength λ of the scattered radiation, the **size parameter**:

$$\alpha_S = \frac{2\pi r}{\lambda}$$

However, Mie theory (for spherical particles) is well developed and a number of numerical models exist to calculate scattering phase functions and extinction coefficients for given aerosol types and particle size distributions (van de Hulst 1980; Wiscombe 1980). Figure 4.2 gives a few examples for single particles or mono-disperse aerosols.

The computational effort is substantially reduced by the introduction of an analytical expression for the scattering phase function, which only depends on a few observable parameters. Most common is the Henyey–Greenstein parameterisation (Henyey and Greenstein 1941):

$$P(\cos\vartheta) = \frac{\left(1-g^2\right)}{4\pi\left(1+g^2-2g\cos\vartheta\right)^{3/2}} \tag{4.23}$$

which depends only on the asymmetry factor g (average cosine of the scattering function):

$$g = \langle\cos\vartheta\rangle = \frac{1}{2}\int_{-1}^{1} P(\cos\vartheta)\cdot\cos\vartheta\, d\cos\vartheta \tag{4.24}$$

For isotropic scattering, $P(\cos\vartheta) = \text{constant}$ and the asymmetry factor is $g = 0$. For complete forward scattering $g = 1$. Typical values for tropospheric aerosol are $g \approx 0.6 - 0.7$.

Tropospheric aerosol is either emitted from the surface (sea salt, mineral dust, and biomass burning) or forms in the gas phase by condensation of chemically formed hygroscopic species (primarily sulphate, nitrates, or oxidised organic material). The aerosol load of the atmosphere, i.e. particle number density and size distribution, depends on the aerosol origin and history.

Parameters for typical aerosol scenarios (urban, rural, maritime, and background) can be found in various data bases and overview publications which list refractive

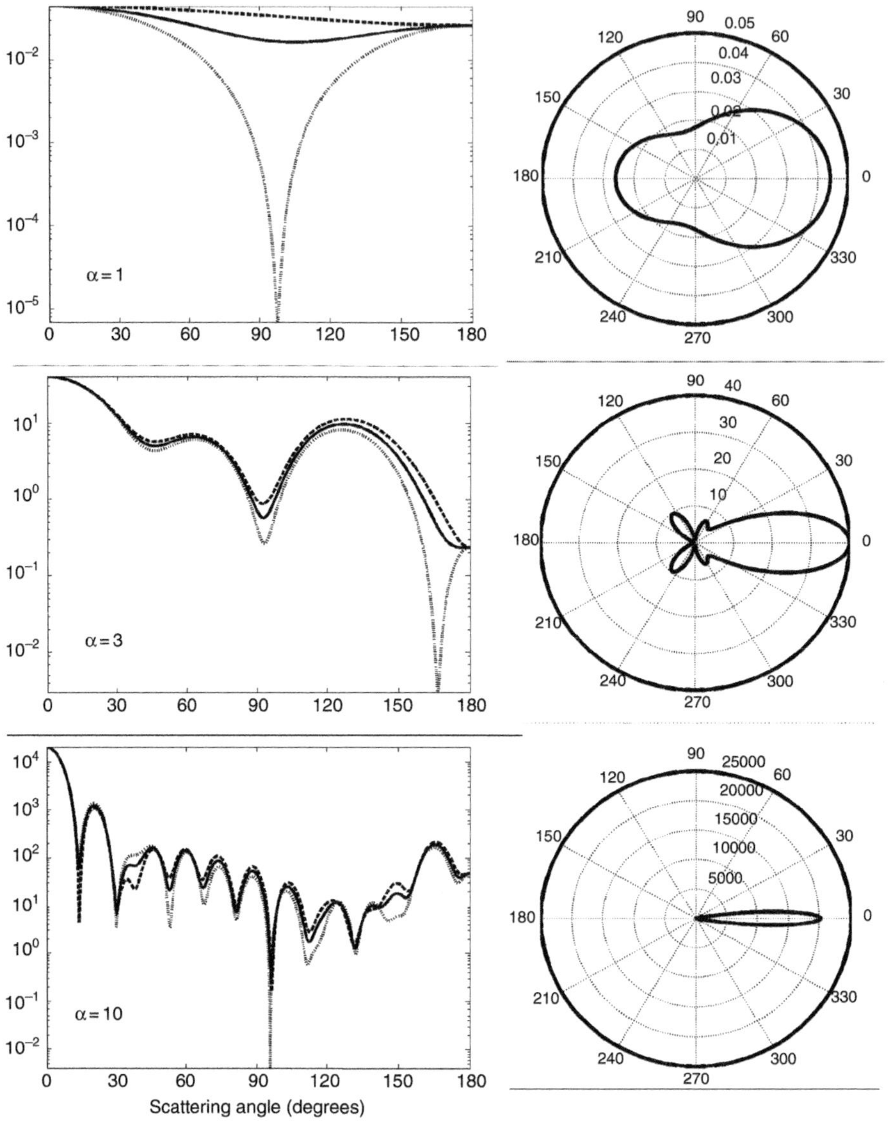

Fig. 4.2 Sample phase function for Mie scattering from individual particles (or mono-disperse aerosol) with size parameters $\alpha_S = 1$, 3, and 10. **Left column**: Phase function on log scale; **Right column**: Polar diagrams of the same phase functions on linear scale. It becomes apparent that for larger values of α_S essentially all energy is scattered in forward direction. Reprinted from Shangavi, Diploma Thesis, University of Heidelberg, 2003

Table 4.6 Typical Mie scattering coefficients, calculated Rayleigh scattering coefficients, corresponding extinction lengths and visibility range

Wavelength (nm)	Rayleigh scattering cross-section σ_R[a] (10^{-26} cm^2)	Rayleigh scattering coefficient ε_R[b] (10^{-5} m^{-1})	Typical Mie scattering coefficient ε_M (10^{-5} m^{-1})	Rayleigh extinction length (km)	Total extinction length L_E (km)	Visibility range[c] $L_V(C = 0.02)$ (km)
300	5.653	13.524	5.8	7.39	5.1	–
400	1.672	4.013	4	24.9	12.5	49
500	0.672	1.612	3	62.1	21.7	85
600	0.317	0.760	2.4	131.6	31.6	123
700	0.170	0.408	1.9	245.4	43.3	169

[a] According to (4.9)
[b] At 1 atm and 20 °C
[c] For definition of visibility range see Sect. 4.7

indices and their spectral dependence, size distributions, single scattering albedos, assymetry factors, etc. for different types of aerosol (Isaacs et al. 1987; Levoni et al. 1997, Dubovik et al. 2002).

Mie scattering is partly an absorption process, but by similar arguments as in the case of Rayleigh scattering, for narrow beams it can be described by Lambert-Beer's law with the extinction coefficient:

$$\varepsilon_M(\lambda) = \varepsilon_{M0} \cdot (\lambda/\lambda_0)^{-\alpha}$$

with the parameters ε_{M0} and λ_0. The Angström exponent α is inversely related to the mean aerosol particle radius and a constant ε_{M0}. Typically α is found in the range 0.5–2.5, with an 'average' value of $\alpha = 1.3$ (Angström 1930). For the ideal case of an exponential aerosol size distribution (with parameter r_0):

$$\frac{\Delta N}{\Delta r} = (r/r_0)^{-(\nu_J+1)}$$

The Angström exponent is related to the Junge index ν_J by $\nu_J = \alpha + 2$ (Junge 1963). Thus, an Angström exponent of 1.3 would correspond to a Junge index of 3.3.

In summary, a more comprehensive description of atmospheric extinction (in the presence of a single trace gas species) can be expressed as:

$$I(\lambda) = I_0(\lambda) \exp[-L((\sigma(\lambda) \cdot c) + \varepsilon_R(\lambda) + \varepsilon_M(\lambda))] \quad (4.25)$$

Typical extinctions due to Rayleigh and Mie scatterings at 300 nm are 1.3×10^{-6} and $1–10 \times 10^{-6}$ cm^{-1}, respectively (see Table 4.6).

4.3 The Radiation Transport Equation

The propagation of radiation (radiation transport) in absorbing and scattering media, such as the atmosphere, is a complex process. This is mostly due to the fact that scattered light can propagate in any direction and can undergo further scattering and absorption processes. In addition, at long wavelengths, thermal emission from gas molecules and aerosol can play a role. Although the contribution of thermal emission is negligible compared to the solar radiation in the near IR or even in the visible or UV spectral regions, we include it here for completeness. In this section, we present the basic equations describing the radiation transport in absorbing and scattering media as described earlier by the elementary processes. Analytical solutions of the equations can only be derived for greatly simplified cases, for instance for a collimated light beam travelling a finite distance through the atmosphere or a horizontally infinite atmosphere (see Sect. 4.4.1). In general, solutions have to be calculated by numerical methods, which can be divided into two groups. First, numerical solutions for the resulting radiation field can be calculated. Second, in a statistical approach the paths of an ensemble of individual photons, with randomly chosen parameters, through the atmosphere are calculated for the desired conditions. The radiation field can then be derived from the photon density. Examples of the former approach are the 'Discrete Ordinate Method' (e.g. DISORT and LIDORT (Stamnes et al. 1988; Spurr 2004; Spurr and Christi 2019)) of the latter Monte Carlo Methods (e.g. McArtim (Deutschmann et al. 2011)).

4.3.1 Absorption of Radiation

The radiance $I(\lambda)$ will be reduced by the amount $dI_a(\lambda)$ after traversing an absorbing layer of the (infinitesimal) thickness ds:

$$dI_a = \frac{d\Phi}{\Delta\Omega \cdot A_s} = -I(\lambda) \cdot \varepsilon_a(\lambda)\, ds = -I(\lambda) \cdot \sigma_a(\lambda) \cdot N\, ds \tag{4.26}$$

where ε_a is the absorption coefficient, $\sigma_a(\lambda)$ is the absorption cross-section of the absorber (molecule or aerosol), and N is the number of absorbing molecules (or aerosol) per unit volume.

Integration of (4.26) yields [with $I_0(\lambda)$ denoting the initial radiance]:

$$\ln\left(\frac{I_0(\lambda)}{I(\lambda)}\right) = \sigma_a(\lambda) \cdot \int_0^L N ds = \sigma_a(\lambda) \cdot S = \tau(\lambda) \tag{4.27}$$

where

$$S = \int_0^L N\, ds \tag{4.28}$$

is the column density of the absorber in which L denotes the thickness of the layer penetrated by the radiation. D is the optical density, which can also be written as:

$$D = \int_0^L \varepsilon_a(\lambda)\, ds \text{ or } dD = \varepsilon_a(\lambda)\, ds \tag{4.29}$$

Thus, the radiance after traversing the absorbing layer becomes:

$$I(\lambda)_L = I_0(\lambda) \cdot e^{-\sigma_a(\lambda) \cdot S} = I_0(\lambda) \cdot e^{-D} \tag{4.30}$$

This is the well-known Lambert–Beer's law (or Bouguer–Lambert law, see Sect. 6.1).

4.3.2 Scattering of Radiation

The scattered radiance is (in analogy to the absorption) given by:

$$dI_s = \frac{d\Phi_s}{\Delta\Omega \cdot A_s} = I(\lambda) \cdot \varepsilon_s(\lambda)\, ds = I(\lambda) \cdot \frac{d\sigma_s(\lambda)}{d\Omega} \cdot N\, d\Omega\, ds \tag{4.31}$$

where ε_s is the scattering coefficient, $\sigma_s(\lambda)$ is the scattering cross-section of a scattering centre, and N is the number of scattering centres (e.g. molecules or particles) per unit volume.

The scattered radiant flux is given by:

$$d\Phi_s = I(\lambda) \cdot \Delta\Omega \cdot \frac{d\sigma_s(\lambda)}{d\Omega} \cdot N\, d\Omega dV \tag{4.32}$$

where $d\sigma_s(\lambda)/d\Omega$ denotes the differential scattering cross-section (depending on scattering angle ϑ and polar angle φ).

The total scattering cross-section is thus given by:

$$\sigma_s(\lambda) = \int_{4\pi} \frac{d\sigma_s(\lambda)}{d\Omega} \cdot d\Omega \tag{4.33}$$

Frequently, the dimensionless phase function, $P_S(\vartheta, \varphi)$ is used to describe the angular dependence of the scattered radiation:

$$P_S(\vartheta, \varphi) = \frac{4\pi}{\sigma_s} \cdot \frac{d\sigma_s(\lambda)}{d\Omega} \tag{4.34}$$

The total extinction experienced by the radiation after traversing a layer of the medium of thickness *ds* is thus given by the sum of absorption and scattering:

$$dI = -dI_a - dI_s = -I(\lambda) \cdot (\varepsilon_a(\lambda) + \varepsilon_s(\lambda))\, ds \tag{4.35}$$

Radiance added by scattering is given by:

$$dI_S^* (\lambda) = \varepsilon_s (\lambda)\, ds \int_0^{\pi}\int_0^{2\pi} I^* \left(\lambda, \vartheta^*, \varphi^*\right) \cdot \frac{P_S (\vartheta^*, \varphi^*)}{4\pi} d\varphi^* \cdot \sin \vartheta^* d\vartheta^* \quad (4.36)$$

4.3.3 Thermal Emission

The thermal emission $[dI_{th}(\lambda, T)]$ from a volume element $(dV = Asds)$ is given as:

$$dI_{\text{th}} (\lambda, T) = \varepsilon_a (\lambda) \cdot I_p(\lambda, T)\, ds \quad (4.37)$$

where ε_a denotes the absorption coefficient as in (4.29) and $I_p(\lambda, T)$ denotes the Planck function (see also Sect. 7.2.1):

$$dI_p (\lambda, T) = \frac{2hc^2}{\lambda^5} \cdot \frac{d\lambda}{e^{hc/\lambda kT} - 1} \quad (4.38)$$

4.3.4 The Radiation Transport Equation and Simplifications

The three processes discussed above, i.e. absorption, scattering, and thermal emission are combined in the radiation transport Eq. (4.39):

$$\frac{dI (\lambda)}{ds} = -\left(\varepsilon_a (\lambda) + \varepsilon_s (\lambda)\right) \cdot I (\lambda) + \varepsilon_a (\lambda) \cdot I_p (\lambda, T) + \varepsilon_s (\lambda) \cdot \int_0^{\pi}\int_0^{2\pi} I^* \left(\lambda, \vartheta^*, \varphi^*\right) \cdot \frac{P_S (\vartheta^*, \varphi^*)}{4\pi} d\varphi^* \cdot \sin \vartheta^* d\vartheta^* \quad (4.39)$$

Simplifications of the radiation transport equation are possible, if only partial systems are of interest.

For instance, at short wavelengths (UV and visible), the Planck term can usually be neglected and (4.39) reduces to:

$$\frac{dI (\lambda)}{ds} = -\left(\varepsilon_a (\lambda) + \varepsilon_s (\lambda)\right) \cdot I (\lambda) + \varepsilon_s (\lambda) \int_0^{\pi}\int_0^{2\pi} I (\lambda, \vartheta, \varphi) \cdot \frac{P_S (\vartheta, \varphi)}{4\pi} d\varphi \cdot \sin \vartheta d\vartheta \quad (4.40)$$

If the thermal radiation (from the atmosphere) is of interest due to its long wavelength, Rayleigh scattering and Mie scattering at aerosol particles (but not at cloud droplets) can often be neglected and given by:

$$\frac{dI (\lambda)}{ds} = \varepsilon_a (\lambda) \cdot \left(I_p (\lambda, T) - I (\lambda)\right) \quad (4.41)$$

With the optical density [see definition in (4.30)]:

$$dD = \varepsilon_a(\lambda)\ ds \text{ and } ds = \frac{1}{\varepsilon_a(\lambda)} dD$$

The above equation becomes:

$$\frac{dI(\lambda)}{dD} = I_p(\lambda, T) - I(\lambda) \tag{4.42}$$

This latter equation is also known as Schwarzschild equation (van de Hulst 1980).

4.4 Light Attenuation in the Atmosphere

When considering the question of the attenuation of radiation in the atmosphere, two extreme cases can be distinguished, as illustrated in Fig. 4.3: Wide beam (as e.g. the illumination of the Earth's atmosphere by the Sun) and narrow beam (as e.g. the light beam emitted by a search-light type DOAS light source).

4.4.1 Wide Beams in the Atmosphere, the Two-Stream Model

In this section, we consider the transport of thermal radiation in the atmosphere. Compared to the general form of (4.39), the problem is simplified by assuming a flat, horizontally homogeneous, infinite atmosphere. Therefore, only the vertical component of the radiance has to be considered, as sketched in Fig. 4.4.

The attenuation dI of the radiance I in a layer of thickness dz is given by (4.26):

$$dI = -I \cdot dD = -I \cdot \varepsilon_a\, dz \tag{4.43}$$

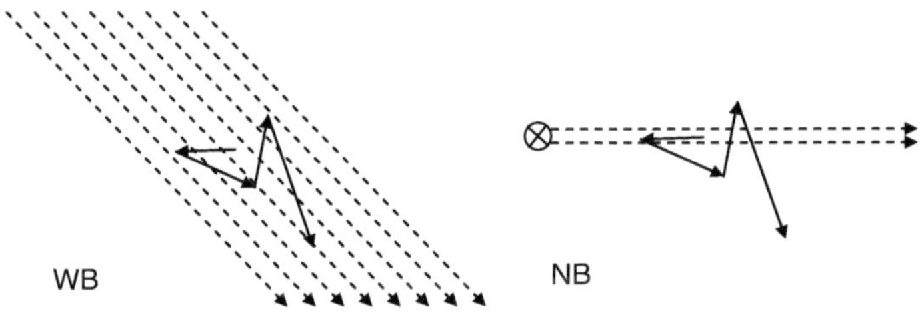

Fig. 4.3 Difference between wide beams, WB (**left**) as e.g. the illumination of the Earth's atmosphere by the Sun, and narrow beams, NB (**right**) as e.g. the light beam emitted by a search-light type DOAS light source. In the narrow beam case, the probability of a photon being scattered back into the beam after being scattered out of the beam is negligible. In the case of a wide beam, the lateral (i.e. perpendicular to the propagation of the incident radiation flux) radiation flux can be neglected. Therefore, the scattering has only the effect of reflecting some of the incoming light, as can be calculated by, e.g., a two-stream model

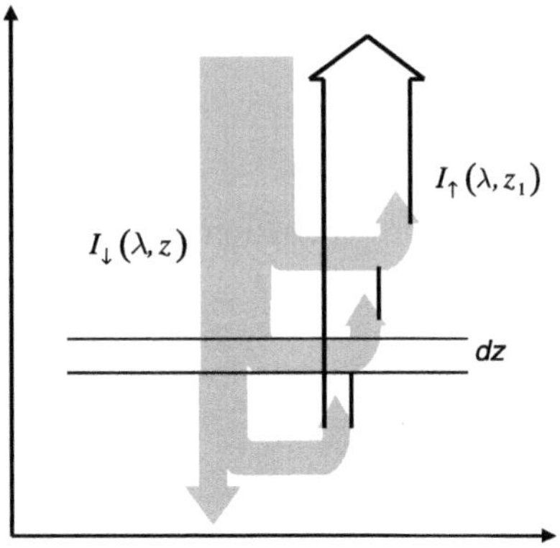

Fig. 4.4 The two-stream model: In a horizontally homogeneous atmosphere, radiation transport can be represented by just two 'streams of radiation', the downwelling flux and the upwelling flux

This is, however, only true for photons traversing the layer in vertical direction. In the general case, where the angle of incidence $\vartheta \neq 0\,°$, the transmission factor T_z of the layer becomes:

$$T_z = \frac{I(\lambda, z_2)}{I(\lambda, z_1)} = e^{-D/\cos\vartheta} \tag{4.44}$$

Since the thermal radiation impinges onto the layer from all directions (Φ denotes the azimuth angle), the average transmission factor is given by:

$$T_z(\lambda, z_1, z_2) = \frac{1}{\pi}\int_0^{2\pi}\int_0^{\pi/2} \exp\left(\frac{\tau(z_1, z_2)}{\cos\vartheta}\right)\cos\vartheta\cdot\sin\vartheta\, d\vartheta\, d\Phi \tag{4.45}$$

Only the layers above the reference height z contribute (according to their temperature and absorptivity or emissivity) to the downwelling radiance $I_\downarrow(\lambda, z)$:

$$I_p(\lambda, T(z'))\cdot\varepsilon_a(\lambda, z')\, dz \tag{4.46}$$

The contribution from the height z' must be weighted by the transmission factor T_z (λ, z, z'):

$$dI_\downarrow(\lambda) = I_p(\lambda, T(z'))\cdot\varepsilon_a(\lambda, z')\cdot T_z(z, z')\cdot dz' \tag{4.47}$$

Thus, the total downwelling radiance is given by the integral:

$$I_\downarrow(\lambda, z) = \int_z^\infty I_p(\lambda, T(z'))\cdot\varepsilon_a(\lambda, z')\cdot T_z(z, z')\cdot dz' \tag{4.48}$$

In an analogous manner we obtain the upwelling radiance $F_\uparrow(\lambda, z)$:

$$I_\uparrow(\lambda, z) = I_p(\lambda, T_{\text{ground}})\cdot\varepsilon_{\text{ground}}\cdot T_z(0, z) + \int_0^z I_p(\lambda, T(z'))\cdot\varepsilon_a(\lambda, z')\cdot T_z(z, z')\cdot dz \tag{4.49}$$

where T_{ground} and $\varepsilon_{\text{ground}}$ denote temperature and emissivity of the ground surface, respectively. The net radiance is then composed of both components (up- and down-welling radiances):

$$I_{\text{n}}(\lambda, z) = I_{\uparrow}(\lambda, z) - I_{\downarrow}(\lambda, z) \tag{4.50}$$

4.4.2 Narrow Beams in the Atmosphere

In the case of narrow beams the probability of a photon being scattered back into the beam after being scattered out of the beam is generally negligible (see Fig. 4.3); therefore, extinction can be treated like absorption, which is given by Lambert–Beer's law, (4.30) with the optical density:

$$D = \int_0^L (\varepsilon_{\text{a}}(\lambda) + \varepsilon_{\text{s}}(\lambda))\, ds \tag{4.51}$$

4.5 The Effect of Atmospheric Refraction (El-Mirage Effects)

Due to the atmospheric pressure gradient, the air density, and thus the index of refraction, decreases with altitude (see Chap. 2). This effect is somewhat counteracted by the decrease in temperature with altitude. For example, in the lowest 100 m of the atmosphere, the pressure gradient alone would lead to a reduction in air density of about 1.25% (assuming a scale height of 8000 m), while the dry adiabatic temperature gradient alone would increase the air density by about 0.34%, thus leading to a net decrease in air density of about 0.9%.

The corresponding change in the index of refraction of air at 550 nm would be from $n_0(z = 0) = 1.0002930$ to $n(z = 100) = 1.0002904$. Since the optical path length is $L_{\text{opt}}(n) = L_0/n$, this leads to a radius of curvature, where L_0 denotes the length of any horizontal path and ΔL the optical path difference at the two altitudes $\Delta L_{\text{opt}} = L(z = 0, n_0) - L(z = 100,\ \text{n})$:

$$R = \frac{\Delta z \cdot L}{\Delta L_{\text{opt}}} \approx \frac{10^2 \times 10^4}{0.03} \approx 3.3 \times 10^7\ \text{m} \tag{4.52}$$

This radius of curvature is about five times the Earth's radius. In other words, the deviation of an actual light path from the straight line on a 10-km path is about 1.5 m downwards, while a 'straight' line always following the curvature of earth would deviate by 7.8 m.

For normal operation of active DOAS instruments, this poses no difficulties. However, problems can arise if the deviation from a straight line changes due to anomalous profiles of index of refraction in the atmosphere. This can happen due to strong heating by solar irradiation near the ground, which causes the well-known 'El-Mirage' effect. On the other hand, strong nighttime inversions in the lower atmosphere can offset or even reverse the adiabatic temperature gradient (for details, see Platt et al. 2007). In summary, the apparent position of the light source in an active, bistatic DOAS system can change by several metres in the vertical, depending on the local temperature gradient. Since the image of the light source is usually roughly the size of the entrance aperture (see Chap. 7), realignment of the receiving telescope pointing direction is needed under conditions of strong and changing temperature gradients. Unattended operation under these conditions will require automatic aligning systems, even if the mechanical mounting of the optics is perfectly stable. For a system involving retro-reflectors, the effect on the returned light should be less severe, as outlined in Chap. 7 (Sect. 7.3.10).

4.6 The Effect of Atmospheric Turbulence

Atmospheric turbulence can influence the propagation of light beams in the atmosphere (e.g. Wolf 1979). Turbulence can be caused either by mechanical effects, i.e. wind, or by thermal convection. Typical relative density fluctuations ΔN/N are about $(3 \times 10^{-3}\,\mathrm{K}^{-1}) \times \mathrm{T}$, while velocity fluctuations of $1\,\mathrm{m\,s}^{-1}$ would correspond to relative density fluctuations on the order of 10^{-5}. In either case, turbulent motion can be characterised by the velocity structure function originally introduced by Kolmogorov (1941), which is defined as (see Andrews 2004):

$$D_\mathrm{v}(\Delta x) = \left\langle (v(x) - v(x + \Delta x))^2 \right\rangle \approx C_\mathrm{v}^2 \cdot \Delta x^{2/3} \tag{4.53}$$

where $v(x)$ denotes the velocity of the flow at point x, and ${C_\mathrm{v}}^2$ the velocity structure constant. The angle brackets indicate the ensemble average over many pairs of points in space separated by the distance Δx. Similarly, a temperature structure function D_T can be defined:

$$D_\mathrm{T}(\Delta x) = \left\langle (T(x) - T(x + \Delta x))^2 \right\rangle \approx C_\mathrm{T}^2 \cdot \Delta x^{2/3} \tag{4.54}$$

Since n − 1 (denoting the refractive index of air) is inversely proportional to the air density, a refractive index structure function D_n can be derived from D_T:

$$D_\mathrm{n}(\Delta x) = \left\langle (n(x) - n(x + \Delta x))^2 \right\rangle \approx C_\mathrm{n}^2 \cdot \Delta x^{2/3} \tag{4.55}$$

where $n(x)$ denotes the index of refraction at point x in space. The refractive-index structure constant ${C_\mathrm{n}}^2$ ranges from $10^{-17}\,\mathrm{m}^{-2/3}$ (weak turbulence) to $10^{-13}\,\mathrm{m}^{-2/3}$ (strong turbulence). For two points in the atmosphere separated by e.g. $\Delta x = 1\,\mathrm{m}$, this translates into average differences in the refractive index $\Delta n = n(x) - n(x + \Delta x)$ around $\Delta n \approx 3 \times 10^{-9}$ to 3×10^{-7}. While these differences are very small, there are a large number of these turbulence elements diffracting the beam, as sketched in Fig. 4.5. This leads to 'beam wander', i.e. random deflections of a beam having propagated a distance L through a turbulent atmosphere. The instantaneous

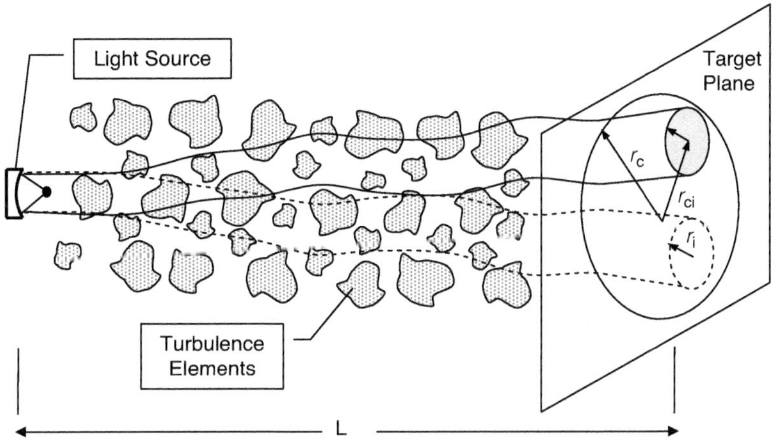

Fig. 4.5 Sketch of the beam wandering process caused by turbulence. At any given time, the 'instantaneous beam' (with an instantaneous radius r_i) arrives at a different position having the distance r_{ci} from the centre of a circle with radius r_c encompassing the long-term (i.e. for times $t >> t_0$) average positions

beam radius r_i is seen when observing for a short time $t << t_0$, where the time constant t_0 is on the order of:

$$t_0 = \frac{r_i}{v}$$

The square of the average deflection of the beam centre is given by the following empirical formula (Andrews 2004):

$$\langle r_c^2 \rangle = 2.87 \cdot \frac{C_n^2 \cdot L^3}{r_i^{1/3}} \quad \text{or} \quad r_c \approx \sqrt{\langle r_c^2 \rangle} = \sqrt{2.87 \cdot C_n^2} \cdot \frac{L^{3/2}}{r_i^{1/6}} \tag{4.56}$$

This would lead to average deflections of a $r_i = 10$ m beam over a 10-km path of 3.6 mm to 0.36 m for weak or strong turbulence, respectively. While this is very small, it should be noted that r_c does not vary much with the beam diameter [see (4.56)]; thus, the relative wander would be larger for smaller beams, but even more importantly the coaxial DOAS optics (see Sect. 7.8.3) relies on turbulence to scatter light from the transmitted beam into the field of view of the receiving telescope (besides the lateral beam offset by the retro reflector).

4.7 Visibility in the Atmosphere

There are many effects of radiation transport that are visible in the atmosphere. For instance, the observation that the sky is blue and the clouds are white is a direct consequence of the different wavelength dependence of Rayleigh scattering compared to Mie scattering. Another important consequence of light scattering and extinction in the atmosphere is the reduction of visibility by haze, smog, dust, or fog. It is an everyday experience that distant objects cannot be seen under conditions of increased scattering.

A quantitative estimate can be made by considering the view of a black object (e.g. a building or a mountain) of area A_0 at a distance L in a homogeneous atmosphere with the scattering coefficient ε. The contrast, i.e. the difference in the brightness, between the object and the background is then given by the following calculation.

When looking at a black object (area A_0), the scattered radiation intensity I_R received by the observer (see Fig. 4.6) is given by:

$$d I_{\mathrm{R}}(x) = d I_{\mathrm{S}}(x) \cdot \frac{A_{\mathrm{E}}}{x^2} \cdot e^{-\varepsilon x} \qquad (A_{\mathrm{E}} = \text{area of the eye})$$

With the intensity I_{S} scattered in a volume element with area $A(x)$ and thickness dx at distance x from the observer:

$$d I_{\mathrm{S}}(x) = \alpha \cdot A(x)\, dx \cdot \varepsilon = \alpha\varepsilon \cdot A_0 \left(\frac{x}{L}\right)^2 dx \quad (\text{where } \alpha \text{ is a constant}) \tag{4.57}$$

we obtain:

$$d I_{\mathrm{R}}(x) = \alpha\varepsilon\, A_0 \left(\frac{x}{L}\right)^2 dx \cdot \frac{A_{\mathrm{E}}}{x^2} \cdot e^{-\varepsilon x} = \alpha\varepsilon \cdot \frac{A_0 A_{\mathrm{E}}}{L^2} \cdot e^{-\varepsilon x} dx \tag{4.58}$$

Integration yields the flux received by the eye:

$$I_R = \int_0^L d I_R(x) = \alpha \cdot \frac{A_0 A_E}{L^3} \cdot \left(1 - e^{-\varepsilon L}\right) \tag{4.59}$$

For $L \to \infty$, the received intensity I_{R} must be equal to the background intensity I_0, so we obtain:

$$I_{\mathrm{R}} = I_0 \cdot \left(1 - e^{-\varepsilon L}\right) \text{ with } I_0 = \alpha \cdot \frac{A_0 A_{\mathrm{E}}}{L^3} \tag{4.60}$$

The relative difference between the radiation intensity received from the black object I_{R} compared to the flux I_0 from its surrounding, the contrast C, is given by:

$$C = \frac{I_{\mathrm{R}} - I_0}{I_0} = \frac{I_0 \cdot \left(1 - e^{-\varepsilon L}\right) - I_0}{I_0} = e^{-\varepsilon L} \tag{4.61}$$

Typically, objects are still visible at a contrast between 0.01 and 0.02; thus, the visibility range L_{V} becomes:

$$C = e^{-\varepsilon L_V} \Rightarrow L_{\mathrm{V}} = -\frac{\ln C}{\varepsilon} \approx (3.9 \ldots 4.6) \frac{1}{\varepsilon} = (3.9 \ldots 4.6) \cdot L_{\mathrm{E}} \tag{4.62}$$

where $L_{\mathrm{E}} = 1/\varepsilon$ and $\varepsilon = \varepsilon_{\mathrm{R}} + \varepsilon_{\mathrm{M}}$, i.e. the sum of the extinction coefficients due to Rayleigh and Mie scattering, as described in Sect. 4.2, where some typical values for

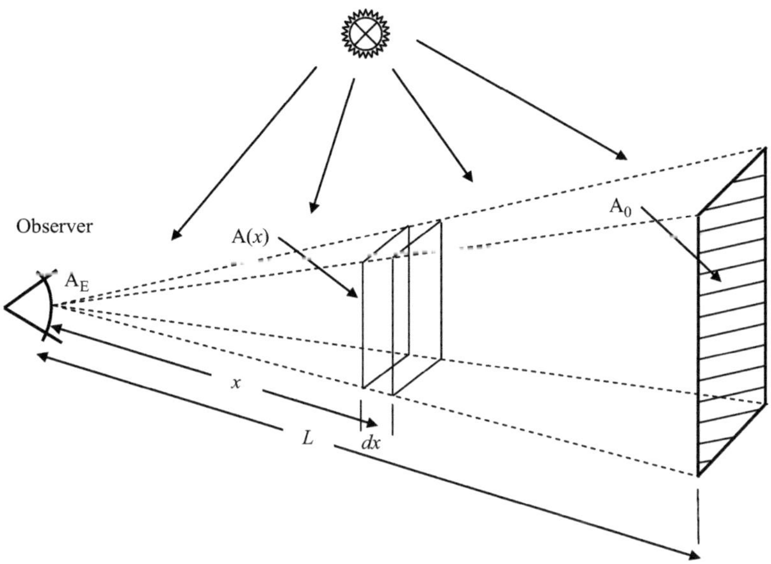

Fig. 4.6 Propagation of radiation in the atmosphere and the visibility of a black surface A_0

L_V are listed in Table 4.6. Of course this is a quite idealised calculation. In reality, the minimum discernible contrast might be lower or higher than assumed. Nevertheless, the estimate of L_E from the visibility of several objects at different distances is quite robust, since only the logarithm of C enters in the calculation (4.60).

Measurement Techniques for Atmospheric Trace Gas Concentrations and Other Parameters

5

Our knowledge about the physical and chemical processes in the atmosphere relies to a large extent on measurements of relevant parameters, such as temperature or ozone mixing ratio, throughout the atmosphere. For instance, the 'meteorological parameters' such as temperature, humidity, pressure, radiation intensities, or wind speed and direction have routinely been measured in the atmosphere. In this book, we focus on measurements of trace gas concentrations and quantities related to chemical processes in the atmosphere, including the intensity of the solar radiation field and chlorophyll fluorescence. These are the experimental prerequisites for understanding the physico-chemical and biological processes on Earth and in the earth's atmosphere. In addition, measurements are made in model atmospheres, e.g. smog chambers or environmental chambers. As the name 'trace constituents' suggests, the species to be investigated here are typically present at minute quantities only (see Chap. 2). Therefore, the determination of trace gas concentrations in the atmosphere constitutes a challenge for the analytical techniques employed. In summary, measurements of atmospheric parameters are done for a variety of purposes:

- Weather forecast and climatological studies
- Meteorological and atmospheric physics research
- Climate studies
- Pollution monitoring and control
- Monitoring of the state of atmosphere
- Atmospheric chemistry research
- Ecology (chlorophyll fluorescence).

In recent years, the term 'chemical weather' was coined to encompass information on the trace gas composition of the atmosphere and air pollution.

U. Platt and J. Stutz, *Differential Optical Absorption Spectroscopy*, Geophysics and Environmental Physics, https://doi.org/10.1007/978-3-662-68639-3_5

5.1 Early Measurement Techniques

After the discovery of a strong, characteristic smell near electric discharges by van Marum in 1785 and the identification and naming of ozone by Christian Friedrich Schönbein in 1839, this species was one of the first atmospheric trace gases to be investigated. An early technique relied on the use of 'Schönbein Paper' (Schönbein 1840), a paper impregnated with iodide (e.g. KI) and starch. Upon exposure to ambient air and hence to the ozone dispersed in it, the following reactions produced a blue colour:

$$2\,H^+ + 2\,I^- + O_3 \rightarrow I_2 + O_2 + H_2O \qquad \text{(R5.1)}$$

$$I_2 + \text{starch} \rightarrow \text{blue colour} \qquad \text{(R5.2)}$$

Together with a suitable colour scale, semi-quantitative measurements could be made. The amount of blue colour strongly depends on relative humidity and wind speed, which were not adequately recorded. Also a negative interference due to SO_2 was found. Consequently, these extensive measurements from the 19th century could not be used for quantitative studies.

These problems were avoided by another technique, which was also introduced during the 19th century. Here, air is bubbled through an arsenite-containing solution, which is oxidised to arsenate (Volz and Kley 1988):

$$O_3 + AsO_3^{3-} \xrightarrow{I^-/I_2} O_2 + AsO_4^{3-} \qquad \text{(R5.3)}$$

Measurements near Paris during 19th century (1876–1910) [the so-called Montsouris series (Albert-Levy 1886)] allowed Volz and Kley (1988) to reconstruct the much lower pre-industrial ozone content of the troposphere.

Besides the 'wet chemical' techniques, optical absorption spectroscopy was very closely linked to early research on atmospheric composition. For instance, Alfred Cornu in 1878 (Cornu 1879) and Sir Walter Noel Hartley (Hartley 1880) showed that a sharp drop in solar intensity below 300 nm was due to atmospheric ozone. This was later used by several researchers to determine the atmospheric O_3 column density (Fabry 1913; Dobson and Harrison 1926; Roscoe and Clemitshaw 1997).

5.2 The Role of Measurements in Atmospheric Chemistry

Measurements of atmospheric parameters can result in a better understanding of our atmospheric environment. Atmospheric parameters can be characterised by the time scale of measurements, i.e. there may be specialised 'measurement campaigns' aimed at investigating certain phenomena or 'long-term studies' to follow the evolution of chemical changes. In addition, we might look at the spatial scale of investigation, as there may be interest in local, regional or global scale phenomena. Apart from measurements made primarily for scientific purposes, a large number of measurements of atmospheric trace gases are made every day to document the pollution level of our atmosphere and to alert the public of extreme pollution events. In trying to group the measurement efforts, we can use the following criteria.

5.2.1 Global and Long-Term Observations

There is great interest in changes in the composition of our atmospheric environment. Some well-known examples include the rise of CO_2 mixing ratios directly observed since the 1950s (e.g. Keeling et al. 1970). Since then, many other trace gases have been followed, e.g. chlorofluorocarbon species (CFC), methane, or the thickness of the stratospheric ozone layer (i.e. the total atmospheric O_3 column density).

Long-term observations are frequently aimed at monitoring gradual changes in the trace gas composition of atmosphere. They include observations of:

- Trends in stratospheric ozone
- Change of stratospheric chemistry, as e.g. realised in the Network for the Detection of Stratospheric Change, (NDSC)
- Evolution of species supplying halogens to the stratosphere (e.g. CFC and HCFC species)
- Trend of the tropospheric ozone mixing ratio (e.g. by the Global Atmospheric Watch project, GAW)
- Trends of greenhouse gas concentrations (including CO_2, CH_4, or N_2O)
- Trends of gases indicating the atmospheric oxidation capacity (i.e. the ability of the atmosphere to remove trace gases). For instance, ozone, CH_3CCl_3 and ^{14}CO are monitored for this purpose.
- Emission of volcanic gases (e.g. by satellite or by the Network for the Observation of Volcanic and Atmospheric Change, NOVAC).

5.2.2 Regional and Episodic Studies

Regional and episodic studies seek to investigate the causes, extent, and consequences of air pollution. Routine monitoring is often performed to follow the progress of air quality regulations designed to protect human health. In addition, many fundamental scientific questions can be investigated only by observations made on a regional scale. Typical measurement tasks in this context are:

- Monitoring of air pollutants at the surface, such as: O_3, SO_2, NO, NO_2, hydrocarbons, particulate matter, for regulatory purposes.
- Investigation of urban and continental plume evolution (e.g. with respect to O_3 and particle formation downwind of source regions)
- Impact of secondary aerosol formation on regional climates

- Observation of the Antarctic Stratospheric Ozone Hole
- Investigation of polar boundary-layer ozone loss events (the 'tropospheric ozone hole').

5.2.3 Local and Rapid In-Situ (Photo)Chemistry

Investigation of fast (time scale of seconds to minutes) in-situ chemical and photochemical processes (see Sect. 2.5) in the open atmosphere allows us to largely neglect the effect of transport, since transport takes place only at longer time scales. Thus, it is possible to study the chemical mechanisms directly in the atmosphere. This is particularly true for free radical (OH/HO_2) photochemistry, where the lifetime of the reactive species is on the order of seconds.

Studies in 'Smog'-Chambers (frequently called 'Reaction Chambers' or 'Photo-Reactors') allow the suppression of transport processes; thus, the effect of chemistry alone can be investigated. In fact, the phenomenon of tropospheric ozone formation was observed in smog-chambers long before the chemical mechanism (see Chap. 2) was discovered. The main disadvantage of smog chambers is the presence of surfaces; thus, care should be taken to avoid artefacts that may arise from chemical processes at the chamber walls. In order to minimise these problems, very large smog chambers with volumes exceeding 100 m^3 have been built; these facilities offer surface/volume ratios below 1 m^{-1}.

5.3 Requirements for Measurement Techniques

Useful measurement techniques for atmospheric trace species should fulfill two main requirements. First, they must be sufficiently sensitive to detect the species under consideration at their ambient concentration levels. This can be a very demanding criterion. For instance, species present at mixing ratios ranging from as low as 0.1 ppt (mixing ratio of 10^{-13}, equivalent to about 2×10^6 molecules/cm^3 at the surface) to several parts per billion (ppb) (1 ppb corresponds to a mixing ratio of 10^{-9}) can have a significant influence on the chemical processes in the atmosphere (Logan et al. 1981; Perner et al. 1987a; Seinfeld and Pandis 2016). Thus, detection limits from below 0.1 ppt up to the ppb range are required, depending on the application.

Second, it is equally important for measurement techniques to be specific (or selective), which means that the result of the measurement of a particular species must be neither positively nor negatively influenced by any other trace species simultaneously present in the probed volume of air. Given the large number of different molecules present at ppt and ppb levels, even in clean air (see Chap. 2), this is not a trivial condition.

In practice, there are further desirable properties of a measurement technique, including simplicity of design and use of instruments. In addition, the capability of real-time operation (as opposed to taking samples for later analysis) and the possibility of unattended operations are important issues. Further considerations are weight, portability, and dependence of the measurements on ambient conditions.

To date, no single measurement technique can fulfill all the diverse requirements for trace gas measurements in the atmosphere. Therefore, many different techniques have been developed (e.g. Clemitshaw 2004; Heard 2006; Heard et al. 2021).

5.4 Measurement Technique Categories

An important criterion of the available techniques is the degree of specialisation; for instance, there are instruments for the measurement of a single species or parameter. Simply speaking, we distinguish between **specialised techniques** and **universal techniques** capable of determining many parameters with a single instrument.

Another fundamental property of instruments is their ability to make **in-situ** or **remote sensing** measurements. While in-situ measurements determine trace gas concentrations in a 'spot' in space that is usually very close to the instrument, remote sensing techniques allow measurements from a large distance, as far as from a satellite instrument in the earth's orbit. Remote sensing instruments usually average the trace gas concentrations over a large volume of air, and always rely on the sensing of electro-magnetic radiation, i.e. they are primarily spectroscopic methods.

To date, a large variety of measurement techniques for atmospheric trace gases (and other atmospheric parameters) have been developed. A few examples of instruments belonging to these categories include:
Specialized techniques:

- Chemiluminescence (e.g. for the detection of NO or O_3, in-situ))
- Chemical amplifiers for the detection of peroxy radicals (e.g. Cantrell et al. 1984, 1993; Clemitshaw et al. 1997, in-situ)
- Matrix isolation—electron spin resonance (MI-ESR) (e.g. Mihelcic et al. 1985, in-situ)
- Derivatisation—HPLC or Hantzsch reaction (e.g. for determination of CH_2O) (Fung and Grosjean 1981; Nash 1953, in-situ)

Universal techniques:

- Gas chromatography (GC, in-situ)
- Optical spectroscopy (in-situ, remote sensing)
- (Chemical Ionisation-)Mass spectrometry ((CI)MS, in-situ)
- Photoacoustic detection (e.g. Sigrist 1994b, in-situ)
- Electrochemical techniques (in-situ)
- 'Bubbler' combined with wet chemistry, colorimetry (e.g. West and Gaeke 1956; Saltzman 1954), or ion-chromatography (IC, in-situ)

Table 5.1 Overview of species of relevance to atmospheric chemistry research and measurement techniques

Species	UV/vis	FT-IR	TDLS (IR)	GC	MS (CIMS)	Fluoresc. Chemolum.	Other
NO	O	O				+	
NO_2	+	O	+			+	MI-ESR[a]
NO_3	+						MI-ESR, LIF
HNO_2	+				+		Denuder, Derivat.
HNO_3		O	O		+		Denuder
OH	+				+		LIF
HO_2/RO_2					+		LIF, Ch. A[b]
H_2O_2		O	+		O	+	
O_3	+	O	O		+	+	Electroch.[c]
HCHO	+	O	+		+		Derivat.[d]
RCHO	+				+		Derivat.[d]
Alkanes		+		+			
Olefines		+		+	O		
Aromatic	+	O		+	+		
CO				+		+	
DMS				+	+		
SO_2	+					+	
N_2O			+	+			
CFC's		+	+	+			
HX[e]		+			+		Wet. Chem.
XO[e]	+				+	+	
HOX[e]					+		

Symbols denote: well measurable(+), measurable(O), not measurable (empty field)
Abbreviations: UV/vis, UV/visible absorption spectroscopy; FT-IR, Fourier-Transform IR Spectroscopy; TDLS, tunable diode laser spectroscopy; GC, gas chromatography; MS (CIMS), mass spectrometry (Chemical Ionisation MS)
[a]Matrix isolation—electron-spin-resonance
[b]Chemical amplifier
[c]Electrochemical cell
[d]Derivatisation + HPLC or absorption photometry
[e]X = halogen atom (F, Cl, Br, I)

Today, atmospheric chemistry has a comprehensive arsenal of measurement techniques at its disposal, Table 5.1 gives an overview of popular trace gas measurement techniques for a series of key species relevant for studies of atmospheric chemistry. A review of a series of in-situ techniques (but largely neglecting spectroscopic techniques) can be found at Hidy et al. (2017) with comments by Kleinman et al. (2017). Among a large number of specialised techniques (like the gas-phase chemoluminescence detection of NO), universal techniques are of great interest due to their relative simplicity—a single instrument can measure a large range of species.

Spectroscopic techniques are a powerful variety of universal methods. They are highly sensitive, very specific, universally usable, give absolute results, and have the potential for remote sensing. It is, therefore, not surprising that spectroscopic techniques assume a unique role among many other methods that are in use today.

5.4.1 In-Situ Versus Remote Sensing Techniques

Remote sensing techniques allow detection of the properties of an object from a distance. Using these techniques, the trace gas composition in atmosphere can be measured at a point that is remote from the probing instrument. Examples include LIDAR instruments (see Sects. 5.7.3 to 5.7.5) and observations of trace gas distributions from space. Many DOAS applications also fall in the remote sensing category. In contrast to that, in-situ instrumentation measures trace gas concentrations (or other parameters) at a particular location—ideally at a point in space. There are a series of applications where localised measurements are very desirable, e.g. the determination of spatial gradients. For example, many trace gases have a vertical gradient close to the ground, and the observation of these gradients requires measurements at different altitudes. It should be noted that in-situ measurements frequently require relatively long averaging times, (t_m). In the presences of wind with a speed v_w, they will thus average over a distance d_m:

$$d_m = t_m \cdot v_w$$

Thus, an integration time of 5 min at 2 m s^{-1} wind speed already translates to a spatial average of $d_m = 600$ m.

5.5 Experimental Evidence for the Presence of Radicals in the Atmosphere

As an example of the combination of measurements by different technologies to solve a difficult problem in atmospheric chemistry, we briefly discuss both direct and indirect evidence for the presence of free radicals in the atmosphere.

Direct detection of free radicals has proven notoriously difficult because of the high reactivity and low concentration of these species in the atmosphere. For instance, although OH-initiated reaction chains are central to our understanding of atmospheric chemistry (see Sect. 2.4), direct experimental verification of this theory took a long time. From the suggestion of the central role of OH in the degradation of trace gases and ozone formation (see Chap. 2) in the early 1970s, it took more than a decade before the first reliable direct observations of atmospheric OH became available (see Table 5.2). Only since the 1990s have tropospheric OH measurements of high quality been reported. Although these data match the model predictions, they do not yet constitute a data set that covers all the conditions of the atmosphere. On the other hand, there has been a substantial amount of indirect evidence for the presence of OH radicals in the atmosphere. In addition, a number of techniques

Table 5.2 Evidence for the presence of OH radicals in the troposphere

Evidence	Literature (not intended to be complete)
Indirect	
Photochemical ozone formation	e.g. Crutzen (1974)
Pattern of VOC—degradation	e.g. Calvert (1976)
NO_X—dependence of peroxide formation	Tremmel et al. (1993)
Global budgets of CO, ^{14}CO, CH_3CCl_3	Volz et al. (1981), Brenninkmeijer et al. (1992), Prinn et al. (1987)
Semi-direct	
Detection by Chemical Ionisation Mass-Spectrometry	Eisele and Tanner (1991), Eisele et al. (1994, 1994), Berresheim et al. (2000)
Detection by in-situ oxidation of ^{14}CO	Campbell et al. (1979), Felton et al. (1990)
Direct	
Long-Path Differential Optical Absorption Spectroscopy (DOAS) in the UV	Perner et al. (1976, 1987b), Platt et al. (1988), Mount and Eisele (1992), Mount (1992), Dorn et al. (1996), Brandenburger et al. (1998), Fuchs et al. (2012), Kanaya et al. (2012)
Laser-Induced Fluorescence (LIF)	Hard et al. (1992a, 1995), Holland et al. (1995, 1998, 2003), Hofzumahaus et al. (1996, 1998), Brauers et al. (1996), Dorn et al. (1996), Brune et al. (1995, 1998), Creasey et al. (1997), Hausmann et al. (1997), Carslaw et al. (1999), George et al. (1999), Kanaya et al. (2000, 2001),Heard and Pilling (2003), Hard et al. (2002), Bloss et al. (2003), Holland et al. (2003), Rohrer and Berresheim (2006), Schlosser et al. (2007), Kanaya et al. (2012), Griffith et al. (2016), Brune et al. (2020)

have been developed that, while not identifying the OH molecule directly, provided a quantitative measurement of OH abundance [like CO, ^{14}CO (Volz et al. 1981; Brenninkmeijer et al. 1992), CH_3Cl (e.g. Prinn et al. 1987)]. In the following section, these methods shall be referred to as semi-direct.

Table 5.2 summarises the available evidence for the presence of OH radicals in the troposphere. It can be concluded that, although direct, unequivocal identification of these species in the atmosphere can only be performed in a few cases; there is an enormous amount of indirect evidence of their presence. Thus, the central importance of HO_X for atmospheric chemistry has been proven.

In the case of peroxy radicals, the situation is less satisfactory. To date, there are very few direct measurements of HO_2 and some other peroxy radicals by matrix isolation—electron spin resonance detection (see Table 5.3). Infrared spectroscopic detection of HO_2 has not yet been achieved in the troposphere. On the other hand, there are techniques for indirect detection of HO_2 available that rely on the conversion to OH via reaction with NO (Table 5.3). Hydroxyl radicals can then be detected by the well-developed LIF technique. Another widely used ‘semi direct’ technique employs

Table 5.3 Evidence for the presence of HO_2/RO_2 radicals in the troposphere

Evidence	Literature (not intended to be complete)
Indirect	
Ozone formation	Crutzen (1974)
Pattern of H_2O_2 (org. peroxide) formation	Tremmel et al. (1993)
Change in the Leighton Ratio ('Missing oxidant')	Parrish et al. (1986)
Semi-direct	
OH–LIF after titration by NO	Hard et al. (1995), Brune et al. (1995, 1998), George et al. (1999), Creasey et al. (1997), Kanaya et al. (2000), Hofzumahaus et al. (1996), Holland et al. (2003), Bloss et al. (2003), Griffith et al. (2016); Brune et al. (2020)
Chemical Amplifier (ROX-Box) Measurements	Stedman and Cantrell (1981), Cantrell et al. (1984, 1993), Hastie et al. (1991), Monks et al. (1996), Clemitshaw et al. (1997), Carslaw et al. (1997), Perner et al. (1999), Mihelcic et al. (2003), Wood et al. (2017), Duncianu et al. (2020)
Direct	
Matrix-Isolation Electron-Spin-Resonance Detection	Mihelcic et al. (1985, 1993, 2003)
Tuneable Diode-Laser Spectroscopy (TDLS)	Werle et al. (1991)

a chemical amplifier (ROX-Box) to convert NO added to the sampled airflow into NO_2. Although this simple technique appears to have severe limitations, it can be used to investigate the tropospheric HO_X system, especially in clean air (e.g. Cantrell et al. 1984; Monks et al. 1996).

The first direct detection of NO_3 radicals (by DOAS) in the troposphere dates back to 1979 (see Table 5.4), but the idea of considerable influence of NO_3 on tropospheric chemical cycles only slowly gained acceptance. Since there appears to be no simple way to derive the global (or regional) average NO_3 concentration, this quantity must be determined from long-term measurement series (e.g. Heintz et al. 1996; Plane and Nien 1991). Again, there is considerable indirect evidence for the importance of NO_3 radical reactions in the troposphere; these include (1) the lack of seasonal dependence of the NO_Y deposition (e.g. Calvert et al. 1985), suggesting non-photochemical NO_X–NO_Y conversion via NO_3–N_2O_5 during winter; (2) observation of nighttime radical formation (e.g. Mihelcic et al. 1993; Gölz et al. 2001); (3) degradation patterns of hydrocarbons (Penkett et al. 1993), details are discussed in Sect. 2.6.

A substantial amount of direct and indirect evidence has been accumulated for the presence of halogen atoms, halogen oxide radicals, and other reactive halogen species in the troposphere (e.g. Von Glasow et al. 2007; Saiz-Lopez et al. 2012; Simpson et al. 2015; see also Table 5.5). Early observations were often made using indirect

Table 5.4 Evidence for the presence of NO_3 radicals in the troposphere

Evidence	Literature (not intended to be complete)
Indirect	
Seasonal variation of the NO_Y deposition	Calvert et al. (1985)
Nighttime observation of RO_2 formation	Platt et al. (1990), Mihelcic et al. (1985), Gölz et al. (1997)
Nighttime degradation pattern of VOCs	Penkett et al. (1993)
Semi-direct	
Conversion to NO + chemoluminescence	–
Direct	
Differential Long-Path Absorption Spectroscopy (LP-DOAS)	Noxon et al. (1980), Noxon (1983), Platt et al. (1979, 1981, 1984), Plane and Nien (1991), Heintz et al. (1996); Aliwell and Jones (1998); Allan et al. (1999, 2000a), Martinez et al. (2000), Geyer et al. (2001a, 2002, 2003a), Asaf et al. (2009), Brown and Stutz (2012), Tsai et al. (2014), Lu et al. (2022), Zhu et al. (2022)
Cavity Ringdown Spectroscopy & CE-DOAS	King et al. (2000), Brown et al. (2001–2003), Brown and Stutz (2012), Dorn et al. (2013)
Matrix-Isolation Electron-Spin-Resonance Detection	Mihelcic et al. (1985, 1993, 2003)
Laser-Induced Fluorescence (LIF)	Wood et al. (2003), Matsumoto et al. (2005)

methods, such as the 'Hydrocarbon clock', or semi-direct methods, such as atomic fluorescence and chemical amplifiers (Table 5.5). In parallel, longpath and multiaxis DOAS lead to the direct observation of the halogen oxides, ClO, BrO, and IO in polar regions, marine areas, over salt lakes, and in the free troposphere (see Table 5.5 for references). The advent of chemical ionisation mass spectrometry added the direct observation of BrO by this technique, as well as the measurements of other reactive halogen species that are halogen radical precursor or radical reaction products, such as Cl_2, Br_2, I_2, HOCl, $ClNO_2$, HOBr, and HOI (see Table 5.5 for references).

In summary, we have a considerable body of indirect as well as direct evidence for the presence of several types of free radicals in the atmosphere.

5.6 Spectroscopic Techniques

To date, no single measurement technique can fulfill the necessary requirements for all applications, therefore, the selection of a technique is based on specific requirements: What species are to be measured? Is simultaneous determination of several species necessary? What is the required accuracy, time resolution, and spatial resolution? Other aspects to be considered include logistic requirements such as power consumption, mounting of light sources or retro-reflectors (see below), or accommodation of the instrument on mobile platforms.

Table 5.5 Evidence for the presence of halogen radicals in the troposphere

Evidence	Literature (not intended to be complete)
Indirect	
'Hydrocarbon clock' observations (Cl, Br-atoms),	Solberg et al. (1996); Jobson et al. (1994), Rudolph et al. (1999), Singh et al. (1996), Wingenter et al. (1996), Ramacher et al. (1997, 1999), Röckmann et al. (1999), Platt et al. (2004)
Sudden loss of ozone in the (polar) boundary layer	e.g. Barrie et al. (1988), Platt and Lehrer (1996b)
Semi-direct	
Atomic fluorescence after titration by NO (BrO, ClO)	Toohey et al. (1996)
Chemical Amplifier (ROX-Box) measurements	Perner et al. (1999)
Direct	
Long-Path Differential Absorption Spectroscopy (LP-DOAS), detection of BrO, ClO, IO	Hausmann and Platt (1994), Unold (1995), Tuckermann et al. (1997), Platt and Lehrer (1996b), Kreher et al. (1997), Hegels et al. (1998), Wagner and Platt (1998), Richter et al. (1998), Martínez et al. (1999), Alicke et al. (1999), Allan et al. (2000a, 2001), Wittrock et al. (2000a), Frieß et al. (2001), Wagner et al. (2001a), Matveev et al. (2001), Leser et al. (2002), Stutz et al. (2002b), Hönninger and Platt (2002), Roozendael et al. (2002), Pundt et al. (2002), Bobrowski et al. (2003), Hollwedel et al. (2004), Hönninger et al. (2004b, c), Frieß et al. (2004), Saiz-Lopez (2004a, 2004a), Bobrowski (2005), Zingler and Platt (2005), Saiz-Lopez et al. (2007), Read et al. (2008), Mahajan et al. (2009), Seitz et al. (2010), Stutz et al. (2011), Gómez Martín et al. (2013), Tuite et al. (2018)
Multi Axis Differential Absorption Spectroscopy (MAX-DOAS), detection of BrO, IO	Platt and Hönninger (2003), Leser et al. (2003), Wagner et al. (2007c), Martin et al. (2009), Seitz et al. (2010), Mahajan et al. (2012), Großmann et al. (2013), Gómez Martín et al. (2013), Mahajan et al. (2019), Prados-Roman et al. (2015), Simpson et al. (2017), Takashima et al. (2022)
Chemical Ionisation Mass Spectrometry (CIMS), detection of BrO, ClO, IO, Cl_2, Br_2, I_2 and $ClNO_2$	Osthoff et al. (2008), Liao et al. (2011), Liao et al. (2012), Custard et al. (2016), Wang and Pratt (2017), Raso et al. (2017), Riedel et al. (2013)

Spectroscopic techniques can be broadly divided into methods relying on absorption of radiation directed to the sample from some source, or into spectroscopic analysis of radiation emitted by the sample itself. Further divisions can be made according to the wavelength range used and thus the kind of internal state of the sample molecules the radiation interacts with. Figure 5.1 illustrates these different categories by giving a 'family tree' of spectroscopic techniques. In this section we give a brief overview of some spectroscopic techniques used to analyse atmospheric composition.

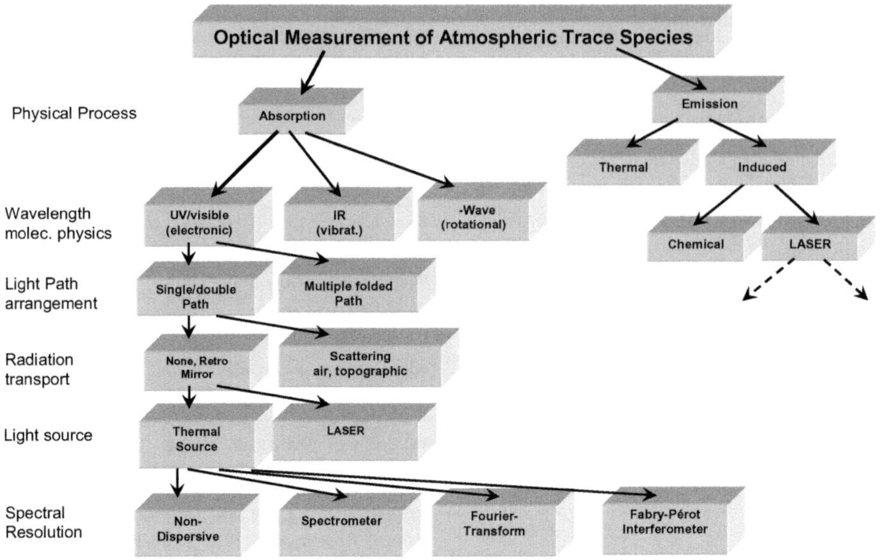

Fig. 5.1 'Family tree' of spectroscopic measurement techniques for atmospheric trace species. Absorption spectroscopy can be categorised by the type of light source—natural (e.g. sunlight) or artificial (e.g. incandescent-, arc lamps, or lasers), the arrangement of the absorption path, the method of obtaining spectral information (non-dispersive, dispersive techniques, type of spectrometer, etc.), or other characteristics of the particular technique

5.6.1 Microwave Spectroscopy

Spectroscopy in the microwave and sub-mm wavelength range can, in principle, be used in active and passive configurations. To our knowledge, no detection of atmospheric gases with active microwave instruments (i.e. with instruments employing their own radiation sources) has been made. However, the measurements of atmospheric parameters (clouds, precipitation, and turbulence) with RADAR techniques are common.

Passive measurements of stratospheric species by microwave and sub-mm wave emission are state of the art (e.g. Janssen 1993). This technique registers the thermal emission due to rotational transitions of the atmospheric molecules. For instance, the ClO-molecule radiates at $\nu_o = 649.448$ GHz, corresponding to 0.46 mm wavelength ($18_{1/2} \rightarrow 17_{1/2}$ transition of ^{35}ClO). Other halogen species are also detectable by this technique. The line width is dominated by collisional (pressure) broadening, which dominates over Doppler-broadening ($\Delta\nu_D/\nu_0 \approx v_{\text{molec}}/c \approx 10^{-6}$), as described in Sect. 3.6. While this strong variation in line width with pressure, and thus altitude of the absorbing molecule, allows the retrieval of vertical profiles from the recorded line shape, it also limits the detection sensitivity at an atmospheric pressure exceeding a few mbars. Thus the best sensitivity is reached in the upper stratosphere, while detection at lower stratospheric or even tropospheric altitudes (i.e. below about 15–25 km) is usually difficult (e.g. Janssen 1993; Klein et al. 2002; Livesey et al. 2003). The microwave and sub-mm wavelength range is presently not used for measurements in the troposphere due to the relatively large pressure broadening of the lines.

5.6.2 IR Spectroscopy

IR spectroscopy has been in use for several decades, and was initially developed for the detection of atmospheric CO_2 by non-dispersive instruments ['Ultra-Rot AbSorption' (German for 'ultra-red absorption'), URAS]. More modern instruments are based on Fourier transform (FT) techniques or Tunable Laser Diode Spectroscopy (TDLS) to measure HNO_3, CH_2O, HCOOH, H_2O_2, and many other species on km path length using atmospheric paths or multiple-reflection cells (e.g. Pitts et al. 1977; Tuazon et al. 1980). The sensitivity is in the low ppb range; thus, those instruments appear to be best suited for studies of polluted air. IR spectroscopy has been applied in four modes of operation:

1. Active operation, where an artificial broadband light source, usually a thermal radiator, is used (e.g. Pitts et al. 1977; Tuazon et al. 1980; Galle et al. 1994; Griffith et al. 2018).
2. Active operation, where a tuneable diode laser serves as light source. (e.g. Harris et al. 1989; Schiff et al. 1990; Li et al. 2016; Genner et al. 2020; see also Sect. 5.7.1).
3. Passive operation using solar occultation FT spectroscopy, where the sun serves as a light source (Mellqvist et al. 2010; Johansson et al. 2014a; Kille et al. 2022)
4. Passive operation using thermal emission from the trace gases under consideration. A particular example is the Michelson Interferometer for Passive Atmospheric Sounding (MIPAS) (Fischer, 1993; Fischer and Oelhaff 1996; Notholt et al. 1997; Dinelli et al. 2021).

5.6.3 UV/Visible Absorption Spectroscopy

At ambient temperatures, thermal emission of UV/visible radiation is completely negligible; nevertheless, active and passive arrangements can be used, the latter technique relying on radiation from naturally hot objects, e.g. the sun.

The most common optical arrangement is shown in Fig. 6.1; the light source and the receiving system are typically separated by several kilometres. UV/visible absorption spectroscopy can be used in both 'active' and 'passive' modes; in the active mode, an artificial light source provides the radiation, while in the passive mode natural light sources, usually the sun or stars, act as source of radiation (see Chap. 6). The strength of these techniques lies in good specificity, the potential for real-time measurements, and the absence of wall losses. The first property makes spectroscopic techniques more suitable for the detection of unstable species such as OH radicals or nitrate radicals (see Chap. 10). Limitations of some systems using a separated light source and receiving system are due to logistic requirements, for example the need for two suitable locations with a line of sight. Some of these logistical challenges can be avoided if the light path is folded into a small multi-reflection cell or a cavity. However, these folded light path arrangements suffer from some of the same challenges as in-situ observations. Common to all UV-vis absorption techniques is poor performance under conditions of poor atmospheric visibility (see Sect. 4.7).

LIDAR techniques combine the absence of wall losses and good specificity with somewhat smaller logistic requirements and the capability to make range-resolved measurements. In contrast to LIDAR, most of the above systems can only make point or path-averaged measurements. Unfortunately, this advantage of LIDAR is usually obtained at the expense of reduced sensitivity (see Sect. 5.7.3).

5.7 Selection Criteria for Spectroscopic Techniques

The important technical criteria of spectroscopic instruments are the wavelength region used (see Fig. 5.1), the physical principle (i.e. absorption or emission spectroscopy), the arrangement of the light path (path in the open atmosphere or enclosed—frequently folded—path), and the type of light source. The following techniques are presently employed to measure atmospheric trace gases:

- Tunable Diode Laser Spectroscopy (TDLS)
- Photo Acoustic Spectroscopy (PAS)
- Light Detection And Ranging (LIDAR)
- Differential Absorption LIDAR (DIAL)
- Laser-Induced Fluorescence (LIF)
- Cavity-Ringdown Spectroscopy (CRD)
- Mask Correlation Spectroscopy
- Differential Optical Absorption Spectroscopy (DOAS).

While, in principle, most techniques (i.e. TDLS, LIDAR, DIAL, LIF, and DOAS) allow light paths (or sensitive volumina, in the case of LIF) in the open atmosphere, some techniques lend themselves to enclosed light paths.

5.7.1 Tuneable Diode Laser Spectroscopy (TDLS)

Tunable diode laser spectroscopy uses a narrow-band, frequency-variable laser source to scan a suitable trace gas absorption structure. This laser can be frequency tuned e.g. by varying the diode current. In practice, a sawtooth-like waveform of the laser current results essentially in a linear scan across a certain spectral interval, typically a small fraction of a wave number. In principle, no further wavelength selective elements are required. Early applications relied on lead–salt (Harris et al. 1989; Schiff et al. 1990; Sigrist 1994a) diode lasers operating in the mid-infrared region (about 3–30 μm), thus covering the fundamental vibrations of many molecules of atmospheric interest. The disadvantages of lead–salt diode lasers are their limited commercial availability and the requirement for cooling to temperatures around (and frequently below) the boiling point of liquid nitrogen. Today, tuneable diode lasers operating at room temperature in the near IR (about 1.6–2 μm) are available, such as communication laser diodes, which allow very compact and lightweight instruments (e.g. Durry et al. 1999; Gurlit et al. 2005). In the near IR, weaker overtone-bands are available. Quantum cascade lasers (QCL) have expanded this range into the mid-IR (e.g. Kosterev and Tittel 2002; Jimenez et al. 2004).

TDLS have measured HNO_3, NO, NO_2, N_2O, CH_2O, CO, and H_2O_2 at ppb and sub-ppb levels (Harris et al. 1989; Schiff et al. 1990, Li et al. 2016; Genner et al. 2020). When coupled to a multiple reflection cell, the strength of TDLS lies in the compact design and low power consumption, which results in high mobility of the instrument, allowing measurements on board ships and airplanes (Genner et al. 2020). Limitations include the necessity to operate under low pressure (in many applications), thus introducing possible losses at the walls of the closed measurement cells. Diode-lasers are available for non-scientific applications, such as telecommunication, but diodes optimised for spectroscopic purposes, especially in the visible and ultraviolet spectral ranges, are still difficult to obtain.

5.7.2 Photo Acoustic Spectroscopy (PAS)

Photo Acoustic Spectroscopy is based on the detection of the pressure change occurring as a consequence of the temperature rise due to the energy absorbed by trace gas molecules illuminated by radiation at suitable frequency. The absorption of radiation by atmospheric trace gases is very minute, thus giving rise to only small temperature and pressure changes. Extremely sensitive devices for detecting periodic atmospheric pressure changes are available in the form of microphones for the detection of sound waves (e.g. Sigrist 1994a).

A typical PAS instrument consists of an intensity-modulated light source, usually a laser, illuminating the interior of a cell (e.g. a tube of a few millimetres of inner diameter and a fraction of a metre length) equipped with a microphone 'listening' to the periodic pressure changes occurring at the frequency of the modulation. The weak signal is sometimes enhanced by tuning the modulation frequency to an acoustic resonance of the absorption cell, leading to a signal increase by the 'quality factor' Q of the acoustic resonator (Sigrist 1994a). A particular example of employing acoustic resonance is the Quartz Enhanced Photoacoustic Spectroscopy (QEPAS) technique (Sampaolo et al. 2022), which uses a high-Q quartz resonator as detector.

Note that photoacoustic detectors directly measure the difference $I_0 - I$ (approximately proportional to $D \cdot I_0$ and thus the trace gas concentration) rather than determining I and I_0 separately.

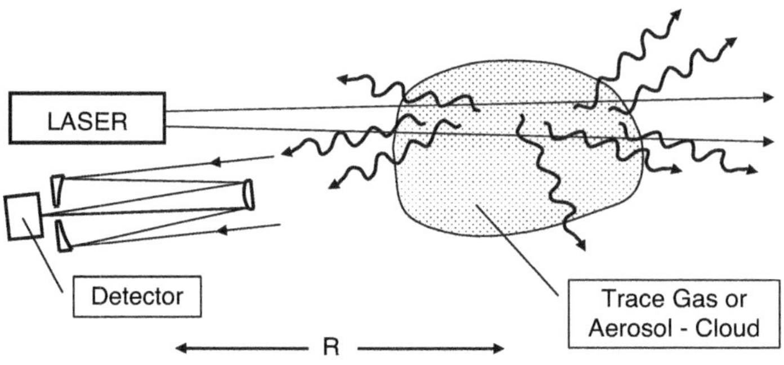

Fig. 5.2 The principle of the LIDAR technique

5.7.3 Light Detection And Ranging (LIDAR)

The term LIDAR was coined to resemble the well-known RADAR (RAdiowave Detection And Ranging) (e.g. Rothe et al. 1974; Hinkley 1976; Svanberg 1992; Sigrist 1994a). With this technique short pulses of a strong, collimated light source, typically a pulsed laser, are emitted into the atmosphere. By analysing the temporal evolution of the intensity backscattered from the atmosphere, the spatial distribution of scattering and extinction along the direction of the emitted (and received) radiation can be deduced, as illustrated in Fig. 5.2.

The distance from which the radiation intensity is scattered back is given by:

$$R = \frac{c \cdot t}{2}$$

where c is the speed of light, and t is time after emission of the laser pulse at which the signal reaches the detector.

The general LIDAR equation is:

$$E(\lambda, R, \Delta R) = K \cdot E_0 \cdot c_S(R)\,\sigma_{SR} \frac{\Delta R}{R^2} \cdot \\ \times \exp\left(-2 \int_0^R \left[\sigma_A(r) \cdot c_A(r) + \sigma_S(r) \cdot c_S(r)\right] dr\right) \quad (5.1)$$

where:

$E(\lambda, R, \Delta R)$ denotes radiation energy received from a volume of air between R, $R + \Delta R$;
ΔR is the distance interval to be averaged over (corresponds to $\Delta t = \frac{2\Delta R}{c}$);
K is a constant of the system (describes the influence and size of the receiving system);
E_0 is the radiation energy emitted by the laser;

$c_s(R)$ is the concentration of backscattering centres (molecules, aerosol particles);
σ_{SR} is the backscatter cross-section;
σ_A is the absorption cross-section (aerosol or gas);
$c_A(R)$ is the concentration of the absorbers (i.e. gas molecules);
σ_S is the total scattering cross-section (note that $\sigma_S \neq \sigma_{SR}$!).

The LIDAR techniques combine the absence of wall losses with the unique capability to make range-resolved measurements along the line of sight. In contrast to that, the systems discussed earlier can only make point measurements or path-averaged measurements. LIDAR systems have been successfully used to map aerosol distributions. There are, however, two central problems with LIDAR systems:

1. The signal [see (5.1)] depends on backscattering as well as on total scattering (or extinction) properties of the atmosphere, which are, in the case of Mie scattering, difficult to separate. The ratio σ_S/σ_{SR} is usually unknown and must be estimated from a-priori information unless the Raman–LIDAR technology is used.
2. The backscattered signal is usually very weak, and thus even when high-energy lasers are used, only a relatively small number of photons is received. Therefore, the above advantages are usually obtained at the expense of high sensitivity.

5.7.4 Differential Absorption LIDAR (DIAL)

In order to measure the distribution of trace gases, at least two different wavelengths are required (Svanberg 1992; Sigrist 1994a). The DIAL equation is obtained by dividing two general LIDAR equations (5.1) for two different wavelengths λ_1 and λ_2, respectively:

$$\frac{E(\lambda_2, R)}{E(\lambda_1, R)} = \exp\left(-2\left(\sigma_A(\lambda_2) - \sigma_A(\lambda_1)\right) \cdot \int_0^R c_A(r)\,dr\right) \tag{5.2}$$

It is assumed that σ_{SR} and σ_S are the same for λ_1 and λ_2, respectively. This assumption appears justified, as long as $\Delta\lambda = \lambda_2 - \lambda_1$ is sufficiently small (a few nanometres). The two wavelengths are chosen so that, while their difference is as small as possible, the difference in the absorption cross-sections $\Delta\sigma = \sigma_A(\lambda_2) - \sigma_A(\lambda_1)$ is as large as possible, i.e. one wavelength is at the centre of an absorption line, the other besides the line.

5.7.5 White Light LIDAR and 'Scheimpflug LIDAR'

Recently, experiments of three different LIDAR techniques were performed using broadband 'white light' sources in conjunction with wavelength dispersing detectors making use of the DOAS principle, see Sect. 5.7.9. If these techniques prove viable, they could combine the advantages of LIDAR and DOAS:

1. A LIDAR principle that combines the differential UV-visible absorption spectroscopy and the classical LIDAR technique. The novel element of the system is the use of an imaging spectrometer in conjunction with a two-dimensional CCD detector array to simultaneously spectrally and temporally resolve backscattered radiation. The 'white' light (i.e. broadband radiation) required for this LIDAR approach is provided either by a flash-lamp or a broadband dye laser of 10–20 nm full width at half maximum and tuneable across the UV-visible spectral region. Simultaneous range-resolved measurement of multiple molecular species is possible by the DOAS technique. A first application demonstrated absorption studies in the spectral regions where NO_3 and H_2O absorb using both elastic and inelastic (Raman) backscattered radiation (Povey et al. 1998; South et al. 1998). In addition, the technique has applicability for a wide range of molecules, including O_3, NO_2, and other spectrally structured absorbers, and for atmospheric temperature sounding, which may be derived from either rotational Raman return or temperature-dependent absorptions such as those of O_2.
2. In this approach, ultra-short (femtosecond) laser pulses produce a 'beam' of white light in the atmosphere, which can then be observed with a separate wavelength-dispersing detector system. Atmospheric constituents will leave their spectroscopic fingerprint on the distance between the 'beam' and the detector, which can then be detected by DOAS techniques (e.g. Rairoux et al. 2000; Kasparian et al. 2003).
3. A completely different approach uses spatially separated (by e.g. a dozen metres) transmitter and receiver. The range resolution is achieved by tilting e.g. receiver field of view along the transmitted light beam (see e.g. Barnes et al. 2007). This technique, also known as 'Scheimpflug LIDAR', is described in Sect. 7.8.5.

5.7.6 Laser-Induced Fluorescence (LIF)

Laser-Induced Fluorescence is a very sensitive and specific technique for the detection of atmospheric trace gases. It relies on the excitation of trace gas molecules by absorption of (laser) radiation, the frequency of which corresponds to a suitable transition from the molecule's ground state to an (electronic) excited state (Fig. 3.6). The number of fluorescence photons detected is then proportional to the atmospheric trace gas concentration.

A central problem with LIF is the separation of Rayleigh scattered exciting radiation from the fluorescence signal. While Rayleigh scattering cross-sections of air molecules (i.e. O_2 and N_2) are typically six to eight orders of magnitude smaller than trace gas absorption cross-sections, typical trace gas mixing ratios are 10^{-12} to 10^{-9}. Combined with the fact that fluorescence efficiencies can be orders of magnitude below unity, scattered excitation radiation must be suppressed by many orders of magnitude. Several techniques have been developed to achieve this goal, which include:

1. Excitation at higher (vibrational) levels above the excited state in which fluorescence can be observed. The detected wavelengths (red) are thus shifted from the excitation wavelength, and can be separated by suitable filters (wavelength filtering).
2. Use of short laser pulses for excitation. The fluorescence signal can be recorded after the excitation signal has decayed and thus separated. This technique frequently requires low pressure in order to avoid too rapid a decay of the fluorescence signal by collision quenching and hence its denomination 'temporal' or 'baric' filtering (Hard et al. 1979).

Other potential problems in the technique are photochemical formation of species to be detected by excitation radiation [e.g. OH by O_3 photolysis (Ortgies et al. 1980; Davis et al. 1981; Shirinzadeh et al. 1987)], or saturation of the transition under consideration. In addition, absolute calibration from 'first principles' has been found to be extremely difficult (e.g. Hofzumahaus et al. 1996, 1998; Holland et al. 2003), and so experimental calibration is necessary (e.g. Hofzumahaus et al. 1996, 1997, 1998; Schultz et al. 1995).

5.7.7 Cavity-Ringdown (CRDS) and Cavity Enhanced Spectroscopy (CEAS)

Cavity-Ringdown Spectroscopy (CRDS) and Cavity Enhanced Spectroscopy (CEAS make use of the effect of atmospheric absorbers on the quality of a passive optical resonator (a 'cavity'), see Sect. 7.7.3.5 (e.g. Paldus and Zare 1999; Brown 2003). After initially exciting the resonator with a laser and then switching off the laser, the decay (or 'ringdown') of the resonator is observed, which is influenced by the extinction of light due to trace gases present in the resonator cavity. If excited with continuous radiation, the resonator will exhibit certain attenuation, which can be monitored. The energy loss per 'round trip' of the radiation in an optical resonator with no absorber is given by the reflectivity of the two mirrors, see Sect. 7.7.3.5. In the case of very weak absorbers, the quality, Q, of the cavity is only determined by mirror reflectivity, R_M, which can reach 0.9999. Thus, effective light path lengths on the order of many kilometres can be reached under favourable conditions with resonator lengths below 1 m. The limitations of CRDS and CEAS are similar to the limitations of DOAS in that the effective light path will be limited by visibility, unless particle filters are used in the inlet. CRDS and CEAS have developed into a useful tool for laboratory measurements and several measurements in the atmosphere, such as NO_3 radicals, NO_2, etc., have been reported (e.g. King et al. 2000; Ball and Jones 2003; Brown et al. 2001–2003; Simpson 2003; Ball et al. 2004; Dorn et al. 2013; Coburn et al. 2014; Wu et al. 2014; Aalto et al. 2015; Min et al. 2016).

5.7.8 Mask Correlation Spectroscopy (COSPEC)

Mask Correlation Spectroscopy was originally conceived as a tool for oil exploration by the detection of atmospheric iodine (I_2) vapour (Davies 1970), which was frequently associated with oil deposits. The fundamental idea of COSPEC is to analyse the absorption features (i.e. absorption bands) of atmospheric constituents imprinted on scattered sunlight. In modern nomenclature, it is a 'passive' instrument (see Chap. 6). The instrument allowed some degree of remote sensing of atmospheric trace gases. While the original idea of I_2-vapour detection did not prove practical, other applications were found, including the observation of tropospheric column densities of SO_2 and NO_2 in industrial emission or volcanic plumes (Giovanelli et al. 1979; Beilke et al. 1981; Redemann Fischer 1985; Hoff 1992).

The COSPEC instrument was originally designed in 1960s (Barringer et al. 1970; Davies 1970; Davies et al. 1975); it uses an innovative opto-mechanical correlator (e.g. Davies 1970; Millan and Hoff 1977; Millán and Hoff 1978; Millan 1972, 1978, 1980) to identify and quantify the spectra of atmospheric trace gases. In this way, a remarkably compact and reliable instrument could be built, since spectral recognition of the molecules to be measured was done by special opto-mechanics (Fig. 5.3) with relatively simple electronics and without a computer. Since neither fast-scanning, multi-channel detectors nor powerful computers were available at that time, the mask correlation technique was an attractive way to measure total columns.

By making the instrument mobile and horizontally traversing emission plumes, a total integrated concentration cross-section of the plumes can be obtained (Giovanelli et al. 1979). After multiplication with the concentration weighted wind component perpendicular to the cross-section, the total emission from the source (e.g. in kg s^{-1}) can be deduced.

For around 30 years, the COSPEC instrument has been the principal tool for remote surveillance of volcanic plumes. Although it was not originally designed for volcanological research, it had a major impact on the discipline, not only in volcano monitoring and eruption forecasting but also in estimating the global contribution of volcanic volatiles to the atmosphere (e.g. Weibring et al. 1998). By allowing the measurements of SO_2 and NO_2 flux in airborne plumes, it has fulfilled a vital role in the management of many volcanic crises, notably at Pinatubo in 1991 (Hoff 1992), and at Soufrière Hills Volcano. The COSPEC instrument was a major technological innovation at its time, but is now outdated in several important respects. Problematic aspects of its concept are interferences from other gases, non-linearity, solar Fraunhofer lines, polarisation effects, and multiple scattering in aerosol and clouds. In particular, each compound to be measured required its individual mask, and the system is sensitive to wavelength shifts due to mechanical distortions and temperature variations.

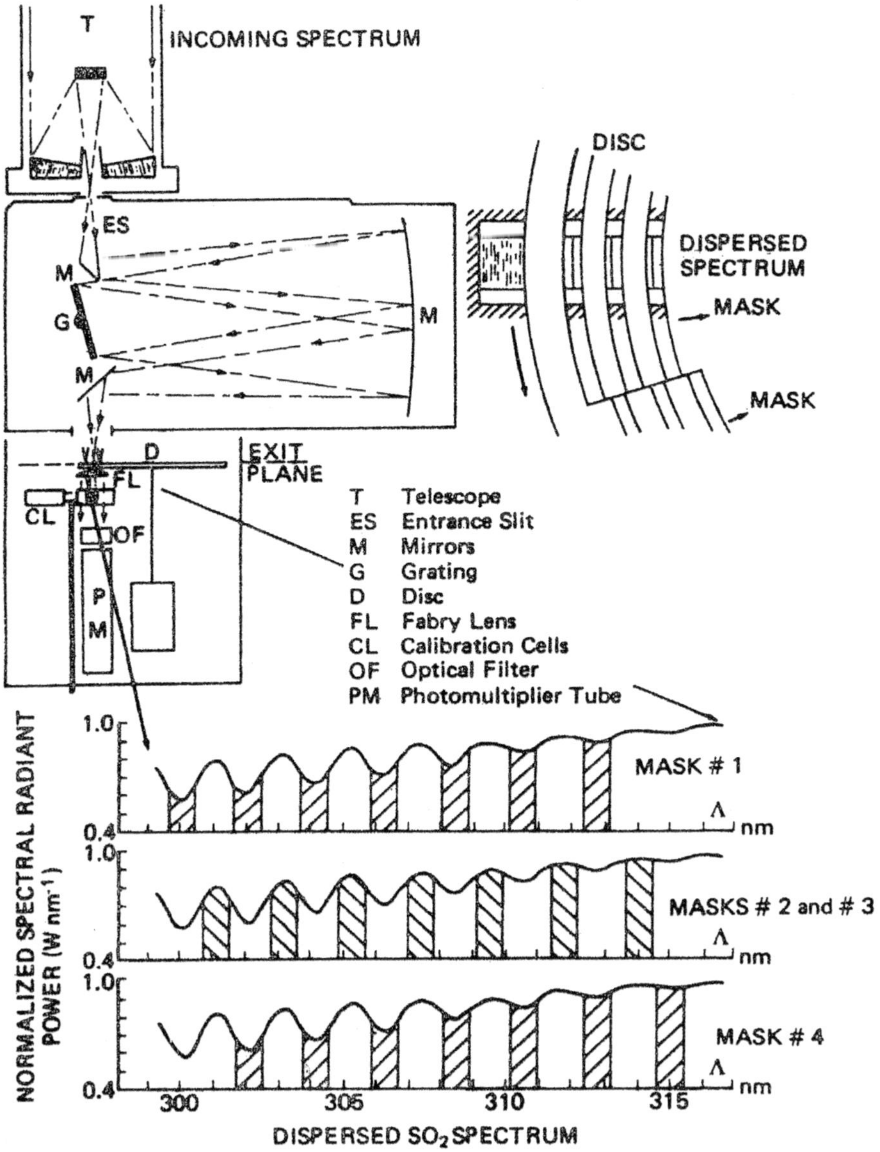

Fig. 5.3 The principle of a COSPEC instrument. Reprinted with permission from Millán and Hoff 1978, ©Elsevier. All rights reserved

5.7.9 Differential Optical Absorption Spectroscopy (DOAS)

Many of the advances of UV-vis absorption spectroscopy as an analytical technique discussed in this and previous chapters can be attributed to the development of Differential Optical Absorption Spectroscopy (DOAS). DOAS uses structured absorption of many trace gases of atmospheric interest while ignoring the rather 'smooth' extinction features due to aerosol or instrumental effects. The DOAS principle and its technical realisation builds on some of the earlier developments, superseding them in many respects and supplementing other spectroscopic techniques (e.g. Platt and Perner 1980; Platt 1994b, 2000; Plane and Smith 1995; Roscoe and Clemitshaw 1997; Russwurm 1999; Finlayson-Pitts and Pitts 2000; Clemitshaw 2004).

The remainder of this book is dedicated to the theoretical and practical aspects of DOAS, as well as its application to monitor and study the atmosphere.

Differential Absorption Spectroscopy 6

Absorption spectroscopy is a well-established tool for the analysis of the chemical composition of gases. As such, it has played a prominent role in the discovery of the physical and chemical properties of the earth's atmosphere.

This chapter focuses on a modern method to quantitatively measure a large variety of trace gases in the atmosphere. DOAS is now one of the most commonly used spectroscopic techniques to measure trace gases in the open atmosphere. This chapter includes a general introduction to absorption spectroscopy and DOAS, an overview of different experimental approaches of DOAS, a discussion of the precision and accuracy of this method, and a rigorous mathematical description of the various DOAS applications.

6.1 The History of Absorption Spectroscopy

Spectroscopic studies of the earth's atmosphere date back more than 100 years. Some milestones in the investigation of atmospheric composition through spectroscopy include:

1879 Marie Alfred Cornu concludes from the change of the edge of the intensity decay in the UV that a trace species in the earth's atmosphere must be causing this UV-absorption (Cornu 1879).

1880 Sir Walter Noel Hartley discovers the absorption of UV-radiation by ozone. This led to the name Hartley-bands for ozone absorption below 300 nm (Hartley 1880, 1881).

1880 M. J. Chappuis discovers the absorption of visible light by ozone, which is today called the Chappuis-band. Chappuis also speculates that light absorption by ozone is the reason for the blue colour of the sky (Chappuis 1880).

U. Platt and J. Stutz, *Differential Optical Absorption Spectroscopy*, Geophysics and Environmental Physics, https://doi.org/10.1007/978-3-662-68639-3_6

1890 Sir William Huggins discovers a new group of lines in the spectrum of Sirius, which are later explained by Fowler and Strutt as absorption of terrestrial ozone. These long wavelength UV-bands of ozone are referred to as Huggins-bands.
1904 Discovery of the infrared absorption of ozone near 4.8, 5.8, and 9.1–10 μm by Knut Johan Ångström.
1913 Balloon measurements of the UV absorption of ozone up to 10 km altitude by Albert Wigand showed essentially no change with altitude.
1918 John William Strutt (better known as Lord Rayleigh) concludes that atmospheric ozone must reside in a layer above 10 km altitude above the surface.
1920 First ozone column measurements were made by Charles Fabry and Henri Buisson, who determine a column of about 3 mm (at atmospheric pressure), with large variations.
1925 First application of a dedicated ozone spectrometer by Gordon Miller Bourne Dobson (Dobson and Harrison 1926).
1926 Paul Götz confirms the theory of an ozone layer by observing the so-called 'Umkehr' effect, and determines its altitude to be about 25 km.
1934 Direct observation of the ozone layer by UV-spectroscopy by Erich Regener (TH Stuttgart).
1948 Marcel Migeotte (Ohio State University) discovers methane and carbon monoxide in the earth atmosphere by near-infrared absorption spectroscopy (Migeotte 1948, 1949).
1950 Discovery of the emission bands of the hydroxyl radical (OH) in the nightglow (the Meinel bands of the OH-radical). As a consequence, HO_x-chemistry is viewed in connection with ozone chemistry by David R. Bates and Marcel Nicolet (1950), and Bates and Witherspoon (1952).
1975 First detection of OH in the atmosphere by Dieter Perner and colleagues using differential optical absorption spectroscopy (Perner et al. 1976).

These studies illustrate the role that spectroscopy has played in the measurement of reactive trace gases in the atmosphere, most notably ozone. The identification and quantification of gases was primarily accomplished by the analysis of atmospheric absorptions. This is still the case in most current applications of atmospheric spectroscopy. The use of emission bands is restricted to the thermal infrared wavelength region (see Chap. 5) or to the excited gas molecules in the upper atmosphere, which emit light at higher energies, i.e. shorter wavelength. Both applications are in use today, but are not the topic of this book.

The initial use of spectroscopy in the atmosphere concentrated on the identification of various gases. Soon, however, this method was put in use to quantify the concentrations (or column densities) of these species. The contributions of Dobson, who constructed the first instrument for the regular measurement of atmospheric ozone, should be highlighted (Dobson and Harrison 1926).

6.2 Classical Absorption Spectroscopy

The basis of the early spectroscopic measurements, and many present quantitative trace gas analytical methods in the atmosphere and the laboratory, is Lambert–Beer's law, often also referred to as Bouguer–Lambert or Beer's law. The law was presented in various forms by Pierre Bouguer in 1729, Johann Heinrich Lambert in 1760, and August Beer in 1852. Bouguer first described that, 'In a medium of uniform transparency the light remaining in a collimated beam is an exponential function of the length of the path in the medium'. There is disagreement in the naming of this law. In this book, we refer to it as Lambert–Beer's law.

A variety of spectroscopic techniques make use of the absorption of electromagnetic radiation by matter (Fig. 6.1). In a formulation suitable for the analysis of gaseous or liquid absorbers, Lambert–Beer's law can be written as:

$$\mathrm{I}(\lambda) = \mathrm{I}_0(\lambda) \cdot \exp\left(-\sigma(\lambda) \cdot \mathrm{c} \cdot \mathrm{L}\right) \tag{6.1}$$

Here, $I_0(\lambda)$ denotes the initial intensity of a light beam emitted by a suitable source of radiation, while $I(\lambda)$ is the radiation intensity of the beam after passing through a layer of thickness L, where the absorber is present at a uniform concentration of c. The quantity $\sigma(\lambda)$ denotes the absorption cross-section at wavelength λ. The absorption cross-section as a function of wavelength is a characteristic property of any species. The determination of the light path length, L, is usually trivial for active DOAS applications (see Chap. 4). Once those quantities are known, the average trace gas concentration, c, can be calculated from the measured ratio $I_0(\lambda)/I(\lambda)$:

$$\mathrm{c} = \frac{\ln\left(\frac{\mathrm{I}_0(\lambda)}{\mathrm{I}(\lambda)}\right)}{\sigma(\lambda) \cdot \mathrm{L}} = \frac{D}{\sigma(\lambda) \cdot \mathrm{L}} \tag{6.2}$$

The expression

$$\mathrm{D} = \ln\left(\frac{\mathrm{I}_0(\lambda)}{\mathrm{I}(\lambda)}\right) \tag{6.3}$$

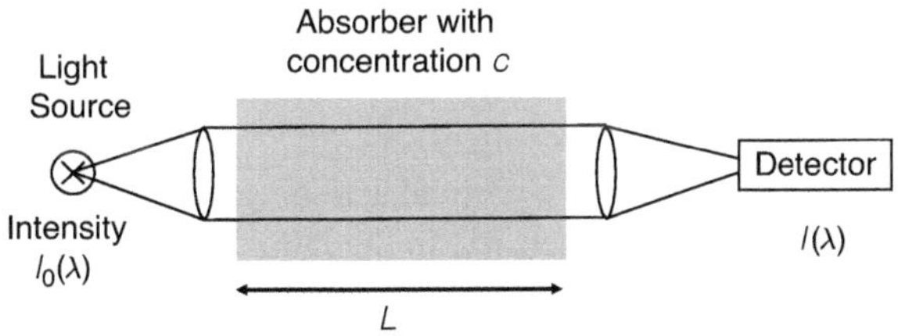

Fig. 6.1 The basic principle of absorption spectroscopic trace gas detection. A beam of light passes through a volume of length L containing the absorber with concentration c. At the end of the light path the intensity is measured by a suitable detector

is called the optical density of a layer of a given absorber. In the literature, the decadal as well as the natural logarithm is used in the definition of optical density. In this book, we exclusively use the natural logarithm.

Equation (6.2) is the basis of most absorption spectroscopic applications in the laboratory, where the intensities $I(\lambda)$ and $I_0(\lambda)$ are determined by measurements with and without the absorber in the light beam.

However, the application of Lambert–Beer's law is more challenging in the open atmosphere. Here, the true intensity $I_0(\lambda)$, as it would be received from the light source in the absence of any atmospheric absorber, is difficult to determine. It would involve removing the air, or more precisely the absorbing gas, from the atmosphere. While this may seem to present a dilemma rendering atmospheric absorption spectroscopy useless, the solution lies in measuring the so-called 'differential' absorption, i.e. the difference between the absorptions at two different wavelengths. This principle was used by Dobson in the 1930s to determine the total column of atmospheric ozone. In an ingenious experimental setup, the Dobson spectrometer compares the intensity of direct solar light of two wavelengths-λ_1, λ_2-with different ozone absorption cross-sections, $\sigma_1 = \sigma(\lambda_1)$, $\sigma_2 = \sigma(\lambda_2)$ (Dobson and Harrison 1926).

6.3 The DOAS Principle

A schematic setup of an experiment to measure trace gas absorptions in the open atmosphere is shown in Fig. 6.2. Similar to Fig. 6.1, light emitted by a suitable spectral broadband source with an intensity $I_0(\lambda)$ passes through a volume with absorbers (here the open atmosphere) and is collected at the end of the light path. As the light travels through the atmosphere, its intensity is reduced through the absorption of a specific trace gas. It also undergoes extinction due to absorption by other trace gases, scattering by air molecules, and extinction by aerosol particles. The transmissivity of the instrument (mirrors, grating, retro-reflectors, etc.) will also decrease the light intensity, as will the light beam widening by turbulence. By expanding Lambert–Beer's law, one can consider the various factors that influence the light intensity using an equation that includes the absorption of various trace gases with concentration c_j and absorption cross-sections $\sigma_j(\lambda)$, Rayleigh and Mie extinction, $\varepsilon_R(\lambda)$ and $\varepsilon_M(\lambda)$, and instrumental effects and turbulence, summarised in $A(\lambda)$:

$$\mathrm{I}(\lambda) = \mathrm{I}_0(\lambda) \cdot \exp\left[-\mathrm{L} \cdot \left(\sum \left(\sigma_\mathrm{j}(\lambda) \cdot \mathrm{c}_\mathrm{j}\right) + \epsilon_\mathrm{R}(\lambda) + \epsilon_\mathrm{M}(\lambda)\right)\right] \cdot \mathrm{A}(\lambda) \quad (6.4)$$

To determine the concentration of a particular trace gas, it is necessary to quantify all other factors influencing the intensity. In the laboratory, this can be achieved by removing the absorber from the light path. In the atmosphere, however, where this is impossible, the multiple factors influencing the intensity pose a dilemma.

Differential optical absorption spectroscopy overcomes this challenge by using the fact that aerosol extinction processes, the effect of turbulence, and many trace gas absorptions show very broad or even smooth spectral characteristics. Certain trace gases, however, exhibit narrowband absorption structures. The foundation of

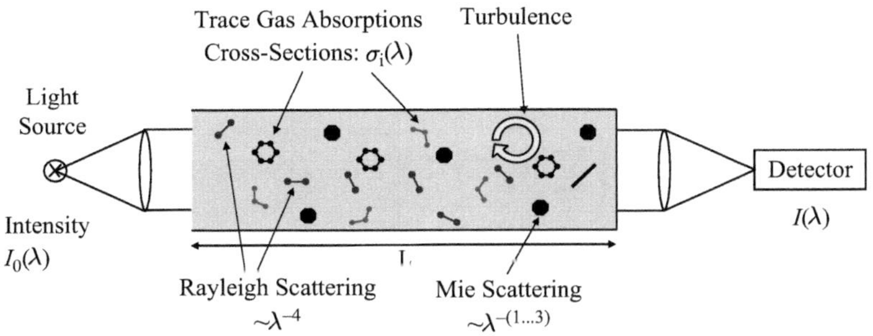

Fig. 6.2 Sketch of an experiment to measure trace gas absorptions in the open atmosphere

Fig. 6.3 Principle of DOAS: I_0 and σ are separated by an adequate filtering procedure into a narrow (D' and σ') and broad band part (I_0' and σ_b)

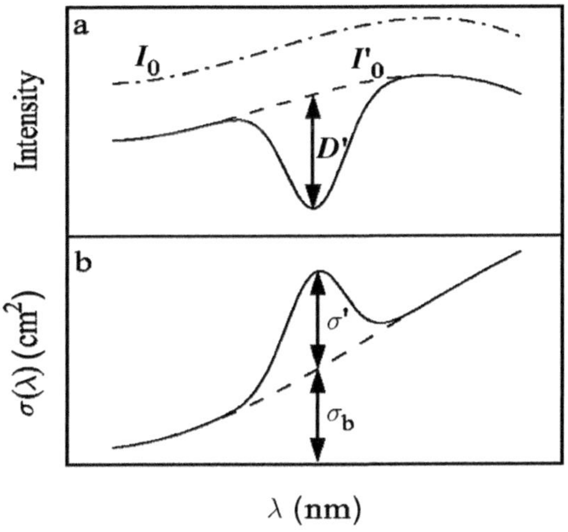

DOAS is thus to separate broad- and narrowband spectral structures in an absorption spectrum in order to isolate these narrow trace gas absorptions (Fig. 6.3). The broad spectrum is then used as a new intensity spectrum $I_0'(\lambda)$, and Lambert–Beer's law can again be applied to the narrowband trace gas absorptions.

Figure 6.3 illustrates the separation of the narrow- and broadband structures for one absorption band, both for the intensity and the absorption cross-section:

$$\sigma_j(\lambda) = \sigma_{j0}(\lambda) + \sigma_j'(\lambda) \tag{6.5}$$

σ_{j0} in (6.5) varies 'slowly' with the wavelength λ, for instance describing a general 'slope', such as that caused by Rayleigh and Mie scattering, while $\sigma_j'(\lambda)$ shows rapid variations with λ, for instance due to an absorption band (see Fig. 6.3). The meaning of 'rapid' and 'slow' variation of the absorption cross-section as a function of wavelength depends on the observed wavelength interval and the width of the absorption bands to be detected. Inserting (6.5) into (6.4), we obtain:

$$\mathrm{I}(\lambda) = \mathrm{I_0}(\lambda) \cdot \exp\left[-\mathrm{L} \cdot \left(\sum_j \left(\sigma'_j(\lambda) \cdot c_j\right)\right)\right] \cdot$$
$$\exp\left[-\mathrm{L} \cdot \left(\sum_j \left(\sigma_{j0}(\lambda) \cdot c_j\right) + \epsilon_R(\lambda) + \epsilon_M(\lambda)\right)\right] \cdot \mathrm{A}(\lambda) \qquad (6.6)$$

where the first exponential function describes the effect of the structured 'differential' absorption of a trace species, while the second exponential function constitutes the slowly varying absorptions as well as the influence of Rayleigh and Mie scattering. The attenuation factor $A(\lambda)$ describes the broad wavelength-dependent transmission of the optical system used and turbulence. A quantity $\mathrm{I'_0}$ can thus be defined as the intensity in the absence of differential absorption:

$$\mathrm{I}'_0(\lambda) = \mathrm{I_0}(\lambda) \cdot \exp\left[-\mathrm{L} \cdot \left(\sum_j \left(\sigma_{j0}(\lambda) \cdot c_j\right) + \epsilon_R(\lambda) + \epsilon_M(\lambda)\right)\right] \cdot \mathrm{A}(\lambda) \qquad (6.7)$$

The corresponding differential absorption cross-section $\sigma'_j(\lambda)$ is then substituted for $\sigma_j(\lambda)$ in (6.1) and (6.2). $\sigma'_j(\lambda)$ is determined in the laboratory or taken from literature data, just like $\sigma_j(\lambda)$. A differential optical density, D', can be defined in analogy to (6.3) as the logarithm of the quotient of the intensities I'_0 and I_0 (as defined in (6.7) and (6.6), respectively):

$$\mathrm{D}' = \ln \frac{\mathrm{I'_0}(\lambda)}{\mathrm{I}(\lambda)} = \mathrm{L} \cdot \sum_j \sigma'_j(\lambda) \cdot c_j \qquad (6.8)$$

Atmospheric trace gas concentrations can then be calculated according to (6.2), with differential quantities D' and $\sigma'(\lambda)$ substituted for D and $\sigma(\lambda)$, respectively. A separation of the different absorptions in the sum of (6.8) is possible because the structures of the trace gases are unique, like a fingerprint (see Sect. 6.5).

Both the separation of broad, and narrow spectral structures, and the separation of the various absorbers in (6.8) require the measurement of the radiation intensity at multiple wavelengths. DOAS measurements usually observe the intensity at 500–2000 individual wavelengths to accurately determine the concentrations of the various absorbing trace gases. The use of multiple wavelengths is an expansion of the principles used by Dobson, who employed two or four wavelengths in his ozone measurements.

The use of differential absorptions over an extended wavelength range has a number of major advantages. Because the transmission of optical instruments typically shows broad spectral characteristics, no calibration of the optical properties or their change with time is necessary. This often makes the instrumentation much simpler and less expensive. The use of a multitude of wavelengths allows the unique identification of trace gas absorptions. A further major advantage of this approach is the opportunity to observe and quantify extremely weak absorptions corresponding to optical densities around $D' = 10^{-4}$. In particular, the ability to use very long light paths in the atmosphere increases the sensitivity of DOAS and, at the same time, provides spatially averaged values. For example, light path lengths of active DOAS applications can be 10–20 km long, while path lengths of passive DOAS applications can reach 1000 km.

6.4 Experimental Setups of DOAS Measurements

The DOAS principle can be applied in a wide variety of light path arrangements and observation modes as described below. To provide a general overview of different setups, we introduce a classification system that will later be used in the description of the different analysis methods (see Sect. 6.7 and Chap. 8) and the underlying technology (Chap. 7).

We distinguish between **active** and **passive** DOAS by light source. Active DOAS uses artificial light, while passive DOAS relies on natural light sources, i.e. solar, lunar, or stars. An overview of the most common active (Fig. 6.4) and passive (Fig. 6.5) setups illustrates the breadth of DOAS applications that are in use today.

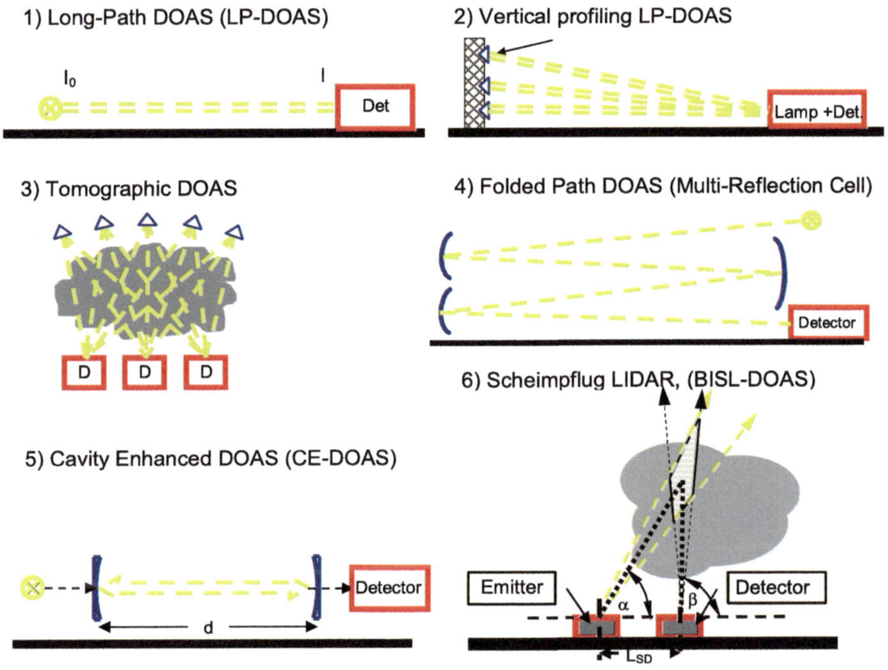

Fig. 6.4 The DOAS principle can be applied in a wide variety of light path arrangements and observation modes using artificial light sources

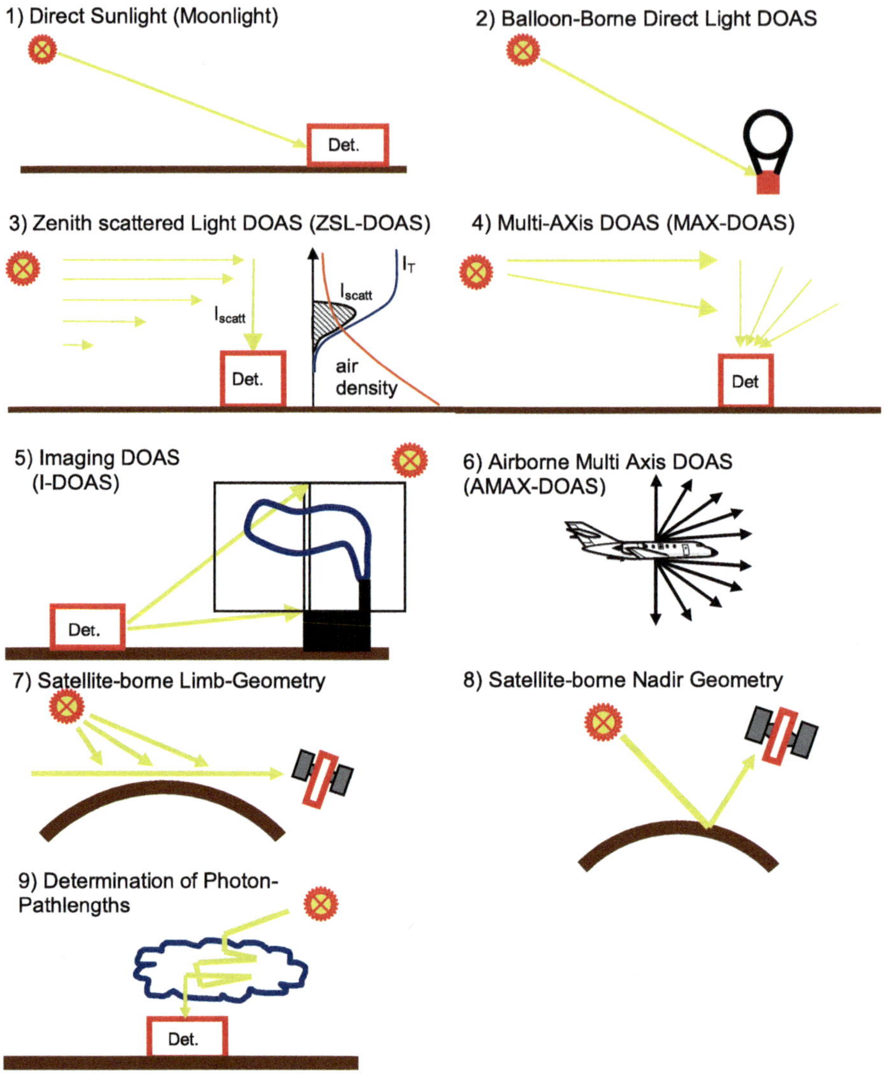

Fig. 6.5 The DOAS principle can be applied in a wide variety of light path arrangements and observation modes using natural light sources

6.4.1 Active DOAS

Active DOAS applications have in common, that they rely on an artificial light source coupled to an optical setup that is used to send and receive light in the atmosphere. Spectroscopic detection is achieved by a spectrometer at the end of the light path. Active DOAS is very similar to classical absorption spectroscopy, as employed in laboratory spectrophotometers. However, the low trace gas concentrations in the atmosphere require very long light paths (up to tens of kilometres in length), making the implementation of these instruments challenging (see Chap. 7 for details). Active DOAS applications are typically employed to study or monitor tropospheric composition and chemistry. Active DOAS systems are also used in smog- and aerosol chamber experiments.

As summarised in Fig. 6.4 the following active DOAS configurations are used for measurements:

1. Long-path systems where the absorption path extents from a (searchlight-type) light source to a receiving system, alternatively, the light path can be 'folded' by a reflector so that light source and receiving system are located at one end of the path (Axelsson et al. 1990; Merten et al. 2011). This was the earliest active DOAS setup, for example used by Perner et al. (1976) to measure OH radicals for the first time. While Perner et al. (1976) used a laser as the 'broad band' light source, configurations using xenon-arc lamps, to measure trace gases such as O_3, NO_2, SO_2, etc., are currently most common (e.g. Stutz and Platt 1997a, b).
2. Vertical profiles of trace gas concentrations can be measured by arranging several light paths in the vertical. This is typically achieved using one sending/receiving telescope and multiple reflectors mounted on a tower or on a mountainside (e.g. Platt 1978; Stutz et al. 2004b; Tsai et al. 2014).
3. Installing a number of light paths across a certain area can create a system able to determine the 2-dimensional distribution of trace gases. If multiple light paths are crossed, tomographic measurements can be performed (see e.g. Hartl et al. 2006; Pöhler 2010; Stutz et al. 2016; Casaballe et al. 2020).
4. Optical multipath systems allow sensitive measurements of trace gas concentrations in small air volumes (see Sect. 7.7.) (e.g. Ritz et al. 1992). Because the light can pass the multiple reflection cells around 100 times, long light paths can be achieved. This setup has been used in the laboratory and the atmosphere (e.g. Alicke et al. 2003; Kurtenbach et al. 2002).
5. Optical resonators are slowly replacing multiple reflection systems (see e.g. Fiedler et al. 2003; Platt et al. 2009; Zheng et al. 2018). Because resonators allow between 1000 and 10000 traverses of the measurement volume, very compact setups with high sensitivities can be realised (see Sect. 7.7.3).
6. Backscattered radiation form an artificial light source allows the determination of vertical concentration profiles with a Scheimpflug-LIDAR arrangement (see Sect. 7.8.5)

Active DOAS measurements have contributed to the discovery and quantification of a number of important atmospheric trace species, most notably the radicals OH, NO_3, BrO, and IO (e.g. Perner et al. 1976; Platt et al. 1979). The elegance of active DOAS is that the expanded Lambert–Beer's law (6.4) can be directly applied to the calculation of trace gas concentrations, based only on the absorption cross-section, without the need for calibration of the instrument in the field. This gives active longpath DOAS high accuracy and excellent sensitivity.

6.4.2 Passive DOAS

Passive DOAS utilises light from natural sources. The two most important sources are the sun and the moon. However, the use of light from other stars has also been reported. While the measurement of light directly from moon and stars is possible, sunlight offers two alternatives: direct sunlight and sunlight scattered in the atmosphere by air molecules and particles. We will further subdivide passive DOAS applications into **direct** and **scattered light** techniques (see also Chap. 11).

As summarised in Fig. 6.5 the following passive DOAS configurations are in use:

1. Direct light measurements use the sun, moon, or stars as light sources, and thus share the advantage of active DOAS of directly applying Lambert–Beer's law (Fig. 6.5, Panels 1 and 2). Since the light crosses the entire vertical extent of the atmosphere, a direct conversion of the absorption signal to trace gas concentrations is not possible. Instead, the total column density, i.e. the concentration integrated along the path, is the result of these measurements. Only by using geometric and radiative transfer calculations can these measurements be converted into vertically integrated column densities (VCD) or vertical concentration profiles. The most common example for VCDs is the total ozone column, which is measured in Dobson units (Dobson and Harrison 1926).
2. Besides direct measurements of sun, moon, and star light from the ground, balloon-borne solar measurements have been very successful (Fig. 6.5, Panel 2). Measurement series during ascent or descent provide vertical concentration profiles of trace gases.
3. The measurement of scattered light from the zenith (Fig. 6.5, Panel 3) was one of the earliest applications of passive DOAS (see Sect. 11.2). Inversion of derived column-density time series during sunset or sunrise (by the Umkehr Method, see Götz et al. 1934) allows the derivation of vertical trace gas concentration profiles and has contributed considerably to our understanding of stratospheric chemistry (see e.g. Mount et al. 1987; Solomon et al. 1987, 1988, 1989).
4. A widespread application of passive scattered DOAS is the use of multiple viewing geometries (Fig. 6.5, Panel 4). This multi-axis DOAS (MAX-DOAS) uses the fact that, at low viewing elevations, the length of the light path in the lower troposphere is considerably elongated (e.g. Hönninger et al. 2004). It is thus possible to probe the lower troposphere very sensitively. In addition, vertical profiles can be derived by suitable inversion techniques, provided that a sufficient number of elevation angles are probed (see e.g. Frieß et al. (2019)).
5. Another passive DOAS approach is aimed at imaging trace gas distributions in the atmosphere, and allows visualisation of emission plumes from volcanoes, power stations, or industrial plants (Fig. 6.5, Panel 5). Imaging DOAS (I-DOAS) can also be used to determine gas emission fluxes (see Sect. 11.4.5).
6. MAX-DOAS can also be employed from airborne platforms, allowing the measurements below, at, and above the flight altitude (Fig. 6.5, Panel 6).
7. DOAS is being used for satellite-borne measurements (Fig. 6.5, Panels 7 and 8), which use sunlight scattered either by the atmosphere, the ground, or both (see

Sect. 11.6). Three viewing geometries are common (for details and examples, see Chap. 11). In the nadir geometry, the DOAS system looks down towards the earth's surface. Satellite instruments provide global concentration fields of trace gases, such as O_3, NO_2, SO_2, and HCHO. Some instruments also employ measurements in limb geometry or use solar occultation, which allows the determination of vertical trace gas profiles with high resolution.

8. Zenith scattered light has also been used to study the radiative transport in clouds (Fig. 6.5, Panel 9), which is an important topic in climate research (e.g. Wagner et al. 1998a; Pfeilsticker et al. 1998b, 1999).

The advantage of passive DOAS applications is the relatively simple experimental setup. For example, scattered light measurements require only small telescopes. In addition, no artificial light source is needed. However, a number of additional challenges have to be addressed in passive DOAS applications. Because solar and lunar light is spectrally highly structured, special care must be taken in the analysis of these observations. To detect very small trace gas absorptions, the strong Fraunhofer bands must be accurately measured. In addition, the fact that the light source spectrum contains narrow and deep absorptions also makes the application of the DOAS technique, which was outlined in Sect. 6.3, more difficult. This will be discussed in more detail below. The largest challenge in using passive DOAS is the conversion of the observed column densities to vertical column densities, concentrations, and vertical profiles. This is, in particular, the case for scattered light setups, where the length of the light path is difficult to determine. The interpretation of these measurements must be based on detailed radiative transfer calculations (see Chap. 9).

6.5 Trace Gases Measured by DOAS

The separation of broad and narrow spectral structures (Sect. 6.3) makes absorption spectroscopy usable in the atmosphere but restricts DOAS measurements to trace gases that have narrow band absorption structures with widths narrower than $\sim$10 nm. In theory, all gases that have these narrow absorption bands in the UV, visible, or near IR can be measured. However, the concentrations of these compounds in the atmosphere, and the detection limits of today's DOAS instruments, restrict the number of trace gases that can be detected. As DOAS instruments improve in the future, the number of measurable trace gases will most likely increase.

Figures 6.6 and 6.7 show the absorption cross-sections of a number of trace gases that are regularly measured by DOAS. Several features about these cross-sections are significant. First and most importantly, each trace gas spectrum has a unique shape. Most of the trace gases only absorb in certain wavelength intervals. However, many spectral regions can contain a large number of simultaneous absorbers. For example, between 300 and 400 nm, the following trace gases will show absorption features if they are present at high enough concentrations: O_3, SO_2, NO_2, HONO, HCHO, and BrO. A separation of the absorptions is possible, due to their unique spectral structure. From (6.8), it is clear that spectral regions with higher σ' will

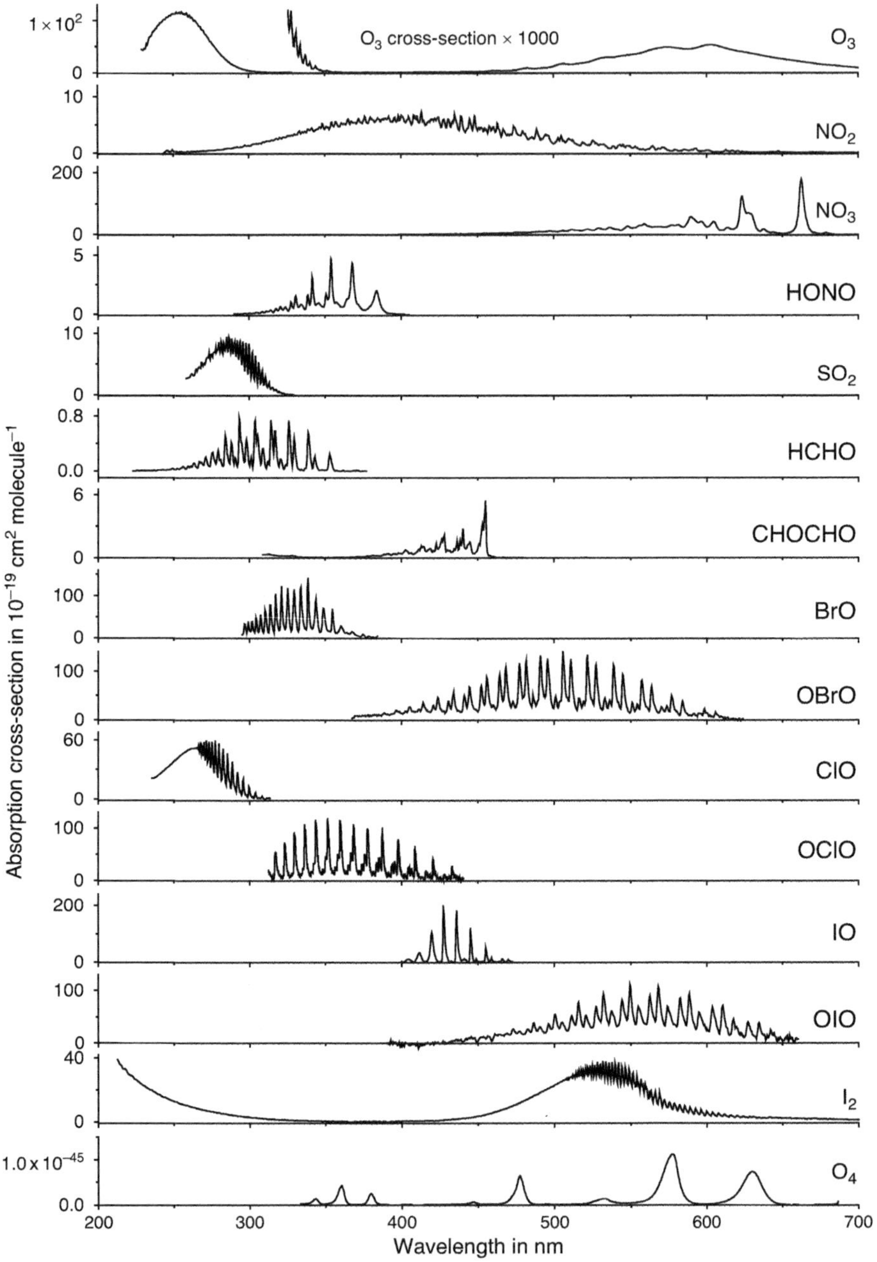

Fig. 6.6 Details of the absorption cross-section features of a number of species of atmospheric interest as a function of wavelength (in nm). Note the 'fingerprint' nature of the different spectra

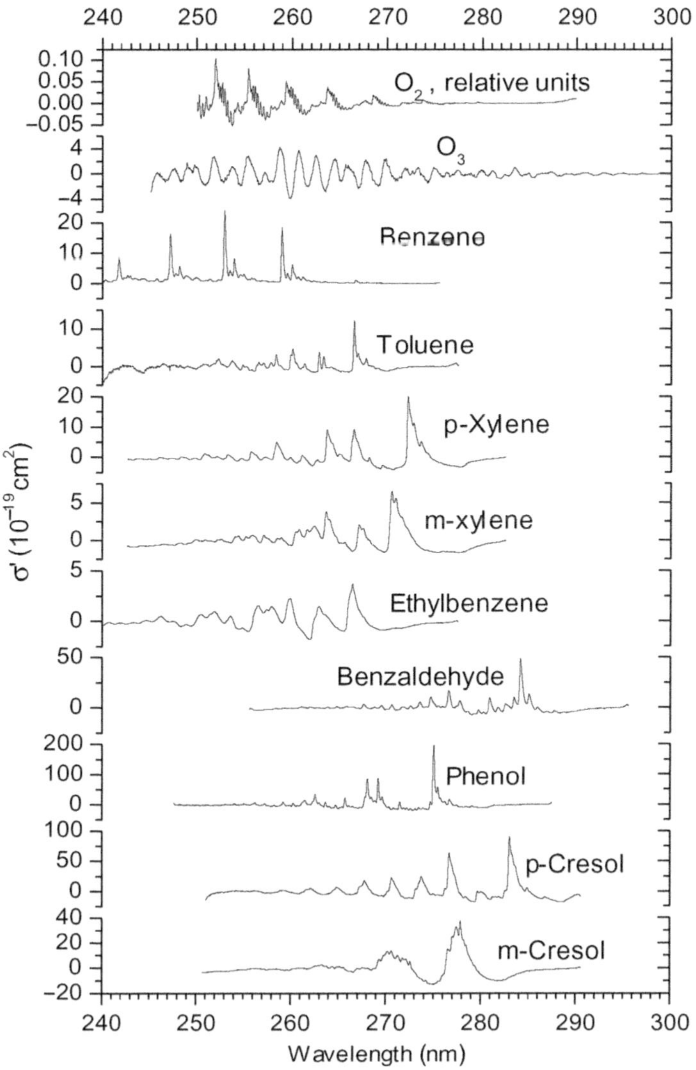

Fig. 6.7 Details of the differential absorption cross-section features of a number of monocyclic aromatic species, O_2, and O_3 as a function of wavelength (in nm). Note the 'fingerprint' nature of the different spectra

show the largest optical densities. These spectral intervals are thus preferable for DOAS measurements since the sensitivity improves in these wavelength regions. In principle, each trace gas has an optimal wavelength interval. In practice, however, one often has to compromise in the choice of the wavelength interval to measure more than one trace gas simultaneously. Because expanding the wavelength window is in conflict with high spectral resolution of typical grating spectrometers, this can reduce the sensitivity.

The wavelength window thus has to be carefully chosen to include all targeted trace gases. Figure 6.8 shows the detection limits of an extended set of trace gases for long-path applications in the troposphere, which can be useful in the selection of the best wavelength region for specific DOAS applications.

A number of trace gases, besides those shown in Figs. 6.6 and 6.7, can be measured. Table 6.1 gives an overview of the various trace gases measured by DOAS, including stratospheric trace gases. At shorter wavelengths, the usable spectral range of DOAS is limited by rapidly increasing Rayleigh scattering and O_2 absorption (Volkamer et al. 1998). Those effects limit the maximum light path length to ∼200 m in the wavelength range from 200–230 nm, where, for instance, the sole usable absorption features of species such as NO (Tajime et al. 1978) and NH_3 are located (see Fig. 6.8 and Table 6.1).

6.6 Precision and Accuracy of DOAS

Selectivity, precision, and accuracy are important aspects of an analytical method. We thus give a brief overview of the different sources of random and systematic errors that must be considered in DOAS. A detailed mathematical treatment of these errors is provided in Chap. 8.

Selectivity describes the ability of an analytical method to clearly distinguish different trace gases in a measurement. In wet-chemical methods, for example, it is often challenging to identify species with similar chemical properties. Spectroscopic methods, on the other hand, provide excellent selectivity, since the absorption structures of different trace gases are a unique property of each compound. This property has been the basis of spectroscopic studies in the laboratory for many years and transfers directly to DOAS. As illustrated in Fig. 6.6, each differential absorption cross-section is unique. The numerical separation procedures described in Chap. 8 allow separation of 10 or more overlaying absorptions without creating cross-sensitivities. A condition for the successful differentiation of the trace gases is that the absorption cross-section of a certain gas does not accidentally contain absorption structures of other trace gases, for example due to impurities in the target gas when the cross-section was recorded. Such absorptions could cause interferences in the numerical analysis procedure. Careful analysis of the cross-sections can prevent this problem, and with today's high-quality laboratory techniques, cross-sections are typically pure. DOAS is also able to distinguish different isomers of a species, for example, of aromatic hydrocarbons, such as the various xylenes.

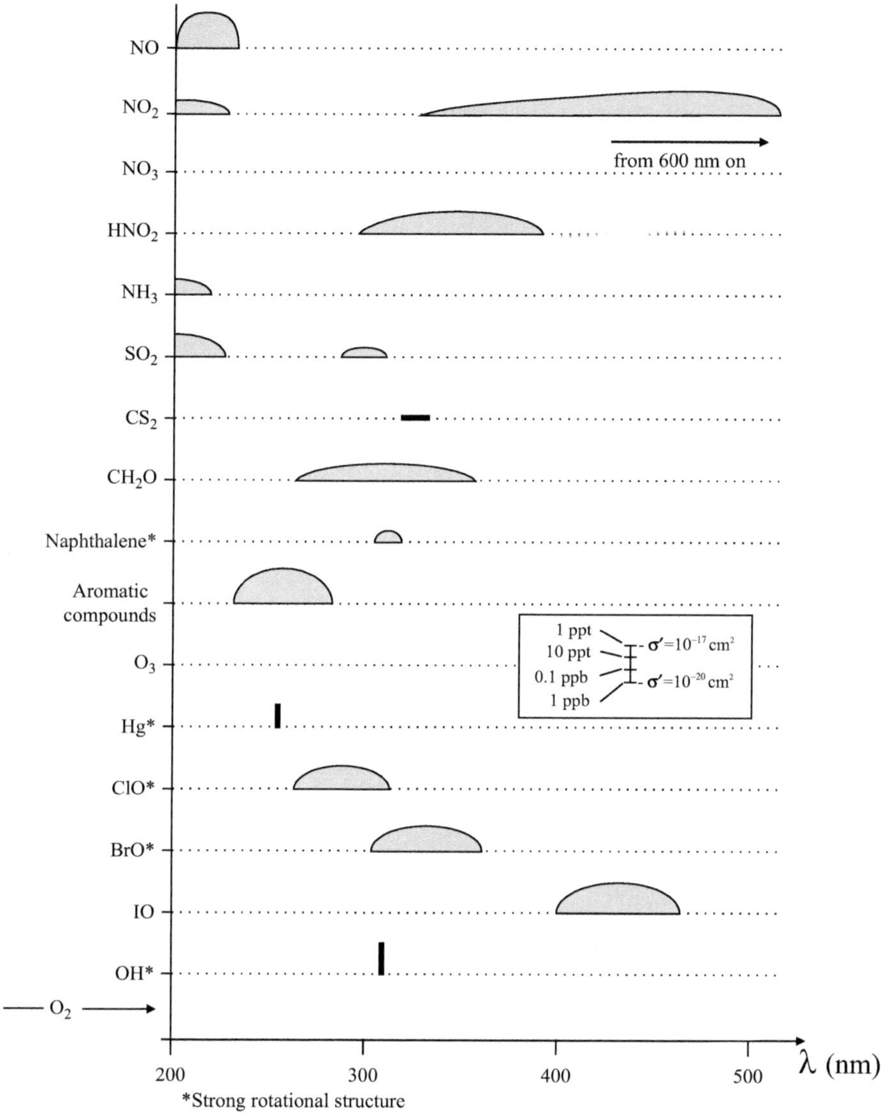

Fig. 6.8 Overview of UV-visible spectral ranges usable for the detection of atmospheric trace gases. Vertical scale: Log of absorption cross-section of the molecule (10^{-20} to 10^{-17} cm^2/molec.), as well as an approximate detection limit at 10 km light path length (1 ppt to 1 ppb, see insert). Molecules exhibiting strong rotational structure at atmospheric pressure are noted by an *asterisk* (*)

Table 6.1 Substances detectable by active UV/visible absorption spectroscopy. Detection limits were conservatively calculated for a minimum detectable optical density of 5×10^{-4}

Species	Wavelength interval (nm)	Approximate differential absorption cross-section 10^{-19} cm^2/molec.	Column density detection limit 10^{15} molec./cm	Detection limit (5 km light path) ppt
SO_2	200–230	65	0.077	62[a]
	290–310	5.7	0.88	70
CS_2	320–340	0.4	13	1000
NO	200–230	24	0.21	167[a]
NO_2	330–500	2.5	2.0	160
NO_3	600–670	200	0.025	2
NH_3	200–230	180	0.028	22[a]
HNO_2	330–380	5.1	0.98	78
O_3	300–330	0.1	50	4000
H_2O	above 430	$3 \cdot 10^{-4}$	$1.67 \cdot 10^4$	$1.33 \cdot 10^4$
CH_2O	300–360	0.48	10	830
ClO	260–300	35	0.14	11
OClO	300–440	107	0.047	3.5
BrO	300–360	104	0.048	4
OBrO	400–600	113	0.044	3.5
I_2	500–630	18	0.28	22
IO	400–470	170	0.029	2
OIO	480–600	110	0.045	3.6
Benzene	240–270	21.9	0.23	180[a]
Toluene	250–280	12.8	0.39	310[a]
Xylene (o/m/p)	250–280	2.1/6.6/20.3	2.4/0.76/0.25	2000/650/210[a]
Phenol	260–290	198	0.0025	20[a]
Cresol (o/m/p)	250–280	20.1/31.8/87.2	0.25/0.16/0.06	200/135/50[a]
Benzaldehyde	280–290	44	0.11	90[a]
Glyoxal	400–480	10	0.5	40

[a] 500-m light path

The precision of DOAS measurements is mostly determined by the quality of the instrument and atmospheric conditions. Different sources of errors that influence the precision must be considered for the various DOAS applications (see Table 6.2). A common and often the dominating source of error in all applications is random noise in the spectra. The most important noise source originates from photon statistics, and is thus unavoidable. In general, the precision increases as more photons are collected. In addition, the noise of the detector can play a role when light levels are low. With today's technology, however, detector noise is often a minor problem. In many DOAS applications, unexplained random spectral structures are encountered. These structures have a different origin than pure photon or detector noise. In most cases,

Table 6.2 Errors that influence the precision and accuracy of DOAS

Type of instrument	Precision	Accuracy
Active DOAS	• Noise	• Noise
	• Unexplained spectral structures	• Unexplained spectral structures
	• Insignificant: path length	• Insignificant: path length
		• Accuracy of absorption cross-section
Passive direct light DOAS	• Noise	• Noise
	• Removal of Fraunhofer bands	• Removal of Fraunhofer bands
	• Temperature-dependent absorption cross-section	• Temperature-dependent absorption cross-section
	• Unexplained spectral structures	• Unexplained spectral structures
		• Accuracy of absorption cross-section
Passive scattered light DOAS	• Noise	• Noise
	• Removal of Fraunhofer bands	• Removal of Fraunhofer bands
	• Temperature-dependent absorption cross-section	• Temperature-dependent absorption cross-section
	• Unexplained spectral structures	• Unexplained spectral structures
	• Path length/radiative transfer	• Path length/radiative transfer
		• Accuracy of absorption cross-section

their spectra show variations that act simultaneously on a number of neighbouring detector channels. For active DOAS applications, the random spectral interferences (also known as ‘optical noise’) are often the primary limitation of the precision. The influence in passive DOAS measurements is less severe. A detailed discussion of the treatment of these structures is given in Chap. 8. Both noise and unexplained spectral structures can be determined for each individual absorption spectrum during the analysis routine. Variations in the environmental conditions, such as changes in atmospheric transmission, will automatically be reflected in these uncertainties.

The precision of DOAS can also be influenced by uncertainties in the determination of the absorption path length. In the case of active and passive direct light DOAS, path lengths can usually be determined with high accuracy and thus contribute little to the errors in these applications. In contrast, absorption path length determination in passive scattered light DOAS applications is a major challenge. Only through detailed radiative transfer calculations is it possible to convert observed column densities to vertical column densities or concentrations. While the radiative transfer models used in this conversion are well tested, they rely on correct input data. Information on the vertical distribution of the trace gases and aerosol, as well as the optical properties of aerosol particles, is required. Since this information is often not available with sufficient detail, the results of the radiative transfer calculations are inherently uncertain. While we treat this uncertainty as a random effect due to its dependence on random temporal changes in the atmosphere, one can argue that this uncertainty may contribute more to systematic uncertainties that influence the accuracy. DOAS offers the opportunity to validate the radiative transfer and determine the uncertainty of this calculation by analysing trace gas absorptions from gases with known concentrations, such as O_2 and O_4. More information on the radiative transfer calculations, input parameters, and results is given in Chap. 9.

Two other uncertainties are specific for passive DOAS applications. In all passive DOAS applications, the spectral structure of sunlight, the Fraunhofer bands, needs to be accurately removed. Because the optical densities of these bands are, in most cases, much larger than those of the trace gas absorptions, the uncertainty in the removal or modelling of the Fraunhofer structure directly influences the uncertainty of the trace gas measurements. Moreover, due to Raman scattering (see Sect. 4.2.3) in the atmosphere, the Fraunhofer lines are distorted to varying degrees (the 'Ring Effect', see Sect. 9.8.1). This also needs to be compensated for. Another error in passive DOAS applications stems from the temperature dependence of the various absorption cross-sections. Even if this dependence is known, the often unknown temperature structure of the atmosphere, the location of the trace gas, and, in case of scattered light applications, the uncertain light path leads to problems in the interpretation of the data.

The determination of the accuracy of any atmospheric measurement is challenging since, in principle, it would require the knowledge of true concentration of the respective trace gas. However, in many cases, it is possible to estimate the various sources of errors, such as sampling artifacts in in-situ methods, from controlled experiments. Fortunately, this is not necessary for DOAS. Sampling artifacts and losses, as well as chemical transformations occurring after sampling that are often encountered by other methods, do not have to be considered in DOAS, since the measurements are made in the open atmosphere without disturbing or influencing the trace gases. Noise, unexplained spectral structures, Fraunhofer band removal, and temperature dependent cross-section influence accuracy in the same way they influence precision. However, the radiative transfer methods used to interpret passive scattered measurements may introduce additional non-random errors that can influence the accuracy, but not the precision, of a measurement. The accuracy of the cross-sections, which is often the largest factor influencing the accuracy of DOAS, is typically in the range of 1–10% (see Appendix B). These numbers will most likely improve in the future, as more laboratory measurements become available.

One of the most important properties of analytical techniques is the detection limit, which gives the smallest possible trace gas amount that can be detected. It is difficult to give this number for all possible DOAS applications, in particular with different path lengths between the setups shown in Fig. 6.4. To provide an overview of the detection limits of DOAS, we have thus listed two different values in Table 6.1. The first is the lowest column density in units of molecules/cm^2 that can be measured, assuming a minimum detectable optical density of 5×10^{-4}. To determine the detection limit for a specific application, this number must be divided by the path length in units of centimetres. The last column in Table 6.1 shows the detection limits that can be achieved with an active DOAS system with a path length of $L = 5$ km for trace gases absorbing above 300 nm, or $L = 500$ m for trace gases absorbing below 300 nm. Figure 6.8 offers a more graphical view on the same data, and also facilitates the choice of the wavelength with the lowest detection limits.

In summary, DOAS is a highly versatile, selective, and accurate technique. The accuracy is primarily determined by the known uncertainties of the absorption cross-sections used. One of the main advantages of DOAS is the ability to determine the precision of a single measurement based on the analysis of the absorption spectrum. In principle, a DOAS measurement is a "spectral photograph" of the atmospheric composition that allows the identification and quantification of its components and the determination of the uncertainty of this measurement.

6.7 Mathematical Description of the DOAS Approach

The following section provides a mathematical description of DOAS, including the influence of the actual measurement on the shape of the absorption structures. This description is the basis of most other topics discussed in this book. While the basic principles in Sect. 6.3 still apply, their application to the analysis of DOAS measurements is limited due to the omission of some important aspects of the DOAS process. An expanded discussion of the principle of DOAS illustrates the approach that needs to be taken to overcome some of the challenges faced in today's wide variety of DOAS applications.

6.7.1 Fundamentals of the DOAS Approach

An idealised experimental setup of a DOAS instrument, as shown in Fig. 6.9, serves as a model to describe the different applications shown in Fig. 6.4.

Light of intensity $I_0(\lambda)$ emitted by a suitable source passes through the open atmosphere and is collected by a telescope. As it passes through the atmosphere, the light undergoes extinction due to absorption by different trace gases, and scattering by air molecules and aerosol particles (see Chap. 4). The intensity $I(\lambda, L)$ at the end of the light path is given by (6.9), using Lambert–Beer's law. The absorption of a trace species j is characterised by its absorption cross-section $\sigma_j(\lambda, p, T)$, which depends on the wavelength λ, pressure p, and temperature T, and by its number concentration $c_j(l)$ at the position l along the light path. The Rayleigh extinction and Mie extinction by aerosols are described by the extinction coefficients $\varepsilon_R(\lambda, l)$ and $\varepsilon_M(\lambda, l)$, respectively. $N(\lambda)$ is the photon noise, which depends on $I(\lambda, L)$. For simplicity, we have omitted the influence of the instrument spectral characteristics and atmospheric turbulence, A(λ), described in Sect. 6.3. The spectrum I(λ, L) at the entrance of the spectrograph (Fig. 6.9a) arises from light that traverses the atmosphere with several absorbers over length L.

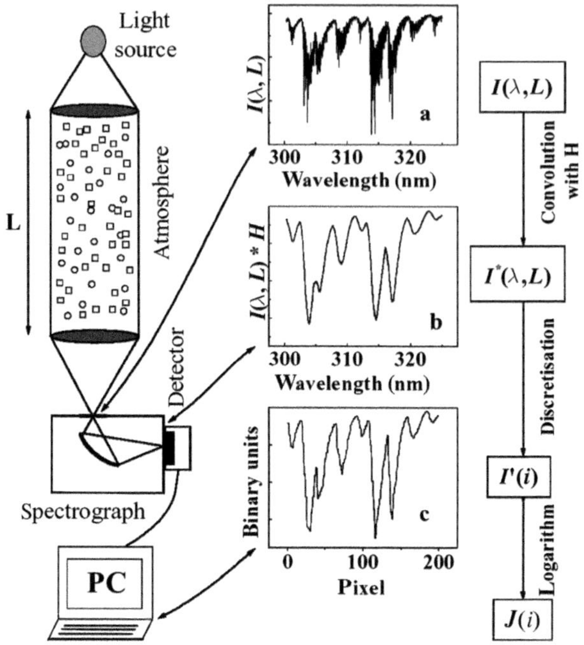

Fig. 6.9 Schematic view of a DOAS instrument used to measure trace gas concentrations. Collimated light undergoes absorption processes on its way through the atmosphere. In **a**, an example of this light entering the spectrograph is given, where HCHO is assumed to be the only absorber and the lights source has smooth spectral characteristics. This absorption spectrum shows the vibrational–rotational structure of the absorption bands. **b** The same spectrum convoluted by the spectrograph instrumental function reaches the detector. In the detector, the wavelength is mapped to discrete pixels. This spectrum **c** is then stored in the computer and can be analysed numerically. Reprinted with permission from Stutz and Platt, 1996, ©Optical Society of America. All rights reserved

$$\mathrm{I}(\lambda, \mathrm{L}) = \mathrm{I_0}(\lambda, \mathrm{L}) \cdot \exp\left[-\left(\int_{o}^{L} \sum_{j} \left(\sigma_j(\lambda, p, T) \cdot c_j(l)\right) + \epsilon_R(\lambda, l) + \epsilon_M(\lambda, l)\, dl\right)\right] + N(\lambda) \tag{6.9}$$

Special consideration must be given to passive scattered light DOAS. The light observed in these applications follows complicated paths in the atmosphere, which are determined by the solar position in the sky, the viewing direction of the telescope, and most importantly the spatial distribution of air molecules and aerosol particles. The measured intensity is the sum over intensities of different light beams. Each of these beams has travelled on an individual path with a specific length through the atmosphere. One can introduce a simplified mathematical formulation of this phenomenon by describing the measured total intensity as an integral over a light path distribution function and a path length dependent $I_0(\lambda, L)$. It is then possible to show the validity of an approximation for cases with weak absorptions, which describes the measured intensity with (6.9). In this approximation, $I_0(\lambda, L)$ represents the total intensity, and L represents the intensity-weighted average path length. The average path length will be discussed in the context of radiative transfer calculations in Chap. 9. A number of approaches to solve the challenges with scattered light measurements will be described in Chaps. 9 and 11.

In most DOAS instruments, light of intensity $I(\lambda, L)$ is focused on the entrance of a grating spectrograph, with a detector recording the spectrum. Due to the typically limited resolution of the spectroscopic instruments, the shape of spectrum $I(\lambda, L)$ changes. The mathematical description of this process is a convolution of $I(\lambda, L)$, with the instrument function $H(\lambda)$ of the spectrograph: $I^*(\lambda, L) = I(\lambda, L)*H(\lambda)$. Figure 6.9b shows the spectrum I after convolution with a typical instrument function H.

$$I^*(\lambda, L) = I(\lambda, L)*H = \int I(\lambda - \lambda', L) \cdot H(\lambda')d\lambda' \tag{6.10}$$

During the recording by a detector, the wavelength range is mapped to n discrete pixels/channels, numbered by i, each integrating the light in a wavelength interval from $\lambda(i)$ to $\lambda(i+1)$. This interval is given by the wavelength-pixel-mapping Γ_I of the instrument. In the case of a linear dispersion, $\Gamma_I : \lambda(i) = \gamma_0 + \gamma_1 \times i$, the spectral width of a pixel is constant $\Delta\lambda(i) = \lambda(i+1) - \lambda(i) = \gamma_1$. The signal $I'(i)$ seen by a pixel i (omitting the sensitivity variations of individual pixels) is given by:

$$I'(i) = \int_{\lambda(i)}^{\lambda(i+1)} I^*\left(\lambda'\right) d\lambda' \tag{6.11}$$

In general, the wavelength-pixel-mapping Γ_I of the instrument can be approximated by a polynomial:

$$\Gamma_I : \quad \lambda(i) = \sum_{k=0}^{q} \gamma_k \cdot i^k \tag{6.12}$$

The parameter vector (γ_k) determines the mapping of pixel i to the wavelength $\lambda(i)$. A change in parameter γ_0 describes a spectral shift of the spectrum. Changing γ_1 squeezes or stretches the spectrum linearly. Parameters γ_k of higher k describe a distortion of the wavelength scale of higher order. Changes in the parameter vector (γ_k) can be caused by different measurement conditions of the spectra, as grating spectrometers usually show a temperature drift of 1/10 of a pixel per K. A variation in air pressure, seen in particular in aircraft measurements, also changes the wavelength alignment due to a change in the index of refraction of air. It is, therefore, necessary to correct these effects in the analysis procedure.

Figure 6.9c shows the discrete spectrum $I'(i)$ as it was recorded. The analysis of $I'(i)$ with respect to the different absorbers in the spectrum is one of the most important aspects of DOAS.

6.7.2 Application of the DOAS Approach in Practical Situations

Equations (6.9) and (6.10) can be combined into one equation:

$$\begin{aligned}I^*(\lambda, L) &= I(\lambda, L)*H \\ &= \int_{-\Delta\lambda}^{\Delta\lambda} I_0(\lambda-\lambda', L)\exp\left(-\int_o^L \sum_j \left(\sigma_j(\lambda-\lambda', p, T)\cdot c_j(l)\right) + \varepsilon_R\left(\lambda-\lambda', l\right)\right. \\ &\quad \left. + \varepsilon_M\left(\lambda-\lambda', l\right)\, dl\right)\cdot H(\lambda')d\lambda' \end{aligned} \tag{6.13}$$

We have omitted the noise term in (6.13), since the following discussion focuses on a mathematical description of the DOAS method that does not consider statistical uncertainties. Noise and its influence on the results of DOAS are presented in Chap. 8.

In order to simplify the mathematical treatment of the DOAS approach, we neglect the effect of wavelength discretisation and concentrate on the continuous spectra arriving at the detector, $I^*(\lambda, L)$. The omission of the discretisation step (6.11) has no influence on the general results of our discussion. When the width of the instrument function H is much larger than the wavelength interval of one pixel, the approximation that the intensity at the centre wavelength of a pixel $I^*(\lambda(i), L)$ is very similar to the integral over the entire pixel i is quite good. If this is not the case, other effects such as aliasing have to be considered. However, since the discussion of these effects does not aid in the understanding of DOAS, we will not consider them in this chapter, and we will assume that the integral over the pixel can be approximated by the intensity of the centre wavelength of the pixel.

Although, from a purely mathematical point of view, the integration of the first integral in (6.13) should extend from $-\infty$ to ∞, in practice it is sufficient to constrain it to a small interval around the instrument function with a width of about $2\Delta\lambda$. Assuming a Gaussian function $H(\lambda') = C\cdot\exp\left[-\left(\lambda'/\lambda_H\right)^2\right]$ as an approximation of the instrument function, $\Delta\lambda$ can be chosen as a small multiple of λ_H. For example, choosing $\Delta\lambda = 4\cdot\lambda_H$ would result in an error $< 10^{-4}$. Following the general DOAS approach (see Sect. 6.3), we can simplify (6.13) by including all broadband terms in a new $I_0'(\lambda, L)$:

$$I^*(\lambda, L) = \int_{-\Delta\lambda}^{\Delta\lambda} I_0'(\lambda-\lambda', L)\exp\left(-\int_o^L \sum_j \left(\sigma_j'(\lambda-\lambda', p, T)\cdot\rho_j(l)\right) dl\right)\cdot H(\lambda')d\lambda' \tag{6.14}$$

It should be noted that the separation of spectrally broad and narrow terms leading to (6.14) is not an a-priori defined procedure, and there are many implementations of this separation, such as regressions and Fourier filters (see Chap. 8). Here, we treat this separation in the most general term, without giving a quantitative definition of what we consider as broad or narrow. Any separation is applied to the intensity spectra $I(\lambda)$, and not to the absorption cross-section $\sigma(\lambda)$. This may, at first glance, contradict the approach in Eq. 6.5. However, because we are free to choose the procedure to separate broad and narrow structures, we can generalise Eq. 6.5 to: $\exp(-\sigma(\lambda)) = \exp(-\sigma_0(\lambda)) \cdot \exp(-\sigma'(\lambda))$, where any separation procedure is applied to $\exp(-\sigma(\lambda))$. The choice of separation procedure, which will be discussed in more detail in Chap. 8, does not impact the following discussion.

The DOAS problem can, in principle, be solved by a numerical model of equation (6.14), where the trace gas concentrations $c_j(l)$ are adjusted to optimise the agreement between $I^*(\lambda, L)$ and the measured spectrum. However, in practice, this is difficult and time consuming, since it requires a non-linear optimisation procedure. A variety of simplifications and adaptations have thus been developed.

A number of these implementations for different DOAS applications are presented below. As an initial simplification, we assume that the differential absorption cross-section $\sigma'_j(\lambda)$ is independent of temperature and pressure. For active DOAS applications, which typically measure a very small altitude interval, this restriction is applicable. For certain passive DOAS applications, this assumption is difficult to sustain and would, in principle, require a rigorous numerical solution of equation (6.14). There is, however, an approximate solution, which splits $\sigma'_j(\lambda, T, p)$ into a small number of different cross-sections $\sigma'_k(\lambda, T_k, p_k)$ for individual temperature/pressure combinations (note that there is a correlation between pressure and temperature in the atmosphere): $\sigma'_j(\lambda, T, p) = \sum b_k \cdot \sigma'_k(\lambda, T_k, p_k)$. With this approach, $\sigma'_j(\lambda)$ becomes independent of the integration over the path length, and the integration over dl only has to be applied to $c_j(l)$.

Based on our assumption that $\sigma'_j(\lambda)$ is independent of temperature and pressure, or can be split into independent parts, we can now introduce the path-averaged gas concentration:

$$\bar{c}_j = \frac{1}{L} \int_0^L c_j(l) dl \tag{6.15}$$

Equation (6.14) thus becomes:

$$I^*(\lambda, L) = \int_{-\Delta\lambda}^{\Delta\lambda} I'_0(\lambda - \lambda', L) \cdot \exp\left(-\sum_j \left(\sigma'_j(\lambda - \lambda') \cdot \bar{c}_j \right) \cdot L \right) \cdot H(\lambda') d\lambda' \tag{6.16}$$

Equation (6.16) is the simplest description of a DOAS measurement containing differential absorption of multiple trace gases, a wavelength-dependent I'_0, and the convolution process representing the measurement.

Based on (6.16), we can define a number of parameters that influence the choice of DOAS implementation (see Table 6.3):

- *The differential optical density of the trace gas absorption.* In general, differential optical densities $D' = L \cdot \sum \sigma'_j(\lambda) \cdot \bar{c}_j$ that are below ~0.1 are denoted as 'small'.

Table 6.3 Overview of the different cases for DOAS applications

Case	Diff. OD of absorber	Resolution of spectrometer	Spectrum of source	Approach	Remark
1	Small	High and low	Smooth	Linearise Lambert–Beer's law	Classical case
2	Large	High	Smooth	Lambert–Beer's Law	Classical case
3	Large	Low	Smooth	Nonlinear, modelling of entire equation system	Saturated absorber
4	Small or large	High[a]	Structured	Divide by $I_0(\lambda)$	
5	Small	Low	Structured	High res. Model + Lambert–Beer's Law	
6	Large	Low	Structured	Non-linear, modelling of entire equation system	

[a] $2\Delta\lambda << \lambda_I$ and $2\Delta\lambda << \lambda_B$

It will depend on the mathematical treatment of the DOAS implementation if this limit applies to the original or convoluted absorption band. The change in differential optical density due to the convolution depends on the width of the absorption band relative to the instrument function width in the convolution. The change is more pronounced in narrow bands.

- *The spectral resolution or instrument function of the spectrograph–detector in comparison to the spectral width of the absorption bands.* In general, we can distinguish between high-resolution cases, in which the natural width of the absorption band, $\Delta\lambda_B$, is spectrally resolved, and low-resolution cases, in which the spectral width of the bands is not resolved by the measurements. In (6.16), this translates into the condition that $2\Delta\lambda << \Delta\lambda_B$ for spectrally resolved bands and $2\Delta\lambda \geq \Delta\lambda_B$ for unresolved bands. Note that a band that looks resolved after a measurement may consist of a number of narrower absorption lines that are not resolved. Thus, the measurement would only show the envelope of this line system.
- *The spectral structure of the light source, $I_0'(\lambda)$.* 'Smooth' light sources have structures broader than the resolution of the instrument. For example, incandescence or high-pressure Xe lamps have this property. 'Structured' light sources show spectral features that are narrower than the instrument resolution. The most important example is the sun, with its narrow Fraunhofer lines. In (6.16), this leads to the condition that $2\Delta\lambda$ is much smaller than the bandwidth, λ_I, of $I_0'(\lambda)$ for a smooth $I_0'(\lambda)$, and larger for structured $I_0'(\lambda)$.

Six different cases can be identified based on these three parameters (Table 6.3). We will now discuss in more detail how Eq. 6.16 has to be applied to determine the desired trace gas column densities for the different situations. The individual cases will be discussed in a general way. However, we will also give the mathematical derivation for the different implementations, wherever possible. For more details see the 'operator representation of DOAS' (Wenig 2001; Wenig et al. 2005).

Case 1: Weak Low- or High-Resolution Absorbers and a Smooth Light Source

The most basic approach of DOAS applies to measurements using a smooth light source, i.e. active DOAS, where weak absorption features are measured with low or high instrumental resolution.

In this case, Lambert–Beer's law can be linearised by taking the logarithm of the ratio of $I^*(\lambda, L)$ and $I_0'(\lambda, L)$, resulting in the following representation of (6.16):

$$\ln\left(\frac{I^*(\lambda, L)}{I_0'(\lambda, L)}\right) = \sum_j \frac{\bar{c}_j \cdot L}{\alpha_j} \cdot \ln\left[\int_{-\Delta\lambda}^{\Delta\lambda} \exp\left(-\sigma_j'(\lambda - \lambda') \cdot \alpha_j\right) \cdot H(\lambda')d\lambda'\right] \tag{6.17}$$

Equation (6.17) states that $\ln(I^*(\lambda, L)/I'_0(\lambda, L))$ can be described by a sum of spectra of pure trace gas absorptions that are scaled by the average number concentration $\bar{c}_j$ and the path length L (the factor α_j will be discussed later). This mathematical approach is far more common than one may expect, and some of the other cases discussed later also rely on it.

The linearisation of Lambert–Beer's law is the most common form of DOAS implementation. The following equation is loosely based on the discrete form of (6.17). The logarithm of the discreet form of $I^*(\lambda, L)$: $J(i) = \ln(I'(i))$ is described by:

$$J(i) = J_0(i) + \sum_{j=1}^{m} a_j' \cdot S_j'(i) \tag{6.18}$$

$J_0(i)$ is the logarithm of the discretisation of $I'_0(i)$, which was defined in (6.11). The differential absorption structures of the trace gases are described by individual 'reference spectra' $S_j'(i)$, which were also discretised with (6.11). a_j' are the scaling factors for the individual discretised reference spectra. We will see that, for different DOAS cases, the reference spectra calculation is different, while (6.18) can still be used. For smooth light sources, the spectra $S_j'(i)$ are determined through the following three equations:

$$S_j^*(\lambda) = \int_{-\Delta\lambda}^{\Delta\lambda} \exp\left(-\sigma_j'(\lambda - \lambda') \cdot \alpha_j\right) \cdot H(\lambda')d\lambda'$$

$$S_j^*(i) = \int_{\lambda(i)}^{\lambda(i+1)} S_j^*(\lambda)d\lambda'$$

$$S_j'(i) = \frac{1}{\alpha_j}\ln(S_j^*(i)) \tag{6.19}$$

We have introduced a scaling parameter α_j that will play a role for Case 3. This parameter can be interpreted as the product of concentration and path length used in the measurement or simulation of S_j^*. For Case 1, where the differential optical densities are small, α_j can be set to unity. In the case of small differential optical density ($\alpha_j = 1$), the first equation in (6.19) is taken from the sum of (6.17), and describes the convolution of the exponential of the negative differential absorption cross-section $\sigma_j'(\lambda)$. The second equation is the discretisation according to (6.11). The last equation calculates the logarithm of the discretised spectrum.

However, if the differential optical densities are not small, (6.18) and (6.19) can only be used if values for $\alpha_j \neq 1$ are chosen. The scaling factors

$$a_j' = \frac{\bar{c}_j \cdot L}{\alpha_j} \tag{6.20}$$

in (6.18) are the product of the average number densities and the path length, divided by the reference spectrum scaling factor α_j.

The solution of (6.18) requires the calculation of the $S'(i)$ through (6.19). For this calculation, the knowledge of instrument function $H(\lambda)$ and the highly resolved absorption cross-section $\sigma_j'(\lambda)$ is required. Both the mathematical approach to solve (6.18) with the goal of retrieving the a_j' factors, and the details of the calculation in (6.18) are presented in Chap. 8. Figure 6.10 shows an example of (6.18), for a real atmospheric measurement. The top spectrum represents $J'(i)$ while the other spectra represent the products of the absorption cross-sections and the scaling factors: $a_j' \cdot S_j'(i)$ (in this case $\alpha_j = 1$).

After describing this 'classical' DOAS approach, we now discuss how this implementation can be derived mathematically from (6.16). We will use two related approximations that are directly valid for small differential optical density of the absorption structures:

$$\ln(x) \approx x - 1 \quad \text{for} \quad 1 - \varepsilon < x < 1 + \varepsilon \tag{6.21}$$

and

$$\exp(x) \approx 1 + x \quad \text{for} \quad -\varepsilon < x < \varepsilon \tag{6.22}$$

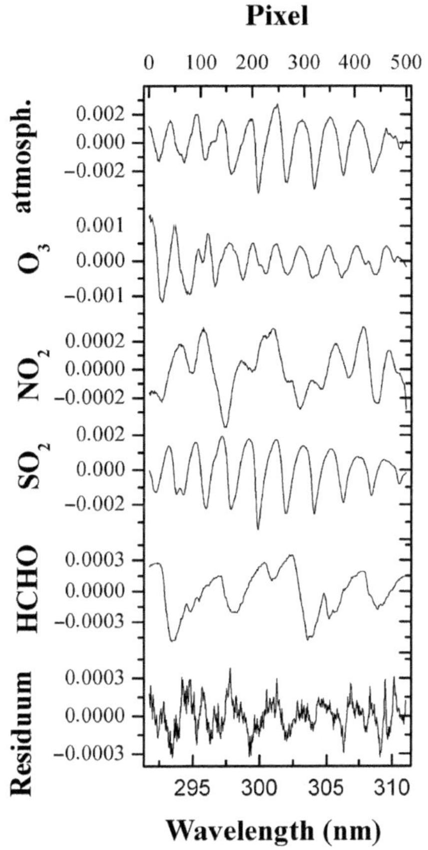

Fig. 6.10 Sample of an atmospheric spectrum (*uppermost trace*) recorded in Heidelberg on August 27, 1994, with overlapping absorptions due to O_3, NO_2, SO_2, and HCHO (traces 2–5 from top, note different scales). The broad spectral structures were described by a fifth order polynomial. The remaining 'residual spectrum' after removal of the absorption structures is shown as the *bottom trace*. Reprinted with permission from Stutz and Platt, 1996,

These approximations are good for ε close to zero. At $\varepsilon = 0.1$, the error imposed by (6.21) and (6.22) is $\sim 5\%$.

We assume that the spectra are measured using a smooth light source. This condition leads to the approximation that $I'_0(\lambda - \lambda', L) \approx I'_0(\lambda, L)$ in the interval from $\lambda - \Delta\lambda$ to $\lambda + \Delta\lambda$. Finally, we use the condition that the integral of the instrument function is scaled to unity. This condition can always be met by dividing the integral over $H(\lambda)$ by a constant:

$$\int_{-\Delta\lambda}^{\Delta\lambda} H(\lambda')d\lambda' = 1 \tag{6.23}$$

Mathematically, this is only true if $\Delta\lambda = \infty$. In all practical cases, however, the integral can be limited to an interval around H.

We begin the derivation of (6.18) by taking the logarithm of the ratio of $I^*(\lambda, L)$ and $I'_0(\lambda, L)$ in (6.16):

$$\ln\left(\frac{I^*(\lambda, L)}{I'_0(\lambda, L)}\right) = \ln\left[\int_{-\Delta\lambda}^{\Delta\lambda} \exp\left(-\sum_j \left(\sigma'_j(\lambda - \lambda') \cdot \bar{c}_j\right) \cdot L\right) \cdot H(\lambda')d\lambda'\right] \tag{6.24}$$

In the case of weak absorbers, the integral in (6.24) is close to unity and we can employ (6.21) to approximate the logarithm:

$$\ln\left(\frac{I^*(\lambda, L)}{I'_0(\lambda, L)}\right) \approx \int_{-\Delta\lambda}^{\Delta\lambda} \exp\left(-\sum_j \left(\sigma'_j(\lambda - \lambda') \cdot \bar{c}_j\right) \cdot L\right) \cdot H(\lambda')d\lambda' - 1 \tag{6.25}$$

This approximation applies to the convoluted absorption structures, which are smaller than the original absorptions before convolution. Using (6.23), this can now be transformed into:

$$\begin{aligned}\ln\left(\frac{\mathrm{I}^*(\lambda, L)}{\mathrm{I}'_0(\lambda, L)}\right) &\approx \int_{-\Delta\lambda}^{\Delta\lambda} \exp\left(-\sum_{\mathrm{j}} \left(\sigma'_{\mathrm{j}}(\lambda - \lambda') \cdot \bar{c}_{\mathrm{j}}\right) \cdot L\right) \cdot H(\lambda')d\lambda' - \int_{-\Delta\lambda}^{\Delta\lambda} H(\lambda')d\lambda' \\ &= \int_{-\Delta\lambda}^{\Delta\lambda} \left[\exp\left(-\sum_{\mathrm{j}} \left(\sigma'_{\mathrm{j}}(\lambda - \lambda') \cdot \bar{c}_{\mathrm{j}}\right) \cdot L\right) - 1\right] \cdot H(\lambda')d\lambda'\end{aligned} \tag{6.26}$$

Further applying the approximation for the exponential function (6.22):

$$\ln\left(\frac{\mathrm{I}^*(\lambda, L)}{\mathrm{I}'_0(\lambda, L)}\right) \approx \int_{-\Delta\lambda}^{\Delta\lambda} \left[-\sum_{\mathrm{j}} \left(\sigma'_{\mathrm{j}}(\lambda - \lambda') \cdot \bar{c}_{\mathrm{j}}\right) \cdot L\right] \cdot H(\lambda')d\lambda' \tag{6.27}$$

This approximation applies to the absorption structures before convolution, and is thus more restrictive than the approximation used in (6.25). We can now exchange the sum and the integral. In addition, we introduce the factor α_j from (6.20):

$$\ln\left(\frac{\mathrm{I}^*(\lambda, L)}{\mathrm{I}'_0(\lambda, L)}\right) \approx \sum_{\mathrm{j}} \frac{\bar{c}_{\mathrm{j}} \cdot L}{\alpha_{\mathrm{j}}} \cdot \int_{-\Delta\lambda}^{\Delta\lambda} \left[\sigma'_{\mathrm{j}}(\lambda - \lambda') \cdot \alpha_j\right] \cdot H(\lambda')d\lambda' \tag{6.28}$$

In the next step of derivation, the approximation for the exponential function (6.22) is applied again, this time in reverse direction:

$$\ln\left(\frac{I^*(\lambda, L)}{I'_0(\lambda, L)}\right) \approx \sum_j \frac{\bar{c}_j \cdot L}{\alpha_j} \cdot \int_{-\Delta\lambda}^{\Delta\lambda} \left[\exp\left(-\sigma'_j(\lambda-\lambda')\cdot\alpha_j\right) - 1\right] \cdot H(\lambda')d\lambda' \tag{6.29}$$

It is important to note here that the errors due to approximations in (6.27) and (6.29) tend to cancel each other. This counterbalance is best if $\alpha_j \approx \bar{c}_j \cdot L$, and actually lifts the restriction to weak absorptions before convolution for the steps from (6.25) to (6.29). The derivation is finalised again using (6.23) and (6.21). The approximation of the logarithm (6.21) is again applied to the convoluted spectral structures.

$$\begin{aligned}\ln\left(\frac{I^*(\lambda, L)}{I'_0(\lambda, L)}\right) &\approx \sum_j \frac{\bar{c}_j \cdot L}{\alpha_j} \cdot \int_{-\Delta\lambda}^{\Delta\lambda} \left[\exp\left(-\sigma'_j(\lambda-\lambda')\cdot\alpha_j\right)\right] \cdot H(\lambda')d\lambda' - \int_{-\Delta\lambda}^{\Delta\lambda} H(\lambda')d\lambda' \\ &\approx \sum_j \frac{\bar{c}_j \cdot L}{\alpha_j} \cdot \ln\left[\int_{-\Delta\lambda}^{\Delta\lambda} \exp\left(-\sigma'_j(\lambda-\lambda')\cdot\alpha_j\right) \cdot H(\lambda')d\lambda'\right]\end{aligned} \tag{6.30}$$

Equation (6.30) is equivalent to (6.17) and states that in the case of small optical densities after convolution, the DOAS approach can be solved by linearising Lambert–Beer's law. The reference spectra used in this fit must be calculated as the logarithm of the convolution of the exponential function of high-resolution absorption cross-sections (6.19).

There are various approximations in the derivation of (6.30). Equations (6.21) and (6.22) were applied twice, in forward and backward directions. While the approximations were applied to different arguments, there is a high degree of cancellation by the twofold use, which makes the approximation better than one would expect from the single application. In particular, in the case that $\alpha_j = \bar{c}_j \cdot L$, the twofold approximations are quite good. A detailed calculation, which uses a Taylor expansion to the second order, shows that the difference between (6.27) and (6.29) is approximately $\frac{1}{2}\sum_{j\neq k} \sigma_j\sigma_k\bar{c}_j\bar{c}_k L^2$, if $\alpha_j = \bar{c}_j \cdot L$. The derivation shown above is, therefore, more widely applicable than just for Case 1, and we will come back to it in Case 3. For small absorptions before the convolution, we can set $\alpha_j = 1$, since all the approximations apply in this case without restrictions. The preceding derivation is the basis of most DOAS applications today. Most measurements of O_3, NO_2, HCHO, NO_3, HONO, halogen oxides, etc. in the troposphere have used this approach (see Fig. 6.10).

Case 2: Strong Absorber at High Resolution and a Smooth Light Source

The case of strong absorbers measured with a spectral resolution that is better than the width of the absorption bands is encountered in a number of DOAS applications. The approach described in Case 1 (6.17–6.19) can also be used in this case. However, the mathematical proof is not as obvious, since the approximations of the logarithm and exponential function (6.21 and 6.22) do not apply in the case of strong absorbers.

We therefore use an approximation that is based on the fact that the spectral interval covering the instrument function $H(\lambda)$, $2\Delta\lambda$, is smaller than the width of the absorption band. Within this interval, we can approximate the exponential function in (6.16) by a linear function, using the derivative of the exponential function as the slope. In addition, this approximation can also be made by the exponential function of the pure scaled absorption cross-section (6.30). Two new functions, $g(\lambda)$ and $h(\lambda)$, for these exponential functions are introduced:

$$\exp\left(-\sum_{\mathrm{j}}\left(\sigma_{\mathrm{j}}'(\lambda-\lambda')\cdot c_{\mathrm{j}}\cdot L\right)\right)=g(\lambda-\lambda')\approx g(\lambda)+\left.\frac{dg}{d\lambda}\right|_{\lambda}\cdot\lambda' \tag{6.31}$$

$$\exp\left(-\sigma_{\mathrm{j}}'(\lambda-\lambda')\cdot\alpha_{\mathrm{j}}\right)=h_j(\lambda-\lambda')\approx h_j(\lambda)+\left.\frac{dh_j}{d\lambda}\right|_{\lambda}\cdot\lambda' \tag{6.32}$$

We again consider $I'_0(\lambda, L)$ as being constant within $\lambda \pm \Delta\lambda$. This is certainly true because the resolution of the instrument is very high and, compared to the interval $2\Delta\lambda$, the change of $I'_0(\lambda, L)$ is small. Applying the approximation in (6.31), the equivalent of (6.24) now becomes:

$$\begin{aligned}\ln\left(\frac{\mathrm{I}^*(\lambda, L)}{\mathrm{I}'_0(\lambda, L)}\right)&=\ln\left(\int_{-\Delta\lambda}^{\Delta\lambda} g(\lambda-\lambda')\cdot H(\lambda')d\lambda'\right)\\&\approx\ln\left(\int_{-\Delta\lambda}^{\Delta\lambda}\left(g(\lambda)+\left.\frac{dg}{d\lambda}\right|_{\lambda}\cdot\lambda'\right)\cdot H(\lambda')d\lambda'\right)\end{aligned} \tag{6.33}$$

Equation (6.33) is now reorganised by using the fact that some of the terms under the integral are independent of λ':

$$\ln\left(\frac{\mathrm{I}^*(\lambda, L)}{\mathrm{I}_0'(\lambda, L)}\right)\approx\ln\left(g(\lambda)\cdot\int_{-\Delta\lambda}^{\Delta\lambda}H(\lambda')d\lambda'+\left.\frac{dg}{d\lambda}\right|_{\lambda}\cdot\int_{-\Delta\lambda}^{\Delta\lambda}\lambda'\cdot H(\lambda')d\lambda'\right) \tag{6.34}$$

The derivative of $g(\lambda)$ can now be calculated as:

$$\left.\frac{dg}{d\lambda}\right|_{\lambda}=-g(\lambda)\cdot\frac{d\left(\sum_j\sigma'(\lambda)\cdot c_j\cdot L\right)}{d\lambda}=-g(\lambda)\cdot\sum_j\frac{d\sigma'(\lambda)}{d\lambda}\cdot c_j\cdot L \tag{6.35}$$

Equation (6.34) then becomes:

$$\ln\left(\frac{\mathrm{I}^*(\lambda, L)}{\mathrm{I}_0'(\lambda, L)}\right) \approx \ln\left(g(\lambda)\cdot\int_{-\Delta\lambda}^{\Delta\lambda} H(\lambda')d\lambda' - g(\lambda)\cdot\sum_j \frac{d\sigma'(\lambda)}{d\lambda}\cdot\bar{c}_j\cdot L\cdot \int_{-\Delta\lambda}^{\Delta\lambda} \lambda'\cdot H(\lambda')d\lambda'\right)$$

and after rearranging:

$$\ln\left(\frac{\mathrm{I}^*(\lambda, L)}{\mathrm{I}_0'(\lambda, L)}\right) \approx \ln\left(g(\lambda)\cdot\int_{-\Delta\lambda}^{\Delta\lambda} H(\lambda')d\lambda'\right) + \ln\left(1-\sum_j\left(\left.\frac{d\sigma_j'(\lambda)}{d\lambda}\right|_\lambda\cdot\bar{c}_j\cdot L\cdot\int_{-\Delta\lambda}^{\Delta\lambda}\lambda'\cdot H(\lambda')d\lambda'\right)\right) \tag{6.36}$$

The approximation for the logarithm from (6.21) and the fact that $H(\lambda)$ is normalised to unity can now be used. The application of the approximation is justified because the sum over j in the above equation is small, and thus the argument of the second ln in the equation is near unity.

$$\ln\left(\frac{\mathrm{I}^*(\lambda, L)}{\mathrm{I}_0'(\lambda, L)}\right) \approx \ln(g(\lambda)) - \sum_j\left(\left.\frac{d\sigma_j'(\lambda)}{d\lambda}\right|_\lambda\cdot\bar{c}_j\cdot L\cdot\int_{-\Delta\lambda}^{\Delta\lambda}\lambda'\cdot H(\lambda')d\lambda'\right) \tag{6.37}$$

After the following transformations,

$$\begin{aligned}\ln(g(\lambda)) &= -\sum_{\mathrm{j}}\left(\sigma_{\mathrm{j}}'(\lambda)\cdot\bar{c}_{\mathrm{j}}\cdot L\right) = -\sum_{\mathrm{j}}\left(\alpha_j\cdot\sigma_{\mathrm{j}}'(\lambda)\cdot\frac{\bar{c}_{\mathrm{j}}\cdot L}{\alpha_j}\right)\\ &= \sum_{\mathrm{j}}\left(\frac{\bar{c}_{\mathrm{j}}\cdot L}{\alpha_j}\cdot\ln\left(\exp\left(-\alpha_j\cdot\sigma_{\mathrm{j}}'(\lambda)\right)\right)\right) = \sum_{\mathrm{j}}\left(\frac{\bar{c}_{\mathrm{j}}\cdot L}{\alpha_j}\cdot\ln\left(h_j(\lambda)\right)\right)\end{aligned} \tag{6.38}$$

one derives:

$$\ln\left(\frac{\mathrm{I}^*(\lambda, L)}{\mathrm{I}_0'(\lambda, L)}\right) \approx \sum_{\mathrm{j}} \left(\frac{\bar{c}_{\mathrm{j}} \cdot L}{\alpha_j} \cdot \ln\left(h_j(\lambda)\right)\right)$$
$$-\sum_{j} \left(\frac{\bar{c}_j \cdot L}{\alpha_j} \cdot \left.\frac{d\sigma_j(\lambda)}{d\lambda}\right|_{\lambda} \cdot \alpha_j \cdot \int_{-\Delta\lambda}^{\Delta\lambda} \lambda' \cdot H(\lambda')d\lambda'\right)$$
$$= \sum_{\mathrm{j}} \frac{\bar{c}_{\mathrm{j}} \cdot L}{\alpha_j} \cdot \left(\ln\left(h_j(\lambda)\right) - \left.\frac{d\sigma_j(\lambda)}{d\lambda}\right|_{\lambda} \cdot \alpha_j \cdot \int_{-\Delta\lambda}^{\Delta\lambda} \lambda' \cdot H(\lambda')d\lambda'\right) \tag{6.39}$$

This equation can be further transformed by again employing the approximation of the logarithm (6.21) in the reverse direction. This step is followed by a number of transformations based on the properties of the logarithm in (6.23):

$$\ln\left(\frac{\mathrm{I}^*(\lambda, L)}{\mathrm{I}_0'(\lambda, L)}\right) \approx \sum_{\mathrm{j}} \frac{\bar{c}_{\mathrm{j}} \cdot L}{\alpha_j} \cdot \left(\ln\left(h_j(\lambda)\right) - \int_{-\Delta\lambda}^{\Delta\lambda} \left.\frac{d\sigma_j(\lambda)}{d\lambda}\right|_{\lambda} \cdot \alpha_j \cdot \lambda' \cdot H(\lambda')d\lambda'\right)$$
$$\approx \sum_{\mathrm{j}} \frac{\bar{c}_{\mathrm{j}} \cdot L}{\alpha_j} \cdot \left(\ln\left(h_j(\lambda)\right) + \ln\left(1 - \int_{-\Delta\lambda}^{\Delta\lambda} \left.\frac{d\sigma_j(\lambda)}{d\lambda}\right|_{\lambda} \cdot \alpha_j \cdot \lambda' \cdot H(\lambda')d\lambda'\right)\right)$$
$$\approx \sum_{\mathrm{j}} \frac{\bar{c}_{\mathrm{j}} \cdot L}{\alpha_j} \cdot \left(\ln\left(h_j(\lambda) \cdot \int_{-\Delta\lambda}^{\Delta\lambda} H(\lambda')d\lambda' - h_j(\lambda) \cdot \int_{-\Delta\lambda}^{\Delta\lambda} \left.\frac{d\sigma_j(\lambda)}{d\lambda}\right|_{\lambda} \cdot \alpha_j \cdot \lambda' \cdot H(\lambda')d\lambda'\right)\right)$$
$$\approx \sum_{\mathrm{j}} \frac{\bar{c}_{\mathrm{j}} \cdot L}{\alpha_j} \cdot \ln\left(\int_{-\Delta\lambda}^{\Delta\lambda} h_j(\lambda) \cdot H(\lambda')d\lambda' - \int_{-\Delta\lambda}^{\Delta\lambda} \left(h_j(\lambda) \cdot \left.\frac{d\sigma_j(\lambda)}{d\lambda}\right|_{\lambda} \cdot \alpha_j \cdot \lambda' \cdot H(\lambda')\right)d\lambda'\right) \tag{6.40}$$

Finally, the derivative of $h_j(\lambda)$:

$$\left.\frac{\mathrm{dh_j}(\lambda)}{\mathrm{d}\lambda}\right|_{\lambda} = \exp\left(-\sigma_{\mathrm{j}}'(\lambda) \cdot \alpha_{\mathrm{j}}\right) \cdot -\frac{\mathrm{d}\sigma_{\mathrm{j}}'(\lambda)}{\mathrm{d}\lambda} \cdot \alpha_{\mathrm{j}} = -\mathrm{h_j}(\lambda) \cdot \frac{\mathrm{d}\sigma_{\mathrm{j}}'(\lambda)}{\mathrm{d}\lambda} \cdot \alpha_{\mathrm{j}}$$

and the definition of $h_j(\lambda)$ (6.33) are introduced:

$$
\begin{aligned}
\ln\left(\frac{I^*(\lambda, L)}{I_0'(\lambda, L)}\right) &\approx \sum_j \frac{\bar{c}_j \cdot L}{\alpha_j} \cdot \ln\left(\int_{-\Delta\lambda}^{\Delta\lambda} \left(h_j(\lambda) - h_j(\lambda) \cdot \left.\frac{d\sigma_j(\lambda)}{d\lambda}\right|_{\lambda} \cdot \alpha_j \cdot \lambda'\right) \cdot H(\lambda')d\lambda'\right) \\
&\approx \sum_j \frac{\bar{c}_j \cdot L}{\alpha_j} \cdot \ln\left(\int_{-\Delta\lambda}^{\Delta\lambda} \left(h_j(\lambda) + \left.\frac{h_j(\lambda)}{d\lambda}\right|_{\lambda} \cdot \lambda'\right) \cdot H(\lambda')d\lambda'\right) \\
&\approx \sum_j \frac{\bar{c}_j \cdot L}{\alpha_j} \cdot \ln\left(\int_{-\Delta\lambda}^{\Delta\lambda} h_j(\lambda - \lambda') \cdot H(\lambda')d\lambda'\right) \\
&\approx \sum_j \frac{\bar{c}_j \cdot L}{\alpha_j \cdot} \ln\left[\int_{-\Delta\lambda}^{\Delta\lambda} \exp\left(-\sigma_j'(\lambda - \lambda') \cdot \alpha_j\right) \cdot H(\lambda')d\lambda'\right]
\end{aligned}
\tag{6.41}
$$

This equation is now equivalent to (6.30), showing that the approach of Case 1 is applicable. It should be emphasised that, in this derivation, the approximations were based on the fact that the convolution would change the absorption structure only slightly. In contrast to Case 1, we did not make any assumption about the strength of the absorptions.

Case 3: Strong Absorbers at Low Resolution and a Smooth Light Source

The case that the optical density of an absorber before the convolution in (6.16) exceeds unity is quite common in low-resolution DOAS applications. Several molecules (e.g. O_2, NO, H_2O, halogen monoxides, some aromatics, and glyoxal) have highly resolved rotational structures in the near UV and visible spectral ranges. It is thus possible that the optical density at the centre of these lines becomes very high. If such an absorption line is measured by a low-resolution instrument, the optical density of the absorption band after convolution is not proportional to the one before convolution.

Figure 6.11 illustrates how the optical density (OD), in this case simply calculated as the logarithm of the ratio of the intensities at the band minimum and its border, for a single narrow absorption line changes after it is convoluted with an instrument function (here assumed to be of Gaussian shape) that is broader than the initial band (see also the bands in the inserts in Fig. 6.11). At small values, the optical densities of the convoluted line depend linearly on the optical density of the narrow line. This is the situation described in Case 1. At increasing values, the two optical densities start to deviate, with the low-resolution OD becoming increasingly smaller than the high-resolution OD. The two optical densities are thus no longer linearly dependent on one another.

The straightforward solution to this problem is to model the entire function (6.16). However, the introduction of the parameter α_j in (6.19) expands the applicability of Case 1 to high optical densities. This is most easily seen for one absorber $j = 1$, when $\alpha_j = \bar{c}_j \cdot L$. In this case, (6.24) and (6.30) become identical. In cases where the

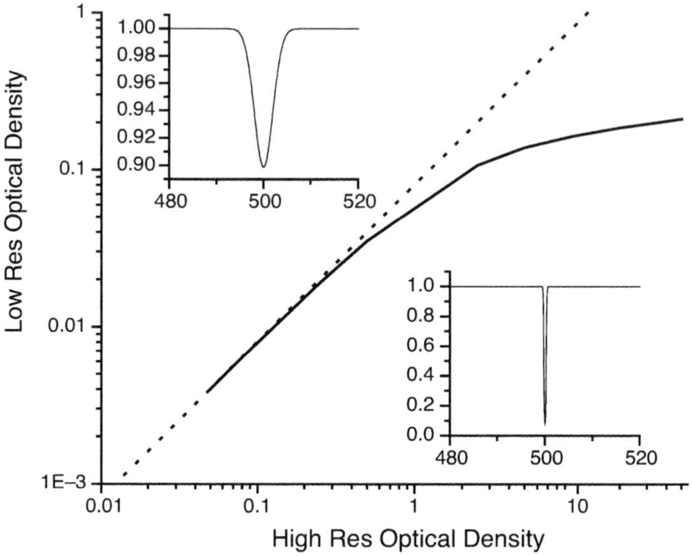

Fig. 6.11 Deviation of the low resolution optical density on the optical density of a narrow absorption band after convolution with a broad instrument resolution

value of α_j is close to $\bar{c}_j \cdot L$, (6.18) and (6.19) can still be used. However, the initial value of $\alpha_j \approx \bar{c}_j \cdot L$ has to be guessed. Alternatively, a lookup table for reference spectra for different α_j can be used.

While it is clear that, for one strong absorber the linearisation of Lambert–Beer's law can be applied if α_j is guessed correctly, one can show that, in the case of one strong absorber and various weak absorbers, (6.18) and (6.19) still apply. However, α_j for the strong absorber must be guessed correctly. When strong differential absorptions of several absorbers overlay each other, this approach leads to systematic errors in the analysis. This situation is, however, rather uncommon in DOAS applications.

Another effect helps in the application of (6.18) and (6.19) for strong absorbers. For many trace gases, the width of the absorption bands is larger than the width of the instrument function. In this case, the convolution does not change the optical density dramatically, as illustrated in Fig. 6.11. For larger optical densities, one operates somewhere between Case 1 and Case 2. By using a good initial guess for α_j, it is then possible to use (6.18) and (6.19), even for fairly strong absorbers. This is, for example, the case for NO_2 and O_3 absorptions, which are dominated by absorption bands that are resolved by typical DOAS instruments.

Because the numerical solution of (6.16) is slow, most DOAS applications rely on (6.18) and (6.19) in their analysis routines. Strong absorbers are taken into account by lookup tables of simulated reference spectra or some other numerical schemes to derive appropriate correction factors (Volkamer et al. 1998; Maurellis et al. 2000; Buchwitz et al. 2000; Frankenberg et al. 2004).

Case 4: Weak Absorbers, High Resolution, and a Structured Light Source

Our fourth case is that of a highly structured light source spectrum combined with weak trace gas absorptions and a high spectral resolution. The case of a structured light source is characteristic for the measurement of solar light, either direct or scattered, which shows narrow spectral structures due to Fraunhofer absorption bands in the photosphere of the sun. One can distinguish two situations here: If the spectral width of the instrument function is smaller than the width of the structures of $I_0(\lambda)$, we can apply Case 1. In the case that the width of the instrument function is between the width of $I_0(\lambda)$ and the bandwidth of the absorbers, Case 5 can be applied.

Case 5: Weak Absorbers, Low Resolution, and a Structured Light Source

In the case of weak absorbers, a low spectral resolution, and a structured light source, the solution shown in Case 1 has to be adopted. In contrast to Case 1, however, where the convolution of the exponential of the scaled absorption cross-section can be used to describe the absorption spectrum, (6.19) does not directly apply. The following equation must be used to calculate the absorption cross-section instead of (6.19):

$$S_j^*(\lambda)\bigg|_{case5} = \frac{\int\limits_{-\Delta\lambda}^{\Delta\lambda} I_0'(\lambda-\lambda', L) \times \exp\left(\sum\limits_j \sigma_j'(\lambda-\lambda') \times \alpha_j\right) \times H(\lambda')d\lambda'}{\int\limits_{-\Delta\lambda}^{\Delta\lambda} I_0'(\lambda-\lambda', L) \times H(\lambda')d\lambda'} \tag{6.42}$$

In principle, the calculation of (6.42) is very similar to that of (6.19). The only difference is that the convolution now includes $I_0'(\lambda, L)$ and the result, therefore, has to be normalised to the convolved $I_0'(\lambda, L)$. In the case of a smooth light source where $I_0'(\lambda-\lambda', L) \approx I_0'(\lambda, L)$, (6.42) is equivalent to (6.19). Because $I_0'(\lambda, L)$ is used in (6.42), the change in the absorption cross-section is often also referred to as the 'I_0 effect' (Platt et al. 1997; Alliwell et al. 2002; Wagner et al. 2002c).

Before discussing the derivation of (6.42), we introduce the convolution of $I_0'(\lambda, L)$ with the instrument function, which helps to simplify the equations below:

$$I'^*_0(\lambda, L) = I_0'(\lambda, L) * H = \int I_0'(\lambda-\lambda', L) \cdot H(\lambda')d\lambda' \tag{6.43}$$

Mathematically, (6.42) can be determined from (6.16). The derivation follows the approach taken in Case 1 [using the approximations from (6.21) and (6.22)], except that the assumption of a smooth light source is dropped, i.e. $I_0'(\lambda-\lambda', L)$ cannot be taken out of the integral. We begin with transforming (6.16) in the same way as in Case 1:

$$\ln\left(\frac{I^*(\lambda, L)}{I'^*_0(\lambda, L)}\right) = \ln\left(\frac{\int\limits_{-\Delta\lambda}^{\Delta\lambda} I'_0(\lambda-\lambda', L)\cdot \exp\left(-\sum_j\left(\sigma'_j(\lambda-\lambda')\cdot c_j\right)\cdot L\right)\cdot H(\lambda')d\lambda'}{I'^*_0(\lambda, L)}\right) \tag{6.44}$$

As in the above examples, (6.21) can now be applied to approximate the logarithm. The resulting equation is further reorganised to extract a factor $\left[I'^*_0(\lambda, L)\right]^{-1}$:

$$\ln\left(\frac{I^*(\lambda, L)}{I'^*_0(\lambda, L)}\right) \approx \left[\int\limits_{-\Delta\lambda}^{\Delta\lambda} I'_0(\lambda-\lambda', L)\cdot \exp\left(-\sum_j\left(\sigma'_j(\lambda-\lambda')\cdot c_j\right)\cdot L\right)\cdot H(\lambda')d\lambda'\cdot\left[I'^*_0(\lambda, L)\right]^{-1} - 1\right]$$

$$= \left[I'^*_0(\lambda, L)\right]^{-1}\cdot\left[\int\limits_{-\Delta\lambda}^{\Delta\lambda} I'_0(\lambda-\lambda', L)\cdot \exp\left(-\sum_j\left(\sigma'_j(\lambda-\lambda')\cdot c_j\right)\cdot L\right)\cdot \times H(\lambda')d\lambda' - I'^*_0(\lambda, L)\right]$$

$$= \left[I'^*_0(\lambda, L)\right]^{-1}\cdot\left[\int\limits_{-\Delta\lambda}^{\Delta\lambda} I'_0(\lambda-\lambda', L)\cdot\left[\exp\left(-\sum_j\left(\sigma'_j(\lambda-\lambda')\cdot c_j\right)\cdot L\right) - 1\right]\cdot H(\lambda')d\lambda'\right] \tag{6.45}$$

As in Case 1, the exponential function is then approximated using (6.22):

$$\ln\left(\frac{I^*(\lambda, L)}{I'^*_0(\lambda, L)}\right) \approx \left[I'^*_0(\lambda, L)\right]^{-1}\cdot\left[\int\limits_{-\Delta\lambda}^{\Delta\lambda} I'_0(\lambda-\lambda', L)\cdot\left(-\sum_j\left(\sigma'_j(\lambda-\lambda')\cdot c_j\right)\cdot L\right)\cdot H(\lambda')d\lambda'\right] \tag{6.46}$$

This equation is further rearranged, and (6.22) is applied again, this time in reverse order. The application of (6.22) in forward and backward directions lifts the restriction of small absorptions to a certain extent, as discussed in Case 1.

$$\ln\left(\frac{I^*(\lambda, L)}{I_0'^*(\lambda, L)}\right) \approx \left[I_0'^*(\lambda, L)\right]^{-1} \cdot \sum_j \frac{c_j \cdot L}{\alpha_j} \cdot \left[\int_{-\Delta\lambda}^{\Delta\lambda} I_0'(\lambda - \lambda', L) \cdot \left[\left(\sigma_j'(\lambda - \lambda')\right) \cdot \alpha_j\right] \cdot H(\lambda')d\lambda'\right]$$

$$\approx \left[I_0'^*(\lambda, L)\right]^{-1} \cdot \sum_j c_j \cdot L \cdot \left[\int_{-\Delta\lambda}^{\Delta\lambda} I_0'(\lambda - \lambda', L) \cdot \left[\exp\left(\sigma_j'(\lambda - \lambda') \cdot \alpha_j\right) - 1\right] \cdot H(\lambda')d\lambda'\right] \quad (6.47)$$

In the next step, this equation is again rearranged (remembering the normalisation of $H(\lambda)$ to unity):

$$\ln\left(\frac{I^*(\lambda, L)}{I'^*_0(\lambda, L)}\right) \approx \left[I'^*_0(\lambda, L)\right]^{-1} \cdot \sum_j c_j \cdot L \cdot \left[\int_{-\Delta\lambda}^{\Delta\lambda} I_0'(\lambda - \lambda', L) \cdot \left[\exp\left(\sigma_j'(\lambda - \lambda')\right) \cdot \alpha_j\right] \cdot H(\lambda')d\lambda' - 1\right]$$

$$\approx \sum_j c_j \cdot L \cdot \left[\left[I'^*_0(\lambda, L)\right]^{-1} \cdot \int_{-\Delta\lambda}^{\Delta\lambda} I_0'(\lambda - \lambda', L) \cdot \left[\exp\left(\sigma_j'(\lambda - \lambda')\right) \cdot \alpha_j\right] \cdot H(\lambda')d\lambda' - 1\right] \quad (6.48)$$

Finally, (6.21) is applied to transform the last term in the sum to a logarithm:

$$\ln\left(\frac{I^*(\lambda, L)}{I_0'^*(\lambda, L)}\right) \approx \sum_j c_j \cdot L \cdot \ln\left[\frac{\int_{-\Delta\lambda}^{\Delta\lambda} I_0'(\lambda - \lambda', L) \cdot \left[\exp\left(\sigma_j'(\lambda - \lambda')\right)\right] \cdot H(\lambda')d\lambda'}{\int_{-\Delta\lambda}^{\Delta\lambda} I_0'(\lambda - \lambda', L) \cdot H(\lambda')d\lambda'}\right] \quad (6.49)$$

It is clear from (6.49) that $I^*(\lambda, \mathrm{L})$ can be analysed by a classical DOAS approach (Case 1) if the differential absorption cross-sections are calculated by (6.42). The discussion about the restrictions imposed by the multiple applications of the approximations of the logarithm and the exponential functions applied here, follows the arguments of Case 1 (see above).

Case 6: Strong Absorbers, Low Resolution, and a Structured Light Source

Case 6 applies to similar trace gases as those described in Case 3, however measured with passive DOAS instruments. As in Case 3, a mathematically correct solution can be achieved in many cases by using (6.18) and (6.42), if α_j are chosen close to $\bar{c}_j \cdot L$. A lookup table approach, as well as a number of other numerical approximations, can also be used (e.g. Maurellis et al. 2000).

The discussion above shows that, in most cases, the DOAS problem can be solved by a linearisation of Lambert–Beer's law, using (6.19) in the case of active DOAS applications and (6.42) in the case of passive DOAS applications. The non-linearity that is imposed by the convolution of strong absorbers can, in many cases, be overcome if the factor α_j is close to the product $\bar{c}_j \cdot L$. The approach shown in (6.18), (6.19), and (6.42) is thus the most important implementation of DOAS. Its numerical solution is discussed in detail in Chap. 8.

The Design of DOAS Instruments

7

7.1 DOAS Instruments

A large variety of designs of active and passive DOAS systems, as well as of its optical, mechanical, electronic, and software subsystems, have been developed in response to the requirements of atmospheric research and monitoring (see below and Chaps. 6, 10, and 11). Many of the characteristics, design considerations, and trade-offs are common to several or all varieties of DOAS systems. In this chapter, we describe the options and design criteria for DOAS instruments, as well as for the required subsystems. In particular, we describe the various decisions that have to be made in design and operating procedures of a DOAS system.

7.1.1 Design Considerations of DOAS Instruments

The design of a particular DOAS system must be based on various requirements, such as spectral properties (resolution and spectral range), and practical aspects (cost, weight, and power consumption). To achieve high sensitivity the instrument must have the capability to detect very weak absorption features, with optical densities often below 10^{-3} (Chap. 6). Depending on the intended application, we can identify the following performance requirements:

- Scientific studies of atmospheric chemistry are frequently performed with ground-based active DOAS systems, but passive, in particular, multi-axis DOAS (MAX-DOAS) systems have become more important. Many scientific studies require high sensitivity (low detection limit), specific detection, and a large degree of flexibility (different light path lengths, wide range of species to be measured and thus a large accessible spectral range). Small size and simplicity of design is usually not a priority in this case.

U. Platt and J. Stutz, *Differential Optical Absorption Spectroscopy*, Geophysics and Environmental Physics, https://doi.org/10.1007/978-3-662-68639-3_7

- Air pollution and emission monitoring, and in particular long-term observations are typically performed with passive MAX-DOAS and active DOAS systems. Here, automated design, simple operation, and low cost are requirements. To some extent, this can be realised at the expense of high sensitivity and flexibility of the instrument.
- In-situ and laboratory observations require folding long absorption paths of active DOAS instruments into a small volume to achieve high sensitivities. This can be achieved through multi-reflection cells or passive resonators.
- Long-term (ground-based) measurements of atmospheric gases require an automated design which needs little supervision. Thus, for the monitoring of stratospheric gases, zenith scattered light (ZSL) passive DOAS is the method of choice. For tropospheric species active as well as passive instruments (e.g. MAX-DOAS) can be considered.
- Global and long-term observation of stratospheric and tropospheric trace gas distributions is achieved by satellite measurements, where the DOAS technique led to considerable progress, in particular with respect to the quantitative measurement of tropospheric species. Naturally, unattended operation is a prerequisite for satellite instruments. But at the large data rates, rapid spectral evaluation can also be a challenge.

From a technical standpoint, the central aspects of any DOAS system can be summarised in the following groups:

1. Active or passive operation: Active systems work in a wider spectral range (i.e. below 300 nm) and also in the absence of sunlight at night, but are relatively complex, both in technical design and logistic requirements. Passive instruments require much simpler hardware (no light source, smaller telescope, etc.), but are usually restricted to operation during daylight hours.
2. The light throughput, i.e. the amount of light passing through an instrument without being lost is an important requirement. High throughput reduces noise, allowing detection of low optical densities of trace gas absorption structures. For active systems, the light throughput is determined by both, light source and receiving optics. In passive systems only the latter is of importance.
3. The spectral range determines the number of species to be observed: While a large spectral range is certainly of advantage, its selection will usually be a compromise because increasing the spectral range may affect spectral resolution. It may also be difficult to obtain a uniform signal throughout a large spectral range.
4. The spectral resolution should be sufficient to resolve the targeted trace gas absorption structures: A better spectral resolution can improve the detection limit and the ability of the instrument to de-convolute overlapping bands of different species (see Chap. 8). On the other hand, improvement of the resolution comes at the expense of light throughput and size (and therefore cost) of the instrumentation.

5. The spectral stability of the instrument: Since DOAS relies on the detection of weak absorptions, the instruments must have the capability to eliminate the effect of overlying, stronger structures (due to Fraunhofer spectra, lamp emission spectra, or other atmospheric absorbers with strong, overlapping spectra) in the same wavelength range. If the spectral characteristics change between the recording of the measurement spectrum and overlapping structures, e.g. the Fraunhofer reference, the performance is degraded.

7.1.2 Key Components of DOAS Systems

Depending on the particular type of application and desired measurement, a DOAS system includes the following key components:

- Receiving optics (telescope)
- Spectrometer and input optics
- Detector, analogue-to-digital converter (ADC) and electronics
- Computer and auxiliary electronics
- Software for evaluation
- Database for absorption cross-sections of species (atoms or molecules) to be measured

In the case of artificial light source (active) DOAS, additional components are required:

- Artificial light source
- Optics to arrange the light path (transmitting telescope, retro-reflector, or multi-reflection optics)

The components can be arranged in many different ways. For instance, a search-light type arrangement consisting of an artificial light source coupled to a transmitting telescope can project a light beam across a distance in the open atmosphere. The other end of the light path would be defined by a receiving telescope collecting part of the transmitted beam and transferring the light to a spectrometer coupled to a suitable detection system. Details of the various possibilities are provided in Sect. 7.8.

Passive DOAS instruments can be operated from the ground, on balloons and aircraft, as well as on satellites in space. These environments and their specific viewing strategies have led to different realisations of passive DOAS instruments that are presented in Sect. 7.10.

7.2 Light Sources for DOAS

To be suitable for DOAS applications, light sources must fulfill several important requirements:

1. Spectral intensities must cover a wavelength interval that is considerably wider than the width of molecular vibrational bands (i.e. around $\approx 0.5 - 5$ nm, see Chap. 6). For active DOAS, a smoothly varying spectral intensity is highly desirable.
2. High intensity of the light source is required to allow a large number of photons to be collected in a short time and improve the sensitivity and time resolution of the instrument.
3. In active DOAS systems, the light has to be projected over a long distance; therefore, a good beam collimation is required. For thermal or LED light sources, this is equivalent to a need for high radiance L_R (or brightness), i.e. high emitted power per unit area of the radiator, solid angle, and wavelength interval.
4. For long-term applications (e.g. in monitoring applications), long lifetime and low cost of light sources are also important.

There is no ideal light source and active DOAS measurements are made using a variety of artificial light source types. Passive DOAS systems rely on light from the sun or stars. Depending on the application, sunlight can be measured directly, or as light scattered by the atmosphere or reflected from the ground.

7.2.1 Characteristics of Artificial Light Sources

The advantage of artificial light sources is that they are always available, while sunlight, moonlight, or starlight can only be used under certain circumstances (day, full-moon, no cloud cover, etc.) and during certain periods of time. In addition, the useable spectral range of artificial light sources can extend further into the UV, as well as into the IR. Light from extra-terrestrial sources that reaches the earth's surface has to penetrate the ozone layer (see Chap. 2); thus, there will be virtually no intensity below 280 nm. Measurements with UV radiation at wavelengths as short as 200 nm can be made in the troposphere with suitable artificial light sources. On the other hand, the intensity of Rayleigh scattered sunlight, for example used in MAX-DOAS observations, becomes rather small at near-IR wavelengths; thus, artificial light sources or direct sunlight can offer advantages in these situations. A summary of available light sources which have been used for DOAS applications and some of their properties is given in Table 7.1.

A basic distinction can be made between thermal light sources, such as incandescent lamps or arc lamps, and non-thermal light sources, such as light emitting diodes (LEDs), atomic emission lamps, fluorescent lamps, and laser light sources. Criteria for the selection of light sources are spectral brightness or radiance (W per

Table 7.1 Comparison of different light sources used for active DOAS applications

Type of lamp	Brightness	Typical input power W	Spectral range nm	Typical lifetime hours
High pressure Xe-Arc	High	75 to several 1000	<200–3000	200–2000
Laser Driven Light Source (LDLS)	High	100	<200–3000	2000–20000
Deuterium-Arc	Medium	25	180–300	
Incandescent lamps	Low	10–500	300–3000	50–1000 depending on voltage
(Broadband) Laser	Very high	0.01–10	<0.3 nm in a 200–3000 nm, interval	Dependent on type
Light emitting diodes (LED)	Low-medium	0.05–4.0	240[a] to several μm, certain wavelengths	>1000–100000

[a]Technological advances towards lower wavelengths LEDs are being made

unit of radiating area, unit of solid angle, and wavelength interval) and spectral range interval.

The radiation energy density $\rho(\nu, T)$ (in J/m^3) as a function of frequency ν and temperature T in a cavity (Hohlraum) is given by the Planck function:

$$\rho\ (\nu, \mathrm{T}) = \frac{8\pi\ \mathrm{h}\nu^3}{\mathrm{c}^3} \cdot \frac{\mathrm{d}\nu}{\mathrm{e}^{\mathrm{h}\nu/\mathrm{kT}} - 1} \tag{7.1}$$

where h, k, and c denote Planck and Boltzmann constants and the speed of light, respectively. By substituting $\nu = \frac{c}{\lambda}$ and $d\nu = \frac{d\nu}{d\lambda} d\lambda = -\frac{c}{\lambda^2} d\lambda$, (7.1) can be converted into the corresponding wavelength-dependent expression:

$$\rho\ (\lambda, \mathrm{T}) = \frac{8\pi\ \mathrm{hc}}{\lambda^5} \cdot \frac{\mathrm{d}\lambda}{\mathrm{e}^{\mathrm{hc}/\lambda\mathrm{kT}} - 1} \tag{7.2}$$

Figure 7.1 shows some examples of $\rho(\lambda)$ for cavities at different temperatures. For thermal emitters, the former quantity is essentially given by the Planck function, and thus the temperature T and the emissivity ε_L of the radiating area. The emissivity is unity for a perfect 'black body' i.e. an object that absorbs all incident light. Kirchhoff's law states:

$$\begin{aligned} \alpha_\mathrm{L}\ (\lambda) &= \frac{\text{Radiation absorbed by the object}}{\text{Radiation incident on the object}} \\ &= \frac{\text{Radiation emitted by the object}}{\text{Radiation emitted by a black body}} = \varepsilon_\mathrm{L}\ (\lambda) \end{aligned} \tag{7.3}$$

Typical temperatures and emissivities are: $T \approx 3000\,\mathrm{K}$, $\varepsilon_\mathrm{L} = \varepsilon_\mathrm{tung} \approx 0.35$ for the tungsten filament of incandescent lamps and $T \approx 6000$–$10000\,\mathrm{K}$, $\varepsilon_\mathrm{Xe} \approx 0.03 \ldots 0.4$

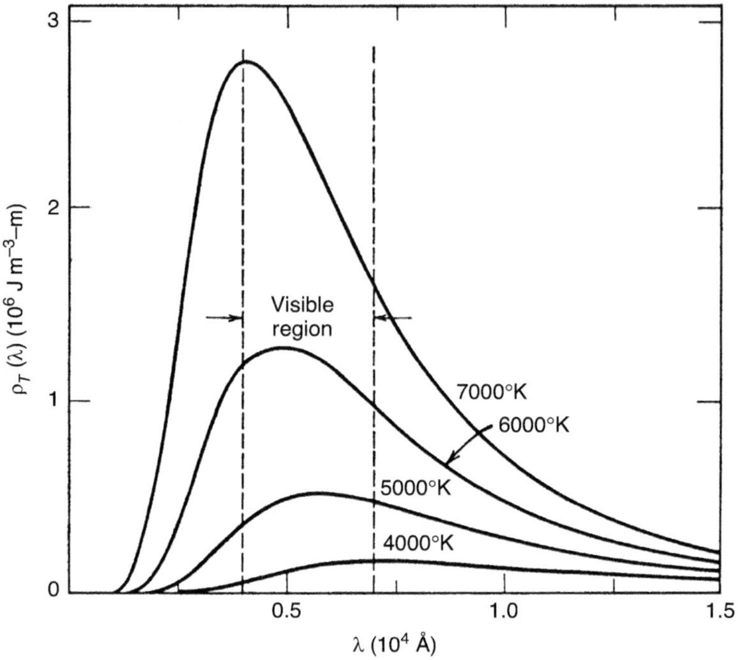

Fig. 7.1 Spectral intensity distributions of black bodies at 4000–7000 K. Reprinted with permission from Eisberg-Resnick 1985, ©John Wiley & Sons. All rights reserved

for high pressure Xe arc lamps. Radiation energy density ρ and radiance (or brightness) L_R from a surface are related by (see also Eq. 4.2):

$$L_R\ (\lambda, T) = \frac{\varepsilon_{\mathrm{L}} \cdot c}{4\pi} \cdot \rho\ (\lambda, T) \tag{7.4}$$

This leads to the radiance of a radiating surface with the emissivity ε (from a lamp filament or arc) in units of $\mathrm{W/m^2}$, per steradian and wavelength interval $d\lambda$:

$$L_R\ (\lambda, T) = \varepsilon_L \cdot \frac{2hc^2}{\lambda^5} \cdot \frac{d\lambda}{e^{hc/\lambda kT} - 1} \tag{7.5}$$

or expressed as a function of ν in the frequency interval $d\nu$:

$$L_R\ (\nu, T) = \varepsilon_L \cdot \frac{2h\nu^3}{c^2} \cdot \frac{d\nu}{e^{h\nu/kT} - 1} \tag{7.6}$$

Incandescent lamps are inexpensive, easy, and relatively safe to operate. The spectrum of incandescent lamps is very similar to that of a black-body radiator at 2800–3000 K, with a maximum in the near IR (1.0–2 μm, see Fig. 7.2). There is little emission in the UV below 400 nm. In practice, quartz–iodine lamps ('halogen lamps') are used because they reach somewhat higher filament temperatures and last

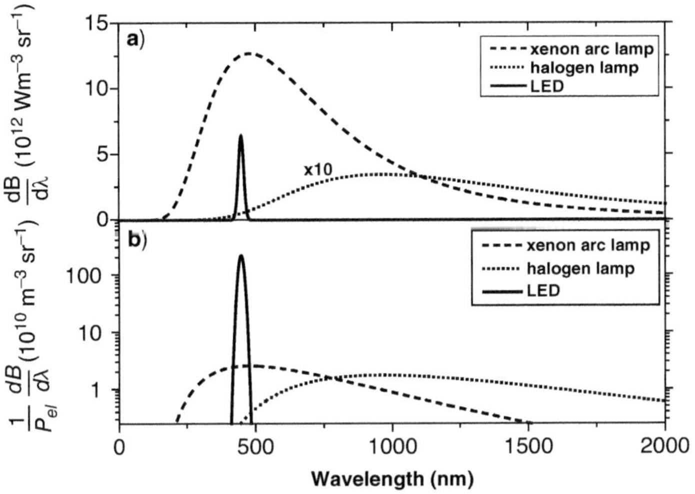

Fig. 7.2 Approximate intensity distributions of light sources. The **top panel** shows the spectral radiance (here denoted by B) distributions of three light sources: (1) An incandescent lamp (represented by a Planck's spectrum at $T = 3000$ K and an emissivity of $\varepsilon_{\text{tung}} = 0.35$). For clarity the intensity is multiplied by 10, (2) a Xenon-arc lamp (Osram XBO 450 W/2, represented by a Planck's spectrum at $T = 6000$ K and $\varepsilon_{\text{Xe}} = 0.4$), and (3) a LED (Luxeon LXHL-LR3C royal blue LED operated at 2.6 W electrical input power). **Bottom panel**: Same figures as above, intensity on a log scale and normalised to the electrical input power of the light sources. Reprinted with permission from Kern et al. 2006, ©Optical Society of America. All rights reserved

longer than traditional incandescent lamps. However, the main advantage of quartz–iodine lamps for spectroscopy in the UV lies in the fact that their envelope is made of UV-transmitting quartz. Quartz–iodine lamps of the type widely used for home illumination purposes can be used. If UV output is desired, care must taken that the lamps are not of the 'UV-protected' type. The output can be greatly increased by operating the lamps at 20–50% higher voltages than nominal at the expense of a shortened lifetime (which normally is in the 50–1000 h range). In some cases, absorption lines due to I_2 were observed in quartz–iodine lamps (Senzig 1995).

Since brightness is more important than total light flux, lamps of higher power than 20–50 W will be of no advantage. Higher-power lamps have larger radiating areas (i.e. filament sizes), but the emitted radiation per unit area stays the same.

Arc Lamps (Xenon and Deuterium) rely on light emission from a plasma generated by an electric discharge current or by Laser heating. Due to their arc temperature ($T \approx 6000$–10000 K, $\varepsilon \approx 0.03$–0.4) being much higher than that of any filament, xenon-arc lamps (Fig. 7.3) are much brighter compared to incandescent lamps. They emit an essentially smooth spectrum below 800 nm, with the exception of several weak lines and a group of emission lines between 400 and 450 nm (see Fig. 7.4; Larche 1953). Xe-arc lamps are available with power ratings from about 75 W to 25 kW and more. As in the case of incandescent lamps, the brightness of the arc of a specific lamp design is largely independent of the absolute power of the lamp. More

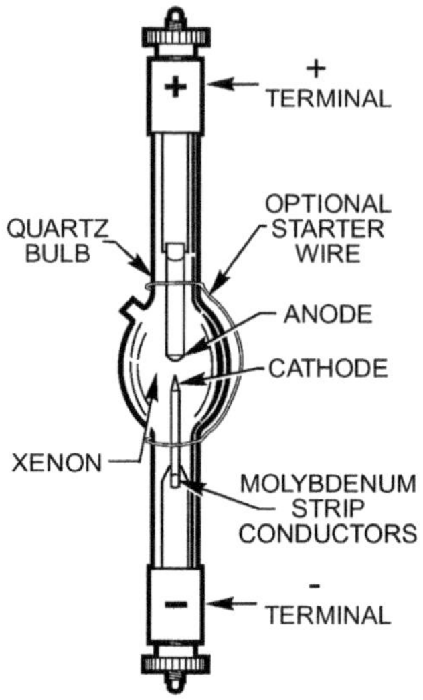

Fig. 7.3 Mechanical outline of a standard high-pressure Xe-arc lamp. The bulb of the lamp is made of quartz, the electrode terminals of metal. A very small arc (around a millimetre long, see text) develops at the centre of the 'bulb' (see Fig. 7.5). Figure courtesy of Hamamatsu Photonics, ©Hamamatsu Photonics. All rights reserved

details of the technology of high-pressure Xe-arc lamps can be found in the Osram Brochure 'XBO—theater lamps' (2024).

For special purposes other types of arc lamps are employed. For instance, good UV intensity is obtained from **Deuterium lamps**.

Arc lamps, in particular Xe-arc lamps, are relatively difficult to operate. While running at about 18 V DC, they require high voltage (about 25 kV) for ignition, and (in some models) about 60 V during the initial seconds of warm-up. As in any discharge lamp, there is a need to stabilise the lamp current. The high-voltage burst during ignition, typically provided by a Tesla transformer in the 'lamp igniter' circuit, leads to emission of a strong pulse of electromagnetic radiation, which can be dangerous for other electronic parts of the DOAS systems, such as nearby personal computers. In addition, the high internal pressure of 30 atm and more in cold lamps and two to three times as much during operation poses an explosion hazard. Therefore, precautions must be taken during operation, as well as when changing a burned out bulb. Typically, Xe-arc lamps are mounted in a lamp housing that provides electromagnetic shielding against the electromagnetic pulse generated during lamp ignition, as well as protection of the operators against UV radiation and consequences of possible lamp explosions.

Warning: Most arc lamps (including some atomic line calibration lamps) are powerful sources of UV light, with wavelengths in the UVA to UVC range. This radiation represents a severe hazard to eyes and can cause sunburn and ultimately skin cancer. Thus appropriate protection (goggles, radiation tight clothing, sun lotion) must always be used when operating these light sources. Xe arc lamps are operated at very high internal pressure, thus there is a severe hazard of explosion with both operating and non-operating lamps. Appropriate protection (safety goggles, protective clothing, etc.) must always be used when handling these light sources.

Xe-arc lamps come in several varieties (see, for example, Table 7.2):

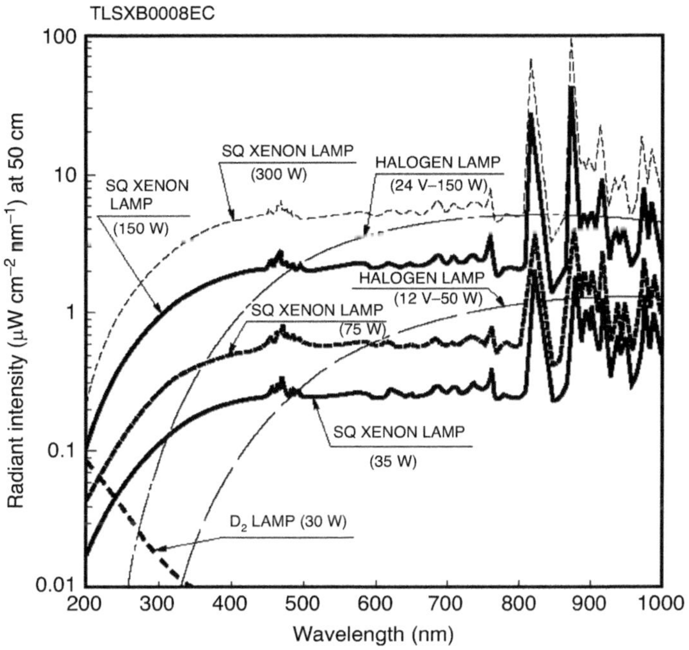

Fig. 7.4 Spectral intensity distribution (expressed as irradiance at 50 cm distance) of high pressure Xe-arc lamps (Hamamatsu) in comparison to Deuterium and incandescent lamps. Note that the radiant intensity at 0.5 m distance is plotted. The luminous intensity of high-pressure short-arc Xe lamps is far superior to incandescent lamps throughout the visible and near IR spectrum. Figure courtesy of Hamamatsu Photonics,

Table 7.2 Characteristics of several high pressure Xenon short-arc lamps (from Hermes 1999)

Lamp manufacturer and type	Size of Arc (50% point) horiz. × vertical μm	Figure	Size of arc (data sheet) horiz. × vertical mm	Intensity in 200 μm circle around brightest spot, 10^3 cd cm^{-2}
Osram XBO 450	770 × 530	7.6	0.9 × 2.7	5.910
Hanovia 959C 1980	380 × 220	7.7	0.3 × 0.3	87.500
Narva XBO 301	300 × 180	7.8	0.2 × 0.3	131.450

- Standard lamps (e.g. Osram XBO 450 or Hamamatsu SQ—lamps, Fig. 7.3) emit a Planck spectrum corresponding to a black body of about 6000–8000 K superimposed with some emission lines (Fig. 7.4) which, due to their width (on the order of 1 nm), can interfere with the DOAS data evaluation. Typical bulb shapes are cylindrical with electrode connections at both ends. Electrode gaps are on the order of millimetres, for instance 4.5 mm in the case of a 450-W lamp. Cross-sections of the intensity distribution across the arc are given in Figs. 7.5 and 7.6.

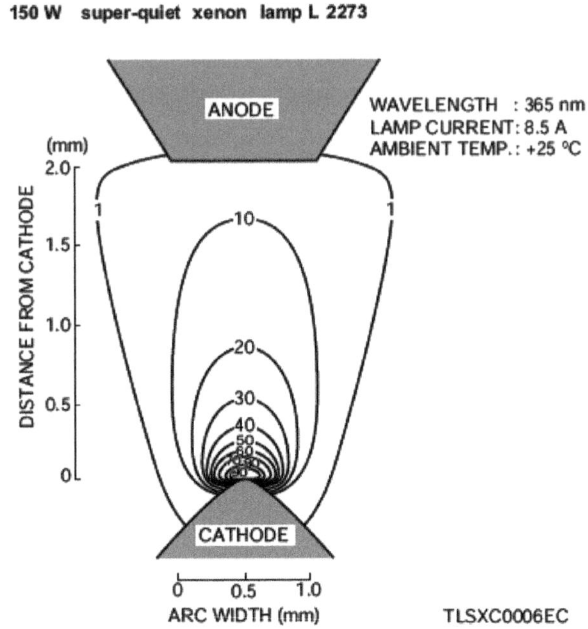

Fig. 7.5 Spatial intensity distribution of a high-pressure Xe-arc lamp. Lines indicate locations of equal luminance (in 10^3 cd cm^{-2}). Figure courtesy of Hamamatsu Photonics, ©Hamamatsu Photonics. All rights reserved

Normally, the lamps can only operate in vertical orientation with the anode at the top. Typical lifetimes are on the order of 1000–2000 h.

- High-intensity lamps (e.g. Osram XBO 500, Narva XBO 301) typically have a much smaller electrode gap than standard Xe-arc lamps, for instance <1 mm for a 500-W lamp. They also emit a Planck spectrum corresponding to a black body of about 8,000–10,000 K. Compared to standard lamps, the radiation is coming from a much smaller area. This can be seen from cross-sections of the intensity distribution across the arc in Figs. 7.7 and 7.8. In addition, the intensity of the emission lines superimposed on the continuum is typically much smaller, and their width is larger; thus, their interference with the DOAS data evaluation is greatly reduced. The lifetime of high intensity lamps is generally much shorter (on the order of 200 h) compared to standard lamps.
- In ozone-free lamps the bulb of the lamp is made of a specially doped type of quartz that blocks radiation below 242 nm, the threshold wavelength for O_2 photolysis. Depending on the type of lamp radiation, output may be degraded at wavelengths as long as 300 nm. Ozone-free lamps are readily available since they are the standard in movie projectors. Absence of ozone pollution is also an advantage in the operation of DOAS instruments.
- Ozone-producing lamps are sometimes offered with different qualities with respect to radiation absorbing impurities of quartz (e.g. Suprasil and Infrasil).

Arc Lamp Stability: Short-term stability is measured over seconds, while long-term stability is measured over minutes, hours, or even days. Short-term stability is affected by arc 'wander', 'flare', and 'flutter'. Arc wander is the movement of

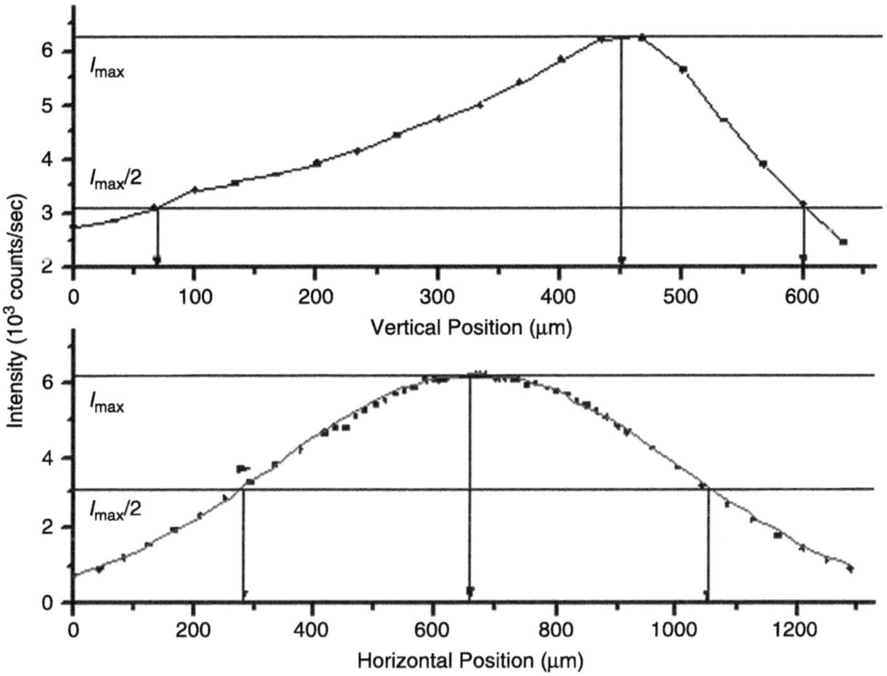

Fig. 7.6 Cross-sections (vertical: in the direction of the current, horizontal: perpendicular to current flow at the plane of highest brightness) of the intensity distribution of a standard 450 W high-pressure Xe-arc lamp (Osram XBO 450). Reprinted from Hermes, Diploma Thesis, 1999, University of Heidelberg

the attachment point of the arc on the cathode surface. Typically, the arc circles around the conical cathode tip (see Fig. 7.5) and takes several seconds to complete a full circle. If arc flares occur, there are momentary changes in brightness as the arc moves to an area on the cathode with a better emissive quality than the previous attachment point. Arc flutter is the rapid side-to-side displacement of the arc column as it is buffeted by convection currents in the xenon gas, which are caused by the gas being heated by the arc and cooled by the envelope walls (Fischer, 1987). Arc wander and flare can sometimes be reduced by a slight decrease in the operating current.

Xe-arc Lamp Lifetime: Lamp life varies dependent on the type. Lifetimes usually range from 200 to 2000 h. The useful life of compact arc lamps is determined by several factors: (1) The decrease of luminous flux due to deposit of evaporated electrode material at the inner wall of the envelope (bulb). The lamp bulb visibly 'blackens', (2) the increase of arc instability, and/or (3) failure of the lamp to ignite or burn within specified parameters.

Frequent ignition accelerates electrode wear and blackening of the envelope. Usually, the lamp life specified by the manufacturer is based on approximately 20 min of operation for each ignition. The end of the lamp life is the point at which the UV output has decreased by approximately 25%, the arc instability has increased beyond

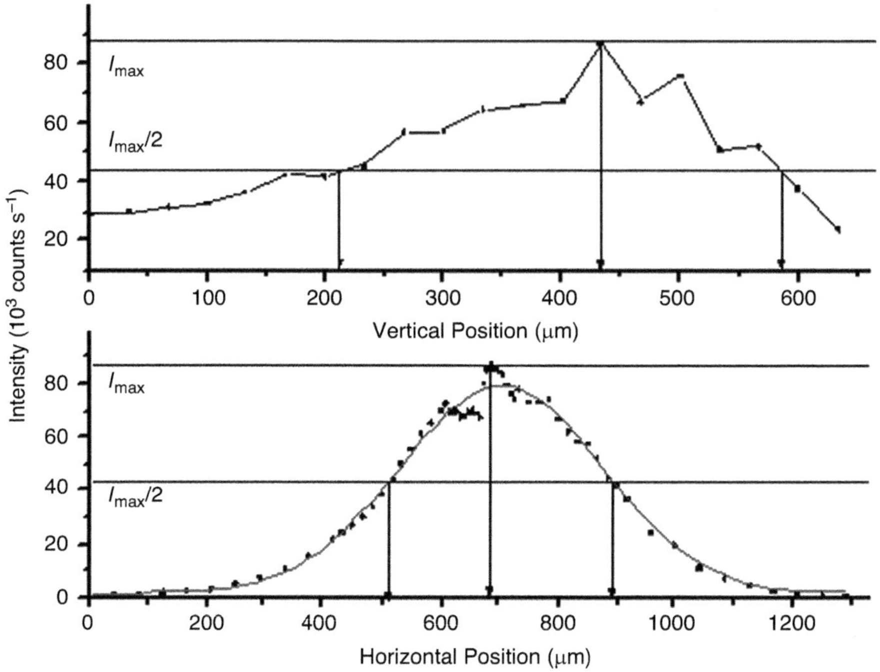

Fig. 7.7 Cross-sections (vertical: in the direction of the current, horizontal: perpendicular to current flow at the plane of highest brightness) of the intensity distribution of a 'high brightness' 500-W high-pressure Xe-arc lamp (Hanovia 959C 1980). Reprinted from Hermes, Diploma Thesis, 1999, University of Heidelberg

10%, or the lamp has ceased to operate. Lamps should be replaced when the average lamp life has been exceeded by 25%.

As the lamp ages, the operating voltage will increase. Lamp current should be decreased to maintain output until the minimum operating current is reached. At this time the lamp should be replaced.

LASER Driven Light Source, LDLS: This type of light source is fundamentally also a Xe-lamp. However, in an LDSL the energy is supplied to the xenon plasma by an infrared laser, rather than by an electrical discharge. During normal operation the plasma is heated by an IR laser focus. Electrodes are usually present, but are only used for ignition, i.e. the initial plasma generation during lamp-startup. The absorption of the laser radiation takes place via inverse bremsstrahlung on the plasma electrons (Horne et al. 2010). The absorbed energy is then transferred to the gas by electron-atom collision, heating the gas to very high temperatures (>20,000K). Since electrode erosion, the main cause for Xe-discharge lamp failure, is largely eliminated the lifetime can become very long (10,000 h or more, see below) and the light emitting area is very small. The hot spot of the lamp is defined by the laser optics. Therefore arc wandering is absent, leading to a much more stable light output.

The LDLS combines the advantages of a high power xenon lamp with long lifetime and high spatio-temporal stability of LEDs at a modest power consumption

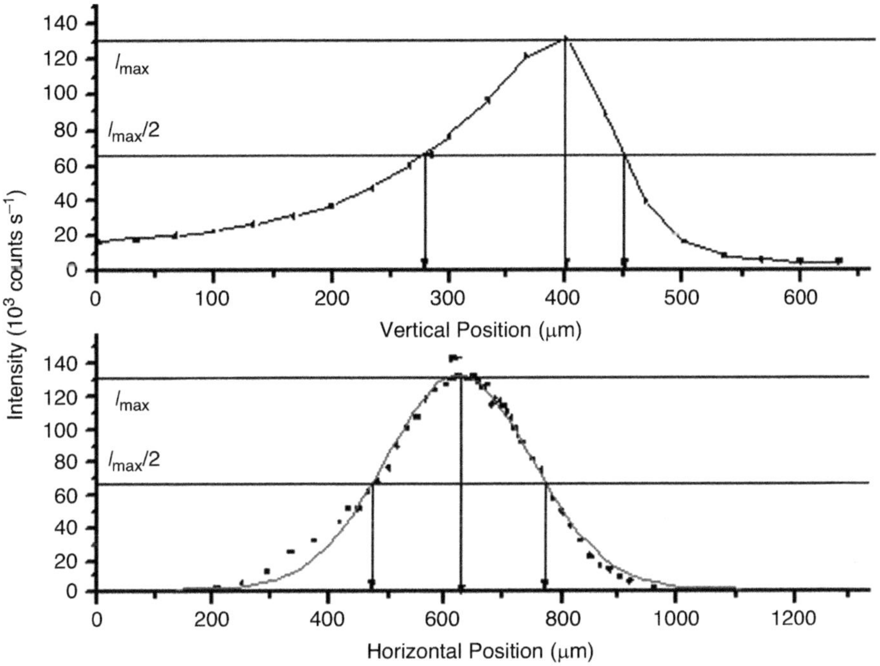

Fig. 7.8 Cross-sections (vertical: in the direction of the current, horizontal: perpendicular to current flow at the plane of highest brightness) of the intensity distribution of a 'high brightness' 300-W high-pressure Xe-arc lamp (Narva XBO 301). Reprinted from Hermes, Diploma Thesis, 1999, University of Heidelberg

(typically 75 to 150W). Similar to conventional xenon lamps, xenon emission lines are visible, the differential nature of which can limit sensitivity in DOAS applications, in particular around 450nm, a spectral window in which e.g. IO or glyoxal can be detected. The lines are broadened by a high Xe pressure in the bulb. Details on LDLS can be found in Zhu and Blackborow (2011a, b), Horne et al. (2010), or Islam et al. (2013). A DOAS application in laboratory tests and different field measurements is described by Nasse et al. (2019). The authors report a lifetime of 22,500 h for such a device.

Light Emitting diodes, LEDs, emit a smooth spectrum with an emission bandwidth of 10–60 nm or larger. There is a steady improvement of LED brightness and an extension of the spectral range from the original near-IR towards shorter wavelengths (e.g. marked by the introduction of GaN devices) (Hermes 1999; Kern 2004; Kern et al. 2006).

LED's are commercially available from the UV (about 240 nm) throughout the visible spectral range and well into the near- and mid-IR (up to about 4.8μm). Typical spectral intensity distributions are given in Fig. 7.9. While the emitted total power is frequently orders of magnitude smaller than that of thermal light sources and most laser types, emission only occurs in a relatively narrow spectral band and, more importantly, from a very small area on the order of 100 μm diameter (Kern 2004; Kern et al. 2006; Sihler et al. 2009). In fact, the power emitted per unit area from many

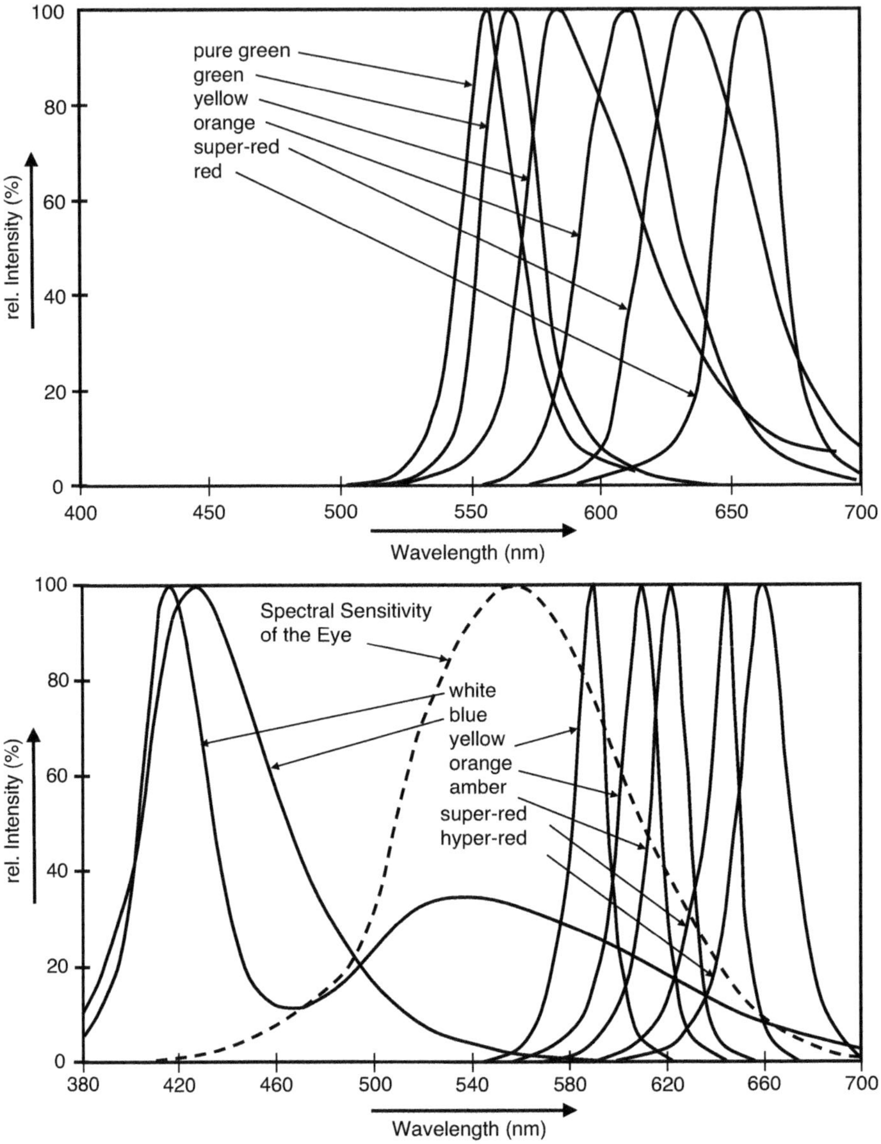

Fig. 7.9 Spectral intensity distribution of a series of LEDs. **Upper panel**: Standard LEDs. **Lower panel**: High brightness GaN LEDs. The dashed line indicates the eye sensitivity

modern LEDs can be comparable to at least that of incandescent lamps. Compared to the electrical input power, LEDs give the best spectral intensity (Fig. 7.2) of any light source, with the exception of lasers. Some LEDs exhibit strong periodic (etalon) structures in their emission spectra, these can be largely eliminated by tilting the LED optical axis relative to the axis of the tranfer optics (see Sihler et al. 2009).

Lasers (broadband) give a far lower beam divergence and higher spectral intensity than thermal light sources, but are usually much more complex devices. The spectral emission bandwidths and tuning ranges of lasers are usually quite narrow. Since the standard DOAS operation requires 'white light' sources, it is usually difficult to observe molecular absorption bands with laser light sources. Therefore, lasers have rarely been used as DOAS light sources. A notable exception is the measurement of OH radicals (together with SO_2, CH_2O, and naphthalene, see Chap. 6), where several rotational lines of 1–2 pm spectral width are observed in a spectral interval of about 0.2–0.3 nm, which can be covered by a 'broadband' laser.

7.2.2 Natural Light Sources for Passive DOAS

Natural light sources include light sources outside the atmosphere i.e. sun, moon, or even starlight. These light sources provide a light path through the entire atmosphere.

The largest problem with these natural light sources, in comparison to most artificial light sources, is the presence of strong absorption features due to the solar (or stellar) atmosphere, the so-called Fraunhofer lines (see Fig. 7.10 and Chap. 9). Fraunhofer lines are highly structured, and the spectral width of the structures is smaller than the resolution of most DOAS instruments. The structures are more prominent in the UV, where they have large optical densities. In addition, the brightness of natural light sources other than direct sunlight, is less than that of good artificial light sources. Other limitations come from the fact that sun, moon, or starlight can only be used under certain circumstances (day, full moon, no cloud cover, etc.) and during certain periods of time (see Sect. 7.2.1).

Direct sunlight allows the determination of the total column density of atmospheric trace gases. Due to the high brightness of the solar disk, the solar intensity is very high. Sun-following optics with moving parts are required to direct the sunlight into the instrument. Measurements are only possible if the direct sunlight is not blocked by clouds.

Complications arise from the fact that the optical densities of the Fraunhofer lines are not constant across the solar disk. This effect is known as 'centre-limb darkening' (see e.g. Bösch 2002; Bösch et al. 2003). Therefore, care has to be taken to always observe the full solar disk. This can constitute a problem on mobile platforms or in occultation studies where radiation originating from the lower rim of the solar disk is passing through denser parts of the atmosphere, and thus will be more strongly attenuated (Bösch 2002; Bösch et al. 2003).

Since the contribution of scattered light in direct sunlight is very small, there is no noticeable filling in of the lines due to Raman scattering (i.e. the 'ring effect', see Chap. 9) in the spectra.

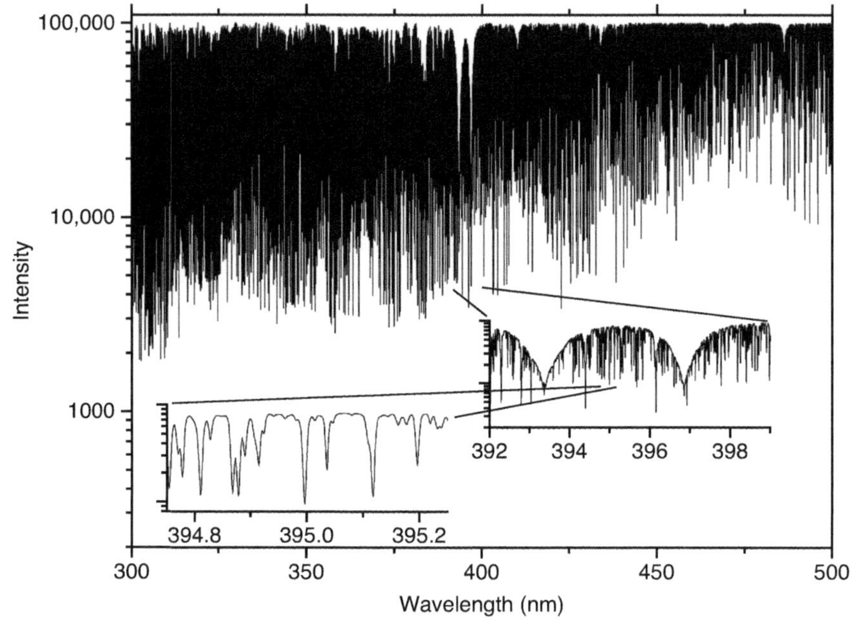

Fig. 7.10 Section of a high-resolution solar spectrum between 300 and 500 nm (from Kurucz et al. 1995)

Direct moonlight observations are subject to essentially the same advantages and restrictions as direct sunlight. However, compared to direct sunlight, the intensity of moonlight is by about five orders of magnitude lower. Moon-following optics are required. In addition, moonlight can only be used around the time of full moon. If the moon is not full, brightness, in addition to the smaller visible area, rapidly diminishes because of the large illumination and observation angles under which the lunar surface is seen. On the other hand, moonlight measurements offer the advantage of nighttime measurements that allow the study of the abundance of photo-labile species, which cannot be observed with direct or scattered sunlight. Since moonlight is reflected sunlight, the spectral features with respect to Fraunhofer lines are the same as in direct sunlight. Since the moon reflects light from all parts of the solar disk, the centre-limb darkening effect (Bösch et al. 2003) should be absent.

Sunlight scattered in the atmosphere is weaker in intensity than direct sunlight by about four orders of magnitude. Nevertheless, DOAS measurements using scattered sunlight have several advantages. No sun-following optics and hence no moving parts are required, which makes unattended operation very simple. In addition, measurements are also possible in the presence of clouds. Disadvantages include much more complicated calculation of vertical trace gas column densities from the primary 'slant-' (or in this case rather 'apparent') column densities, particularly if Mie scattering, due to aerosol or clouds, is involved. These conversions typically involve detailed radiation transport calculations (Chap. 9). In addition, the observed spectra are distorted due to (rotational) Raman scattering, leading to a 'filling in' effect of solar Fraunhofer lines at large SZA, the ‘ring effect’ (Grainger and Ring 1962), see Sect. 9.6.

Since the atmosphere scatters light very uniformly from all parts of the solar disk in the same way, the centre-limb darkening effect should not be large.

In principle, scattered light from each part of the sky can be used. However, certain arrangements have proven to be particularly useful. Zenith measurements have long been employed to monitor stratospheric composition. Observing scattered sun light at shallow angles above the horizon is most sensitive to the lower troposphere, and is the basis of MAX-DOAS. More details on these techniques are given in Chaps. 9 and 11.

7.2.3 Calibration Light Sources

Wavelength calibration of DOAS spectrometers and determination of their instrument function (Sect. 7.4) are usually performed by recording light with known, usually narrow, spectral features. This can either be accomplished by subjecting white light to structured and spectrally stable absorption by (sufficiently inert) trace gases in absorption cells (cuvettes), or by suitable glasses, e.g. holmium glass. Alternatively, light sources emitting spectrally structured radiation, such as lasers, atomic emission lamps, or the Fraunhofer structure of sunlight, can be used.

Lasers, in particular gas lasers and some solid-state lasers, can have well-defined emission wavelengths, and are ideally suited for precise wavelength calibration. In addition, the narrow spectral bandwidth ($<10^{-3}$ nm in the visible) is a good approximation of a single wavelength, the response to which directly gives the instrument function of the spectrograph–detector combination. Some small and inexpensive laser types, such as semiconductor lasers, show a large variation of the emitted wavelength with temperature and therefore can only be used for determining the instrument function, not the wavelength calibration. For the latter purpose, He–Ne lasers or some solid-state lasers (Nd:Yag) are useful, since their wavelength is determined by atomic transitions. The main disadvantage of using lasers for calibrations is their high cost and large size.

Atomic emission lamps are a very inexpensive tool for both, wavelength calibration, and instrument function determination of DOAS spectrometers. They consist of a small glass or quartz envelope filled with a small amount of the particular element. Two electrodes allow flow of current in a low-pressure discharge. In the case of noble gases (He, Ne, Ar, Kr and Xe), a few millibars of the gas are the only component. For lamps showing metal lines (e.g. Hg, Na, and Cd), Ne is also usually added in order to sustain an initial discharge current to heat the lamp until the metal exhibits sufficient vapour pressure to show emission lines. Depending on the active filling, the lamps emit many discrete emission lines of well-known (usually to a few pm or better) spectral positions (see Table 7.3). Line widths are typically on the order of 1 pm, but depend somewhat on operating conditions (they tend to be wider if the lamp is operated at higher temperature). Some care has to be taken to avoid using lines that are actually multiplets, and which might not be resolved at the spectral

Table 7.3 Recommended wavelengths (air) and wave numbers (vacuum) for selected Hg spectral lines emitted by pencil-type lamps (adapted from Sansonetti et al. 1996)

Line intensity[a]	Wavelength[b] (nm) in air	Wave number (cm^{-1}) in vacuum
3000000	253.6521	39412.236
160	289.3601	34548.888
2600	296.7283	33691.025
280	302.1504	33086.464
2800	312.5674	31983.828
1900	313.1555	31923.765
2800	313.1844	31920.819
160	334.1484	29918.220
5300	365.0158	27388.271
970	365.4842	27353.171
110	366.2887	27293.096
650	366.3284	27290.138
4400	404.6565	24705.339
270	407.7837	24515.883
34	434.7506[b]	22995.229
10000	435.8335	22938.095
10000	546.0750	18307.415
1100	576.9610	17327.389
1200	579.0670	17264.372

[a]Intensities are relative values based on irradiance values from Reader et al. (1996) with the intensity of 436 nm set arbitrarily to 10,000
[b]The wavelength uncertainty is 0.0001 nm, with the exception of that of the 434.7506-nm line

resolution of typical DOAS instruments (typically 0.1–1 nm). It should also be noted that the relative intensities of the various lines can vary considerably from lamp to lamp and also with the operating conditions of a particular lamp. Atomic emission lamps are available in many different configurations and gas mixtures.

Most widely used are mercury, neon, and cadmium lamps. Spectral positions of selected lines of cadmium, mercury, hydrogen, neon, and zinc are given in Appendix A. Some mercury line positions, together with approximate intensities (Sansonetti et al. 1996), are summarised in Table 7.3. More line positions may be found in the literature e.g. in the CD-ROM and internet database by Kurucz and Bell (1995).

Broadband light sources plus monochromators can also produce narrow-band radiation usable for calibration. While it may be difficult to produce radiation with very well known wavelengths, the radiation can be used to determine the instrument function (see Sect. 8.5.2) of the spectrograph at arbitrary wavelengths, as opposed to particular wavelengths dictated by the availability of atomic- or Laser emission lines. A grating monochromator, such as a spectrograph equipped with an exit slit instead, with the required spectral resolution (few pm) is a rather bulky instrument, and thus has been rarely used. However, recent advances in Fabry-Pérot interferometers (see Sect. 7.4.3) allow construction of much more compact monochromator-based light sources.

7.3 Optical Elements for DOAS Systems

DOAS instruments require various optical elements to guide the light from its source into a spectrometer/detector. In the case of active DOAS systems, optical elements are needed to collimate the light from the source, send it though air on a single or multiple paths, and receive the light at the other end of the path. For passive systems only receiving optics are needed. These optical setups generally consist of optical elements such as plane or curved mirrors, lenses, and prisms. This section provides a brief overview of these elements, their principles of operation, and some properties. A full treatment of the subject is beyond the scope of this book, and the reader is referred to textbooks of optics such as Smith and King (2000), Hecht (2002), and Bergman (1966).

7.3.1 Principles of Optics

Electromagnetic radiation, propagating as electromagnetic waves, may conveniently be represented as rays, which are geometric lines giving the direction along which the electromagnetic energy flows. Optical elements change the direction of these rays. Alternatively, electromagnetic waves may be characterised by their wave fronts, which are perpendicular to the direction of the rays. Optical elements act to change the shape of the wave fronts. Visualisation, design, and calculation of optical systems consisting of several elements are typically based on 'ray tracing' techniques.

When optical elements are large compared to the wavelength of radiation, we can use a 'geometrical optics' approach. Radiation propagates along straight lines until reflected by mirrors or refracted by dense, but transparent material. Accordingly, important optical elements are plane and curved surface mirrors, prisms, and lenses.

Radiation incident on a surface is either absorbed, reflected, or transmitted. Neglecting absorption and transmission, the reflection from a flat surface is said to be either specular from a very smooth mirror-like surface, or diffuse from a matte-finish surface. The terms specular or diffuse denote extreme cases, which can only be approximated in real reflectors, as shown in Fig. 7.11.

In the case of specular reflection (Fig. 7.11, left diagram), the light is reflected at the surface like an elastic sphere bouncing back from a hard surface. Incoming and reflected rays are in a plane with angle of incidence α between the normal of the surface and the incident ray. α is equal to the exit angle α' between the normal of the surface and the exiting ray.

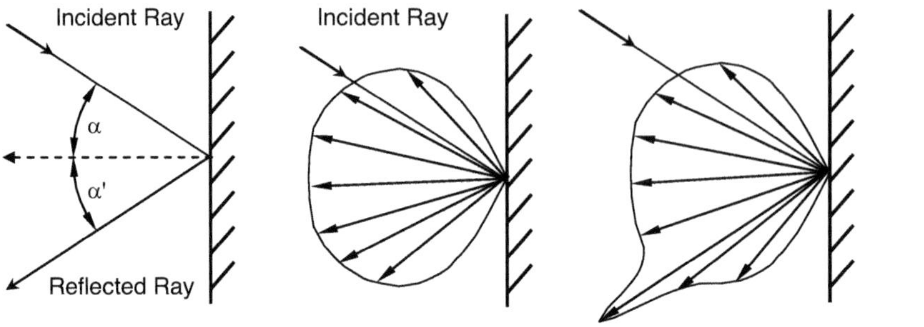

Fig. 7.11 Purely specular (**left**) and purely diffuse (**centre**) reflection from a flat surface are idealised cases. In reality, these cases can only be approximated and usually a combination is observed (**right**)

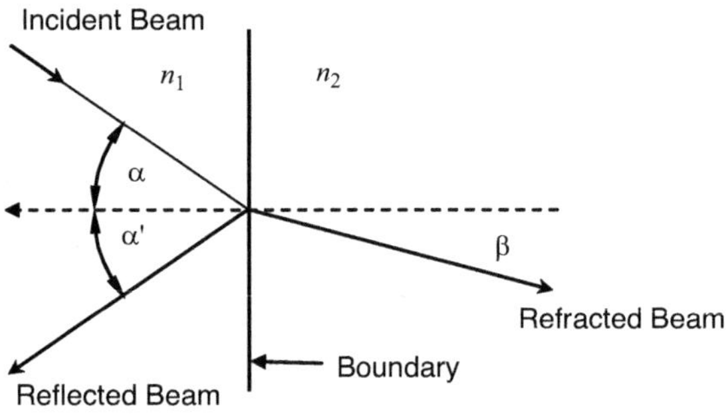

Fig. 7.12 Refraction and (specular) reflection from transparent surfaces. In this example, $n_1 < n_2$; for instance, the ray may enter from air (**left**) through the boundary between the media into glass (**right**)

Radiation incident on transparent surfaces is only partially reflected. Another part of the radiation intensity enters the medium and its direction of propagation is refracted. The direction of the refracted beam (see Fig. 7.12) is given by Snell's law:

$$\frac{\sin\alpha}{\sin\beta} = \frac{n_2}{n_1} \quad \text{or} \quad \sin\alpha = \sin\beta \cdot \frac{n_2}{n_1} \tag{7.7}$$

The degree of refraction is determined by the indices of refraction (refractive indices) n_1 and n_2 of the two materials. A ray approaching the surface from within the material with the higher index of refraction (n_2 in our example) will exit at $\alpha \geq 90°$, and therefore not exit the material at all if:

$$\sin\alpha = \sin\beta \cdot \frac{n_2}{n_1} \geq 1 \quad or \quad \sin\beta \geq \frac{n_1}{n_2} \tag{7.8}$$

The ray will undergo total internal reflection. Total internal reflection is essentially loss free, if the reflecting surface (boundary between materials with different index of refraction) is clean. It is used in retro-reflector prisms (see Sect. 7.3.10).

Reflected Intensity: The reflectance $R(\lambda) = I_{reflected}/I_{incident}$ is different for the two polarisation directions; parallel to the plane of incidence and perpendicular to the plane of incidence. It is given by the Fresnel equations:

$$R(\alpha, \beta, \lambda)_{perpendicular} = \left(\frac{n_1 \cos\alpha - n_2 \cos\beta}{n_1 \cos\alpha + n_2 \cos\beta}\right)^2 = \left(\frac{\sin(\alpha - \beta)}{\sin(\alpha + \beta)}\right)^2 \quad (7.9a)$$

$$R(\alpha, \beta, \lambda)_{parallel} = \left(\frac{n_2 \cos\alpha - n_1 \cos\beta}{n_1 \cos\beta + n_2 \cos\alpha}\right)^2 = \left(\frac{\tan(\alpha - \beta)}{\tan(\alpha + \beta)}\right)^2 \quad (7.9b)$$

where the wavelength dependence is derived from the wavelength dependence of the index of refraction $n(\lambda)$. The last term in (7.9a, b) is derived by introducing Snell's law (7.7) into the preceding term. The total reflectance (for unpolarised light) is given by:

$$R(\alpha, \beta, \lambda) = \frac{1}{2}(R_{parallel} + R_{perpendicular}) = \frac{1}{2}\left(\frac{\sin^2(\alpha - \beta)}{\sin^2(\alpha + \beta)} + \frac{\tan^2(\alpha - \beta)}{\tan^2(\alpha + \beta)}\right) \quad (7.9c)$$

The dependence on the indices of refraction in (7.9a–7.9c) is implicit, since the equations depend on both angles α and β. The dependence on β may be eliminated by introducing $\cos\beta = \sqrt{1 - \sin^2\beta} = \sqrt{1 - \frac{n_1^2}{n_2^2} \cdot \sin^2\alpha}$ into (7.9a–7.9c).

The transmittance $T(\lambda) = I_{transmitted}/I_{incident} = 1 - R(\lambda)$ is given by:

$$T(\lambda)_{perpendicular} = \left(\frac{2n_1 \cos\alpha}{n_1 \cos\alpha + n_2 \cos\beta}\right)^2 = \left(\frac{2\sin\beta \cdot \cos\alpha}{\sin(\alpha + \beta)}\right)^2 \quad (7.9d)$$

$$T(\lambda)_{parallel} = \left(\frac{2n_1 \cos\alpha}{n_1 \cos\beta + n_2 \cos\alpha}\right)^2 = \left(\frac{2\sin\beta \cdot \cos\alpha}{\sin(\alpha + \beta) \cdot \cos(\alpha - \beta)}\right)^2 \quad (7.9e)$$

At normal incidence, i.e. $\alpha = 0$, (7.9a–c) simplify to:

$$R(n_1, n_2) = \left(\frac{n_1 - n_2}{n_1 + n_2}\right)^2 \quad (7.9f)$$

For typical indices of refraction of $n_1 = 1.0$, $n_2 = 1.5$ (e.g. refraction at an air–glass interface), and normal incidence ($\alpha = 0$), a reflectance of $R = 0.04$ is obtained. Figure 7.13 shows an example of the variation of $R_{parallel}(\alpha)$, $R_{perpendicular}(\alpha)$, and $R(\alpha)$ with α for the above indices of refraction. At small angles of incidence (up to about 20°), the reflectance stays near 4%, sharply increasing to unity as α approaches 90°. Note that $R_{parallel}$ has a minimum with $R_{parallel} = 0$ at some angle of incidence α_B. This angle is called the Brewster angle.

The index of refraction of (transparent) materials varies as a function of wavelength. Some examples of various glass and quartz types are shown in Fig. 7.14.

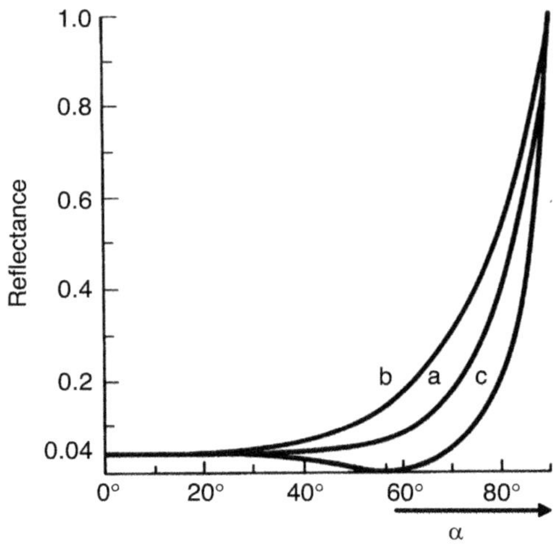

Fig. 7.13 Intensity reflected from a flat, dielectric surface. Curve (**a**) shows total reflectance $R(\alpha)$ for the unpolarised incident light. Curves (**b**) and (**c**) show the reflectances for polarised light with perpendicular and parallel polarisation, respectively. Note the zero reflectance, and thus 100.00% transmission, for parallel-polarised light at the Brewster angle (curve c at about 55°)

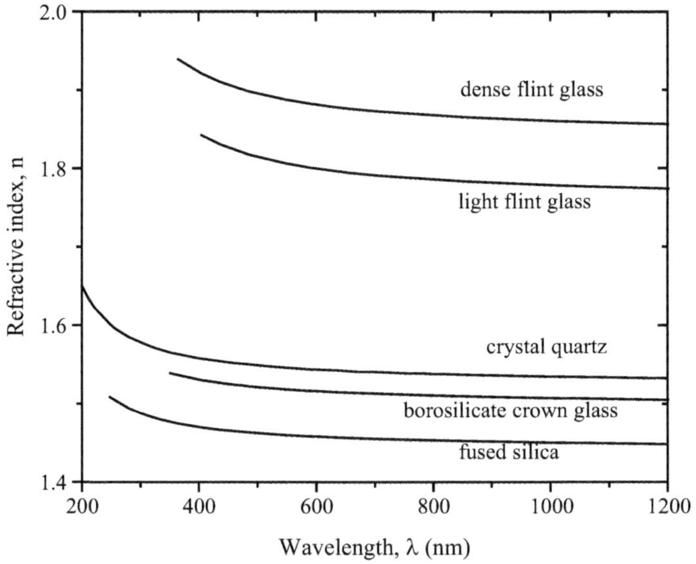

Fig. 7.14 Index of refraction as a function of wavelength for various glasses and quartz (data courtesy of Schott, Mainz Schott (2022))

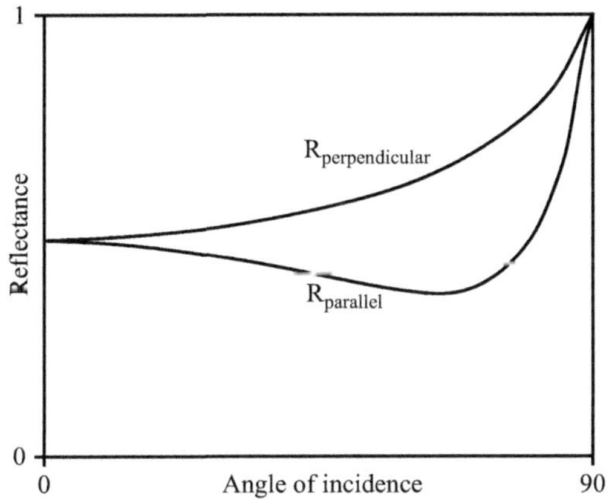

Fig. 7.15 Typical reflectance ($R_{perpendicular} = R_{\perp}$) for linearly polarised beams of radiation incident on a metal as a function of angle of incidence. $R_{parallel}$ drops to a minimum at the so-called 'principal angle of incidence'

The index of refraction of gases is a function of their density, which in turn are functions of temperature T and pressure p.

$$n\,(p,T) = \frac{p}{p_0} \cdot \frac{T_0}{T} \cdot (n_0 - 1) + 1 \tag{7.10}$$

For dry air, $n_0 \approx 1.000293$ denotes the index of refraction at 589 nm, standard pressure and temperature, $p_0 = 1$ atm (101325 Pa) and $T_0 = 273.15$ K (0 °C), respectively.

Reflection on Metals: Metals can be treated as materials having a given index of refraction, but also an attenuation coefficient due to the electric conductivity of metals. In the case of a zero attenuation coefficient and a very high index of refraction, Snell's law (7.7) would be applicable and the reflectivity would become very high. The sketch of a typical metal reflectivity as a function of polarisation (parallel or perpendicular to the plane of incidence) in Fig. 7.15 shows that the reflectance of metals behaves qualitatively similarly to the reflectance of dielectric materials.

Figure 7.16 shows reflectance curves R as a function of wavelength for the popular mirror coatings silver, gold, copper, and aluminium. In the red and infrared spectral regions, gold has superior reflectance (R ≈ 0.99, hence its reddish colour). Silver is almost as good a reflector (R ≈ 0.98) in the visible and near-IR spectral ranges. Aluminium is much less reflective than the noble metals, but superior in the UV. Note that optical mirrors are usually first surface mirrors, i.e. the reflective coating is on the surface oriented towards the incident radiation. Therefore, the metal surfaces, e.g. silver and to an even larger extent aluminium mirrors, are susceptible to corrosion and normally require protective layers, usually a thin overcoat of SiO or MgO.

7.3.2 Mirrors

The reflection by plane mirror surfaces is shown in Fig. 7.11, with the angle of incidence equal to the angle of reflection. Curved mirror surfaces, however, can have

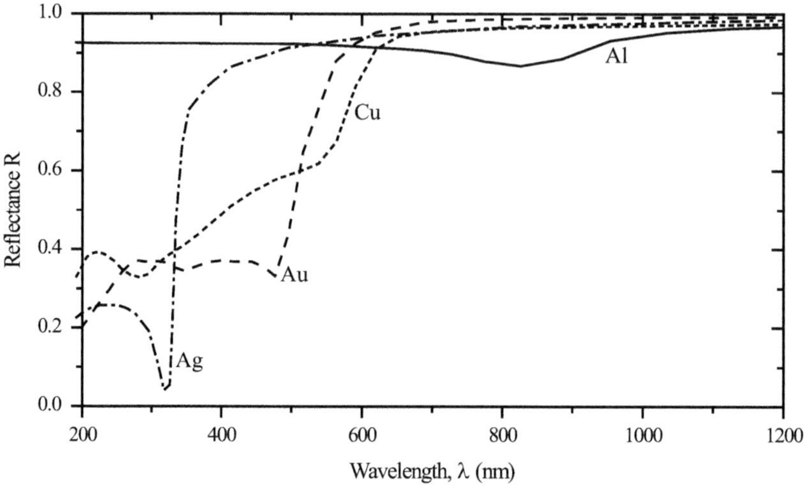

Fig. 7.16 Reflectances of silver, gold, copper, and aluminium as a function of wavelength. Reprinted with permission from CRC (2008),

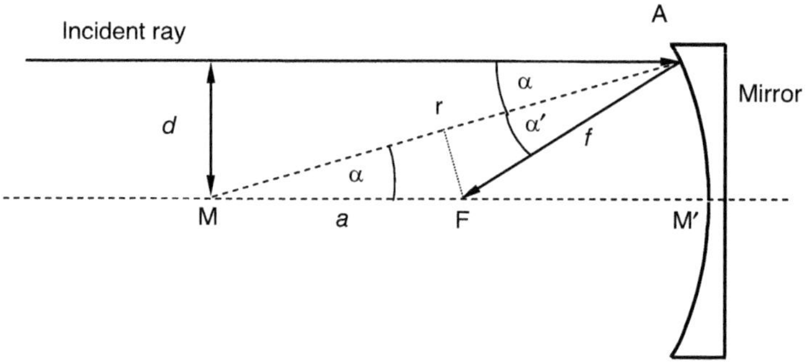

Fig. 7.17 Imaging by a (concave) spherical mirror, with radius of curvature r. Line AM denotes the normal to spherical surface at point A. The angle of incidence is α. In the paraxial approximation ($\cos\alpha \approx 1$), all rays incident parallel to the optical axis (MM′) of the mirror (at distance d) will be reflected such that they cross the focal point F at the focal length (AF = FM′ = f) of $f = r/2$

imaging properties. Here, we discuss the imaging properties of a spherical, concave mirror shown in Fig. 7.17.

A sufficiently small surface element of any curved mirror can be considered plane. A ray entering parallel to the mirror axis (solid line in Fig. 7.17) at distance d will hit a mirror with a curvature of radius r, at point A. It is incident at an angle α with the normal to the mirror surface at A (dashed line AM in Fig. 7.17). The ray is reflected at an angle of α and intersects the mirror axis at point F. The reflected ray forms an angle of $\alpha' = \alpha$ with the normal intersecting the mirror axis at point M. Thus, AMF is an isosceles triangle with $a = f$ and:

$$\cos\alpha = \frac{r/2}{a}, \quad a = f = \frac{r}{2\cos\alpha} \quad \text{and also } \sin\alpha = \frac{d}{r} \tag{7.11}$$

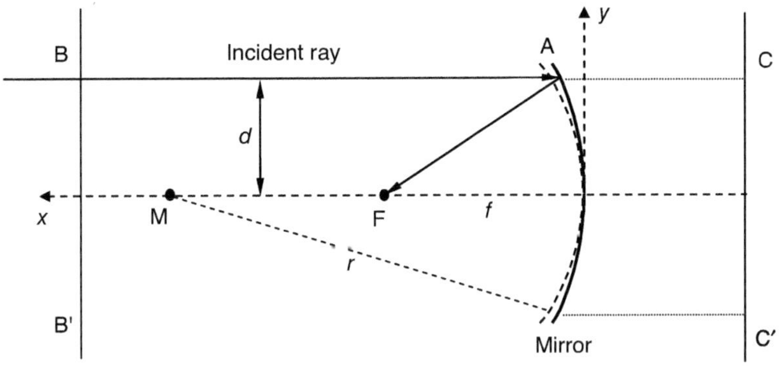

Fig. 7.18 Imaging by a parabolic mirror. For any ray incident parallel to the optical axis the distance BA + AF must be equal to BC i.e. constant. A parabolic mirror can be aproximated by a spherical surface as shown by the dashed line

For rays entering very close to the optical axis ('paraxial' rays), α will become very small, and thus $\cos\alpha \approx 1$. In this 'paraxial approximation', a spherical mirror has the focal length $f = r/2$.

As a more general case, we consider a concave mirror, as shown in Fig. 7.18. All rays incident at an arbitrary distance d from the optical axis must travel the same distance from points B on plane BB′ via A to F. Thus, the distances BA + AF must be equal to BC (where C is on plane CC′) for all values of d. A curve containing all points with the same distance between a line (i.e. CC′) and a point (F in this case) is a parabola. The points of the parabola are related through the formula:

$$x = \frac{1}{4f} \cdot y^2; \quad y = \sqrt{4f\,x} = 2\sqrt{f\,x} \tag{7.12}$$

A parabola can be approximated by a sphere given by $y^2 + (x - r)^2 = r^2$ (centre of the sphere M is shifted from the origin by r) or $x^2 - 2rx + y^2 = 0$. Solving for x gives:

$$x = r \pm \sqrt{(r^2 - y^2)} \tag{7.13}$$

Expansion in a binominal series gives:

$$x = \frac{y^2}{2r} + \frac{y^4}{2^2 2! r^3} + \frac{3y^6}{2^3 3! r^5} + \cdots \tag{7.14}$$

The first term of the expansion corresponds to the equation of a parabola (7.12) with $f = r/2$; the remaining terms describe the deviation between a sphere and a parabola:

$$\Delta x = \frac{y^4}{8r^3} + \frac{3y^6}{48r^5} + \cdots \tag{7.15}$$

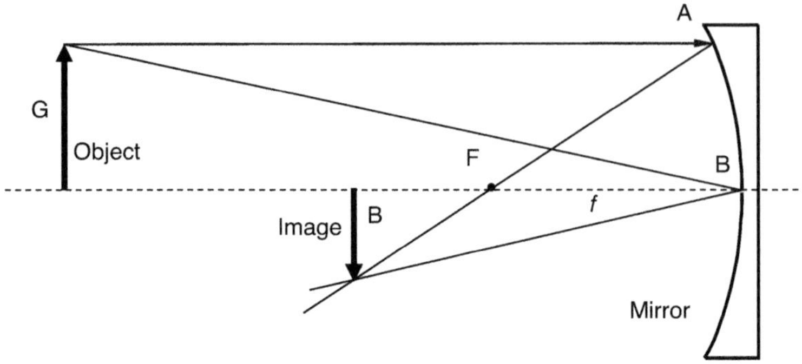

Fig. 7.19 Imaging by a parabolic mirror. An object (G) further away from the surface than $2f$ will be imaged in reduced size (B) at a distance between f and $2f$

For small values of y (i.e. paraxial rays), the deviation will be relatively small and spherical, and parabolic curves will be indistinguishable.

A reflecting parabolic (concave) mirror with the focal length f images any object (size: G) further away from the surface than $2f$ in reduced size (B) at a distance between f and $2f$ (see Fig. 7.19). The relationship between focal length f, distance of object and image from the vertex of the parabola g and b, respectively, is given by:

$$\frac{1}{f} = \frac{1}{g} + \frac{1}{b} \tag{7.16}$$

7.3.3 Prisms

Prisms are optical elements made out of glass or quartz with two refracting surfaces oriented at an apex angle γ (see Fig. 7.20). This arrangement allows the spatial separation of different wavelengths of light using the wavelength dependence of the refractive index of the prism material. Prisms can also be used to change the direction of parallel rays.

A ray entering at an angle of incidence β_1, i.e. the angle between the direction of the ray and the normal on the surface at the left surface, is refracted and continues at an angle β_2 inside the prism, until exiting at angle β_3 at the right surface, thereby being refracted again outside of the prism at an angle β_4. The different angles are connected by Snell's law of refraction (7.7):

$$n_1 \sin \beta_1 = n_2 \sin \beta_2 \text{ and } n_2 \sin \beta_3 = n_1 \sin \beta_4$$

where n_1 and n_2 denote the indices of refraction outside and inside the prism, respectively. The total deviation β between the incoming and outgoing beam is $\beta = \beta_1 - \beta_2 - \beta_3 + \beta_4$. Since in the triangle OAB, $\gamma = \beta_2 + \beta_3$, we obtain $\beta = \beta_1 + \beta_4 - \gamma$.

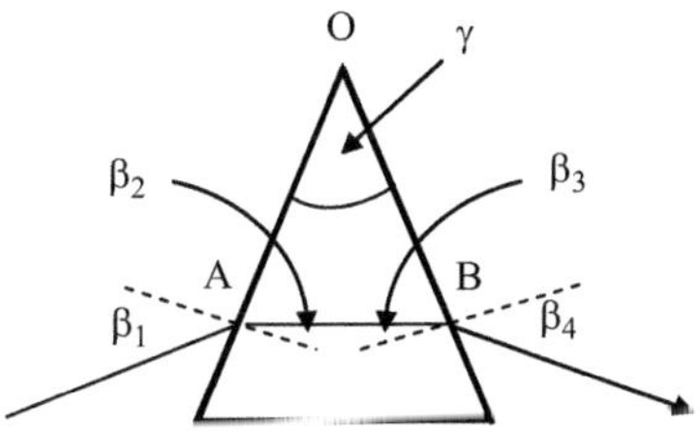

Fig. 7.20 Sketch of ray propagation through a prism

For small angles γ we have:

$$\beta = (n - 1) \cdot \gamma$$

Prisms can be used to separate radiation of different wavelengths in spectrometers. Putting polychromatic light through a prism will result in the rays being refracted in different directions with the total angular deviation:

$$\frac{d\beta}{d\lambda} = \gamma \cdot \frac{dn}{d\lambda} \tag{7.17}$$

where $dn/d\lambda$ denotes the change of the refractive index with wavelength, the dispersion (see Fig. 7.14), which for most glass types is on the order of 10^{-4} nm^{-1}.

7.3.4 Lenses

Treating a lens as a series of thin prisms leads to an expression relating the radii of curvature of both lens surfaces to the focal length of the lens, the 'lensmaker's formula':

$$\frac{1}{f} = (n - 1) \cdot \left(\frac{1}{r_1} - \frac{1}{r_2}\right) \tag{7.18}$$

Note that the difference in the indices of refraction n of the lens material and the material around the lens (here set to unity) is included in the equation. For lenses, the Cartesian sign convention of r_1 and r_2 is such that the radii of surfaces, for which the centre of curvature is on the opposite side of the light source, are defined as positive. If the centre of curvature is on the same side as the light source, r is defined as negative. For instance, for the lens in Fig. 7.21, r_1 would be defined as positive, while r_2 would be negative. In the case of a biconvex lens, the focal length would be:

$$\frac{1}{f} = (n - 1) \cdot \left(\left|\frac{1}{r_1}\right| + \left|\frac{1}{r_2}\right|\right)$$

Equation (7.18) can also be applied to biconcave or concave–convex lenses. An important property of lenses is the relatively strong dependence of the focal length on the index of refraction, with $f \propto 1/(n - 1)$, and thus on the wavelength of the

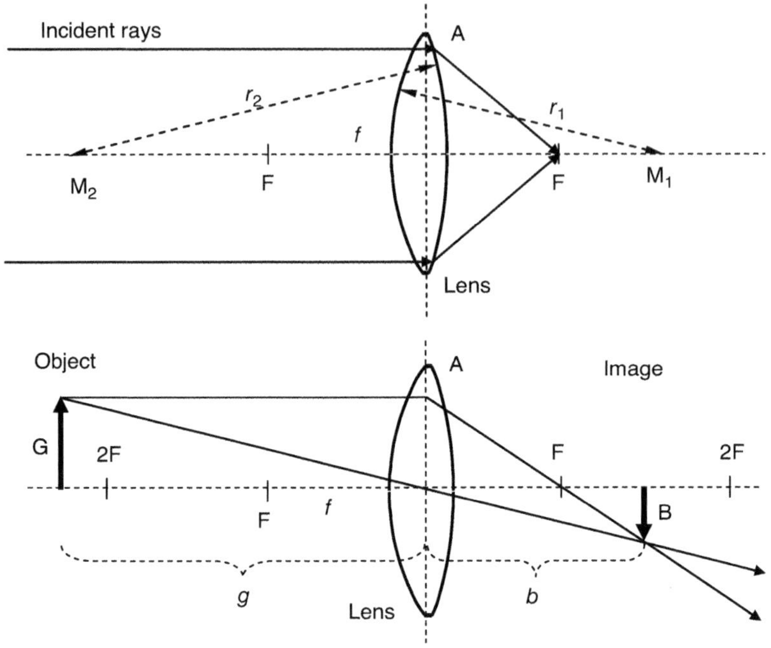

Fig. 7.21 Imaging by a thin biconvex lens. Upper panel: relationship between radii of curvature and 'focal length (see 'lensmaker's formula'). Lower panel: Imaging of object with extension G into image with extension B. The vertical dashed lines indicate the 'principal plane' of the lens i.e. the position of an idealised, infinitely thin lens

radiation, since n varies considerably with the wavelength, in particular in the UV (see Fig. 7.14).

A convex lens can produce an image of an object. The Gaussian lens formula relates the distance g between the object and the lens to the distance b between lens and image:

$$\frac{1}{f} = \frac{1}{g} + \frac{1}{b} \tag{7.19}$$

where f denotes the focal length of the lens. An example is outlined in Fig. 7.21. The formula is the same as (7.16), but object and image are on opposite sides of the optical element (lens). The size B of the image can be calculated from the relationship:

$$\frac{\mathrm{B}}{\mathrm{G}} = \frac{b}{g} = \mathrm{M} \tag{7.20}$$

with the magnification M. The angular extents of object and image (i.e. the angle under which the object or image are seen from the position of the lens) are given by:

$$\tan\beta = \frac{\mathrm{B}}{b} \quad \text{and} \quad \tan\gamma = \frac{\mathrm{G}}{g} \tag{7.21}$$

7.3.5 Apertures, Entendue, Lagrange Invariant

The amount of radiation power gathered by an optical system (e.g. consisting of mirrors or lenses) will be directly proportional to its entrance area. A large clear entrance area will usually gather more light than a smaller one. Of course, this will only be true if all of the entrance area is illuminated by the source. The size (linear extent) of the image produced by an optical system will vary proportionally to its focal length f. Since a given amount of light enters through the entrance area, the irradiance in the focal plane of an optical system of given entrance area will be inversely proportional to the image area, and hence f^2.

Most optical elements are circular in shape, and thus their area is proportional to the square of their diameter $D = 2R$ (R: radius of the element). The ability to produce a bright image is therefore proportional to $(D/f)^2$. A customary way to specify this is the relative aperture D/f or its inverse, the focal ratio or **f-number** $F = f/D$. Frequently, the f-number is written as $f/\#$. Optical systems with a smaller f-number will produce a brighter image. Comparing a system with a given $(f/\#)$ to one having $(f/\#)/2$ will result in four times the brightness of the image in the system with $(f/\#)/2$. Another way to specify the ability of an optical system is its **numerical aperture** (NA), which is defined as R/f. Both quantities are related by:

$$F = f/\# = \frac{f}{D} = \frac{1}{2\,(\mathrm{NA})} \tag{7.22}$$

A further useful quantity to specify the light throughput of an optical system is its **entendue**, which is defined as the product of the entrance area A times the aperture solid angle Ω: $E = A \cdot \Omega$. The entendue is a constant of optical systems, i.e. the entendue of the light entering an optical system is as large as the entendue of the light leaving it. For example, reduction of the size B (and thus the area in proportion to B^2) of an image by a lens will increase the aperture solid angle, thus keeping the entendue constant.

Object and image, respectively, see the lens in Fig. 7.21 under the angles α_G and α_B. Using the lens radius R_L we derive:

$$\tan \alpha_\mathrm{G} = \frac{R_\mathrm{L}}{g} \quad \text{and} \quad \tan \alpha_\mathrm{B} = \frac{R_\mathrm{L}}{b}$$

Considering $g \cdot \tan \alpha_\mathrm{G} = R_\mathrm{L} = b \cdot \tan \alpha_\mathrm{B}$ and (7.20) we get:

$$G \cdot \tan \alpha_\mathrm{G} = B \cdot \tan \alpha_\mathrm{B} = \mathrm{L_I} \tag{7.23}$$

where L_I denotes the **Lagrange invariant**. L_I is another conserved quantity.

7.3.6 Diffraction at Apertures

Radiation passing through apertures of finite size or grazing edges will undergo diffraction, meaning that part of the incident radiation deviates from its original direction.

In the case of a parallel beam passing through a **circular aperture** (i.e. a circular hole in a piece of non-transparent matter oriented perpendicular to the propagation direction of the radiation, or a circular mirror) with radius R, there will be concentric rings of diffracted radiation around the centre beam. The angle ϑ between the diffracted radiation and the original beam direction and its intensity $I(\vartheta)$ is given by:

$$I(\vartheta) = I_0 \left(\frac{2J_1(x)}{x} \right)^2 \quad \text{with } x = \frac{2\pi R}{\lambda} \cdot \sin\vartheta \tag{7.24}$$

$J_1(x)$ denotes the Bessel function of first degree, λ the wavelength, and I_0 the original intensity. The diffraction angles and intensities of the first few minima and maxima are provided in Table 7.4

For a very long rectangular aperture with width w_S (i.e. a slit with a height $h_S >> w_S$), the diffracted intensity is given by:

$$I(\vartheta) = I_0 \frac{\sin^2 x}{x^2} \quad \text{with } x = \frac{\pi w_S}{\lambda} \cdot \sin\vartheta \tag{7.25}$$

Thus, the first minimum is at $x = \pi$ with $\sin\vartheta = \lambda / w_S$.

7.3.7 Quartz Fibres, Mode Mixers, and Cross-Section Shaping

Quartz (or glass) fibres rely on total internal reflection (see Sect. 7.3.1) to conduct radiation. In practice, the total reflection does not occur between the quartz (or glass) and the surrounding air, but rather between the (relatively) high index of refraction (n_f) core of a fibre clad with a outer layer of material with a lower index of refraction (n_c) (see Fig. 7.22). Here we focus on most commonly used fibres with circular cross-sections. It should be noted that fibres with rectangular cross-sections are also available, see e.g. Amann et al. 2019.

Table 7.4 First maxima and minima of radiation diffracted at a circular aperture

Maximum/minimum no.	I/I_0	$\frac{\sin\vartheta}{\lambda/R} = \sin\vartheta \cdot \frac{R}{\lambda}$
Min. 1	0	0.61
Max. 1	0.0175	0.815
Min. 2	0	1.08
Max. 2	0.00415	1.32
Max. 3	0.00160	1.85

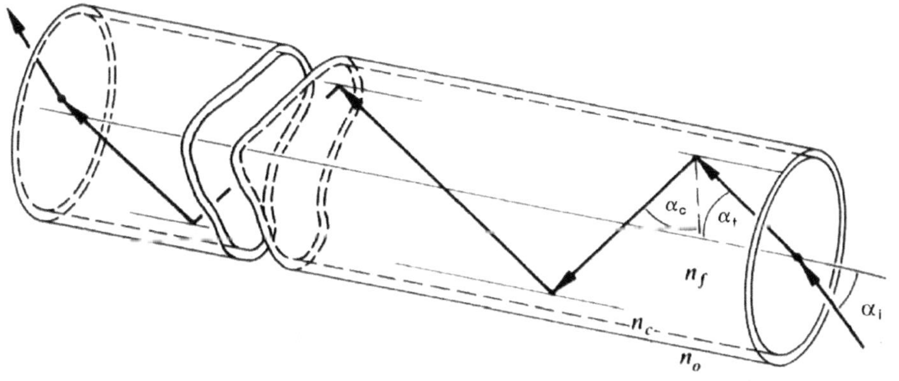

Fig. 7.22 Propagation of radiation in a cylindrical optical fibre. Total internal reflection occurs between the core (index of refraction = n_f) and the cladding (index of refraction = n_c) of the fibre, not at its outer surface

A maximum angle of incidence $\alpha_i = \alpha_{max}$ exists, below which the light entering the fibre is transmitted by total internal reflection. The external α_i corresponds to an internal α_t due to refraction at the boundary between the ambient medium (e.g. air, index of refraction $n_0 \simeq 1$) and the core of the fibre (index of refraction n_f). A detailed calculation yields:

$$\sin \alpha_{max} = \sqrt{n_f^2 - n_c^2}$$

The NA and f-number of a glass fibre are given by:

$$\mathrm{NA} = \sqrt{n_f^2 - n_c^2} \qquad f/\# = \frac{1}{2\sqrt{n_f^2 - n_c^2}} \tag{7.26}$$

Fibres are conveniently employed to connect optical elements within DOAS systems, for example the receiving telescope to the spectrometer. This arrangement gives freedom in the placement of the components; for instance, the telescope can be placed outside, while the spectrometer is sheltered. The spectrometer does not need to be physically attached to the telescope (see also Sect. 7.8.3).

Beyond their convenience in setting up optical assemblies, optical fibres can be important optical components, improving the overall sensitivity of a DOAS system considerably. This is due to several properties of optical fibres:

1. Non-uniform illumination of the field of view of the spectrometer can cause residual structures in the spectra, which may limit the minimum detectable optical density to $2 - 10 \times 10^{-3}$ (Stutz and Platt 1997b). Introducing 'mode mixers' into the quartz fibre connecting the receiving telescope to the spectrometer can greatly reduce this effect. Figure 7.23 shows a sketch of a mode mixer arrangement.

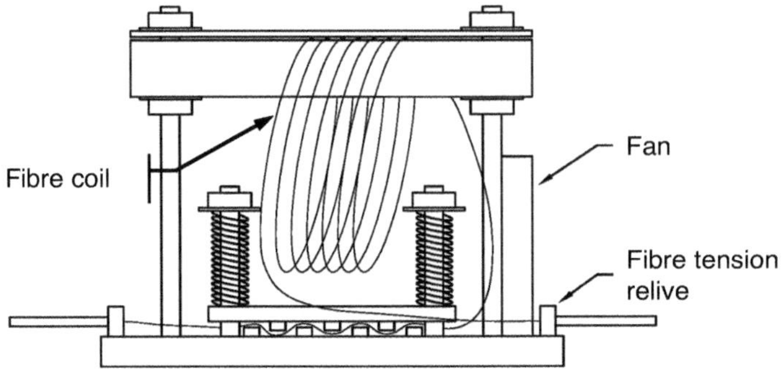

Fig. 7.23 Sketch of a fibre-mode mixer. The device consists of two elements: (1) A fibre coil (top part of the figure), which is slightly 'shaken' by an air stream generated by the fan and (2) a section of the fibre that is put under mechanical stress by a spring loaded plate introducing several sharp bends in the fibre (bottom part of the figure) (redrawn after Stutz and Platt 1997a)

Fig. 7.24 Cross-section change from a circular entry to a column at the exit by an optical fibre bundle

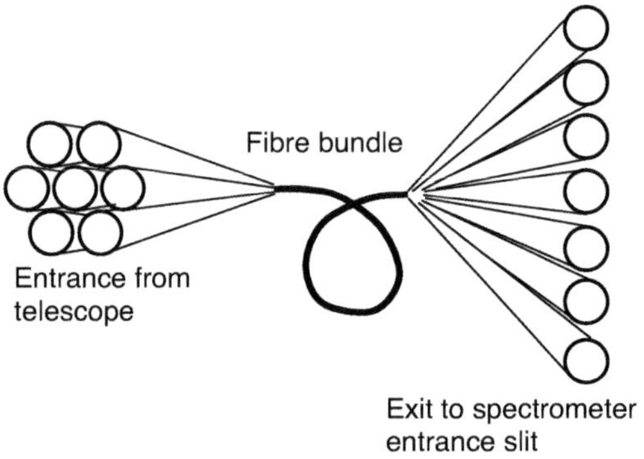

2. Most fibres scramble the polarisation state of the transmitted radiation, thus effectively removing the polarisation sensitivity of DOAS instruments. Polarisation sensitivity can be a problem in DOAS instruments, in particular in passive, scattered light applications, where the incoming radiation is, in general, partially polarised (see Chap. 4). Optical fibres of modern production appear to be less effective in polarisation scrambling than fibres produced around the year 2000 (see Tirpitz 2021). Studies on optical fibres and their depolarisation efficiency can also be found in Amann et al. (2019).
3. Fibre bundles can be used to change the cross-section of a light bundle. This property can be used to match the approximately circular (see Sect. 7.2.1) shape of e.g. a Xe-arc light source to the rectangular aperture of the spectrometer entrance slit. Figure 7.24 shows an example in which a bundle of fibres is arranged in an approximately circular shape at the entrance and a column at the exit.

7.3.8 Slit Homogenizers

In the theory of spectrographs (see Sect. 7.4) it is assumed that the entrance slit of the instrument is homogeneously illuminated. However, this might not be the case in many DOAS applications, in particular for passive DOAS instruments looking at topographic targets such as nadir-looking satellite instruments and ToTaL-DOAS applications (Frins et al. 2008). MAX-DOAS applications or satellite limb scanning might also suffer from non-uniform illumination of the field of view. The common problem is that part of the field of view receives more intensity than the rest. Consider a nadir-looking satellite instrument observing an instantaneous field of view consisting of two parallel stripes aligned parallel to the entrance slit (see Fig. 7.25, left panel A). Assume one of the stripes has a very high albedo, the other a low albedo (e.g. a cloud edge over the ocean or a coastline). Most of the light reaching the entrance slit will come from the high albedo (bright) stripe, which, in this example, is imaged onto right half of the slit. The effective slit-width is only about half as wide in this example as the physical slit width w_S (for a detailed calculation see Hummel et al. 2021). Consequently, there will be a considerably different instrument function compared to the uniform slit illumination case, leading to poor fit results, high residuals, and possible measurement errors.

In order to reduce or eliminate this effect Slit Homogenizer Devices (SHD) have been developed for satellite instruments (see e.g. Caron et al. 2019; Amann et al. 2019), for instance for the Sentinel-5/UVNS (see e.g. Hummel et al. 2021; Crowell et al. 2023). The principle is shown in Fig. 7.25, panel B. The spectrograph entrance slit is replaced by two parallel mirrors facing each other and mounted at a distance equal to the slit width w_S. The height of the mirrors corresponds to the slit height h_S. Their length is 10–100 times larger than the slit width (i.e. a few mm). Before entering the spectrograph, the light is reflected many times, thus scrambling the intensity distribution across the slit and smoothing (homogenizing) it. Challenges include (1) many reflections of the incoming light are necessary, resulting in reflection losses. (2) Strong polarisation effects can occur due to the rather large angles of incidence on the reflecting surfaces (see Fig. 7.25, panel B and Fig. 7.15) as pointed out by Crowell et al. (2023). The required high mirror reflectivity is most likely the reason why present designs are only for near-IR applications, where gold surfaces provide 99% reflectivity.

Other designs may provide solutions to these challenges. For instance the reflectivity and polarisation problems may be solved by the use of total internal reflection, i.e. by replacing the two parallel mirrors with a glass plate with mirror dimensions and a thickness equal to slit width. SHDs based on fibre bundles have also been proposed (see. e.g. Caron et al. 2019; Amann et al. 2019, and Hummel et al. 2022).

SHD technology is still under development, in particular for short wavelengths. Nevertheless, given the potential problems caused by non-uniform slit illumination, the use of this technology should be considered for almost all DOAS applications.

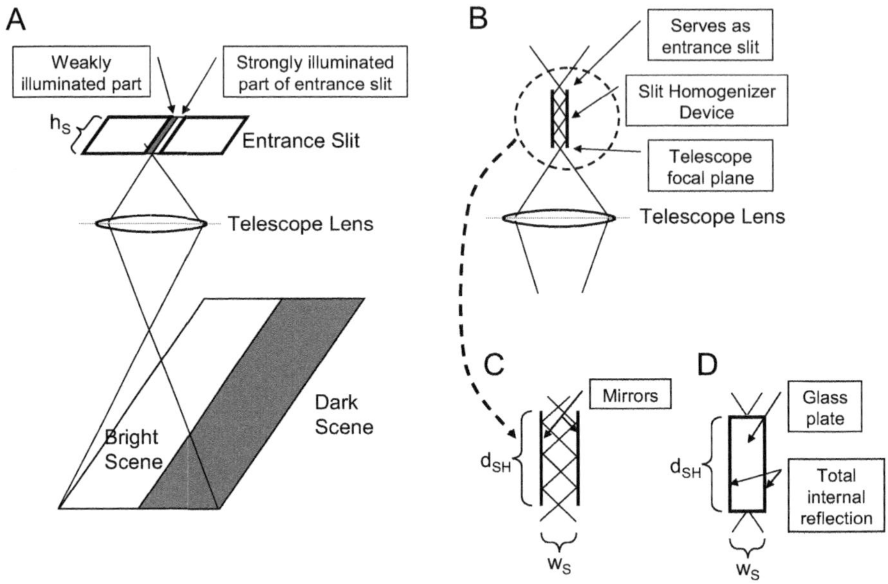

Fig. 7.25 Origin of non-uniform slit illumination and schematics of a Slit Homogenizer Device (SHD). **Panel A**: Imaging of a (ground) scene consisting of parallel stripes with high and low albedo, leads to non uniform illumination of the entrance slit. In this example with stripes of equal width essentially only half the slit width is illuminated, resulting in approximately half the effective slit width with corresponding change of the instrument function (see text). **Panel B**: Sketch of a SHD (inside dashed circle) consisting of two parallel mirrors mounted at distance w_S in front of the entrance slit. Multiple reflections between the mirrors scramble the light and reduce non-uniformity of slit illumination. **Insert C**: Details of mirror-based SHD. Insert D: Alternative SHD design using a glass (or quartz) plate

7.3.9 Filters

In DOAS systems spectral filters are usually required for several purposes. These include blocking unwanted spectral orders (see Sect. 7.4.1) or reducing the amount of stray light in the spectrometer (see Sect. 7.4.5) by limiting the spectral bandwidth of the radiation entering the instrument. In addition, it is desireable to eliminate the visible part of the light from the light source in order to avoid blinding humans and to make the radiation emitted by the DOAS instrument less visible.

Filters can either be 'colour glass' type (e.g. Schott UG5), thin film partially reflecting filters, or thin-film interference filters. Figure 7.26 shows the transmission curve for several UG-type UV-bandpass colour glass filters used to block visible light and thus reduce spectrometer stray light.

Under certain circumstances the radiation intensity may be too large for a given detector, for example when observing direct sun signals by an instrument optimised for scattered light operation. In these cases, optical attenuation is required. While this can be achieved by several measures, such as defocusing or reducing the aperture,

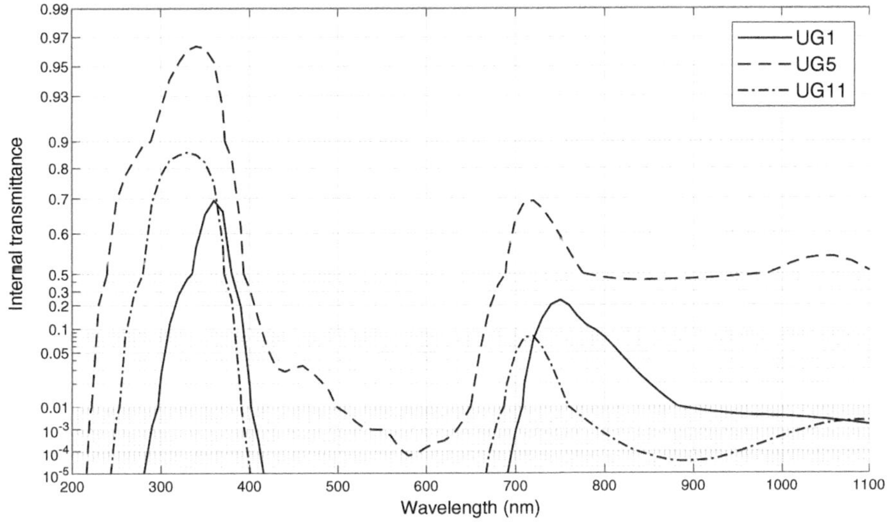

Fig. 7.26 Sample transmission curve of 2 mm thick colour glass filters (UG-1, UG-5, UG-11 from Schott 2022). These filters are used to reduce spectrometer stray light in DOAS applications. Note that the second transmission maximum in the red/near-IR region is where the sensitivity of silicon detectors usually reaches its maximum (data courtesy of Schott, Mainz Schott 2022)

grey filters offer a simple and clean way to reduce the signal level essentially without changing other properties of the optical system. Grey filters are usually characterised by an attenuation factor $A = \mathrm{I}/\mathrm{I}_0$ (I_0 and I denoting incident and transmitted intensities, respectively), frequently expressed as optical density base 10 ($\log^{10}(A)$ or $\lg A$). The attenuation A should ideally be independent of the wavelength. There are three popular types of optical attenuators: Grey glass or quartz filters, thin-film interference filters, and mesh-type filters. A potential difficulty of interference filters for DOAS applications is the presence of a Fabry–Pérot etalon effect (see Sect. 7.4.3).

7.3.10 Retro-Reflectors

The purpose of retro-reflectors is to return a beam of light back to the sending-receiving telescope of a DOAS instrument. While, in principle, plane mirrors could be used (see e.g. Perner et al. 1976 and Hübler et al. 1984), there are several types of optical arrangements that return the incident light exactly into the direction of incidence (Eckhardt 1971; Sugimoto and Minato 1994). The most popular design is the 'corner cube' retro-reflector, also called 'trihedral' retro-reflector or triple retro-reflector (Fig. 7.27). It consists of three plane reflecting surfaces arranged at right angles (Fig. 7.27). The reflecting surfaces can be viewed as being arranged like the three surfaces that meet in the corner of a cube. The radiation enters from the direction of the cube's main diagonal. In a typical active DOAS instrument, a light beam from a search light-type light source would be directed towards one or several

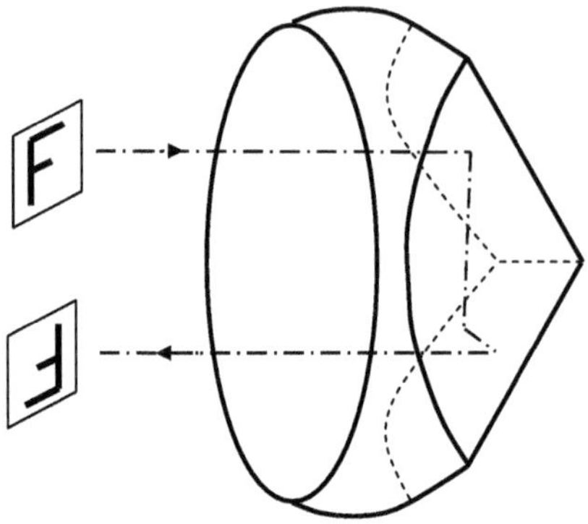

Fig. 7.27 Reflections in a corner cube retro-reflector prism

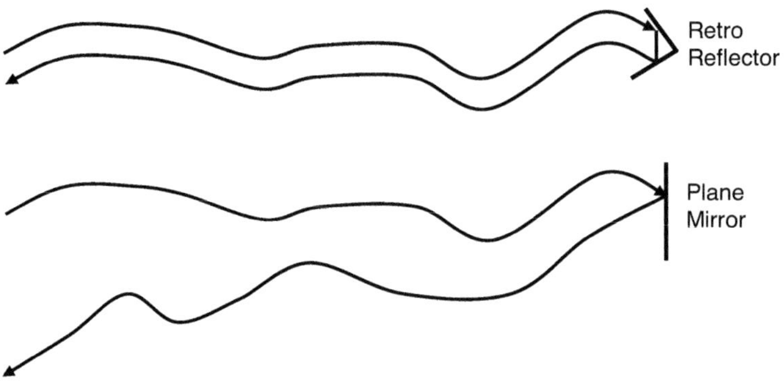

Fig. 7.28 Compensation of turbulent beam dispersion by a (corner cube) retro-reflector arrangement (**upper panel**), in comparison to reflection by a plane mirror (**lower panel**)

retro-reflectors mounted at a distance equal to half the total desired light path (see Fig. 6.2). The necessary angular precision of the retroreflectors, measured as the angle between incoming and outgoing beams, is on the order of a few arc seconds (about 10^{-5} radians), which usually requires specially made high-precision devices. An additional property of retro-reflectors is reduction of the effect of atmospheric turbulence on the light beam. As illustrated in Fig. 7.28, this is due to the property of corner cube retro-reflectors to return the incident light exactly into the direction of incidence, although with a lateral offset.
There are several varieties of the 'corner cube' retro-reflector designs (Table 7.5):

- Hollow corner cubes made of three flat mirrors mounted at very precise 90° angles. In order to protect the mirror surfaces, which are difficult to re-coat without taking the assembly apart, quartz front panes are sometimes added. These front panes represent four additional surfaces to the light returned by the retro-reflector leading to additional light loss due to Fresnel reflection (see Eq. 7.9).
- Prism-type retro-reflectors need no additional protection of front surfaces. The reflection at the rear surfaces of prism-type retro-reflector elements can either be due to total internal reflection or by a coating applied to the outside of the flat corner cube surfaces of the prism.

Table 7.5 Characteristics of different corner cube retro-reflector designs

Type of retro-reflector	Reflectance (%)	Acceptance angle range degrees	Comment
Hollow cube corner	R^3 typ. 50	≈32–56	Least expensive type for a given aperture. Vulnerable to dirt
Hollow cube corner with protective pane	$R_S{}^4 \cdot R^3$ typ. 40	≈32–56	Protected, but more expensive, while being less efficient
Corner cube prism, coated	$R_S^2 \cdot R^3$ typ. 45	60–90	Type of choice for wide angle applications
Corner cube prism, total internal reflection	R_S^2 typ. 92	20–60	Good compromise between protection and economy, most widely used type

R = reflectance of one of the corner cube surfaces, R_S = reflectance of a surface perpendicular to the light beam, e.g. of a protective (quartz) cover pane

Both approaches have advantages and drawbacks. Prism based corner cubes with internal reflection have the highest efficiency of returning light. The only loss is the reflection off the front surface of the prism. The useable range of acceptance angles is smallest for this type of prism because the limiting angle of total reflection is exceeded at higher incidence angles (Rityn 1967). In contrast, the coated corner cube prism has relatively large reflectance losses, in particular in the UV wavelength range. Typical reflectivities are on the order of R ≈ 0.8–0.9 (see Fig. 7.16). For three reflections, this amounts to an overall efficiency of 40–50% including the 4% losses ($R_S \approx 0.96$) occurring twice at the surface of the prism, see Table 7.5. The coated corner cube prism features the largest acceptance angle range (much larger than in the internal reflectance case, but also larger than the hollow cube corner), since all incident rays are diffracted towards the symmetry axis, i.e. their angel of incidence becomes smaller inside the prism (Rityn 1967).

In order to save weight, an array of small retro-reflector elements is typically used, rather than a single large unit. Typically, reflector prism sizes are on the order of 50–60 mm diameter. Special design variants have also been explored (Minato et al. 1992; Minato and Sugimoto 1998).

There are several limitations for the use of retro-reflectors: For most applications the reflectors must be of very high precision. For instance, if a deviation of the returning beam of a typical retro-reflector diameter (60 mm, see above) over a distance of 5 km is desired, this corresponds to an angle between incoming and outgoing light beams of about $\delta = 0.06\,\text{m}/5000\,\text{m} \approx 1.2 \times 10^{-5}$ radian ($\approx 7 \times 10^{-4}$ degrees or about 2 arc seconds).

Another limitation is due to diffraction at the retro-reflector aperture. The angle of diffraction for the first maximum of a circular aperture, see Table 7.4, is $\alpha \approx \sin\alpha \approx 0.815 \cdot \lambda/R$. This results in $\alpha \approx 1.25 \times 10^{-6}$ radians at a wavelength of $\lambda = 500$ nm, and a diameter of a retro-reflector element of $D = 2R = 63$ mm (2.5 in.). Thus, for the customary size of retro-reflector elements, the diffraction limitation is already close to the tolerable limits for long light paths.

7.4 Spectrographs/Interferometers for DOAS Systems

A central and critical component of all DOAS systems is a device which separates measured light into different wavelengths, so that the intensity of each wavelength, or to be more precise multiple narrow wavelength intervals, can be quantified. There are two fundamentally different approaches in use, spectrometers and interferometers.

Spectrometers separate radiation of different wavelengths into different spatial directions, where the spectrum can be measured by a spatially resolving detector (see Sect. 7.5). Spectrometers used for DOAS systems are based on three basic physical principles:

(1) Refraction by prisms (see Sect. 7.3.3), which rely on dispersion in transparent optical materials such as quartz, glass, or plastic. Advantages of prism spectrometers are the absence of overlapping orders, as in the case of grating spectrometers. In addition, their wavelength separation (linear dispersion and hence spectral resolution) is larger at shorter wavelengths. Because this matches the narrower trace gas absorption features of many trace gases, a wider spectral range is possible. Disadvantages include limited resolution ($\lambda/\Delta\lambda$ < a few 10,000 for practical purposes) and a non-linear wavelength scale. Today the non-linearity can be readily compensated in digitised spectra. Currently, prism spectrometers are rarely used, although there may be many advantages in DOAS applications, including higher light throughput (efficiency near 100%), absence of overlapping orders, and larger spectral range.

(2) Diffraction by gratings makes use of spatial interference of light separated by small distances introduced by geometric reflection (or transmission) patterns (Sect. 7.4.1). Advantages of gratings include the option of very high resolution ($\lambda/\Delta\lambda$ in excess of 100,000 is possible), approximately linear spatial separation of wavelengths (dispersion), and the ability to build robust instruments with no moving parts. It is thus not surprising that most DOAS instruments are based on grating spectrometers. One disadvantage is that interference is the same for light with wavelengths that are full multiples, i.e. wavelengths of $N \cdot \lambda$ (where $N = 1, 2, 3, \ldots$ denotes the order of the spectrum) are dispersed in the same direction.

(3) Spatial or temporal interference effects are the basis of Michelson and Fabry-Pérot interferometers. Michelson or Fourier Transform Interfometers produce (and record) an interference pattern which is the Fourier transform (FT) of the spectral intensity distribution of the received radiation. Although, in principle, spatial as well as temporal interferograms can be produced, in practice only the latter principle is in widespread use, see Sect. 7.4.4.
Fabry-Pérot interferometers use interferences due to multiple reflections between two plates and can be used with a single detector by scanning over a range of wavelengths, or combined with two-dimensional detectors as spectrometers (Sect. 7.4.3).

7.4.1 Diffraction Gratings

A grating can be viewed as a series of equally spaced slits (distance between slits = w_G), which transmit or reflect light, while the area in between does not. The angles β at which the rays, incident under angle α, from the slits interfere constructively are given by:

$$w_G\,(\sin\alpha + \sin\beta) = N\lambda \tag{7.27}$$

The left side of (7.27) gives the optical phase difference (see Fig. 7.29) which, for constructive interference, must be equal to an integer (N) multiple of the wavelength λ. The integer N is called the order of the spectrum. In practice, reflective gratings are used instead of transmission gratings, as illustrated in Fig. 7.29. Orders with $N\lambda/w_G < 2$ can be observed, i.e. in first order w_G should be larger than $\lambda/2$.

The angular dispersive power is defined as the rate of change $\partial\beta/\partial\lambda$ of the angle of emergence β versus wavelength. For a given angle of incidence α and a given order N, we obtain:

$$\frac{\partial\beta}{\partial\lambda} = \frac{1}{w_G}\cdot\frac{N}{\cos\beta} = g\cdot\frac{N}{\cos\beta} \tag{7.28}$$

The number of grooves per unit length (e.g. grooves per millimetre) $1/w_G$ is also known as grating constant g. $\partial\beta/\partial\lambda$ is independent of the angle of incidence, but increases with order and grating constant. It also becomes larger at large angles

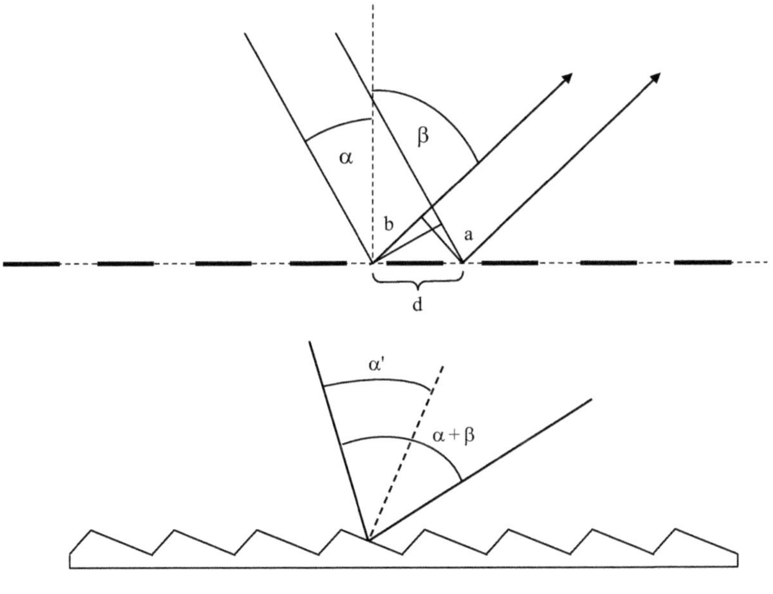

Fig. 7.29 Reflective diffraction grating. **Upper panel**: The optical path differences for the incident ray are $a = d\cdot\sin\alpha$, for the exiting ray $b = d\cdot\sin\beta$. The total path difference is $a + b$. **Lower panel**: The blaze angle of a grating is defined as the angle $\alpha + \beta$ where the refracted beam has the same direction as a specular reflected beam would have

of emergence, where $\cos\beta$ is small. For instance, for an angle of emergence $\beta = 60°$ ($\cos\beta = 0.5$), $g = 1200$ grooves per millimetre ($w_G = 0.8333\,\mu\text{m}$ or $1/w_G = 1.2 \times 10^6\,\text{m}^{-1}$), and for first order one obtains $\partial\beta/\partial\lambda = 2.4 \times 10^6$ radians m^{-1} or 2.4×10^{-3} radians nm^{-1}. Thus, two parallel bundles of radiation with wavelength 1 nm apart will emerge as two bundles propagating at an angle of 2.4 mradians.

Ruled Versus Holographic Gratings: Depending on the manufacturing process, there are two popular types of gratings: ruled gratings and holographic gratings. The former are produced by mechanically drawing grooves into a suitable substrate. Since this is a lengthy process, the gratings commonly used in spectrometers are replicated from a ruled 'master' grating. Ruled gratings are easily blazed to improve their efficiency in certain wavelength regions, see below. In contrast, holographic gratings (interference gratings) are produced by recording the interference pattern of two (laser) light beams (e.g. Lerner and Thevenon 1988). Since all grooves are formed simultaneously, holographic gratings are virtually free of random and periodic deviations in line positions, and thus they have no 'ghosts' (see below) and low levels of stray light. In addition, holographic techniques allow more freedom in producing gratings with uneven, or continuously varying, line spacing.

Gratings can be produced such that the reflecting surfaces are inclined with respect to the plane of the grating, as shown in the lower panel of Fig. 7.29. These gratings are known as 'blazed' gratings. The blaze angle of a grating is defined as the angle $\alpha + \beta$ where the refracted beam has the same direction as a beam, which is specular reflected from the sawtooth-shaped grooves of the grating (see Fig. 7.29, lower panel). The highest grating efficiency is achieved when the angle between incident and refracted beam equals the blaze angle.

In principle, a grating distributes the incident radiation energy into an infinite number of orders. In practice, most of the energy is diffracted in a few or only one order. The fraction of the incident energy reaching the strongest order is called the efficiency of the grating. Generally, the grating efficiency is a function of wavelength and the polarisation of the light. Typical maximum grating efficiencies are on the order of 20–80%. For a particular application, it must be ensured that the grating efficiency is sufficient in the desired spectral range. Note that a low grating efficiency not only reduces the light throughput of the spectrometer, but may also considerably increase its stray light level, since low efficiency usually means that the missing radiation is diffracted in other orders.

Another complication, in particular in scattered light applications where the incident radiation is usually partially polarised, is the polarisation dependence of the grating efficiency.

Grating Anomalies: Real gratings can exhibit a number of prominent deviations from the ideal behaviour described earlier. In particular, individual grooves (or groups of grooves) can be misplaced randomly or periodically. The latter imperfection gives rise to grating ghosts. An example for grating ghosts is the Rowland ghost (named after one of its discoverers, Hernry A. Rowland) caused by periodic misplacement of lines in the ruled grating. If we assume that each nth line is displaced by an amount $\Delta d_G << d_G$ from its nominal distance d_G to its neighbouring line, then the resulting grating can be viewed as an overlay of three gratings with slightly different line spacing $d_G - \Delta d_G$, d_G, and $d_G + \Delta d_G$. Such a grating will produce two additional spectra, or 'ghosts', i.e. each spectral line will have two satellites.

An anomaly of a different kind is the Wood's anomaly (Wood 1902). Many gratings show sharp changes of the diffracted intensity within narrow spectral regions. Typical reflection gratings produce a series of orders of diffracted light. At some critical wavelength, the diffracted light falls back into the plane of the grating. The light which would be sent into the forbidden region is redistributed back into the allowed orders, and appears as an addition to the spectral response, with a sharp increase at the critical wavelength and a steep decline towards higher wavelengths. This effect acts as an enhancement of the grating efficiency, as if the light from two orders are combined. A Wood's anomaly is almost entirely polarised perpendicular to the grating grooves.

A limitation of the resolving power of gratings is that the grating spacing $w_G = 1/g$ cannot be much smaller than the wavelength λ. In IR applications, this can lead to the preference for interferometers (see Sect. 7.4.4). This limitation can be overcome to some extent by using Immersed Gratings. This type of grating (see e.g. Larsson and Neuhaus 1968) makes use of the fact that the wavelength in materials λ_M with index of refraction n is reduced by the factor $\lambda_M = \lambda_{air}/n$. Thus the grating is not at the surface of the substrate and illuminated through air, but rather illuminated from the back through the substrate (e.g. glass, diamond or, in the short-wave IR, silicon). A spectrometer based on an immersed grating has been successfully used in the near IR channel of the TROPOMI satellite (Van Amerongen et al. 2010).

7.4.2 Spectrographs

The purpose of a spectrograph is to separate and re-image radiation of different wavelengths in order to record spectral signatures. In principle, this can be done in several ways, depending on the purpose of the instrument and the particular requirements regarding spectral resolution, light throughput, etc. (see e.g. Bingham 1979). A scanning **monochromator** selects and transmits a narrow spectral interval of the incoming radiation, in connection with a suitable detector. A monochromator can be thought of as an optical band-pass filter. Alternatively, a **spectrograph**, which images a range of wavelengths at the exit focal plane for detection by multi-channel detector or photographic film, can be used.

Traditionally, the spectrograph of a DOAS instrument produces a 'spectrum' i.e. spatially separates the radiation into wavelength intervals, so that the intensity in the different wavelength intervals can be recorded by a suitable detector capable of resolving spatial differences in radiation intensity. The latter task can be achieved by detectors using moving parts to convert spatial separation in temporal separation, such as a moving exit slit, or by a collection of many sensors that simultaneously record spatial variations in intensities, i.e. detector arrays.

As mentioned earlier, the 'classic' spectrograph design rests on the two optical principles of prism spectrographs and grating spectrographs, which are used in several design variants (Fig. 7.30 and 7.31).

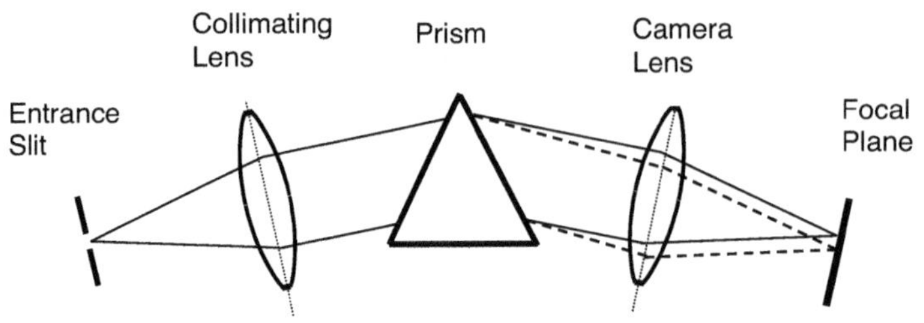

Fig. 7.30 Optical outline of a typical prism spectrograph

The Prism Spectrograph: In the early days of optical spectroscopy (e.g. Dobson, 1926), prism spectrographs (see Fig. 7.30) were the instruments of choice. The typical setup consists of an entrance slit, a collimator lens that produces a parallel light beam that illuminates the prism. The light dispersed by the prism is focused by another lens onto the focal plane of the spectrograph, where it can be fed through an exit slit to isolate a specific wavelength or be projected onto an area detector to record full spectra. The wavelength in the focal plane can be adjusted by rotating the prism in its axis with or without rotating the focusing assembly. The angular dispersion is given by (7.17). While rarely used in DOAS applications, the prism spectrograph has some merits which may make it worthwhile to consider its future use in DOAS systems. For example, the non-linearity of the dispersion can be corrected with modern software. In fact, higher resolution would usually be required in the UV than in the visible spectral range. The absence of higher orders is also an advantage.

Plane Grating Spectrographs: In recent decades, the grating spectrograph has become the dominating spectrograph. This is due to its inherent advantages, such as the almost linear dispersion curve (see 7.28), the higher resolution at a given size of the instrument, and advances in manufacturing of diffraction gratings. The Czerny–Turner-type spectrograph (see Fig. 7.31a) and variants have traditionally been selected for DOAS instruments (e.g. Platt et al. 1979; Platt and Perner, 1983). Besides the grating, this spectrograph requires at least two additional reflecting surfaces, i.e. a collimating and a camera mirror. The additional reflection losses at these surfaces are usually offset by high-efficiency (60–80%) plane blazed gratings. The wavelength range is readily selected by rotating the grating.

The linear dispersion $\partial s/\partial\lambda$ of a grating spectrograph is given by $\partial\beta/\partial\lambda$ (7.28) multiplied by the focal length of the camera mirror:

$$\frac{\partial s}{\partial \lambda} = f\frac{\partial \beta}{\partial \lambda} = \frac{f}{d}\cdot\frac{N}{\cos\beta} = fg\cdot\frac{N}{\cos\beta} \tag{7.29}$$

In special applications, for example if only a small aperture is required, Fastie–Ebert-type (Fig. 7.31c) or Littrow-type spectrographs (Fig. 7.31b) are also often used. While the design of these spectrograph types appears to be simpler, since a single mirror doubles as collimating mirror and camera mirror, the disadvantages include larger size (in the case of Fastie–Ebert) and the requirement to work at a wavelength where

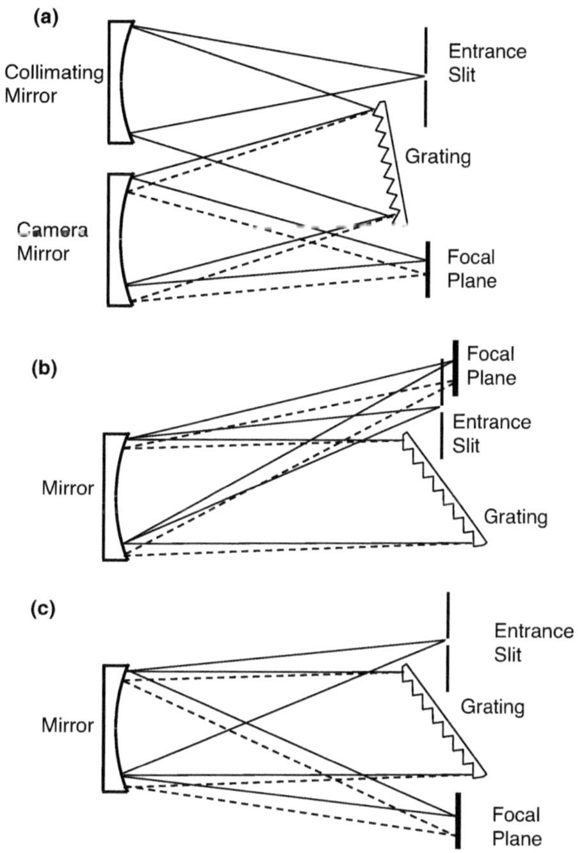

Fig. 7.31 Optical outlines of different variants of grating spectrographs. **a** Czerny–Turner; **b** Littrow; **c** Fastie–Ebert

the angle of incidence and the angle of refraction of the grating are equal (Littrow-type spectrograph).

Immersed Grating Spectrographs: As mentioned in Sect. 7.4.1, a common limitation of optical gratings is the fact that the efficiency diminishes when the groove spacing becomes comparable to or smaller than the wavelength of the radiation to be diffracted. This problem can be overcome to some extent by using an immersed grating.

Concave (Holographic) Grating Spectrographs: Modern holographic grating technology (Sect. 7.4.1) allows the design of spectrographs only consisting of entrance slit, concave grating, and detector. In contrast to the traditional Rowland mounting that also uses a concave grating as the only imaging element in the spectrograph, modern concave holographic gratings can produce a flat spectrum over a certain range (hence the name flat-field spectrographs).

The lower efficiency of most holographic gratings (20–40%, Sect. 7.4.1) is compensated by the reduced losses due to the smaller number of optical surfaces. While the stray light produced by holographic gratings is lower compared to ruled gratings, the higher intensity of unwanted orders may increase the stray light level in the spectrometer. The unused orders are also focused and can thus be removed relatively easily by light traps. A disadvantage of most flat field gratings is their limitation to a relatively small spectral range that normally cannot be changed.

Imaging Spectrographs: The term 'imaging spectrograph' refers to spectrographs that can image a position along the entrance slit onto a corresponding position in the focal plane. The imaging direction is perpendicular to the dispersion direction. Thus, two-dimensional detectors (typically 2D CCD or CMOS detectors) in the focal plane record multiple spectra and each point along the entrance slit corresponds to a line encompassing an individual spectrum in the focal plane. As illustrated in Fig. 7.32, there are n_x pixels in the 'dispersion direction' determining the spectral sampling, similar to a one-dimensional diode array (or CCD detector). In the 'slit direction', n_y rows of n_x pixels each provide spatial resolution. Theoretically, the imaging spectrographs behaves like n_y individual spectrographs, each with a spectral sampling of n_x pixels. The imaging quality of modern imaging spectrometers perpendicular to the dispersion direction are quite good in the centre of the focal plane, but degrades towards the edges.

Imaging spectrographs are used in several applications:

- MAX-DOAS (see also Sect. 7.10.3). In this application, a column of fibres illuminate the entrance slit. Each fibre (or set of fibres) is coupled to a telescope observing the sky under an individual angle.
- One-dimensional imaging in order to resolve e.g. vertical trace gas profiles.
- Two-dimensional spectral imaging. Since only one dimension is covered by the slit, other scanning mechanisms (e.g. mechanical; Lohberger et al. 2004) must be employed to cover the other dimension. Examples are satellite or aircraft, the movement of which provides the second scanning dimension (e.g. Laan et al. 2000; Levelt et al. 2000).
- Two-dimensional imaging of emission plumes (e.g. volcanic or industrial), see for instance (Kern et al. 2010; Pikelnaya et al. 2013; Kuhn et al. 2019; Fuchs et al. 2021).

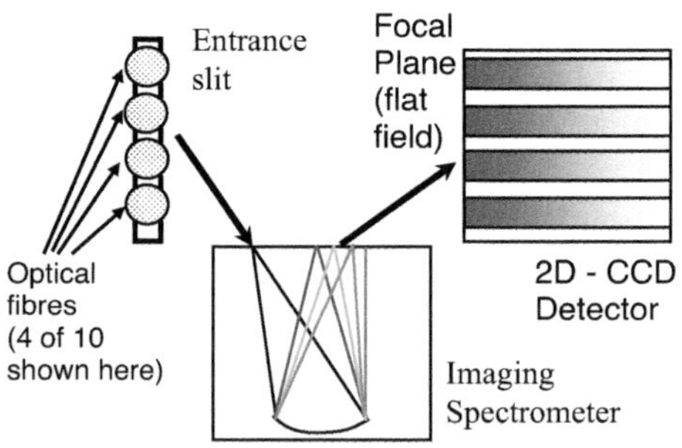

Fig. 7.32 Sketch of an imaging spectrometer. Here, a concave grating spectrometer is shown. But, in principle, any type of spectrometer can be used. In this example, $n_y = 4$ optical fibres could connect to four individual telescopes looking at different viewing angles. In reality, frequently larger numbers of fibres (e.g. 10; Heismann 1996) are present (figure courtesy of T. Wagner)

Imaging devices play an increasingly important role in atmospheric measurements, including global monitoring of trace gases by satellite instruments or regional observations of industrial, urban, or volcanic emission sources. More examples are given in Chap. 11.

7.4.3 Fabry-Pérot Interferometers

The Fabry-Pérot Interferometer (FPI) is a fundamentally very simple optical device, the principle of which has been known for more than a century (e.g. Perot and Fabry 1899; Vaughan 1989). It consists of two plane parallel surfaces, each with reflectance R, separated by a distance, d, which are illuminated by a parallel beam of light inciding under an angle α (Fig. 7.33). In practice the setup can be a single plate of transparent material where the two outer surfaces form the active FPI surfaces as shown in Fig. 7.33. Alternatively, when very small plate distances are required, as in the case of trace gas imaging FPI-instruments, two separate, parallel plates can be used, where only the inner surfaces (facing each other) are active, while the outer surfaces are made inactive by either an anti-reflective coating or by applying a small wedge angle.

In either case the medium between the plates is assumed to have an index of refraction n. Light incident on the device under an angle α is split up in partial beams, with different optical path lengths between the two reflecting surfaces. Due to interference of the transmitted partial beams, the spectral transmission of the FPI is characterised by periodic transmission peaks at wavelengths where constructive interference between the partial beams occurs. For sufficiently high orders of interference, the free spectral range (FSR) Δ_λ between two transmission peaks in units of wavelength λ is approximately given by:

$$\Delta_\lambda(\lambda) = \frac{\lambda^2}{2 \cdot n \cdot d \cdot cos(\alpha)} \tag{7.30}$$

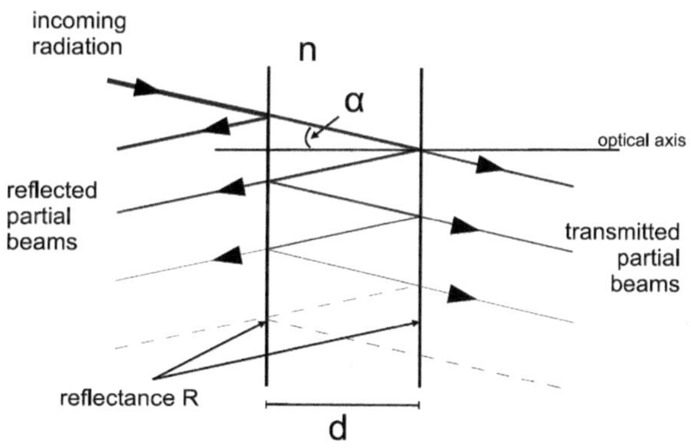

Fig. 7.33 Sketch of an Fabry–Pérot interferometer: Incoming radiation (from upper left) undergoes multiple reflections between two plane-parallel surfaces of reflectance R, mounted at distance d. Interference of the incoming with the twice reflected partial beams leads to a periodic transmission structure (see Eq. 7.34)

The finesse F of a FPI is defined as the ratio of FSR to the full width half maximum (FWHM) δ_λ of a transmission peak:

$$F = \frac{\Delta_\lambda(\lambda)}{\delta_\lambda(\lambda)} \tag{7.31}$$

F is basically a measure of the number of effectively interfering partial beams. Thus F increases with the reflectance R of the FPI surfaces. In an ideal case F is only a function of R and is (for R > 0.5) given by:

$$F = \frac{\pi}{2\arcsin(\frac{1}{\sqrt{C_F}})} \approx \frac{\pi\sqrt{R}}{1-R} \tag{7.32}$$

with the fines coefficient C_F:

$$C_F = \frac{4R}{(1-R)^2} \approx \frac{4F^2}{\pi^2} \tag{7.33}$$

In practice, F also depends on the alignment and quality (in particular flatness and parallelity) of the surfaces. Superposition of all transmitted partial beams with their respective phase shifts (neglecting absorption effects) yield the spectral transmission of the FPI as a function of λ and the FPI's instrument parameters d, n, α, and R (Perot and Fabry 1899):

$$T_{FPI}(\lambda, d, n, \alpha, R) = \frac{1}{1 + C_F \cdot sin^2(\frac{2nd\pi}{\lambda}\cos(\alpha))} \tag{7.34}$$

The etendue of an FPI is given by the area of its active surface (typically circular with an area around 100 mm^2) times its acceptance angle, which depends on the desired resolution and can vary between a few degrees and a fraction of a degree. FPIs can be used for DOAS applications in several ways:

1. FPIs can be used in lieu of customary medium spectral resolution (fraction of a nm) grating spectrometers or Michelson interferometers.
2. While replacement of medium-resolution grating spectrometers by FPIs is possible, more interesting is the capability of FPIs to realise high spectral resolution (around 1 pm) in a very compact setup (see e.g. Kuhn et al. 2021).
3. An entirely different application is the use of the periodic nature of T_{FPI} to identify atmospheric trace gases (see e.g. Kuhn et al. 2014).

The first option has, to our knowledge, not been used for DOAS applications.

In contrast, the second option, a DOAS-device reaching high spectral resolution simultaneously with superior light throughput compared to grating spectrographs, has been realised recently by Kuhn et al. (2021) and Kuhn et al. (2023). Applications are described in Sect. 7.8.2. High spectral resolution (e.g. $\delta\lambda \approx 2pm$ or $\lambda/\delta\lambda \approx 10^5$)

is obtained by FPIs with high finesse (around 100 or larger), resulting in a rather small free spectral range Δ_λ (around 0.2 nm at 300 nm in this example). Therefore the FPI has to be combined with an order sorting device, which can be a narrow-band filter or a suitable medium spectral resolution (grating) spectrometer. In the latter case, large spectral ranges can be realised simultaneously with high resolution. A sketch of a high resolution FPI instrument (Kuhn et al. 2023) employing a grating spectrometer for order sorting is shown in Fig. 7.34 A and B. Figure 7.34 D illustrates the combined dispersion of order sorting spectrometer and FPI. The (high) FPI dispersion leads to separation of wavelengths within a particular order in a radial direction (indicated by concentric circles) while the (low) dispersion of the order sorting spectrometer separates the orders in left-to-right direction. A two-dimensional detector records the resulting pattern, which can be converted to an 'ordinary' spectrum with high resolution $\delta\lambda$ (2 pm in this example) covering many free spectral ranges Δ_λ (i.e. a multiple of 0.2 nm).

The third type of FPI application, identification of trace gases by FPI-filters with their transmission structure matched to the periodicity of the gas under investigation, has been known for decades (see e.g. Barrett and Myers 1971 or Vargas-Rodríguez and Rutt 2009) but was applied mostly in the IR spectral range. The use of FPIs for imaging volcanic gas emissions has been reported only recently (see Kuhn et al. 2014 and references therein). This application is illustrated in Fig. 7.35: One key element is a FPI where the design parameters ($d \approx 22\mu\text{m}$, $R \approx 0.65$) are chosen such that the transmission maxima under a certain incidence angle α_A coincide with the absorption maxima of the trace gas (in this case SO_2), while the finesse (in this case $F \approx 7$) approximately matches the width of the absorption maxima (see Fig. 7.35b). Changing the incidence angle to α_B will shift the transmission maxima to the minima of the trace gas absorption, as shown in Fig. 7.35c. Changing the FPI angles (or alternatively d or n) by a suitable amount (typically a few degrees) will lead to minimal (A) or maximal (B) transmission of the FPI in the presence of the trace gas. In the absence of the trace gas, the intensity changes will be very small since the wavelength shift is small (half the wavelength difference between two consecutive absorption maxima). The effect of using an FPI in this way is equivalent to determining the correlation between the wavelength-dependent absorption cross-section $\sigma(\lambda)$ and the FPI transmission function $T_{FPI}(\lambda)$. The correlation is determined for each point within the FPI aperture. This allows imaging of trace gases. Devices and measurements are described by a number of authors, for instance with the demonstration of SO_2 measurements (Kuhn et al. 2014; Fuchs et al. 2021), calculations for several other species including O_3, HCHO, IO or OClO (Kuhn et al. 2019), or measurements of SO_2 and BrO (Nies et al. 2023), see also Section 11.4.5.

7.4.4 Michelson Interferometers (FT Spectrometry)

Interferometers also make use of the wave properties of electromagnetic radiation in order to effectively produce (and record) the Fourier Transform (FT) of the spectral

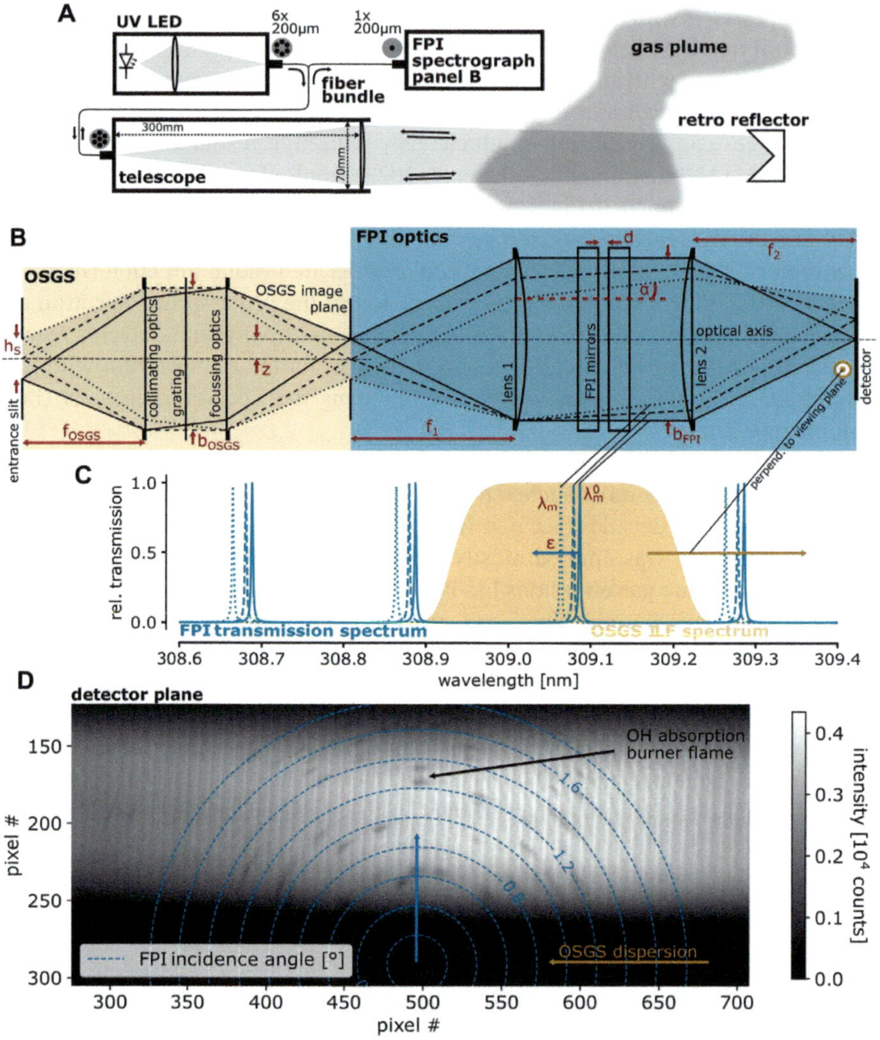

Fig. 7.34 (**A**) Open-path set-up as used for the measurement at Nyiragongo: The light of a UV LED is coupled into 6 "transmitting" optical fibres, collimated by the telescope, sent across the volcanic plume, reflected back by a retro reflector, coupled into a single 'receiving' fibre by the same telescope, finally reaching the FPI spectrograph. (**B**) Optical schematic of the FPI spectrograph with grating order sorting mechanism, OSM: The entrance slit of the order sorting grating spectrometer, OSGS is imaged onto the FPI spectrographs detector. The superposition of linear grating dispersion and radially symmetric FPI transmission is indicated in (**C**), illustrating how the FPI's (blue) and the GS's (brown) spectral transmission change across the detector. (**D**) Spectrum of a UV LED after traversing a burner flame. Absorption lines of OH are visible, aligned along the indicated lines of equal FPI incidence angle. Reprinted under CC BY 4.0 license from Kuhn et al. (2023). ©The Author(s)

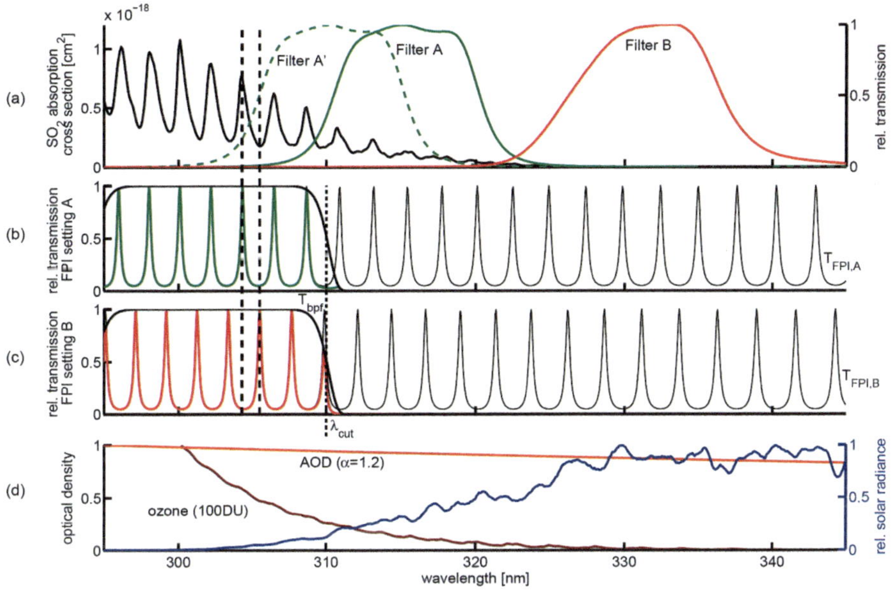

Fig. 7.35 Examples of trace gas (in this case SO_2) imaging by filter camera and FPI camera. Panel (**a**): SO_2 absorption cross-section (black drawn line, left ordinate axis, (data from Bogumil et al. 2003) as a function of wavelength, and transmission curves of filters A, A' and B (right ordinate axis). Panels (**b**) and (**c**): FPI transmission curves for FPI settings A and B; the dashed vertical lines mark the positions at which FPI transmission maxima coincide with maximum and minimum trace gas absorptions, respectively. Panel (**d**): Incident solar radiance (blue drawn line, right ordinate axis in relative units), optical density of ozone absorption (100 DU, left ordinate axis), and aerosol extinction (for AOD = 1 at 295 nm, Ångström exponent α = 1.2, left ordinate axis). Reprinted with permission from Kuhn et al. (2014), ©CC 3.0. All rights reserved

intensity distribution in the received radiation. Therefore, interferometers are frequently called FT spectrometers. Although, in principle, spatial as well as temporal interferograms can be produced, in practice essentially all instruments are based on the latter principle. The most popular temporal interferometer design is the Michelson interferometer (Michelson 1882) depicted in Fig. 7.36. However, there are also other designs (see e.g. Fischer et al. 2008). A Michelson interferometer consists of two mirrors, one fixed (top) and one moving (left), and a beam splitter plate. The collimated light beam to be analyzed enters from the right, is split into two partial beams at the beam splitter, and recombined there after being reflected at the respective mirror. The interferogram is then converted to an electric signal by the detector, digitised and Fourier transformed to the desired spectrum.

FT spectrometers have been in use for IR absorption spectroscopy for a long time (FT-IR) (e.g. Tuazon et al. 1980; Galle et al. 1994; Notholt et al. 1997; Hanst and Hanst 1994). Technological advances made their application in the UV/visible spectral range (now called FT-UV), and thus for DOAS, possible (Colin et al. 1991; Vandaele et al. 1992).

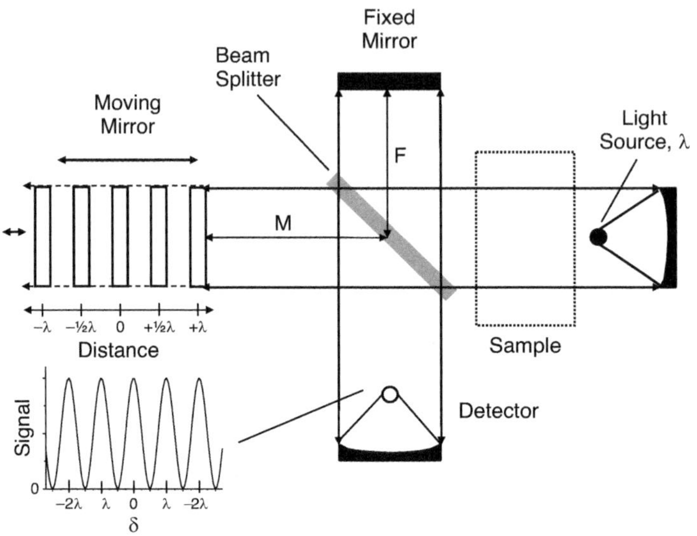

Fig. 7.36 Optical outline of a Michelson-type FT spectrometer

7.4.5 Characteristics of Spectrographs

As shown earlier, there are a large variety of spectrograph designs with widely varying characteristics, such as:

Spectral dispersion—The spatial separation of radiation of different wavelengths in the focal plane of the instrument is referred to as spectral dispersion. This dispersion is given by $ds/d\lambda$ measured in, e.g., mm/nm, where s denotes the linear distance in the focal plane and λ the wavelength, see (7.29).

Spectral resolution—The capability of the instrument to separately measure the radiation intensity at different wavelengths is characterised by the spectral resolution, $\Delta\lambda_R$. There are several definitions of the spectral resolution, usually based on the capability to separate the intensity of two monochromatic light sources (atomic emission lines) simultaneously observed by the instrument. A more general and meaningful term is the instrument or slit function, $H(\lambda–\lambda_0)$ at the wavelength λ_0, which is defined as the instrument's response to a delta function $\delta(\lambda–\lambda_0)$ as input intensity distribution. The spectral resolution $\Delta\lambda_R$ at the wavelength λ_0 can be seen as the full width at half maximum (FWHM) of $H(\lambda–\lambda_0)$.

Flatness—Flatness of the focal plane determines how the spectral characteristics change in the focal plane. This determines how well a 1D or 2D detector of a given size can be accommodated. Typically the focal plane is reasonably flat in the dispersion direction (i.e. over ~25 mm or more). Imaging spectrographs are necessary to achieve a flat field perpendicular to the dispersion direction.

Thermal Stability—Experience and theoretical calculations show that even slight variations in the position of a recorded atmospheric spectrum with respect to a reference spectrum can lead to relatively large errors in the spectral fit (see Chap. 8). Changes in dispersion have the same effect, since they lead to spectral shift of one portion of the spectrum compared to another. In practice, shifts or dispersion changes as small as 0.01 spectral pixels can lead to noticeable degradation of detection limits and accuracy of DOAS instruments. Although shift and change in dispersion can be compensated for by evaluation software (see Chap. 8), occurrence of these effects will always degrade the result. It is, therefore, of great advantage to minimise spectral shift and change in dispersion. In practice, these effects are largely caused by a change in the temperature of the spectrometer. Thus, either the susceptibility of spectrographs to temperature changes should be minimised, e.g. by use of material with low thermal expansion, and/or the temperature changes must be kept at a minimum. Air density changes due to changes in temperature also lead to a small drift.

Although efforts are being made to design spectrographs with low thermal effects (e.g. made from Zerodur), the most widely used approach is to actively stabilise the temperature of the entire spectrograph subsystem. This is usually accomplished by insulating the instrument and heating it to a temperature above the highest expected ambient temperature. Typical temperature stability reached in field experiments are on the order of ± 0.1–1 K. Typical wavelength shifts are 0.05 nm/degree. Thermally stabilizing the detector is usually accomplished as a by-product of the necessary cooling. In some designs, both functions are combined by cooling and stabilizing the entire spectrograph–detector assembly at a low temperature (Hönninger 2002; Bobrowski et al. 2003).

Pressure (Air Density) Effects—Changes in ambient atmospheric pressure and temperature will usually also lead to a wavelength shift since the index of refraction of air $n_{air}(p, T)$, the speed of light, and thus the wavelength of the radiation changes, see (7.10).

$$\lambda = \frac{c}{\nu} = \frac{c_0}{\nu} \cdot \frac{1}{n_{air}} \tag{7.35}$$

In a grating spectrograph, this will alter the dispersion of the instrument. In a Fabry–Pérot interferometer essentially the spacing of the transmission maxima is changed.

Stray Light—Stray light requires careful consideration when selecting or designing a spectrograph. Sources of stray light include: Light scattered from optical elements (grating and mirrors), reflection of unused diffraction orders off the spectrograph walls, reflection of unused portions of the spectrum from walls near the focal plane, and reflections from the detector surface (e.g. Pierson and Goldstein 1989). Stray light can be reduced by coating the inner walls of the spectrograph with superblack material like "Vantablack" or "Dark Chamaleon Dimers".

A particular form of stray light is due to 'grating ghosts', i.e. portions of radiation diffracted in unwanted directions (see Sect. 7.4.1). An avoidable source of stray light is due to incorrect illumination of the spectrometer. If the F-number of the illumination exceeds that of the spectrometer, radiation will overfill the collimating mirror and hit interior walls of the spectrograph, where it can be reflected to the detector.

Stray light tends to be comparatively high in spectrographs filtering a relatively broad wavelength interval from a continuous spectrum, as in DOAS applications. To illustrate this point, we consider a typical stray light level of 10^{-5} for a Czerny–Turner spectrometer (Pierson and Goldstein 1989). This number gives the fractional light intensity found anywhere in the spectrum when a single (laser) line of very narrow spectral width enters the spectrometer. Continuous light entering the spectrometer can be thought of as being composed of a series of lines spaced at centre-to-centre distances equal to their width (i.e. the spectral resolution of the instrument). Since the total spectral range of light entering the spectrometer (for instance from 300 to 600 nm) is on the order of 1000 times larger than the spectral resolution (typically 0.3 nm for a low resolution DOAS instrument), the level of stray light (per wavelength interval) is roughly three orders of magnitude larger than would be expected from the single line definition usually considered. Thus, in DOAS applications, stray light levels I_{SL}/I can be expected to be closer to 10^{-2} than to 10^{-5}. Stray light reduces the recorded optical density D, as shown in Sect. 8.8.2.

Reentrant Light: In many spectrometers where a linear or matrix solid-state detector array is used, there might be reflection of radiation from the detector surface back to the grating. The solution is to either tilt the array up to the point that the resolution begins to degrade or, if the system is being designed for the first time, to work out of plane.

In some Czerny–Turner spectrometers, a diffracted wavelength other than that at which the instrument is set may hit the collimating mirror (see Fig. 7.31) and be reflected back to the grating, where it may be rediffracted and directed to the detector. If this problem is serious, a good solution is to place a mask perpendicular to the grooves across the centre of the grating. The mask should be somewhat higher than the entrance slit and detector array. If the wavelength is known, it is possible to calculate the point on the grating where the radiation of the reflected wavelength hits. In this case, the only masking necessary is at that point.

Light Throughput of Spectrometers—An important characteristic of any spectrometer is its light throughput, L_t. It is defined as the product of acceptance solid angle Ω, effective entrance aperture A (i.e. the area of the entrance slit), and the light transmission T of the instrument:

$$L_t = \Omega \cdot A \cdot T = E \cdot T \; (E = \text{Entendue, see Sect. 7.3.5})$$

In principle, L_t is affected by many parameters including:

- The aperture ratio ($f/\#$) of the instrument
- The entrance slit dimensions
- The light losses inside the instrument.

The light throughput is approximately inversely proportional to the square of the spectral resolution, which scales with the width, and thus the areas A of both the entrance slit and the detector pixels. Thus, $L_t \propto 1/\Delta\lambda_R$.

7.5 Detectors for UV/Visual Spectrometers

The basic function of detectors for UV/vis spectrometers is to convert spatial radiation intensity differences $I(\lambda)$ into a signal $I_D(\lambda)$ that can be transferred to a computer for further evaluation.

$$I_D(\lambda) = \alpha \cdot I(\lambda) \tag{7.36}$$

where α denotes the detector sensitivity. Detectors must rapidly capture the entire spectrum in order to minimise additional noise due to atmospheric turbulence. Good quantum efficiency is also desirable. However, the most significant requirement is the capability to reliably detect relative intensity differences within the spectrum on the order of 10^{-4}. This latter requirement implies low noise and excellent linearity, as well as sufficient sampling of the information. Presently, several different detector designs are in use for DOAS spectrometers. Optomechanical scanning devices, OSDs, which record the spectra sequentially, make use of only a small fraction of the light, and thus can be regarded as outdated for scientific applications. However OSDs are still found in several commercial DOAS instrument designs. Recent scientific instruments employ solid-state detector arrays known as photodiode array (PDA) detectors, CCD array detectors, and complementary metal oxide semiconductor (CMOS) array detectors.

7.5.1 Geometrical Focal Plane Sampling Requirements

The requirements for DOAS sensors can be described by a number of properties:

Geometry—For spectroscopic purposes, pixels with an aspect ratio that is narrow in the direction of the scan (i.e. dispersion direction of the spectrometer) and wide in the direction perpendicular to the scan are required. Typically, a pixel width of a fraction of the width of the instrument function is desirable, to ensure that the instrument function is adequately sampled. The spatial instrument function is the resulting spatial intensity distribution in the focal plane of a spectrometer as a response to a delta peak of monochromatic light at the entrance of the spectrometer. Taking an entrance slit width of 50 μm corresponding to, for example, a width of the spatial instrument function of 75 μm, a detector pixel width on the order of $w_D = 15\,\mu m$ would be required.

On the other hand, the 'height' (extension perpendicular to scan direction or dispersion direction) of each pixel should be at least as large as the height h_S of the illuminated area in the focal plane. For a typical case of a one-to-one imaging spectrometer, h_S is at least as large as the illuminated height of the entrance slit. Active DOAS systems often employ an optical fibre or fibre bundle with heights at the entrance slit of h_s between 100 μm to 500 μm. The detector height, h_D thus has to be in the same range (see Sect. 7.3.7). For passive systems observing the scattered sky light, the imaged object (i.e. the sky) is very large and the light throughput is limited by the entrance area of the spectrometer and its aperture ratio, which we shall not consider here. Since the slit width is determined by the required spectral

Table 7.6 The characteristics of PDAs versus CCD and CMOS sensors

Type	PDA	CCD	CMOS
Characteristic Geometry	1-D only	Available in 2-D (usually) and 1-D	Available in 2-D (usually) and 1-D
Readout Noise (electrons)	High readout noise (several 10^3 electrons)	Low readout noise ($\approx 1 - 20$ electrons)	Low readout noise (≈ 10 electrons)
Typical pixel capacity (electrons)	10^8	10^5	10^5
Typical noise due to photoelectron statistics	10^4	300	300
Relative S/N	10^{-4}	3×10^{-3}	3×10^{-3}
Dominant noise source	Photoelectron statistics, if full well capacity can be used	Usually photoelectron statistics	Usually photoelectron statistics
Dark current	100–1000 s to saturate pixel	100–1000 s to saturate pixel	100–1000 s to saturate pixel
Signal-dependent dark current	Yes	Yes	Unknown
Typical readout time	20 ms	10 ms to 1 s (2-D)	10 ms to 1 s (2-D)
Spectral range	NIR (Silicon limit) to UV (200 nm)	NIR (Silicon limit) to UV, many types	NIR (silicon limit) to UV
Quantum efficiency	Good	Excellent (back-thinned arrays)	Good
On-chip ADC available	No	No	Yes

resolution, the only free parameter is the slit height, which should be maximised. Thus, large pixel heights h_D of the order of several millimetres are of advantage.

Readout Time—In active DOAS applications with large optics (Sect. 7.8.3) a signal level in the spectrometer focal plane of 10^9 photoelectrons per second and 25 μm resolution element can be reached. For passive DOAS systems observing scattered sunlight (zenith or off-axis geometry), typical signal levels are about two orders of magnitude lower. The detector readout has to be fast enough to keep up with the signal accumulation. In fact it would be desirable for the actual readout time to be short compared to the exposure time required to fill the detector pixels near full-well capacity. The full-well capacitance of a typical (one-dimensional) CCD detector is of the order of 10^5 electrons (see Table 7.6), this means that in active DOAS applications up to 10^4 complete readout processes per second may be required. For a 2000-element linear CCD sensor, this figure would correspond to about 20 MHz pixel frequency. This might be a factor of 4 better for the CCD's listed in Table 7.6, because their pixel width, and thus signal intensity, is a factor of 2 lower than assumed (13 or 14 μm instead of 25 μm as assumed earlier), but still some CCDs fall short by a factor of 5 from the above requirement. For 2D-CCD arrays, which have comparable full well capacities, the readout time can be a limitation.

Photodiode arrays have a much larger full well capacity of $6 \times 10^7 - 1.2 \times 10^8$ electrons per pixel and the read out times can be considerably lower. However, due to their electronic layout, the readout frequency is typically limited to ~100 kHz.

Co-adding Readouts and Exposure Time—In order to achieve the desired signal-to-noise ratios (e.g. $SNR = 10^4$), a sufficiently large number N of photoelectrons needs to be recorded so that $N > SNR^2$ and $N > B (B =$ combined effects of all fixed noise sources, e.g. detector readout noise and preamplifier noise). Frequently, the desired number of photoelectrons cannot be accumulated in a single scan, and thus several scans have to be co-added (see Sects. 7.5.2 and 7.5.3). For a good N/B ratio of a single scan, a signal level not too much below the maximum detector signal is desirable. On the other hand, overexposure, e.g. a signal larger than the maximum allowable detector signal (e.g. due to pixel saturation or ADC overload), must be avoided. Various schemes for exposure control have been developed, mostly based on predicting the future signal on the basis of the already recorded signal. The magnitude of the signal can be controlled through the high voltage in the photo-multipliers, PMT, or exposure time for solid-state detectors. If radiation signals are fluctuating in intensity, as can occur in many active and passive DOAS applications, overexposure cannot always be avoided. In this case, overexposed spectra must be discarded. Note that each individual spectrum must be checked for overexposure, since it may not be detectable after a few overexposed spectra are buried under a large number of properly exposed co-added spectra.

Spectral Sensitivity—A detector must be sufficiently sensitive in the spectral range of interest. For most classical DOAS applications, a spectral range from about 200 to 800 nm is required (see Chap. 6 and Appendix B). Recent applications, however, extend this range into the near-infrared wavelength region. Silicon allows measurements from below 200 nm up to about 1100 nm, where the upper wavelength end is set by the band gap of silicon. For measurements at longer wavelengths, detectors based on other semiconductor materials are used. A popular material for wavelengths, up to 1600 nm or even 2400 nm is InGaAs. Single elements and linear arrays based on this material are available.

For PMTs, the long wavelength end of the sensitivity range is given by the requirement that the photon energy must exceed the work function, i.e. work to release an electron from the cathode material. Modern cathode materials (e.g. multi-alkali) give sensitivities to about 600 nm. In contrast, the short wavelength end of the sensitivity range is set by the window material. Glass windows will give about 400 nm, quartz windows <200 nm range.

The detector spectral sensitivity is frequently expressed in terms of the quantum efficiency Q:

$$Q = \frac{\text{Number of photoelectrons}}{\text{Number of photons}} \tag{7.37}$$

Spectral Sampling—In order to obtain a true digital representation of the spectrum projected into the focal plane of the spectrometer, sufficiently fine spectral sampling is required. Obviously, the spatial frequency (e.g. density of detector pixels) should match the spectral resolution of the dispersing instrument. In other words, there must be a sufficient density of pixels compared to the spectral resolution. A rule of thumb is that the spectral width of a detector pixel should be about 10–20% of the FWHM of the instrument function (see Chap. 8 and Hofmann et al. 1995; Stutz 1996; Roscoe et al. 1996; Chance et al. 2005).

Aliasing problems—Aliasing problems are well known from many areas of electronic signal evaluation. Aliasing will arise in cases where the spectral resolution is too high for a given number of detector pixels in the focal plane of the spectrometer, or if there are too few pixels in the focal plane sensor in comparison with the spectral resolution of the spectrometer. Figure 7.37 illustrates this problem for the case of discrete absorption lines. In this example, the position of an emission line (smooth Gaussian in (a) is shifted by the half-width of a pixel (b). As shown, there can be a drastic change in line shape of the digitised representation, shown as a histogram-like curve in panels (a) and (b), which actually describe the same line shape.

It is tempting to make 'best possible use' of a given number of pixels and thus to specify only the minimal twofold oversampling (Nyquist criterion). However, practical experience and numerical modelling show there must be a sufficiently large number of signal samples $I(i)$ (or detector pixels) across one 'resolution interval'. Sample calculations demonstrate that on the order of 5–10 samples per resolution interval lead to acceptable errors (Hofmann et al. 1995; Stutz 1996; Roscoe et al. 1996; Platt et al. 1997). In principle, advanced interpolation techniques and other procedures can be used to correct for this problem (e.g. DOAS evaluation of GOME spectra, Chance et al. 2005).

Detector Linearity—Detector linearity is a prerequisite for good sensitivity and measurement precision of DOAS systems. It can be described as:

$$I_D(i) = \alpha \cdot I(i) \quad or \quad \frac{dI_D(i)}{dI(i)} = \alpha = \text{constant} \tag{7.38}$$

where $I_D(i)$ denotes the signal generated by the ith detector pixel, while $I(i)$ is the radiation intensity reaching this pixel (see Chap. 8). The effect of non-linearity, i.e. the sensitivity α varying as a function of the radiation intensity I, is illustrated by the following consideration. In the limit of small optical densities, the (differential) optical density D of a spectral structure is given by:

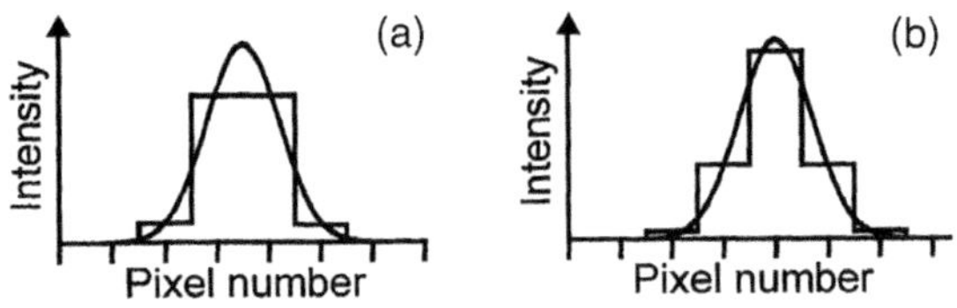

Fig. 7.37 Aliasing problems occur if there are too few pixels in the focal plane sensor in comparison with the spectral resolution of the spectrometer. If in this example, the position of an emission line **a** is shifted by a half width of a pixel **b**, there can be a drastic change in line shape of the digitised representation. The histogram-like curve in panels (**a**) and (**b**) actually describe the same line shape

$$D = \ln\frac{I_0}{I} \approx \frac{I_0}{I} - 1 = \frac{I_0 - 1}{I} = \frac{\Delta I}{I} \tag{7.39}$$

In a practical measurement, however, I can only be estimated from the detector signal I_D:

$$D_D \approx \frac{\Delta I_D}{I_D}$$

Assuming for the sake of simplicity, two different sensitivities α_1 and α_2 for low and high signals, respectively (see Fig. 7.38), we obtain:

$$D_D \approx \frac{\Delta S_D}{S_D} = \frac{\alpha_2 \Delta I}{\alpha_1 I} = \frac{\alpha_2}{\alpha_1} \cdot D \tag{7.40}$$

Thus, the measured optical density (and consequently trace gas level) would deviate from the true value by the factor α_2/α_1. This could be interpreted as the following requirement: If a 1% precision of the trace gas level is desired, the sensitivity $\alpha(I)$ cannot deviate more than 1% from its average value at any I. While this may be sufficient if there is only one dominating absorber present, there might be problems if weak spectral structures are to be detected in the presence of strong ones. For instance, in passive DOAS systems most absorbers are weak compared to the Fraunhofer lines. Here, distortion of the strong spectral structure due to non-linearity of the detector can introduce large errors in the determination of weak spectral absorption structures. Consequently, a much better detector linearity than estimated above might be required. A more detailed analysis and mathematical treatment of the effects of detector non-linearity can be found at Horbanski (2015) and Chap. 10.

7.5.2 Optomechanical Scanning Devices and Photomultiplier Tube

Optomechanical scanning devices (OSD) scan a section of the dispersed spectrum, typically several nanometres to 100 nm, in the focal plane of the spectrograph by a moving exit slit, as shown in Fig. 7.39. In the most common design, radial slits are etched in a thin metal disk ('slotted disk') rotating in the focal plane (see Fig. 7.40). At a given time, one particular slit serves as an exit slit. The start of a scan is

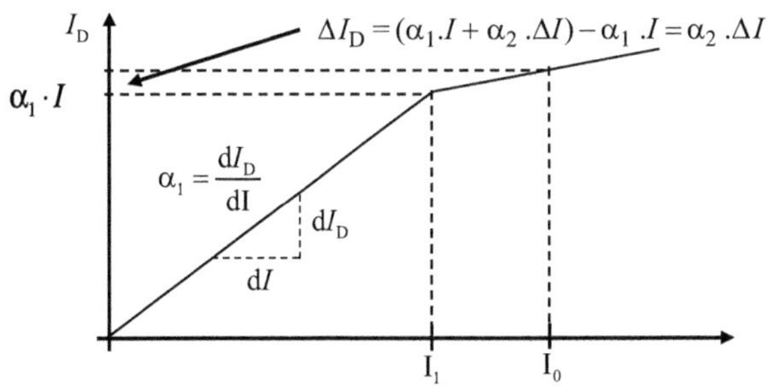

Fig. 7.38 Insufficient linearity of a detector can cause incorrect determinations of the optical density

determined by an infrared light barrier. The light passing through this slit is received by a photomultiplier tube (PMT), the output signal of which is digitised by a high-speed ADC and recorded by a computer. The signal is adapted to the ADC range by variable preamplifier gain and/or changing the PMT high voltage. During one scan (i.e., one sweep of an exit slit over the spectral interval of interest), several hundred digitised signal samples are taken. Consecutive scans are performed at a rate of typically 100 scans per second, and are digitised into 500–1000 'channels', which are signal averaged by the software (thus individual samples of the signal are recorded at 50–100 kHz frequency), thereby meeting the requirement of detection of very small signal differences.

Alternatively, vibrating exit slits can be used. However, disadvantages include less uniform scanning speed and possible influences of the vibration on other parts of the optics.

Since a single scan takes ≈10 ms, the effect of atmospheric scintillations is negligible, because the frequency spectrum of atmospheric turbulence close to the ground peaks around 0.1–1 Hz and contains very little energy at frequencies above 10 Hz. In addition, typical spectra obtained during several minutes of integration time represent an average over 10,000–40,000 individual scans. Thus, effects of noise and temporal signal variations are very effectively suppressed. In fact, even momentarily

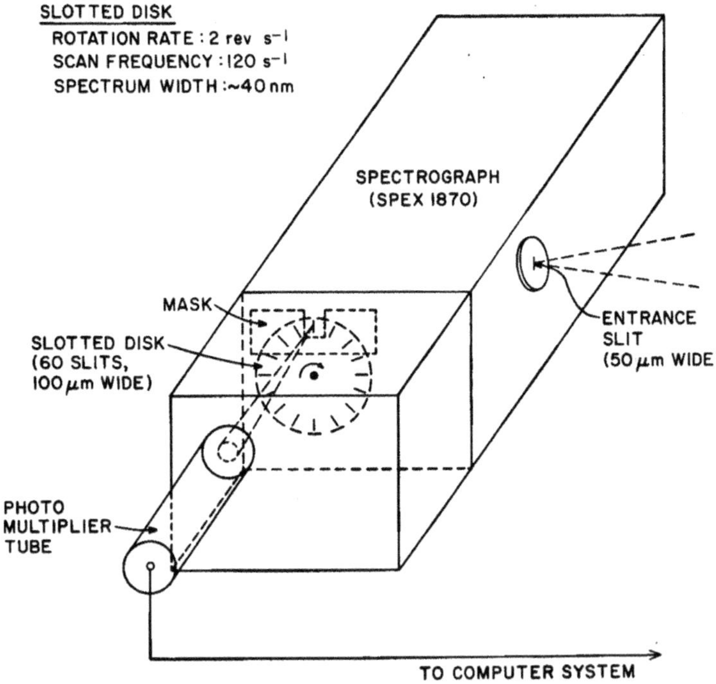

Fig. 7.39 Sketch of an OSD, the 'slotted disk' design. A series of (in this case 60) radial slits are etched in a thin metal disk (*'slotted disk'*) rotating in the focal plane

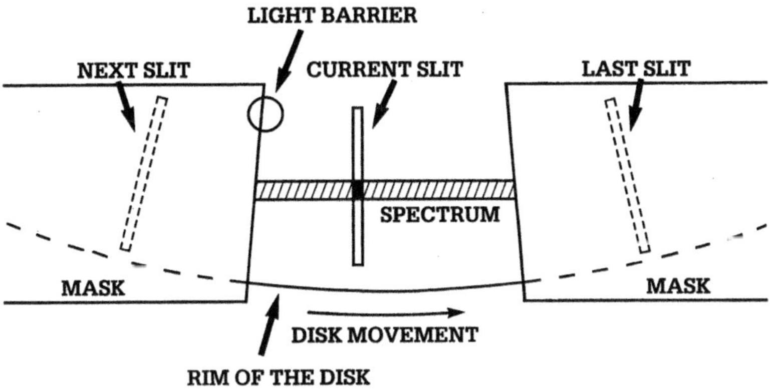

Fig. 7.40 OSD, the 'slotted disk'. A mask allows only one slit to serve as an exit slit at any given time. The start of a scan is signalled by an infrared light barrier. The light passing through this slit is received by a PMT. Reprinted with permission from Platt and Perner, 1983, ©Springer. All rights reserved

blocking the light beam entirely (e.g., due to a vehicle driving through the beam) has no noticeable effect on the spectrum.

The main disadvantage of the optomechanical scanning of absorption spectra is the large multiplex loss, since only the section of the spectrum passing through the exit slit reaches the detector. Only $\sim 1\%$ of the light is utilised, making relatively long integration times necessary.

The conversion of the radiation intensity to a detector signal is performed by the PMT. The design of the voltage divider chain supplying the voltages for the dynodes of the PMT is important. Since DOAS detectors are 'high anode current' applications (anode currents can reach $I_A = 100\,\mu A$), relatively high currents in the voltage divider (at least $10 \cdot I_A$) are required in order to not compromise detector linearity (see Sect. 7.5.1). Alternatively 'active voltage dividers' using transistors can be employed. Another important aspect is the response of the PMT electronics to transients. Commercial voltage dividers frequently employ capacitors between dynodes to improve response for light pulses; this is not needed or desirable for DOAS applications. However, PMT detectors are inherently fast response devices, so a low-pass filter is usually required before or within the preamplifier, supplying a time constant roughly equivalent to the time to record one detector pixel (typically $\tau \approx 20\,\mu s$). Finally, PMT photocathodes must be chosen according to the desired long wavelength end of the response. In general, cathodes with higher sensitivity at longer wavelength have higher dark currents. Available cathodes (e.g. multialkali, S20) limit the spectral range to the long wavelength end of the visible range.

7.5.3 Solid-State Array Detectors and Characteristics

Solid-state detector arrays consist of a large number, typically several hundred or more, of individual photodetectors (i.e. photodiodes) arranged in a linear row or in a two-dimensional, matrix-like fashion. They are, in principle, far superior to OSD in recording absorption spectra, since radiation of all wavelengths is recorded simultaneously rather than sequentially. This so-called 'multiplex advantage' amounts to about two orders of magnitude more photons recorded in typical systems. Accordingly, the measurement time can be shortened by the same factor or, alternatively the photon noise be reduced by about one order of magnitude. Additional advantages include the absence of moving parts, and nearly simultaneous sampling of all wavelength intervals. In addition, no high voltage is required and the quantum efficiency (at least in the red and near-infrared spectral regions) is better than that of most PMT cathodes (see e.g. Yates and Kuwana 1976).

There are three basic designs of solid-state multi-element photodetectors, which are sketched in Fig. 7.41: PDA, CCD, and CMOS devices (see also Table 7.6). Either design principle allows construction of linear or two-dimensional arrangements of hundreds to many thousands of individual photosensors, which can be read out sequentially.

In the case of **PDAs**, individual photodiodes are arranged on a silicon chip, and are read out by sequentially connecting them, via a field effect transistor, to a common readout line that conducts the accumulated charge to a readout amplifier. There may be more than one readout line (e.g. one for all even numbered, one for all odd-numbered diodes), and thus the readout time is reduced accordingly. PDAs are usually only available as one-dimensional (linear) arrays.

In contrast, **CCDs** have no readout line. Rather, the charge of each sensor is sequentially transferred through all pixels until it reaches a readout amplifier connected to the very last pixel in the row. Advantages of CCDs over PDAs include a simpler design, since no readout transistors and associated driving circuits are required, combined with much smaller output capacitance. The output capacitance is equivalent to one pixel, while the capacitance of a typical PDA readout line is equivalent to dozens or hundreds of pixels. As a consequence, the readout noise of a typical CCD is about two orders of magnitude lower than that of a PDA (see Table 7.6). On the other hand, the charge transfer efficiency of CCDs is limited to values of typically 99.999%, leading to some cross-talk between pixels, which is absent in PDAs. CCDs are usually produced as linear (1D) or area (2D) sensors. In the latter case, they provide additional information, which can be utilised e.g. to simultaneously record several spectra. However, the capacitance of the individual elements is much larger in PDAs, as outlined in Table 7.6. Therefore, the total noise can be limited by photoelectron shot noise, despite the higher readout noise of PDAs.

CMOS sensors are similar to PDAs, in that they are read out by sequentially connecting the individual pixels (via a field-effect transistors) one by one to a common readout line, which conducts the accumulated signal to a readout terminal. In contrast to PDAs, amplifiers (and sometimes reset gates) are frequently integrated directly at each pixel. Another advantage is that the CMOS process allows integration of other devices such as amplifiers and ADC directly on the sensor chip. There is also typically random access to each pixel. If amplifiers are attached to each pixel, non-destructive readout is possible, allowing simple 'exposure control'.

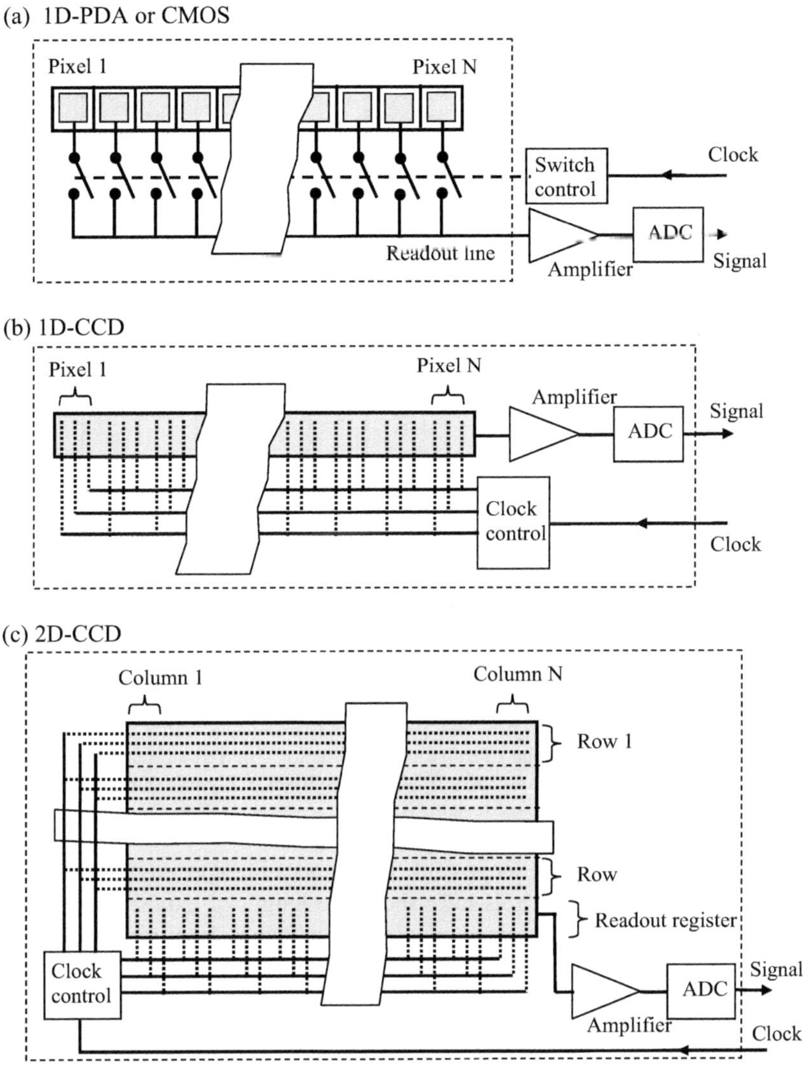

Fig. 7.41 Comparison of linear and two-dimensional solid-state photodetector array design principles. Shown are **a** layout of PDA or CMOS, **b** One-dimensional CCD, **c** Two-dimensional CCD devices

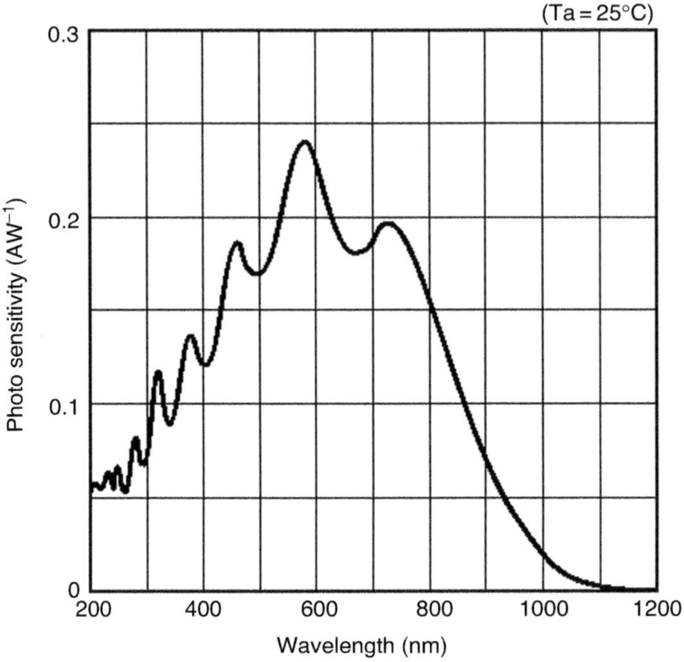

Fig. 7.42 Typical spectral response (in ampere photocurrent per watt of incident radiation) of a PDA detector (Hamamatsu Model S3904-1024). Note strong modulation of response due to the Fabry–Pérot etalon effect of the overcoat of the detector chip (from Hamamatsu data sheet NMOS Linear Image Sensors S3900, 3901-1024Q, 3904-2048Q). Figure courtesy of Hamamatsu Photonics, ©Hamamatsu Photonics. All rights reserved

A more subtle point concerns possible differences in exposure times for the different pixels. In PDAs, all pixels have the same length of exposure, but the time intervals do not quite overlap since exposure takes place from readout to readout. Readout is thus earlier for the first diode than the last, and thus the exposure time interval for the last diode is shifted in time (later) by the readout time. This may be of consequence at relatively short exposure times (compared to the readout time) and varying light levels. On the other hand, linear CCD detectors usually have a readout gate that transfers the charge of all pixels simultaneously into the readout CCD register, which is not exposed to light. Therefore, exposure time for all pixels covers exactly the same interval. In particular, in 2D sensors exposure to radiation during readout can be a problem. These problems can be overcome by using a shutter to darken the detector during readout.

Although there are clear advantages of solid-state array detectors, these devices pose some problems when used to record weak absorption features in DOAS instruments. First, the signal dependence of the dark current of the photodiodes has to be considered. Another problem is the presence of Fabry–Pérot etalon structures in the recorded spectra, which are caused by the protective overcoat of many commercially available detector arrays and also by possible vapour deposits on the array (Mount et al. 1992; Stutz and Platt 1992, 1996), see Fig. 7.42 for an example. Particular problems are due to changes in the thickness of the layer of vapour deposit causing continuous changes in the 'etalon' modulation of the array sensitivity. In any case, diode sensitivity (and dark current) varies $>\pm 1\%$ between individual diodes. This usually constant fixed pattern must be compensated for.

Solid-state array detectors have clear advantages, in particular in applications where only low-light levels are available (as in ZSL observation of stratospheric species). For the observation of low optical densities ($D\approx 10^{-3}$–10^{-4}) OSDs appear to give comparable results, but at inferior light utilisation and thus much longer exposure times. Solid-state array detectors can be described by a number of characteristic features:

Spectral Sensitivity (Quantum Efficiency)—The most popular semiconductor material for solid-state detectors is silicon. Its spectral response extends from <200 nm to about 1100 nm, see Fig. 7.42. The long wavelength (or lower photon energy) end of the range is set by the band gap of silicon, while there is no hard limitation at the short wavelength end. In practice, thin active Si layers are required to obtain good UV sensitivity. While UV-optimised silicon devices (in particular back-illuminated CCDs) can have >20% quantum efficiency below 300 nm, many detectors show little sensitivity below 400 nm. A simple way to obtain UV sensitivity consists of applying a UV-fluorescent coating (such as Metachrome II or Lumogene) on the surface of the device that converts the UV to visible (largely green) light at a wavelength near the maximum of the Si response. Disadvantages of this approach include the relatively low UV quantum efficiency of these coatings (around 10%), their limited lifetime, and the reduction of the sensitivity at longer wavelengths. It should be noted that, in some cases, a glass window with little transmission below ≈370 nm can be the limiting factor in the UV sensitivity.

Memory Effect—Illuminating the PDA with light of high intensity generates a signal which can sometimes also be observed (commonly at much lower intensity) in subsequently recorded spectra, although the detector is completely darkened. This effect is called memory effect. It was e.g. studied by Stutz (1996), but its origin is not quite clear. By waiting a few seconds and/or recording some spectra with the shortest possible integration time, it is usually possible to remove this effect (Alicke 1997; Geyer 2000).

Dark Current—As the name indicates, the 'dark current' of an optical detector is the current (or more generally speaking its output signal) which is still present if no radiation (within the detectors spectral sensitivity range) is admitted to the device. The dark current can be minimised by detector design and, for a given detector, by reducing its temperature. The dark current decreases exponentially with temperature, and can always be reduced to negligible levels by cooling the array to low temperatures (−40° to −80°C). The remaining dark signal can, in principle, be determined by observing the detector signal without illumination and then be subtracted from the signal.

Signal-dependent Dark Current—Naive subtraction of the diode array dark signal determined by recording the signal without illumination leads to wrong results. The origin of this observation can be found by inspection of the (simplified) circuit diagram of an individual diode (Fig. 7.43). It shows that the average voltage across the p-n junction is largest (approximately equal to U_o) if the diode is not illuminated. Under illumination (i.e. during an actual measurement), the voltage across the junction necessarily decreases. This apparent paradox is due to the different average voltages across the diode junctions during exposures, when the array is darkened or irradiated. At an exposure close to saturation, the average voltage during the signal integration period will be about $U_o/2$. Thus the 'dark signal' obtained by a measurement with no light admitted to the detector will overestimate the actual dark signal by up to a factor of two (Stutz and Platt 1992). In fact, the effect can be even larger, since the diode capacitance increases with decreasing voltage. Note that this effect occurs with linear diode arrays using readout switches, as well as with CCD arrays. A detailed explanation of this effect and procedures for its elimination is given by Stutz and Platt (1992).

Fixed Pattern 'Noise': Another potential problem when using solid-state detectors is the high variability of the sensitivity from diode to diode. The diode sensitivity depends on the wavelength of the incoming light and can vary from 1% up to several percent between the diodes of one array (Stutz 1996). Usually, however, this sensitivity variation is systematic, (the popular term 'fixed pattern noise' is a misnomer), and can be readily corrected. Since DOAS is intended to observe optical densities down to well below 10^{-3}, it is necessary to accurately remove the diode sensitivity structure from a spectrum.

An easy way would be the division of each spectrum by a 'white light' lamp reference spectrum scanned just before each spectrum. However, this procedure can lead to additional lamp structures in the spectrum, in particular when Xe high-pressure lamps are used, since Xe emission lines differ in both spectra, due to differences

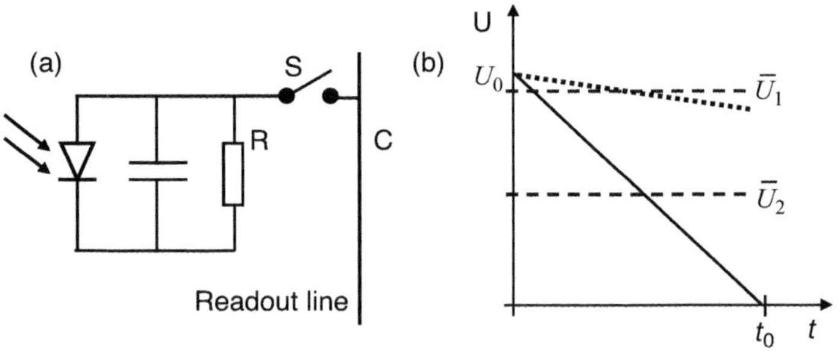

Fig. 7.43 **a** Simplified circuit representation of a single pixel in a diode array (or CCD) detector. During readout switch S is closed; (**b**) Temporal evolution of the voltage of a pixel: During acquisition of a dark spectrum the average voltage $\bar{U}_1$ across the capacitance C of the diode is close to the initial voltage U_0, and the dark current $\bar{I}_{DC} = U/R$ is relatively high. However, in an actual measurement the average voltage $\bar{U}_2$ is always lower (in fact $\bar{U}_2 \approx U_0/2$) if the signal is close to saturation. Therefore, the dark current during a typical measurement will be considerably lower than indicated by a measurement with an un-illuminated detector

of the illumination of the detector. Note that, in passive DOAS systems, the diode sensitivity structure is usually removed while eliminating the Fraunhofer structure (see Sect. 7.13.3). A common method to eliminate diode sensitivity structures in active long path DOAS spectra is the 'multi-channel scanning technique' (MCST) introduced by Knoll et al. (1990). The basic idea of the MCST is the combination of a multi-channel detection system (e.g. a PDA) with the scanning technique generally used to cover a larger spectral region with a single channel detection system. This technique is described in detail in Sect. 7.13.2.

7.5.4 PDA Detectors

Figure 7.44 shows the layout and cross-section of a typical silicon PDA detector (S3904-1024 by Hamamatsu) encompassing 1024 diodes, each measuring 0.025 by 2.5 mm, integrated on a single silicon chip. Thus, the total sensitive area measures 25 mm (in dispersion direction) by 2.5 mm. Note that there are two sets of diodes, 'active diodes' and 'dummy diodes'. The latter are read out separately. Taking the difference of 'active' and 'dummy' signals reduces the fixed pattern noise of the device. Data of some popular PDA types are summarised in Table 7.7.

7.5.5 CCD Array Detectors

Another type of solid-state photodetector is the charge coupled device (CCD) detector. Its name is derived from the readout mechanism. The charge generated by the radiation impinging on the detector is stored in potential wells generated by electrodes on the surface of the semiconductor (silicon). By applying suitable waveforms to the electrodes, the charge can be moved across the semiconductor, usually towards one edge of the device, where it is fed to an electrode connected to a readout amplifier. While PDAs are usually one-dimensional (i.e. the individual photodiodes

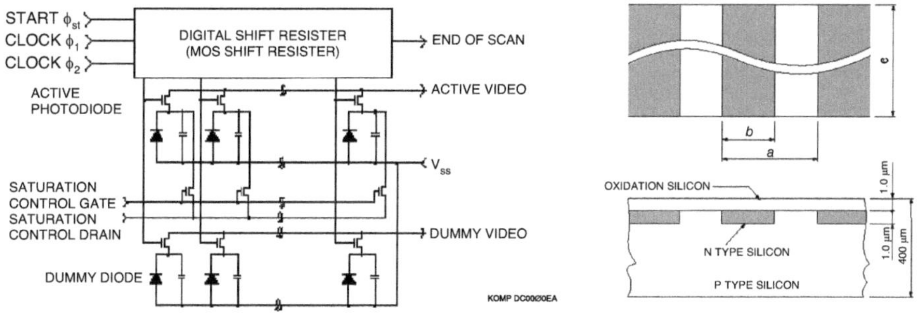

Fig. 7.44 Circuit schematic and layout of a PDA detector (Hamamatsu Model S3904-1024) (from Hamamatsu data sheet NMOS Linear Image Sensors S3900, 3901-1024Q, 3904-2048Q). Figure courtesy of Hamamatsu Photonics, ©Hamamatsu Photonics. All rights reserved

Table 7.7 Characteristics of some linear Photo-Diode Arrays (PDA) for spectroscopic applications

Type (Manufacturer)	Pixel aspect ratio ($h_D \times w_D$)μm	No. of elements N	Sensitive Area mm × mm	Saturation Charge electrons	Max. pixel rate MHz	Dark current pA
S3900-1024, NMOS (Hamamatsu)	2500 × 25	1024	25.6 × 2.5	10^8	2	0.1
S3904-2048, (Hamamatsu)	2500 × 25	2048	51.2 × 2.5	1.6×10^8	2	0.1
S3904-1024Q (Hamamatsu)	5000 × 50	1024	51.2 × 5	6.3×10^8	2	0.4

Table 7.8 Some CCD sensors for linear spectroscopic applications

Type (Manufacturer)	Pixel aspect ratio ($h_D \times w_D$) μm	No. of elements N	Sensitive area mm × mm	Saturation Charge electrons	Max. pixel rate MHz	Separate readout shift register
ILX 511 (SONY)	200 × 14	2048	28.67 × 0.2	1.6×10^5	2	Yes
TCD1205D (Toshiba)	200 × 14	2048	28.67 × 0.2	$\approx 2 \times 10^5$	2	Yes
TCD1304AP (Toshiba)	200 × 8	3648	29.18 × 0.2	?	0.6	Yes
IL-C6 (DALSA)	500 × 13	2048	26.6 × 0.5	$\approx 2 \times 10^5$	15	Yes
IT-F6 TDI[a] (DALSA)	13 × 1248(13 × 96 × 13)	2048	26.6 × 1.3	4.6×10^5	8 × 20	Yes
S7030-1007 (Hamamatsu)	24 × 24	1024 × 122	24.58 × 2.93	3×10^5		
CD30-11 (E2V)	26 × 26	1024 × 256	26.6 × 6.7	5×10^5	2	Yes
eXcelon (Teledyne/E2V)	20 × 20	1340 × 100/400	26.8 × 2.0/8.0	10×10^5	2	Yes

[a]UV sensitive to 250 nm

are arranged in a linear row), CCD's are typically two-dimensional. However, one-dimensional devices are also available (e.g. ILX-511 from SONY, see Table 7.8).

Geometry of Linear CCD Detectors: In principle, linear CCDs are in widespread use in the industry, for instance in fax machines, computer scanners, bar code scanners, copiers, etc. However, in most of these applications two-dimensional originals are scanned, so the CCD line sensor pixels must be, at least approximately, quadratic. Typical pixel dimensions (h_D by w_D) are, therefore, 15 μm by 15 μm or 7 μm by 7 μm (see Table 7.8). For special purposes, some linear CCDs with aspect ratios (h_D by w_D) of about 10 are available (see Table 7.8).

Table 7.9 Some CMOS sensors for spectroscopic applications

Type (Manufacturer)	Pixel aspect ($h_D \times w_D$) μm	No. of elements N × M	Sensitive area mm × mm	Saturation charge electrons	Dark current electrons/s	Max. pixel rate MHz	Readout noise electrons
Star250[a] (fill factory[b])	25 × 25	512 × 512	12.8 × 12.8		4750	8	74
Star1000[a] (fill factory[b])	15 × 15	1024 × 1024	15.4 × 15.4	120,000	3500	12	39
ACS-1024[a] (PVS[c])	13 × 13	1026 × 1026	13.21 × 13.21	300,000	< 3000	30 fps	≈100
ELIS-1024[d] (PVS[c])	125 × 7.8125 × 64.4	1024 × 128	0.125 × 7.998	800,000–6.4 × 10^6	12,000	30	1600
LIS-1024[d] (PVS[c])	125 × 7.8	1024 × 1	0.125 × 7.998	8 × 10^6	?	20	≈500
LNL2048R[e] (pixel devices[f])	200 × 7	2048 × 1	0.2 × 14.34	187,500	8000	32	17
S8378-1024Q[d,e] (Hamamatsu)	500 × 25	1024 × 1	25.6 × 0.5	3.9 × 10^7	2.5 × 10^5	0.5	≈6000

fps = frames per second
[a]On chip 10-bit ADC
[b]Fill factory: http://www.fillfactory.com/index2.htm
[c]Photon vision systems: http://www.photon-vision.com/p-lis.htm
[d]UV sensitive
[e]On chip amplifier(s)
[f]Pixel devices

Geometry of Area CCDs: The problem of insufficient pixel height can be overcome by using two-dimensional CCDs and summing up all pixels of a column.

Readout Speed: Readout speed is on the order of 10^6–10^7 pixels per second (1–10 MHz, see Table 7.8). While a linear CCD can be read out in 2–0.2 ms, readouts of 2D CCD arrays may take up to 0.25 s. Considering the requirement for a complete readout in less than 0.1 s for active DOAS systems, only fast detectors can be used. Using CCD sensors with multiple readout taps (e.g. 8 as in the case of DALSA IT-F6 TDI, see Table 7.8) can provide the required speed, although at the expense of additional electronics (8 ADCs plus associated electronics are required). Readout can also be sped up by 'binning' pixel in the vertical, i.e. combining the charge of several or all pixel of a column in the readout shift register.

There is a large and growing variety of CCD detectors available. While linear CCDs are ideally suited for DOAS, a number of crucial parameters have to be evaluated.

1. The pixel width w_D and height h_D of most standard scanner CCDs is appropriate for practically all DOAS applications. For passive DOAS applications higher h_D would be desirable.

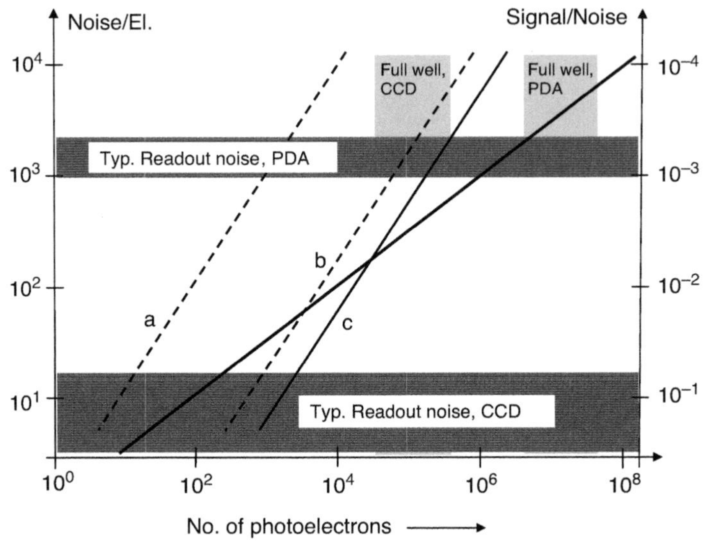

Fig. 7.45 Sketch of signal-to-noise (S/N) ratios as a function of the number of photoelectrons and characteristics of typical CCD and PDA detectors. Thick line: S/N ratio from photoelectron statistics alone; Lines a–c: contribution of the ADC to the S/N ratio of the detector system. **a** 16-bit ADC for a typical CCD detector (6×10^4 photoelectrons full well); **b** 12-bit ADC for a typical CCD detector; **c** 16-bit ADC for a typical PDA (6×10^7 photoelectrons full well)

2. The UV sensitivity of linear CCDs is frequently not sufficient. Linear CCDs are not available in back-illuminated versions. This can partly be circumvented by adding a UV fluorescent coating (such as Metachrome II or Lumogene) that converts the UV to visible (largely green) light at a wavelength near the maximum of the Si response. Limitations are due to relatively low quantum efficiency (near 10%) compared to >20% for special UV Si devices or back illuminated (two-dimensional) CCDs.
3. Readout time can be a problem, in particular in active DOAS systems. The fastest linear CCD systems or multiple tap CCDs appear to be just fast enough for active DOAS applications. Two-dimensional CCDs are frequently too slow by about two orders of magnitude for active DOAS applications.
4. The charge transfer efficiency has to be considered. While stated efficiencies >99.999% may sound impressive, the charge generated in e.g. the first pixel is shifted through about 1000 pixels. This may lead to a charge loss of the order of 1%. This charge is distributed across other pixels, thus adding to the cross-talk (Fig. 7.45).

A comparison of PDA and CCD detector characteristics is illustrated in Fig. 7.45. In particular the margins for shot noise limited detection are much wider for CCDs.

7.5.6 CMOS Detectors

Complementary metal–oxide–semiconductor (CMOS) arrays are also common spectroscopic detectors. The fact that the detector elements can be integrated on the chip with other electronic components, such as amplifiers and ADC converters, makes them an interesting option. CMOS sensors typically exhibit less 'blooming' (i.e. bleeding of photo-charge from an over-exposed pixel into neighbouring pixels). Noise levels of CMOS detectors match or surpass those of CCD detectors (Fig. 7.45). Table 7.9 gives details of currently available CMOS sensors.

7.6 Telescope Designs

In DOAS systems, telescopes are used for two purposes: to collect light from natural or artificial light sources and, in active DOAS applications, to transmit a collimated beam of radiation from the light source through the open atmosphere. The telescope designs either employ refracting elements, i.e. lenses (refractors) or reflecting elements, i.e. mirrors (reflectors). For DOAS applications, reflectors are often chosen due to absence of chromatic aberrations. Reflectors also need fewer large optical surfaces, which are expensive to manufacture. Traditionally, there are several design choices for reflecting telescopes, which are known as Newtonian, Cassegrain, or Gregor arrangements, as shown in Fig. 7.46. A popular design is the Newtonian, which combines compact and simple optics with easy adjustment. Also, Newtonian telescopes usually form the basis of a transmitting–receiving telescope combination (see Fig. 7.55, Sect. 7.8.3). In contrast, Cassegrain-type telescopes are much more compact, but difficult to manufacture and require strict tolerances in adjustment. Nevertheless Cassegrain-type telescopes have been used for DOAS instruments, see e.g. Volten et al. (2012). A newer fibre-based design for longpath DOAS telescopes will be discussed in Sect. 7.8.3

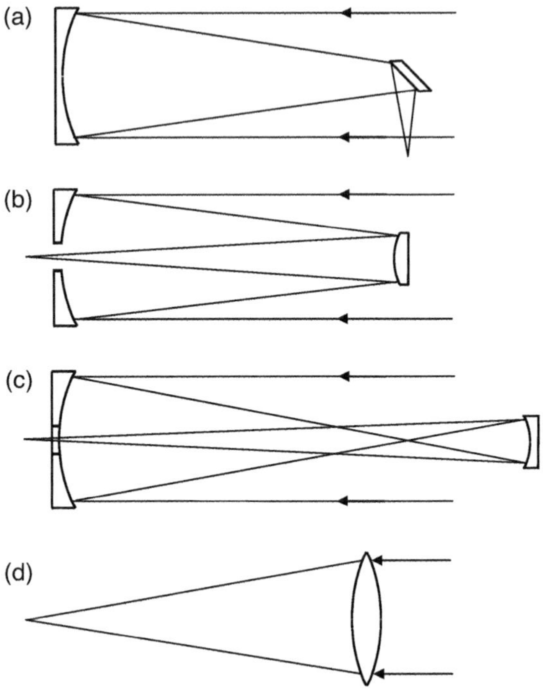

Fig. 7.46 Different telescope designs in use for DOAS applications. Reflectors: **a** Newtonian, **b** Cassegrain, **c** Gregor type telescope, **d** Refractor

7.7 Optical Multi-pass Systems

An attractive alternative light path arrangement is to fold the absorption light path in a multiple pass (or multiple reflection) cell (MRC), where the probing beam of radiation traverses the same volume (the 'base path') many times. There are several optical arrangements that allow the multiple folding of light beams. The most widely used are:

- White Cells (White 1942, 1976)
- Herriot Cells (Herriot et al. 1964; Herriot and Schulte 1965)
- Passive resonators (Engeln et al. 1998; Ball et al. 2004)

For each design, a number of variants were developed (Horn and Pimentel 1971; Schulz-DuBois 1973; Chernin and Barskaya 1991; Ritz et al. 1992; MacManus et al. 1995; Vitushkin and Vitushkin 1998; Grassi et al. 2001).

Multi-reflection systems have several advantages. The measurement volume is greatly reduced, thus allowing atmospheric in-situ observations and laboratory measurements. For White Cells, the length of the light path can be changed increments of small multiples of the base path, depending on the particular design of the cell.

Therefore, it is possible to keep the length of the light path close to its optimum (see Sect. 7.12) for the respective atmospheric conditions. In closed cells reference spectra of pure air can be measured, effectively allowing measurement of I_0 with an atmosphere where all trace gases are removed.

Disadvantages of multi-reflection cells include the relatively complicated optical design, the requirement for extreme mechanical stability, and small reflection losses of the mirrors. Typically, high reflectivity ($R > 0.99$–0.9999) of the cell mirrors over spectral intervals of several 10 nm is required, but difficult to obtain and to maintain in the field (e.g. due to dust settling on the surfaces), in particular for large area mirrors and in the UV. In some applications, the high-radiation intensity inside the folded light path can cause photochemical reactions that may change the concentration of the species to be measured. In addition, the level of stray light originating from the light source is difficult to measure in multi-reflection cells. Another problem can be due to chemical reactions of the trace gases with the surfaces of optical or mechanical parts of the cell, although 'open path' cells can reduce this effect.

7.7.1 White Multi-reflection Cells

The basic White multi-reflection system (Fig. 7.47; White 1942) consists of three spherical, concave mirrors of identical radius of curvature, and thus focal length. The distance between the mirror sets is twice the focal length. The front mirror A faces the two side by side back mirrors B and C. Radiation from the light source is focused in point F_0 such that it illuminates back mirror B, which images point F_0 into F_1 on mirror A. Mirror A is oriented such that the radiation originating from F_1 illuminates back mirror C, which in turn images F_1 into F_2. This completes a round trip of the radiation traversing the system four times. F_2 can be thought of acting as a new entrance to the system, with the next round trip producing foci F_3 and F_4. Since F_2 is closer to the axis of the system, F_3 and F_4 will be as well. Ultimately, there will be two sets of foci $F_0, F_2, F_4, \ldots$ and $F_1, F_3, \ldots$ across the surface of 'nesting mirror' A. Usually, mirrors B and C will be tilted slightly such that even and odd numbered reflections of mirror C form two separate rows on mirror A. Finally, one

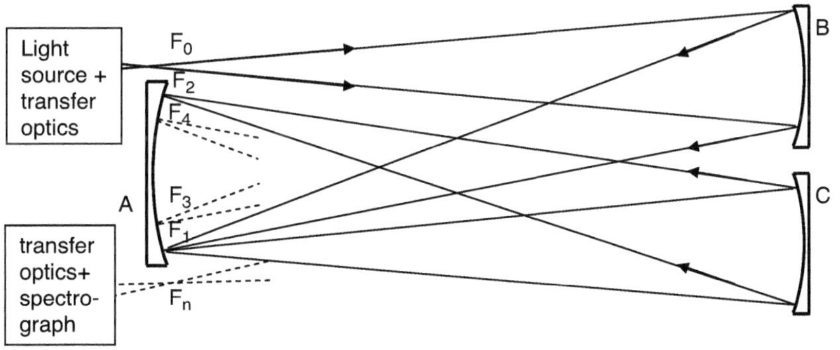

Fig. 7.47 Schematic diagram of a basic White system (redrawn after J.U. White 1942)

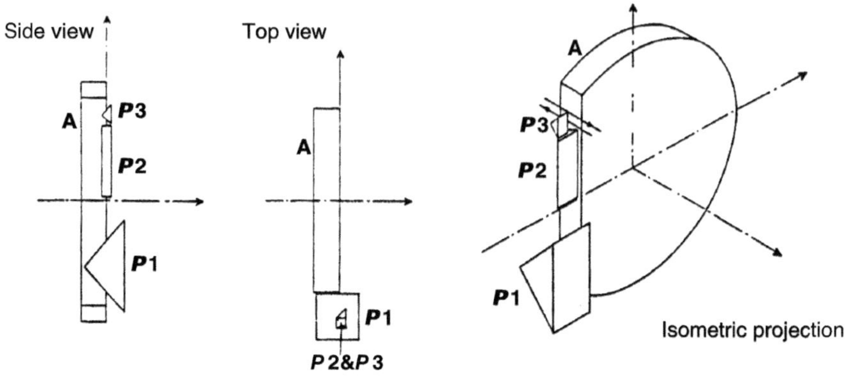

Fig. 7.48 Schematic diagram showing the nesting mirror (**A**) and the prisms P1, P2 and P3. Reprinted with permission from Ritz et al. 1992, ©Society of Photo-Optical Instrumentation Engineers (SPIE). All rights reserved

of the images F_N (not shown) falls off the surface of mirror A, leaving the system. The number of passes can be adjusted in steps of 4.

In the more advanced White system (White 1976), two or three quartz prisms are added near the front mirror (Fig. 7.48). In its basic function, the system functions as the classic White system (White 1942) described earlier. However, the radiation leaving the mirror system after N traverses near the bottom of mirror A is reflected back by prism P1. After making another N traverses, it is returned again by prism P2, giving rise to another $2N$ traverses. Finally, prism P3 reverses the entire set of light paths for a total of $8N$ traverses. Thus, the number of passes can be adjusted in steps of 32 traverses. This scheme not only multiplies the number of passes by 8, but more importantly also considerably increases the stability against misalignment of the system. This is due to the fact that each prism acts as retro-reflector, returning the radiation nearly exactly back to its original direction in one plane. However, the planes of P1 and P2 are perpendicular to each other, thus compensating for misalignments within the system only as long as they do not become too large.

Two prisms would be enough to accomplish this effect, as originally described by White (1976). The vertex angle of the prisms is actually slightly smaller than 90° (Ritz et al. 1992). Further optimisations of the design were suggested by Hausmann et al. (1997) and Grassi et al. (2002). White-type multi reflection cells employing more than two back mirrors (see Fig. 7.47) have become known as Chernin-type systems (Chernin and Barskaya 1991; Steyert et al. 2001).

The transmission of radiation through a multi-reflection system is given by the transmission of the transfer optics, the reflectivity R of the mirrors, and the extinction of the air inside the system. The transfer optics consists of two parts. First, optics matching the aperture of the light source to that of the entrance of the multi-reflection cell, and second, optics matching the exit aperture of the multi-reflection cell to the entrance of the DOAS spectrometer. Typically, the desired aperture from a Xe-arc lamp or LED might correspond to an f-number of $F_1 \approx 1$, while a White cell might have $F_2 = 100$ (i.e. $2f$ divided by the diameter of the mirrors B and C). Transfer optics would then image the Xe-arc at a magnification of $F_2/F_1 \approx 100$. At the exit side (assuming the exit $f/\#$ of the white cell to be the same as the entrance $f/\#$), matching the white cell to the entrance of a spectrometer with $F_3 = 5$ would require a reduction in image size by $F_3/F_2 \approx 20$. Given a diameter $d \approx 0.2$ mm of the brightest spot in the arc (see Sect. 7.2.1), the images on nesting mirror A of the White cell (Fig. 7.47) would be 20 mm in diameter. Thus, in order to accommodate many reflection mirrors, A has to be quite large. Alternatively, the aperture of the light source (F_1) could be reduced or that of the White cell (F_2) increased.

Overall, the (broadband) transmission of a White multi-reflection cell is given by:

$$T = \frac{I}{I_0} = G \cdot R_M^N \cdot e^{-\varepsilon_B} \tag{7.41}$$

where G denotes the transmission of the transfer optics, R_M the mirror reflectivity, N the number of reflections, and ε_B the broadband extinction of the air (see Chap. 6).

7.7.2 Herriott Multi-reflection Cells

The optical design of a basic Herriott-type multi-reflection cell (e.g. Herriott et al. 1964; Herriott and Schulte 1965; McManus and Kebabian 1990; McManus et al. 1995; Zimmermann 2003) is shown in Fig. 7.49. A nearly collimated beam of radiation enters through a hole in circular mirror A at a certain distance R_0 from the centre (marked by a cross in Fig. 7.49) of the mirror. The radiation is then reflected N times between mirrors A and B and leaves through the same hole in mirror A. The reflections are located on either mirror on circles, or more generally on ellipses, with radius R_0. The vertex angle θ in a triangle formed by the centre of mirror A and the locations of two consecutive reflections is given by $\theta = 2\pi M/N$, where M denotes the number of times the reflections circle around the optical axis of the system. The

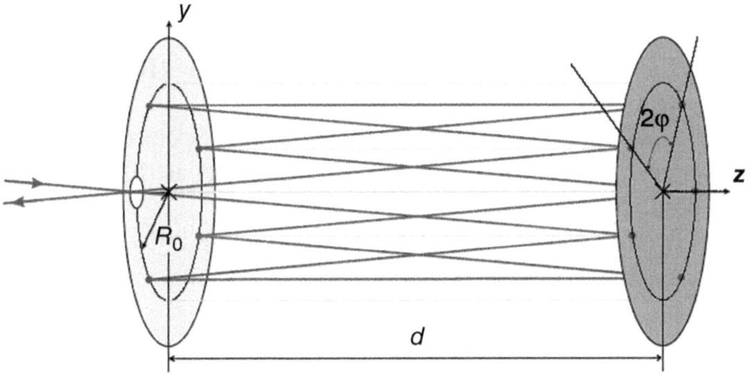

Fig. 7.49 Schematic diagram of a Herriott cell. The vertex angle $\theta = 2\phi$ is marked on the right side mirror. Reprinted from Zimmermann, Ph.D. Thesis, 2003, University of Heidelberg

angle θ is determined by the ratio of the distance d between the mirrors and their radius of curvature R:

$$\cos\theta = 1 - \frac{d}{R} \tag{7.42}$$

The total absorption path L_H is given by $L_H = d \cdot N$. Advantages of Herriott cells are their simple optical setup combined with a good stability of the beam profile due to the refocusing properties of the optics. In addition, the cell is insensitive to tilt of the mirrors against each other and lateral translation of the mirrors. A major disadvantage is the need for collimated beams. Typically F-numbers (f/#) around 30–50 are required. More details on the theory of Herriott cells can be found in Herriott and Schulte (1965), and some details for applications in Zimmermann (2003) and McManus and Kebabian (1990).

Improvements of the Herriot cell principle employ astigmatic off-axis resonator cells (see e.g. Herriott and Schulte 1965 and McManus et al. 1995). In this type of multi-reflection cell each of the two mirrors have two different radii of curvature (sagittal and meridional). By rotating the mirrors around their optical axis a variety of non-circular and non-elliptic spot patterns on mirror B can be obtained. A further improved version of Herriot-type cells is described by Robert (2007). It consists of three mirrors in basically the same arrangement as shown in Fig. 7.49. However mirror A is cut along its diameter in two halves (mirrors A and A'), which are slightly rotated against each other with the axis of rotation laying in the plane of mirrors A, A'. A total of 600 traverses through the cell has been demonstrated.

7.7.3 Passive Resonators (CEAS, CE-DOAS, CRDS)

Passive resonators can provide long optical paths. This technique has become known under a number of names:

- Cavity Enhanced Absorption Spectroscopy (CEAS) (e.g. Engeln et al. 1998; Brown 2003; Ball et al. 2004; Fiedler 2005)
- Broad-Band Cavity Enhanced Absorption Spectroscopy (BB-CEAS), (e.g. Aalto et al. 2015; Min et al. 2016; Chen and Venables 2021)
- Incoherent Broad-Band Cavity Enhanced Absorption Spectroscopy (IBB-CEAS), (e.g. Fiedler et al. 2003; Wu et al. 2014; Nakashima et al. 2017; Tang et al. 2019, 2020; Liu et al. 2019)
- Cavity Enhanced DOAS (CE-DOAS), (e.g. Platt et al. 2009; Meinen et al. 2010; Coburn et al. 2014; Min et al. 2016)

The basic idea is to introduce white light into an optical resonator (examples are shown in Fig. 7.50) consisting of two nearly plane mirrors M_1, M_2 with high reflectivity R (typically $R > 0.999$). Alternatively, a ring resonator consisting of several (typically 2–6) prisms can be used. In this case the light is reflected by total internal reflection (R $=$ 1.000) inside the prisms and the losses when entering and leaving the prisms are minimised by anti-reflection coatings or by orienting the prisms such

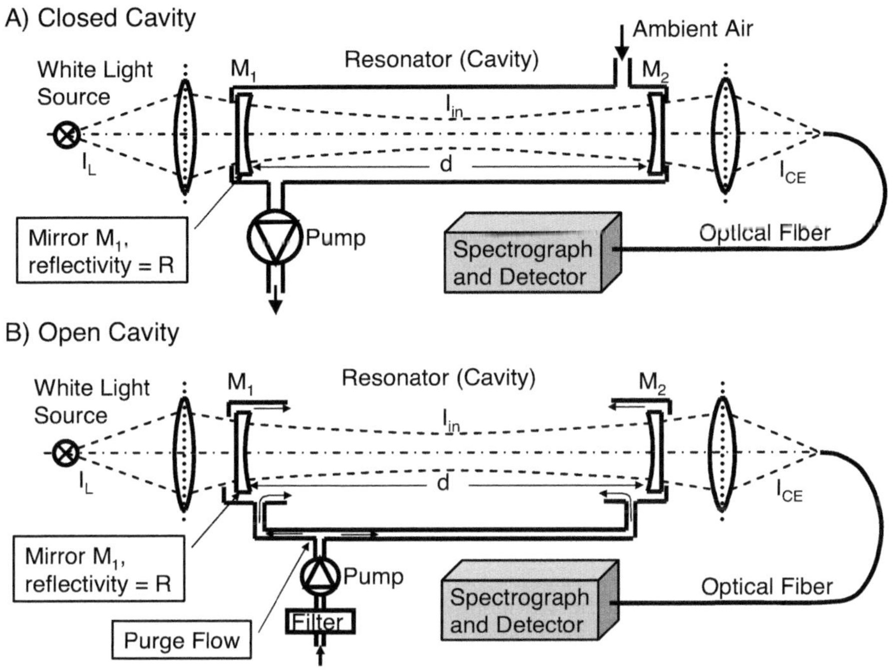

Fig. 7.50 Sketch of cavity-enhanced absorption-based DOAS systems (CE-DOAS, e.g. Ball et al. 2004; Fiedler 2005, Platt et al. 2009; Meinen et al. 2010). A) Enclosed cavity, ambient air is drawn through the cell by a pump. B) Open cavity system, the resonator is set up in the open atmosphere. The mirrors are protected by a filtered air purge flow system. Here care has to be taken that possible wind-driven back-flow of purge air is avoided. Note that the curvature of the cavity mirrors and the beam waist are exaggerated

that the prism surfaces are under the Brewster angle. Remaining losses are due to surface imperfection and possibly due to birefringency of the prism material (see e.g. Lehmann et al. 2009; Lee et al. 2014).

There are two basic approaches for cavity enhanced spectrometry relying on a pumped optical resonator: (1) Cavity Ring-Down Spectroscopy (CRDS) measures the decay time (ringdown time) of light intensity leaking from the resonator after switching off the pump light source. Thus it is a time dependent measurement. (2) Cavity Enhanced Absorption Spectroscopy (CEAS, CE-DOAS, etc.) measures the transmittance of the resonator by recording the wavelength-dependent leak-out intensity in a continuous measurement. Thus, it is in principle time independent. Both techniques have been reviewed by a number of authors (e.g. by Ball et al. 2003; Brown 2003; Zheng et al. 2018; Chen and Venables 2021).

Approach (1) requires the knowledge of the ringdown time of the empty cavity, the determination of which is possible, but difficult. Moreover it is a monochromatic measurement, with the disadvantages of this type of measurement (see Chap. 6). Therefore we concentrate on approach (2), which, in principle, has all advantages of a DOAS measurement.

7.7.3.1 Theory of CE-DOAS (CEAS)

In CE-DOAS, incoherent broadband light of intensity I_L is introduced into the optical resonator (Fig. 7.50) consisting of two mirrors with reflectivity $R_1 = R_2 = R$ mounted at a distance d_0. In the case of different reflectivities of the two mirrors R_1, R_2 the overall reflectivity is the harmonic average $R = (R_1 R_2)^{1/2}$. The two mirrors are concave with long focal lengths $f_1 = f_2 > d_0$. They are mounted with the focus of M_1 to the right of M_2 and the focus of M_2 to the left of M_1 (see Fig. 7.50). The light path can be in the open atmosphere or within a cell which is flushed by the measurement air.

Initially only the fraction $\varrho_R = 1 - R$ of the radiation emitted by the light source, I_L (see Fig. 7.50), will enter the resonator. However, once inside the resonator the radiation will be reflected about $1/(1 - R)$ times on average (neglecting other losses for the moment). Finally, in the absence of any extinction in the cavity, half the radiation will leave the resonator through Mirror 1, the other half through Mirror 2, and each fraction will have the intensity I_0:

$$I_0(\lambda) = I_L(\lambda) \cdot \frac{1 - R(\lambda)}{2} = I_L(\lambda) \cdot \frac{\varrho_R(\lambda)}{2} \tag{7.43}$$

In the following we omit the wavelength dependence of the variables I_0, I_L, R, ρ_R, ϵ_R, L for simplicity. Usually, 'empty cavity' measurements are made, i.e. measurements with a purified air filled cavity, therefore it is convenient to include the Rayleigh extinction ϵ_R of pure air in an effective mirror reflectivity R_0:

$$R_0 = R - \epsilon_R d_0 \quad \text{or} \quad \varrho_0 = 1 - R_0 = 1R + \epsilon_R d_0 = \varrho_R + \epsilon_R d_0 \tag{7.44}$$

Where the index 0 always corresponds to the cavity without absorbers. A detailed discussion of the relationship between reflectance, extinction, and differential absorption is given by e.g. Fiedler (2005), Fiedler et al. (2005, 2007), and Platt et al. (2009).

During a single traverse of the resonator the intensity will change by the factor $(1 - TR_0)$, where T denotes the transmissivity of the cavity (see Fig. 7.50). We then derive the decay of the intensity of a light pulse inside the cavity I_{in} after n traverses:

$$-\frac{dI_{in}(0)}{dn} = I_{in}(n) \cdot (1 - TR_0) \tag{7.45}$$

The number of reflections n is used as a continuous variable here, although in practice n can only assume positive integer values. Equation (7.45) can be integrated with the boundary condition that the initial light intensity before the first traverse through the cavity is $I_{in}(n = 0) = I_{in}(0)$:

$$I_{in}(n) = I_{in}(0) \cdot \exp\left[n \cdot (TR_0 - 1)\right] \tag{7.46}$$

or

$$\ln\left(\frac{I_{in}(0)}{I_{in}(n)}\right) = n \cdot (TR_0 - 1) \tag{7.47}$$

Where $I_{in}(n)$ denotes the intensity inside the cavity after n reflections (or traverses through the cavity).

Re-writing the cavity transmission $T = 1 - \tau$ and the mirror reflectance $R = 1 - \varrho$ with $\tau, \varrho \ll 1$ (Note: here we use the symbol $\tau = 1-$ Transmission, not to be confused with the ring-down time or optical density) we obtain:

$$T R_0 = (1 - \tau)(1 - \varrho_R) = 1 - \tau - \varrho_R + \tau \varrho_R \approx 1 - \tau - \varrho_R \tag{7.48}$$

Introducing TR_0 from above in the expression for $\ln[I_{in}(0)/I_{in}(n)]$, Eq. 7.47, and replacing ρ_R by ρ_0 (i.e. considering an air-filled cavity, see Eq. 7.44) we obtain:

$$\ln\left(\frac{I_{in}(0)}{I_{in}(n)}\right) = n \cdot (T R_0 - 1) \approx n \cdot (1 - (1 - \tau - \rho_0) = n \cdot (\tau + \rho_0) \tag{7.49}$$

Note that according to Lambert-Beer's law τ can be written as:

$$\tau = \sigma \cdot \overline{c} \cdot d_0 \tag{7.50}$$

7.7.3.2 The CE-DOAS Light Path Length

The average number of passes through the cavity, $\overline{n}$, which the photons travel and the average light path, $\overline{L}$, results in the decay of the initial intensity inside the cavity $I_{in}(0)$ to the remaining intensity $I_{in}(\overline{n}) = 1/e \cdot I_{in}(0)$. Thus, we derive from Eq. 7.49:

$$\ln\left(\frac{I_{in}(0)}{I_{in}(\overline{n})}\right) = \ln e = 1 \approx \overline{n} \cdot (\tau + \rho_0) \tag{7.51}$$

With the average number of traverses of the photons through the cavity and associated average length of the light path we finally derive for the length of the average light path $\overline{L}$:

$$\overline{n} \approx \frac{1}{\tau + \rho_0} \text{ and } \overline{L} = d_0 \cdot \overline{n} \approx \frac{d_0}{\tau + \rho_0} = \frac{d_0}{\sigma \cdot \overline{c} \cdot d_0 + \rho_0} \tag{7.52}$$

Note that τ includes all extinction processes in the cavity i.e. aerosol absorption and Mie scattering as well as absorption by trace gases, except those by pure air due to absorption and Rayleigh scattering, which are taken into account by using ϱ_0 instead of ϱ (see above). For a cavity filled with pure air we therefore have:

$$\overline{L}_0 = \frac{d_0}{\rho_0} \tag{7.53}$$

Here the index 0 always corresponds to the cavity without absorbers as stated above. If additional (broadband) extinction is assumed, the average length of the absorption path $\overline{L}$ will be approximately:

$$\overline{L}_0 = \frac{d_0}{\rho_0 + \epsilon_B d_0} \tag{7.54}$$

where ϵ_B denotes all broadband extinctions except Rayleigh scattering, e.g. due to Mie scattering and broadband trace gas absorption.

In practice, the length of the average 'pure air' light path ($\overline{L}_0$) can be determined by a number of different techniques, e.g. by measuring the ring-down time of the cavity only filled with pure air (Meinen et al. 2008), or by addition of a different gas with known Rayleigh scattering cross-section (Washenfelder et al. 2008). The following examples illustrate the consequences of the above considerations:

1. In a cavity with base path length $d_0 = 0.5m$, mirrors with $R_0 = 0.99995$ ($\varrho = 5 \cdot 10^{-5}$)), and without trace gases ($\tau = 0$), an average light path of $\overline{L}_0(\tau = 0) = 10^4\, m$ or 10 km is obtained according to Eq. 7.53 (neglecting the contribution from Rayleigh scattering, which would amount to $\epsilon_R d_0 \approx 2.55 \cdot 10^{-6}$ at 622 nm (Penndorf 1957) or $\approx 5\%$ of ϱ). This simplification amounts to assigning $\varrho = \varrho_0$ here.
2. Adding an absorber with $\tau = 5 \cdot 10^{-5}$ to the same cavity, we obtain an average light path of $\overline{L}_0(\tau = 5 \cdot 10^{-5}) = 5 \cdot 10^3\, m$ or 5 km.
   In a high finesse resonator (i.e. one with high mirror reflectivity), a rather weak absorber with $\tau = d_0 \cdot \epsilon = 5 \cdot 10^{-5}$ or $\epsilon = 10^{-4}$ significantly reduces the effective average light path. CRDS measurements are based on this light path reduction.

For broad band cavities and structured absorptions, i.e. $\tau = \tau(\lambda_{max})$, this implies that the average light path length varies for different wavelengths. Considering an absorption band, as sketched in Fig. 7.51, and neglecting any broadband absorption, i.e. $I_{in0}(n) = I'_{in0}(n)$, the intensity in the absence of the absorption band after n traverses, $I_{in0}(n)$, is:

$$I_{in0}(n) = I_{in0}(n) \cdot e^{-\sigma \cdot \overline{c} \cdot d_0 \cdot n} \tag{7.55}$$

According to DOAS principles, $I_{in0}(n)$ ($I'_{in0}(n)$ in Fig. 7.51) can be interpolated from the intensity outside the absorption band.

Only if the number of round-trips, n, is the same with and without absorber, the optical density (from Lambert—Beer absorption) can be written as:

$$\ln\left(\frac{I_{in0}(n)}{I_{in}(n)}\right) = \sigma \cdot \overline{c} \cdot d_0 \cdot n \tag{7.56}$$

7.7.3.3 The DOAS Optical Density in a Cavity, D_{CE}

From Eq. 7.56, the trace gas concentration inside the cavity, $\overline{c}$, can be calculated once the absorption cross-section, σ, is known, d_0 is constant and $\overline{L} = d_0 \cdot n$, or the number of round trips n, is known. Outside the cavity, i.e. at the detector, all intensities are lower by a factor $1 - R_0 = \varrho_0$:

$$\varrho_0 \cdot I_{in}(n) = \varrho_0 \cdot I_{in0}(n) \cdot e^{-\sigma \cdot \overline{c} \cdot \overline{L}} \tag{7.57}$$

leading to the same equation for D as for usual DOAS applications (Chap. 6).

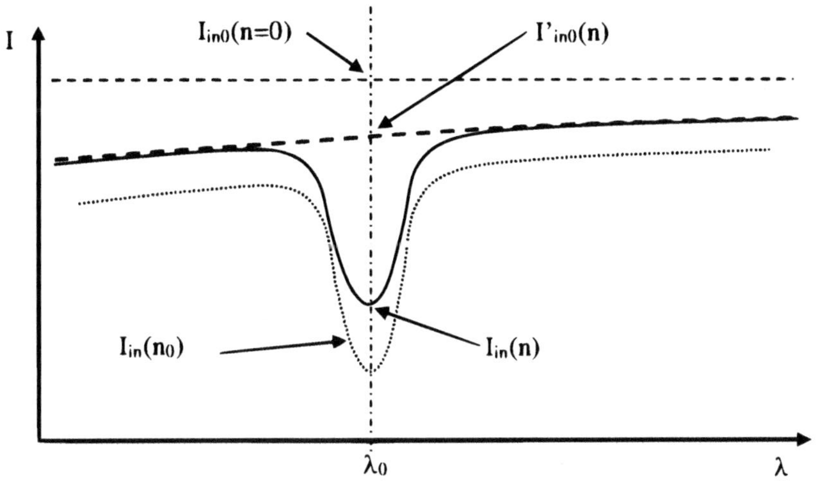

Fig. 7.51 Sketch of intensities versus wavelength: $I_{in0}(n = 0)$: Intensity (during first pass through the cavity) in the pure air filled cavity without any absorbers, $I'_{in0}(n)$: Intensity (after n passes through the cavity) after any continuous absorption (due to gases or aerosol) has taken place, $I_{in}(n)$: Intensity after n passes through the cavity including also differential absorptions (note that n varies with wavelength, since the trace gas absorption cross-section varies with wavelength), $I_{in}(n_0)$: Theoretical intensity for the same absorptions if the number of traverses were not reduced

Another consequence of the reduced light path due to absorbers in the cavity is a different measured optical density D_{CE} compared to a classical long-path measurement. The optical density seen by the detector through the second mirror of the cavity, D_{CE}, after n traverses is given by:

$$D_{CE} = \ln\left(\frac{I_{tot0}}{I_{tot}}\right) = \ln\left(\frac{\int_0^\infty I_{in0}(n)\,dn}{\int_0^\infty I_{in}(n)\,dn}\right) \tag{7.58}$$

Where I_{tot}, I_{tot0} denote the intensities at the detector with and without an absorbing species in the resonator, respectively. Using Lambert-Beer's law, and the fact that $I_{in}(0)$ is a non-zero constant in this context, we can rewrite the above equation as:

$$D_{CE} = \ln\left(\frac{\int_0^\infty I_{in}(0)\cdot e^{-\varrho_0 n}\,dn}{\int_0^\infty I_{in}(0)\cdot e^{-(\varrho_0+\tau)n}\,dn}\right) = \ln\left(\frac{\varrho_0+\tau}{\varrho_0}\right) = \ln\left(1+\frac{\tau}{\varrho_0}\right) \tag{7.59}$$

This is the trace gas optical density, or simply the DOAS signal, from the CE-DOAS measurement. We illustrate the consequences of Eq. 7.59 for two extreme, but realistic, cases:

1. For values of $\varrho_0 \ll \tau$, equivalent to effective mirror reflectivities $R = 1 - \varrho_0$ close to unity, (i.e. very good mirrors) and/or large values of τ, we obtain for the optical density due to extinction in the cavity (above Eq. 7.59):

$$D_{CE} = \ln\left(\frac{\varrho_0 + \tau}{\varrho_0}\right) \approx \ln\left(\frac{\tau}{\varrho_0}\right) \propto \ln \tau = \ln\left(\alpha_{CE} \cdot d_0\right) \tag{7.60}$$

   This equation shows $D_{CE} \propto \ln \tau$ is nearly constant. In other words, the trace gas optical density D_{CE} will only be weakly dependent on the extinction and thus the trace gas concentration $\overline{c}$ in the cavity. The proportionality factor α_{CE} will be further discussed below (Eq. 7.62).

2. For large values of $\varrho_0 \gg \tau$, i.e. relatively poor mirror reflectivities and/or small trace gas absorption, we have:

$$D_{CE} = \ln\left(\frac{\varrho_0 + \tau}{\varrho_0}\right) \approx \frac{\varrho_0 + \tau}{\varrho_0} - 1 = \frac{\tau}{\varrho_0} \propto \tau \tag{7.61}$$

   Since ϱ_0 is constant for a given set-up, D_{CE} will be proportional to $\tau = \sigma \cdot \overline{c} \cdot d_0$. Thus the measured optical density D_{CE} will be proportional to the trace gas absorption τ and concentration $\overline{c}$.

From Eq. 7.61 we can calculate the effective (i.e. intensity weighted average) CE-DOAS light path length $\overline{L}_{eff} = \overline{L}_{eff}(\tau, \varrho)$, which corresponds to an absorption light path of a Lambert Beer absorption with the same optical density:

$$\alpha_{CE} = \frac{D_{CE}}{\overline{L}_{eff}} = \frac{\ln\left(\frac{\varrho_0+\tau}{\varrho_0}\right)}{\overline{L}_{eff}} \approx \frac{\tau}{d_0} = \sigma \cdot c \text{ and } \overline{L}_{eff} = \frac{\ln\left(\frac{\varrho_0+\tau}{\varrho_0}\right)}{\tau} d_0 \tag{7.62}$$

The effective light path length $\overline{L}_{eff}$ corresponds to an effective number of reflections n_{eff} since $\overline{L}_{eff} = n_{eff} \cdot d_0$:

$$n_{eff} = \frac{\overline{L}_{eff}}{d_0} = \frac{\ln\left(\frac{\varrho_0+\tau}{\varrho_0}\right)}{\tau} \tag{7.63}$$

For small absorptions, i.e. $\tau \to 0$ equivalent to the cavity transmission approaching $T \to 1$, we derive the average light path for an empty cavity, L_0:

$$\overline{L}_{eff}(\tau \to 0) = \overline{L}_0 = \frac{d_0}{\varrho_0} \text{ or } \varrho_0 = \frac{d_0}{\overline{L}_0} \tag{7.64}$$

With the above relationship for n_{eff} (7.63) and Eq. 7.61 we can write the effective light path length $\overline{L}_{eff}$ as function of the light path length of the empty (i.e. air filled) cavity L_0:

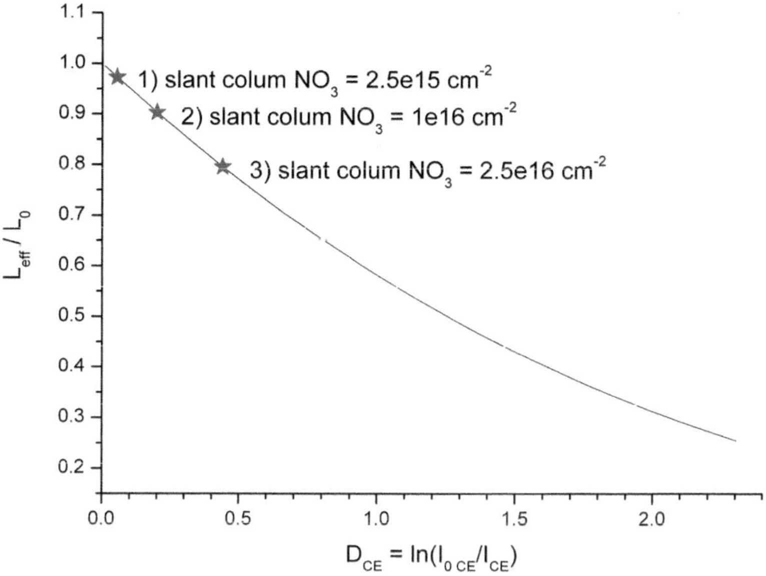

Fig. 7.52 Reduction of the light path L_{eff} in a cavity due to extinction in the cavity expressed as optical density D_{CE}. Plotted is the ratio L_{eff}/L_0, where L_0 denotes the average light path in an empty (air filled) cavity as described in the text. For illustration the effective light paths at 662nm (peak of the NO_3 absorption band) for three realistic NO_3 total column densities (2.5, 10, $25 \cdot 10^{15}$cm^{-2}) in the cavity are marked by stars. Reprinted with permission from Platt et al. (2009),

$$\overline{L}_{eff} = \frac{D_{CE}}{(e^{D_{CE}} - 1)\varrho_0} \cdot d_0 = \frac{D_{CE}}{(e^{D_{CE}} - 1)} \cdot \overline{L}_0 \tag{7.65}$$

The dependence of L_{eff} on D_{CE} is illustrated in Fig. 7.52.

For illustration, the effective light paths at 662 nm (peak of the NO_3 absorption band) for three realistic NO_3 total column densities in the cavity are also shown (see Meinen et al. 2008). Note that the ratio L_{eff}/L_0 only depends on the trace gas column density in the cavity, which is a function of c, d_0, and R.

It is useful to define an effective optical density D_{eff}:

$$D_{eff} = \alpha \cdot \overline{L}_0 \tag{7.66}$$

Using the expression for $\overline{L}_{eff}$ from Eq. 7.62 we obtain:

$$D_{eff} = \alpha \cdot \overline{L}_0 = \frac{D_{CE}}{\overline{L}_{eff}} = D_{CE} \cdot \frac{(e^{D_{CE}} - 1)}{D_{CE} \cdot \overline{L}_0} \cdot \overline{L}_0 = e^{D_{CE}} - 1 = \frac{I_{0tot}}{I_{tot}} - 1 \tag{7.67}$$

and:

$$\alpha(\lambda) = \left(\frac{I_{0tot}}{I_{tot}} - 1\right) \cdot \frac{1 - R(\lambda)}{d_0} = \left(\frac{I_{0tot}}{I_{tot}} - 1\right) \cdot \frac{1}{\overline{L}_0} = \sigma(\lambda) \cdot \overline{c} + \epsilon_{Mie} \tag{7.68}$$

Thus, CE-DOAS contains the presented correction, but has to rely on stable I_0 conditions.
Rewriting D_{eff} leads to an alternative form of the effective light path:

$$D_{eff} = \overline{c} \cdot \sigma \cdot \overline{L}_0 = \left(\frac{I_{in0}}{I_{in}} - 1 \right) = e^{-\overline{c} \cdot \sigma \cdot \overline{L}_{eff}} - 1 \tag{7.69}$$

Solving Eq. (7.69) for $\overline{L}_{eff}$ yields:

$$\overline{L}_{eff} = \frac{\ln\left(\overline{c} \cdot \sigma \cdot \overline{L}_0 + 1\right)}{\overline{c} \cdot \sigma} \tag{7.70}$$

For small optical densities, i.e. $\overline{c} \cdot \sigma \cdot \overline{L}_0 \ll 1$, L_{eff} approaches L_0:

$$\overline{L}_{eff}(\overline{c} \cdot \sigma \cdot \overline{L}_0 \ll 1) \approx \frac{\overline{c} \cdot \sigma \cdot \overline{L}_0 - 1 + 1}{\overline{c} \cdot \sigma} = \overline{L}_0 \tag{7.71}$$

A consequence of this result is that a very high mirror reflectivity (compared to the extinction losses in the cavity) can cause problems. On the other hand, a high mirror reflectivity is a prerequisite for long light paths and thus high sensitivity. Therefore, the reflectivity R_0 cannot usually be made sufficiently small (or ϱ_0 so large) that Case 2 (Eq. 7.61) becomes valid. An optimal strategy thus involves a combination of a high reflectivity cavity and a correction of the observed optical densities for the effect of light path reduction.

7.7.3.4 Correction for Light Path Reduction Due to Extinction

There are several possible solutions for determining the reduction of the effective average light path length $\overline{L}_{eff}$ in the presence of absorbing gases in the cavity:

1. Determine the average light path $\overline{L}_0$ with an empty, i.e. air filled, cavity (where replacing R by R_0 accounts for the known Rayleigh extinction of air as explained above) and use the corrected effective light path length $\overline{L}_{eff}$ (Eq. 7.65) instead of $\overline{L}_0$ for the calculation of the average trace gas concentration $\overline{c}$ from the measured D_{CE}.
2. Use the 'empty cavity' L_0 to calculate a concentration c_0 from D_{CE} and then apply a correction factor on the concentration, i.e. taking the ratio of L_0/L_{eff} from the following equation:

$$\overline{c}_0 = \frac{D_{CE}}{\sigma \cdot \overline{L}_0}, \; \overline{c} = \frac{\overline{L}_0}{\overline{L}_{eff}} \cdot c_0 = \frac{e^{D_{CE}} - 1}{D_{CE}} \cdot c_0 \tag{7.72}$$

3. Convert the trace gas absorption cross-section σ to a corrected value σ_{eff}, which is then used in the DOAS fit and in conjunction with L_0 to calculate the concentration from D_{CE} (again taking the ratio of L_0/L_{eff} from Eq. 7.65):

$$D_{CE}(\lambda) = \sigma(\lambda) \cdot \overline{c} \cdot \overline{L}_{eff}(\lambda) = \sigma_{eff}(\lambda) \cdot \overline{c} \cdot \overline{L}_0 \tag{7.73}$$

$$\sigma_{eff}(\lambda) = \sigma(\lambda) \cdot \frac{\overline{L}_{eff}(\lambda)}{\overline{L}_0} = \sigma(\lambda) \cdot \frac{D_{CE}}{e^{D_{CE}} - 1} \tag{7.74}$$

4. The measured optical density D_{CE} can be corrected to obtain a corrected optical density D_{eff} according to Eq. 7.69.

$$D_{eff} = e^{D_{CE}} - 1 = \frac{I_{0tot}}{I_{tot}} - 1 \tag{7.75}$$

   This procedure corresponds to a linearisation of the problem, since $D_{eff}(\lambda)$ is proportional to the absorption cross-section $\sigma(\lambda)$, thus a "normal" DOAS fit of $D_{eff}(\lambda)$ to $\sigma(\lambda)$ can be performed as in a classic DOAS approach with constant light path length.

All four correction approaches depend on the mirror reflectivity, R_0, and the extinction in the cavity, including the extinction due to the trace gas optical density. Since the true trace gas optical density is a priori unknown (as the true I_0 is unknown), a theoretical value for I_0 has to be assumed. If for instance, the light source and the adjustment of the optics are constant, the I_0 measurement from a pure air-filled cavity or an iterative procedure can be used (for details see Platt et al. 2009).

7.7.3.5 Determination of the Effective Light Path Length

The determination of the effective average light path length in the cavity, L_{eff}, in the presence of absorbing gases is complicated by the dependence of the data evaluation on the stability of the light intensity. Not only is the stability of the light source itself critical, but also the stability of the light transmission through all optical components between light source and the detector (e.g. fibre, lenses, filters, and the spectrograph). In order to overcome these stability problems a relatively rigid and temperature-stabilised setup is required, which makes sensitive instruments heavy. In addition, regular maintenance, such as frequent zero-air reference measurements, is required. Solutions have been described by Horbanski et al. (2019) who introduced an iterative CE-DOAS technique (ICAD), which is robust against fluctuations of the absolute intensity of the light source, as well as against vibrations of the optical set-up. The iterative evaluation scheme starts with all trace gas concentrations set to zero, i.e. assuming Rayleigh scattering as the only extinction process. Then a first estimate of the trace gas concentrations c_{j1} is calculated from the spectrum. These concentrations are used to calculate the wavelength dependent light path, $L_{eff}(\lambda)_1$. This quantity is then used to calculate effective absorption cross-sections (Eq. 7.74), which are used to calculate a set of c_{j2} values. In the next iteration step these are used to calculate $L_{eff}(\lambda)_2$ and so forth until the relative changes between two successive values of all c_j fall below pre-determined limits (e.g. the DOAS fit error). Alternatively, changes in $L_{eff}(\lambda)$ may be used as stopping criterion (see e.g. Tang et al. 2020).

Another approach, the integrated calibration by means of optical modulation (ICOM), developed by Finkenzeller et al. (2023) uses additional hardware to determine $L_{eff}(\lambda)$. The light source is on-off modulated, while the transmission of an optical modulator (e.g. a chopper wheel or acoustic optical modulator), located between the resonator output and the spectrograph entrance slit, oscillates between blocking and transmitting the light with the same frequency as the modulated light source and with a distinct and adjustable phase relation. By adjusting the chopper phase, the spectrum of the light $I(\lambda, t)$ leaving the resonator can be determined as function of the ringdown time t (for details see Finkenzeller et al. 2023). The derived data can be used to calculate $L_{eff}(\lambda)$ and the mirror reflectivity $R(\lambda)$. Since the ringdown time is on the order of µs, the design of the chopper is quite challenging. For this reason, new approaches use clocked CCD-detectors to determine $I(\lambda, t)$, eliminating the need for a chopper.

7.8 Active Long Path DOAS Systems

Active DOAS systems employ their own light source. Therefore, they have to include suitable optics to transmit light through the atmosphere to the spectrometer. The most basic arrangement of a light path in the open atmosphere (see Fig. 7.53) consists of a light source defining one end of the light path and a receiving telescope coupled to a spectrometer defining the other. Earlier DOAS instruments (e.g. Perner and Platt 1979; Platt et al. 1979) were based on this arrangement and they are still in use, in particular in commercial instruments. However, a series of variants have been developed which have clear advantages over the original active DOAS setup.

7.8.1 Classic Active Long-Path System

The original design of active DOAS systems (see Fig. 7.53) consisted of a receiving telescope, coupled to a dispersing device, e.g. a spectrometer and a telescope, 'looking' into the light beam emitted from a light source, e.g. a search light. This type of setup has been used in many investigations since the 1980s (Perner and Platt 1980; Platt et al. 1979, 1981; Harris et al. 1982). Disadvantages of this arrangement include the requirement to align the optical elements at both ends of the light path, and the requirement for power at two locations, which frequently complicates logistics in field campaigns.

In active long-path DOAS systems, a receiving telescope of sufficiently large aperture, usually 150–300 mm, is required to collect the light. The telescope has to be precisely pointed at the light source or retro-reflector array (see Sect. 7.3.10). This can usually only be accomplished by active control of the telescope. Even a very

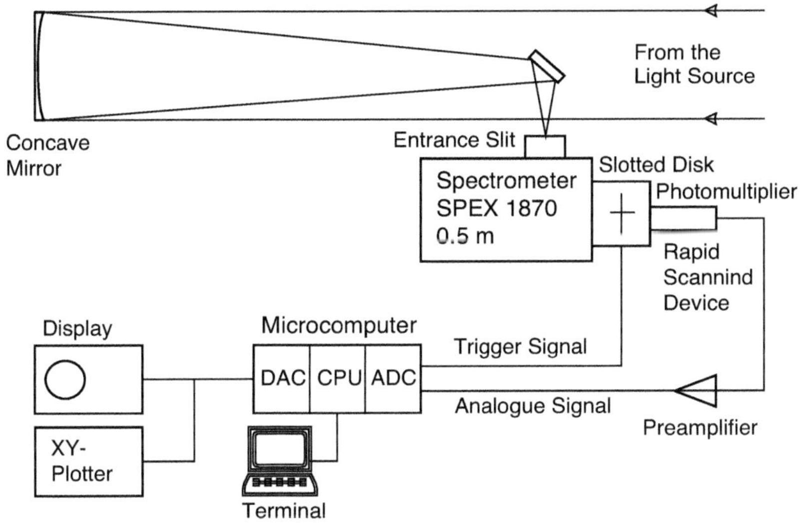

Fig. 7.53 Schematic setup of a 'classic' DOAS system using separate transmitting and receiving telescopes and an optomechanical (rotating 'slotted disk' + photomultiplier) scanning system

Table 7.10 Data pertinent to high resolution and low resolution DOAS systems for the spectroscopic detection of OH radicals and NO_3 radicals

Radical parameter	OH	NO_3
Wavelength of strongest band or line l/nm	308	662
Line width Γ/nm	0.0017	3.2
Effective abs. cross-section σ/cm^2	1.3×10^{-16}	1.8×10^{-17}
Minimum detectable optical density D_0	10^{-4}	10^{-4}
Detection limit C_0 molecules/cm^3 ($L = 5$ km)	1.5×10^6	1.1×10^7
Interfering molecules	SO_2, CH_2O, CS_2, $C_{10}H_8$	H_2O

rigid mounting of the optical building blocks of the system will usually not solve the problem. Due to an 'El Mirage' effect, the light path will 'bend' as a consequence of changing thermal gradients in the atmosphere.

The aperture ratio of the telescope must match that of the spectrometer, or of the connecting quartz fibre (see Fig. 7.53).

7.8.2 High-Resolution Long-Path DOAS Systems

Some species, in particular atoms and certain small molecules, exhibit narrow absorption lines (on the order of one or a few picometers, see Table 7.10), which dictate the use of high spectral resolution spectrometers in combination with high spectral intensity light sources. A prominent case is the OH radical, which was reliably detected

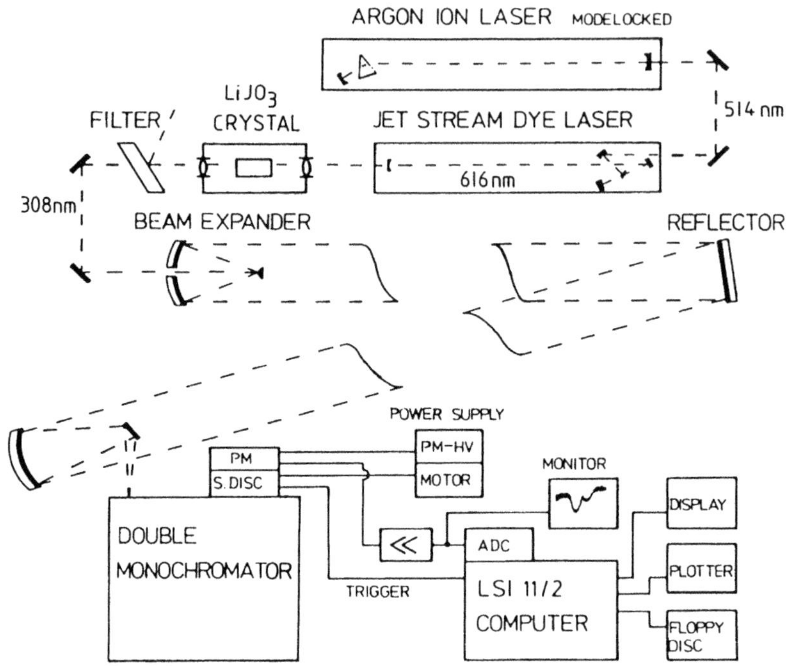

Fig. 7.54 Optical setup of a high-resolution long-path spectrometer. A laser system supplies the high-spectral intensity required due to the high resolution of 0.002 nm (instead of about 0.5 nm for a low-resolution DOAS). Reprinted with permission from Platt et al. 1987,

in the atmosphere for the first time by DOAS (Perner et al. 1976). The required high-spectral energy density can be supplied by lasers and LEDs. Its emission bandwidths can either be broad or narrow compared to the OH lines. The former approach has a fixed laser emission wavelength, but requires combination with a high-resolution spectrometer (Perner et al. 1976, 1987a, 1987b; Hübler et al. 1984; Platt et al. 1987, 1988; Dorn et al. 1988; Hofzumahaus et al. 1991; Mount 1992). The latter variety of the technique uses a narrowband laser scanning the OH line (Zellner and Hägele 1985; Amerding et al. 1990, 1992).

An experiment following the broadband approach is shown in Fig. 7.54. A frequency-doubled dye laser coupled to a beam-expanding telescope serves as a light source. To keep interference at negligible levels, the beam is expanded to about 0.2 m diameter (Perner et al. 1976). The laser emission bandwidth (FWHM) of about 0.15 nm is large compared to the width of the OH line (≈0.0017 nm). The received light is dispersed by a 0.85-m double spectrometer with a resolution comparable to the OH line width (Perner et al. 1987a; Hofzumahaus et al. 1991). Table 7.10 lists some key parameters of a long-path OH spectrometer in comparison to a 'classical' low-resolution DOAS instrument. An important consideration is the amount of self-produced OH, since the intense UV radiation of the laser beam photolyses ozone, yielding O^1D atoms that can react with H_2O to produce OH in the beam.

Later designs of this type of instrument coupled the high-resolution spectrometer to a multi-reflection cell (see Sect. 7.7) to allow more localised measurements and better control of the optical path (Brauers et al. 1996; Dorn et al. 1993, 1996, 1998; Hausmann et al. 1997; Hofzumahaus et al. 1991, 1998; Brandenburger et al. 1998).

In addition, the bulky high-resolution grating spectrometer can be replaced by compact Fabry-Pérot instruments (see Sect. 7.4.3), as described by Kuhn et al. (2021) and Kuhn et al. (2023).

7.8.3 Designs of Active Long-path DOAS Systems

There are several variations of the basic active DOAS-system design. Most use a reflector (usually a corner cube retro-reflector as indicated in Fig. 7.55) to return the light of a source located next to the spectrometer. Thus the total light path length is twice the distance between light source/spectrometer and reflector. A coaxial merge of transmitting and receiving telescope as shown in Fig. 7.55 is often used. This setup was first described by Axelsson et al. (1990b). Advantages and limitations with regard to atmospheric visibility are similar to the classic arrangement described in Sect. 7.8.1. However, power and delicate alignment is only necessary at the spec-

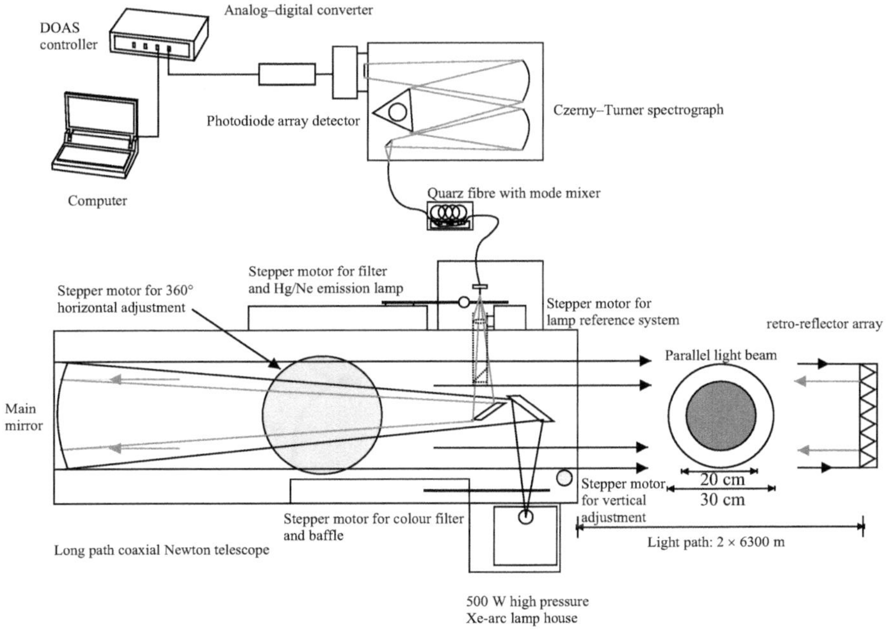

Fig. 7.55 Schematic setup of a DOAS system using a coaxial arrangement of transmitting and receiving telescope in conjunction with a retro-reflector array (Geyer et al. 2001a; 2001b). This type of setup, pioneered by Axelsson et al.(1990a), is the standard for active DOAS systems for research but is gradually replaced by forked fibre bundle telescopes introduced by Merten et al. (2011), see Fig. 7.56. Reprinted from Geyer (2000), Ph.D. Thesis, 2001, University of Heidelberg

trometer/light source site (Plane and Nien 1992; Edner et al. 1993a; Hausmann and Platt 1994; Heintz et al. 1996; Alicke et al. 2002; Geyer et al. 1999; Allan et al. 1999, 2000a, 2000b). The introduction of retro-reflectors also has the advantage of reducing the impact of atmospheric turbulence on the light beam (see Sect. 7.3.10). This is due to the capability of corner cube retro-reflectors to return the incident light exactly (although with some lateral offset) into the direction of incidence. In addition, the light collected by the receiving section of the telescope is transmitted to the spectrometer via quartz fibre, allowing the introduction of a 'mode mixer' (see Sect. 7.3.7; Stutz and Platt 1997b), which further reduces the effect of atmospheric turbulence. Because of these advantages, retro-reflector arrangements such as the one shown in Fig. 7.55 have become the standard setup for active long-path DOAS instruments.

An important improvement of the folded-path design of active DOAS instruments was described by Merten et al. (2011). It employs a forked fibre bundle, the single entrance of which is mounted in the focal plane of a telescope while the other (split) ends are connected to (1) the light source and (2) the spectrometer entrance, as shown in Fig. 7.56. In principle, the light emitted by the fibers from the light source would be returned to exactly these fibers of the bundle E and returned to the light source. In practice, the lateral offset of the returned light together with the slight deflection of the beam by the atmospheric turbulence cause a considerable fraction of the returned light to hit the receiving fibre. In addition, it was found that a slight de-focusing of

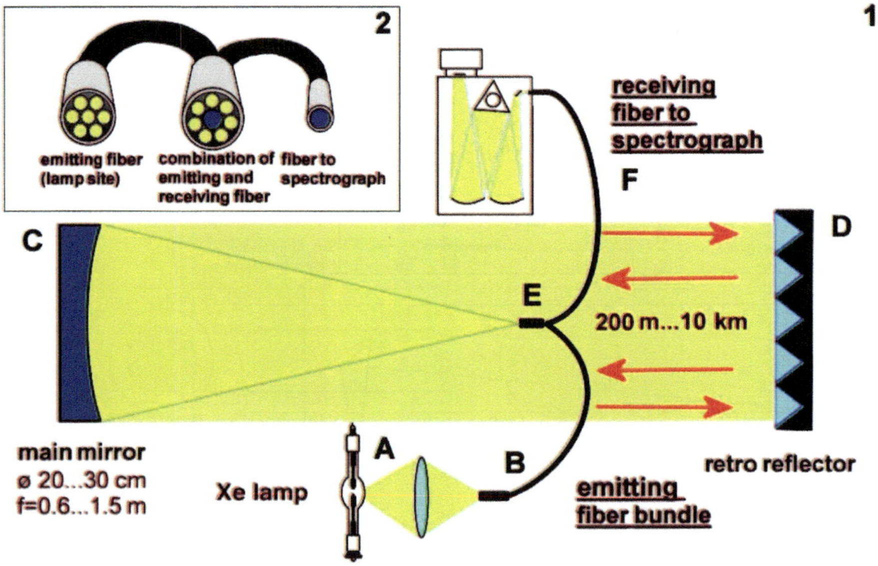

Fig. 7.56 Large figure (1): Sketch of a fibre LP-DOAS instrument with emitting and receiving fibers combined into a single bundle (E). The inset (2) shows the ends of the fibre bundle (left to right): Bundle from light source (B), emitting and receiving bundle in the telescope focal point (E), and fibre to spectrograph (F)). Reprinted with permission from Merten et al. (2011), ©Optical Society of America. All rights reserved

the telescope can enhance the returned light intensity. A detailed model study by Merten et al. (2011) showed that a setup with a single receiving fibre in the centre surrounded by 6 transmitting fibers can yield a maximum transmission of 36% for a 30cm diameter main mirror of an F/5 telescope and 100μm diameter fibers, compared to a maximum transmission of 11.5% for an equivalent coaxial setup. A potential problem associated with de-focusing the telescope is a somewhat larger acceptance angle and thus étendue of the instrument for scattered sunlight.

The main advantage of this design is the very robust setup, as the fibers in the transmitting/receiving bundle in the telescope focus are rigidly mounted together (E in Fig. 7.56). Rotating a diffuse reflector in front of the bundle end E at a short distance (typically around 1mm) allows recording the emitted spectrum of the light source.

A series of "Fibre DOAS" design applications are reported by Sihler et al. (2009); Pöhler et al. (2010); Liao et al. (2011); Vita et al. (2014); Stutz et al. (2016) and a number of improvements of the design are described by Nasse et al. (2019).

7.8.4 DOAS Systems with Optical Multi-pass Systems

The absorption light path may be folded in a multiple reflection cell. The optical arrangements are described in Sect. 7.7. Figure 7.57 shows a typical setup of a DOAS system using a White type open multiple reflection cell. The primary elements are the light source and entrance transfer optics to match the large aperture of the light

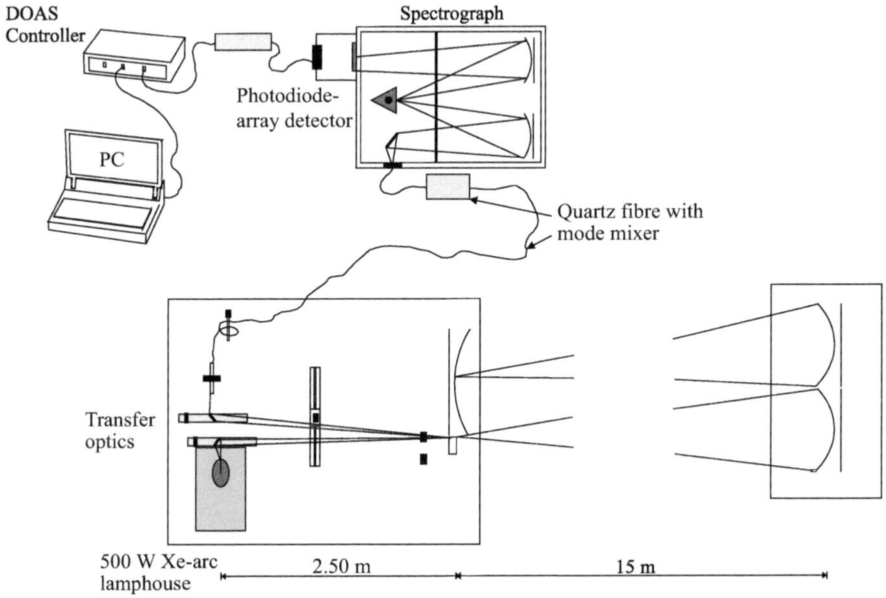

Fig. 7.57 Schematic optical setup of the 'White' multi-pass system with a base path of 15 m, as used during a field campaign in Pabstthum/Germany

source to the small aperture of the White multiple reflection system. After traversing the multiple reflection system, secondary transfer optics match the aperture to that of the fibre and spectrometer. The spectrometer, mode-mixer, and detector are of the same design as in the long-path system described in Sect. 7.8.3.

7.8.5 DOAS-LIDAR Systems (Scheimpflug LIDAR)

An interesting expansion of active DOAS is its use as a LIDAR. In contrast to normal mono-static LIDAR systems, where light source and detector are co-located and the return time of a light pulse is used to determine the distance of a scattering event, a bi-static LIDAR separates light source (emitter) and receiver by a distance $L_{SD} \sim 10m$ (Fig. 7.58). Range resolution is accomplished by changing either the radiation source elevation angle, α, or the radiation detecting telescope elevation angle, β (e.g. Barnes et al. 2003, 2007; Flock 2012; Mei and Brydegaard 2015). The magnitude of these angles is a factor in determining the spatial extent of the overlap region (OR), indicated by the light grey area in Fig. 7.58, between emitted beam and the field of view of the receiving system. From this overlap volume the light is scattered back to the receiving system. The imaging principle of the emitted light beam follows the rules of Theodor Scheimpflug.

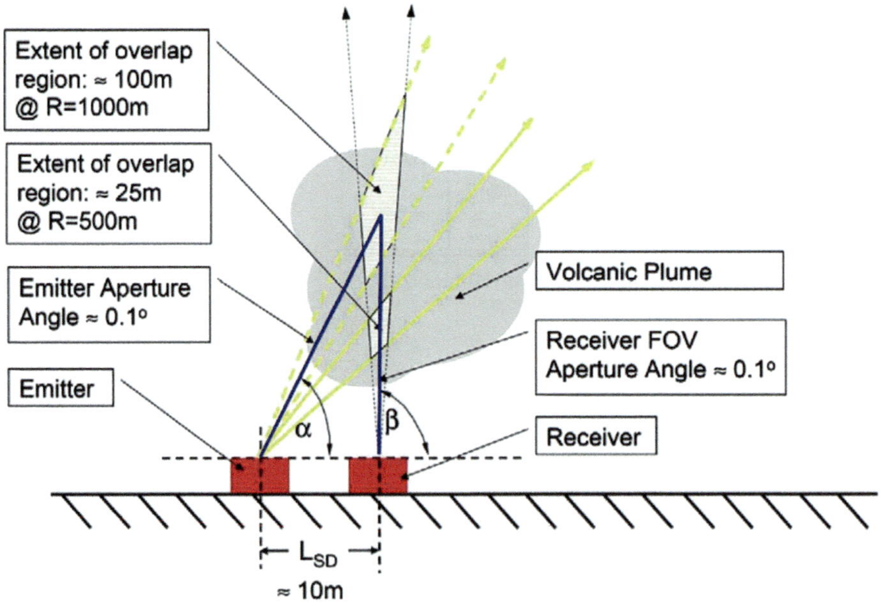

Fig. 7.58 Schematics of a Bi-Static LIDAR-DOAS (BISL-DOAS) instrument. Receiver FOV-elevation angle is arbitrarily assumed at $\beta = 90°$, in reality emitter elevation angle α would only be slightly less than 90°, thus an extended overlap region ensures large signal return. Note that the angle $(\beta - \alpha)$ between receiver FOV and emitter elevation angle is exaggerated

There are several advantages of the Scheimpflug LIDAR:

1. There is no need for a nanosecond pulsed LASER light source, in fact continuous waveform (CW) light emitting diodes, in combination with suitable optics, can be used (see Flock 2012 and Platt et al. 2015).
2. A high time-resolution detector is not required, thus radiation can be received by a 1D imaging telescope–spectrometer combination. Consequently, trace-gas differential absorptions can be measured if broad-band radiation sources are used.
3. The vertical extent of the OR between transmitted beam and detector field of view grows approximately with the square of the distance, R, of the overlap region from the receiver (see Fig. 7.58). Since the amount of radiation received by the detector increases with the volume of the OR, this R^2-dependence (Barnes et al. 2003) approximately compensates the $1/R^2$ decay of signal intensity with increasing distance from the measurement volume. This solves the common LIDAR problem of extreme dynamics of the returned signal, but at the expense of diminishing spatial resolution at larger distances R.

Calculations by Flock (2012) and Platt et al. (2015) and measurements by Flock (2012) show that the approach is feasible and can achieve moderate sensitivity (e.g. a few ppb for NO_2 and several 100 ppb for SO_2) at a range of several kilometres and a range resolution of tens of metres. Therefore DOAS/Scheimpflug-LIDAR applications appear promising for the study of emission plumes from smoke stacks, volcanoes, or other point sources, where elevated trace gas concentrations prevail.

7.9 Light Utilisation in an Active Long-path Spectrometer

The spectral noise, and thus detection limit and accuracy of DOAS instruments, usually depends on the amount of light reaching the detector (see Sect. 7.5). Thus, it is important to maximise the fraction of light provided by the light source that reaches the detector. This section describes light throughput of a typical long-path spectrometer. For simplicity, we take the dimension of a single-path DOAS system, as sketched in Fig. 7.59. The results are also valid for folded-path systems (see Sect. 7.8.3).

The fraction x_1 of light emitted from the radiating surface of the light source, for example from the filament of an incandescent lamp or arc of a Xe high-pressure lamp (see Sect. 7.2.1), transformed into a light beam, can be approximated as follows:

$$x_1 \approx \frac{r^2\pi}{4\pi f^2} = \frac{1}{4} \cdot \left(\frac{r}{f}\right)^2 \tag{7.76}$$

where r denotes the radius of the light-source mirror and f its focal length (see Fig. 7.59). Assuming the radiating surface of the lamp to be circular with the radius r_1 (and thus the surface $r_1^2\pi$), the divergence of the beam (half angle) is given by:

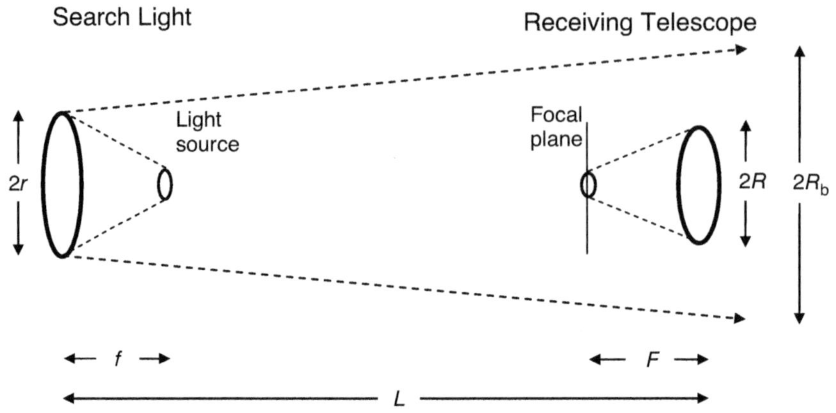

Fig. 7.59 Optical sketch of a single-path active DOAS system. The search light mirror of the transmitting system has radius r and focal length f, the corresponding quantities for the mirror of the receiving telescope are R and F

$$\alpha \approx \frac{r_l}{f} \tag{7.77}$$

and the radius R_b of the light beam at the receiving telescope located at the distance L from the search light (assuming $R_\mathrm{b} >> r$ and setting $\alpha \simeq \tan \alpha$) is:

$$R_b \approx \alpha \cdot L = \frac{r_l \cdot L}{f} \tag{7.78}$$

We obtain the fraction x_2 of the emitted light beam actually collected by the telescope:

$$x_2 = \left(\frac{R}{R_b}\right)^2 = \left(\frac{R \cdot f}{r_l \cdot L}\right)^2 \tag{7.79}$$

Combining (7.79) with (7.76) gives the fraction x_e of the total lamp output received by the telescope:

$$x_e = x_1 \cdot x_2 = \frac{1}{4} \cdot \left(\frac{R \cdot r}{r_l \cdot L}\right)^2 \tag{7.80}$$

Note that for a given radius r, x_e is independent of the focal length f of the search light mirror and thus its aperture. On the other hand, x_e is proportional to the product of the areas of the two mirrors and inversely proportional to the radiating area of the lamp, and thus its luminous intensity (radiated energy per unit of arc area). The illumination intensity D_b in the focal plane of the receiving telescope (i.e. the brightness of the focal spot) is given by:

$$D_b = \frac{P_\mathrm{l} \cdot x_e}{(r \cdot F/L)^2 \pi} \tag{7.81}$$

where P_l denotes the lamp output power and $(r\ F/L)^2\pi = R_f^2\pi$ the area of the focal spot. Substituting x_e from (7.80) yields:

$$D_b = \frac{P_l}{4\pi} \cdot \left(\frac{R}{r_l \cdot F}\right)^2 \tag{7.82}$$

The illumination intensity of the focal spot depends neither on the length of the light path L (if light extinction in the atmosphere is neglected) nor on the properties of the search light optics. The size of the focal spot (i.e. its radius R_f) will be proportional to the size (r) of the search light mirror, and inversely proportional to the light path length:

$$R_f = \frac{r \cdot F}{L} \tag{7.83}$$

From (7.82), it appears as if the light throughput of the system could be maximised by choosing the largest possible aperture, $A_r = 2R/F$, of the receiving telescope. However, aperture A_r, must not be larger than aperture A_s of the spectrometer. If $A_r > A_s$, a portion of the light received by the telescope will simply not be used. Thus, for a given value of A_s, the brightness of the focal spot has a maximum value given by substituting $R = A_s F/2$ into (7.82):

$$D_b = \frac{P_l}{16\pi} \cdot (A_s/r_l)^2 \tag{7.84}$$

For a long-path spectrometer the size of the optical system does not matter, provided the aperture of the receiving system matches that of the spectrometer and the dimensions of the optical components exceed certain 'minimum sizes' given by (7.83).

Depending on the relative sizes of the focus R_f and the entrance slit $(w_S \cdot h_S)$, three regimes can be distinguished (see Fig. 7.60):

1. $w_S, h_S < R_f$: The amount of light entering the spectrometer will be independent of L (except for extinction). This is due to the independence of the illumination intensity from L stated in (7.82).
2. $w_S < R_f < h_S$: The amount of light entering the spectrometer will vary nearly in proportion to L^{-1}.
3. $R_f < w_S, h_S$: The amount of light entering the spectrometer will vary proportional to L^{-2}

It can be concluded that the product of the size of the search light mirror and the focal length of the receiving system (the aperture of which matches the aperture of the spectrometer used) should be sufficiently large as to allow the system to be in Regime 1. Table 7.11 gives examples of the quantities derived earlier. Data are evaluated for typical DOAS systems with radii of search light and receiving telescope mirrors, $r = R = 0.15$ m, focal length of the search light mirror, $f = 0.3$ m, and focal length

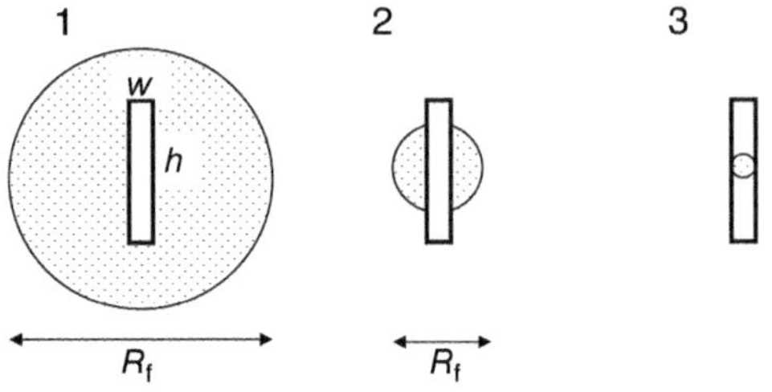

Fig. 7.60 Focusing the image of a retro-reflector or lamp on the spectrometer entrance slit there are three regimes: (1) For $w_S, h_S < R_f$, the radiation flux entering the spectrometer will be essentially independent of L (except for extinction); (2) For $w_S < R_f < h_S$ the radiation flux is approximately $\propto L^{-1}$; (3) If $R_f < w_S, h_S$, the radiation flux is approximately $\propto L^{-2}$

Table 7.11 Light losses in a long path spectrometer system (10-km light path) using a 'slotted disk'-type optomechanical scanner

Optical element/process		Loss factor (optomechanical scanner)	Loss factor (multi-element solid state detector)
Light source into beam	x_1	0.08	0.08
Geometric losses ($\alpha = 10^{-3}$ radian, $R_b = 17$ m)	x_2	3×10^{-4}	3×10^{-4}
Atmospheric attenuation (Mie and Rayleigh scattering)		0.2	0.2
Entrance slit		1.0	1.0
Spectrograph losses		0.2	0.2
Dispersion (fraction of total intensity received by individual pixel)		5.0×10^{-3}	5.0×10^{-3}
Multiplex losses (400 channels)		2.5×10^{-3}	1.0
Detector quantum efficiency		0.2	0.4
Total loss factor	I_{tot}	1.2×10^{-12}	10^{-9}
Total number N of photoelectrons recorded per pixel and second		$\approx 10^5$	$\approx 10^8$

of the telescope mirror, $F = 1.5$ m. Further assuming entrance slit dimensions (width w_S by height h_S) of 0.05 by 0.2 mm, the light paths for Regimes 1–3 are obtained:

Regime 1: $L <$ about 2000 m
Regime 2: $2000 < L <$ about 9000 m
Regime 3: $L >$ about 9000 m

As seen in Table 7.11, the total loss factor from lamp output to the detector is close to 10^{-12} in the case of an optomechanic scanning device with a photomultiplier detector. The total number of photons per second and spectral interval (assuming a division of the spectrum in 400 intervals) reaching the detector can be estimated as follows: A 450-W Xe-arc lamp emits about 2.5 W, corresponding to $dN_0/dt \approx 6 \times 10^{18}$ photons per second, in the form of light in the spectral range 440–460 nm (Canrad-Hanovia). Taking the loss factor from Table 7.11, about $dN/dt \approx 6 \times 10^6$ photoelectrons per second and spectral interval of 0.05 nm width are registered. This figure corresponds to a shot noise limit of about $4 \times 10^{-4}\,\mathrm{s}^{-1/2}$ (1σ).

7.10 Passive DOAS Systems

Passive DOAS systems measure atmospheric trace gases by using extraterrestrial light sources, such as sunlight or moonlight (see also Sect. 7.2.2). There are numerous observation geometries tailored to observe different parts of the atmosphere.

1. Direct sunlight or moonlight DOAS.
2. Zenith Scattered Light DOAS (ZSL-DOAS).
3. Multi-AXis DOAS (MAX-DOAS).
4. Imaging DOAS (I-DOAS).
5. Airborne DOAS (A-DOAS).
6. Satellite DOAS (S-DOAS).

The direct light geometry usually allows simple calculation of the light path based on the elevation of the celestial body above the horizon. In all other cases scattered (solar) radiation is used, where the effective light path depends on the observation geometry and the scattering properties of the atmosphere. Details of the light path calculations are given in Chap. 9. Some sample applications and results are described in Chap. 11.

We can also classify passive instruments by location or operating platform:

1. Ground-based Static DOAS
2. Ground-based Mobile DOAS
3. Airborne DOAS (A-DOAS).
4. Satellite DOAS (S-DOAS).

All passive DOAS instrument consist of a telescope, transfer optics or fibers, and a spectrometer-detector combination. Compared to active DOAS instruments, a relatively large aperture angle of the telescope can be tolerated for passive DOAS systems (on the order of 0.1–1 degrees instead of 10^{-3}–10^{-2} degrees). Therefore, short focal lengths are sufficient. Consequently, much smaller telescopes (aperture 25–50 mm) compared to typical active instruments are usually sufficient. A number of spectrometer designs are in use for passive DOAS, although small spectrographs are most popular (Sect. 7.4.2). Because of their multiplex advantage, solid state detectors (Sect. 7.5.3) are favored. In this section we will discuss the experimental setups for various passive DOAS observational geometries and platforms.

7.10.1 Direct Sunlight or Moonlight Setups

Measurements of direct sun or moon light provide information on the trace gas concentration integrated along the light path. The instruments typically consist of a spectrometer-detector combination coupled to a telescope. An optical and mechanical setup is needed to follow the apparent movement of these celestial bodies in the sky. These solar trackers can be quite complex, depending on their pointing accuracy, in particular when they are used on mobile platforms, such as cars, ships, and aircraft. Intermittent clouds can also cause problems. Care must be taken that the instrument always observes the full solar disk, since the partial solar disk exhibits very different Fraunhofer structures due to centre to limb darkening (see e.g. Bösch et al. 2003 and Scheffler and Elsässer 1974). Usually an opaque disk is inserted to ensure that radiation from all parts of the solar disk is recorded. This element also helps to reduce the solar radiation intensity which can damage the instrument.

7.10.2 Zenith Scattered Light DOAS

An early scattered light DOAS technique relies on scattered solar radiation observed with a telescope pointed to the zenith (Fig. 7.61). From the SCDs, trace gas concentrations in different layers of the atmosphere can be inferred. This technique has also become known as "Umkehr Technique" (see Götz et al. 1934).

ZSL-DOAS has been applied by many authors to the determination of vertical column densities (VCDs) and the vertical distributions of O_3, NO_2, NO_3, OClO, BrO, and $(O_2)_2$ (e.g. Götz et al. 1934; Brewer et al. 1973; Noxon 1975; Pommereau 1982; Mount et al. 1986; Pommereau and Goutail 1987; Wahner et al. 1990a; 1990b; Johnston et al. 1992; Sanders et al. 1987; Solomon et al. 1989a; 1989b; Solomon et al. 1987b; Schiller et al. 1990; Perner et al. 1991; Kreher et al. 1996; Otten et al. 1998; Solomon et al. 1989b; Kreher et al. 1997; Sarkissian et al. 1991; Erle et al. 1995). Examples can be found in Chap. 11.

Compared to direct sunlight observations a fixed telescope is sufficient and stratospheric species can be detected, even in the presence of cloud cover. These advantages come at the expense of relatively difficult calculations of the effective light path length. In addition, several effects of light scattering in the atmosphere have to be carefully compensated for, as outlined in Chap. 9.

Figure 7.61 shows the schematic setup of a ZSL-DOAS system. In this example, two telescopes, each coupled to its own spectrometer, were used to cover different wavelength ranges. Quartz fibres conduct the light from the telescopes to the spectrometers. The quartz fibres also remove the polarisation sensitivity of the instrument.

7.10.3 Multi AXis DOAS (MAX-DOAS) Instruments

Multi AXis DOAS has become a popular technique for the measurement of tropospheric trace gases. The basic principle is to perform a series of off-axis observa-

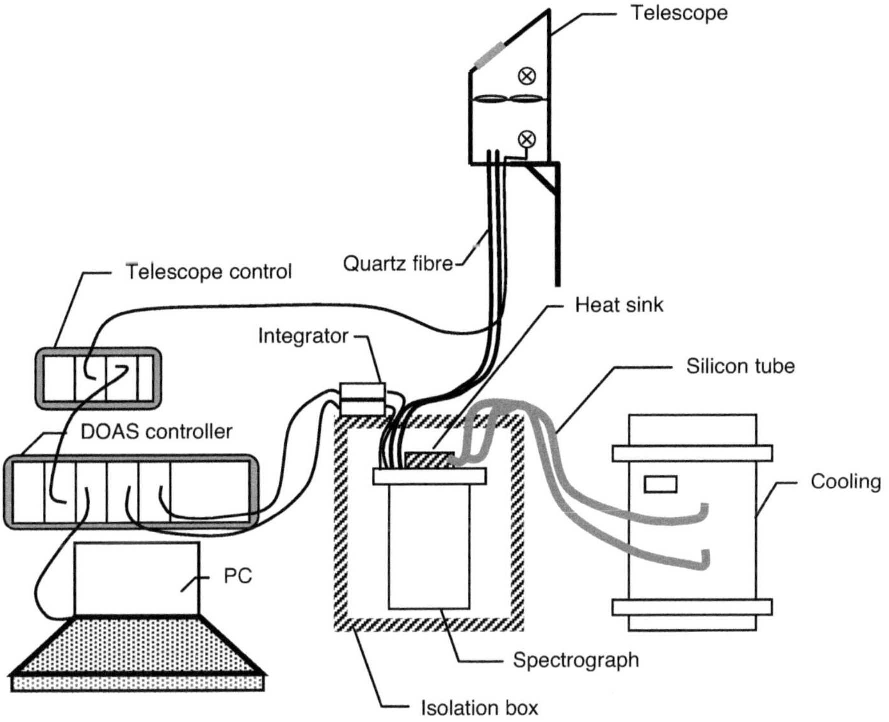

Fig. 7.61 Schematic setup of a ZSL-DOAS system. Reprinted from Hönninger, Ph.D. Thesis, 2002, University of Heidelberg

tions at different observation elevation angles $\alpha_1 \ldots \alpha_n$. The change of the observed trace gas slant column as a function of increasing α gives information about the vertical distribution of the trace gas (see Chap. 9). MAX-DOAS was pioneered by Hönninger 2002 (see also Hönninger and Platt 2002; Leser et al. 2003; Wagner et al. 2004; Bruns et al. 2004; Heckel et al. 2005) and has since been used in a large and growing number of applications. Numerous examples are given in Chap. 11.

There are different technical approaches for MAX-DOAS observations.
(1) Sequential observation at different elevation angles, as shown in Fig. 7.62. This approach has the advantage of a relatively simple setup, requiring only one spectrometer and a mechanism for pointing the telescope in different directions. It has therefore been used in a large number of studies (Chap. 11). The availability of very compact, low-cost spectrometer/detector combinations has led to the development of the "Mini-MAX-DOAS" instrument, shown in Fig. 7.63. In this setup the entire telescope detector assembly is pointed to the desired elevation by a motor. One disadvantage is that the measurements at different elevation angles are not simultaneous. A complete cycle encompassing several observation directions may take several minutes (e.g. Hönninger and Platt 2002). This can be a problem during periods of rapidly varying radiation transport conditions in the atmosphere, e.g. varying cloud cover, aerosol load, or during sunrise/sunset. In addition, the instrument contains

mechanically moving parts, which may be a disadvantage in adverse environments and during long-term operation at remote sites.

(2) Simultaneous observation at different elevation angles solves this problem - although at the expense of other complications. One disadvantage is the increased instrumental requirement, since a spectrometer and a telescope for each of the observation directions is needed. Some simplification can come from using 'two-dimensional' spectrometry, where a single spectrometer is equipped with a two-dimensional (CCD) detector. In these systems one dimension is used for dispersion and the other for spatial resolution, i.e. different viewing angles (Fig. 7.32, Sect. 7.10.4, and Heismann 1996; Wagner et al. 2002c; Wang et al. 2003, 2005; Heue et al. 2005).

A more severe problem lies in the fact that the different instruments, or different parts of a 2D CCD-detector, have to be ratioed against each other in order to eliminate the solar Fraunhofer structure (see Chap. 9). This proves difficult if different instrument functions are involved (e.g. Bossmeyer 2002). Although the situation can be somewhat improved by 'cross convolution' (i.e. convoluting each measured spectrum with the instrumental line shape of the respective spectrum it is to be compared with), it appears that, with present instrumentation, the highest sensitivity is reached with sequential observation.

A solution could lie in the combination of techniques (1) and (2), i.e. the use of multiple spectrometers (one per observation direction) and moving telescopes

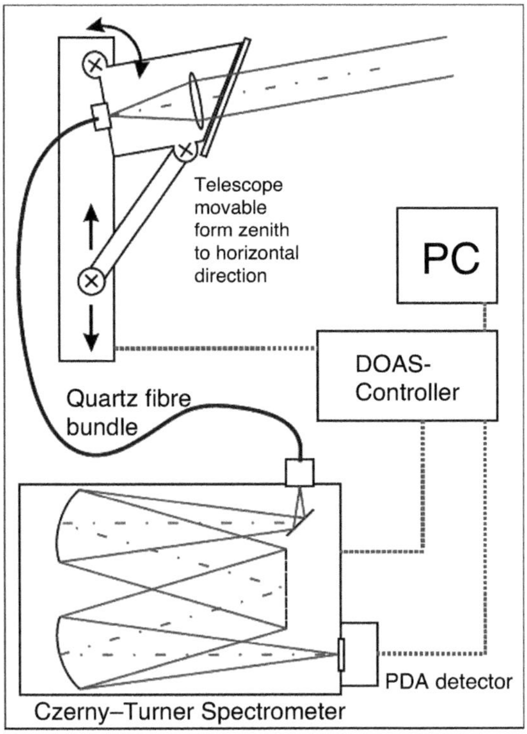

Fig. 7.62 Schematic setup of a MAX-DOAS system. A small telescope coupled to a spectrometer can be pointed to any elevation between horizon and zenith. Reprinted from Hönninger, Ph.D. Thesis, 2002, University of Heidelberg

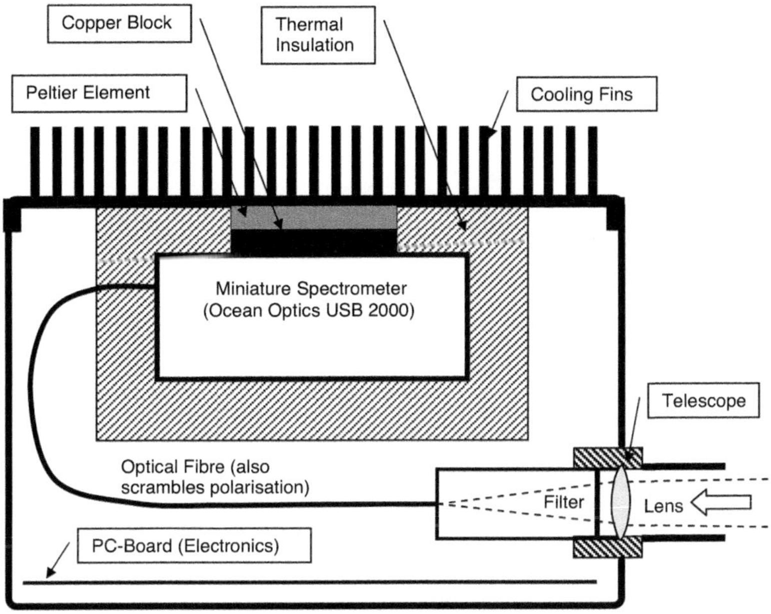

Fig. 7.63 Schematic cross-section through a miniature MAX-DOAS system (Bobrowski 2005). The entire instrument, including the telescope, can be pointed to any elevation between horizon and zenith

(or spectrometer–telescope assemblies). While this approach appears to combine the disadvantages of the two above setups, it also combines their advantages. If each individual instrument sequentially observes all elevation angles, including the zenith, and the observation are phased in such a way that at any time each angle is observed by one instrument, simultaneous observation of all angles is achieved. Instruments based on this approach have been used (see Leser et al. 2003; Bossmeyer 2002).

7.10.4 Imaging DOAS (I-DOAS) Instruments

Imaging DOAS can be thought of as an expansion of MAX-DOAS, where many elevation angles are measured simultaneously or quasi-simultaneously. By adding a scanner and/or using an imaging spectrograph, absorption spectra for a two-dimensional array of directions is measured. Thus, the measurement results in a three-dimensional data product with two spatial dimensions and one wavelength dimension, which can be called a 'super colour' or 'hyperspectral' image, in contrast to ordinary colour images containing intensity information of only three wavelength ranges.

There are three commonly used techniques to obtain this information, distinguished by the temporal sequence of acquiring data of the three dimensions (Fig. 7.64).

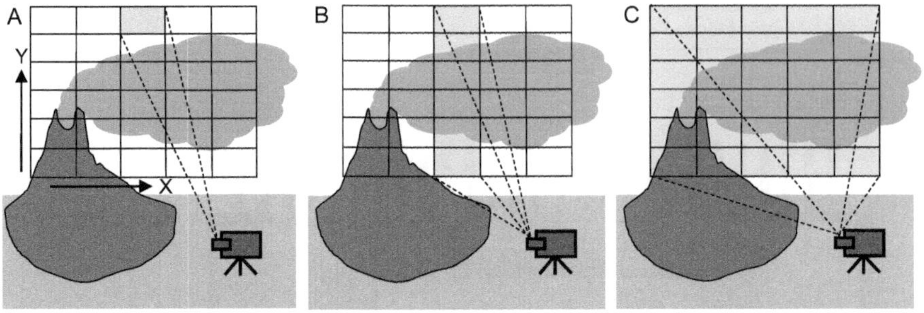

Dim.	Whiskbroom	Pushbroom	Full Frame
X	Time	Time	Space
Y	Time	Space (FP)	Space
λ	Space (FP)	Space (FP)	Time

Fig. 7.64 Principles of DOAS imaging: **A** a "Whiskbroom" imaging system sequentially scans all pixles of the scene. **B** a "Pushbroom" imaging system scans a column at a time. **C** a "Full Frame" imaging system views the entire frame at once, but wavelength scanning is done sequentially

1. **Whiskbroom** imaging sensors acquire the spectrum of one spatial pixel at a time (Fig. 7.64A). Hence, each of the two spatial dimensions needs to be scanned consecutively. Extending non-imaging sensors with a mechanical (two axis) scanning mechanism leads to this type of instrument. While preserving the major advantage of relatively simple design of the diffractive optics and photon detector elements, the disadvantage of the whiskbroom scanning principle is a linear dependence of total measurement time on the total number of spatial pixels, i.e. the time increases proportional to the square of the desired image resolution.
2. **Pushbroom** imaging sensors simultaneously image one spatial direction (Fig. 7.64B) and spectrally disperse this light in order to obtain a spectrum of each pixel of the investigated object. Doing this simultaneously requires a two-dimensional detector. Since scanning is needed only in the one spatial direction (horizontal in Fig. 7.65), the scanner is simplified. In addition, the scanner speed can be much lower (see e.g. Lohberger et al. 2004; Kern and Kelly 2023; General et al. 2014).
3. **Full frame** imagers record the entire image at once using a spectral camera (Fig. 7.64C). While both spatial dimensions are sampled simultaneously, wavelength scanning is done sequentially.

We note that the light throughput and thus measurement time of pushbroom instruments are not inherently superior to whiskbroom instruments if both instruments encompass a spectrograph with the same étendue. For whiskbroom scanning all radiation from a single pixel is directed to the spectrograph while in pushbroom configuration the étendue is shared among all pixels of an image column. Therefore, it takes much longer (by a factor equal to the number n_y of pixels in a column) to record a column for a pushbroom- than for a whiskbroom instrument. Overall the imaging

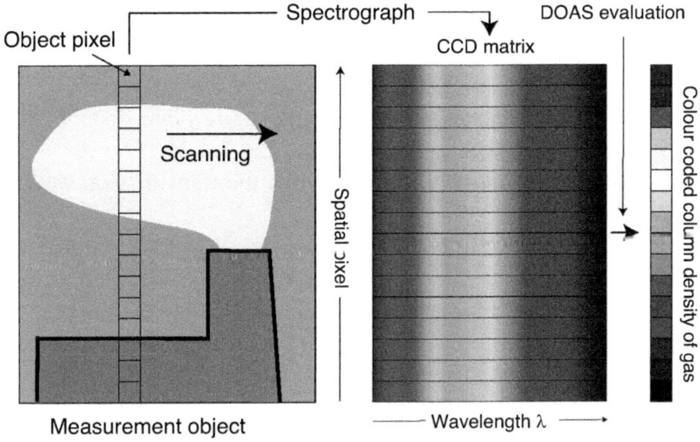

Fig. 7.65 Measurement principle of an I-DOAS instrument. The instrument utilises pushbroom imaging to simultaneously acquire spectra of one spatial direction (i.e. vertical). Spectral information along the second spatial dimension (horizontal) is obtained by scanning the scene using a scanning mirror. The spectra are evaluated using the DOAS technique. Reprinted with permission from Lohberger et al. 2004, ©Optical Society of America. All rights reserved

time will be equal for both systems, but the scanner of a whiskbroom instrument is much simpler.

The term 'Imaging Spectrograph' is used in several ways: Traditionally it refers to a spectrograph that can image a position along the entrance slit onto a corresponding position in the focal plane. The term Imaging Spectrograph is also used in a more general sense for any instrument capable of producing 2D-images with some spectral information, i.e. full-frame instruments also known as "Spectral Cameras" (Sect. 7.10.5). In principle this definition would include colour cameras. However cameras with defined spectroscopic properties are usually implied.

The most common application of I-DOAS is the visualisation of emission plumes, for example from volcanoes or smoke stacks. Radiative transfer is thus characterised by small distances between the instrument and the object of investigation, compared to scattering lengths by atmospheric Rayleigh and Mie scattering. Typical extinctions due to Rayleigh and Mie scattering (at e.g. 400 nm) are $4.6 \times 10^{-7}\,\text{cm}^{-1}$ and $4 \times 10^{-7}\,\text{cm}^{-1}$, which result in scattering lengths of 22 km and 25 km, respectively (see Sect. 4.2). In contrast, the typical range of gas plumes being visualised is a few hundred metres to a few km.

7.10.5 Spectral Cameras

A modern manifestation of imaging DOAS is spectral cameras. These instruments contain wavelength sensitive elements, modifying the intensity of the radiation recorded by the detector. Technical solutions for the wavelength sensitive elements include (Fig. 7.66 and Platt et al. 2015):

1. A set of two or more interference filters, brought into the beam sequentially. This approach is also known as a 'UV-Camera' or 'SO_2-Camera' (see e.g. Mori and Burton 2006 and Kern et al. 2010).
2. A Fabry-Pérot interferometer (FPI) with adjustable plate distance d (by either physically changing d or tilting the FPI, see Sect. 7.4.3)
3. An interference filter, which is tilted to change the transmission wavelength (see e.g. Platt et al. 2015)
4. A cuvette containing the gas species to be measured. This technique is also known as 'Gas Correlation Spectroscopy' (see e.g. Kuhn et al. 2022)
5. A tunable filter, for instance an Acousto Optical Tunable Filter (AOTF) (e.g. Dekemper et al. 2016)

Overall the listed arrangements sample the spatial information (i.e. the image) simultaneously, but the spectral information sequentially. The time to acquire an individual image is a compromise between the amount of spectral information sampled and temporal resolution. For instance, in Case 1 (set of interference filters sampled) two images are typically required, which leads to rapid acquisition of column

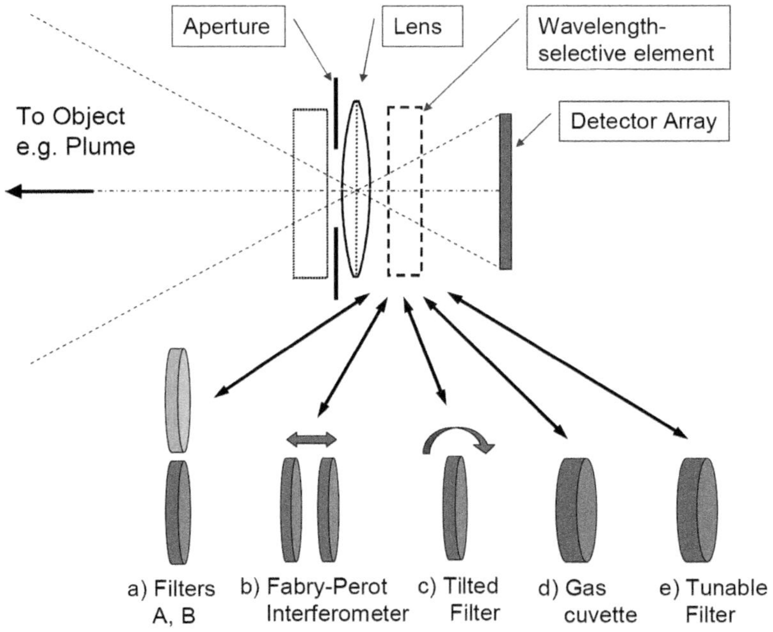

Fig. 7.66 Trace gas camera schemes: The basic set-up consists of a (UV) lens, aperture, wavelength selective element (WSE) and suitable 2-D detector. The WSE can either be mounted in front of the lens (dotted box) or between lens and detector (dashed box), the relative merits of the two approaches are discussed by Kern et al. (2010). The different approaches are distinguished by their WSE, which can be one of the following: (1) two narrow band filters alternatingly being brought into the beam. (2) a Fabry–Pérot interferometer with adjustable transmission wavelength (3) a narrow-band interference filter, which can be rotated. (4) a cell (cuvette) containing the gas to be measured, periodically introduced into the light beam (5) a tunable filter (e.g. AOTF)

density images at the expense of coarse wavelength information (essentially the intensity at two wavelength intervals with strong and weak absorption by the trace gas, respectively). This increases the possibility of spectral interferences. Similar relationships hold for Cases 3 and 5, where the compromise between spatial and spectral information can be changed more easily, i.e. several images with different filter settings can be combined to one column density image. Exceptions are Cases 2 and 4 (Fabry-Pérot interferometer, FPI, and Gas Correlation Spectroscopy, GCS). For the FPI the comb-like transmissions structure of the device can be aligned with regular absorption structures, for example of SO_2, to implement good interference rejection. Nevertheless, Case 2 also requires tuning the interferometer on and off the absorption structure. For GCS the absorption structure of the gas to be measured acts as filter.

7.10.6 Aircraft-Based DOAS (A-DOAS)

Passive DOAS instruments have been used on aircrafts in many applications. There are several possible viewing strategies for these instruments that impact their design:

- ZSL geometry, where one upwards viewing direction is used, follows the general setup of ground-based ZSL systems (e.g. Wahner et al. 1989a; 1989b, 1990a; 1990b; Schiller et al. 1990; Pfeilsticker and Platt 1994; Pfeilsticker et al. 1997a).
- MAX-DOAS geometries, where multiple upwards and downwards angles are implemented. As in the case of ground-based MAX-DOAS, elevation viewing angels can be changed using an active scanner or by using multiple spectrometers or an imaging type spectrometer-detector combination. For airborne MAX-DOAS with upward and downward viewing angles, the telescope has to be mounted on the side of the fuselage or in a pylon where the view is not obstructed. Separate telescope for upward and downward viewing can also be used (e.g. Wang et al. 2003, 2005; Heue et al. 2004a, b; Kreycy et al. 2013; General et al. 2014; Berthet et al. 2017).
- Nadir imaging strategies where a whiskbroom or pushbroom system is used to map the area below the aircraft. Both types of instruments provide spatially resolved measurements perpendicular to flight direction, while the movement of the aircraft is used to provide the other direction (Fig. 7.67). Since these systems are often mounted at the bottom of the aircraft, the measurement of a solar reference spectrum needs special attention.
- Limb and limb-scanning strategies take advantage of the very long light paths that can be achieved when looking horizontally or perpendicular to the flight direction (Fig. 7.67). In this approach the ability of the aircraft to change altitude is used to derive vertical concentration profiles. Many limb instruments also have the ability to point into the zenith to measure solar reference spectra.

Figure 7.67 shows the scanning strategy of the Univ. Heidelberg HAIDI instrument, which combines a nadir pushbroom and a limb-scanning approach. Here the

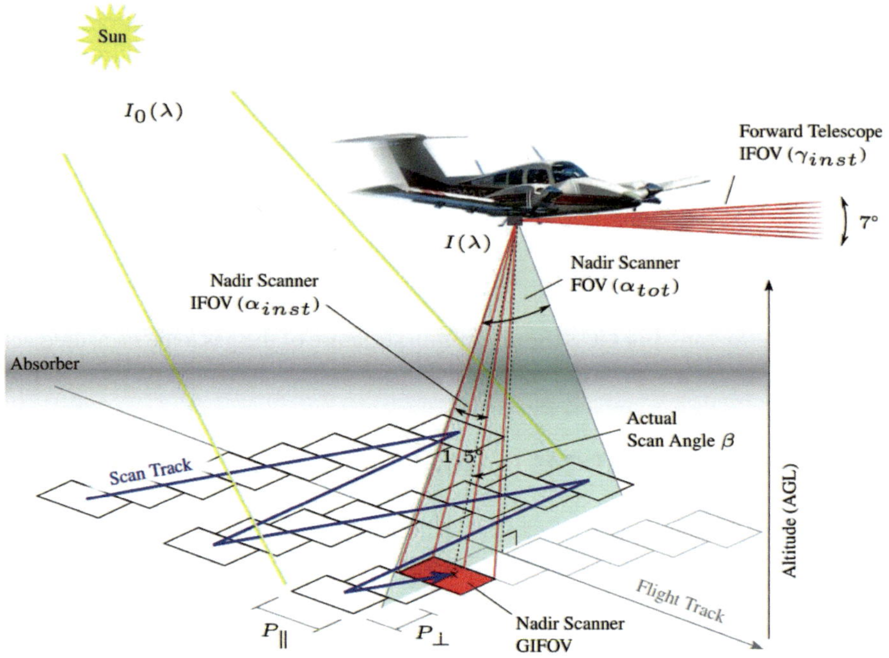

Fig. 7.67 Schematic setup of an airborne MAX-DOAS system (AMAX-DOAS). Shown is the Heidelberg Airborne Imaging DOAS Instrument (HAIDI) system in the configuration for low flying aircraft, which uses a whiskbroom scanner in nadir direction and a pushbroom scanner in forward direction. The instantaneous field of view of the two telescopes (α_{inst}, γ_{inst}) is marked in red. The whiskbroom nadir scanner uses a prism to rotate the viewing direction of the nadir telescope by angle β across track direction, resulting in the total FOV α_{tot} (green), the forward-looking telescope observes multiple viewing angles at once. The light is transmitted via quartz fibre bundles to two spectrographs. Reprinted with permission from General et al. (2014), ©CC 3.0. All rights reserved

nadir pushbroom is realised by a small telescope with a one-dimensional prism-based scanner. The limb direction scanning is achieved by using a telescope and a spectrometer with a two-dimensional CCD detector where different viewing angles are imaged at different heights on the detector (General et al. 2014). Examples of observations with the different viewing strategies are presented in Sect. 11.1.

Aircraft DOAS comes with a number of advantages and challenges. Advantages include the ability to make measurements at many places in a short period of time and in inaccessible locations. The ability of the aircraft to change altitude allows measurement of trace gas profiles during ascents and descents. This is particularly useful for limb geometries. Nadir mapping gives unique insights into the distribution of trace gases and is particularly useful for the validation of satellite instruments.

There are, however, some challenges when operating airborne DOAS systems. Because the orientation of the aircraft changes during flight the viewing angles can vary. Instruments thus have to either implement active control of the telescopes, i.e. correct aircraft pitch and roll in real-time, or correct for the varying viewing direction during the DOAS retrievals. The change in cabin or outdoor pressure poses another challenge because the spectrometer dispersion depends on the refractive index of air and thus pressure and temperature. There are two approaches to address the shift of the spectra with changing pressure: (1) Correct this effect during the spectral analysis, or (2) Maintain constant pressure in the spectrometer, for example by mounting the spectrometers in a vacuum chamber. Finally, airborne DOAS instruments have to be built for high mechanical stability to withstand the vibrations and accelerations encountered during the flight.

7.10.7 Balloon-Borne Instruments

Balloon-borne DOAS derives vertical trace gas concentration profiles by taking a succession of spectra from a balloon-borne spectrometer. Most balloon observations have focused on the stratosphere where low air density makes Rayleigh scattering inefficient. Direct sunlight observations are thus the most common viewing strategies for these instruments, although MAX-DOAS observations have also been performed. Balloon borne observations can be used to derive trace gas vertical profiles in different ways:

1. A series of slant column densities $S(z)$ above the balloon altitude z can be derived from spectra recorded during ascent. The concentration profile, $c(z)$, can be derived, by conversion of $S(z)$ to a vertical column density profile, followed by differentiation of this profile (e.g. Pommereau and Piquard 1994a; Pundt et al. 1996; Camy-Peyret et al. 1995, 1999; Ferlemann et al. 1998; Fitzenberger et al. 2000; Bösch 2002; Bösch et al. 2003).
2. From a balloon floating at high altitude, a series of spectra is taken during sunset (or sunrise). From the derived trace gas column densities $S(\vartheta)$, as a function of SZA and ϑ, the concentration profile $c(z)$ can be derived by an 'onion peeling' approach (Weidner et al. 2005; Dorf et al. 2005).
3. MAX-DOAS techniques, i.e. observation of trace gas SCDs at different elevation angles, can be applied.

An example of the required hardware is shown in Figs. 7.68 and 7.69. Figure 7.68 illustrates the mechanical setup of a balloon-borne direct-sunlight DOAS system, which is housed together with an FT-IR instrument on a three-axis stabilised balloon gondola (Camy-Peyret et al. 1997). A moveable mirror directs the sunlight into the spectrometers.

Balloon DOAS instruments come with similar challenges as airborne DOAS systems. In particular, the large change in pressure as the balloon reaches the stratosphere has to be addressed. Figure 7.69 depicts a cross-section of the pressurised dual DOAS spectrometer module that overcomes the pressure sensitivity.

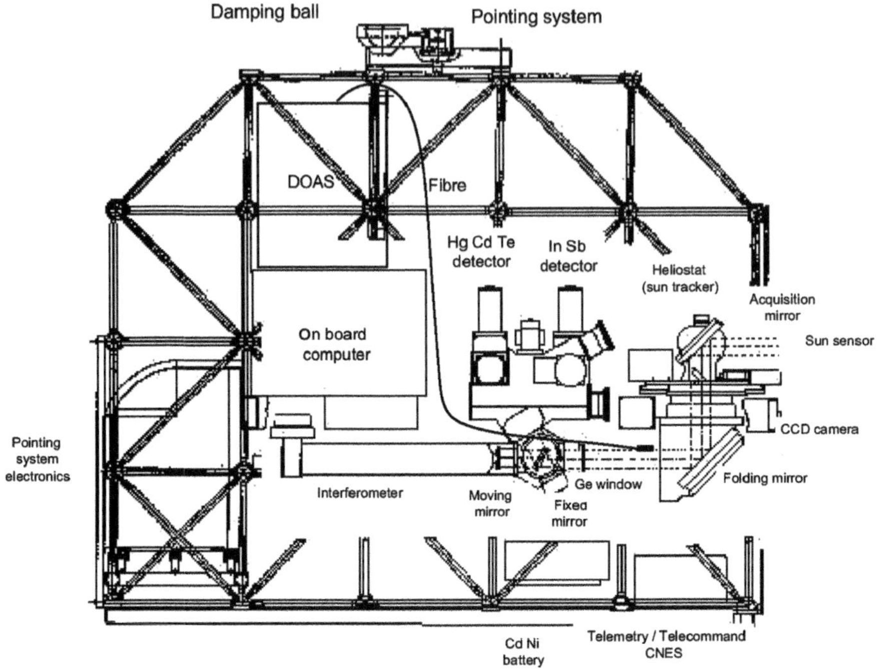

Fig. 7.68 Setup of a balloon-borne direct-sunlight DOAS system. The DOAS instrument is housed together with an FT-IR instrument on a three-axis stabilised balloon gondola. Reprinted with permission from Camy-Peyret et al. (1997), ©CNRS. All rights reserved

7.10.8 Satellite Instruments

The desire to observe the global distribution of trace gases motivated the development of absorption based satellite instruments. Earlier instruments such as the Solar Backscatter UV (SBUV) (Dave and Mateer 1967) measured the ozone column density by taking the ratios of a few wavelength pairs, essentially bringing the concept of the Dobson spectrometer (Dobson and Harrison 1926; Dobson 1968) to space.

Beginning with the launch of the Global Ozone Monitoring Experiment (GOME) on the ESA satellite ERS-2 (GOME Users Manual 1995) in 1995, DOAS principles have been applied to satellite instruments with great success. Since GOME, several additional DOAS-type instruments have been put into orbit. The list includes:

The SCanning Imaging Absorption spectroMeter for Atmospheric CHartographY (SCIAMACHY) (e.g. Burrows et al. 1991; Noël et al. 1999) on the ESA ENVIronmental SATellite (ENVISAT), OSIRIS on ODIN (Murtagh et al. 2002), Global Ozone Monitoring by Occultation of Stars (GOMOS), Ozone Dynamics Ultraviolet Spectrometer (ODUS), the Ozone Monitoring Experiment (OMI, Levelt et al. 2018) on the NASA Earth Observing Satellite Aura (EOS-Aura), the TROPOspheric Monitoring Instrument (TROPOMI, Sentinel-5 precursor mission, Veefkind et al. 2012), the

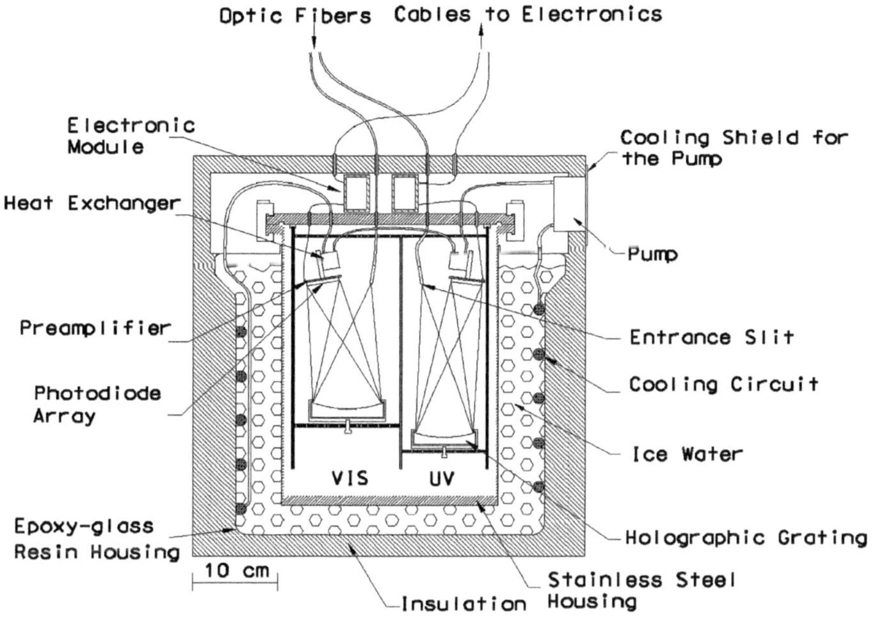

Fig. 7.69 Spectrometer of a balloon-borne direct-sunlight DOAS system (Bösch 2002). Two spectrometers (one for the visible (VIS) and the other for the ultraviolet (UV) spectral range) receive direct sunlight through a heliostat (see Fig. 7.68) and quartz fibres. The spectrometers are sealed in a stainless steel housing and thermostated to 0°C by a ice-water package inside a Styrofoam insulated housing. Reprinted from Bösch, Ph.D. Thesis, 2002, University of Heidelberg

Environmental Trace Gases Monitoring Instrument (EMI, on the Chinese GaoFen-5 satellite, Zhang et al. 2020).

Instruments on satellites in geostationary orbits were also launched: Geostationary Environment Monitoring Spectrometer (GEMS) (GEO-KOMPSAT-2B, Korea Kim et al. 2020), Tropospheric Emissions: Monitoring of Pollution (TEMPO) (Intelsat 40c, NASA, USA, Zoogman et al. 2017). Sentinel-4 (on Copernicus, ESA planned for 2024). Details on launch date, orbit and species measured can be found in Table 11.1.

Three basic viewing geometries are utilised for DOAS measurements (see Sect. 11.7):

1. Nadir view: Looking down from space towards the nadir direction, the sunlight reflected from the earth's surface or atmosphere (called the 'earthshine' in analogy to 'moonshine') can be utilised to obtain information on atmospheric trace gas column densities. Examples of satellite instruments using this geometry include GOME, ILAS, SCIAMACHY, ODIN, and OMI, as shown in Table 11.1.

2. Limb view: In this geometry, light scattered from the earth's terminator is analysed. By observing the limb at different tangent altitudes, vertical trace gas concentration profiles can be inferred.
3. Occultation: By observing the light of the rising or setting sun, moon, or stars through the atmosphere, vertical trace gas concentration profiles can be deduced.

In addition, there are several types of orbits which are particularly suited for probing the atmospheric composition from satellite:

1. Low earth orbit (LEO), where the satellite circles earth at an altitude around 800 km. Particularly advantageous are sun-synchronous polar orbits, from where the satellite instrument can observe the whole earth each day or within a few days. In the latter case, the time for one orbit can be adjusted such that the instrument will see a specific point on earth at essentially the same local time (e.g. 10:30 in the morning in the case of ERS-2)
2. Geostationary orbits, where the satellite circles earth in synchronicity with the rotation of earth at an altitude of 36000 km.

An example of the LEO is GOME on board the second European Research Satellite (ERS-2), later followed by three nearly identical instruments GOME-2A through GOME-2C on ESA MetOp satellites. A schematic outline of the instrument is shown in Fig. 7.70. It consists of a wiskbroom scanning mirror mapping the Earth surface in nadir view, a set of four spectrometers, and thermal and electronic sub-systems (Burrows et al. 1991, 1999b). Light backscattered from earth is collected by the scan mirror, then focused on the entrance slit of the spectrometer by an off-axis parabolic telescope. After being collimated, the light passes through a pre-dispersing prism, which produces an intermediate spectrum within the instrument. The pre-dispersing prism is arranged in such a way that internal reflection results in a polarised beam which is directed towards three broadband detectors monitoring the polarisation state of the light, the polarisation monitoring device (PMD). The PMD observes radiation from the atmosphere in the wavelength ranges 300–400, 400–600, and 600–800 nm. The main beam from the pre-dispersing prism is then directed towards the four individual spectrometer gratings of GOME. Each spectrometer uses refractive optics and a 1024-element diode array detector. Thus the entire spectral range between 232 and 793 nm can be observed at a spectral resolution of about 0.2 and 0.33 nm, below and above 400 nm, respectively.

The direct, extraterrestrial solar irradiance (the Fraunhofer reference spectrum, FRS) was measured by GOME over a diffuser plate to reduce its intensity prior to being reflected by the scan mirror into the instrument. The spectra and other signals recorded by GOME were transmitted to the ESA ground station for storage and processing. The resulting Level 1 products, essentially calibrated radiances, and Level 2 total column data products are available to the scientific community. ERS-2 was launched on 20 April 1995 into a nearly polar, sun synchronous orbit, having an equator crossing time of 10:30 am in a descending node. The scan strategy of

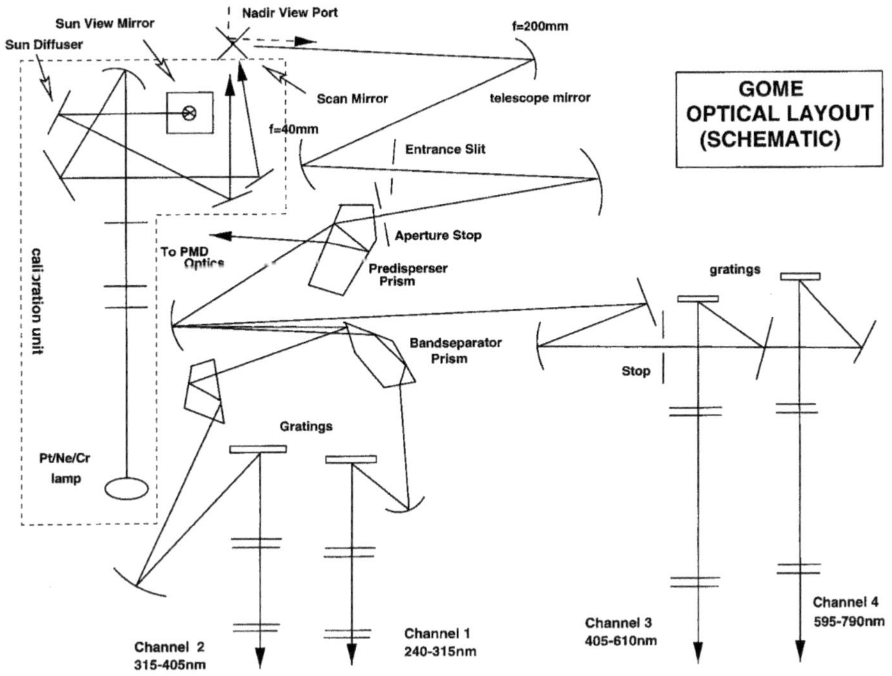

Fig. 7.70 Schematic setup of a satellite DOAS system: GOME on ERS-2. Reprinted with permission from Burrows et al. 1999b, ©American Meteorological Society. All rights reserved

GOME yielded complete global coverage at the equator in 3 d. The swath width of GOME was 960 km, divided into three pixels of $320 \times 40\,\text{km}^2$ for the array detector ($20 \times 40\,\text{km}^2$ for the PMDs, which were read out 16 times faster than the arrays; Burrows et al. 1999a; 1999b, and references therein). The later GOME-2 instruments had a 1920 km swath with $80 \times 40\,\text{km}^2$ ground pixels. Besides ozone, GOME and the later DOAS-type instruments can measure a wide variety of atmospheric trace gases, including bromine monoxide (BrO) (e.g. Hegels et al. 1998; Richter et al. 1998; Wagner and Platt 1998), chlorine dioxide (OClO) (Burrows et al. 1998; Burrows et al. 1999a; 1999b; Wagner et al. 2001b), formaldehyde (HCHO) (Burrows et al. 1999a; 1999b; Chance et al. 2000), sulfur dioxide (SO_2) (Eisinger and Burrows 1998, and references therein), nitrogen dioxide (NO_2) (Burrows et al. 1998; Leue et al. 2001), and water vapour (H_2O) (Noël et al. 1999; Maurellis et al. 2000b). In addition, ozone and aerosol profiles may also be retrieved (e.g. Hoogen et al. 1999, and references therein).

More recent instruments (e.g. OMI, TROPOMI and the geostationary instruments) use pushbroom scanning (Sect. 7.10.4) with 2D-spectrometers and exhibit smaller ground pixels, e.g. around $15 \times 24\,\text{km}^2$ for OMI and $7 \times 3.5\,\text{km}^2$ for TROPOMI (near nadir, nominal values). An example of the scanning scheme of the OMI instrument is shown in Fig. 7.71. A stripe of 2600 km width on the ground is imaged on the 580 pixels in each row of the 2D-CCD detector of the instrument. The radiation is dispersed onto 780 pixels in the column direction. Thus for each of the

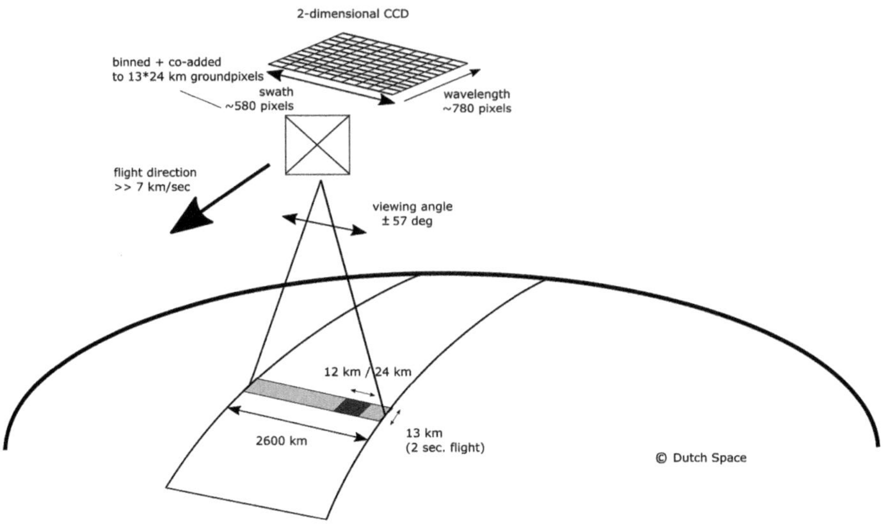

Fig. 7.71 Pushbroom scanning scheme of the Ozone Monitoring Instrument (OMI) on NASA's Aura satellite. The spectrum of a ground pixel is projected on the wavelength dimension (columns) of the charge-coupled device (CCD). The cross-track ground pixels are projected on the swath dimension (rows) of the CCD. The satellite orbital velocity of 7 km/s and an exposure time of 2 s lead to a ground pixel size of 13 km in the flight direction. The viewing angle of 114° leads to a swath width (cross-track extension) on the ground of 2600 km. Reprinted with permission from Schenkeveld et al. (2017),

580 spatial pixels there is a 780 pixel spectrum. All spectra are stored at once and transmitted to the ground for analysis. For instruments in LEO this arrangement eliminates the mechanical scanner found in the earlier instruments (GOME family, SCIAMACHY) using whiskbroom scanning. However, it does not improve the light throughput (Sect. 7.10.4).

Future improvements in light throughput and reduction of ground pixel size (e.g. down to 1 km by 1 km), while also reducing size and weight of the instrument, can be achieved by using arrays of massive parallel miniature spectrometers (see Sect. 7.12.5 and Platt et al. 2021).

7.11 Software Controlling DOAS Instruments

Computers and specialised software for the control of DOAS instruments are an essential part of DOAS system design. In most instances this software is at least comparable in complexity to the hardware. The tasks of DOAS operating software varies with applications, but generally includes:

(a) Record and store the spectra and associated information (meteorological data such as air pressure and temperature required to convert measured concentrations to mixing ratios).

(b) Cycle through measurement sequences (e.g. take atmospheric spectra, background spectra, and dark current, see Sect. 7.13) and calculate the true atmospheric spectra (see Chap. 8). Perform measurements in different wavelength regions, switch to different light paths or observation geometries, etc.
(c) Align optics. For instance, in long-path systems the telescope has to be precisely pointed at the light source or retro-reflector array.
(d) Evaluate spectra and display trace gas concentrations following the principles outlined in Chap. 8. This is the most complex task, and is frequently performed offline.
(e) Perform auxiliary functions, including calculation of sun and moon positions (e.g. SZA). Support wavelength calibration and instrument function determination procedures.

Many software packets have been developed to perform some or all of the above tasks. In many circumstances, and especially in the case of airborne and satellite instruments, the control software is specifically written for the respective system. New software emerges regularly due to changing computer and hardware technology and the reader is referred to the scientific literature to find information on the currently available DOAS control software.

7.12 Optimising DOAS Instruments

During the design of DOAS instruments decisions on the spectral range, spectral resolution, size of the optics, desired measurement time, etc. have to be made. While in many cases extraneous circumstances may dominate the design decisions, it is worth considering what the optimum values for these design criteria are. The central question here is how to optimise the sensitivity of DOAS instruments (typically within a specified measurement time), which in many cases is equivalent to maximizing the amount of light reaching the detector under the given boundary conditions.

7.12.1 Noise and Minimum Detectable Optical Density

The detection limit for a particular substance can be calculated according to Eq. 6.9 if the differential absorption cross-section, the minimum detectable optical density D'_0, and the length of the light path are known. In several applications, however, the goals of maximising light throughput and light path length can contradict each other, as is detailed below.

In general, D'_0 is determined by: (1) Photoelectron statistics (shot noise), which is a function of light intensity and detector quantum efficiency η only, and (2) by any other noise sources including detector noise, induced 'background absorption structures', or 'optical noise' (see Sect. 7.2). This second class of noise sources, summarised in quantity B, is on the order of $B = N/S \approx 10^{-4}$ for typical DOAS systems with solid state detectors or OSDs (see Sect. 7.5). Photoelectron shot noise is proportional to $N^{-1/2}$, where N is the total number of photons recorded around the centre of the absorption line during the time interval t_m of the measurement. The number of photoelectrons N is related to the power P received by a detector pixel.

$$N = \frac{P \cdot t_m \cdot \eta \cdot \lambda}{hc} \tag{7.85}$$

where η denotes the detector quantum efficiency (at the given wavelength λ), c the speed of light, and h Planck's constant. Thus, the minimum detectable optical density D_0 will be proportional to:

$$D_0' \propto \sqrt{\frac{1}{N} + B^2} \tag{7.86}$$

If sufficient light is available (which is rarely the case), D_0' is limited by other noise sources (Platt et al. 1979; Platt and Perner 1983). Using N from Table 7.11, an integration time on the order of minutes for optomechanical detectors and around 1 s for array detectors would be sufficient to make $N^{-1/2} \approx$ B.

While for a given minimum detectable optical density the detection limit improves proportionally to the length of the light path, the actual detection limit, will not always improve with a longer light path. This is because with a longer light path the received light intensity I(λ) tends to be lower, increasing the noise associated with the measurement of I(λ), and thus increasing the minimum detectable optical density.

7.12.2 Optimum Light Path Length in Active DOAS Systems

An important consideration in the practical application of active DOAS is the length (L) of the chosen light path. The value of L has a large influence on the performance of the system:

1. The optical density due to light absorption by trace gases at a given concentration is directly proportional to L.
2. The intensity (I) received from the light source generally decreases with the length of the light path. The decrease dI/dL will depend on the details of the setup and L itself. For folded paths (DOAS systems coupled to multi-reflection cells) I decreases with increasing L.
3. The volume over which the trace gas concentration is averaged for long path systems is proportional to L.
4. Long light paths (folded or unfolded) make the DOAS system more vulnerable to extinction by aerosol, haze, or fog.

A few simple considerations clarify this relationship and the detection limits. For a typical DOAS system there are no geometrical light losses of path lengths up to ~2 km. However, in addition to geometrical light attenuation losses due to atmospheric absorption and scattering affect the magnitude of the received light signal:

$$I_{\text{received}} = I_{\text{source}} \exp[-\mathrm{L}(\Sigma(\sigma_{\mathrm{i}}(\lambda)c_{\mathrm{i}}) + \varepsilon_{\mathrm{R}}(\lambda) + \varepsilon_{\mathrm{M}}(\lambda))] \tag{7.87}$$

or simply:

$$I_{\text{recieved}} = I_{\text{source}} \cdot \exp\left(-\frac{L}{L_0}\right) \tag{7.88}$$

where the absorption length L_0 reflects the combined effects of broadband atmospheric absorption, Mie and Rayleigh scattering, as given in the second term of (6.5), i.e.:

$$L_0 = \frac{1}{\sum_i (\sigma_{\mathrm{i}0}(\lambda)\, c_{\mathrm{i}}) + \varepsilon_{\mathrm{R}}(\lambda) + \varepsilon_{\mathrm{M}}(\lambda)} \tag{7.89}$$

The noise level of a photodetector signal is essentially dependent on the number of photons received and thus is proportional to the square root of the intensity. The signal-to-noise ratio of a DOAS system as a function of the light path length L can be expressed as:

$$\frac{D'}{N} \approx L \cdot G(\lambda) \cdot \exp\left(-\frac{L}{2L_0}\right) \tag{7.90}$$

where $G(\lambda)$ is constant (Regime 1), $G(\lambda) = L^{-1}$ (Regime 2), or $G(\mathrm{i}) = L^{-2}$ (Regime 3), as defined in Sect. 7.9. Relationship (7.90) yields optimum D'/N ratios for light path lengths of $L = 2L_0$ (Regime 1), $L = L_0$ (Regime 2), or a boundary maximum at the shortest light path (Regime 3). While there are no lower limits for L_0 in the atmosphere (consider, for example, fog), the upper limits are given by Rayleigh scattering and the scattering by atmospheric background aerosol, which indicate optimum light path lengths ($2L_0$) in excess of 10 km for wavelengths above 300 nm (see Table 4.6).

Another point to consider is the width G of the absorption line or band. At a given spectral intensity of the light source $N(\lambda)$, the light intensity around the line centre will be $N(\lambda)d\lambda \approx N(\lambda)\,\Gamma$ (in photons per nanometre and second). Thus D_0 becomes:

$$D_0 \propto \sqrt{\frac{1}{N(\lambda) \cdot \Gamma \cdot t} + B^2} \tag{7.91}$$

From (7.91), it becomes clear that the detection of a spectrally narrow absorption line requires a proportionally higher spectral intensity of the light source (see Sect. 7.8.2).

7.12.3 Optimum Spectral Resolution

For a given grating spectrometer, the light throughput varies in proportion to the square of the spectral resolution, as shown in Sect. 7.4.5 (here expressed as the width

Γ_0 of the instrument function in nm); Thus the light intensity at the output becomes:

$$I \approx I_i \cdot \Gamma_0{}^2 \tag{7.92}$$

The 'signal' in the spectrum is the differential optical density D', which is (for a given species, wavelength, and light path) proportional to the concentration.

In the case of shot noise limitation, the minimum detectable optical density D_0 in the spectrum is inversely proportional to the square root of I. Thus:

$$D_0 \sim \frac{1}{\sqrt{I}} \sim \frac{1}{\Gamma_0} \tag{7.93}$$

In general, the signal as a function of the spectral resolution will vary proportional to the differential absorption cross-section as a function of resolution $\sigma'(\Gamma_0)$. Figure 7.72 (upper panel) shows an example for several bands of NO_2 in the visible spectral region. In principle, $\sigma'(\Gamma_0)$ can be an arbitrary function of Γ_0, which decreases with increasing Γ_0. In the example of Fig. 7.72, $\sigma'(\Gamma_0)$ can be approximated as a linear function:

$$\sigma'(\Gamma_0) \approx \sigma'_0\,(1 - a\Gamma_0) \tag{7.94}$$

where a is a constant. In the example of Fig. 7.72, $a \approx 2.4\,\mathrm{nm}^{-1}$.

Thus, in this special case, the signal-to-noise ratio becomes:

$$\frac{D'}{N} = \sigma'_0 \cdot (1 - a\Gamma_0) \cdot \Gamma_0 \propto \Gamma_0 - a \cdot \Gamma_0^2 \tag{7.95}$$

The optimum D'/N is thus obtained at the resolution:

$$\frac{d}{d\Gamma_0}\frac{D'}{N} = 1 - 2a\Gamma_0 = 0 \text{ or } \Gamma_{0,\mathrm{opt}} = \frac{1}{2a} \approx 1.2\,\mathrm{nm}\ (a \approx 2.4\,\mathrm{nm}^{-1},\ \text{see Fig. 7.72}) \tag{7.96}$$

Note that, in general, the instrument function $H(\lambda)$ will not be of Gaussian shape. The spectrum of a given trace species (expressed as optical density), as seen by a particular spectrometer, is obtained by convoluting the true spectrum $D_T(\lambda)$ with the instrument function $H(\lambda)$, as described in Chap. 6.

7.12.4 Optimum Measurement Time

The number of photons recorded in a DOAS spectrum is proportional to the integration time (either of a single detector exposure or as a sum of several co-added spectra). Consequently, the noise in the spectrum, and thus the minimum detectable optical density D_0, will be reduced proportionally to the square root of the integration time, see (7.91). This will not continue indefinitely, but reach a system dependent minimum noise level B, which is determined by other noise sources in the system. An

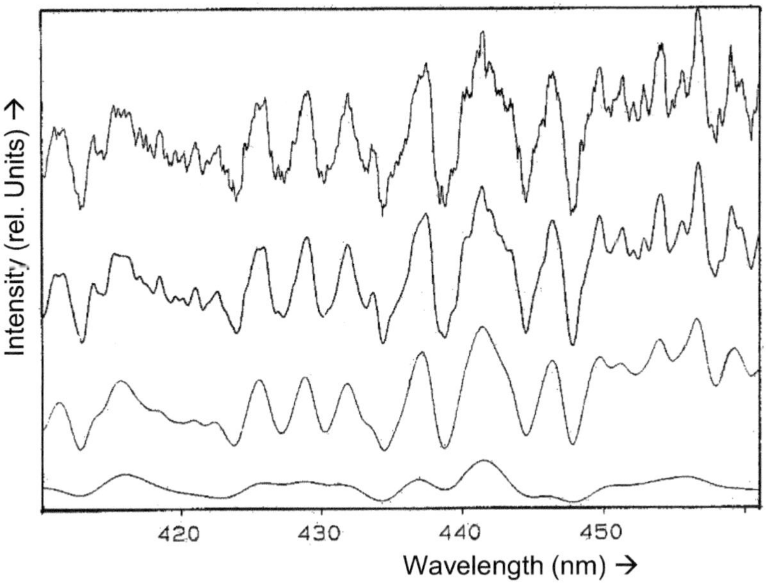

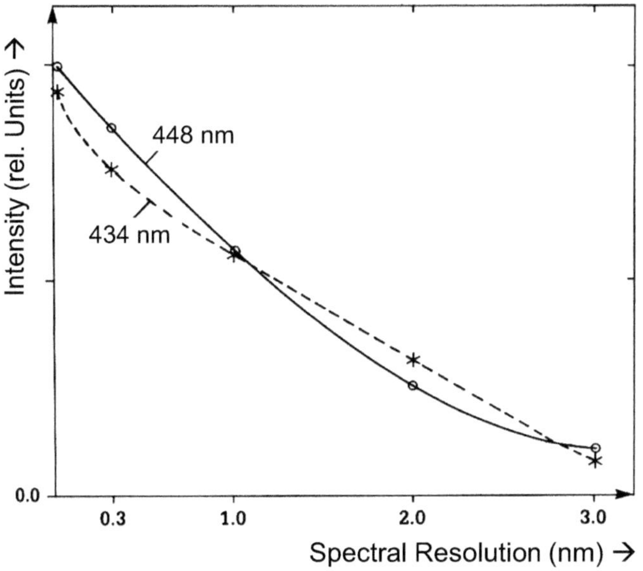

Fig. 7.72 Absorption cross-section of NO_2 as a function of spectral resolution, here expressed as the FWHM Γ_0 of the instrument function $H(\lambda)$. **Upper panel**: Absorption spectrum for different values of the spectral resolution Γ_0, from top to bottom $\Gamma_0 = 0.01, 0.1, 1.0, 3.0\,\text{nm}$; **Lower panel**: Differential optical density for constant NO_2 column density as a function of the spectral resolution Γ_0. There is a nearly linear decay of the optical density as the spectral resolution is reduced, i.e. increasing Γ_0 (adapted from Platt 1994)

example for a typical active DOAS system (Tuckermann et al. 1997) using a diode array detector is shown in Fig. 7.73. This type of diagram is also known as Allan Plot, originally developed by D. W. Allan as a measure of the frequency stability of oscillators (Allan 1966, see also Horbanski et al. 2019). With increasing signal (i.e. photoelectrons accumulated, given as total binary units), the noise levels are reduced proportionally to the square root of the signal. However, from $\sigma^2 \approx 10^{-8}$–10^{-7} the noise will not be further reduced. In other cases it was even found to increase at longer integration times, see e.g. Horbanski et al. 2019. This observation shows that there are additional noise sources which are not signal dependent. Under laboratory conditions with a fixed optical setup, noise levels of at least two orders of magnitude lower are reached at sufficiently long integration times, indicating that the noise source dominating at long integration times is not due to electronic noise, but is more likely due to optical interference effects or processes in the light source (Stutz 1996; Stutz and Platt 1997b). The remaining noise is therefore called 'optical noise'. Obviously, it is of little use to continue a measurement to the point where the photoelectron-shot

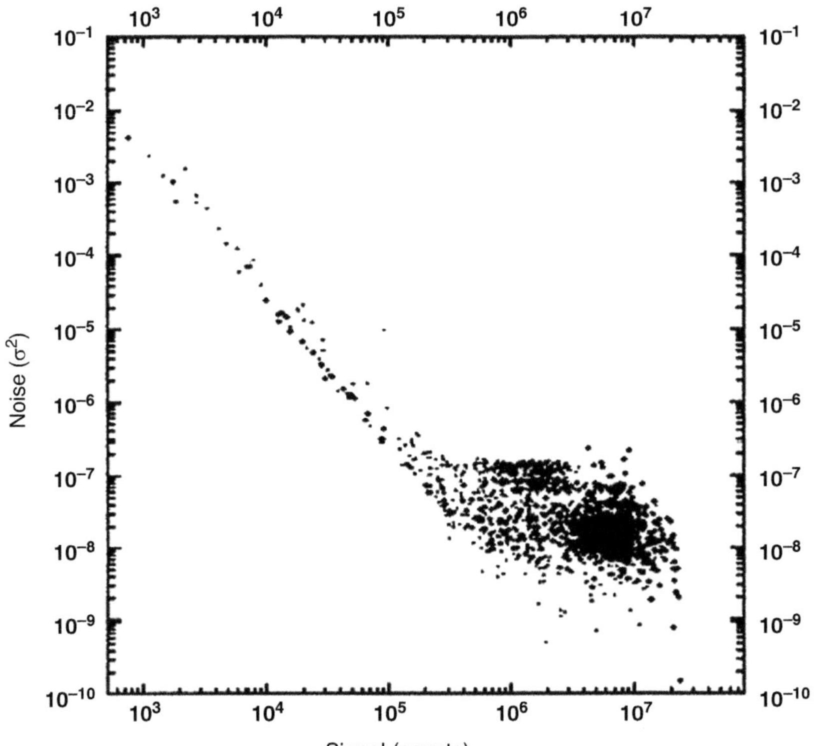

Fig. 7.73 Square of the noise level (σ) as a function of measurement time (log–log scale). This type of analysis is also known as Allan Plot (Allan 1966). Up to a certain limit, there is a linear decrease of σ^2 with the number of photoelectrons recorded (here 1 count signal $\simeq$ 1000 photoelectrons), recording more photoelectrons (or measuring longer) does not further decrease noise, in fact the noise level might increase. (adapted from Tuckermann 1996)

noise becomes much smaller than the combined effect of all other noise sources in the system.

7.12.5 Optimum Spectrograph Size

As illustrated in the three preceding subsections a central parameter for the achievable sensitivity and spatial resolution of a spectroscopic instrument is its light throughput, which is determined by the instrument's étendue, E. The étendue is typically a property of the spectrograph. Measures can be taken to improve the light throughput of spectroscopic (DOAS) instruments (details can be found at Platt et al. 2021). The étendue of an instrument can be enhanced by:

1. Upscaling all instrument dimensions, i.e. lengths L_1, height L_2, and width L_3, and thus the entrance slit area by the same factor, while keeping the acceptance angle (i.e. the spectrograph F number) constant. We refer to this option as "spectrograph size scaling".
2. Increase the étendue while keeping some dimensions of the spectrograph constant. For instance, scale the acceptance angle (i.e. the F number) of the spectrograph while keeping its entrance area A_S constant. We refer to this option as "spectrograph F number scaling".
3. Alternatively, the entrance area A_S may be scaled up while retaining the F number. For instance, the slit height h_S could be made larger while keeping the overall instrument dimensions constant.
4. Operate many instruments of identical design in parallel, and electronically combine the resulting spectra.

In principle, the étendue can be enhanced by arbitrary factors when Options 1 and 4 are applied while the effect of Options 2 and 3 is limited (the acceptance angle can not exceed $2\pi\, sr$, slit height and width are limited for a given spectrograph size).

One may investigate how the spectrograph light throughput changes when the spectrograph size is scaled up or down while keeping the acceptance angle (and thus the spectrograph F-number) constant. For simplicity, it is assumed that the typical dimension of the spectrograph L (e.g. the length of the housing) is scaled from its initial value L_0 to some other value $L = \Gamma \cdot L_0$ such that all other dimensions (including entrance slit dimensions) are scaled by the same factor Γ proportional to L. In particular, the slit dimensions w_{S0} and h_{S0} are scaled to $\Gamma \cdot w_{S0}$ and $\Gamma \cdot h_{S0}$, respectively.

Since the aperture solid angle Ω and F-number will stay constant, the étendue $E(L)$ will change from its initial value $E_0 = E(L_0)$ in proportion to the area A_s of the entrance slit:

$$E(L) = A_s \cdot \Omega \cdot \left(\frac{L}{L_0}\right)^2 = E_0 \cdot \left(\frac{L}{L_0}\right)^2 = E_0 \cdot \Gamma^2 \propto L^2 \qquad (7.97)$$

However, volume and mass of the spectrograph scale with L^3, i.e.:

$$M(L) \propto V(L) = M_0 \cdot \left(\frac{L}{L_0}\right)^3 \propto L^3 \tag{7.98}$$

As a consequence, the volume and thus the mass of a spectrograph scale with its light throughput, as measured by the étendue, as:

$$M(L) \propto V(L) \propto L^3 \propto E^{\frac{3}{2}} \text{ or } E \propto M^{\frac{2}{3}} \tag{7.99}$$

Doubling the linear dimensions of a spectrograph (from $L = L_0$ to $L = 2L_0$ while retaining the aspect ratio) would increase its étendue by a factor of 4, since the area of the entrance slit quadruples. However, its volume and mass would increase by a factor of eight. Thus, the light throughput per liter or kilogram of the instrument is cut in half. For this reason increasing the light throughput of spectrometers by size up-scaling becomes increasingly unattractive with instrument size.

Other measures, such as increasing the F-number, grating efficiency, detector efficiency, slit height, etc. provide limited gains in light throughput given that present day spectrometers are already quite optimised (as e.g. discussed by Platt et al. 2021).

Operating many spectrometers in parallel (Option 4), appears most promising. Clearly the étendue scales proportionally to mass and volume when several identical spectrometers are used instead of a single one. Modern techniques of mass production and microelectronics make this option increasingly attractive.

Combining Options 1 and 4 can lead to great reductions in instrument size and mass while retaining light throughput. Alternatively, one may increase the light throughput while retaining instrument size (or a combination of both parameters). The idea is not to increase the instrument size, but rather to decrease its size, which leads to an improvement of E/M (Eq. 7.99), and to compensate by operating many spectrographs in parallel, which keeps E/M constant. For instance, by scaling down the size (i.e. L) of a given spectrograph by a factor of 10, the light throughput and volume (and thus mass) would be reduced by factors of 100 and 1000, respectively. Running N=100 spectrographs in parallel would yield an instrument with the same light throughput, but only 1/10 of the mass, alternatively at N=1000, one would create an instrument with the same mass as the original but 10-times the light throughput. This may seem surprising, but is actually just a consequence of scaling, as outlined by Haldane (1927).

Obviously, we cannot shrink a spectrograph indefinitely. In addition to possible mechanical constraints, the following phenomena limit the shrinking of spectrographs (Avrutsky et al. 2006):

1. **Light diffraction at the ever shrinking entrance slit**: As discussed in Sect. 7.3.6, diffraction at narrow entrance slits needs to be considered. A more precise calculation yields w_S being slightly smaller than $\lambda \cdot F$, thus for $\lambda = 320$nm and $F = 4$ one obtains w_S around 2μm. A typical entrance slit width is around 50μm, leaving more than an order of magnitude room for down-scaling.

2. **Loss of grating resolving power**: The resolving power $P_R = \lambda/\Delta\lambda$ of a grating is determined by its total number of grooves, N_G, (Sect. 7.4). The smallest spectrographs used in DOAS instruments are equipped with gratings with $N_G \approx$ 36,000, corresponding to a theoretical resolving power P_R =36,000. In practice, because of the relatively wide entrance slits, the typical spectral resolution is just $\Delta\lambda \approx 0.5$nm at 300 nm, corresponding to $P_{pract} \approx 600$. This leaves more than an order of magnitude room for down-scaling.
3. **Small detector pixels size**: Detector arrays of small spectrographs typically have a pixel pitch around 12μm. Down-scaling a spectrograph requires down-scaling of the pixel pitch with the same linear scaling factor. Detectors with pixel pitches below 1 μm are mass produced and are used in many consumer products (smartphones, webcams, etc.). Although these sensors are primarily designed for visible light detection, it has been shown that UV sensing is also possible (Wilkes et al. 2017).

In summary, even rather small 'miniature' spectrographs with focal lengths around $f \approx 50 - 70mm$ could likely be scaled down in size by an order of magnitude or more. For larger spectrographs, such as those used in satellite instruments or active LP-DOAS, even higher scaling factors and corresponding gains in E/M are possible.

7.13 Measurement Process Control

Like most techniques, DOAS measurements require a number of additional parameters besides the recording and evaluation of trace gas absorption spectra. Several supporting systems are needed for DOAS measurements, particularly if automatic operation is desired. These systems include mechanical actuators for optical alignment of the light path and change of measurement modes, recording of auxiliary parameters required for the interpretation of the measurements (length of the light path, air density, etc.), and monitoring of instrument performance.

In this section, we present a series of 'typical' measurement procedures. Generally, these procedures involve taking measurement spectra and a series of auxiliary spectra (background, dark current, direct light source, calibration, and sometimes scattered sunlight spectra).

7.13.1 Active DOAS Systems Standard Approach

A complete DOAS measurement using radiation from an artificial light source includes taking a series of several different spectra. Here, we give an example for a system using a diode array detector (see e.g. Figure 7.55), but other arrangements follow procedures similar to those outlined in Fig. 7.74. For a complete DOAS measurement, the following spectra are usually recorded:

1. Atmospheric raw spectrum [ARS, denoted as $I_M(i)$]. This is the spectrum of atmospheric absorbers (trace gases) superimposed on a light source spectrum. Usually, several individual spectra, N_M, with integration time T_M are co-added.
2. Lamp intensity spectrum [LIS, $I_L(i)$]. This spectrum is recorded by admitting direct light from the lamp to the spectrometer using some optics to bypass the light path.
3. Dark current and electronic offset spectrum [$I_D(i)$]. This spectrum is taken without light on the detector, e.g. by closing a shutter in front of the entrance slit of the spectrometer. By normalising the sum of N_D co-added dark current spectra, we obtain $I_{D1}(i) = I_D(i)/N_D$.
4. A background spectrum [$I_B(i)$] with integration time T_B is sometimes required. Since sunlight scattered into the receiving optics might disturb the measurements, a scattered sunlight or background spectrum is taken by blocking the lamp.

As a result of Steps 1–4, a true atmospheric spectrum I(i) is calculated from the above series of spectra :

$$I(i) = [(I_M - N_M \cdot I_{D1}) - T_M/T_B \cdot (I_B - N_B \cdot I_{D1})] \cdot (I_L - N_L \cdot I_{D1})^{-1} \tag{7.100}$$

where N_B and N_L denote the numbers of co-added background and LIS, respectively. Note that background spectra need not always be taken and compensated for (for instance at night or at wavelengths below 300 nm where scattered sunlight levels are low). Also, while LIS are always required to compensate for diode-to-diode sensitivity variations (see Sect. 7.5.3), it might not be necessary to take one with each ARS.

Once these spectra are recorded, the following steps are taken to analyze the atmospheric spectrum $I(i)$:

1. A non-linear least squares fit of a series of reference spectra is performed. The reference spectra are usually taken from the literature, and must be adapted to the instrument function of the spectrometer used. This procedure is described in Chap. 8.
2. The resulting fit parameters describing the optical density of the reference spectra are used to calculate the respective trace gas column densities (and their uncertainties).
3. The desired trace gas concentrations are derived by dividing by the length of the optical absorption path. If desired, the concentrations can be converted to mixing ratios.
4. The fit parameters for 'shift' and 'squeeze' of the spectra, which are not directly relevant for deriving trace gas column densities or concentrations, can be used to improve the estimate of the uncertainties and to give general information about the quality of the measurement.

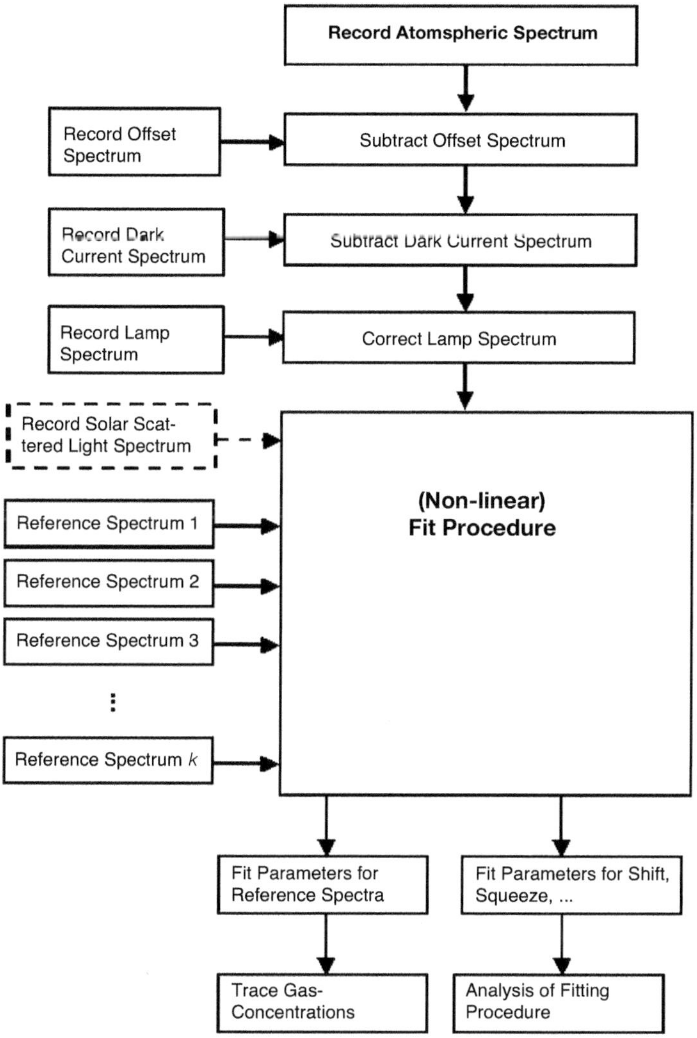

Fig. 7.74 Flowchart of a typical active DOAS measurement and evaluation procedure

7.13.2 Active DOAS Systems MCST

A different approach to removing the diode sensitivity variation is the Multichannel Scanning Technique, MCST, introduced by Knoll et al. (1990) (see also Brauers et al. 1995; Stutz 1996; Alicke 2000). The basic idea of the MCST is the combination of a multi-channel detection system with the scanning technique generally used to cover a larger spectral region with a single-channel detection system. A series of ARS $[I_{MM}(i)]$ is taken and stored. The individual spectra are successively shifted in wavelength, for instance by slightly rotating the grating in such a way that the spectral information is essentially smoothed out when the ARS are co-added, after subtracting

I_D, to give LIS [$I_{LM}(i)$]. Note that the wavelength difference must be smaller than any spectral structure like absorption bands; otherwise this structure not only remains in the sum spectra, but would also be duplicated near itself. Each ARS is divided by the obtained LIS and digitally shifted back to the same wavelength position (for instance of the middle ARS in the series) to obtain a true atmospheric spectrum . The advantage of the MCST is that the diode-to-diode sensitivity variations are even more precisely eliminated, since spectra taken to determine these variations are recorded under precisely the same conditions as the ARS. In addition, measurement time is saved since no separate LIS have to be taken.

7.13.3 Passive DOAS Systems

In passive DOAS systems the primary spectrum consists of absorption features of atmospheric molecules superimposed on a solar or stellar spectrum. Raw spectra of sunlight and moonlight in the UV/visible are dominated by strong Fraunhofer lines with optical densities sometimes exceeding unity in the near UV. Thus, in order to make useful determinations of the weak absorption features of the trace gases in the earth's atmosphere, the very accurate removal of the solar spectral structure (LIS) is an indispensable prerequisite. Unlike the case of the artificial light DOAS, the direct solar spectrum is not accessible for ground-based instruments. Therefore, another approach is used:

In order to remove the Fraunhofer structure, spectra taken at different zenith angles of the observed celestial body, ϑ_1 and ϑ_2, are ratioed. The observed trace gas slant columns S are $S(\vartheta_1) = V \cdot A(\vartheta_1)$ and $S(\vartheta_2) = V \cdot A(\vartheta_2)$, where V is the vertical trace gas column. $A(\vartheta_1)$ and $A(\vartheta_1)$ are airmass factors which describe the enhancement of the light path length compared to the vertical and are dependent on the solar zenith angle. We obtain for the ratio of the measured optical densities:

$$\begin{aligned}\frac{D_2}{D_1} &= \frac{e^{-S(\vartheta_2)\cdot\sigma}}{e^{-S(\vartheta_1)\cdot\sigma}} = e^{-\sigma\cdot[S(\vartheta_2)-S(\vartheta_1)]}\\ &= e^{-\sigma\cdot V\cdot[A(\vartheta_2)-A(\vartheta_1)]}\end{aligned}$$

In other words, the ratio of the optical densities is the optical density based on the difference in the air mass factors (see Chap. 9). For the slant column $S_D = V\ [A(\vartheta_2) - A(\vartheta_1)]$ the term 'differential slant column density' (DSCD) is sometimes used. This should not be confused with the use of the term 'differential' in the context of DOAS evaluation of optical spectra.

A complete scattered light DOAS measurement, for example with a diode array detector, thus consists of taking the following spectra :

1. Two ARS, denoted as $I_{M1}(i)$ and $I_{M2}(i)$, at different (solar) zenith angles. For each spectrum several individual spectra, N_M, with integration times T_M are usually co-added.

2. Dark current and electronic offset spectrum $[I_\mathrm{D}(i)]$. By normalising the sum of N_D co-added dark current spectra, we obtain $I_{\mathrm{D1}}(i) = I_\mathrm{D}(i)/N_\mathrm{D}$.

The procedure is also outlined in Fig. 7.75.

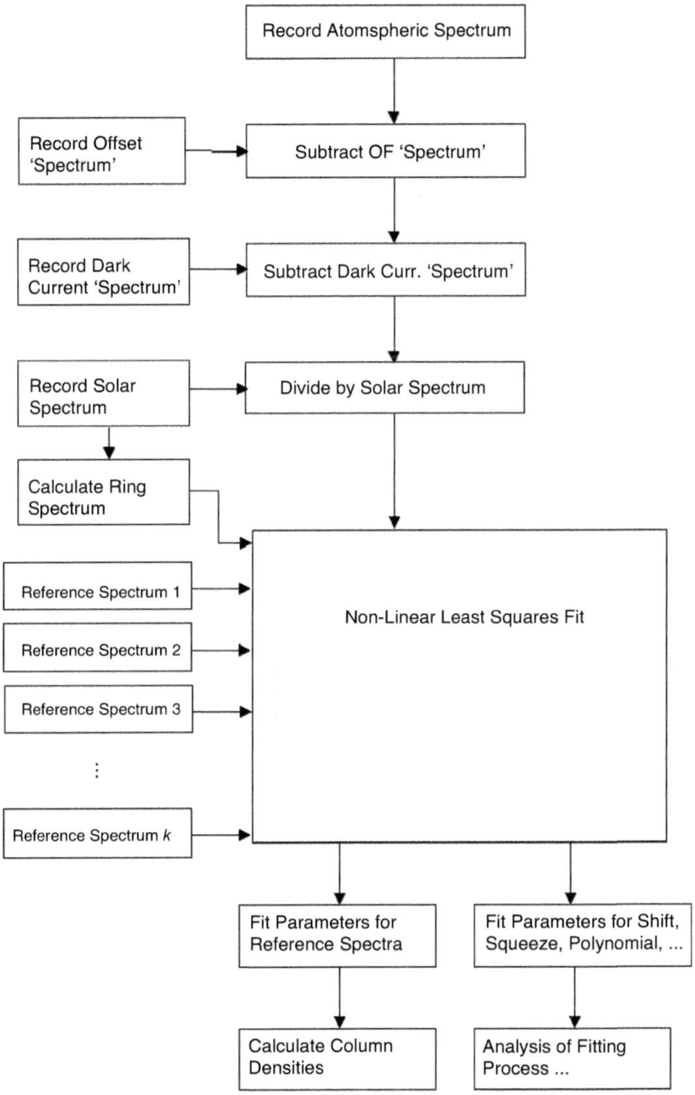

Fig. 7.75 Flowchart of a typical passive DOAS measurement and evaluation procedure

7.13.4 MAX-DOAS and Off-Axis DOAS Systems

MAX-DOAS and Off-axis DOAS measurements make use of scattered light spectra recorded at non-zero angles $\beta_1 \ldots \beta_k$ between zenith and observation direction. In both cases, a zenith spectrum is used to remove the solar Fraunhofer structures. In principle, the approach is very similar to the one described in the previous section. However, now the measurements differ in two angles, β and ϑ. Generally the trace gas absorptions in the stratosphere increase with increasing SZA, ϑ, while the absorption due to gases in the lower atmosphere, i.e. the boundary layer, strongly increases with larger off-axis angles β (see Chap. 9). By comparing measurements with large β to zenith observations $\beta = 0$, that are temporally close, i.e. have similar ϑ, one can probe the lower atmosphere. In this case the procedure is similar to that in Fig. 7.75 where the atmospheric spectrum is recorded at high β and the solar spectrum is measured in the zenith.

7.14 Mechanical Actuators

The DOAS operating procedures described in Sect. 7.13 require a number of supplemental data, such as offset, dark current and other spectra. These data can be measured manually, but will usually be acquired in an automatic sequence under computer control. In order to switch between the different measurement modes e.g. recording atmospheric spectra, dark current, or observing at different elevation angles, mechanical actuators are needed. In addition, in active DOAS systems re-aiming of the receiving telescope is often required, due to mechanical instabilities and refraction of the light path caused by atmospheric temperature gradients. These changes in the optical configuration required for the different measurement modes and the optical alignment tasks are usually performed by electrical motors coupled to appropriate mechanics, such as translation stages. When defined, absolute positions need to be reached. For example, when pointing a telescope to a retro-reflector array. Stepper motor-driven actuators are frequently the devices of choice, although DC-motor–encoder combinations might become important in the future.

For a typical active DOAS system, actuators are used for:

- 'Short circuit' optics can be moved into the light beam to record the lamp emission spectrum.
- A shutter allows the recording of the dark-current spectrum.
- Motorised setting of the spectrometer wavelength allows measurement of spectra in several spectral intervals.
- A filter actuator inserts spectral filters into the optical path, for example to block second order spectra or to reduce stray light.
- Actuators for alignment of the telescope pointing direction (azimuth and elevation) are typically required.
- Alignment of the telescope focus can be motorised in order to accommodate different light path lengths.

In summary, five or more actuators might be needed in a typical long-path DOAS system.

Fewer actuators are typically needed for passive (scattered-sunlight) DOAS systems:

- A shutter allows the recording of the dark-current spectrum.
- Motorised setting of the spectrometer wavelength allows measurement of spectra in several spectral intervals.
- A filter actuator inserts spectral filters into the optical path.
- An actuator for aiming of the telescope in one or two directions (azimuth and elevation), may be required for multi-axis measurements.

7.14.1 Stepper Motors

Stepper motors move in precisely known increments. Typical incremental angles (or step sizes) are 1.8° (or 200 steps per revolution), but a wide range of different step sizes are available. Special drive electronics ensure that the correct current waveforms are supplied to the motor windings. Modern drive electronics also frequently support fractional steps; i.e. by driving the motor windings by different relative currents the steps can be further divided. Today, many drivers allow 'half-step' or 'micro-steps' modes.

Under computer control, such a motor can perform a precisely determined motion pattern. The positions of the motor are repeatable. For instance, different filters mounted on a wheel attached to the motor axis can be selected without further hardware, provided the initial position of the motor is known. This is usually ensured by driving each actuator to a limit switch (and thus to a known mechanical position) during the set-up period of the instrument. In addition, several different telescope pointing directions (i.e. towards several retro-reflectors or light sources) can be selected under software control when azimuth and elevation of the telescope are actuated by stepper motors.

If further mechanical gears are attached to a stepper motor, slackness in the mechanics may be present. This is caused by thread clearance in a translation stage. A consequence is that the actuator will reach a different position when approached from different directions of motion. A possible solution is to always make the final move in the same direction, for instance in 'forward direction', which means that moving in the 'backward' direction consists of two phases: first, move backwards by the desired number of steps plus a number of n extra steps, then move 'forward' by n steps.

Disadvantages of stepper motors include their relatively low speed, high cost, and the possibility of 'losing steps', which is usually not directly detectable by the driving electronics. Furthermore, there might be electrical interference from the rapidly changing currents in the motor windings, for instance with the detector electronics. This problem can usually be avoided by not operating stepper motors during the recording of optical spectra. However, it should be noted that many commercial stepper motor drivers use a 'switch mode' approach to limit the current to the motor windings. Therefore, although the motor is not moving and the currents to the windings should be 'static', there might be high frequency modulated currents on the windings. This is difficult to avoid in micro-step mode, but in full or half-step mode, when currents are off or have the same absolute magnitude, a series resistor to the winding might be chosen to limit the current to a point just below the 'desired' current of the motor driver.

In order to minimise the problems of losing steps, a motor should be selected that has sufficiently large torque for the particular application. Additionally, the motor should be operated at a sufficiently low speed.

Stepper motors have a recommended 'starting speed' (i.e. step frequency) and a 'maximum speed'. These numbers are given by the manufacturer for the motor alone and are obviously smaller in any practical application where the motor shaft is connected to additional mechanical elements, such as lead screws, which increase the inertia of the motor armature. Also the motor has to drive a load. A simple approach involves always operating the motor at the same margin below the starting speed. A more sophisticated approach, allowing faster motor operation, starts the motor at its starting speed, followed by a ramp of the step frequency up to the maximum speed. Towards the end of the motion, the frequency is ramped down to the starting speed again. In either case, care has to be taken to avoid motor resonance frequencies. Motor resonance effects can cause errors that are difficult to diagnose later. Motor resonance frequencies are given by the manufacturer, but they apply to the motor alone and will be changed by added load. Note that motor resonance is frequently less pronounced if a half-step mode rather than full-step mode is used.

7.14.2 Other Actuators

Despite their limitations, stepper motors are most commonly used to drive actuators in DOAS systems. Alternatives include DC motors with or without angular encoders. Given a suitable drive electronics encoder, motors can perform almost like stepper motors, with the advantage of not missing steps since the encoder always allows verification of the motor motion. However, resolution down to a single step is usually not possible.

In some cases, encoders might not be needed; for instance, the position of different filters does not need to be very precise, and thus DC motors with a series of position switches might be sufficient.

For small linear displacements piezo actuators might be useful. This type of actuator has seen rapid technological development in recent years.

7.15 Information Needed for Later Analysis

Besides the trace gas column density calculated from the spectra, additional information is required in order to use the data in an orderly way. This information includes obvious data:

- Site of measurement (i.e. geolocation, typically latitude and longitude of the measurement). This information is always be required for measurements from mobile platforms, but it is also good practice to record this information for fixed sites.
- Time and date of measurement.
- Information on type of measurement (active, passive, long-path DOAS, ZSL, multi-axis, etc.).
- Light-path identification for active systems and length of light path.
- Viewing direction for MAX-DOAS and off-axis DOAS measurements, i.e. viewing azimuth and elevation.
- Solar or lunar zenith angle. This information can also be calculated from time, date, and location of the measurement.
- Information on air density. This information is needed to convert the trace gas concentrations into mixing ratios. The air density can be readily determined by recording atmospheric pressure and temperature.
- Technical information, including number of pixels per spectrum, type of light source, type of spectrometer, grating used, type of detector, optical filter, wavelength, etc.

The DOAS control- and data acquisition software will typically record these data and add them to each spectrum.

8 Evaluation of DOAS Spectra, Sensitivity, and Detection Limits

One of the central aspects of DOAS is the analysis of atmospheric absorption spectra recorded with instruments described in Chap. 7. As discussed in Chap. 6, the analysis serves different purposes. First and most importantly, trace gas columns (the product of concentration and absorption path length) are derived. This process includes the separation of narrowband from broadband absorption features, and the separation of overlying absorption structures. In many DOAS applications, it is also of advantage to perform an automatic spectral alignment of different absorption cross-sections and absorption spectra. The other routine task of a DOAS analysis is the accurate determination of statistical uncertainties of trace gas columns. In contrast to laboratory experiments, the repetition of measurements under the same experimental conditions in the open atmosphere is impossible. Thus atmospheric observations rarely offer the opportunity to derive errors based on the statistical behaviour of measured concentrations. DOAS overcomes this problem by using the spectroscopic information and the quality of analysis of a single spectrum to derive an error for each measured concentration based on sound physical and statistical principles.

This chapter introduces and describes the most common DOAS analysis procedures. Various techniques for the modelling of spectra, the determination of associated measurement errors, and the derivation of detection limits for spectroscopic trace gas measurements will be covered. Most DOAS analysis routines in use today are based on least squares techniques, which will be discussed in the first part of this chapter. Approaches for the determination of the instrument function are presented. The application to the analysis of atmospheric spectra is discussed in the following section, including a brief description of Principal Component Analysis, PCA. A description of the simulation of absorption spectra required in the analysis is given. While this chapter is intended to provide an introduction to DOAS analysis, a certain degree of knowledge in mathematical methods is presumed.

U. Platt and J. Stutz, *Differential Optical Absorption Spectroscopy*, Geophysics and Environmental Physics, https://doi.org/10.1007/978-3-662-68639-3_8

8.1 Linear Fitting Methods

One of the most common methods to analyse experimental data is the linear least squares fitting technique (see e.g. Bevington 1969; Press et al. 1986). The basic idea behind this approach is to minimise the difference between the measured data, $y(x_i)$, where x_i is the independent parameter, and a function, $F(x, a_0, a_1, \ldots)$, which describes the dependence of $y(x_i)$ on the independent parameter x. The model parameters $a_0, a_1, \ldots$ are adapted during the fitting process until the function $F(x, a_0, a_1, \ldots)$ simulates $y(x_i)$ as closely as possible.

To quantify the similarity of $y(x_i)$ and $F(x_i, a_0, a_1, \ldots)$, one introduces χ^2:

$$\chi^2 = \sum_{i=1}^{n} \left(\frac{y(x_i) - F(x_i, a_0, a_1, \ldots)}{\varepsilon(i)} \right)^2 \tag{8.1}$$

where $\varepsilon(i)$ is the statistical error of the n measured data points $y(x_i)$. The meaning of χ^2 becomes clear if one compares it to the statistical definition of the standard deviation of the difference between $y(i)$ and $F(x_i, a_0, a_1, \ldots)$:

$$\sigma = \sqrt{\frac{1}{n} \sum_{i=1}^{n} (y(x_i) - F(x_i, a_0, a_1, \ldots))^2} \tag{8.2}$$

One can associate χ^2 with σ via $\chi^2 = \frac{\sigma^2}{\varepsilon^2} n$ if all the errors $\varepsilon = \varepsilon(i)$ are equal. χ^2 is thus a statistical measure of the difference between measured data and model function. The goal of the fitting routine is to minimise χ^2.

In many cases, the error $\varepsilon(i)$ is omitted from (8.1) by setting $\varepsilon(i) = 1$ for all data points. This is the most commonly used form of linear least square fit, which is called unweighted least squares fit. The approach that includes $\varepsilon(i)$ is called weighted least squares fits. Both unweighted and weighted least squares fit will be discussed in the following section. However, it is first necessary to discuss the general assumptions that the least squares method is based on. Later, it will become clear that these assumptions are not always fulfilled in DOAS analysis routines.

1. The model function $F(x, a_0, a_1, \ldots)$ must be linear in the parameters $a_0, a_1, \ldots$. A common example for such a function is a polynomial $F(x) = \sum_{h=0}^{r} a_h \times x^h$.
2. The model function needs to accurately describe the physical characteristics of the problem, otherwise, systematic biases can be introduced in the results of the fit. It is the task of the user to ensure that this assumption is fulfilled, for example by carefully analysing the residual $R(i) = y(x_i) - F(x_i, a'_0, a'_1, \ldots)$, where $a'_0, a'_1, \ldots$ are the fitting results. $R(i)$ should be statistically distributed around 0 and not show any functional dependence on x. For example, incorrectly using a linear function $F(x, a_0, a_1)$ as a model for a data set that has a quadratic (x^2) component would lead to biases in the results, and $R(i)$ would clearly show a functional dependence.

3. The statistical mean of each error $\varepsilon(i)$ upon repeated measurements has to be 0. In other words, no systematic errors should be imposed on $y(x_i)$. If the mean is not 0, biases may be introduced in the analysis. It is clear that systematic errors are commonly encountered in experimental data. One can, however, soften this requirement by stating that the systematic error of each data point, $y(x_i)$, needs to be much smaller than the respective statistical error, $\varepsilon(i)$. It is crucial that this assumption is tested in a DOAS analysis.
4. The variances of errors $\varepsilon(i)$ need to be known correctly to calculate the error of parameters a_0, $a_1, \ldots$ with the fitting routine. This assumption is often ignored in the application of least squares fits, and also often violated in spectroscopic measurements.

If all the above assumptions are fulfilled, the least squares fit will give the best estimate for the parameters a_0, $a_1, \ldots$ to minimise χ^2.

8.1.1 Unweighted Linear Least Squares Fit

To simplify the mathematical description of linear least squares method, we will use vector-matrix nomenclature to describe the solution of the fitting problem (vectors and matrices are printed in bold face). The n measurements of $y(x_i)$ will be described by a vector $\mathbf{y}$, and the m parameters a_0, $a_1, \ldots a_{m-1}$ with a vector $\boldsymbol{a}$. The physical content of $F(x_i, a_0, a_1, \ldots, a_{m-1})$ is summarised in a $m \times n$ matrix $\mathbf{X}$. The lines in the matrix denote the functional dependence of each parameter a_j in F, i.e. for $F(x, a_0, a_1) = a_0 + a_1 \times x$, the first line in matrix $\mathbf{X}$ would be $(1, x)$.

The parameter vector $\boldsymbol{a}$ that provides the best description for the measured data can then be derived analytically via:

$$
\begin{aligned}
\mathbf{a} &= \left[\mathbf{X}^T\mathbf{X}\right]^{-1}\mathbf{X}^T\mathbf{y} \\
\mathbf{\Theta} &= \hat{\sigma}^2\left[\mathbf{X}^T\mathbf{X}\right]^{-1} \\
\hat{\sigma}^2 &= (n-m)^{-1}\left[\mathbf{y}-\mathbf{Xa}\right]^T\left[\mathbf{y}-\mathbf{Xa}\right]
\end{aligned}
\tag{8.3}
$$

Here, $\mathbf{\Theta}$ represents the covariance matrix. The square root of diagonal elements of $\mathbf{\Theta}$, Θ_{ii}, represents the statistical uncertainty (1σ error) of the parameters a_i:

$$
\Delta a_i = \sqrt{\Theta_{ii}} \tag{8.4}
$$

The non-diagonal elements of $\mathbf{\Theta}$ can be used to determine the correlation between parameters a_i. $\hat{\sigma}$ is the root mean square (rms) of the residuals $R(i)$, and is thus a measure for the average "instrument error" ε, if $n >> m$.

As shown in (8.3), the linear least squares fit is an analytical method. It can be shown that the calculated vector $\boldsymbol{a}$ provides the parameters with which $F(x_i, a_0, a_1, \ldots, a_{m-1})$ describes the data set $y(x_i)$ best (Bevington 1969; Press et al. 1986).

The fact that the linear least squares fit is an analytical method explains its popularity in the analysis of many types of data.

8.1.2 Weighted–Correlated Least Squares Fit

The approach described earlier can be expanded by explicitly including the errors $\varepsilon(i)$ for each measurement $y(x_i)$. Mathematically, this can be achieved by introducing the covariance matrix of the errors $M_{ij} = E(\varepsilon(i) \times \varepsilon(j))$, where E is the expectance value of $\varepsilon(i) \times \varepsilon(j)$ (note that $E(\varepsilon(i) \times \varepsilon(j))$ is different from $\varepsilon(i) \times \varepsilon(j)$). If $\mathbf{M}$ is known, the following solution for the linear least squares fit can be found.

$$\begin{aligned} \mathbf{a} &= \left[\mathbf{X}^T\mathbf{M}^{-1}\mathbf{X}\right]^{-1}\mathbf{X}^T\mathbf{M}^{-1}\mathbf{y} \\ \mathbf{\Theta} &= \hat{\sigma}^2\left[\mathbf{X}^T\mathbf{M}^{-1}\mathbf{X}\right]^{-1} \\ \hat{\sigma}^2 &= (n-(m+r))^{-1}\left[\mathbf{y}-\mathbf{Xa}\right]^{\mathrm{T}}\mathbf{M}^{-1}\left[\mathbf{y}-\mathbf{Xa}\right] \end{aligned} \tag{8.5}$$

If all the errors are statistically independent from one another, $\mathbf{M}$ is a diagonal matrix. We will see in spectroscopic applications that this is not always the case. If all the errors $\varepsilon(i)$ are equal, $\mathbf{M}$ will be a matrix with a diagonal of 1, and the solution will be equal to the one in (8.3). The meaning of $\mathbf{\Theta}$ and the determination of Δa_i are the same as in (8.4).

8.2 Non-linear Fitting Methods

The linear least squares method described earlier offers an elegant way of analysing data. However, the limitation to linearly dependent fit parameters can impose restrictions for certain applications. The problem of spectral alignment cannot be solved by a linear method, and requires non-linear least squares method. It is preferable to use linear methods since they provide an analytical solution to the fitting problem, whereas non-linear routines are iterative and thus often sensitive to the starting conditions. In addition, non-linear methods require long computation times for problems with a large number of parameters.

Here, three different non-linear fitting methods will be presented. In general, one tries to solve a problem where a number of measurements $y(x_i)$ are analysed by fitting a model function, $f(x_i, c_0, c_1, \ldots)$, that is non-linear in the parameters c_0, $c_1, \ldots$. In addition, the approach will rely on minimising χ^2, as defined in (8.1).

8.2.1 Gradient Method

The gradient method is based on finding the minimum of χ^2 by following the steepest descent of χ^2 on the multi-dimensional surface in the $c_0, c_1, \ldots$ space by an iterative process (Press et al. 1986). If one denotes the vector of the parameters c_0, $c_1, \ldots$ during the iteration as $\boldsymbol{\gamma}_k$, where k is the number of iterations of the fit, the mathematical equation allowing the calculation of next iteration is as follows:

$$\boldsymbol{\gamma}_{k+1} = \boldsymbol{\gamma}_k + \text{const.} \times \left[-\nabla\chi^2\left(\boldsymbol{\gamma}_k\right)\right] \tag{8.6}$$

The equation is based on the calculation of the gradient of χ^2 and a step along the steepest descent, with a width that can be chosen through the constant. For small constants, the iteration will converge very slowly, whereas for large constants there is a high risk that the iteration will step over the minimum of χ^2.

The iteration has to be repeated until a certain convergence condition is met. The choice of this condition depends on the problem and the desired accuracy, and it is difficult to guarantee the convergence of this method at all accuracies. The method also depends on the starting conditions $\boldsymbol{\gamma}_0$ for problems that have more than one minimum of χ^2.

8.2.2 Gauß–Newton Method

The Gauß–Newton method is based on the fact that, for a minimum of χ^2, the first derivative of χ^2 in the parameter space has to be zero. By using this fact and applying a Taylor expansion to χ^2, the following expression is derived for an iteration step in the fitting process (Press et al. 1986):

$$\boldsymbol{\gamma}_{k+1} = \boldsymbol{\gamma}_k + -\mathbf{D}^{-1} \times \left[-\frac{1}{2}\nabla\chi^2\left(\boldsymbol{\gamma}_k\right)\right] \tag{8.7}$$

where $\mathbf{D}$ is the Hessian matrix (or simply Hessian) of our model function $f(x_i, c_0, c_1, \ldots)$, which is approximated by the following expression in the Gauß–Newton method:

$$(\mathbf{D})_{rs} = -\frac{1}{2}\frac{\partial\chi^2}{\partial\gamma_r\partial\gamma_s} \approx \sum_{j=1}^{n}\frac{\partial f(\boldsymbol{\gamma}_k)}{\partial\gamma_r}\frac{\partial f(\boldsymbol{\gamma}_k)}{\partial\gamma_s} \tag{8.8}$$

As in the case of the gradient method, a suitable condition for convergence has to be applied to stop the iteration. There is, however, no guarantee that the Gauß–Newton method converges. The method works best close to the minimum of χ^2, where (8.8) is a good approximation.

8.2.3 Levenberg–Marquardt Method

As discussed in the previous sections, the gradient method converges rapidly away from the minimum of χ^2, while the Gauß–Newton method works best close to the minimum. Taking into account the similarity of (8.6) and (8.7), Levenberg

(1944) combined the methods in a single algorithm that is more stable and converges faster than the individual methods. By rewriting the vector step in (8.6) as $\lambda \cdot \mathbf{1} \cdot \left[-\nabla\chi^2(\boldsymbol{\gamma}_k)\right]$ with a parameter λ that stands for the constant, and by including the 1/2 factor into $\mathbf{D}^{-1}$ in (8.7): $\mathbf{D}'^{-1} = 1/2\,\mathbf{D}^{-1}$, he introduced a matrix $\mathbf{A}$:

$$\mathbf{A} = \left(\lambda \cdot \mathbf{1} + \mathbf{D}'^{-1}\right) \tag{8.9}$$

$\mathbf{A}$ can be used to calculate $\boldsymbol{\gamma}_{k+1}$ from $\boldsymbol{\gamma}_k$ in the iteration:

$$\boldsymbol{\gamma}_{k+1} = \boldsymbol{\gamma}_k + \mathbf{A}\left[-\nabla\chi^2\left(\boldsymbol{\gamma}_k\right)\right] \tag{8.10}$$

If λ is large, this method favours the gradient method, where λ is the step size. Small λ leads to the use of the Gauß–Newton method. For medium-sized λ, a linear combination of the two methods is used. Marquardt (1963) describes the following steps for the choice of λ:

1. Choose a starting parameter vector $\boldsymbol{\gamma}_0$, and set λ to a moderate value, for example $\lambda = 0.001$.
2. Perform an iteration step and calculate $\boldsymbol{\gamma}_{k+1}$ from $\boldsymbol{\gamma}_k$ using (8.10). Determine $\chi^2(\boldsymbol{\gamma}_{k+1})$ for the new parameter vector.
3. If $\chi^2(\boldsymbol{\gamma}_{k+1}) > \chi^2(\boldsymbol{\gamma}_k)$, set $\boldsymbol{\gamma}_{k+1} = \boldsymbol{\gamma}_k$, thus rejecting the new parameter vector, and increase λ by a factor of 10. Return to Step 2 to calculate a new parameter vector.
4. If $\chi^2(\boldsymbol{\gamma}_{k+1}) \leq \chi^2(\boldsymbol{\gamma}_k)$, decrease λ by a factor of 10. Return to Step 2 to calculate the next parameter vector.

To stop the iteration, convergence conditions have to be set. For example, one may wish to stop iteration if λ is too large, i.e. when the iteration does not seem to converge, or if λ is too small, when the iteration is already very close to the minimum of χ^2. Another possibility is a limit for the change of χ^2, e.g. if χ^2 does not change more than a certain factor, at which one assumes that the iteration is close to the minimum.

The advantage of the Levenberg–Marquardt method is that it always converges, since the steps that do not reduce χ^2 are rejected while, at the same time, increasing λ. In this case, more weight will be put on the gradient method. The algorithm, on the other hand, will try to make λ smaller and thus give more weight to the Gauß–Newton method, which provides a good approximation close to the minimum of χ^2.

A number of methods have been suggested to calculate the errors of the parameters $c_0, c_1, \ldots$, which are the result of the fit. Press et al. (1986), for example, suggest a method that assumes that the errors of the original data are normally distributed. To derive the errors, they suggest to recalculate $\mathbf{D}'^{-1}$ at the end of iteration, for example by setting $\lambda = 0$ and recalculating $\mathbf{A}$. $\mathbf{D}'^{-1}$ can then be interpreted as the covariance matrix of the non-linear fit.

$$\mathbf{\Theta} = \mathbf{D}'^{-1} = \frac{1}{2}\mathbf{D}^{-1} \tag{8.11}$$

The errors can then be calculated in a similar way as described earlier.

$$\Delta c_i = \sqrt{\Theta_{ii}} \tag{8.12}$$

Cunningham (1993) tested this calculation using Monte Carlo methods and found that it underestimates the error by less than 50%.

8.3 DOAS Analysis Procedure

The DOAS approach was discussed in detail in Chap. 6. It is helpful to recall the most important aspects of the DOAS technique to understand the consequences they have for the design of analysis routines.

- The DOAS technique is based on the separation of slow- and fast-varying spectral structures. A DOAS analysis routine has to be able to either high-pass filter the absorption spectra before the analysis or provide the means to do this during the analysis.
- Atmospheric absorption spectra contain overlaying absorption structures that must be separated by the analysis routine. The typical approach is to use absorption reference spectra or cross-sections that are adapted to the instrument resolution.
- In case the reference spectra are not spectrally aligned, which occurs frequently, the analysis routine has to be able to correct this misalignment.
- Because the absorption spectra always contain noise, which may introduce statistical uncertainties, the analysis routine has to be able to calculate errors of the desired trace gas columns.

As discussed in Chap. 6, the DOAS problem can be summarised by (8.13):

$$I(\lambda, L) = I_0(\lambda) \exp\left(\int_o^L \sum_j \left(-\sigma_j(\lambda, p, T) \times \rho_j(l)\right) - \varepsilon_R(\lambda, l) - \varepsilon_M(\lambda, l)\, dl\right) + N(\lambda) \tag{8.13}$$

where $I_0(\lambda)$ and $I(\lambda, L)$ are the light intensity at the beginning and end of the light path, respectively; $\sigma_j(\lambda, p, T)$ is the absorption cross-section of a trace species j, which depends on the wavelength λ, the pressure p, and the temperature T; $\rho_j(l)$ is the number density at position l along the light path; $\varepsilon_R(\lambda, l)$ and $\varepsilon_M(\lambda, l)$ are the Rayleigh-extinction and Mie-extinction, respectively; and $N(\lambda)$ represents the noise.

The measurement changes $I(\lambda, L)$ in two ways. First, the limited resolution of the spectrograph, which can be described by a convolution with the instrument function

H, $I^*(\lambda, L) = I(\lambda, L) * H$ (see Sect. 8.5.1), increases the width of spectral structures. In addition, the different optical elements can impose a broad spectral structure onto $I^*(\lambda, L)$. Finally, $I^*(\lambda, L)$ is digitised by the detector $I'(i)$.

In Chap. 6, we discussed various implementations of the DOAS technique. In Case 1, a linearisation of (8.13) after convolution and discretisation is possible. This is the most common example of DOAS implementation today, and we will concentrate on this approach in the following section. In cases with strong trace gas absorptions, the modelling of equation $I'(i)$, including convolution and discretisation, is necessary. This is, in principle, possible using the non-linear least squares fits described earlier. However, in practice this approach is slow and cumbersome. It is thus rarely used, and we will not discuss it further.

Most DOAS analysis software relies on the linearisation of $I'(i)$ by calculating the logarithm $J(i) = \ln(I'(i))$:

$$J(i) = J_0(i) + \sum_{j=1}^{m} a'_j \cdot S'_j(i) + B'(i) + R'(i) + A'(i) + N'(i) \tag{8.14}$$

The narrow absorption structures of trace gases are described by their individual differential absorption structure $S'_j(\lambda) = \ln(\exp(-\sigma'_j(\lambda))^* H)$ calculated from the convolution of the cross-section of trace gas j with the instrument function H (see Chap. 6 for the mathematical rationale to calculate $S'_j(\lambda)$ for different DOAS implementations). The scaling factors $a'_j = c_j \times L$ are then the product of average number densities c_j over path-length L. The broad absorptions of trace gases are represented by $B'(i)$. Variations in the spectral sensitivity of the detector or spectrograph are summarised in $A'(i)$ as a function of pixel number. The extinction by Mie and Rayleigh scattering is represented by $R'(i)$. The noise $N'(i) = \ln(N(\lambda))$ is caused by detector noise and photon statistics. The overlaying absorption structures of several trace gases are represented by the sum in (8.14). In practice, the number of absorbers m can be limited to trace gases with absorption structures sufficiently strong to be detectable with DOAS instruments.

8.3.1 The Linear Model

Most DOAS evaluation procedures are based on a model that describes the DOAS spectra according to (8.14). The logarithm of the discrete measured intensity, $J(i)$, is modelled by a function F(i):

$$F(i) = P_r(i) + \sum_{j=1}^{m} a_j \times S_j(d_{j,0}, d_{j,1}, ..)(i) \tag{8.15}$$

The absorption structures of trace gases, S_j, are input data to the procedure. The scaling factors a_j are the result of the fit and can be used to calculate the concentration c_j of the respective trace gases with differential absorption cross-section σ'_j by: $c_j = a_j/(\sigma_j \times \mathrm{L})$.

The polynomial, $P_r(i)$, describes the broad spectral structures caused by the characteristics of the lamp $I_0(i)$, the scattering processes $R'(i)$, the spectral sensitivity $A(i)$, and the broad absorptions by the trace gases $B(i)$:

$$P_r(i) = \sum_{h=0}^{r} c_h \cdot (i - i_c)^h \tag{8.16}$$

The parameter $i_c = \text{int}(n/2)$ represents the centre pixel of the spectral region used for evaluation. The polynomial refers to i_c to maximise the influence of non-linear terms. In this approach, the polynomial is responsible for the separation of broad and narrow spectral structures. The degree of filtering depends on the problem. In case of relatively broad absorption features, for example of NO_3, a low-order polynomial has to be used. If a large number of narrow absorption structures were to be analysed, for example of SO_2, a higher order polynomial can be used. An alternative approach to fitting a higher order polynomial is to high-pass filter the absorption spectra and references before using them in (8.15) (see Sect. 8.3.2).

Among the most important aspects of DOAS analysis using the model function (8.15) is the very accurate description of the reference spectra S_j. As discussed earlier, the least squares methods that are most commonly used in the analysis of DOAS spectra (see Sect. 8.1) require that the model function is an accurate description of the physical characteristics of the problem. Reference spectra of poor quality can introduce systematic biases in the analysis.

8.3.2 High and Low-Pass Filtering

The basis of the DOAS method is the separation of narrow and broad spectral structures. While this can be done directly in the analysis routine, this separation is often also performed before the actual fitting process. While a comprehensive survey of available filtering methods is beyond the scope of this book, a number of commonly used filtering approaches will be briefly described below. In addition to high-pass filters, low-pass filters, which reduce the magnitude of noise in the spectra, are also employed. Examples of the various methods described here are shown in Fig. 8.1.

Low-Pass Filters

Low-pass filters are used to reduce the unwanted high frequency components of absorption spectra. These can be caused by noise, pixel-to-pixel sensitivity changes, or other instrumental problems. Care should be taken when using low-pass filters in a DOAS analysis, as the noise reduction will lead to an unrealistically small fitting error. The following filters are often used in DOAS applications:

Moving average: The moving average approach replaces the content of a channel i by taking the average of the original data point $y(i)$ and the m neighbour channels to the left and right:

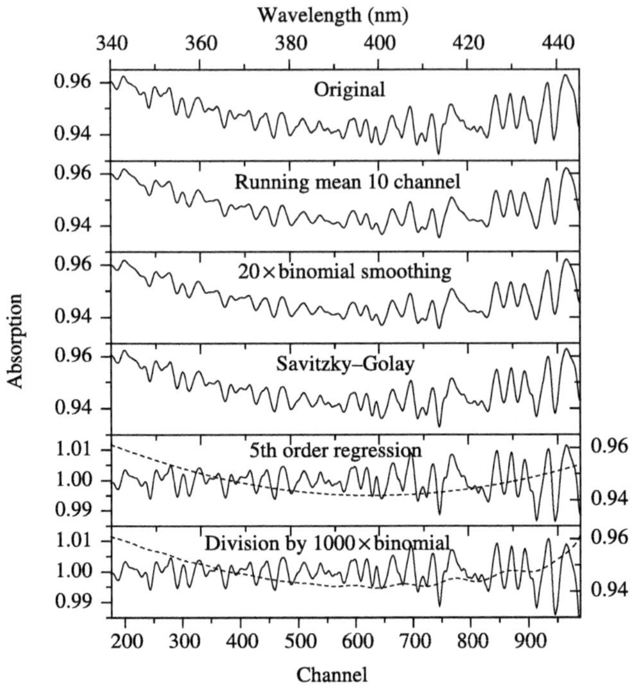

Fig. 8.1 Example of low- and high-pass filtered absorption spectra

$$\hat{y}(i) = \frac{1}{2m + 1} \sum_{j=-m}^{m} y(i + j) \tag{8.17}$$

The width of the box $2m + 1$ determines the degree of smoothing of the spectrum. A wider box leads to the removal of broader structures. Care must be taken at the edges of the spectrum.

Binomial (triangular) smoothing: In this method, each channel is replaced by a value equal to the weighted average of neighbouring channels. Each channel is weighted according to the binomial distribution so that points further from the original channel contribute less. The simplest example of a binomial smoothing approach is a three-point binomial filter with the following weighting coefficients: 0.25, 0.50, 0.25:

$$\hat{y}(i) = \frac{1}{4} y(i - 1) + \frac{1}{2} y(i) + \frac{1}{4} y(i + 1) \tag{8.18}$$

Equation (8.18) can be applied multiple times, which will be case the smoothing will be equivalent to a higher order smoothing.

Polynomial smoothing The basic principle of polynomial smoothing method is the fitting of a polynomial of nth degree to the neighbouring channels surrounding a particular channel i in the dataset. The 'smoothed' value of the channel i is then assigned the value of the polynomial at i. The two parameters influencing the degree of smoothing are the degree of the polynomial and the window over which the polynomial is fitted.

An efficient algorithm for computing the polynomial filter has been described by Savitzky and Golay (1964). It enables rapid calculation of the polynomial coefficients. Another implementation is a cubic-spline, which invoke additional conditions for the calculation of polynomials, e.g. the first and second derivatives of adjacent curves have to be equal, thus avoiding discontinuities at the joints.

High-Pass Filters

A number of filters are used to remove broad (much larger than typical absorption band widths) spectral structures from absorption spectra, including:

Division by low-pass filtered spectrum: Any low-pass filter can be converted into a high-pass filter by dividing the original spectrum by the low-pass filtered spectrum. This is often used in combination with binomial smoothing or running averages.

Regression: Another common high-pass filter method is division by a polynomial of degree g, which is fitted to the entire spectrum with a linear least squares method. The degree can be chosen according to the expected broadband behaviour. This method is ideal if the dominant broadband contribution originates from Raleigh scattering. Regressions are typically performed using polynomials with maximum orders of five to nine. The difference to the polynomial low-pass filters described earlier is that, for regression, a large number of channels are used in the fit.

Fourier filters: The most sophisticated means of filtering absorption spectra are methods that are based on Fourier techniques. Here, the spectrum is first transformed into frequency space, for example, by a fast Fourier transform (FFT) calculation. Certain higher or lower frequencies are removed or reduced. After the transformation back into normal space via another FFT, a smoothed spectrum is obtained. There are many different choices for these types of digital filters, and the reader is advised to refer textbooks on this topic.

A word of warning: If filters are used in a DOAS analysis, they must be applied to all spectra, including atmospheric absorption and reference spectra. This will ensure that the absorptions in the reference and atmospheric spectra have the same structure. For this reason, it has also become less common to use regressions, since these do not guarantee that the absorptions change equally in both types of spectra.

8.3.3 Wavelength Alignment

While it should be theoretically possible to measure or calculate reference spectra that are well aligned with the atmospheric absorption spectrum and each other, small uncertainties in the wavelength position and dispersion can arise due to thermal changes in the spectrometer, inaccuracy of the grating position, and errors in the wavelength calibration of literature absorption cross-sections. It is, therefore, often necessary to perform a spectral alignment during the analysis. It should be noted that small misalignments do not affect the scaling factor/concentration of the respective trace gas much, but influence other weaker absorbers by introducing unwanted residual structures.

The task of the analysis procedure is to align the reference spectra $S_j(i)$ (wavelength-pixel-mapping Γ_j) to the atmospheric absorption spectrum $J(i)$ (wavelength-pixel-mapping Γ_J). The procedure, therefore, must recalculate the reference spectrum $S_j^*(i)$ with the wavelength-pixel-mapping Γ_J (see Chap. 6 for a discussion on wavelength-pixel-mapping). This can be described as 'shifting and stretching/squeezing' the reference spectrum in wavelength. As Γ_j is a strongly monotonous function, its inverse can be described by a polynomial: Γ_j^{-1}: $x(\lambda) = \sum_{k=0}^{q} \beta_k \times \lambda^k$, where $x(\lambda)$ represents the non-integer 'pixel number' that results from this inverse transformation. $S_j(\lambda)$ can now be calculated from the continuous spectrum $S_j(x)$. This spectrum must be approximated using an appropriate interpolation method, such as a cubic spline procedure (Press et al. 1986), on the discrete spectrum $S_j(i)$.

We can now calculate $S_j^*(i)$ with the wavelength-pixel-mapping Γ_J by deriving $S_j(\lambda)$ with Γ_j^{-1} from $S_j(x)$, which is approximated by an interpolation on $S_j(i)$, and then applying Γ_J:

$$S_j(i) \stackrel{\text{interpolation}}{\longrightarrow} S_j(x) \stackrel{\Gamma_j^{-1}}{\longrightarrow} S_j(\lambda) \stackrel{\Gamma_J}{\longrightarrow} S_j^*(i) \tag{8.19}$$

It is possible to combine Γ_j^{-1} and Γ_J into one formula, which links i to x using a polynomial with parameters δ_k:

$$x(i) = x(\lambda(i)) = \sum_{k=0}^{q_S \times q_I} \delta_k \cdot i^k \tag{8.20}$$

It is advantageous to use a slightly modified version of (8.20), where the spectral alignment parameters $d_{j,k}$, determining the transformation are zero if the wavelength-pixel-mappings of J and S_j are equal:

$$x = i + f_j(i) \quad \text{with} \quad f_j(i) = \sum_{k=0}^{p_j} d_{j,k} \cdot (i - i_c)^k \tag{8.21}$$

The spectrum $S_j(d_{j,0}, d_{j,1}, \ldots)(i) = S_j^*(i)$ now has the wavelength-pixel-mapping Γ_J, which was calculated with the parameters $d_{j,k}$ using (8.19) and (8.21) and an interpolation of $S_j(i)$.

The parameters $d_{j,k}$ are derived by performing a non-linear fit of the model F to the spectrum J with fixed parameters a_j and c_h. If $p_j = 0$, the spectrum S_j is shifted by $d_{j,0}$ pixels. If $p_j = 1$, the spectrum is additionally linearly squeezed or stretched according to parameter $d_{j,1}$. Higher values of p_j represent a squeeze or stretch of higher order. To achieve the best physical description of the spectra, the degree of the squeeze process p_j should be selected for every reference spectrum S_j. Alternatively, one set of parameters $d_{j,k}$ for two or more reference spectra is used if the wavelength calibration is identical for these spectra.

To quantify the importance of a spectral alignment, one can set up a test spectrum J, which is calculated as the sum of a single trace gas spectrum $S_g^*(d_{g,0}, d_{g,1}, \ldots)(i) = a_g \times S_g(d_{g,0}, d_{g,1}, \ldots)(i)$, with shift and squeeze represented by the parameters d_{gj} and $(m-1)$ with unaltered reference spectra $S_j^* = a_j \times S_j$:

$$J = 1 + S_1^*(0) + \cdots + S_g^*\left(d_{g,0}, d_{g,1}, \ldots\right) + \cdots + S_m^*(0) \tag{8.22}$$

A linear fit based on the model function $F_{\text{sh}} = 1 + a_1^* \times S_1^*(0) + a_2^* \times S_2^*(0) + \cdots + a_m^* \times S_m^*(0)$ is then used to analyse J for different parameters $d_{g,0}, d_{g,1}, \ldots$ in J. By varying the different parameters and spectra, one can calculate a function:

$$\Phi_{j,g}(d_{g,0}, d_{g,1}, ..) = 1 - a_j^* \tag{8.23}$$

which determines the change in the scaling factors a_j^* due to a shift of spectrum S_g^*. For m reference spectra in J, there are m^2 different $\Phi_{j,g}$. $\Phi_{j,g}$ illustrates the sensitivity of the fit results to misalignments. Figure 8.3 shows 16 different $\Phi_{j,g}$ for the spectra in Fig. 8.2 and an alignment by a shift. It can be seen that a shift of one pixel in spectrum S_1, corresponding to approximately 1/20 of the width of the absorption bands, produces a result for spectrum S_3 that differs by 70% from the real value ($\Phi_{3,1}$; Fig.8.3c). This confirms that, without correcting the wavelength calibration, the linear fit procedure can give incorrect results.

8.3.4 Realisation

There are a number of different implementations of DOAS analysis algorithms. Most of them have in common that they use least squares methods to fit a model function to the absorption spectra. The most popular approach uses linearisation (Sect. 8.3.1), which is shown here as one single-model function:

$$F(i) = \sum_{h=0}^{r} c_h \cdot (i - i_c)^h + \sum_{j=1}^{m} a_j \cdot S_j(d_{j,0}, d_{j,1}, ..)(i) \tag{8.24}$$

The model function is determined by three sets of parameters: (1) $c_0, c_1, \ldots, c_h$ determine an additive polynomial; (2) $a_1, \ldots, a_m$ scale the reference spectra/cross-sections; (3) $d_{j,0}, \ldots$ are the parameters for spectral shift and squeeze for each reference spectrum. The first two sets of parameters are linear in $F(i)$ and can be summarised in a vector $\boldsymbol{\beta} = (c_0, c_1, c_2, \ldots, a_0, a_1, a_2, \ldots)$. The third set of parameters is non-linear in $F(i)$ and can only be fitted through non-linear least squares

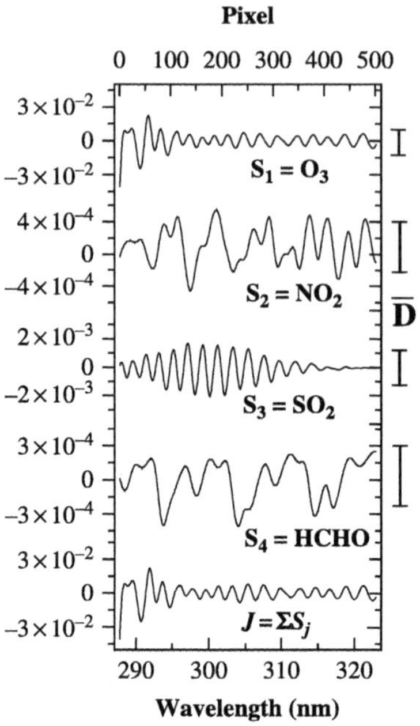

Fig. 8.2 Absorption spectra used for the alignment tests. The spectra were calculated by convoluting high-resolution cross-sections with a Gaussian-shaped instrument function of 1.3 nm half-width. The average optical density $\overline{D}$ of the spectra, 0.0195 for O_3, 0.00066 for NO_2, 0.0023 for SO_2 and 0.00051 for HCHO, is marked with *bars* on the right axis. This corresponds to atmospheric concentrations of 100 ppb for O_3, 2.5 ppb for NO_2, 0.5 ppb for SO_2 and 0.5 ppb for HCHO for a light path of 7.5 km in the troposphere. Reprinted with permission from Stutz and Platt 1996, © Optical Society of America. All rights reserved

methods. While it is possible to use non-linear procedures to optimise all three sets of parameters, we will discuss here an approach that is based on a combination of linear and non-linear methods, which has the advantage of being faster than a purely non-linear approach.

The fitting process starts with the choice of reasonable starting values for the $d_{j,k}$ parameters. In the first step, the parameter vector $\boldsymbol{\beta} = (c_0, c_1, c_2, \ldots, a_0, a_1, a_2, \ldots)$ is determined by a linear least squares method using the measured spectrum $J(i)$ and the coefficient matrix **X**. **X** contains $(r + 1)$ polynomial arguments $(i - i_c)^h$ and the m reference spectra S_j, which are shifted according to the parameters $d_{j,k}$:

$$\mathbf{X} = \begin{pmatrix} 1 & (-i_c)^1 & (-i_c)^2 & \cdots & (-i_c)^r & S_1(0) & S_2(0) & \cdots & S_m(0) \\ 1 & (1-i_c)^1 & (1-i_c)^2 & \cdots & (1-i_c)^r & S_1(1) & S_2(1) & \cdots & S_m(1) \\ 1 & (2-i_c)^1 & (2-i_c)^2 & \cdots & (2-i_c)^r & S_1(2) & S_2(2) & \cdots & S_m(2) \\ \vdots & \vdots & \vdots & \ddots & \vdots & \vdots & \vdots & \ddots & \vdots \\ 1 & (n-i_c)^1 & (n-i_c)^2 & \cdots & (n-i_c)^r & S_1(n) & S_2(n) & \cdots & S_m(n) \end{pmatrix} \tag{8.25}$$

The number of columns in **X** is given by the number of fitted parameters $(r + 1) + m$. The number of lines is equal to the number of pixels $(n + 1)$ of the wavelength interval

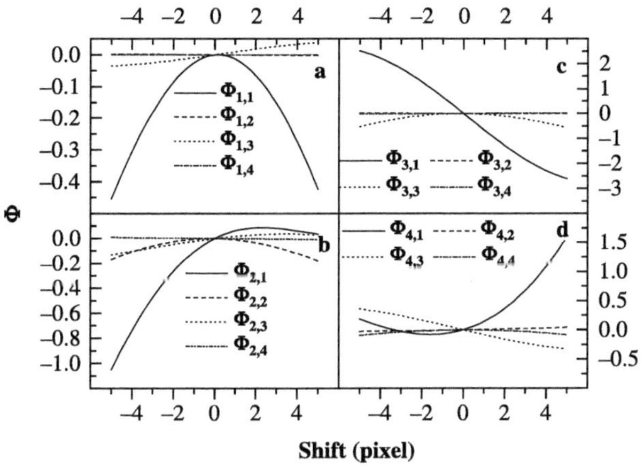

Fig. 8.3 $\Phi_{j,k}$ [see (8.23)] describes the influence on the fit results a_j of the misalignment of spectrum S_k. The $\Phi_{j,k}$ shown in this figure was calculated using the spectra of Fig. 8.2. (**a**) shows the influence of the shift of the four different S_j on a_1 of O_3. (**b**), (**c**), and (**d**) show the influence on NO_2, SO_2 and HCHO, respectively. The highest influence is found for the shift of O_3 on SO_2 ($\Phi_{3,1}$ in (c)), where a shift of 1 pixel changes a_3 by 70%. Reprinted with permission from Stutz and Platt 1996,

of the analysis. Performing the linear least square procedure gives the best estimate for linear parameters, using the given $d_{j,k}$: $\boldsymbol{\beta}^1$.

Next, one iteration of a Levenberg–Marquardt fit is performed to derive the estimate of non-linear parameters $d_{j,k}$. The resulting parameters $d^1_{j,k}$ are used in the successive call of the linear fit, which calculates $\boldsymbol{\beta}^2$. This result is then used in the next call of the non-linear fit. The procedure then continues to alternate the two methods, always using the result of the previous method as values for the next fit method (see also Fig. 8.4). This procedure is repeated until one of several stopping conditions for the non-linear fit is fulfilled.

Normally, the fit is aborted when the relative changes of χ^2 in the last step is smaller than a given value (usually 10^{-6}), and thus the fit has converged. The fit also stops if the number of repetitions of the iteration, determined by the user, is exceeded or if the non-linear method becomes unstable, e.g. the λ parameter of the Levenberg–Marquardt method becomes too large or too small.

The analysis procedure is thus a combination of the well-known non-linear Levenberg–Marquardt method (Levenberg 1944; Marquardt 1963; Press et al. 1986) determining the coefficients $d_{j,k}$ (see Sect. 8.3.3) and a standard linear least squares fit (Press et al. 1986; Bevington 1969; Albritton et al. 1976) to derive the coefficients a_j and c_k.

8.3.5 Linearizing Spectral Misalignment Correction

Compared to the linear fit, the non-linear fit is time consuming and depends on the proper starting conditions. Methods have been developed to linearise the non-linear effects of spectral misalignments (see e.g. Rozanov et al. 2011; Beirle et al.

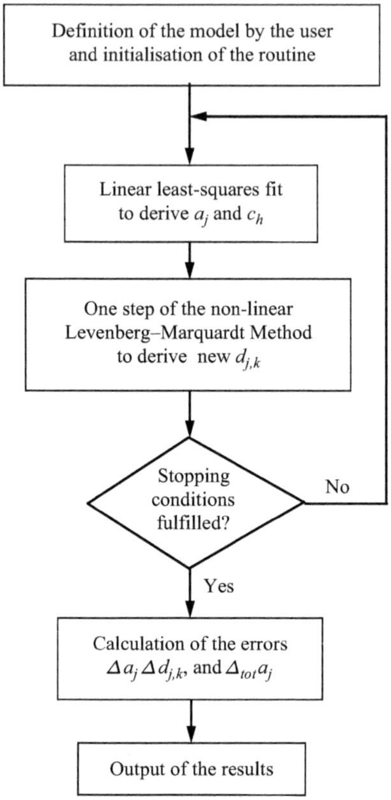

Fig. 8.4 Overview of the linear–non-linear DOAS analysis procedure. Reprinted with permission from Stutz and Platt 1996, ©Optical Society of America. All rights reserved

2013). This approach was found to be computationally faster and more stable, since it excludes local, relative minima of the residual function. In order to investigate the effects of a spectral shift, i.e., a simple offset of $\Delta\lambda$ in the wavelength-pixel allocation between I and I_0, we consider the equation for the optical density (see Chap. 6):

$$\ln\left(\frac{I(\lambda)}{I_0(\lambda)}\right) = \ln(I(\lambda)) - \ln(I_0(\lambda)) = \sum S_i\sigma_i + P(\lambda) \tag{8.26}$$

Typically the cross-sections $\sigma_i(\lambda)$ and $I_0(\lambda)$ (and thus $\ln I_0(\lambda)$) can be assumed to be fixed in their spectral positions. When $I(\lambda)$ is shifted (e.g. because of a drift in the wavelength-pixel mapping of the spectrograph) by $\Delta\lambda$, this can be thought of as artificial spectral structures being created, which can be approximated by a Taylor expansion (Beirle et al. 2013):

$$\begin{aligned}\ln(I(\lambda)+\Delta\lambda) &= \ln(I(\lambda)) + \frac{d\ln(I(\lambda))}{d\lambda}\cdot\Delta\lambda + \mathcal{O}(2)\\ &= \ln(I(\lambda)) + \frac{1}{I(\lambda)}\frac{dI(\lambda)}{d\lambda}\cdot\Delta\lambda + \mathcal{O}(2)\\ &\approx \ln(I(\lambda)) + A_{SH}(\lambda)\cdot\Delta\lambda\end{aligned} \tag{8.27}$$

The term $\mathcal{O}(2)$ denotes second-order and higher order terms. The error introduced by neglecting these terms can be approximated by inspecting the second-order term of the Taylor expansion. Thus, a spectral shift $\Delta\lambda$ of $I(\lambda)$ causes, in first order, an additive spectral structure, which is proportional to $A_{SH}(\lambda) = 1/I(\lambda)\cdot dI(\lambda)/d\lambda$. This "**shift-spectrum**" $A_{SH}(\lambda)$, i.e., the normalised (or 'logarithmic') derivative of $I(\lambda)$, can now be included as an additional 'pseudo-absorber' in the linear part of the DOAS fit. The corresponding fit coefficient directly yields the spectral shift $\Delta\lambda$ of $I(\lambda)$. A spectral 'squeeze' or 'stretch' of the spectrum can be handled by making $\Delta\lambda$ a linear function of λ (Beirle et al. 2013).

8.3.6 Spectral Shift—the 'Tilt Effect'

The shift is often attributed to temporal or other instabilities of the instrument. However, as first noted by Sioris et al. (2003) spectral positions of narrow-band structures like Fraunhofer lines and molecular absorption lines are systematically shifted for different measurement geometries due to an overall slope, or 'tilt', of the measured intensity spectrum. The phenomenon has thus become known as the 'tilt effect' (Sioris et al. 2003). It was first applied for limb satellite observations as a first-order correction.

As pointed out by Lampel et al. (2017) the effect is a consequence of the measurement process itself because atmospheric absorption and convolution with the instrument function do not commute. During the measurement process the highly resolved spectral structures in the spectrum will first be modified by absorption as well as Mie- and Rayleigh-scattering processes in the atmosphere before they are recorded with a spectrometer, which convolutes them with its instrument function. In the subsequent DOAS spectral evaluation process, however, the polynomial (or other function used for this purpose) accounting for broadband absorption is only applied after the convolution is performed. For DOAS evaluations, which allow for a spectral shift of measurement spectrum relative to a reference spectrum, this effect was found by Lampel et al. (2017) to lead to spectral shifts of up to 2 pm at a $\Delta\lambda = 0.6nm$ spectral resolution of the instrument. If no spectral shift is allowed during the evaluation procedure, residual structures of up to $2.5\cdot 10^{-3}$ peak-to-peak were observed. It was shown by Lampel et al. (2017) that the commutator of the DOAS polynomial and the convolution operation can be interpreted as a spectral shift of the measured solar spectrum. Thus, this effect needs to be considered (and corrected) for all DOAS applications aiming at an RMS of the residual of 10^{-3} and below.

8.3.7 Error Analysis

The calculation of the uncertainties of the various fit factors and ultimately the errors of concentrations of the respective trace gases are an important part of the analysis routine. The fit method described in the previous section calculates two sets of errors: Δa_j and Δc_h, which are derived by the linear least squares fit using the alignment parameters $d_{j,k}$; and, in the case of an analysis routine with spectral alignment, $\Delta d_{j,k}$, the errors of the alignment parameters, which can be calculated using the approach suggested by Press et al. (1986). Both sets of errors are calculated independently from one another. The discussion in Sect. 8.3.3 showed that the fit factors a_i depend on the alignment parameters $d_{j,k}$ of the spectra. Following the same arguments, the alignment uncertainties $\Delta d_{j,k}$ will increase the uncertainty Δa_j of the linear fit parameters a_i (we ignore the polynomial parameters here since they are not the main result of the analysis, although the same arguments apply). The following section will present a method to estimate this error propagation.

The residuals of DOAS analyses often include structures that are broader and larger than the expected photon noise. This impacts the accuracy of error calculations in DOAS retrievals. The last part of this section is dedicated to the discussion of various methods that have been developed to address this problem. A discussion of the causes of the residual structures is given in Sect. 8.7.

Influence of Alignment Uncertainty

The impact of spectral alignment errors $\Delta d_{j,k}$ on the total errors of the fitting parameters $\Delta_{\text{tot}} a_j$ depends on the shape of different absorption spectra involved. There is no simple equation that connects the alignment uncertainty with the uncertainty Δa_j derived by the linear fit. Therefore, an approximate method has to be found. The fact that the alignment uncertainty of one trace gas spectrum propagates to the uncertainty Δa_j of another gas complicates this approach. To solve this problem, Stutz and Platt (1996) suggest a method that relies on a similar approach to that described to investigate the dependence of a_j on $d_{j,k}$ in Sect. 8.3.3. A test spectrum is calculated based on the original reference spectra that are shifted and scaled based on the results of the fitting procedure, $S_j^*(i) = a_j \times S_j(d_{j,0}, d_{j,1}, \ldots)(i)$. These reference spectra thus represent the best estimate of the trace gas absorptions contained in the original atmospheric spectrum. In addition, spectrum $S_g^*(i)$ is shifted and squeezed as specified by the set of parameter errors $\Delta d_{g,k}$ according to (8.21):

$$x_{\pm} = i \pm \Delta f_g(i) \quad \text{with} \quad \Delta f_g(i) = \sqrt{\sum_{k=0}^{p_j} \left(\Delta d_{g,k} \times (i - i_c)^k\right)^2} \tag{8.28}$$

Here, $\Delta f_g(i)$ represents the error of the position of pixel i. The test spectrum is then calculated by:

$$J = 1 + S_1^*(0) + \cdots + S_g^*\left(\Delta d_{g,0}, \Delta d_{g,1}, \ldots\right) + \cdots + S_m^*(0) \tag{8.29}$$

A linear fit using the model function $F_{\text{sh}} = 1 + a_1^* \times S_1^*(0) + a_2^* \times S_2^*(0) + \cdots + a_m^* \times S_m^*(0)$ with the references $S_j^*(0)$ is then used to analyse J. In the case of negligible errors $\Delta d_{g,k}$, all the fit parameters a_j^* will be $a_j^* = 1$. If S_g^* in J is shifted and squeezed due to the uncertainty $\Delta d_{g,k}$, the resulting a_j will change. These changes in a_j give an estimate of the propagation of spectral alignment errors of S_i^*, on the fit result for spectra S_j^*. We can use the function $\Phi_{j,g}$ defined above to quantify this dependence. As this error is an absolute value, two spectra $S_i^*(+x)$ and $S_i^*(-x)$ have to be calculated, and two linear fits with the model F_{sh} to the spectrum J have to be performed to derive $\Phi_{j,g}(\pm\Delta f_g)$. The influence of the alignment error of S_g on the error of a_j, $|\Phi_{j,g}|$ can then be calculated by:

$$\left|\Phi_{j,g}\left(\Delta f_g\right)\right| = \frac{1}{2} \times \left(\left|\Phi_{j,g}\left(+\Delta f_g\right)\right| + \left|\Phi_{j,g}\left(-\Delta f_g\right)\right|\right) \tag{8.30}$$

By performing this procedure for all $g = 1, \ldots, m$, one derives a set of $m^2|\Phi_{j,g}|$ values. The set of $|\Phi_{j,g}|$ can now be used to estimate the error $\Delta_{\text{sh}}a_j$ of a_j caused by the uncertainty of the alignment of all spectra included in the fit:

$$\frac{\Delta_{sh}a_j}{a_j} = \sqrt{\sum_{g=1}^{m}\left(\left|\Phi_{j,g}\left(\Delta f_g\right)\right|\right)^2} \tag{8.31}$$

Assuming that this error and the error $\Delta d_{j,k}$ of the linear fit are independent, the total error of a_j is:

$$\Delta_{\text{tot}}a_j = \sqrt{\left(\Delta a_j\right)^2 + \left(\Delta_{\text{sh}}a_j\right)^2} \tag{8.32}$$

This error estimation can be included in the analysis procedure by calculating the $\Phi_{j,g}$ for all reference spectra after the procedure has finished the iteration process to derive $d_{j,k}$.

A Monte Carlo analysis illustrates the behaviour of errors in DOAS analysis (Stutz and Platt 1996). Noise spectra, $N(\sigma)$, with standard deviations (σ) from 10^{-4} to 3×10^{-2} were added to a spectrum similar to J shown in Fig. 8.5. This test spectrum J was calculated as the sum of references S_j (see Fig. 8.2) that were linearly shifted against each other: $J = 1 + S_1\,(0.1) + S_2\,(-1) + S_3\,(1) + S_4\,(-0.1) + N\,(\sigma)$. The individual shifts of the spectra were: O_3, 0.1 pixel; NO_2, -1 pixel; SO_2, 1 pixel; HCHO, -0.1 pixel. To analyse the spectra, a model function allowing individual shifts was used:

$$F = P_4 + a_1 \times S_1\left(d_{1,0}\right) + a_2 \times S_2\left(d_{2,0}\right) + a_3 \times S_3\left(d_{3,0}\right) + a_4 \times S_4\left(d_{4,0}\right)$$

For every level of σ, 800 different artificial noise spectra were calculated. The resulting 800 spectra J were then analysed with the fitting algorithm described earlier. The average results $\overline{a_j}$ and $\overline{d_{j,k}}$ agreed with the values used to calculate J. Figure 8.6a illustrates that $\overline{\Delta d_{j,k}}$ is linearly dependent on noise levels σ, as long as the noise is not too high as compared to the optical density of the reference spectra. It is remarkable that uncertainties in the shift of several pixels to several tens of pixels

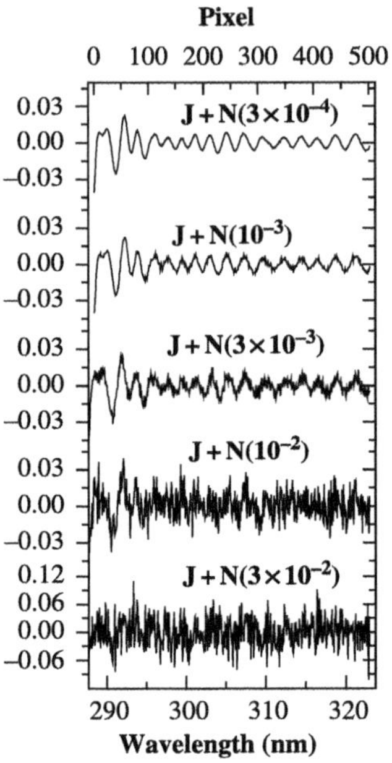

Fig. 8.5 J (Fig. 8.2) with added noise spectra $N(\sigma)$ of different magnitudes from 3×10^{-4} to 3×10^{-2}. The magnitude is defined by the standard deviation σ over the pixel intensities. For high σ, the spectrum J cannot be identified in the noise. Reprinted with permission from Stutz and Platt 1996, © Optical Society of America. All rights reserved

are found due to noise in the spectra. The errors of the shift $\overline{\Delta d_{j,k}}$ agreed well with the true error $\delta(d_{j,k})$, which was calculated from the standard deviation of the fit results $d_{j,k}$ from the true shift.

Figure 8.6b illustrates that the average of the total fit error $\overline{\Delta_{\text{tot}} a_j}$ increases linearly with increasing noise level σ. Figure 8.6c shows how well the error calculation discussed in the previous section describes the statistical fluctuations of the results. The average total errors $\overline{(\Delta a_j)}_{\text{tot}}$ agree with $\delta(a_j)$ within better than 10%, as long as the relative error of a_j is smaller than 0.5. Figure 8.6d shows that the influence of the shift error is, in this case, relatively small compared to the average error $\overline{\Delta^{\text{lin}} a_j}$ of a linear analysis. The influence of the shift error $\Delta d_{j,k}$ is highest for O_3 and SO_2. A shift uncertainty of a few hundredths of a pixel can, in some cases, increase the total error of a_j by 20–50%, thus increasing the derived trace gas concentration error considerably.

The Effects of Residual Structures

A common problem in the application of the DOAS technique is the occurrence of structures other than noise in the residual J–F of the fit. These structures may indicate an unknown absorber, or can be caused by the instrument itself, and occur at random in most of the cases. Possible sources will be further discussed below. Here, only random structures will be discussed. Non-random residual structures

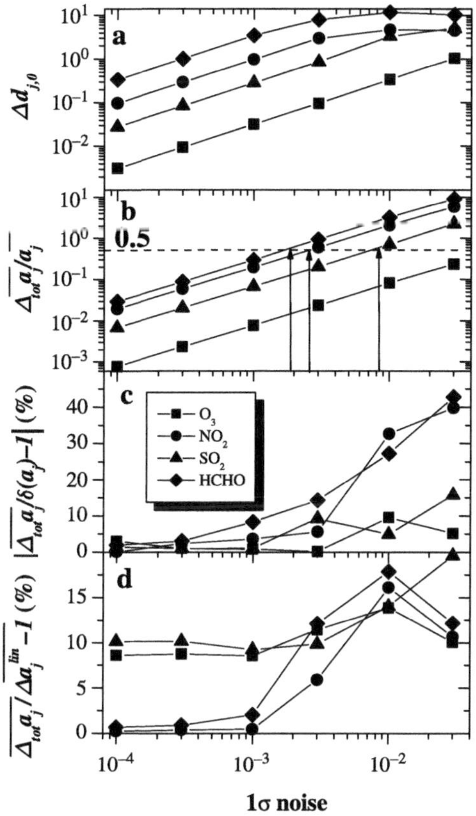

Fig. 8.6 Results of the investigation of the error calculation, including spectral alignment uncertainties. Reprinted with permission from Stutz and Platt 1996, © Optical Society of America. All rights reserved

cause systematic errors in the analysis, which cannot be described by the statistical methods.

In general, the effect of residual structures is to increase the error of the fit parameters a_j and the derived trace gas concentrations. A number of empirical methods have been applied to correct the effect of residuals, most of which are based on the experience of the respective researchers with their instruments. Some of the common methods include:

- The root mean square of the residual structures in the fitting window scaled with the absorption cross-section and path length gives an estimate of the detection limit of the instrument. Smith et al. (1997), for example, used this approach during their NO_3 measurements in California.
- A very conservative approach is the use of peak-to-peak size of the residual structure, scaling it with cross-section and path length.
- The peak-to-peak size of the residual, divided by a weighting function determined by the number of absorption bands in the fitting window.
- The strongest artefact line in the residual closest to the trace gas absorption structure.

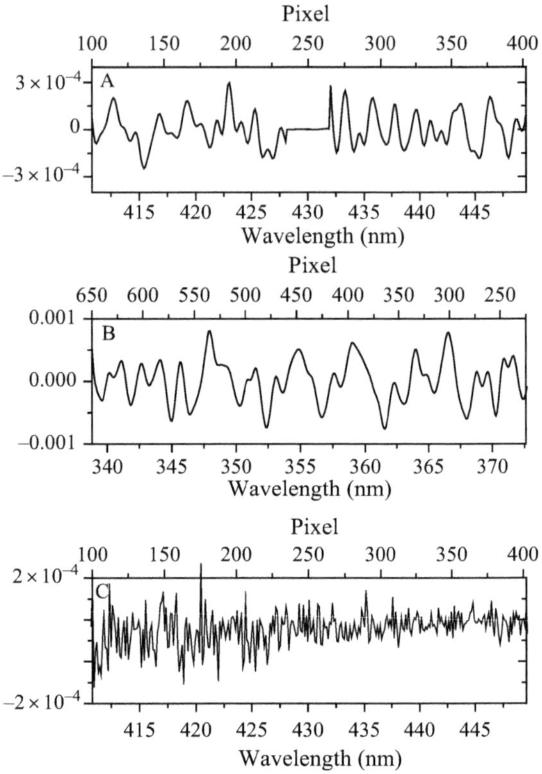

Fig. 8.7 Example of residual structures in different DOAS applications. Plate A is from measurements with a MAX-DOAS system. The flat residual around 430 nm is caused by the exclusion of this interval from the analysis. Plate B shows an example of a LP-DOAS measurement of NO_2 and HONO. Both residuals were filtered to remove random noise and to better illustrate the shape of the residuals. In both cases, the random noise was considerably lower than the residuals. Plate C shows a pure noise spectrum measured using lamp references of a MAX-DOAS system

In recent years, other methods based on more statistical arguments have been suggested in literature. Some of those methods will be discussed below.

Figure 8.7 shows examples of residual structures in different DOAS applications. The general features of these spectra, as compared with pure noise spectra as shown in Fig. 8.7, are structures that are several pixels wide. One can argue that, if these structures are random, the errors of the pixels of such a structure are not independent of each other, i.e. a random event will change all pixels of a structure simultaneously (Stutz and Platt 1996; Hausmann et al. 1999). This argument can also be supported if one recalls that, due to the finite resolution of the spectrometer, a single wavelength is imaged on a number of different pixels determined by the width of the instrument function H. Any random change in the instrument that is strongly wavelength dependent will, therefore, lead to structures that are as wide as, or wider than, the instrument function.

The fact that the errors of individual channels are correlated indicates that an uncorrelated least squares fit is inappropriate in this case. However, all known DOAS analysis programs are based on such an uncorrelated least squares approach.

Stutz and Platt (1996) describe how the correlation between errors may influence the results of a least squares fit. To introduce a correlation between the errors, they used a pure noise spectrum and smoothed it with a moving average. Figure 8.8 com-

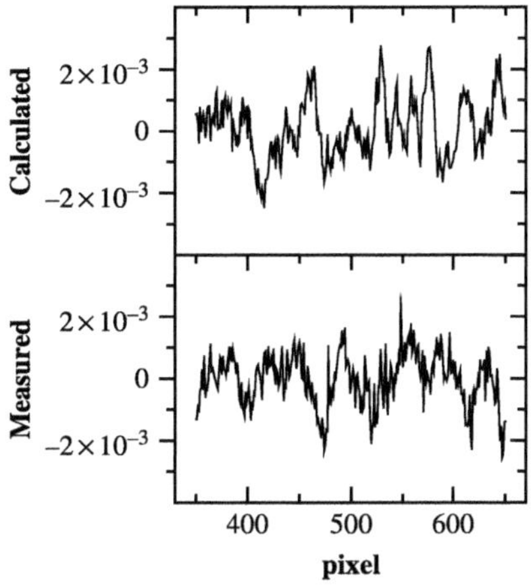

Fig. 8.8 Example of calculated and measured residual structures. Reprinted with permission from Stutz and Platt 1996,

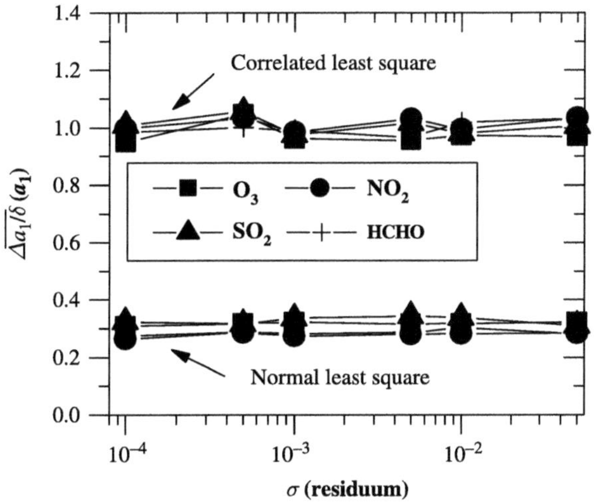

Fig. 8.9 Comparison of the ratio of average fit error and standard deviation of the results of a normal fit and a correlated least squares fit for a spectrum containing residual structures. Reprinted with permission from Stutz and Platt 1996,

pares such a spectrum with a residual from the analysis of an atmospheric absorption spectrum.

The influence of these residuals and the applicability of a correlated least squares fit was tested by Monte Carlo experiments in which several hundred noise spectra were smoothed by a running mean of 9 pixel width. Each of these spectra was added to the test spectrum shown in Fig. 8.2 and analysed with a model function of the sum of the reference spectra, both with an uncorrelated and a correlated linear least square fit without spectral alignment. The changes in the resulting a_j were small, but the calculated errors increased drastically by a factor of 3 (Fig. 8.9) for the uncorrelated least squares fit. The tests also showed that the underestimation of error is independent of the magnitude of the residual.

The correlation of errors was described by the variance–covariance matrix of the errors of the pixel intensity $(\mathbf{M})_{ij} = E\left(\varepsilon_i \times \varepsilon_j\right)$ (8.33), where ε_k is the error of the intensity of pixel k, and E is the expectation value of $\varepsilon_i \times \varepsilon_j$. This matrix can be built for the running mean smoothing, assuming all ε_k to be of equal size 1:

$$M_9 = \frac{1}{81} \times \begin{pmatrix} & & \ddots & \\ \cdots 1\,2\,3\,4\,5\,6\,7 & 8 & 9\,8\,7 & 6 & 5\,4\,3\,2\,1\,0\,0\,0 \cdots \\ \cdots 0\,1\,2\,3\,4\,5\,6 & 7 & 8\,9\,8 & 7 & 6\,5\,4\,3\,2\,1\,0\,0 \cdots \\ \cdots 0\,0\,1\,2\,3\,4\,5 & 6 & 7\,8\,9 & 8 & 7\,6\,5\,4\,3\,2\,1\,0 \cdots \\ & & & \ddots & \end{pmatrix} \tag{8.33}$$

M_9 is a symmetric band matrix with 9/81 in the diagonal. The correlated least squares fit is then performed according to (8.5) on pixel 160 to 260 of the test spectra to reduce calculation time.

Figure 8.9 shows the ratio of the average error calculated from both fitting procedures, $\overline{\Delta a_j}$, and the standard deviation of the fit factors, a_i, $\delta(a_j)$. In the case of a uncorrelated fit, the errors are underestimated by a factor of ~3. The errors derived with the correlated fit agree, on the other hand, within 5% of the statistical scatter of the fit factors.

This suggests that, if the assumption of correlated channel errors during a DOAS analysis is indeed correct, as assumed by several authors (Stutz and Platt 1997a; Hausmann et al. 1999), the uncorrelated least squares fit underestimates the error considerably. A correlated fit will give calculated correct errors if the correlation matrix, **M**, is known.

The most serious limitation in the use of a correlated least squares fit is the unknown error correlation in a DOAS residual spectrum. In the case shown above, the correlation of errors was known, and could thus be incorporated into the analysis. For atmospheric spectra, it is likely that the correlation changes with time, and perhaps even from spectrum to spectrum. This would make the determination of **M** difficult and time consuming. However, this discussion is important since it puts the treatment of residual structures on a mathematical basis, moving away from purely empirical correction methods.

Other methods that have been suggested are based on other statistical methods or approximations.

Stutz and Platt (1996) suggest, based on their use of the running mean in the study of correlated least squares fit, to simulate residuals using this method, and to use Monte Carlo experiments to derive correction factors for the errors of an uncorrelated least squares fit. Experiments were performed with artificial absorption spectra that contained five identical, non-overlapping, Gaussian-shaped, absorption lines with half width, τ, ranging from 2.5 to 50 pixels in the different spectra. These spectra were added to noise spectra $N(W)$ that were smoothed by different running means of width $W = 2$ to 64, and scaled to a standard deviation of 1%. For every combination of τ and W, 500 random noise spectra N were calculated, smoothed, and added to $J = S(\tau) + N(W)$. Then the model $F = P_0 + a_1 \times S(\tau)$ was fitted to the spectrum J. The ratio of the standard deviation of the fit parameter $\delta(a_1)$ and the average error $\overline{\Delta a_1}$ calculated by the fit of 500 different residuals is the desired correction factor

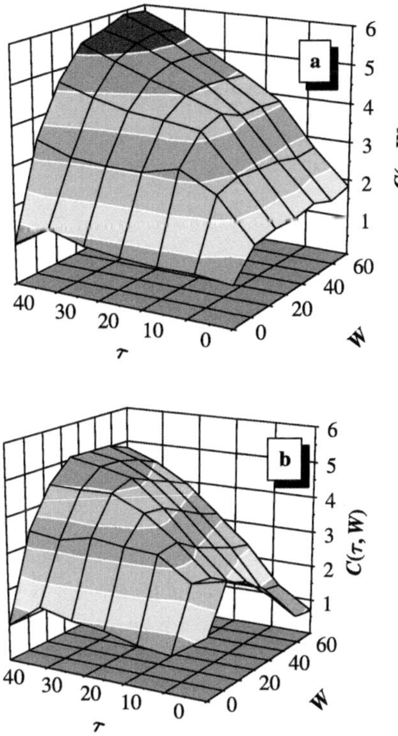

Fig. 8.10 Correction factor $C(\tau,\ W)$ for the linear and non-linear fit. Reprinted with permission from Stutz and Platt 1996, © Optical Society of America. All rights reserved

$C(\tau,\ W) = \delta(a_1)/\overline{\Delta a_1}$. Figure 8.10a shows $C(\tau,\ W)$ for different combinations of W and τ. The correction factor is largest for broad structures of the spectrum and the residual. As the correction factor is independent of the magnitude of the residual, this factor can be applied to correct the uncorrelated fit errors using the width of the residuals and the reference spectra. The same test was repeated with the non-linear procedure for $d_{1,0}$ by analysing the error of the shift after shifting $S(W)$ by 1 pixel against its reference and then adding the residuals $N(W)$. Figure 8.10b illustrates a similar behaviour as in the linear case.

Hausmann et al. (1999) describe two statistical methods to estimate the true error of a DOAS analysis with residual structures. The first method, called Residual Inspection by Cyclic Displacement (RICD), is based on the assumption that artificial residual spectra can be simulated from true residual spectra by a cyclic shift, i.e. shifting a residual spectrum by k pixel, and appending the k pixel that are shifted out of one side of the spectral window to the other side. To derive the true error of the parameters a_j, they suggest to first analyse the atmospheric absorption spectrum. The residual $R(i) = J(i) - F(i)$ of this fit is then shifted cyclically by k pixel, $R^*(i) = R(i+k)$, and all the reference spectra, scaled according to the results of the original fit a_j, are added to this shifted spectrum: $J^*(i) = P(i) + a_1 \times S_1 + a_2 \times S_2 + \cdots + a_m \times S_m + R^*(i)$. Figure 8.11 illustrates the shifting method for an OH radical absorption spectrum. The analysis of $J^*(i)$ with the same procedure gives a new set of fit factors a_j^*. By repeating this procedure with a large number of shift

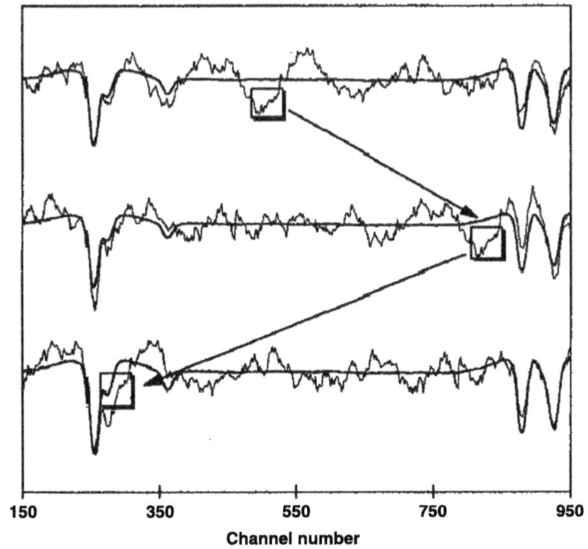

Fig. 8.11 Example of the proposed RICD method. Reprinted with permission from Hausmann et al. 1999,

distances k, a set of fit factors a_j^* is derived. By calculating the standard deviation of this data set for each spectrum S_j, one can estimate the error of fit $\Delta_{\mathrm{RICD}} a_j$. It is obvious that, in the case that the residual spectrum $R(i)$ consists of pure noise, this method will yield similar uncertainties to the least squares fit calculated directly through (8.4). Hausmann et al. (1999) found that this method slightly underestimated the true error, $\Delta_{\mathrm{true}} a_j$, of the fitting parameter, which was calculated from the standard deviation a_j calculated for a large number of different unshifted residual spectra. Values in their experiments with pure noise residuals were 5–30% lower than those calculated by the least squares fit. In the presence of residual structures different from noise, a similar bias between true error and error calculated by the RICD method was found (Table 8.1). The true and the RICD errors were, however, considerably larger than the error that was calculated by the least squares fit alone. The factor between true and pure least squares fit was ~4. The precise value was dependent on the individual reference spectrum (Table 8.1). A limitation in this method is the restriction to residuals that have relatively constant peak-to-peak strength across the spectrum.

The other method is based on the bootstrap approach (BS). In short, the bootstrap method, which was introduced by Efron (1979), is a resampling procedure used to estimate the distributions of statistics based on independent observations. An estimate of the sampling distribution is created by randomly drawing a large number of samples from a population, and calculating for each one an associated statistical parameter.

Hausmann et al. (1999) applied the BS method to DOAS by using the absorption spectra and reference spectra as the original data sets. The statistical estimate is derived by a linear least squares analysis described in Sect. 8.1.1. The BS procedure is performed in three steps.

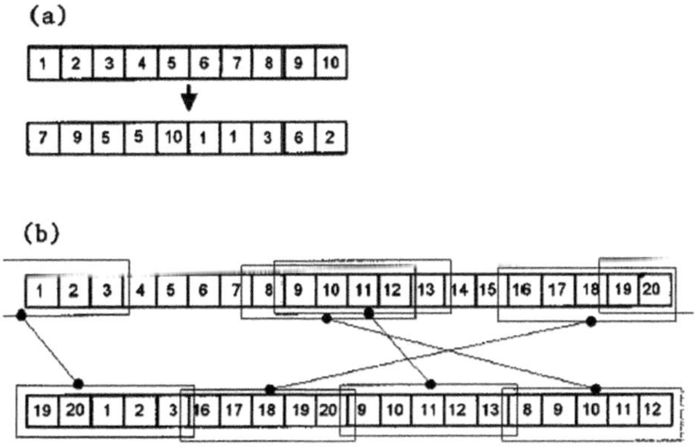

Fig. 8.12 Schematic example of the bootstrap method: (**a**) resampling of single channels, (**b**) block resampling Reprinted with permission from Hausmann et al. 1999,

1. A large number (100–1000) of artificial absorption and reference spectra are created. These spectra have the same number of channels as the original spectra. The content is determined by randomly attributing channels or blocks of channels from the original spectra to channels of the new spectra, as illustrated in Fig. 8.12. This drawing procedure is equally applied (the same pixel of the original is attributed to the artificial spectra) to both atmospheric absorption and reference spectra, yielding new spectra, $J^*(i)$ and $S_j^*(i)$. As illustrated in Fig. 8.12, channels from the original spectrum can occur more than once in the new artificial spectrum, while others may not occur at all.
2. In the second step, the artificial atmospheric absorption spectrum $J^*(i)$ is analysed with a DOAS linear least squares procedure (see Sect. 8.1) that employs the reference spectra with the identical drawing scheme $S_j^*(i)$. This is repeated for each of the 100–1000 different drawings.
3. In the last step, the standard deviation of all the fitting results, i.e. a_j^*, is determined. This is, according to the BS theory, an estimate of the error of the original parameters, a_j.

As illustrated in Fig. 8.12, the block size of resampling can vary from a single channel to as many as hundreds of channels. Hausmann et al. (1999) showed that the results of their calculations are not independent of the block size, and attributed this to the interdependence/correlation between different channels. The dependence on the block size disappears for most spectra when a large block size is used. They recommend a block size of at least 40 channels to account for this effect. At this block size, the error calculated by the BS method, $\Delta_{\mathrm{BS}} a_j$, reaches a plateau for trace gases with multiple absorptions in the spectral window, and larger block sizes do not change the results. However, for absorbers with only one or two absorption bands, this plateau is not reached, and the BS method may not be applicable.

The Monte Carlo test performed by Hausmann et al. (1999) (Table 8.1) shows that, as was the case for the RICD method and the correlated least squares fit (Stutz and Platt, 1996), the true error is about a factor of ~4 higher than that calculated by the normal uncorrelated least squares fit. The BS method underestimates the error by only 30–60%, and is thus a much better estimate of the true error.

Table 8.1 Standard deviation of the ratio between calculated error and statistical scattering of the results of the 100 simulated measurement spectra [a]

Estimation methods	Measurement spectra			
	Ratios for the following errors			
	RICD	Bootstrap	BS ⊕ RICD	err (LSQ)
OH	1.43	1.32	0.91	4.27
$C_{10}H_8$	1.30	1.53	0.96	4.86
SO_2	1.20	1.44	0.91	3.91
HCHO	1.36	1.60	1.00	3.98
Deviation from true error	0.75	0.65	1.05	0.25

Table adapted from Hausmann et al. (1999).
[a]In the case of correct error estimation, the ratio between calculated error and statistical scattering of the results is unity

The authors argue that, because the RICD and BS methods describe different contributions to the spectral evaluation error caused by residual structures, a combination of the two methods, BS ⊕ RCID, is the best estimate. This combination can be calculated by $\Delta_{\text{tot}} a_j = (\Delta_{\text{RCID}} a_j^2 + \Delta_{\text{BS}} a_j^2)^{1/2}$. Since the error of a noise-dominated residual would be overestimated due to the addition of both methods, it is proposed that this error is scaled by a factor of 0.85 to take into account the fact that it is impossible to distinguish between pure noise and residual-dominated spectra.

Validation of Error Analysis Methods

It is challenging to validate the error calculations for a realistic DOAS analysis, i.e. for true atmospheric spectra, due to unknown trace gas amounts in the atmosphere. One older approach is the statistical analysis of fit results for a trace gas that is believed to have much lower concentrations than the detection limit of the DOAS instrument used in its measurement. While this approach can give an approximate check for the correctness of the routine, it is not accurate enough to make precise quantitative statements. The method adopted by Tuckermann et al. (1997) and Alicke et al. (2002) compares the statistical standard deviation of the fit results with the average of errors calculated by the fit for the same data points. Only those data points for which it can be assumed that the concentration of the respective trace gas is much smaller than the detection limit are used. Obviously, this method is limited by the latter assumption which, in many cases, cannot be verified from the DOAS measurements alone.

Figure 8.13 shows the results of such an analysis using HONO measurements in Milan, Italy (Alicke et al. 2002). The histogram of HONO mixing ratios observed by DOAS during times of strong solar radiation scatter resembled a Gaussian distribution, with a maximum of ~0.015 ppb. The width of this distribution was very similar to the average error calculated by the DOAS analysis routine. Based on this analysis, Alicke et al. (2002) concluded that the error calculation based on Stutz and

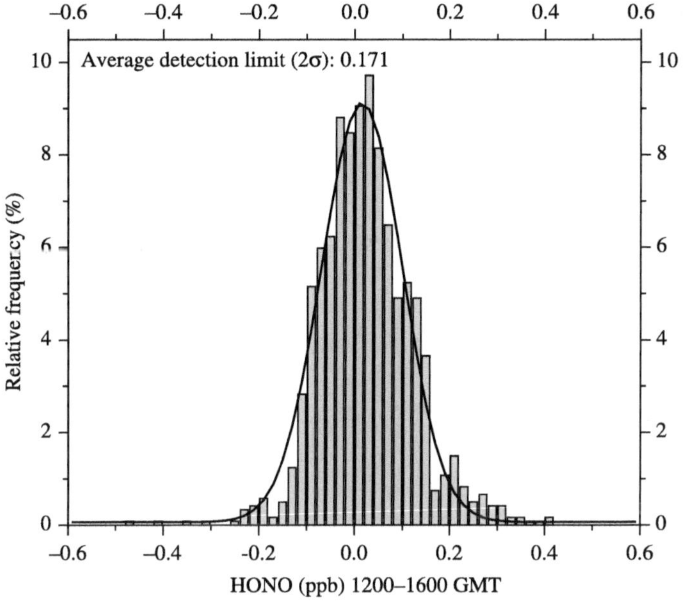

Fig. 8.13 Frequency distribution of HONO mixing ratios during periods of strong actinic flux and expected low HONO concentrations. The relative frequency agrees well with a Gaussian curve (*solid line*, see text) at 15 ppt. The average error of 0.171 ppb agrees well with the width of this distribution. Reprinted with permission from Alicke et al. 2002, © American Geophysical Union. All rights reserved

Platt (1996) does indeed give good estimates of the true error of DOAS results. The results by Tuckermann et al. (1997) show a similar agreement.

Examples of Analysis Procedures

Here we present two examples of the analysis of atmospheric spectra. Figure 8.14 shows an example of the analysis of a spectrum measured using the MAX-DOAS method. The atmospheric spectrum to be analysed, i.e. $J(i)$, is the measurement at 3° elevation displayed in the top trace. The spectra below this trace are the zenith measurement, which serves as Fraunhofer reference and the scaled references of NO_2, H_2O, and O_4, i.e. $a_j \times S'_j(i)$. Also included in the fit is a spectrum describing the Ring effect (see Sect. 9.8). The bottom trace shows the residual of the fit, which was, in this case, larger than the noise one would expect in the respective spectra.

Figure 8.15 shows the example of measurements of NO_2 and HONO by a long-path instrument in Nashville, TN. All spectra were high-pass filtered before analysis. The middle trace shows a comparison of the scaled HONO reference spectrum and the sum of this spectrum and the residual (bottom trace). This comparison allows the visualisation of the quality of the analysis procedure. If the absorptions cannot be identified in the sum spectrum, the trace gas concentration is below or very close to the detection limit of the instrument. If the absorption bands can be identified, the

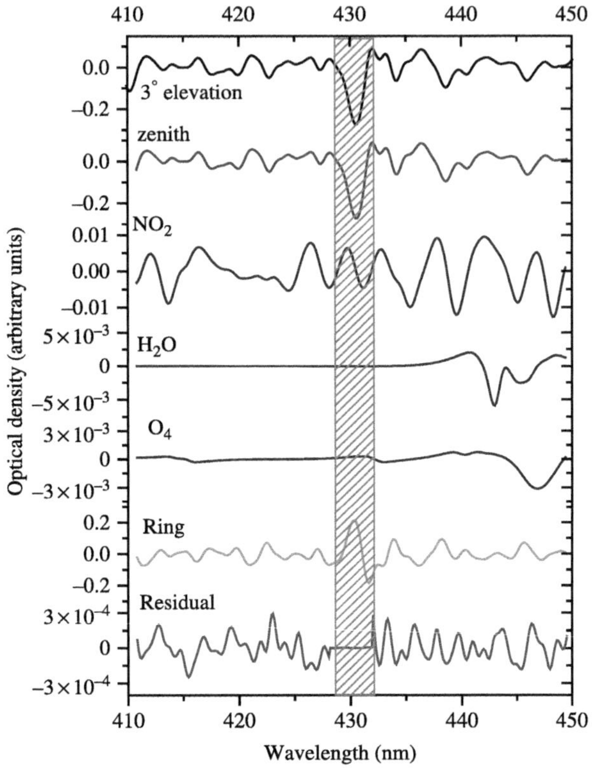

Fig. 8.14 Example of the analysis of a MAX-DOAS spectrum measured in Los Angeles in October, 2003. The *top trace* shows the spectrum measured with a 3° elevation. The *trace below* is the zenith spectrum. The various other traces are the scaled reference spectra of NO_2, H_2O, and the ring-effect spectrum. The *lowest trace* shows the residual structure of this fit. All spectra were high-pass filtered before the analysis

quality of the fit was good and the measurement of the respective trace gas with a small error is possible.

Because of the breadth of DOAS application (see Chap. 6), it is not possible to show examples of all the possible DOAS analysis procedures. Chapters 10 and 11 will, however, show more examples of the implementation of DOAS analysis procedures.

8.4 Estimation of Detection Limits

It is now interesting to derive the detection limit for the spectra S_j at a given noise level σ. We will define the detection limit by the trace gas concentration, which can be determined with a relative error of $\Delta a_j / a_j = 0.5$. For example, Fig. 8.6b can be used to derive the detection limit of the trace gases for a given noise level σ. Unfortunately, this method is very cumbersome, as the Monte Carlo experiments are very time-consuming.

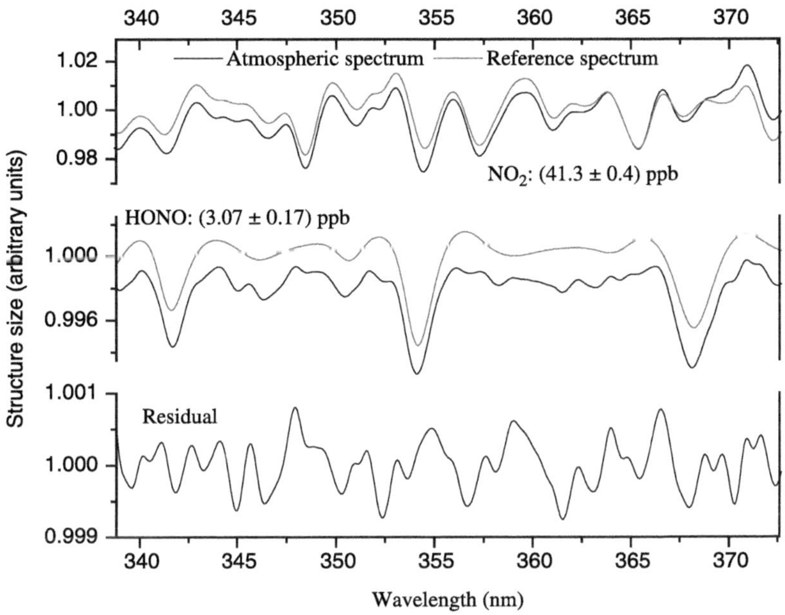

Fig. 8.15 Example of the analysis of a spectrum measured by a long-path DOAS. The two *top traces* show the atmospheric spectrum and the NO_2 reference spectrum after high-pass filtering. The *middle traces* show the HONO reference scaled to the absorption contained in the atmospheric spectrum. Also shown is the sum of this spectrum and the residual, which allows a visual confirmation of the quality of the fit. The *lowest trace* is the residual of the fit. All spectra shown in this figure were filtered to reduce noise in the spectrum

A faster but less exact method is the analysis of the pure linear problem, neglecting the uncertainties of the wavelength-pixel-mapping. This estimation of the detection limits by calculating $\Delta a_j = \sqrt{\Theta_{jj}} = \sigma \times \sqrt{\left(\left(\mathbf{X}^T\mathbf{X}\right)^{-1}\right)_{jj}}$ (8.4) with known σ still requires the calculation of the covariance matrix Θ. Assuming that the absolute value of the linear correlation coefficient between the references is smaller than 0.1–0.2, (8.4) can be simplified further. After this simplification, an expression for the smallest $\overline{D}$ detectable can be derived. For a given noise level σ and a number of pixels n, the detection limit $\overline{D}_{\text{limit}}$ can be approximated by:

$$\begin{aligned} \overline{D}_{\text{limit}} &\approx \sigma \times \frac{6}{\sqrt{n-1}} \quad &\text{(a)} \\ \sigma_{\text{limit}} &\approx \frac{\sqrt{n-1}}{6} \times \overline{D} \quad &\text{(b)} \end{aligned} \tag{8.34}$$

This expression was validated by Stutz and Platt (1996) with their results of the Monte Carlo experiments with a number of pixels $n = 500$. In Fig. 8.6b, the noise levels σ_{limit} given by (8.34b) for the $\overline{D}$ of NO_2, SO_2, and HCHO, respectively, are marked by arrows. The arrows point to the curve of relative error $\Delta a_j / a_j$ of the respective trace gas. The estimated detection limits are within 10–20% of the value derived in the Monte Carlo tests. Admittedly, this is only a rough estimation, but it can be performed on every spectrum without complicated calculations.

From the data in Fig. 8.9b, it can be seen that it is possible to analyse spectra with (1σ) levels of pure uncorrelated noise higher than the typical optical densities of the absorption bands. At the detection limit, where the 1σ error of a_j is $\Delta a_j \approx 0.5 \times a_j$, the noise level can be about a factor 3.5 larger than the average optical density $\overline{D}$ for a wavelength region with 500 pixels. Therefore, the analysis of spectra where absorption structures cannot be identified with the eyes is possible. The detection limits are, in this case, smaller than the intuitively expected values.

In the presence of residual structures, a correction factor of 3–4 (Stutz and Platt 1996; Hausmann et al. 1999) has to be included in (8.34). As a consequence, the detection limits increase by the same factor, and the smallest detectable absorption structure will be of similar size to the residual.

8.5 Determination of Reference Spectra

The most direct, and often most successful, approach to obtaining the correct reference spectra S_j in (8.15) is to measure them with the instrument that is also used to perform atmospheric measurements. For many years, this has been the preferred method for trace gases such as SO_2, NO_2, HCHO, O_3, NH_3, and aromatics in the troposphere. For example, for SO_2 and NO_2 the reference cells have sufficient long-term stability to transport them into the field. While the use of an SO_2 absorption cell is straightforward, NO_2 cells are more difficult to use due to NO_2 photolysis, the temperature-dependent equilibrium of NO_2 with N_2O_4, and other problems which lead to temperature- and time dependent NO_2 concentration in the cell. For a detailed discussion see Platt and Kuhn (2019). The measurement of HCHO in the field is often achieved by heating a cell with para-formaldehyde. Despite the higher temperatures in the cell, this approach has been shown to be quite good. The measurement of O_3 in the field is rarely used anymore due to the substantial experimental effort, and the fact that the O_3 absorption cross-section is temperature dependent in the UV and thus difficult to measure in a temporary field setup. Aromatics can be measured by filling the fresh liquid compound into a short cell and relying on the relatively high vapour pressure of these compounds to provide the concentration needed for cell measurement. The advantage of this method is a very accurate determination of the absorption structure, which is particularly important for the different aromatics, which have very similar absorption spectra.

The direct measurement of pure trace gas absorptions has the advantage of providing the best reference spectra, since the convolution with the instrument function (see Chap. 6) is performed through the measurement. The disadvantage of this approach is that for all trace gases, except perhaps SO_2, an absolute calibration of the absorption cross-section is not easily possible. The absorption spectra must be calibrated after the measurement with literature absorption cross-sections. The very accurate description of S_j, however, outweighs the disadvantage of this calibration.

8.5.1 Theoretical Basis of Reference Spectra Simulation: Convolution

For trace gases that are too reactive to be measured in the field in their pure form and to calibrate the measured reference spectra, one has to rely on the simulation of the absorption spectra for the respective instrument.

The first step in this simulation is the determination of the instrument function of the spectrometer, H (also known as "instrumental spectral response function (ISRF)", "instrument transfer function", or "slit function"). Practical approaches for the determination of H are given in Sect. 8.5.2.

In addition to the emission lines, the accurate determination of the wavelength to pixel mapping is also required in the simulation. This can be determined by using a number of well-known positions of atomic emission lines and/or by using known absorption structures of trace gases or the solar Fraunhofer lines. Most high-quality spectrometers display a nearly linear dependence of channel numbers and wavelengths with a very small non-linear component. Less expensive spectrometers, and in particular miniature spectrometers, can have more complex dependencies. However, in most cases, a polynomial of degree 3 describes the wavelength-pixel mapping sufficiently.

As we discussed earlier, the measurement of an absorption spectrum changes its shape by reducing its spectral resolution. Mathematically, this process can be expressed by a convolution of the absorption spectrum $I(\lambda)$ with the instrument function $H_{\mathrm{sp}}(\lambda)$ of the spectrometer via:

$$\begin{aligned} I^*(\lambda) &= \int I\left(\lambda - \lambda'\right) \cdot H_{sp}\left(\lambda'\right) d\lambda' \\ &= I\left(\lambda'\right) * H\left(\lambda'\right) \end{aligned} \tag{8.35}$$

In addition, most detectors show certain sensitivity overlap (or crosstalk) between adjacent channels. For example, in the case of a photodiode array, ~6.5% of the signal of a pixel stems from each neighbour. This can again be represented by a convolution of, in this case, $I^*(\lambda)$ with the detector response function, which will be called $w(i,\lambda)$, where i is the channel number and λ_i the mid-wavelength of the respective channel.

$$I'(i) = \int I^*\left(\lambda_i - \lambda'\right) \cdot w\left(i, \lambda'\right) d\lambda' \tag{8.36}$$

In reality, it is often impossible to distinguish the convolution by the spectrograph and the detector. By combining (8.35) and (8.36), both processes can be combined to a single instrument function $H = H_{\mathrm{sp}}{}^* w(i,\lambda)$. During the measurement of an atomic emission line, the combined instrument function, H, is measured. The mathematical idea behind this measurement is that an atomic emission line is narrow compared to the resolution imposed by H. Thus, it can be approximated by a δ-function:

$$I^*_{AEL}(\lambda) = \int \delta\left(\lambda - \lambda'\right) \cdot H\left(\lambda'\right) d\lambda' = H(\lambda) \tag{8.37}$$

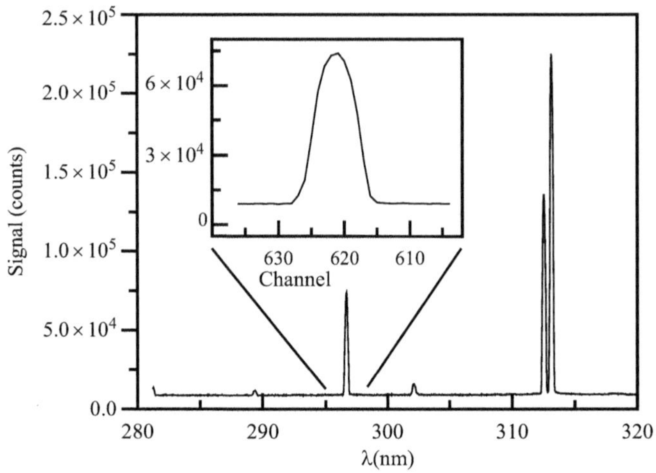

Fig. 8.16 Example of a mercury spectrum measured with a typical spectrometer–detector combination used for DOAS. The insert shows the shape of the mercury atomic emission line (see Appendix A) that is used for a convolution

While in many spectrometers one can assume that H is independent of the wavelength, we have encountered a number of instruments where H is a function of wavelength $H(\lambda)$. This dependence can lead to different emission line widths on both ends of the recorded spectrum.

8.5.2 Determination of the Instrument Function

The instrument function is determined by the optical properties of the spectrometer, including width of entry slit, optics, grating, detector properties etc. It is typically too complex to be accurately calculated by a theoretical model of the instrument. However, it can be measured by two methods:

Illuminate the instrument with (quasi) monochromatic light. Sufficiently monochromatic light sources can be suitable lamps emitting atomic emission lines (see Sect. 7.2.3), broadband light that has passed a monochromator, or a laser light source. For most low-resolution DOAS applications, atomic emission lines from low-pressure lamps provide an effective means to determine H. Because the natural line width of the emission lines is much smaller ($\sim$10 pm) than the typical resolution of DOAS spectrometers ($\sim 0.1 - 1$ nm), the emission line measured with the respective spectrometer–detector system is a very good approximation of H (see Fig. 8.16). In the case of atomic emission lines it is important that only single lines with a good signal-to-noise ratio are used. If possible, several lines at different positions in the spectrum should be chosen to detect, and take into account if necessary, the possible change of H with wavelength.

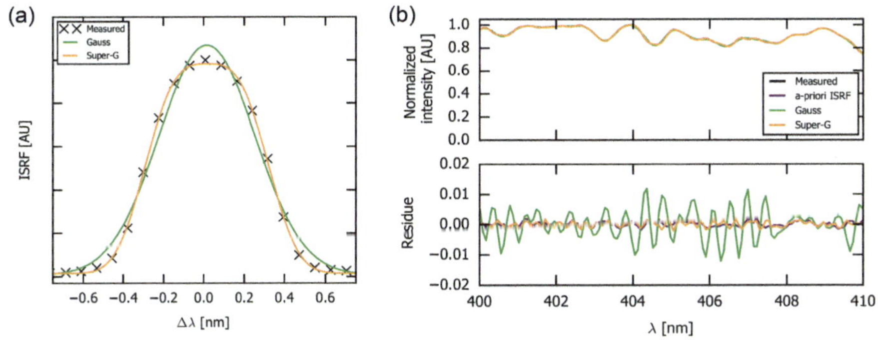

Fig. 8.17 Instrument-function fit results for a small spectrometer (AvaSpec-ULS2048x64). Panel **a** Measured line shape (crosses) using the Hg line at 404.66 nm and best matching Gaussian and SG parameterisations (drawn lines). **b** Least-squares fit of the Instrument-function during wavelength calibration of a zenith-sky measurement in the 400–410 nm interval (top) and corresponding residual (bottom, expanded scale), with the instrument-function derived from the Hg measurement ("a priori") or parameterised as Gaussian and SG. Residuals are small and similar when using Hg-derived or SG fitted instrument functions, while a Gaussian parametrisation of the instrument function results in much larger residuals. Reprinted with permission from Beirle et al. (2017),

Parameterisation of the instrument function shape: Parameters describing the instrument function can be derived by comparing a measured spectrum to a high-resolution spectrum of known structures, frequently the solar Fraunhofer spectrum (see e.g. Chance and Kurucz 2010). During a spectral fit, as performed in a DOAS evaluation (see Chap. 6), additional parameters describing the instrument function can be fitted, as described e.g. by Beirle et al. (2017). Many mathematical functions can describe the instrument function. In practice, the Gaussian function has been chosen most frequently. A better choice is the "Super Gauss" (SG) function introduced for this purpose by Beirle et al. (2017). It is described by:

$$S_G(\lambda) = A(w_{SG}, k) \cdot \exp\left(-\left|\frac{\lambda}{w_{SG}}\right|^k\right) \tag{8.38}$$

The SG function is described by two parameters, the width w_{SG} and the shape parameter k. For $k < 2$, S becomes "peaked" at the top, with long tails on both sides. For k=2 and normalisation factor $A(w_{SG}, 2) = 1/(\sqrt{2\pi} \cdot 0.5 \cdot w_{SG})$ one obtains an 'ordinary' Gauss-distribution with width $\sigma = w_{SG}/2$. For $k > 2$, S becomes "flat-topped", converging to a "box" shape of width $2w_{SG}$ for k approaching infinity. The SG as defined in Eq. (8.38) is symmetric, but it can easily be extended to an asymmetric SG (ASG) by introducing asymmetry parameters for w_{SG} and/or k, as described in Beirle et al. (2017). The example in Fig. 8.17 shows that SG-parametrisation of the instrument function leads to superior results compared to Gaussian parametrisation.

A high quality parameterisation of the instrument function, for example by a SG-function, provides several advantages:

1. The instrument function can be derived by fitting a catalog Fraunhofer spectrum to a Fraunhofer spectrum recorded with the instrument. This is trivial for passive DOAS instruments using direct or scattered sunlight. But it is also possible for active DOAS instruments, which usually record Fraunhofer spectra as daytime-background or can see sunlight, e.g. by introducing a diffuser into the light-path. In this way the sun can be used as a readily available 'spectral lamp'.
2. The dependence of the instrument function on wavelength can be derived by applying this approach to different parts of the spectrum.
3. The instrument function can be determined very frequently in the field.
4. Even if spectral lamps are used it can be of advantage to parametrise the instrument function since, at typical oversampling ratios of DOAS spectrometers around 4–6 pixel per resolution element, the instrument function is not very well sampled. By averaging the parameters derived from several lines (in spectral vicinity) a better overall representation of the instrument function can be obtained.

8.5.3 Practical Implementation of Reference Spectra Simulation

To simulate the absorption spectra, and thus obtain reference spectra for (8.15) for a specific spectrometer–detector combination, the following steps have to be performed:

1. Choose an accurate absorption cross-section $\sigma(\lambda)$ for the respective trace gas as found e.g. in 'The MPI-Mainz UV/VIS Spectral Atlas of Gaseous Molecules of Atmospheric Interest' (Keller-Rudek et al. 2013). Ideally, the spectral resolution of this cross-section should be much better than the spectral resolution of the spectrograph–detector system. While finding a highly resolved absorption cross-section is sometimes challenging, it has been shown that a resolution at least 10–100 times better is necessary to avoid problems with this approach (see below). Because the cross-section is also digitised with a step interval of $\Delta\lambda_\sigma$, we will call it $\sigma(j)$.
2. Determine the wavelength to pixel mapping, for example by using an atomic emission line spectrum or through comparison with a catalog Fraunhofer spectrum. If the dispersion is linear, the distance between two pixels will be $\Delta\lambda_{AEL}$.
3. Determine the atomic emission lines that can be used for convolution or derive an instrument function using the procedure described in Sect. 8.5.2. If possible, choose several lines, or a wavelength dependent parameterisation, to take into account the possible changes of H with the wavelength.
4. Because the convolution applies to the intensity, an absorption spectrum is calculated next:

$$I_{ref}(j) = I_0(j) \exp(-A \cdot \sigma(j)) \tag{8.39}$$

We can distinguish two cases for this calculation. In the case of smooth light sources, such as those provided by artificial lamps, $I_0(j)$ can be set as constant, e.g. $I_0(j) = 1$. If the light source has strong narrow-band spectral features, e.g. the sun, it is often better to use the known $I_0(j)$ of the light source. We will come back to this point later. The parameter A represents the trace gas column (concentration × absorption path length). In general, A should be chosen to be similar to the expected column. If the absorber is weak, the precise value of A is not very important. If the absorber is strong or there is a risk that some of the high-resolution features of the spectrum are in, or close to, saturation, one must choose A as close to the real column as possible to simulate saturation effects accurately. This is, for example, often the case for H_2O and O_2.

5. In the next step, the convolution is performed. To make this step simple, one needs to bring I_{ref} and the mercury lines chosen for the convolution to the same wavelength grid. We have successfully used polynomial extrapolation routines to expand the mercury lines to the $\Delta\lambda_\sigma$ grid. After this expansion, the convolution is performed using a suitably chosen wavelength/channel window from j_{low} to j_{up}:

$$I^*_{ref}(j) = \sum_{j_{low}}^{j_{up}} I_{ref}(j - j')H(j') \tag{8.40}$$

It is also possible to use a Fast Fourier Transform approach to perform the convolution in a more efficient way.

6. Finally, the convoluted spectrum $I_{ref}{}^*(j)$ with a wavelength grid of $\Delta\lambda_\sigma$ needs to be transferred to the grid determined by the wavelength pixel mapping. The content of one pixel can be calculated by integrating $I_{ref}{}^*(j)$ from the lower to the upper wavelength border of the pixel:

$$I^{**}_{ref}(j) = \int_{\lambda(j)-\frac{\Delta\lambda}{2}}^{\lambda(j)+\frac{\Delta\lambda}{2}} I^*_{ref}(\lambda_k)d\lambda \tag{8.41}$$

7. To calculate the spectrum in the form that can be used in the model function of the fit (8.15), the logarithm of the ratio of $I_{ref}{}^{**}(j)$ and $I_0{}^{**}(j)$ has to be calculated. If I_0 is wavelength dependent, the same convolution as in (8.39–8.41) with $A = 0$ needs to be performed before the division.

$$S_{ref}(j) = -\ln\left(I^{**}_{ref}(j)/I^{**}_0(j)\right) \tag{8.42}$$

The reference spectrum $S_{ref}(j)$ calculated in (8.42) should be an accurate representation of the respective trace gas absorption, if all the input data were correct.

Figure 8.18 shows a comparison of the measured and calculated spectrum of SO_2. The calculation was performed with the absorption cross-section of SO_2 and the

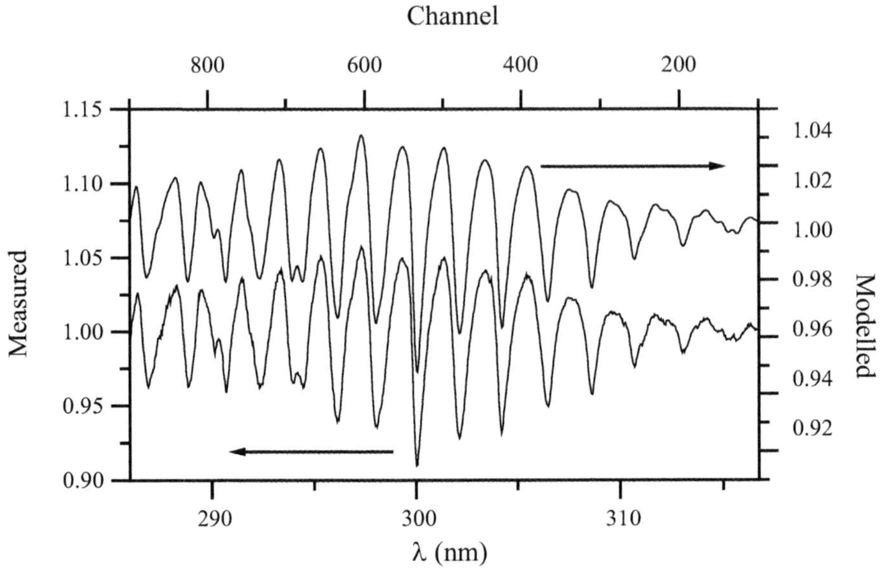

Fig. 8.18 Comparison of a measured and modelled SO_2 absorption spectrum. The mercury line shown in Fig. 8.16 and the absorption cross-section by Vandaele et al. (1994) were used in the simulation. The correlation between the two spectra is 0.9999

mercury line shown in Fig. 8.16. The correlation coefficient between the two spectra is 0.9999, showing the excellent agreement that can be achieved with this method.

8.5.4 Optimum Resolution of Literature Reference Spectra

The method described in the previous section relies on the assumption that the spectral resolution of the literature absorption cross-section can be ignored. It is worthwhile to consider when this assumption is fulfilled. We can perform a simple calculation based on Gaussian-shaped instrument functions, with κ_{AXS} and κ_H defined as the half-width of the Gaussian instrument function for the absorption cross-section and the instrument, respectively. To determine an optical density better than $\varepsilon\%$, the spectral resolution has to fulfill the following condition:

$$\kappa_{AXS} < \frac{\varepsilon}{100\%} \cdot \kappa_H \tag{8.43}$$

To describe the half-width of an absorption band better than $\varepsilon\%$, the following condition has to be met:

$$\kappa_{AXS} < \sqrt{\frac{\varepsilon}{100\%}} \cdot \kappa_H \tag{8.44}$$

These two equations illustrate that the spectral resolution of the literature absorption cross-section used from the convolution process should be at least 10 to 100 times better than the instrument resolution. Fortunately, the data set of high resolution absorption cross-sections has been increasing considerably over recent years (see e.g. the compilation by Keller-Rudek et al. 2013).

Finally, we will discuss what approaches are available if a high-resolution absorption cross-section is not available. If both the instrument function of the absorption cross-section and of the instrument are available, it is possible to convolute the atmospheric absorption spectrum with the instrument function of the absorption cross-section, and to convolute the cross-section with the instrument function of the spectrograph used in the atmospheric measurement following the same strategy as outlined earlier. Both the absorption cross-section and the absorption spectrum will be based on the product of the two instrument functions. Any high-resolution absorption cross-section that is included in the analysis will have to be convoluted with both instrument functions to match the resolution. This also illustrates that it is essential to report absorption cross-sections with the instrument function that was used for their measurement.

8.6 The Additive Polynomial

A common problem in the analysis of DOAS spectra is the presence of an intensity signal I_{add} that is additive to the light intensity I that is the target of a measurement. This phenomenon is most often noticeable in passive DOAS applications that measure scattered sunlight. The presence of such an additive signal can degrade the quality of the spectral retrievals by introducing biases and increasing the retrieval error. The additive signal can originate from instrumental effects, such as uncertainties in the correction of stray light in the spectrometer, or atmospheric effects, such as fluorescence from the surface. Here we discuss how an additive signal can impact a DOAS retrieval and how it can be corrected. We begin by using the formalism we developed in Sect. 6.3, where we consider a known absorption band with an intensity at the band edge of I_0 and the intensity at the band centre of I (Fig. 8.19). When an additive signal, here for simplicity considered a constant I_{add}, is present, the sum of the two signals $I^{sum} = I + I_{add}$ and $I_0^{sum} = I_0 + I_{add}$ will be the measured signals (Fig. 8.19). The change in optical density due to the additive signal can be derived as follows:

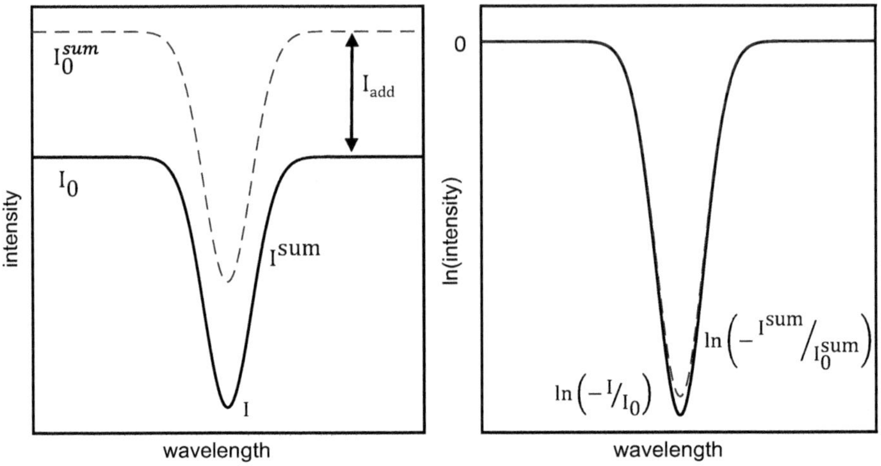

Fig. 8.19 Impact of an additive signal on the optical density of an absorption line

$$
\begin{aligned}
\ln\left(\frac{I^{sum}}{I_0^{sum}}\right) &= \ln\left(\frac{I+I_{\text{add}}}{I_0+I_{\text{add}}}\right)\\
&= \ln(I) + \ln\left(1+\frac{I_{\text{add}}}{I}\right) - \ln(I_0)\\
&\quad - \ln\left(1+\frac{I_{\text{add}}}{I_0}\right), \quad \text{with } \ln(1+x)\approx x\\
&\approx \ln\left(\frac{I}{I_0}\right) + \frac{I_{\text{add}}}{I} - \frac{I_{\text{add}}}{I_0}\\
&= \ln\left(\frac{I}{I_0}\right) + \frac{I_{\text{add}}}{I_0}\cdot\left(\frac{I_0}{I}-1\right)\\
&\quad \text{with } \ln(1+x)\approx x \quad \text{and} \quad x = I_0/I - 1\\
&\approx \ln\left(\frac{I}{I_0}\right) + \frac{I_{\text{add}}}{I_0}\cdot\ln\left(\frac{I_0}{I}\right)\\
&= \ln\left(\frac{I}{I_0}\right)\cdot\left(1-\frac{I_{\text{add}}}{I_0}\right)
\end{aligned}
\tag{8.45}
$$

The effect of an additive signal is thus to decrease the optical density of an absorption band. This effect is particularly obvious during retrievals of scattered sunlight, where one often finds that the retrieved fit scaling factor of the solar reference to the measured spectrum is not equal to the expected value of 1.

8.6.1 Inclusion of Additive Signal in a DOAS Retrieval

To overcome the challenge posed by the presence of an additive signal, it is beneficial to introduce a methodology to include the additive signal in spectral retrievals. Because DOAS retrievals are based on the logarithm of the intensity, we develop a formalism that allows the inclusion of the additive signal as a reference spectrum in a DOAS retrieval.

We begin by assuming that the detected signal is the sum of the original intensity $I(\lambda)$ and an additive component $I_{add}(\lambda)$. As in Eq. 8.45, we apply a logarithm to this sum, $\ln(I^{sum}(\lambda))$:

$$\begin{aligned}\ln(I^{sum}(\lambda)) &= \ln(I(\lambda) + I_{add}(\lambda)) \\ &= \ln(I(\lambda)) + \ln\left(1 + \frac{I_{add}(\lambda)}{I(\lambda)}\right)\end{aligned} \tag{8.46}$$

We will now use a Taylor expansion on the second term in the sum $\ln(1+x) \approx x - \frac{1}{2}x^2 + \mathcal{O}(x^3)$. Applying this to Eq. 8.46 yields:

$$\ln(I^{sum}(\lambda)) \approx \ln(I(\lambda)) + \frac{I_{add}(\lambda)}{I(\lambda)} - \frac{1}{2}\left(\frac{I_{add}(\lambda)}{I(\lambda)}\right)^2 + \mathcal{O}\left(\frac{I_{add}(\lambda)}{I(\lambda)}\right)^3 \tag{8.47}$$

Here the first term, $\ln(I(\lambda))$, can be analyzed using the methods described in Sect. 8.3. The other terms of Eq. 8.47 approximate the impact of the additive polynomial on the shape of $\ln(I^{sum}(\lambda))$. In most cases the additive signal is small compared to $I(\lambda)$. Consequently, only the first term of the Taylor expansion is included in the retrieval. In practice, the additive signal is often not exactly known and its size has to be determined in the fitting procedure by adding a reference spectrum. Following Eq. 8.47 this reference spectrum has the form:

$$S_{add}(\lambda) = \frac{R_{add}(\lambda)}{I(\lambda)} \tag{8.48}$$

where $R_{add}(\lambda)$ is the shape of the additive signal, such as a polynomial. The reference $S_{add}(\lambda)$ and a scaling factor a_{add} is then added to the fit model function (Eq. 8.15). The magnitude of the additive signal is then $a_{add} \cdot R_{add}$.

The challenge of including an additive polynomial in a retrieval, however, is that its effect on trace gas absorptions is similar to the trace gas absorption scaling factor that is optimised in the fitting procedure. It is thus necessary to link the additive spectra to a spectral structure that is not optimised in the fit. In passive DOAS applications this is done by setting the scaling factor for the solar reference to a fixed value of 1, and letting the additive signal describe any change of the optical densities of the solar Fraunhofer lines. This approach offers a physically more correct retrieval, as the optical density of solar Fraunhofer lines do not change in the atmosphere. In addition, because the change of optical density is determined by the ratio of I_{add}

and I (Eq. 8.45), it can be uneven over the spectral range if the either I_{add} or I have broadband variations in the retrieval interval. The inclusion of the additive polynomial thus provides advantages in the DOAS retrieval, but has to be used with care, due to potential impacts on the retrieved trace gas scaling factors.

8.7 Residual Spectra

Residual structures influence the spectral analysis and the general performance of atmospheric DOAS measurements. In particular, the error determination is challenging if structures other than pure noise are present. While a number of approximation methods to determine the errors of the derived trace gas levels have been developed, an accurate statistical description of the effect of residual structures is still not available.

A more serious problem is the obstacle that residual structures pose to further improvement of the detection limits of DOAS. Most instruments, and in particular active DOAS systems, show residual structures that are much larger than the noise levels one would expect from photon statistics. The residual thus determines the smallest optical density that can be analysed. Reducing or removing residual structures is therefore an important undertaking.

Solutions to numerically solve the residual challenge have thus far not been very successful. Often, the reduction of the residuals, for example by including an average residual spectrum or some other estimate of the residual, reduces the random error in the analysis, however, at the price of increasing the systematic uncertainty of the measurement. The solution to the residual problem is thus, most likely, the study of the sources of these residuals and the improvement of DOAS instruments.

While, in many cases, we do not know what causes the residual structures in the DOAS analysis, we will discuss here a number of obvious candidates that could be responsible. We will begin with listing some trivial examples that are often encountered in DOAS applications. These cases can be identified through a careful analysis of the residual. The residual spectrum often offers the opportunity to study the performance of an analysis. For expert DOAS users, the shape of the residual as compared to the reference spectra can give indications of possible mistakes in the analysis.

The following examples of residual structures are often easily fixed:

- The omission of an absorber in the analysis will lead to a residual that is often similar to the spectral structure of the absorber. However, the residual will not be identical to the absorber, since the analysis routine will try to reduce the χ^2 by finding linear combinations of the reference spectra to describe the missing absorber.
- Different resolutions of the atmospheric spectrum and the reference spectra will lead to residuals that often look like the second derivative of the spectra. The same applies if different low-pass filters are used for different spectra.

- Spectral misalignments between the spectra in an analysis will lead to residuals that are similar to the first derivative of one or more of the spectra.
- The presence of inaccurately represented strong absorbers in a classical DOAS analysis can also lead to specific spectral structures.

While these examples are fairly obvious, the random spectral structures often found in DOAS applications are more elusive. In the following paragraphs, we will speculate about the causes of these residuals.

- One of the causes for residual structures that has been discussed in literature (Stutz and Platt 1997) is the direction sensitivity of solid-state detectors. Because in many DOAS applications the illumination of the spectrograph detector system varies with time, the response of different pixels of the detector will also change. This change in response will cause spectral structures that cannot be easily removed during the analysis. Optical setups and measurement methods have been developed to reduce this effect (see Chap. 10). It is, however, unclear whether the detector sensitivity can be entirely removed by these methods.
- Any inaccuracy in the reference spectra can also lead to residual structures. With further improvement of laboratory measurements of absorption cross-sections, it will become clear in future if this can explain some of the residuals.
- Another possible source of residual structures could be deficiencies in the spectrographs that are commonly used in DOAS applications. Stray light in the spectrometer, for example, could cause an uneven additional signal on top of the original spectrum. The magnitude of the stray light signal would only have to be of the size of the residual, i.e. in the range of 0.1% of the original light intensity. Stray light of this magnitude is commonly encountered in grating spectrometers.

These examples illustrate that very small deficiencies in the optical instrumentation or input data for the analysis can lead to residual structures of the size found in DOAS. Although the residuals are a serious problem for DOAS, they are indeed a very small signal, often equivalent to optical densities below 0.1%.

8.8 Systematic Errors in the Analysis

There are many ways in which systematic uncertainties can be introduced to a DOAS analysis. In this section we will present some of the more obvious examples and discuss other potential sources that are more difficult to recognize.

8.8.1 Spectral Interferences

A number of problems are known to cause interferences between different trace gases in the DOAS analysis. The most common is the contamination of absorption cross-sections and reference spectra with other trace gas absorptions. At first glance, this should be easily overcome by carrying out careful laboratory experiments to determine the absorption cross-sections of the respective gases. In a number of cases, however, it is difficult to prepare pure gases. For example, HONO made in the laboratory often contains NO_2, which is either a byproduct of the HONO formation process, or is formed through secondary reactions in the cell. It is also difficult to make pure NO_2, since it is partially transformed to HONO on the walls of the cell.

A simple calculation can show what effect a contamination of a trace gas reference spectrum will have on the results of a DOAS analysis. We will simplify (8.14) to provide a mathematical description of the influence of such an interference by only considering two absorbers:

$$J(i) = J_0(i) + a_1' \times S_1'(i) + a_2' \times S_2'(i) \tag{8.49}$$

We analyse this spectrum with the linear function described earlier (8.15), where the reference spectrum of Absorber 2 is contaminated by absorptions from Absorber 1: $S_2(i) = S_2^*(i) + \varepsilon \times S_1(i)$, where $S_2^*(i)$ is the reference spectrum of the pure Absorber 2. The model function, omitting the polynomial and the shift and squeeze of the reference spectra, is then:

$$F(i) = 1 + a_1 \times S_1(i) + a_2 \times S_2(i) = 1 + a_1 \times S_1(i) + a_2 \times \left(S_2^*(i) + \varepsilon \times S_1(i)\right) \tag{8.50}$$

The least squares analysis of this equation results in the correct scaling factor for a_2', but the result for Absorber 1 would be $a_2' = (a_1' - a_2' \times \varepsilon)$. The true absorption would thus be underestimated, depending on the amount of contamination and the scaling factor for Absorber 2. In certain cases, i.e. NO_2 and HONO, this interference may be difficult to recognise because the two trace gases are also linked to each other through their chemistry in the atmosphere. It is then basically impossible to separate spectral interference from correlation from the atmospheric chemistry of these compounds.

The best solution to this problem is thus to ensure that the trace gas absorption cross-section or reference spectra are not contaminated. This can often be done by comparing different measurements or different published absorption cross-sections. In some instances, and especially for passive DOAS applications, higher-order absorption structures can also lead to residual structures. Puķīte and Wagner (2016) introduced an iterative retrieval algorithm correcting for the higher-order absorption structures in the DOAS fit as well as for the absorption dependence on temperature and scattering processes by introducing additional terms (known as 'Pukite Terms') in the modelled spectrum.

8.8.2 Spectrometer Stray Light and Offsets

A problem common to all grating spectrometers, which leads to systematic uncertainties, is stray light or unremoved offsets in the spectra (see also Sects. 7.4.5 and 8.6). Stray light originates from light entering the spectrometer that is not directly projected onto the detector, i.e. other orders of the diffraction grating or light of undetected wavelengths. This light should ideally be removed by absorption on the spectrometer walls (see Sect. 7.4.5). Because a fraction of this light will eventually fall onto the detector, a small amount of non-dispersed light will be measured. Offsets can originate from drifting electronics or light that enters the spectrometer accidentally, for example due to scattered solar light in active DOAS applications. We will assume here that this light has a smooth spectral characteristic. One can then describe the intensity at a certain wavelength as follows.

$$I_{meas}(\lambda) = I(\lambda) + I_{stray}(\lambda) \tag{8.51}$$

where I_{stray} is the intensity from all effects. The correction of stray light in a DOAS analysis uses the additive polynomial fitting approach described in Sect. 8.6.

Because the ratio of stray light to signal intensity, $I_{stray}(\lambda)/I'_0(\lambda)$, is often less than 1%, the error imposed by stray light is small. However, a number of cases are known, especially for measurements below 300 nm, where $I_{stray}(\lambda)/I_0(\lambda)$ can reach 10% or more. In this case, stray light has to be considered.

Stray light can also influence the shape of an absorption spectrum. If the stray light intensity varies over the spectrum, absorption bands at different wavelengths will be changed differently, i.e. the optical density of bands with higher stray light will be reduced more than those with lower stray light. Consequently, the relative sizes of the absorption bands, and thus the overall shape of the spectrum, will change. In the case that the spectral shape of the stray light is known, an additive stray light spectrum can be included in the analysis (Sect. 8.6). Unfortunately, it is difficult to determine the stray light spectrum, and higher degree polynomials are thus often used to describe the wavelength dependence of stray light.

8.9 Choosing an Optimal Spectral Interval (Vogel-Sihler-Plot)

A particular challenge in setting up DOAS retrievals is optimisation of the retrieval parameters for each specific case and particular trace gas(es) of interest. Of these parameters, the retrieval wavelength range is the most important. For many trace gases there are obvious choices of the evaluation wavelength interval. Clearly, the retrieval wavelength range should include prominent absorption features of the trace gas of interest in order to obtain optimum sensitivity, while excluding strong absorption features of other trace gases. Spectral artefacts of the instrument (typically the spectrometer) should also be excluded. Overall, a balance has to be found between sensitivity and possible cross correlations. A broader wavelength range increases the information available to the algorithm, but also introduces the danger of incorporating the strong absorption structures of other gases. In some cases, a broader wavelength range can increase errors due to insufficient correction of the broad-band terms. It is thus clear that a systematic procedure for identifying an optimal retrieval wavelength range and for a quantitative assessment of its suitability is desirable. An approach to reach this goal was invented by Vogel et al. (2013) and has become known as

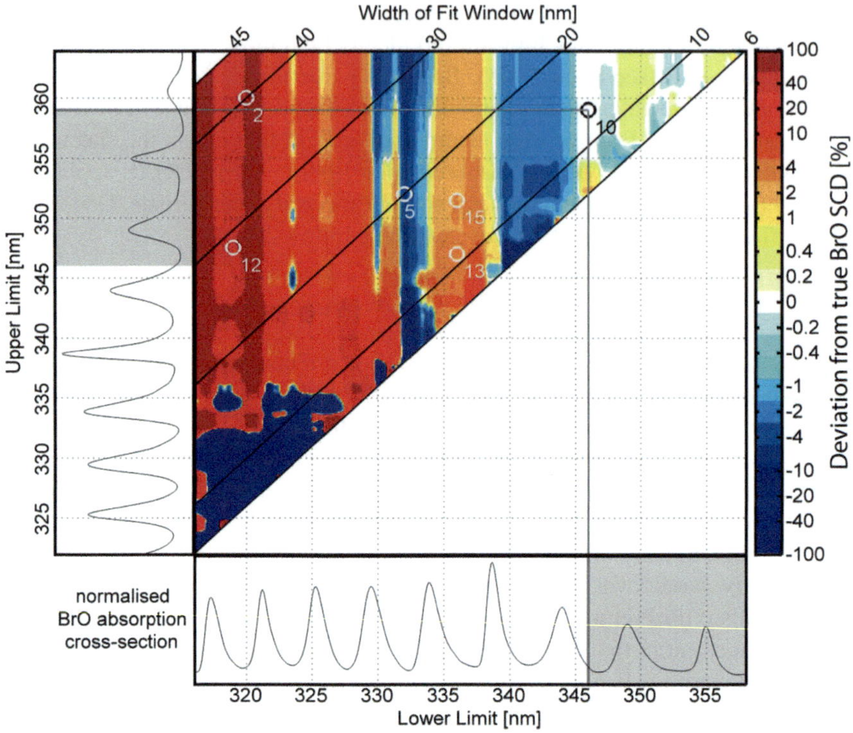

Fig. 8.20 Wavelength interval mapping (VSP) of a synthetic spectrum. Zenith-sky measurements of stratospheric BrO (Table 2) are simulated with a true BrO SCD of $1.5 \cdot 10^{14}$ molec cm^{-2}. The retrieval applies I_0-corrected reference cross-sections. The abscissa axis shows the lower wavelength limit of the retrieval wavelength interval, the ordinate axis the upper wavelength limit. Deviations from the true BrO SCD are displayed on a logarithmic colour coded scale. The circles mark selected retrieval wavelength intervals in the literature, (see text). The shaded wavelength interval (346–359nm, indicated by white circle 10) denotes the optimal retrieval wavelength interval found in an intercomparison exercise (Aliwell et al. 2002). Reprinted with permission from Vogel et al. (2013),

'Vogel-Sihler-Plot' (VSP). The VSP is a tool to visualize the effect of different evaluation wavelength ranges. The central idea is to map retrieved trace-gas column densities in the retrieval wavelength space and thus visualize the consequences of making different choices of spectral retrieval ranges. In addition, the effect of erroneous absorption cross-sections, cross correlations, and instrumental features can be systematically investigated. Based on the information gathered, an optimal retrieval wavelength range can be determined.

In a VSP, relevant features of the retrieval (e.g. deviation from the true value in a simulation or the retrieval error) are displayed in a colour map where the abscissa axis shows the lower wavelength limit of the retrieval wavelength interval, while the ordinate axis shows the upper wavelength limit. Thus each point in the plot corresponds to a wavelength interval and lies above the diagonal, where the upper limit exceeds the lower limit. The example in Fig. 8.20 shows a VSP of the deviation of retrieved and true BrO column density on a logarithmic scale. Wide retrieval intervals are found in the upper left edge of the plot. Narrow intervals close to the diagonal are given by black lines parallel to the diagonal (interval width (10–45 nm)). The numbered white circles indicate a few examples of wavelength intervals used in other publications (2, 5: Bobrowski and Platt 2007; 10: Aliwell et al. 2002; 12: Salawitch et al. 2010; 13: Begoin et al. 2010; 15: De Smedt et al. 2004). As can be seen from the plot, small errors (below 1%) can be reached for a large range of wavelength intervals. On the other hand, the wrong choice of wavelength interval can cause large retrieval errors.

Minimal retrieval errors are not the only criterion for a retrieval, therefore VSPs can be produced to display interferences or other parameters as, for example, discussed by Vogel et al. (2013). Other examples of VSP can be found in Fickel and Delgado Granados (2017), Seo et al. (2019), Davis and McLaren (2020).

8.10 Principal Component Analysis and DOAS Retrievals

The standard DOAS approach models each measured optical density spectrum $D(\lambda)$ by summing the effects of a series of trace gas reference spectra, a polynomial of suitable degree, a Ring spectrum, in the case of scattered sunlight, and possibly other pseudo absorbers. In a linear/nonlinear fitting procedure, the contributions of the different trace gases, polynomial coefficients, strength of Ring spectrum, and scaling factors for pseudo absorbers are calculated.

Another approach relies on performing a Principal Component Analysis (PCA, see Pearson 1901) of a series of measured spectra (e.g. Li et al. 2013). In this particular application, the principal components (PCs) are vectors (i.e. spectra) which "explain" the measured spectra in the sense that the measured spectra (except for a residual spectrum) can be composed of a weighted sum of these PCs. Thus, in principle, the a-priori knowledge of Fraunhofer and trace gas reference spectra (and their convolution with the instrument function) is not necessary. As demonstrated by Li et al. (2013) for satellite observations of volcanic SO_2 PCA can be used to derive background spectra from a series of measurements where the species to be measured is not present. In the evaluation procedure a set of PCs is extracted, which capture most of measurement-to-measurement variation of the intensity spectra in the absence of the signal of interest. Thus the derived PCs will capture physical and measurement details other than those associated with the SO_2 absorption. The PCs are ordered so that the first PC, corresponding to the largest eigenvalue of the covariance matrix, explains most of the variance, the second PC explains the second most (i.e. less) of the variance, and so on. Typically, the first few PCs were found to describe the background spectra sufficiently well (Li et al. 2013, 2017; Theys et al. 2015). The problem is that the PCs can not readily be associated with weaker absorbers. However, a study by Theys et al. 2015 found the DOAS and PCA algorithms for strong absorbers, i.e. the OMI SO_2-retrieval, to be in good agreement. For instance a new algorithm, OMSO2VOLCANO, uses characteristic features extracted directly from OMI radiances (Li et al. 2017).

A further application of PCA is the analysis of residuals with respect to common features in a series of residuals. This was demonstrated by Ferlemann (1998) in analysing spectra from a balloon-borne DOAS instrument.

Scattered-Light DOAS Measurements 9

9.1 Scattered-Light DOAS Measurements

Absorption spectroscopy of sunlight scattered by air molecules and particles has a long tradition for the study of atmospheric composition. Götz et al. (1934) introduced the 'Umkehr' technique, which is based on the observation of a few select wavelengths of scattered sunlight. The analysis of strong absorptions in the ultraviolet wavelength region allowed the retrieval of ozone concentrations in several atmospheric layers, which yielded the first remotely measured vertical profiles of ozone. The COSPEC technique developed in late 1960s was the first attempt to study tropospheric species by analysing scattered sunlight in a wider spectral range with the help of an optomechanical correlator (Millan et al. 1969; Davies 1970), see Sect. 5.7. It has been applied over five decades for measurements of total emissions of SO_2 and NO_2 from various sources, e.g. industrial emissions (Hoff and Millan 1981) and volcanic plumes (Stoiber and Jepsen 1973; Hoff et al. 1992).

Experience with Dobson spectro-photometers and COSPEC instrument, in combination with the development of modern spectrometers and detectors lead to the use of DOAS with direct or scattered sunlight as the light source. Scattered sunlight DOAS is an experimentally simple and very effective technique for the measurement of atmospheric trace gases and aerosols. Since scattered light DOAS instruments analyse radiation from the sun, rather than relying on artificial sources, they are categorised as **passive** DOAS instruments (see also Chap. 6).

All passive DOAS instruments are similar in their optical setup, which essentially consists of a telescope to collect light, coupled to a spectrometer–detector combination (see Chap. 7). However, different types of passive DOAS instruments employ a wide variety of observation geometries for different platforms and measurement objectives. An overview of different scattered light techniques is given in Table 9.1.

The earliest scattered light DOAS applications were ground-based and predominately observed light from the zenith (Noxon 1975; Syed and Harrison 1980;

U. Platt and J. Stutz, *Differential Optical Absorption Spectroscopy*, Geophysics and Environmental Physics, https://doi.org/10.1007/978-3-662-68639-3_9

Table 9.1 Overview and history of the different scattered light passive DOAS applications

Method	Measured quantity	No. of axes, technique	References
COSPEC	NO_2, SO_2, I_2	1, S	Millan et al. 1969, Davies 1970, Hoff and Millan 1981, Stoiber and Jepsen 1973, Hoff et al. 1992
Zenith scattered light DOAS	Stratospheric NO_2, O_3, OClO, BrO, IO	1	Noxon, 1975; Noxon et al. 1979, Harrison 1979, McKenzie and Johnston 1982, Solomon et al. 1987a, Solomon et al. 1987b, McKenzie et al. 1991, Fiedler et al. 1993, Pommereau and Piquard 1994a,b, Kreher et al. 1997, Wittrock et al. 2000a
Off-axis DOAS	Tropospheric and stratospheric BrO, NO_3 profile	1 or 2	Arpaq et al. 1994, Miller et al. 1997, Friess et al. 2001, 2004,
Zenith Aircraft-DOAS	Tropospheric BrO, NO_2	2	McElroy et al. 1999
Limb Balloon DOAS	Stratospheric NO_2 and BrO	1	Weidner et al. 2005
Limb Aircraft-DOAS	Tropospheric BrO, IO, NO_2 glyoxal, HONO	2	McElroy et al. 1999
Nadir Aircraft-DOAS	Tropospheric BrO, IO, NO_2 glyoxal, HONO	2	McElroy et al. 1999
Multi Axis DOAS	Tropospheric concentrations and concentration profiles	3+ S	Hönninger and Platt 2002; Löwe et al. 2002; Oetjen 2002; Heckel 2003; Zingler et al. 2005
Airborne Multi Axis DOAS	Trace gas profiles	3+, M	Wagner et al. 2002, Wang et al. 2003, Heue et al. 2003
Imaging DOAS	Industrial and volcanic plumes	20+, M	Wagner et al. 2002; Wang et al. 2003; Heue et al. 2003
Nadir Satellite DOAS	Stratospheric and Tropospheric O_3, NO_2, HCHO, SO_2, BrO, IO	1, M	Wagner et al. 2002, Wang et al. 2003, Heue et al. 2003
Limb Satellite DOAS	Industrial and volcanic plumes	3+, M	Wagner et al. 2002, Wang et al. 2003, Heue et al. 2003

S = Scanning instrument, M=Multiple telescopes

McMahon and Simmons 1980; Pommereau 982, 1994; McKenzie et al. 1982, 1991; Solomon et al. 1987, 1988, 1993; Perner et al. 1994; Van Roozendael et al. 1994; Slusser et al. 1996). This Zenith Scattered Light–DOAS (ZSL-DOAS) geometry is particularly useful for the observation of stratospheric trace gases, and has made major contributions to the understanding of the chemistry of stratospheric ozone, in particular through the measurement of stratospheric NO_2, OClO, BrO, and O_3 (Pommereau 1982, 1994; McKenzie et al. 1982, 1991; Solomon et al. 1987, 1988, 1993; Perner et al. 1994; Van Roozendael et al. 1994a, b, c; Slusser et al. 1996; Sanders 1996; Sanders et al. 1997).

In an attempt to further increase the sensitivity of scattered DOAS measurements to stratospheric trace gases an off-axis viewing geometry was introduced (Sanders et al. 1993). This approach was particularly successful for trace gases that rapidly photolyze in the atmosphere, such as NO_3 and OClO (e.g. Weaver et al. 1996; Sanders et al. 1993; Smith and Solomon 1990; Smith et al. 1993; von Friedeburg et al. 2002). to stratospheric and upper tropospheric trace gases

Soon after, it was recognised that off-axis viewing strategies also lead to enhanced sensitivities in the troposphere. To overcome the challenge of separating tropospheric and stratospheric contribution to the observed scatter sunlight. Hönniger and Platt (2003), argued that the use of multiple viewing angles, including the zenith, would be needed. This new multi-axis (MAX) DOAS method was soon adopted widely and is now one of the most popular forms of passive DOAS. In short, MAX-DOAS typically employs 3–10 different viewing elevations, which are often observed sequentially through a mechanical scanner-telescope combination. Because the sensitivity to stratospheric and upper tropospheric trace gases is only weakly dependent on the viewing elevation, it is possible to determine trace gas levels in the lower troposphere. MAX-DOAS has become a tool to monitor pollutants, such as NO_2 (see e.g. Brinksma et al. 2008; Chan et al. 2018; Hendrick et al. 2014), to study sources and formation of organics such as HCHO and glyoxal (see e.g. Heckel et al. 2005; Pinardi et al. 2013; Zhang et al. 2021; Sinreich et al. 2007, 2010), to investigate chemistry of the halogen oxides BrO and IO (see e.g. Wagner et al. 2007c; Bognar et al. 2020; Hönninger and Platt 2002; Hönninger 2003; Simpson et al. 2017; Großmann et al. 2013; Gómez Martín et al. 2013; Prados-Roman et al. 2015; Takashima et al. 2022), and to determine the distribution of aerosol in the lower troposphere (see e.g. Wagner et al. 2004; Sinreich et al. 2005; Irie et al. 2008; Chan et al. 2018). MAX-DOAS instruments are most frequently deployed at fixed ground sites, but mobile deployments on vehicles, ships, and even buoys are also common. The ease of operating MAX-DOAS instruments and their high sensitivity has led to large measurement networks that are operational across the globe (Verhoelst et al. 2021; De Smedt et al. 2021; Kanaya et al. 2014). The number of MAX-DOAS publications has grown greatly over recent years and only selected examples will be discussed in this book (see also Chap. 11).

Airborne platforms were explored early on in the development and use of scattered light DOAS. Initially the main purpose of these activities was to reach areas that were otherwise difficult to access, such are polar regions. The first application was the study of stratospheric ozone depletion chemistry using zenith viewing geometries (Schiller et al. 1990; McElroy et al. 1999; Pfeilsticker and Platt 1994). Balloon measurements were performed for the same purpose around the same time (Harder et al. 1997b; Ferlemann et al. 1998; Fitzenberger et al. 2000; Dorf et al. 2008; Butz et al. 2009). These measurements are usually based on a limb (elevation angle 0°) or near limb viewing geometry using the ascent and descent of the balloon to derive vertical concentration profiles of NO_2 and BrO. Similarly, limb viewing aircraft DOAS have also contributed to our understanding of atmospheric chemistry of BrO, IO, and glyoxal (see e.g. Baidar et al. 2013; Prados-Roman et al. 2011; Hüneke et al. 2017; Stutz et al. 2017). Frequently the range of airborne MAX-DOAS extends from zenith viewing to nadir (downwards) viewing, thus covering 180°. This viewing direction arrangement allows the measurement of trace gases below and above the aircraft (e.g. Wang et al. 2003).

A natural extension of MAX-DOAS is Imaging DOAS (see also Sect. 7.4.2 and Fig. 7.64), which is based on more than 10 (up to thousands) close viewing directions. Deployed on the ground, these systems can be used to visualise emission plumes and quantify the emission fluxes of industrial facilities and volcanos (Lohberger et al. 2004; Bobrowski et al. 2006; Louban et al. 2009; Pikelnaya et al. 2013). In addition, airborne Imaging DOAS systems pointed in the nadir can be used in a push broom mode to map spatial concentration distributions, for example of NO_2 and HCHO (Heue et al. 2008; General et al. 2014; Meier et al. 2017; Tack et al. 2019).

One of the most exciting developments of passive DOAS in the last few decades has been the launch of spaceborne DOAS instruments (see Sect. 7.10.8 and Chap. 11). The instruments typically operate in a nadir wisk-broom or push-broom viewing mode that provides spatial coverage. Most of the instruments are orbiting earth in a polar orbit and provide global coverage in one day. Most recently, geostationary instruments have been launched that observe a more limited area, but provide multiple observations per day. Originally, the purpose of these instruments was the operational monitoring of the global ozone layer. In recent years the focus has shifted to provide regional and global data on air pollution, with a focus on NO_2 and HCHO. In addition, these system have provided global distributions of BrO, IO, glyoxal, etc. A few space-borne instruments have also employed limb observations, for example SCHIAMACHY and OSIRIS, with the goal of deriving vertical trace gas profiles.
A common characteristic, that distinguishes scattered light absorption spectroscopy measurements from active DOAS (for examples see Chap. 10) or direct sunlight DOAS is that a single light path cannot be defined as photons travel through the atmosphere along a multitude of routes. In addition, these routes depend on atmospheric conditions and the solar zenith and azimuth angles. Considerable effort has to be invested in converting the observed trace gas absorption strength to quantities, i.e. vertically integrated concentrations or trace gas profiles, that are useful for the interpretation of observations. This usually involves numerically modelling the radiation transport in the atmosphere (see Chap. 3) to determine an effective light path length in the atmosphere.

This chapter provides a general introduction into the methods required to interpret scattered light DOAS measurements. We will begin by introducing the basic concepts that are needed to understand scattered light DOAS, and then discuss the details of radiative transfer calculations and models.

9.2 Air Mass Factors (AMF)

The classical concept of absorption spectroscopy as an analytical method is based on the knowledge of absorption path length and the assumption that the conditions along the light path do not vary (Chap. 6). For scattered and direct sunlight DOAS measurements, in which the light crosses the vertical extent of the atmosphere, this assumption is usually not valid. New concepts thus have to be introduced to interpret these measurements. In this section we will introduce these concepts and the quantities that are necessary to quantitatively analyse DOAS observations. We will use an approach that loosely follows the history and development of the interpretation of spectroscopic observations, beginning with direct sun measurements, followed by zenith and off-axis scattered sunlight applications.

Before discussing the individual aspects of these observational strategies, it is useful to introduce the quantity that is commonly the final result of passive DOAS observations, the vertical column density (VCD). Historically, the vertical column density, V, has been defined as the concentration of a trace gas vertically integrated over the entire extent of the atmosphere:

$$V = \int_0^\infty c(z)\, dz \tag{9.1}$$

Here z denotes the altitude above the ground. This concept can be expanded by varying the limits of integration to cover the stratosphere, troposphere, or height intervals from z_1 to z_2 of the atmosphere. We will, therefore, expand this equation by introducing partial columns:

$$V(z_1, z_2) = \int_{z_1}^{z_2} c(z)\, dz \tag{9.2}$$

9.2.1 Direct Light AMF

The earliest applications of absorption spectroscopy in the atmosphere relied on the measurement of direct sun or moonlight. As a consequence of the movement of the solar or lunar disk in the sky (Fig. 9.1), the path length used in Lambert–Beer's law to convert the observed trace gas absorption changes as a function of the solar or lunar position. It is common to use the angle between the zenith and the sun or moon to quantify this position. This Solar or Lunar Zenith Angle (SZA, LZA), ϑ, is 0° when the sun or moon is in the zenith, and 90° when they are on the horizon. In addition, the Solar or Lunar-Azimuth-Angle (SAZ, LAZ) is used to define the horizontal position. The SAZ (LAZ) is zero by definition when the sun or moon

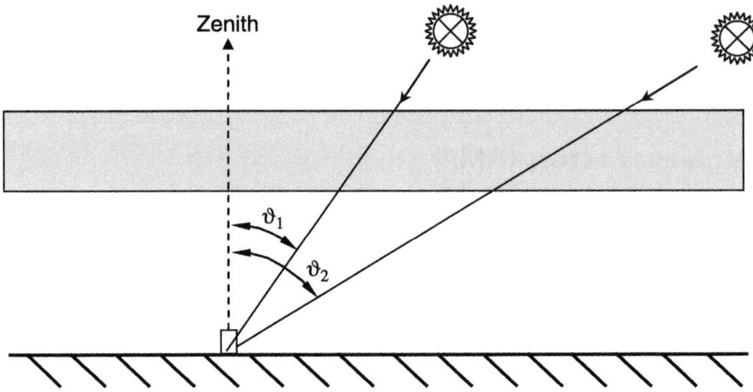

Fig. 9.1 Sketch of direct light observation geometries. In first approximation, the light path through a trace gas layer varies with $1/\cos\vartheta$ (ϑ = zenith angle of celestial body observed)

is in the northern direction and increases clockwise. We can, however, see that the azimuth angle does not play an important role when interpreting direct solar or lunar measurements.

To describe the observations of trace gases, we introduce the 'slant column density' (SCD), S. Historically, the SCD has been defined by integrating the concentration c(s) over the light path in the atmosphere.

$$S = \int_0^\infty c(s)\, ds \tag{9.3}$$

In contrast to the definition of VCD in (9.1), the element of path ds does not need to be vertical. In the case sketched in Fig. 9.1, the SCD can be determined by the geometrical enhancement of a slanted light path in the atmosphere, i.e. $ds = 1/\cos\vartheta\, dz$ for small SZA. This concept of the SCD will, however, lead to problems in the interpretation of scattered light observations, since the column seen by the instrument is an 'apparent' column, which is intensity weighted over an infinite number of different light paths through the atmosphere. We will, therefore, define SCD more generally from the observed slant column density as the ratio of measured differential optical density D' and known differential absorption cross-sections σ', i.e. $S = D'/\sigma'$ (see Chaps. 6 and 8).

To relate the observed SCD to the desired result of the measurement, i.e. the vertical column density, we now introduce the airmass factor (AMF), A, as:

$$A = \frac{S}{V} \tag{9.4}$$

The AMF is the proportionality factor between the observed slant column density and the VCD (see Noxon et al. 1979). The most basic example of an AMF is the direct light AMF for an observation geometry where the instrument looks directly towards a celestial body (e.g. sun, moon, star), which is assumed to be point-like. Neglecting

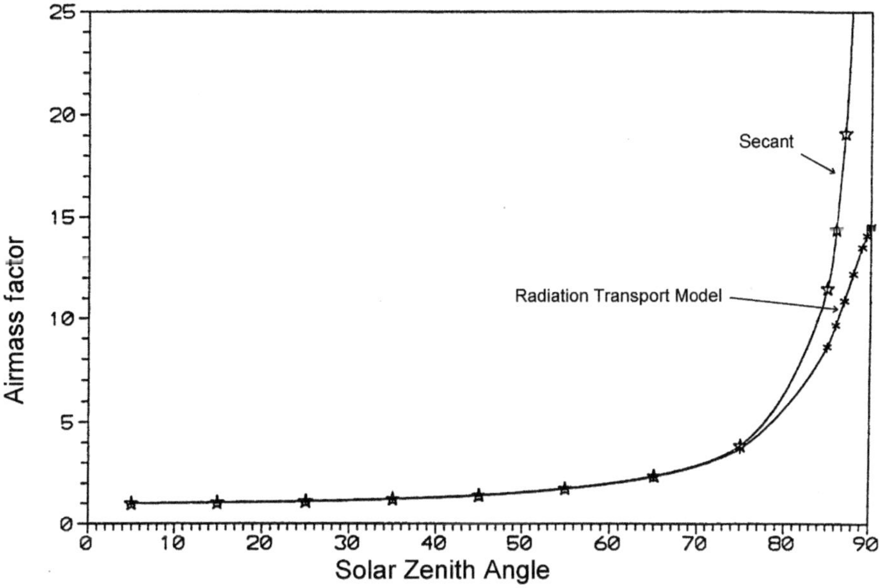

Fig. 9.2 Direct light air mass factors. Model calculations including curvature of earth as well as refraction inside the atmosphere are compared to the simple secants $\vartheta = 1/\cos\vartheta$ approximation. Deviations become apparent at $\vartheta > 70°$. Reprinted from Frank, Diploma Thesis, 1991, University Heidelberg

the curvature of earth and refraction in the atmosphere (i.e. for sufficiently small zenith angles), we obtain for the AMF as A_D (Fig. 9.1):

$$A_D = \frac{\text{length of slant path}}{\text{length of vertical path}} \approx \frac{1}{\cos\vartheta} \tag{9.5}$$

Up to an SZA of ≈75°, (Eq. 9.5) is a good approximation for the direct AMF (Fig. 9.2). Above 75°, effects such as the earth's curvature and atmospheric refraction have to be considered. Refraction in the atmosphere is caused by the dependence of the refractive index of air on temperature, pressure and thus its change with altitude.

9.2.2 Scattered Zenith Light AMF

A multitude of passive DOAS applications use scattered sunlight to measure trace gas absorptions. The telescopes in these scattered sunlight DOAS instruments are aimed at a point in the sky other than the sun or moon. Consequently, we need to consider the viewing direction of the DOAS telescope in addition to the solar position. This viewing direction is again characterised by two angles:

1. The elevation angle, which gives the angle in the vertical between the horizon (for ground based instruments) and the viewing direction. The zenith in this case is at

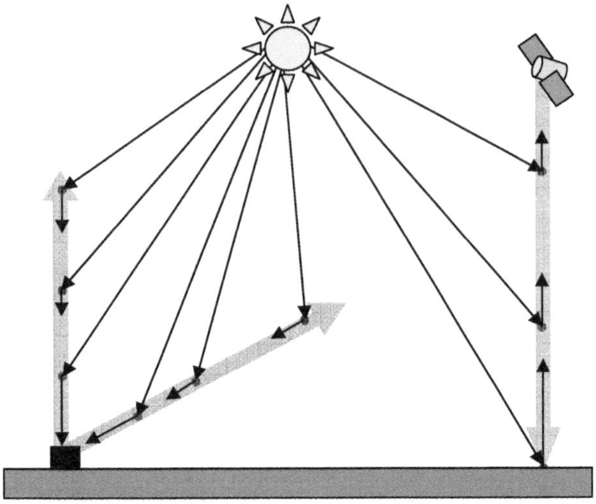

Fig. 9.3 Sketch of ground-based and satellite-borne passive DOAS observations

an elevation of $\alpha = 90°$. For downward-looking geometries, such as from airborne or satellite instruments, we use negative values, i.e. the nadir is at $\alpha = -90°$. The viewing direction can also be measured from the zenith, in which case viewing parallel to the ground is at a zenith angle of $\vartheta = 90°$ and the nadir is at $\vartheta = 180°$.

2. The viewing azimuth angle, which is defined in the same way as the azimuth of the solar position.

The zenith viewing geometry, i.e. $\alpha = 90°$, has historically been one of the most successful applications of passive DOAS. Many of the basic concepts of AMF calculations have been determined for this viewing geometry, mostly in the context of studying stratospheric trace gases and the chemistry leading to the Antarctic ozone hole.

As illustrated in Fig. 9.3, in zenith scattered light DOAS applications the irradiance received by the detector originates from light scattered by the air molecules and particles that are located along the viewing direction of the telescope. Assuming, for now, that only one scattering process occurs between the sun and the detector, one can gain a basic understanding of the zenith sky observations.

Figure 9.4 illustrates that two processes have to be taken into account to understand the measurement of radiances at the detector. First, one has to consider the efficiency by which solar light is scattered from its original direction towards the detector. Secondly, one needs to consider the extinction, either by trace gas absorption or by Rayleigh scattering, along the different light paths. The main process changing the direction of light under clear sky conditions is Rayleigh scattering in the zenith, which depends primarily on air density, i.e. the number of scattering air molecules. The scattering efficiency will thus be highest close to the ground and decrease exponentially with altitude. In a thin layer at height z', the intensity of light scattered towards the detector depends on the intensity reaching the scattering point, I_z, the Rayleigh scattering cross-section, σ_R, and the air density, $\rho(z)$ (Solomon et al. 1987):

$$I_S(\lambda, z') = I_0(\lambda, z') \cdot \sigma_R(\lambda) \cdot \rho(z')\, dz' \tag{9.6}$$

The extinction along the light path will depend on the length of each light path in the atmosphere, the concentration of air molecules and, in the case of strong absorbers, the concentration of these gases.

$$I_S(\lambda, z) = I_0(\lambda, z) \times \exp\left(-\sigma_R(\lambda) \int_z^{\infty} \rho(z') \cdot A(z', \vartheta)\, dz'\right) \tag{9.7}$$

where $A(z, \vartheta)$ is the direct sun AMF for the light before scattering at altitude z'. Because the AMF depends on the solar zenith angle, the length of each light path will also depend on the SZA, as can be seen in Fig. 9.1. This leads to a decrease in intensities I(z) with SZA. The dependence of Rayleigh scattering on air density indicates that the intensity reaching the zenith point at which it is scattered will be lowest close to the ground and increase with altitude.

Combining the two effects, which show opposite altitude dependences, gives rise to a distribution of scattered light intensity that is small at the ground, increases to a maximum, and then decreases again with altitude. The maximum of this scattered light distribution represents the most probable height that the observed light originates from. The contribution of scattering in a thin layer at altitude z to the intensity

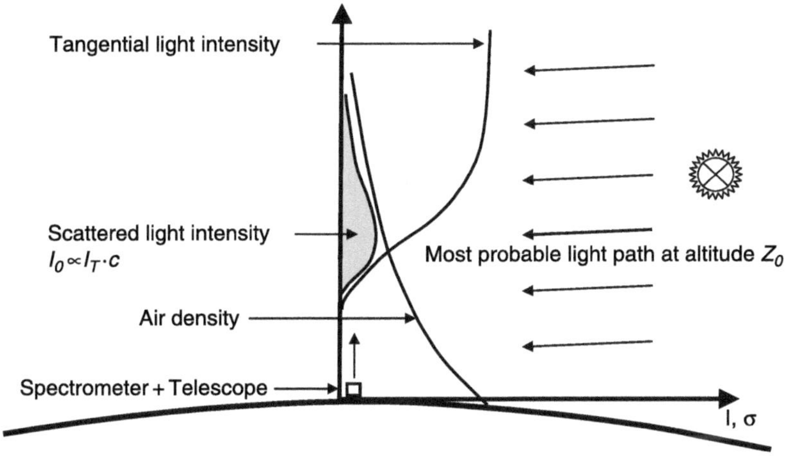

Fig. 9.4 Geometry of Zenith Scattered Light DOAS–radiation transport in the atmosphere. There is an infinite number of possible light paths. However, at solar zenith angles around 90° the light seen by the spectrometer most likely originates from a certain altitude range around Z_0

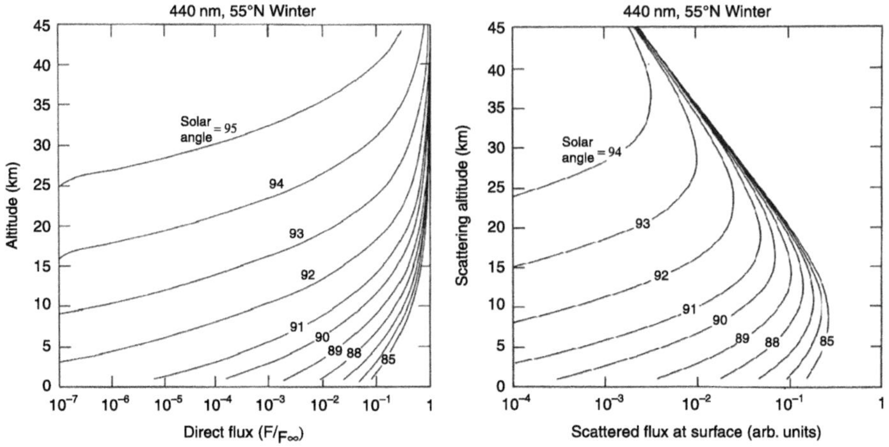

Fig. 9.5 Calculated intensities of the direct (**left panel**) and scattered (**right panel**) radiative fluxes as a function of altitude for various SZA. (Reprinted with permission from Solomon et al. (1987a)a,

observed by the detector is then:

$$I_S(\lambda, z) = \sigma_R(\lambda) \cdot \rho(z) \cdot I_0(\lambda) \cdot \exp\left(-\sigma_R(\lambda) \int_z^{\infty} \rho(z') \cdot A(z', \vartheta)\, dz\right)$$
$$\times \exp\left(-\sigma_R(\lambda) \int_h^{z} \rho(z')\, dz'\right) \tag{9.8}$$

The last term in this equation is the Rayleigh extinction on the light path from the scattering height to the detector, which is assumed to be at an altitude h.

This equation allows the discussion of the dependence of scattering properties with the SZA. At large SZA, the light paths in the atmosphere become large, and at lower altitudes the intensity reaching the scattering height is reduced more than at higher altitudes (Fig. 9.5). The most probable scattering height, therefore, moves upwards as the SZA increases. The most probable scattering height also depends on the wavelength, due to the wavelength dependence of Rayleigh scattering. Z_o typically varies from about 26 km (327 nm) to 11 km (505 nm) at 90° SZA.

In the presence of an absorbing trace gas, (9.8) is expanded by including the absorption cross-section and the trace gas concentration at each height.

$$I_S^A(\lambda, z) = I_S(\lambda, z) \cdot \exp\left(-\sigma(\lambda) \int_z^{\infty} C(z') \cdot A(z', \vartheta)\, dz\right)$$
$$\cdot \exp\left(-\sigma(\lambda) \int_h^{z} C(z')\, dz'\right) \tag{9.9}$$

Because the large direct sunlight AMFs, $A(z', \vartheta)$, are approximately equal to $1/\cos\vartheta$, the absorption is largest at large SZAs. The slant light paths through the atmosphere are quite long under twilight conditions. At 90° zenith angle and 327 nm, the horizontal light path length through the stratosphere is 600 km. The AMFs for the path below the scattering height, i.e. the second exponential term in (9.9), are equal to unity.

Our discussion shows that the light reaching the detector is an average over a multitude of rays, each of which takes a somewhat different route through the atmosphere. The detector, therefore, measures the intensity-weighted average of the absorptions along the different light paths arriving at the telescope. This 'apparent' column density S is, for historic reasons, also called SCD, although it has no resemblance to the slanted column of direct solar measurements.

Based on our discussion above, we can now use the definition of SCD (9.3) to write down a simplified expression for SCD for scattered sunlight:

$$S(\vartheta) = \frac{1}{\sigma(\lambda)} \ln \left[\frac{\int_h^\infty I_S^A(\lambda, z)dz}{\int_h^\infty I_S(\lambda, z)dz} \right] \tag{9.10}$$

This SCD can now be used in (9.4) to calculate the AMF for zenith scattered light. Figure 9.6 shows such an AMF for a stratospheric absorber at 450 nm. As expected, the AMF increases with SZA as the light path in the stratosphere becomes longer and the most probable scattering height moves upwards. The decrease at very large SZA occurs when Z_0 moves above the absorption height. To put it simply, the detected light from the zenith can be represented by a most probable light path through the atmosphere defined by the most likely scattering height Z_O in the zenith.

While our simplified description illustrates the principles of radiative transfer and AMF calculations, it is insufficient to make accurate AMF calculations. Other physical processes such as scattering by aerosol particles, refraction, and multiple scattering need to be considered in the AMF calculations. Interpretation of the SCD, therefore, requires radiation transport modelling. More details on atmospheric radiation transport are available in Chap. 4, and, for example, Solomon et al. (1987), Frank and Platt (1990), or Marquard et al. (2000).

9.2.3 Scattered Off-Axis and Multi-Axis AMF

Scattered-light DOAS viewing geometries, other than the zenith, have become increasingly popular in recent years. The motivation for using smaller elevation angles is twofold. With respect to stratospheric measurements, lower viewing angles can improve the detection limits by increasing the light intensity reaching the detector. The stronger motivation is the ability to achieve long light paths, i.e. larger AMFs, for tropospheric trace gases. To understand these effects, the underlying radiative transfer principles are discussed here.

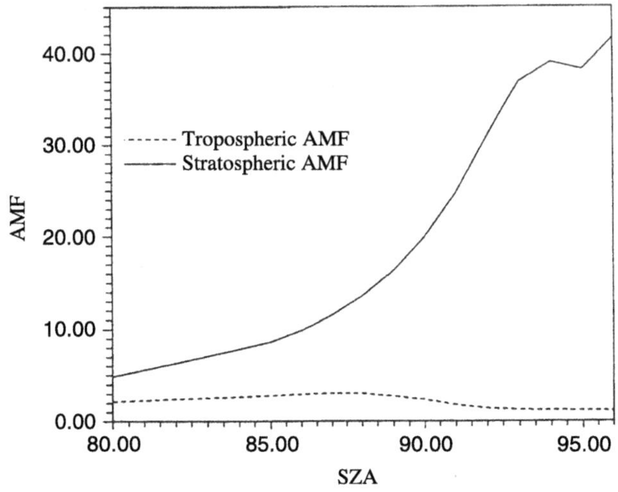

Fig. 9.6 Stratospheric and tropospheric AMF of NO_2 at 445 nm determined using a single scattering radiative transfer model. Reprinted from Stutz, Diploma Thesis, 1992, University of Heidelberg

Our argument follows very closely the approach we have adopted for ZSL-DOAS in Sect. 9.2.2. The light of the detector originates from scattering processes within the line of view of the detector. Because the detector aims at lower elevations than in the ZSL case, i.e. the viewing path crosses through layers with a higher air density, scattering events closer to the ground will contribute more to the detected intensity. Consequently, the most probable scattering height will move downwards in the atmosphere as the viewing elevation angle decreases. It is typically somewhere in the troposphere for all wavelengths. We can now expand (9.8) and (9.9) by including an AMF, $A_T(z, \alpha)$, for the path between the scattering event and the detector. In an approximation based on purely geometrical arguments, $A_T(z, \alpha)$ is equal to $1/\cos\alpha$, i.e. it increases with decreasing viewing elevation angle. This will change the distribution of the scattering term in (9.9), as well as increase the Rayleigh extinction between the scattering event and the detector. It should be noted here that both the SZA and the elevation angle also influence σ_R (see Chap. 4).

$$\begin{aligned} I_S(\lambda, z, \alpha) =& \sigma_R(\lambda) \cdot \rho(z) \cdot A(z', \alpha) \cdot I_0(\lambda) \\ & \cdot \exp\left(-\sigma_R(\lambda) \int_z^\infty \rho(z') \cdot A_S(z', \vartheta)\, dz\right) \\ & \cdot \exp\left(-\sigma_R(\lambda) \int_h^z \rho(z') \cdot A_T(z', \alpha)\, dz'\right) \end{aligned} \tag{9.11}$$

The smaller elevation angles also influence the absorption of trace gases. Equation (9.9) thus has to be expanded to include $A_T(z, \alpha)$ in the second integral:

$$I_S^A(\lambda, z, \alpha) = I_S(\lambda, z, \alpha) \cdot \exp\left(-\sigma(\lambda) \cdot \int_z^\infty C(z') \times A_S(z', \vartheta)\, dz\right) \cdot \exp\left(-\sigma(\lambda) \cdot \int_h^z C(z') \cdot A_T(z', \alpha)\, dz'\right) \tag{9.12}$$

The lower elevation in our simplified model does not change the first integral that describes the light path before the scattering event. The behaviour of stratospheric trace gases, for example, does not depend on the elevation angle, while that of tropospheric trace gases will. The SCD for the off-axis case can now be calculated according to (9.10).

To address the separation of stratospheric and tropospheric absorbers in more detail, we will further simplify our discussion by concentrating on the most probable light path. This eliminates the integrations in (9.10) to calculate the SCD and the AMF. Figure 9.7 illustrates this simplified view of off-axis viewing geometries. In this simplified picture, the AMFs $A_S(z, \vartheta)$ and $A_T(z, \alpha)$ are independent of altitude z and can be approximated as $1/\cos\vartheta$ and $1/\cos\alpha$, respectively. We further introduce the tropospheric and stratospheric vertical column densities, $\mathrm{VCD_T}$ and $\mathrm{VCD_S}$, which integrate the vertical trace gas concentration profile from the ground to the scattering altitude, and from the scattering altitude to the edge of the atmosphere, respectively. After these simplifications, we find that the SCD can be described by:

$$S(\alpha, \vartheta) = V_T \cdot A_T(\alpha) + V_S \cdot A_S(\vartheta) = \frac{V_T}{\cos\alpha} + \frac{V_S}{\cos\vartheta} \tag{9.13}$$

The SCD, therefore, depends both on the SZA, which controls the contribution of the stratospheric column, and the elevation angle, which controls the contribution of the tropospheric column.

While this equation illustrates the dependence of the SCD and the total AMF on the SZA, the elevation angle, and the vertical trace gas profile, it is highly simplified. In the case of low elevations, which are often used to increase the tropospheric path length, multiple scattering events, curvature of the earth, and refraction become significant (see Fig. 9.8). In addition, the higher levels of aerosols in the troposphere make Mie scattering an important process that must be included in the determination of AMFs. Finally, stratospheric AMFs are weakly dependent on the viewing direction. To quantify these effects, a detailed radiative transfer model is required. A short description of such models will be given in Sect. 9.3.

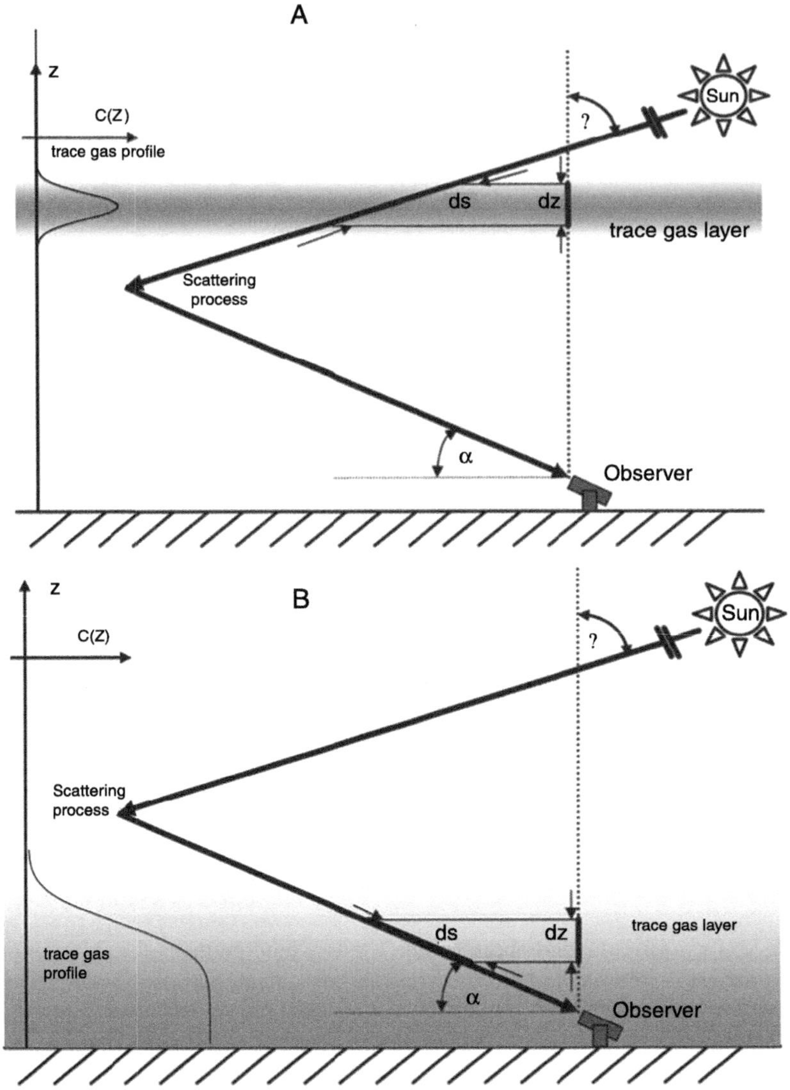

Fig. 9.7 Geometry of Off-axis DOAS and a sketch of the associated radiation transport in the atmosphere. Like in the case of ZSL-DOAS, there is an infinite number of possible light paths. Reprinted with permission from Hönninger et al. 2004a,

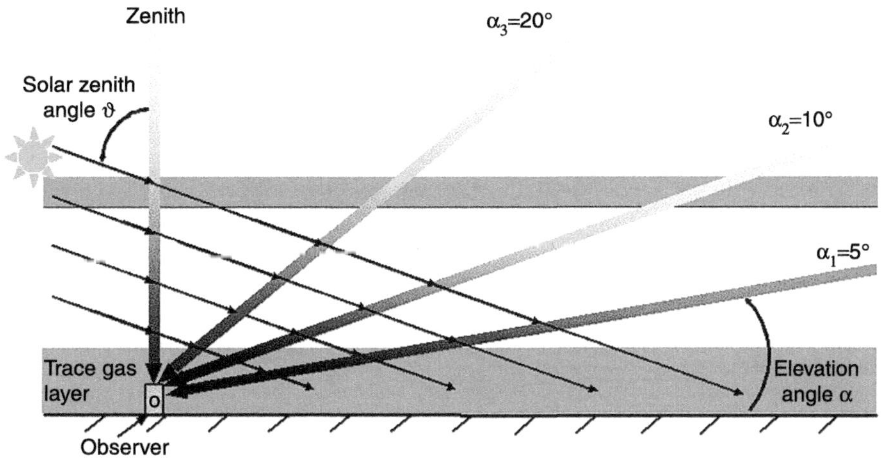

Fig. 9.8 Geometry of Multi-axis DOAS (MAX-DOAS) and a sketch of the associated radiation transport in the atmosphere. As in the case of ZSL-DOAS, there is an infinite number of possible light paths. Reprinted with permission from Hönninger et al. 2004a, © CC 2.5. All rights reserved

9.2.4 AMFs for Airborne and Satellite Measurements

Airborne and satellite DOAS measurements have become an important tool to study atmospheric composition on larger scales. These measurements are based on scattered sunlight detection, and AMFs have to be calculated to interpret the observations. In principle, the approach is very similar to that shown for other scattered light applications.

As with the multi-axis approach, the SZA and the viewing angle have to be considered. In this case, however, the instruments look downwards and can, at least at higher wavelengths, see the ground. Besides the scattering on air molecules and aerosol particles, clouds and the albedo of the ground have to be considered. The ground is typically parameterised by a wavelength-dependent albedo and the assumption that the surface is a Lambertian reflector, or by a bi-directional reflectivity function (BDRF), which parameterises the reflection based on incoming and outgoing reflection angles. Clouds seen from an airplane or a satellite can be parameterised by introducing parallel layers of optically thick scatterers in a multiple scattering model or, in the case of thick clouds, by parameterising them as non-Lambertian reflectors in the model at a certain altitude (Kurosu et al. 1997).

An additional problem, which we will be discussed later, is the fact that downward-looking airborne or satellite instruments often observe areas of the earth's ground that may also be partially covered by clouds. The spatial averaging over the earth surface together with clouds is a challenge for any radiative transfer model.

9.2.5 Correction of Fraunhofer Structures Based on AMFs

A challenge in applying DOAS to the measurement of atmospheric trace gases is the solar Fraunhofer structure, which manifests itself as a strong modulation of $I_0(\lambda)$ due to absorption in the solar atmosphere (see Chap. 6). All passive DOAS spectra are overlaid with a strong Fraunhofer structure, and weak spectral bands can usually only be detected in the ratio of two spectra, one I(λ) with a large AMF A and one $\mathrm{I_{ref}}(\lambda)$ with a small AMF A_{ref} (e.g. a zenith spectrum). The latter is known as the "Fraunhofer reference spectrum". The ratio Q(λ) of the two spectra is then given by:

$$Q(\lambda) = \frac{I(\lambda)}{I_{ref}(\lambda)} = \frac{I_0(\lambda) \cdot e^{-A \cdot V \cdot \sigma(\lambda)}}{I_0(\lambda) \cdot e^{-A_{ref} \cdot V \cdot \sigma(\lambda)}} = e^{-(A-A_{ref}) \cdot V \cdot \sigma(\lambda)} = e^{-(S-S_{ref}) \cdot \sigma(\lambda)} \tag{9.14}$$

This means that the result of the evaluation of the ratio spectrum Q(λ) is the difference of the SCDs of the measurement spectrum and the Fraunhofer reference spectrum. This SCD is also known as 'Differential Slant Column Density', DSCD. The determination of the trace gas content of the reference spectrum is generally a challenge. We will discuss two approaches to overcome this problem, one for the measurement of stratospheric trace gases and one for the measurement of tropospheric gases. Both techniques are based on the choice of suitable Fraunhofer reference spectra that can be used in the analysis procedures described in Chap. 8, or in simple terms by which the observed spectrum is divided. Ideally, one would like to choose a spectrum without any absorption of the respective atmospheric trace gas. Recording a scattered light spectrum with small absorption of the target gas can be possible in cases where strong horizontal gradients of the target gas exist, for instance when observing emission plumes from point sources such as smoke stacks or volcanoes. Satellite instruments often have the ability to measure a direct solar spectrum outside of the atmosphere and can thus also circumvent this problem. When this is not possible (e.g. for stratospheric gases or in the absence of plumes) one has to rely on the choice of a Fraunhofer reference spectrum where the trace gas absorptions are small. Based on our discussion in Sect. 9.2.5, this is equivalent to a small AMF in the reference spectrum as compared to the actual observation.

In the case of ZSL measurements of stratospheric trace gases, our earlier discussions revealed that the AMF increases with the solar zenith angle. The obvious choice for a Fraunhofer reference spectrum is a spectrum measured at small SZA. Ideally, the ratio of a spectrum taken at a higher SZA, for example at sunset, and the Fraunhofer spectrum will consist of pure trace gas absorptions. As discussed above, taking the ratio of spectra recorded at two different SZAs would eliminate the solar intensity, i.e. $I_S(\lambda, z)$. However, this approach poses another challenge since the division also eliminates the absorptions that are originally in the Fraunhofer reference spectrum, which are not known. The analysis of the ratio between the two spectra see Eq. 9.14 results in the DSCD, S', which is the difference between the SCDs of high SZA, ϑ_2, and the reference spectrum measured at low SZA ϑ_1:

$$S' = S(\vartheta_2) - S(\vartheta_1) \tag{9.15}$$

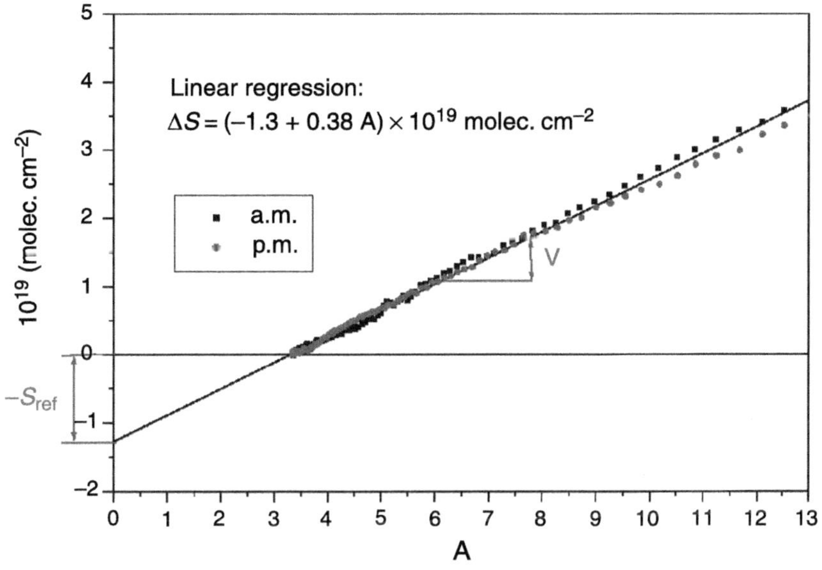

Fig. 9.9 Sample Langley plot: Measured ozone differential slant columns (DSCDs) are plotted as a function of the airmass factor A calculated for the solar zenith angle of the measurement. The slope of the plot indicates the vertical column density V. The ordinate section indicates the slant column density in the solar reference spectrum, S_{ref}

A determination of the VCD is not directly possible from S'. A solution to this problem presents itself if the vertical column density of the absorbing trace gas remains constant with time, i.e. same at ϑ_2 and ϑ_1. Equation (9.15) can then be written as:

$$S' = S'(\mathrm{AMF}) = \mathrm{V} \times \mathrm{A}(\vartheta_2) - \mathrm{S}(\vartheta_1) \tag{9.16}$$

This linear equation can then be used to determine the VCD and $\mathrm{S}(\vartheta_1)$ by plotting the DSCD against the AMF and applying a linear fit to the resulting curve. An example of this so-called Langley Plot is shown in Fig. 9.9. The slope of the curve gives the VCD, while the extrapolation to $A = 0$ results in an ordinate intersection at $-S_{ref}$.

In the case where the interest is more in tropospheric trace gases, the above method will not lead to the desired result since the tropospheric AMF is only weakly dependent on the SZA. However, as shown in Sect. 9.2.3, changing the viewing elevation will considerably influence the tropospheric AMF (9.13). The approach to measure tropospheric gases is thus to use a zenith spectrum ($\alpha = 90°$) as the Fraunhofer reference to analyse spectra measured at lower elevations. To guarantee that the stratospheric AMF is the same for these two spectra, one has to measure the zenith and the low elevation spectrum simultaneously, or at least temporally close together. Using (9.13), this approach describes:

$$S' = S(\alpha, \vartheta) - S(90, \vartheta) = V_T \times A_T(\alpha) \tag{9.17}$$

It should be noted that, despite the fact that we have called V_T the tropospheric vertical column density, the height interval over which the vertical concentration profile is integrated in V_T depends on the radiative transfer. Typically V_T does not cover the entire troposphere, but rather the boundary layer and the free troposphere. A more detailed discussion of this will be given in Sect. 9.4.

9.2.6 Using Catalog Fraunhofer Reference Spectra Derive 'Absolute' SCDs

Spectra of solar Fraunhofer bands, recorded at high spectral resolution (about 0.04 nm with ≈0.001 nm absolute calibration), are available in literature (e.g. Chance and Kurucz 2010). These spectra were measured from high altitudes in the atmosphere (high mountains and from balloons) to reduce atmospheric extinction. The remaining atmospheric signatures are additionally removed from the spectra through various correction methods. These Fraunhofer catalog spectra (also known as 'Kurucz Spectrum' or SAO96 or SAO2010.solref) are used for wavelength calibration and for the determination of the instrument function as described in Chap. 8. Because the Fraunhofer catalog spectra are free from atmospheric absorption structures one could conclude that they could, after convolution with the instrument function, serve as ideal Fraunhofer reference spectra in a DOAS retrieval. However, while most of the Fraunhofer structures are removed from passive DOAS observations, relatively large residuals are typically found. Therefore, the use of current literature Fraunhofer spectra reduces the sensitivities of DOAS retrievals, and they are rarely used for this purpose. One exception, as pointed out by Lübcke et al. (2016), are cases with very strong absorptions, such as SO_2 in volcanic plumes. The advantage of this approach is that absolute slant column densities can be determined, when the 'traditional' approaches of determining Fraunhofer reference spectra yields considerable lower trace gas SO_2 SCDs.

9.3 AMF Calculations

The detailed calculation of AMFs requires consideration of different physical processes influencing the radiative transfer (RT) in the atmosphere. Scattering processes, either Rayleigh or Mie scattering, reflection on the earth's surface, refraction, the curvature of the earth, and the vertical distribution of trace gases play a role in the transfer of solar radiation. Consequently, sophisticated computer models are employed to calculate the RT in the atmosphere and the AMFs needed to retrieve vertical column densities from DOAS observations of scattered sunlight.

Early AMF calculations based on single-scattering radiative transport models used the approximations that the radiation propagating through the atmosphere is scattered only once before it is detected and that the total optical density at the detector position is equivalent to the sum of the optical densities along the single paths weighted by their probabilities (see Sarkissian et al. 1995; and references therein). A linear weighting scheme (see e.g. Solomon et al. 1987) is then used to determine the AMFs. While this approach is computationally very efficient, the omission of multiple scattering leads to inaccurate AMFs, especially in cases with large SZAs and in the troposphere. Consequently, this approach is rarely used today.

The most common computation procedure for the calculation of AMFs for a certain absorber in an atmosphere with multiple scattering is as follows :

1. The radiances $I_S(\lambda, 0)$ and $I_S{}^A(\lambda, \sigma)$ are calculated by an appropriate RT model. This means that two model simulations must be performed: one simulation where the absorber is included and one simulation where the absorber is omitted from the model atmosphere.
2. The slant (or apparent) column density S is calculated according to the DOAS method, i.e. (9.1).
3. The vertical column density is calculated by integrating the number density of the considered absorber, which has been used as an input parameter for the modelling of $I_S{}^A(\lambda, \sigma)$ in Step 1, in the vertical direction over the spatial extent of the model atmosphere.
4. The AMF is then derived according to (9.4).

This computational procedure is used to calculate AMFs when multiple scattering is taken into consideration and has also been used in some single-scattering RT models. It was first described and applied by Perliski and Solomon (1993).

At present, this procedure appears to be implemented in most existing RT models that consider multiple scattering, such as discrete ordinates radiative transfer (DISORT) models (Stamnes et al. 1988; Dahlback and Stamnes 1991; Laszlo et al. 2016), GOMETRAN/SCIATRAN (Rozanov et al. 1997; Rozanov et al. 2005, 2014), LIDORT (Spurr et al. 2001; Spurr and Christi 2019), VLIDORT (Spurr 2006; Spurr et al. 2022), and certain backward Monte Carlo RT models (e.g. Perliski and Solomon 1992; Marquard 1998; Marquard et al. 2000). We refer to this method as the 'traditional' (linear) AMF computation method. In some backward Monte Carlo models, such as McArtim Deutschmann et al. (2011), a more computationally effective approach is used, where a weighted photon path length over a given height interval is calculated and then vertically integrated.

9.3.1 Single-Scattering RT Models

As mentioned above, earlier RT models considered only one scattering event in the atmosphere, and were thus termed 'single-scattering models'. Figure 9.10 illustrates the approach most often taken in these simple RT models. The curved atmosphere is subdivided in distinct layers. Light travels from the direction of the sun, defined by the SZA, through the atmosphere until it reaches the zenith over the DOAS instrument. At the zenith, a parameterisation of Rayleigh scattering and Mie scattering is used to calculate the fraction of light scattered towards the earth surface from this scattering point (Fig. 9.11). Refraction is considered whenever a ray enters a new

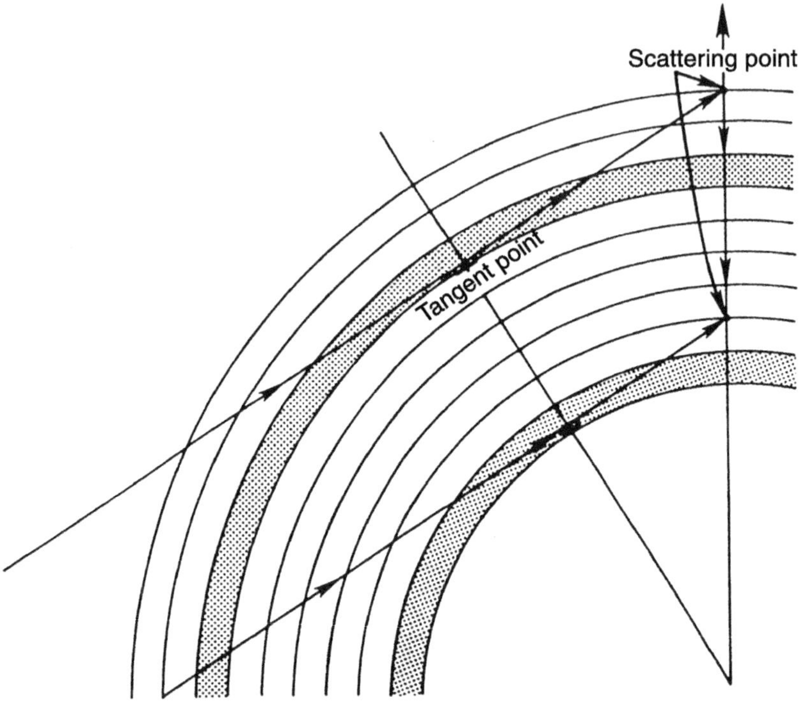

Fig. 9.10 Definition of the tangent point in spherical geometry. Reprinted with permission from Perliski and Solomon 1993, © American Geophysical Union. All rights reserved

layer following Snell's law. Absorption is calculated from the path length in each layer, vertical concentration profile, and absorption cross-section supplied to the model. The model then performs a numeric integration (see Eq. 9.10) to derive the SCD for each SZA and wavelength. The ratio of this SCD with the VCD calculated from the vertical trace profile is then the desired AMF.

A number of single-scattering models have been developed in the past (Frank 1991; Perliski and Solomon 1993; Schofield et al. 2004). The advantage of these models is their simplicity and computational efficiency. However, the omission of multiple scattering events poses a serious challenge. Figure 9.11 compares the contribution of single and multiple scattering to the signal of a ground-based ZSL instrument. In the lower atmosphere the contribution of multiple scattering is substantial. This effect is even more severe if off-axis geometries are used, where the most probable scattering height is in the troposphere. Multiple scattering also has to be considered when high aerosol concentrations are present. For these reasons, single-scattering models are rarely used today.

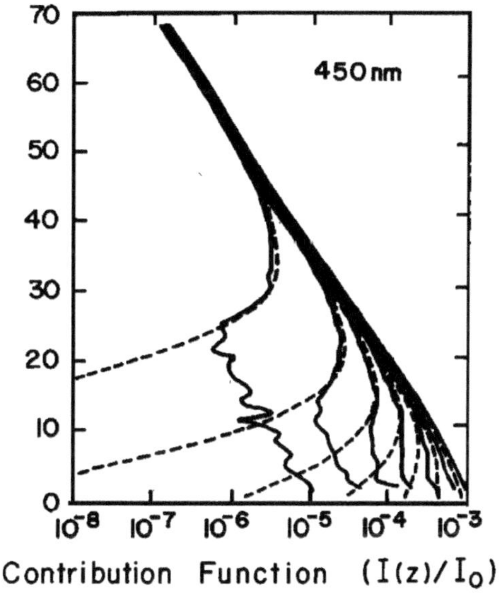

Fig. 9.11 Contribution of different altitudes to the intensity detected by a ground-based ZSL-DOAS instrument for different SZA. The dashed and solid lines show results from a single and a multiple scattering model, respectively. Reprinted with permission from Perliski and Solomon, 1993, © American Geophysical Union. All rights reserved

9.3.2 Multiple-Scattering RT Models

Radiative transfer models that consider multiple scattering events are the standard in current AMF calculations. In general, these model solve the RT equation introduced in Chap. 3 on a vertical grid through the atmosphere, i.e. in one dimension, or sometimes on a three-dimensional grid. Aside from the dimensionality, the models differ in the mathematical approach used to solve the RT equation and the detail in which polarisation is considered. For details on the mathematics underlying the various solution we refer the reader to the cited publications and to textbooks on radiative transfer (e.g. Liou 2002).

Current RT models used in the DOAS community (see Table 9.2) are based either on the analytical discrete-ordinate solution of the RT equation (Lenoble 1985; Stamnes et al. 1988; Spurr 2001) or a numerical Monte Carlo solution (Metropolis and Ulam 1949; Marchuk et al. 1980; Perlisky and Solomon 1993; Marquard and Platt 1997). All modern models consider the earth sphericity and multiple scattering. Several of the analytical models are also full vector models and explicitly consider polarisation, i.e. the solution is based on the full Stokes vector (Spurr 2006; Spurr et al. 2022; Rozanov et al. 2014). In short, the discrete-ordinate method solves the RT equation by expressing the radiation field as a Fourier cosine series in the azimuth (Stamnes et al. 1988). A numerical quadrature scheme is then used to replace the integrals in the RT equation by sums. The RTE equation is, therefore, reduced to a set of coupled linear first-order differential equations, which are consequently solved. There are currently three discrete-ordinate RT models in wider use to calculate AMFs (see Table 9.2): DISORT (web.archive.org/web/20180905023345/lllab.phy.stevens.edu/disort/) , SCIATRAN

Table 9.2 Most commonly used radiative transfer model for DOAS applications

RT Model	Aproach	Source
DISORT	Discrete-ordinate	Lin et al. (2015b); Laszlo et al. (2016); Stamnes et al. (2017)
SCIATRAN	Discrete-ordinate	Rozanov et al. (2005, 2014, 2017)
VLIDORT/LIDORT	discrete-ordinate	Spurr et al. (2001); Spurr (2004, 2006); Spurr and Christi (2019); Spurr et al. (2022)
McArtim	Monte Carlo	Deutschmann et al. (2011)
MYSTIC	Monte Carlo	Emde et al. (2016); Mayer and Kylling (2005); Mayer (2009)
smartMOM	Linearised vector	Sanghavi et al. (2014); Sanghavi (2022); Sanghavi and Frankenberg (2023)

(www.iup.uni-bremen.de/sciatran/), and LIDORT/VLIDORT (avaliable from RT SOLUTIONS Inc., rtsolutions@verizon.net).

The Monte Carlo solution is based on statistical sampling experiments, following a large number of photons as they travel through the atmosphere (Metropolis and Ulam 1949; Marchuk et al. 1980; Lenoble 1985). Photons can undergo random processes such as absorption, Rayleigh scattering, Mie scattering, reflection on the ground, etc. For each of these processes, a probability is determined, which is then used to randomly determine if, at any point in the atmosphere, the photon undergoes a certain process. A statistical analysis of the fate of the photon ensemble provides the desired RT results. The Monte Carlo RT solution can be achieved by tracing photons from the source, i.e. the sun, to the detector, which is often called forward calculations. This is the method implemented in the MYSTIC model (www.libradtran.org, Emde et al. 2016; Mayer and Kylling 2005; Mayer 2009). A more computationally efficient method is the backward Monte Carlo approach, where photons leave the detector and are then traced through the atmosphere (Marquard et al. 2000; von Friedeburg et al. 2003). Several backward Monte Carlo models have been developed over the years (Marquard et al. 2000; von Friedeburg et al. 2003). The latest model, McArtim (www.tim-deutschmann.de/McArtim/, Deutschmann et al. 2011), has been widely used, but is not being further developed at this time. The advantage of Monte Carlo methods is their precise modelling of RT without the need for complex numerical solutions or the simplifications of the underlying RT equation (Chap. 4). In addition, only Monte Carlo models are currently able to perform RT simulations in three dimensions. The disadvantage, however, is that Monte Carlo models are inherently slow due to the large number of single-photon simulations that are needed to determine a statistically significant result.

9.3.3 Model Initialisation

Independent of the RT model used, the accuracy of the AMFs depends crucially on the correct input data. Certain input data is typically included in the model implementation, such as the solar spectrum, Rayleigh scattering coefficients, etc.

However, parameters specific to the measurement conditions have to be supplied by the user. The following list summarises the most common parameters that RT models need to calculate AMFs:

- **Model grid**: The grid spacing of the model, either in 1D or 3D, has to be adapted to the specific DOAS application. For example, for MAX-DOAS measurements a finer grid is typically chosen in the lower atmosphere (~100 m spacing), while a coarser grid is sufficient in the stratosphere (1-2 km spacing).
- **Atmospheric temperature and pressure**: Refraction and Rayleigh scattering within the model domain depend on the vertical profiles of temperature and pressure. In addition, many models include oxygen and oxygen collisional complex absorptions that are calculated based on these profiles. Temperature and pressure profiles are available through operational soundings, for example at airports, or from meteorological models.
- **Solar position and instrument viewing geometry**: The position of the sun, i.e. zenith and azimuth angles, as well as the instrument viewing direction, i.e. viewing azimuth and zenith angles and, in the case of Monte Carlo models the instrument opening angle, are crucial input data for the RT models. For airborne or spaceborne observations the altitude of the instrument above the surface also needs to be supplied.
- **Wavelength**: The wavelength or wavelength interval has to be specified in the model.
- **Absorption cross-sections**: Absorption cross-sections of strongly absorption trace gases at the chosen wavelength, such as O_2, O_3, H_2O, have to be supplied. These gases can have an impact on the RT and thus have to be considered independent of the species for which the AMF will be calculated. The absorption cross-section of the target species is also needed. If available, these cross-sections should be provided as a function of temperature and pressure, i.e. each altitude grid has its specific absorption cross-section.
- **Aerosol Profiles**: Information on the vertical distribution of the aerosol is needed to correctly model particle scattering throughout the atmosphere. This is most often achieved by providing vertical profiles of aerosol extinction either from climatology, atmospheric chemistry models, or, when available, from LIDAR observations. For balloon or aircraft observations in-situ observations of aerosol extinction profiles or size distributions are often available. In the latter case a Mie model can be used to calculate aerosol extinction.
- **Aerosol optical properties**: Aerosol optical properties are often the most demanding input for RT models. In all models, scattering and absorption coefficients, or the single extinction and scattering albedo (Sect. 4.2.5) are required. Information on the angular distribution of scattering, either in the form of an asymmetry parameter for the Henyey-Greenstein function or as a phase function (Sect. 4.2.5), is also needed. Because these parameters are often not available, climatological data (Levy et al. 2007; Li et al. 2014; Dubovik et al. 2002) are often used. In some cases, satellite or ground-based networks can supply information on optical properties of aerosol at the time of the DOAS measurement. With the emergence of powerful

analytical methods for aerosol size and composition it is often possible to directly calculate scattering and absorption coefficients and phase functions through Mie calculations. Refractive indices of specific aerosol types/sizes, which are often taken from literature, have to be used in this calculation (Sect. 4.2.5, Massie and Hervig 2013; Gordon et al. 2017). In fact, several of the RT models listed above are linked to Mie codes and allow the performance of these calculations.

- **Albedo or BRDF**: Reflection at the surface impacts AMF calculations. Models thus either need the albedo at the respective wavelength or, if available, the bidirectional reflectance distribution function, BRDF. Both have been determined for specific surfaces and can be found in literature (e.g. Wanner et al. 1997; Schaaf et al. 2002, 2011; Shuai et al. 2011; Liu et al. 2017).
- **Clouds**: The presence of clouds can have profound impacts on the RT and therefore AMFs. In the case of satellite AMF calculations the presence of clouds is often characterised by a fractional cloud cover and cloud top heights, which are corrected for outside of RT models. The inclusion of clouds, and in particular broken cloud cover, requires 3D RT models and at this point in time is an focus of ongoing research (Emde et al. 2022).

The accuracy of these input parameters is often more important than the selection of a specific model. For example a model intercomparison study by (Hendrick et al. 2006) found that four RT models using the same input data agreed better than $\pm 5\%$ for NO_2 and HCHO AMFs. The selection of the various input parameters is therefore often the most challenging task when calculating AMFs.

9.3.4 Applications and Limitations of the 'Traditional' DOAS Method for Scattered Light Applications

The classical DOAS approach has been widely and successfully applied to measurements of scattered sun-light. It uses the same tools as described in Chaps. 6 and 8. In short, based on Lambert–Beer's law:

$$I(\lambda) = I_0(\lambda) \cdot \exp\left(-\sigma(\lambda) \cdot c \cdot L\right) \tag{9.18}$$

one can calculate the product of path length and concentration of a trace gas, i.e. the column density, by measuring $I(\lambda)$ and $I_0(\lambda)$ and using the known absorption cross-section of the gas $\sigma(\lambda)$.

$$c \cdot L = -\frac{1}{\sigma(\lambda)} \ln\left(\frac{I(\lambda)}{I_0(\lambda)}\right) \tag{9.19}$$

Most applications of scattered light DOAS follow this approach by using the measurement in the zenith or at a low elevation angle as $I(\lambda)$ and, as described above, a spectrum with small SZA or different elevation angle for $I_0(\lambda)$. In this case, the fitting of one or more absorption cross-sections modified to the instrument resolution will yield the SCD, or more precisely the DSCD. It is also clear that the SCD is independent of the wavelength. This appears to be trivial here. However, we will see below that this may not be the case for certain scattered light measurements.

In this section we argue that this classical DOAS approach can lead to problems in the analysis of scattered light applications. To simplify the discussion, we assume that $I_0(\lambda)$ is the solar scattered light without the presence of an absorber, as assumed in Sect. 9.2. The discussion can easily be expanded to the case where a different solar reference spectrum is used (see Sect. 9.2.5).

We will use a simplified version of (9.10), where we replace the integrals over height with integrals over all possible light paths that reach the detector. In addition, we simplify this equation further by replacing the integrals with a SCD for each path, S'. As assumed in the classical approach, the left side of the equation is the desired VCD times the AMF, the SCD determined by DOAS measurements.

$$V \cdot A = \frac{1}{\sigma(\lambda)} \ln \left[\frac{\int\limits_{all\ paths} I_S(\lambda, z) \cdot \exp\left(-\sigma(\lambda) \cdot S'\right) dz}{\int\limits_{all\ paths} I_S(\lambda, z) dz} \right] \tag{9.20}$$

The comparison with (9.10) reveals that the application of the logarithm is not as straightforward as in the case of Lambert–Beer's law. The logarithm has to be applied on the integrals over the intensities of the light passing the atmosphere on different light paths. We can now distinguish three cases to further interpret this equation:

1. In the case that SCD S' is the same for all paths, i.e. in case of direct solar measurements, the exponential function in the numerator can be moved in front of the integral and we have the classical Lambert–Beer's law. In this case, the methods described in Chap. 8, i.e. fitting of absorption cross-section, can be applied.
2. If we assume that the exponential function in the numerator of (9.20) can be approximated using $\exp(x) \approx 1 + x$, for $-\varepsilon < x < \varepsilon$, which is the case for a weak absorber, we can perform the following approximation:

$$\int_{all\ paths} I_S(\lambda)\cdot\exp\left(-\sigma(\lambda)\cdot S'\right) \approx \int_{all\ paths} I_S(\lambda)\cdot\left(1-\sigma(\lambda)\cdot S'\right) =$$
$$= \int_{all\ paths} I_S(\lambda) - \sigma(\lambda)\cdot \int_{all\ paths} I_S(\lambda)\cdot S'$$
$$= \left[\int_{all\ paths} I_S(\lambda)\right]\cdot\left(1-\sigma(\lambda)\cdot\bar{S}\right) \quad with \quad \bar{S} = \frac{\int_{all\ paths} I_S(\lambda)\cdot S'}{\int_{all\ paths} I_S(\lambda)}$$
$$\approx \left[\int_{all\ paths} I_S(\lambda)\right]\cdot\exp\left(-\sigma(\lambda)\cdot\bar{S}\right)$$

In the case of a weak absorber, i.e. typically with an optical density below 0.1, the classical approach can still be employed since the absorption cross-section $\sigma(\lambda)$ is now outside of the integral. The SCD $\bar{S}$ is the intensity-weighted average of all slant columns on different paths. One can, therefore, use the fitting of an absorption cross-section and a classical AMF for the analysis of the data.

3. In the case of a strong absorber (OD $>$ 0.1), such as ozone in the ultraviolet wavelength region, the approximation used above cannot be employed. The RT in the atmosphere cannot be separated from the trace gas absorption. The integral in the numerator of (9.20) now becomes dependent on $\sigma(\lambda)$ through absorptions along each path, as well as the weighing of each path during the integration, i.e. paths with stronger absorptions have a smaller intensity and thus contribute less to the integral than paths that have weaker absorptions.

Richter (1997) investigated this effect and found that the classical approach introduces small VCD errors of ~2% for ozone absorptions in the ultraviolet for SZAs below 90° in ZSL applications. However, the VCD error can reach 15% if the classical DOAS approach is used for SZAs above 90° in this wavelength region. For ozone in the visible and NO_2, the error generally remains below 2% for all SZA. Richter (1997) also showed that in the UV above 90° SZA, the residual of the fit increases due to this effect.

A number of methods to correct for this problem have been proposed in the literature, primarily for ozone in the UV or trace species that absorb in this wavelength range, such as BrO and SO_2. Coldewey-Egbers et al. (2004, 2005) introduced the weighting function WF-DOAS method in which a wavelength dependent weighting function is fitted instead of the absorption cross-section. The weighting function describes the relative intensity change caused by changing atmospheric concentrations. Similarly, Suleiman et al. (2019) multiply their BrO absorption cross-section with a wavelength dependent AMF prior to the spectral retrieval to take into account the impact of changing ozone absorptions in their fitting window. A different approach was adopted by Puķīte et al. 2010 who apply a Taylor expansions on the wavelength dependent AMF or SZA and include the first terms of the Taylor series in their DOAS fit. An additional problem in the use of the DOAS approach for solar measurements is the temperature dependence of absorption sections. Because the light reaching the detector crosses the entire atmosphere, absorption occurs at different temperatures found at different altitudes. Equation (9.20), therefore, needs to be expanded by introducing a temperature dependent $\sigma(\lambda, T)$. The light path through the atmosphere, i.e. the RT, now plays an important role since regions with different temperatures are weighted differently. Most RT models today consider this effect.

9.4 AMFs for Scattered Light Ground-Based DOAS Measurements

The analysis of scattered light DOAS observations relies on the principles outlined earlier. However, the different observational setups require analysis strategies adapted to the particular geometry of each application. Central problems include the dependence of the radiation transport, and thus the AMFs, on the a-priori unknown amount of Mie scattering and the location of trace gases in the atmosphere. Here we discuss how AMFs depend on various atmospheric parameters, such as the solar position, the trace gas profile, etc., and how this impacts the analysis of DOAS observations.

9.4.1 ZSL-DOAS Measurements

Zenith scattered light applications have been and still are widely used to study stratospheric chemistry. The AMF can vary widely as a function of wavelength, trace gas absorption, vertical trace gas profile, and stratospheric aerosol loading (e.g. Solomon et al. 1987; Perliski and Solomon 1993; Fiedler et al. 1994). On the other hand, tropospheric aerosol and clouds are of comparatively little influence, thus making ground-based measurements by this technique possible when the SZA is smaller than 95°.

Dependence on SZA

For SZAs smaller than 75°, the AMF can be approximated by $1/\cos\vartheta$. At larger SZA, a RT model yields results as shown in Fig. 9.12 for ozone. The AMF increases continuously, reaching a value of 6–20 at 90° SZA, depending on the wavelength.

A small dip around 93° is caused by the most likely scattering height passing above the altitude of the stratospheric absorption layer, leading to a reduction of the effective path length. Wavelengths above 550 nm do not show this effect since the influence of scattering and absorption processes in the atmosphere decrease as the wavelength increases.

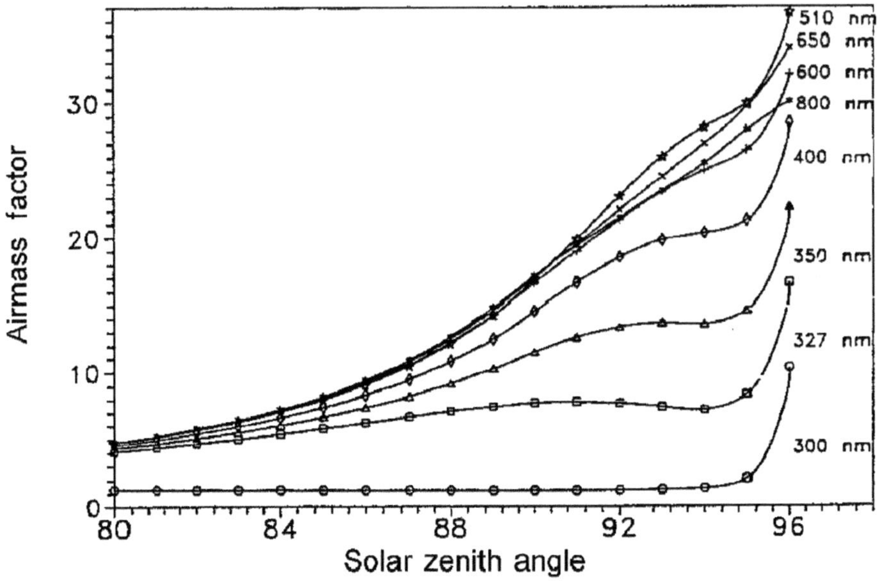

Fig. 9.12 Examples of Zenith Scattered Light (ZSL) airmass factors (AMFs) for stratospheric ozone and different wavelength as a function of solar zenith angles (in degrees) Reprinted from Frank, Diploma Thesis, 1991, University of Heidelberg

Dependence on Solar Azimuth

The dependence of the AMF on the solar azimuth is generally small for ZSL appications. The only exception is when the stratospheric trace gas or aerosol is not homogeneously distributed, and the changing effective viewing direction causes a change in the parameters influencing the RT.

Dependence on Wavelength

The ZSL-AMF depends on wavelength in the same way as Rayleigh and Mie scattering, as well as certain trace gas absorptions, are wavelength dependent. Figure 9.13 shows the dependence of the ozone AMF on the wavelength. The most notable feature is a steep increase of the AMF at lower wavelengths. This is predominantely caused by the wavelength dependence of Rayleigh scattering, which influences the weighing of the absorption layer.

A minimum around 570 nm is caused by the Chappuis ozone absorption band, which increases in strength at increasing SZA.

Dependence on Trace Gas Profile

The AMF also depends on the vertical profile of the respective trace gas. As illustrated in Fig. 9.4, the light collected at the ground is weighted towards a distribution around the most probable scattering angle. A trace gas profile with a maximum at this altitude will lead to a larger AMF than a profile with a maximum above or below the most probable scattering altitude because the maximum of the profile is more

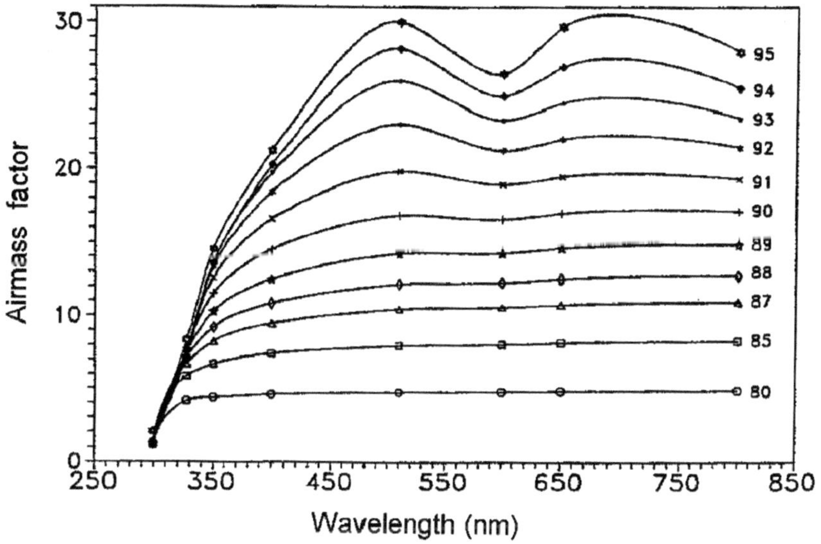

Fig. 9.13 Examples of Zenith Scattered Light (ZSL) airmass factors (AMFs) for stratospheric ozone and different solar zenith angles (in degrees) as a function of wavelength Reprinted from Frank, Diploma Thesis, 1991, University of Heidelberg

heavily weighted. An extreme example for the dependence of the vertical profile is a comparison of the ZSL AMF of a trace gas located in the troposphere and the stratosphere. Figure 9.6 shows that the stratospheric AMF, for example for NO_2 around 445 nm, increases from 6 at 80° SZA to 20 at 90° SZA. In contrast, the tropospheric AMF is much smaller, with a value of 2 at 80° SZA and ~1 at 90° SZA. The much smaller values of the tropospheric AMF are due to the relatively small amount of light being scattered in the troposphere and the short path on which light scattered in the stratosphere passes through the troposphere. Above 90° SZA, less and less light is scattered in the troposphere, and the tropospheric AMF approaches 1 as stratospheric light passes the troposphere vertically.

Dependence on the Aerosol Profile

Aerosol particles influence the RT since they efficiently scatter solar light into the receiving instrument. The impact on the AMF depends on the vertical distribution of aerosol and its scattering coefficient. For example, a stratospheric aerosol layer located below the maximum concentration of a stratospheric trace gas will lead to a reduction of the AMF since the most probable scattering altitude is shifted downwards. On the other hand, the AMF can, theoretically, be increased if an aerosol layer and a trace gas layer are collocated. The most prominent example of the impact of aerosol on AMF occured during the 1992 volcanic eruption of Mount Pinatubo. Since the volcanic aerosol was located below the ozone layer, the AMF were changed up to 40% relative to the pre-eruption case (e.g. Dahlback et al. 1994).

Tropospheric aerosol has little influence on the ZSL-AMF of a stratospheric trace gas, in particular at high SZA, since most of the scattering events occur in the stratosphere. Similarly, tropospheric clouds have little influence.

Chemical Enhancement

An additional problem in the AMF determination is found when measuring photoreactive species that change their concentration according to solar radiation (e.g. Roscoe and Pyle 1987). Examples of such species are NO_2 and BrO. Because their concentration and vertical profile change during sunrise and sunset, the temporal change in these parameters have to be considered when calculating the AMF dependence on the SZA. This is typically achieved by using correction parameters derived from a photchemical model of stratospheric chemistry.

Accuracy of AMF Determinations

The accuracy of the AMF calculations directly affects the accuracy of the VCDs derived by ZSL–DOAS instruments. Much effort has thus gone into comparing RT models (Sarkissian et al. 1995; Hendrick et al. 2004, 2006).

Hendrick et al. (2006), for example, compared six RT models to determine the systematic difference between different solutions to the ZSL-DOAS retrieval. The models included single-scattering models, multiple-scattering models based on DISORT, CDIPI, or similar analytical approaches, and one Monte Carlo model. All models were constrained with the same boundary conditions, which included time-dependent profiles of the photoreactive trace gases BrO, NO_2, and OClO to describe chemical enhancement. Figure 9.14 shows a comparison of the models run in single scattering (SS) and multiple scattering (MS) mode. Note that the figure shows the SCD calculated by the models rather than the AMF. For BrO and OClO, the different models agreed better than $\pm 5\%$. The agreement for NO_2 is $\pm 2\%$ for all except one model. There is a systematic difference between SS and MS models, in particular for OClO, for which the altitude of the aerosol is similar to that of the OClO layer. These results show the typical systematic uncertainty of current RT models for ZSL-DOAS interpretation. It should be noted that this intercomparison does not take into account uncertainties introduced by the errors in the aerosol profile, trace gas profile, and chemical enhancement used for the retrieval of real ZSL-DOAS observations.

9.4.2 MAX-DOAS Measurements

The main difference between ZSL–DOAS and 'off-axis' or 'multi-axis'-DOAS is a viewing elevation angle different from 90°. This leads to an increased effective path length in the troposphere and relatively small changes in stratospheric trace gas AMFs as compared to the zenith viewing geometry. Consequently, tropospheric absorbers are more heavily weighed in low elevation observations. This property is one of the main motivations to use low elevation viewing angles. However, it also increases the number of parameters that have to be considered in the RT calculations.

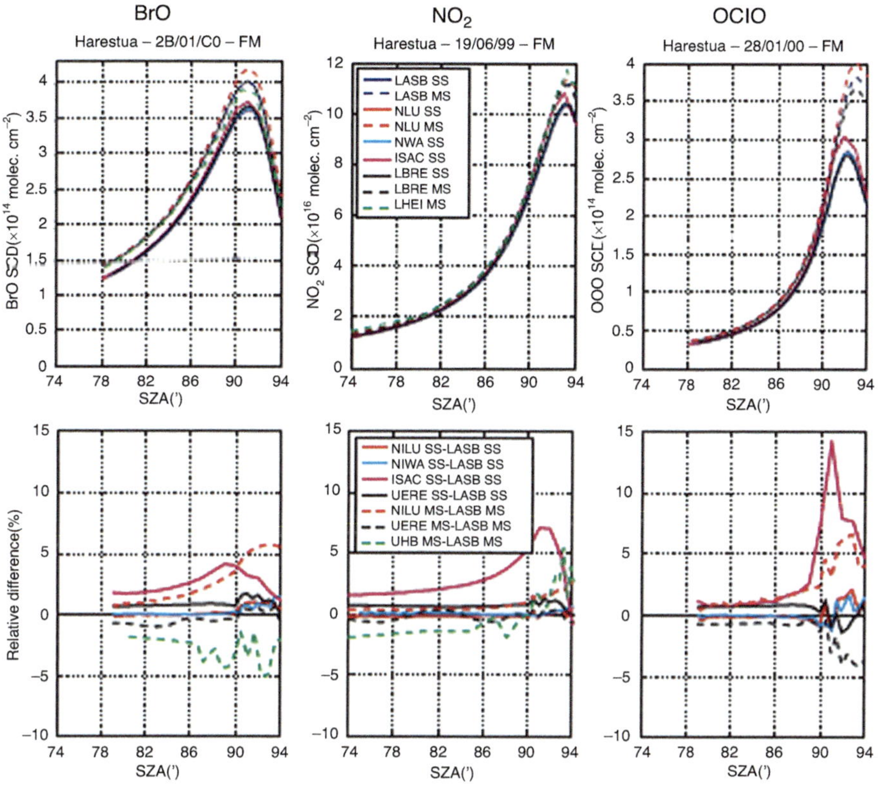

Fig. 9.14 Comparison of stratospheric and tropospheric SCDs of NO_2 at 445 nm from various RT models. Reprinted with permission from Hendrick et al. 2006,

In particular, the vertical profiles of tropospheric trace gases and aerosol play a more important role. The RT is also more dependent on the albedo and the solar azimuth than in a ZSL case.

The behaviour of the AMF under various conditions has been discussed by Hönninger et al. (2003) using the Monte Carlo radiative transfer model "Tracy" (v. Friedeburg 2003), which includes multiple Rayleigh and Mie scattering, the effect of surface albedo, refraction, and full spherical geometry. To investigate the dependence of the AMF on the vertical distribution of an absorbing trace gas, Hönninger et al. (2003) considered a number of artificial profiles (Fig. 9.15). These were used together with a number of different aerosol extinction profiles and phase functions (Fig. 9.15). Calculations were performed at a wavelength of $\lambda = 352$ nm, and a standard atmospheric scenario for temperature, pressure, and ozone. The vertical grid size in the horizontal was 100 m in the lowest 3 km of the atmosphere, 500 m between 3 and 5 km, and 1 km from 5 km up to the top of the model atmosphere at 70 km.

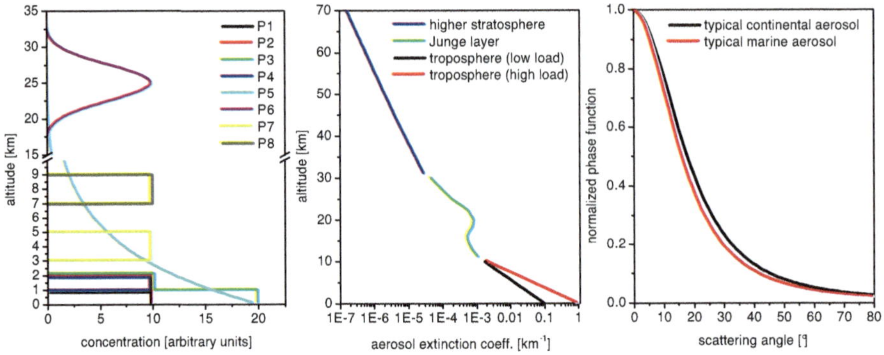

Fig. 9.15 Profile shapes of an atmospheric absorber (**left**), atmospheric aerosol (**middle**) and the aerosol scattering phase functions (**right**) used for MAX-DOAS radiation transport studies. The profiles P1–P4 assume a constant trace gas concentration in the 0–1 km and 0–2 km layers of the atmosphere. Profile P5 is that of the oxygen dimer, O_4. P5 is a purely a stratospheric profile centred at 25 km with a FWHM of 10 km. Reprinted with permission from Hönninger et al. 2004a, © CC 2.5. All rights reserved

SZA Dependence of the AMF/Stratospheric AMF

The change of the AMF with SZA depends strongly on the vertical distribution of the trace gas. As in the ZSL geometry, the AMF for a stratospheric trace gas depends strongly on the SZA, in particular at large SZA (left panel in Fig. 9.16). The dependence of tropospheric AMF is much smaller and only becomes significant above a SZA of ~75° (middle panel in Fig. 9.16). Above 75°, a small dependence on SZA can be observed. For trace gases that are present both in the troposphere and the stratosphere the total AMF will be a mixture of the terms.

The functional dependence of the AMF on the SZA can be best understood by analysing the altitude of the first and last scattering events between the sun and the detector (Fig. 9.17). In the model atmosphere investigated by Hönninger et al. (2003), the first scattering altitude (FSA) for $\alpha = 2°$ and SZA $< 75°$ is approximately 6 km, while the last scattering altitude, i.e. the altitude of the last scattering event before a photon reaches the MAX-DOAS instrument, is ~0.6 km.

At larger SZA, the FSA slowly moves upwards in the atmosphere into the stratosphere. This is in agreement with the concept of an upward-moving most probable light path as the sun sets (see Sect. 9.2.2), and explains the SZA dependence of the stratospheric AMF. The LSA is largely independent on the SZA. Therefore, the tropospheric AMF changes little.

Dependence of AMF on Viewing Elevation

The dependence of the AMF on the viewing elevation angle, α, is strongly influenced by the vertical profile of the trace gas. Stratospheric AMFs show little dependence on α (left panel in Fig. 9.16) at low SZA because the first scattering event occurs below the stratosphere, and the light path in the stratosphere is approximately geometric, i.e. proportional to $1/\cos(\vartheta)$. Only at larger SZA does the stratospheric AMF show a weak dependence on α.

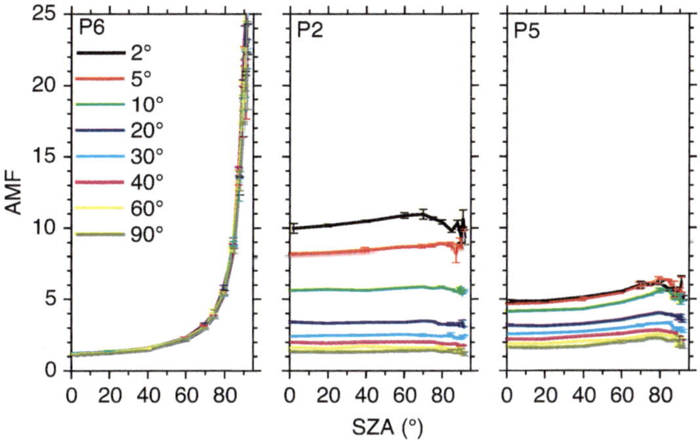

Fig. 9.16 SZA-dependence of the AMF for a typical stratospheric profile (P6) and a boundary layer profile (P2) as well as the O_4 profile (P5) for comparison (for description of profiles see caption of Fig. 9.15 and Hönninger et al. 2004a). The expected strong SZA dependence is observed for the stratospheric absorber, with no significant dependence on the viewing elevation angle listed in the figure legend. In contrast, for the tropospheric profiles P2 and P5 significant differences for the various viewing elevation angle can be seen, while the SZA dependence is significant only at higher SZA. Reprinted with permission from Hönninger et al. 2004a,

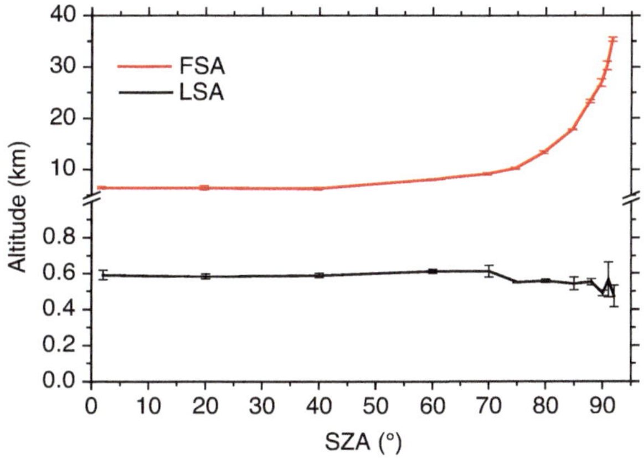

Fig. 9.17 Average altitude of first and last scattering event (FSA, LSA) for observation at 2° elevation angle. The altitude of the FSA strongly depends on the SZA, whereas the LSA altitude is largely independent of the SZA. Note the axis break at 1 km altitude and the expanded altitude scale below. Reprinted with permission from Hönninger et al. 2004a,

For a trace gas located in the lowest kilometre of the atmosphere, the dependence of the AMF on α is strong (middle panel in Fig. 9.16). Since most of the scattering events occur above the trace gas absorptions, the dependence is close to geometric, i.e. proportional to 1/sin (α) (see also Sect. 9.2.3). Deviations from this dependence are observed at SZA larger than 75°.

For trace gases extending above 1 km or located in the free troposphere, the dependence on α is more complicated than for the lower tropospheric case. In general, one finds that the dependence on α decreases as the altitude of the trace gas increases above the last scattering altitude. The dependence nearly disappears at the altitude of the first scattering event.

The dependence of the AMF on α is the basis of MAX-DOAS. If simultaneous (or temporally close) measurements are made at different elevation angles α, there is essentially no change in ϑ and thus in the stratospheric part of the AMF. Thus, the stratospheric contribution to the total absorption can be regarded essentially a constant offset to the observed SCD.

Influence of the Trace Gas Profile Shapes

In the previous section, we indicated that the AMF depends on the vertical profile of the trace gas. Hönninger et al. (2003) calculated AMFs for six different profiles, P1–P6 in Fig. 9.18, for a pure Rayleigh scattering atmosphere, i.e. no aerosol. The dependence on the elevation angle is strongest for trace gases located close to the ground (profile P1) and decreases as the gases are located higher in the atmosphere, reaching AMFs of about 15 for very small α. A comparison with the geometric AMF shows how well 1/sin(α) approximates the AMF in this case. As the profiles extend further aloft (P2, P3, P5), the AMF decreases. Profile P5 deserves special attention since it describes the exponential decreasing concentration of atmospheric O_4. Because the O_4 levels and the vertical profile of O_4 do not change, they can be used to validate RT calculations (see Sect. 9.4.4).

For elevated trace gas layers, i.e. profile P4, the AMF peaks at $\alpha = 5$, because at lower viewing elevation angles a DOAS instrument would predominately see the air below the layer. The AMF is almost independent of the viewing direction for the stratospheric profile P6.

Dependence on Surface Albedo

Figure 9.19 shows that tropospheric AMFs (P1 – P5) are also influenced by the surface albedo. In general, higher albedos lead to larger AMF, because the reflection at the ground and the upwelling radiation will increase the effective absorption path length in the troposphere. Light that enters a DOAS instrument after being reflected at the ground may have passed parts through an absorption layer in the lower troposphere twice, thus increasing the trace gas absorption as it enters the detector. Since the effect of higher albedo increases all AMFs similarly, it has no significant effect on the elevation angle dependencies.

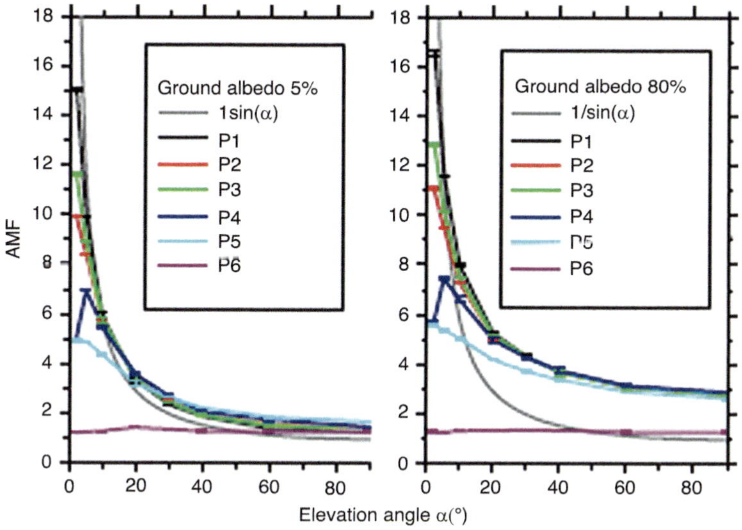

Fig. 9.18 AMF dependence on the viewing direction (elevation angle α) for the profiles P1–P6 (for description of profiles see caption of Fig. 9.15) calculated for 5% ground albedo (**left**) and 80%, albedo (**right**), respectively. A pure Rayleigh case was assumed. Reprinted with permission from Hönninger et al. 2004a, © CC 2.5.

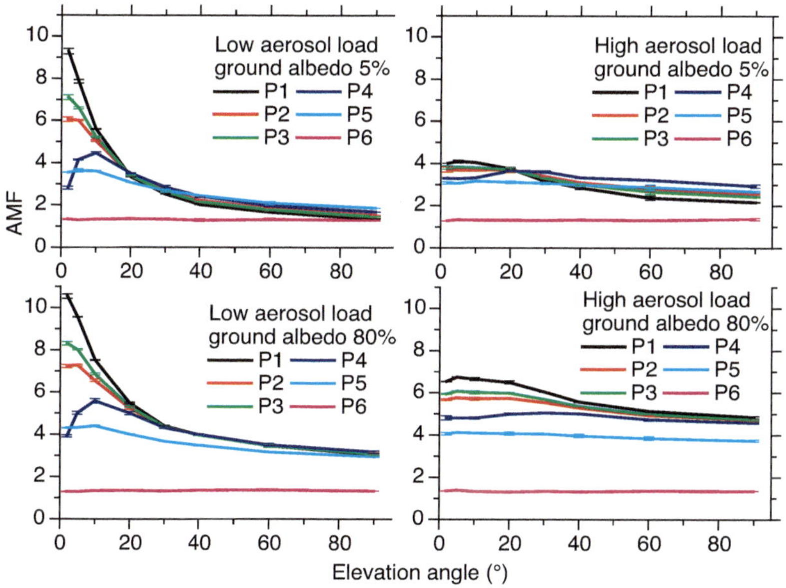

Fig. 9.19 AMF as a function of viewing direction, i.e. elevation angle α for the profiles P1–P6 (see caption of Fig. 9.15 and Hönninger et al. 2004a) calculated for low (**left column**) and high (**right column**) aerosol load (Fig. 9.15) and for 5% (**top**) and 80% (**bottom**) ground albedo, respectively. Reprinted with permission from Hönninger et al. 2004a, © CC 2.5.

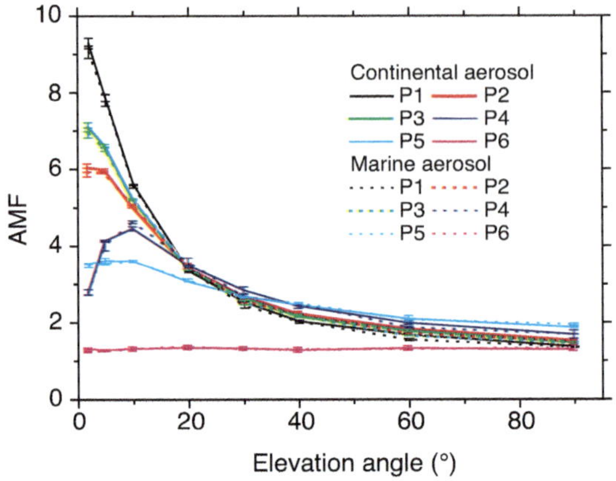

Fig. 9.20 The effect of different aerosol types and different profiles (see caption of Fig. 9.15 and Hönninger et al. 2004a) on the AMF. Only small differences result from different scattering phase functions, Reprinted with permission from Hönninger et al. 2004a, © CC 2.5. All rights reserved

Dependence on Aerosol Profile

The presence of aerosol particles in the atmosphere significantly enhances the scattering of radiation. Figure 9.19 illustrates how the AMFs for different viewing directions change for two different aerosol scenarios (high and low tropospheric aerosol load, Fig. 9.19) as well as for large and small surface albedo (see above). The AMF for a low aerosol load shows a similar shape to the Rayleigh case (Fig. 9.18). However, the absolute values of the AMFs at elevation angles below 20° are considerably reduced as compared to the Rayleigh case. Consequently, the geometric approximation is not good at low elevation angles, even at low aerosol loads. At high tropospheric aerosol loads, the absolute AMFs are further reduced and the dependence on α becomes much weaker.

The comparison of Rayleigh, low aerosol, and high aerosol cases also show how the dependence of the AMF on α changes with aerosol load. For trace gases in the lower part of the troposphere (P1–P3), the difference between AMF at low and high elevations slowly decreases. An extreme case of this dependence is dense fog, for which the α dependence of the AMF completely disappears. The high aerosol load and high albedo case in Fig. 9.19 also shows that, in this case, the AMF at very small α, i.e. 2°, can indeed be smaller than those at slightly larger α, e.g. 5°. This effect can be explained by the reduction of effective light path length at low elevation angles under these conditions (see below). The shift of the AMF maximum to larger α is even more pronounced for an elevated trace gas layer (P4), for which the maximum is at $\alpha = 5°$ for the Rayleigh case, $\alpha = 10°$ for a low aerosol load, and $\alpha = 40°$ for a high aerosol load.

The optical properties of aerosol, i.e. the scattering phase function (Fig. 9.15) and the single scattering albedo, influence the AMF little, as shown by the comparison of a continental and marine aerosol in Fig. 9.20.

The fact that the tropospheric AMF decreases with increasing tropospheric aerosol load can be explained by a shorter mean free path of photons due to aerosol extinction.

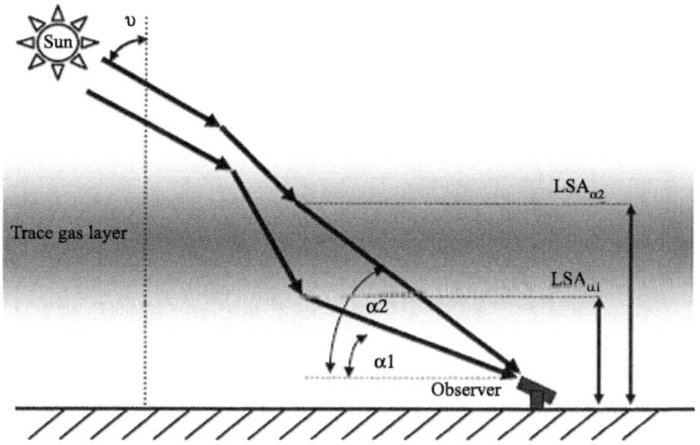

Fig. 9.21 The last scattering altitude (LSA): for low elevation angles, the mean free path in the viewing direction is shorter due to higher density and/or aerosol load. This can result in the slant path through absorbing layers at higher levels being shorter for lower elevation angles than for higher ones. Reprinted with permission from Hönninger et al. 2004a, © CC 2.5. All rights reserved

In particular, at low α, the light path in the lowest atmospheric layers is shorter, and a less slanted path in the higher layers is probable as illustrated in Fig. 9.21.

The last scattering altitude (LSA), is a key parameter to understand the sensitivity of MAX-DOAS measurements towards different vertical profiles of gases and aerosol. The LSA dependence on the viewing elevation angle for different aerosol profiles is shown in Fig. 9.22. The LSA moves downward in the atmosphere as α decreases. In addition, the LSA decreases for increasing aerosol load and increasing albedo. In general, the geometric approximation for the AMF can only hold for trace gases below the LSA. Once the LSA reaches the trace gas layer, the AMF begins to decrease. Due to the α dependence of the LSA, this decrease occurs first at small α, leading to smaller AMF at very low elevation angles.

Dependence of the AMF on the Solar Azimuth Angle

In contrast to the ZSL measurements, low elevation angle measurements can show a dependence of the AMF on the solar azimuth angle. In general, this dependence is determined by the relative azimuth angle between the sun and the viewing direction. The effect arises due to the shape of scattering phase functions for the atmospheric scattering processes. Light paths taken by photons at different relative azimuth angles may be different, thus impacting the AMF. Figure 9.23 shows that, overall, the relative azimuth angle, here calculated for a SZA of 30° and the profile P4, has only a small influence on the AMF and its α dependence. However, this effect is found to increase with ground albedo and tropospheric aerosol load.

9.4.3 Box Air Mass Factors (BAMFs)

To understand the sensitivity of MAX-DOAS measurements and RT calculations on the vertical distribution, the concept of the Box Air Mass Factor was introduced.

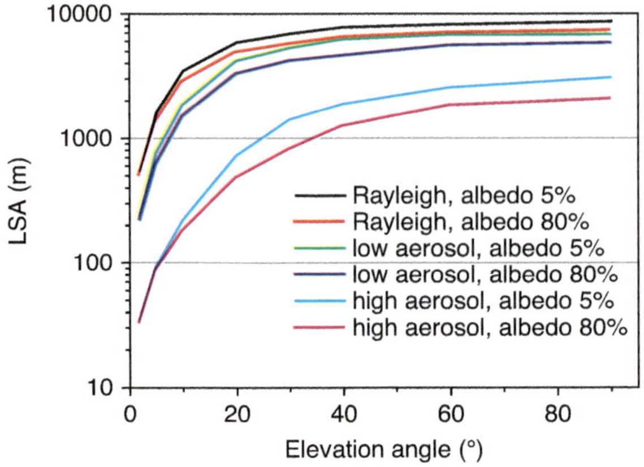

Fig. 9.22 Last scattering altitude (LSA) for pure Rayleigh, low and high aerosol load scenarios and both 5% and 80% albedo. The LSA is generally below 1 km for the lowest elevation angle and above 1 km for the highest elevation angles, especially for zenith viewing direction. Reprinted with permission from Hönninger et al. 2004a, © CC 2.5. All rights reserved

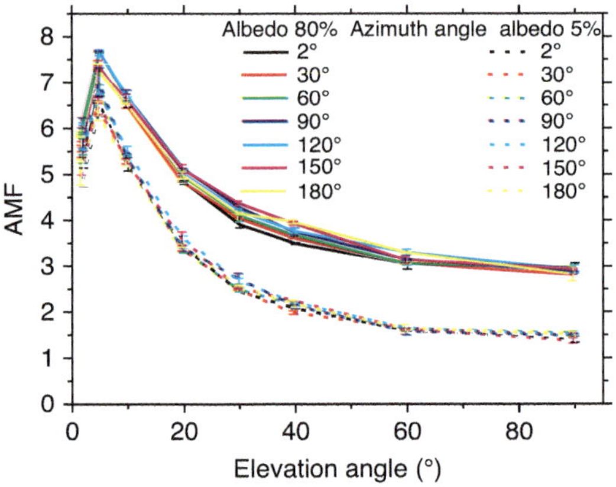

Fig. 9.23 Azimuth-dependence of the AMF for 30° SZA and profile P4 (trace gas layer at 1–2 km altitude). For relative azimuth angles between 2° (looking almost towards the sun) and 180° (looking away from the sun) only a small effect can be seen. Reprinted with permission from Hönninger et al. 2004a, © CC 2.5. All rights reserved

The Box-AMF represents the contribution of a trace gas located at a certain altitude interval, or vertical 'box', to the overall AMF under given atmospheric conditions. It is thus a measure of the sensitivity of a particular viewing direction towards an absorber being present in a specific altitude interval. The SCD of a trace gas, and thus

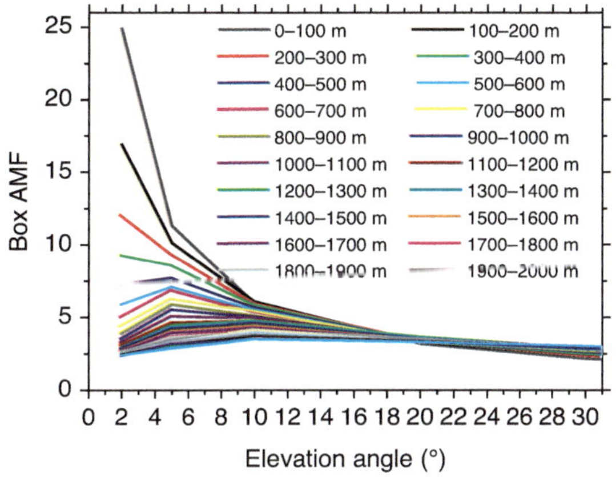

Fig. 9.24 Box AMF for the model layers below 2 km altitude. The sensitivity towards the lowest layers changes strongly for elevation angles below 10°, for elevation angles smaller than 5° the sensitivity already decreases for layers above 400 m altitude. Reprinted with permission from Hönninger et al. 2004a,

indirectly the AMF, at a given elevation viewing angle, α_i, is related to the Box-AMF through the following equation:

$$S(\alpha_i) = \sum_{j=1}^{m} \left[(A_{Box})_{ij} \cdot \Delta h_j \cdot c_j \right] \tag{9.21}$$

In this equation, Δh_j is the height of altitude interval j, c_j the concentration, and $(A_{\mathrm{Box}})_{ij}$ the Box-AMF in this interval for elevation viewing angle, α_{i}.
Figure 9.24 shows modelled Box-AMFs for trace gas layers of $\Delta h_j = 100$ m thickness from the ground up to 2 km altitude. The Box-AMF is largest near the surface for very low elevation angles. For α smaller than 10°, the Box-AMFs vary strongly with altitude. Consequently, MAX-DOAS observations at small α are very sensitive to the shape of the vertical profile in the lower troposphere. This sensitivity can be used to retrieve trace gas profiles from MAX-DOAS observations (see e.g. Frieß et al. 2019). By measuring trace gas SCDs at different elevation angles below 20°, one can set up a linear equation system based on (9.21), which can then be solved for the concentrations in each layer.

9.4.4 Accuracy of MAX-DOAS AMF Calculations

Determining the accuracy of RT calculations of AMFs is a challenging task. As in the ZSL case, intercomparisons of different RT models allow the estimate of the uncertainties introduced by numerical calculations in the models. In addition, MAX-DOAS measurements of tropospheric trace gases offer another opportunity to constrain or validate RT calculations: The use of observations of stable atmospheric gases such as O_2 and O_4. Both of these methods will be discussed in this section.

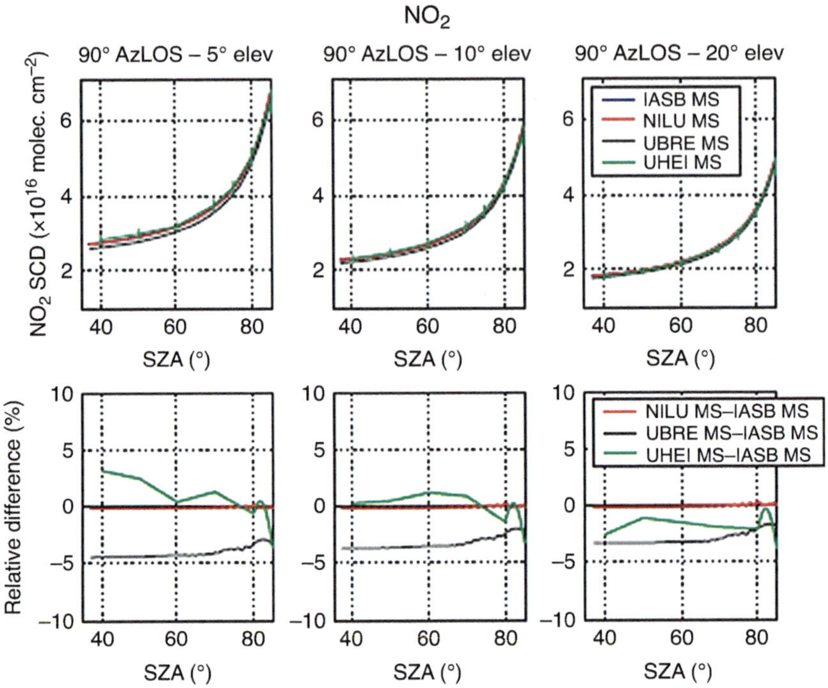

Fig. 9.25 Intercomparison of low viewing elevation slant column densities for NO_2 calculated by four different multiple scattering models. Reprinted with permission from Hendrick et al. 2006, © CC 2.5. All rights reserved from Hendrick et al. (2006)

Model Intercomparison/Model Accuracy

To support the increasing number of MAX-DOAS observations, AMF calculations for low viewing elevations have been implemented in all RT models. Successful models include spherical geometry, refraction, and multiple-scattering. Figure 9.25 illustrates an intercomparison of four RT models, which were previously described in Sect. 9.4.1 (Hendrick et al. 2006). All models used the same boundary conditions, i.e. trace gas and aerosol profiles. Comparisons were made for viewing elevation angles of 5°, 10°, and 20°. In general, the agreement between the models was better than ±5% for the two trace gases investigated: NO_2 (Fig. 9.25) and HCHO (not shown). Hendrick et al. (2006) conclude that the largest discrepancies between the models are caused by the different treatments of aerosol scattering in each model.

The good agreement between RT models was also confirmed by Wagner et al. (2007b), who reports on a RT model intercomparison exercise for radiances and BAMFs for MAX-DOAS applications. In most cases BAMFs agreed to within 5%. Larger differences in cases with low optical depth were attributed to inclusion of the Earth's sphericity in some models while others only used a plane parallel atmosphere. Based on past studies, one can conclude that most modern RT models agree to better than 5% when initialised with the same atmospheric and aerosol parameters (Hendrick et al. 2006; Wagner et al. 2007b).

Validation of RT Calculations with O_2 and O_4 Measurements

One of the most challenging aspects of RT calculations is to determine the accuracy of AMFs for atmospheric measurements. The best way of achieving such validations is through the use of the known absorbers, such as O_2 or the oxygen dimer, O_4 (Wagner et al. 2019; Beirle et al. 2022). The atmospheric O_2 and O_4 concentration profiles can be accurately determined based on knowledge of temperature and pressure profiles, i.e. the O_4 profile $c_{O4}(z) \propto (0.21 \cdot c_{air}(z))^2$ (Wagner et al. 2021a). The observed O_2 and O_4 SCDs are thus solely a function of the radiative transfer in the atmosphere, which is impacted by various atmospheric parameters, such as aerosol profiles, aerosol optical properties, etc. (see Sect. 9.3.3). A comparison of observed O_2 and O_4 SCDs with RT model calculations initialised with knowledge of the atmospheric parameters will allow the determination of the uncertainty of the RT model calculations. However, it should be noted that highly accuracy O_4 comparisons have been elusive, i.e. even under condition where input parameters are well known, observed, and modeled, O_4 columns often do not agree (Wagner et al. 2019; Beirle et al. 2022). While the validation of RT models is a worthwhile undertaking, the use of observation/model comparisons of O_4 SCDs is more valuable by using them to retrieve atmospheric parameters, such as aerosol extinction profiles (see Sect. 9.5, below).

9.5 MAX-DOAS Vertical Profile Retrievals

The development of MAX-DOAS has opened new ways of using DOAS to study the vertical distribution of aerosol and trace gases in the troposphere. As mentioned in the previous section, the RT, and therefore the interpretation of observed DSCD of tropospheric gases, depends strongly on aerosol scattering, or more precisely on the aerosol extinction profile, as well as the vertical distribution of the measured trace gases. Here will briefly review how MAX-DOAS observations of O_4 and trace gas DSCDs at various elevation angles can be used to retrieve aerosol and trace gas profiles.

It was recognised relatively early that O_4 slant column densities and intensities provide valuable information on the vertical distribution of aerosols (Wagner et al. 2004; Friess et al. 2006). The underlying idea is that O_4 has a well known vertical profile and that variations of O_4 are controlled by the atmospheric RT and the viewing direction. Similarly, the intensity at a given direction in the sky is determined by the solar illumination and the viewing geometry, well-known Rayleigh scattering, as well as by the spatial distribution and optical properties of the aerosol. If the aerosol extinction profile is known, a set of trace gas DSCDs at different elevation angles can be used to retrieve the vertical concentration profile of the respective gas using a framework similar to that introduced in Sect. 9.4.3 for box air mass factors.

Aerosol and trace gas profiles are most often retrieved sequentially (e.g. Sinreich et al. 2005; Bösch et al. 2018). In the first step, O_4 DSCD observations at a range of viewing elevation angles are compared to RT model calculations, varying the aerosol extinction profile until a best match of modeled and observed DSCDs is found. In the

second step the retrieved extinction profile is used to derive concentrations profiles based on the trace gas DSCDs in the measurement sequence.
While the general principle of the first step is straightforward, the realisation of such an optimisation is difficult due to the underlying non-linearity of the problem and the fact that only limited information, i.e. a limited number O_4 DSCDs at selected elevation angles, is available. Additional information thus has to be included in the retrieval.
The problem of under-constrained retrievals of atmospheric parameters is well-known in the remote sensing community and we refer the reader to extensive literature describing the underlying mathematical principles (for example Rodgers, 2000). In short, these methods (known as Optimal Estimation Methods, OEM) invert the the following equation describing the observations $\mathbf{y}$ as a function of the atmospheric parameters $\mathbf{x}$

$$\mathbf{y} = \mathbf{F}(\mathbf{x}, \mathbf{b}) + \boldsymbol{\epsilon} \tag{9.22}$$

For MAX-DOAS $\mathbf{y}$ is a vector of the measured O_4 DSCDs, and $\mathbf{x}$ is a vector of state variables that should be retrieved. For aerosol retrievals, this is most commonly the aerosol extinction profile. Vector $\mathbf{b}$ contains other variables describing the atmosphere and aerosol, such as temperature, pressure profiles, aerosol optical properties, albedo, etc. $\mathbf{F}(\mathbf{x}, \mathbf{b})$ is the forward model that describes how the state variables $\mathbf{x}$ and $\mathbf{b}$ lead to the observations. For MAX-DOAS retrievals the RT models described in Sect. 9.3 or look-up tables from RT runs typically serve as the forward model. $\boldsymbol{\epsilon}$ describes measurement noise and forward model errors.
The problem tackled by OEM is to invert this equation and derive the best estimate of the state vector $\hat{\mathbf{x}}$ from the data $\mathbf{y}$. This is achieved through minimizing a cost function:

$$\chi^2 = (\mathbf{y} - \mathbf{F}(\mathbf{x}, \mathbf{b}))^T \mathbf{S}_\epsilon^{-1} (\mathbf{y} - \mathbf{F}(\mathbf{x}, \mathbf{b}) + (\hat{\mathbf{x}} - \mathbf{x}_a)^T \mathbf{S}_a^{-1} (\hat{\mathbf{x}} - \mathbf{x}_a) \tag{9.23}$$

Here $\mathbf{x}_a$ is the a-priori state vector that provides prior knowledge to overcome the under-constrained nature of the problem. In the case of aerosol retievals, a climatological extinction profile or an exponentially decaying profile can serve as a-priori profiles. The covariance matrix $\mathbf{S}_a$ describes the uncertainty in the knowledge of $\mathbf{x}_a$. $\mathbf{S}_\epsilon$ is the covariance matrix of $\boldsymbol{\epsilon}$. In the case where only measurement errors are considered, this is a diagonal matrix where each element is determined by square of the measurement error of $\mathbf{y}$.
The OEM algorithm minimises the cost function, often using linear least squares methods similar to those described in Chap. 8. The result of this step is $\hat{\mathbf{x}}$ and an averaging kernel $\mathbf{A} = \frac{\partial \hat{\mathbf{x}}}{\partial \mathbf{x}}$, which describes the sensitivity of the retrieved state vector $\hat{\mathbf{x}}$ as a function of the true state vector. The trace of the averaging kernel matrix quantifies the degrees of freedom, i.e. number of independent pieces of information, that have been gained by the retrieval.
There are currently two different approaches for the aerosol retrieval based on MAX-DOAS data. The first uses the full OEM approach and cost function by providing an a-priori state vector, which in most cases is a vertical extinction profile. The second

approach uses a parameterisation of the forward model, for example by describing the aerosol profile using a small number of parameters, such as a boundary layer height and aerosol optical depth. For this approach the second term in Eq. 9.23 can be omitted.
The retrieval of trace gas concentration profiles follow the general approach described for aerosol. However, the measurement vector **y** represents the trace gas DSCDs, ideally determined from the same observations as O_4, i.e. the same MAX-DOAS spectra. The forward model is an RT model constrained by the aerosol extinction profiles determined in step one, i.e. the aerosol extinction profile is part of **b**. The a-posteriori results $\hat{\mathbf{x}}$ are the trace gas concentrations.
A number of implementations and improvements of aerosol retrieval algorithms have emerged in past years (Beirle et al. 2019b; Bösch et al. 2018; Irie et al. 2011; Wagner et al. 2011). Several intercomparisons of these algorithms have been performed to assess the differences between the specific retrieval approach. Frieß et al. (2016) provide a direct comparison of some of these algorithms based on field observations. Figure 9.26 illustrates that the different algorithms can retrieve the vertical distribution of aerosol quite well on certain days. Overall, Frieß et al. (2016) found that correlation coefficients between aerosol optical thickness from a colocated sun-photometer and the MAX-DOAS retrievals are better than 0.8 for all algorithms. However, the algorithms underestimate aerosol optical thickness (AOT) by factors between 0.6 and 0.9, with mean AOT differences below 0.07. Analysis of the averaging kernel of one of the models (Fig. 9.27) shows that information on aerosol extinction can only be retrieved in the lowest 2000 m of the atmosphere, i.e. averaging kernels become very small above this altitude, and that the MAX-DOAS is most sensitive at the ground. The degrees of freedom obtained by the retrievals is in the range of 1–2.
A comparison confirms the ability of a more extensive set of retrieval algorithms to retrieve both aerosol and trace gas profiles based on synthetic data (Frieß et al. 2019). Overall the root-mean-square difference between original and retrieved data was in the range of $(0.05 - 0.1)\,\text{km}^{-1}$ for aerosol extinction and $(2.5 - 5.0) \times 10^{10}\,\text{molec.cm}^{-3}$ for concentration profiles of HCHO and NO_2.

9.6 Aircraft Observed Scattered Light (AMAX-DOAS)

Airborne platforms give high mobility to DOAS applications, allowing rapid mapping of trace gas distributions over large areas and accessing regions of the atmosphere that are otherwise difficult to reach, such as the remote arctic or the upper troposphere. There are three different observational geometries for aircraft DOAS. Early observations, which were focused on stratospheric chemistry in polar regions, were based on zenith viewing geometries (Wahner et al. 1989a, b, 1990a, b; Schiller et al. 1990; Brandjen et al. 1994; Pfeilsticker and Platt 1994, 1997; Erle et al. 1998; McElroy et al. 1999). The RT calculation for these applications are very similar to those described in Sect. 9.4.1, and will not be further discussed here.
To take advantage of the mapping capability offered by aicraft, downward (i.e. nadir)

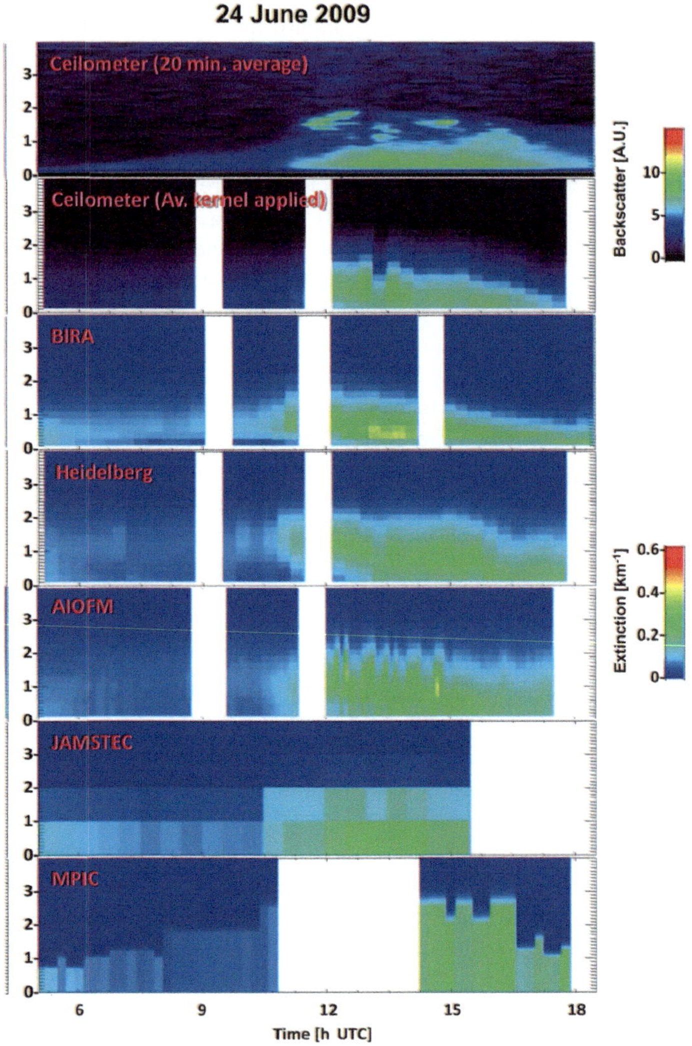

Fig. 9.26 Extinction profiles retrieved from MAX-DOAS observations using different algorithms compared to ceilometer backscatter profiles 24 June, 2009 (top panel). Backscatter signal after application of the University of Heidelberg averaging kernels is shown in the second panel. The five panels below show retrievals from BIRA (Brussels, Belgium), Univ. Heidelberg (Germany), AIOFM (Hefei, china), JAMSTEC (Yokosuka, Japan), and MPIC (Mainz, Germany). Reprinted with permission from Frieß et al. (2016), © CC 3.0. All rights reserved

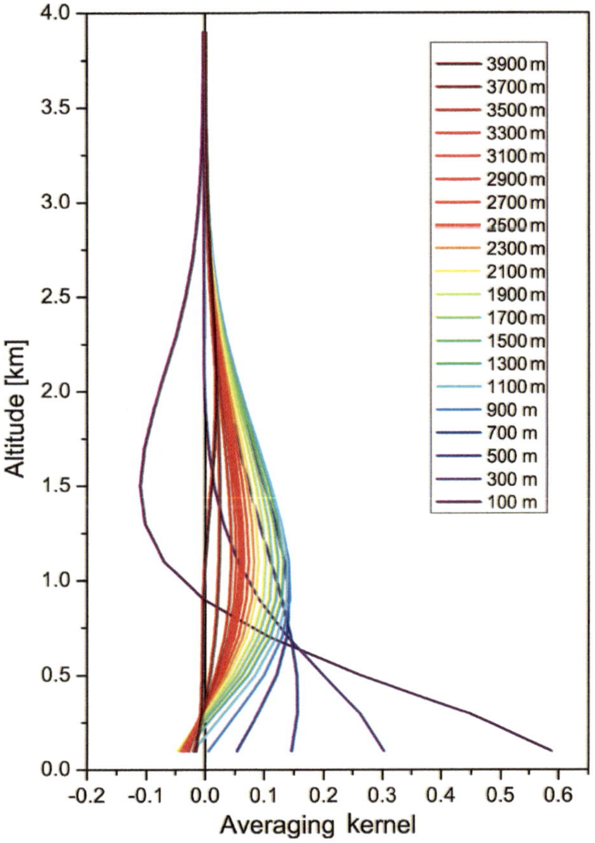

Fig. 9.27 Aerosol extinction averaging kernels from the Heidelberg retrieval algorithm for 2 July 2009, 12:00 UTC. Reprinted with permission from Frieß et al. (2016), © CC 3.0. All rights reserved

viewing DOAS systems have become one of the most common airborne DOAS applications. Here single or multiple downward viewing directions provide insight into trace gas columns below the aircraft. The focus of these observations has predominately been to investigate pollution below the aircraft, which is typically concentrated in the boundary layer. The task of RT calculations in this context is to convert the observed nadir DSCD to a VCD below the aircraft. This conversion is achieved by calculating the AMF or Box-AMFs below the aircraft using the models introduced in Sect. 9.3. The main differences between ground-based RT calculations and those from an airborne nadir DOAS system are:

- **Flight altitude, topography, and aircraft orientation**: AMFs and BAMFs below the aircraft depend on the altitude of the aircraft (see Fig. 9.28). All RT models allow calculations for altitudes other than the surface. The challenge, however, is to accurately determine the distance between aircraft and ground (see e.g. Lawrence et al. 2015). If the aircraft is equipped with a radar altimeter, this altitude data can be used when the topography changes slowly. In other cases flight altitude and surface elevation models are employed to determine the true altitude for each

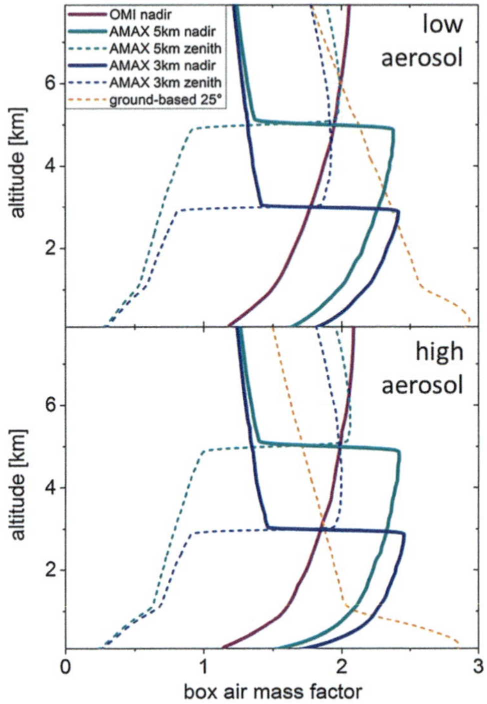

Fig. 9.28 Examples of 435 nm BAMFs for aircraft observations at 3 and 5 km altitude, AOD of 0.3 (low) and 0.5 (high) and a SZA of 17°. Also shown are 446 nm BAMFs for the OMI satellite observation geometry and for 435 nm ground based observations. It should be noted that nadir BAMFs have a non-zero contribution from above the aircraft. Reprinted with permission from Oetjen et al. (2013), © American Geophysical Union. All rights reserved

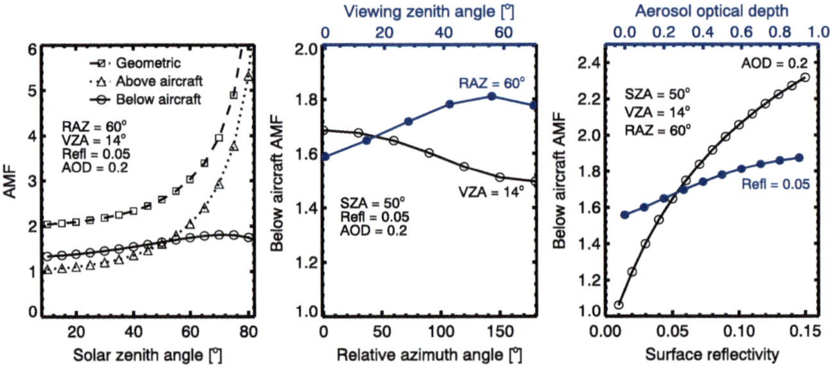

Fig. 9.29 Sensitivity aircraft AMFs at 8 km altitude. (**left**) Above and below aircraft AMFs calculated with the VLIDORT RT model as a function of SZA. Also shown are a corresponding geometric AMF. (**middle**) Dependence of below aircraft AMF on relative azimuth angle (RAZ) and viewing zenith angle (VZA). (**right**) Below aircraft AMF as a function of surface reflectivity at AOD of 0.2 (black) and versus AOD at an albedo of 0.05 (blue). Reprinted with permission from Lamsal et al. (2017), © American Geophysical Union. All rights reserved

flight segment (Lawrence et al. 2015). Aircraft orientation relative to the sun and the ground permanently changes during flights. In many cases this has a profound impact on the RT calculations (Fig. 9.29). It is therefore common to perform RT calculations for each observation during a flight using the respective aircraft parameters.

- **Albedo or BRDF**: The impact of albedo/BRDF of the ground is much larger in airborne nadir observations than in other DOAS applications (see Wang et al. 2005; Lamsal et al. 2017; Nowlan et al. 2016). Larger albedos/BRDFs increase the sensitivity, i.e. BAMFs, of the nadir observations to trace gases at the surface. For example, the right panel in Fig. 9.29 illustrates how the below aircraft AMF increases from 0.1 to 2.3 when the albedo increases from 0 to 0.15. Because trace gas levels are often highest near the surface, an accurate knowledge of the albedo/BRDF is therefore important. In situations with inhomogeneous surfaces, such as urban areas, this poses a considerable challenge for RT calculations. The AMF varies strongly over short distances as a consequence of albedo variations (Fig. 9.30). Many DOAS observations do not directly provide albedo/BRDF. Consequently, these values are often taken from external sources, such as available satellite data bases (Lamsal et al. 2017; Nowlan et al. 2016). Alternatively, methods to determine the albedo from uncalibrated observations and RT models have been developed (see e.g. Lawrence et al. 2015). The most direct approach is to radiometrically calibrate the airborne DOAS instrument and determine surface reflectivity directly from the observations (Popp et al. 2012).
- **Aerosol and clouds**: Airborne nadir DOAS is sensitive to the presence of aerosol (Fig. 9.29, right panel) (see e.g. Wang et al. 2006; Lamsal et al. 2017). Aerosol concentrations are often highest in the boundary layer, thus directly impacting the sensitivity to the lowest part of the atmosphere, i.e. BAMFs at the surface decrease with increasing aerosol extinction (see Fig. 9.28). Knowledge of the boundary layer aerosol extinction, or retrieval of this extinction from simultaneous O_4 observations, is thus crucial. Elevated aerosol layers and clouds are another challenge that require good prior knowledge of extinction and altitude to be considered accurately.
- **Trace gas profile**: As illustrated in Fig. 9.28, BAMFs and therefore the sensitivity of nadir observations are altitude dependent. Consequently, nadir VCDs depend on the trace gas profile used in the AMF calculations. Trace gas profiles are most often derived from climatological information or from atmospheric chemistry models.

There are currently two different approaches to derive nadir viewing below aircraft AMFs. A geometric approximation calculates the AMF as $AMF_{belowaircraft} = 1 + 1/cos(\theta)$ (Oetjen et al. 2013). This approximation works well for flight altitude below 4km if the SZA $< 50°$, while more detailed RT calculations are needed for aircraft flying at higher altitudes (see Lamsal et al. 2017). Generally BAMF and sensitivity of nadir observations are highest just below the flight altitude, with BAMF values above 2. The BAMFs decrease towards the surface where, depending on the albedo, they can reach values between 0 for very low albedo and 2 for a very high albedo.

It should be noted that nadir observations are also sensitive to trace gases above the flight altitude (Fig. 9.28). An additional temporally closed zenith measurement is thus often performed. However, the AMF of the zenith observations is different than the sum of the nadir BAMFs above the aircraft, and RT calculations have to be performed for an accurate correction of the above aircraft component of nadir observations.

The third application of airborne DOAS is the use of the very long light path lengths in and around the limb of the aircraft (Dix et al. 2009; Merlaud et al. 2011; Prados-Roman et al. 2011; Baidar et al. 2013; Volkamer et al. 2015; Dix et al. 2016). Figure 9.31 shows the weighing functions, i.e. a scaled representation of the BAMFs, for different viewing angles from an aircraft at 10km altitude (Bruns et al. 2004). The near limb directions, i.e. 88° and 92° (nadir = 0°, zenith = 180°) have the highest sensitivities/BAMFs. While it is possible to use various viewing angles for retrievals around the flight altitude, the overlapping sensitivities below 5 km and above 13km in Fig. 9.31 illustrate the difficulty in the retrieval of information on the trace gas over a larger altitude range in the atmosphere (Bruns et al. 2004). It has therefore become more common to make observations in the limb and use the ability of the aircraft to change altitude to derive atmospheric profiles.

RT calculations in the limb are uniquely sensitive to the trace gas absorption in a ~2km wide altitude interval around the flight altitude. While the BAMFs in these intervals can reach 50–100, BAMFs do not drop to zero outside of this interval. Consequently, limb measurements are still, albeit slightly, sensitive to traces gases below and above of the aircraft. To overcome this sensitivity and to derive vertical trace gas profiles, most limb observations are performed during an ascent or descent of the aircraft. The measurement vector, **y**, is then a set of DSCDs at different altitudes. This measurement vector is then used in OEM retrievals (see Sect. 9.5) to derive aerosol extinction profiles using O_4 DSCDs, and for trace gas profile retrievals. An example of an aerosol extinction profile is shown in Fig. 9.32. The averaging kernels illustrate how each measurement is sensitive to the O_4 extinction in a narrow interval around the aircraft altitude. The retrieval is capable of describing the RT within an error of $\sim \pm 10\%$ (right panel, Fig. 9.32). As in the case of MAX-DOAS retrievals (Sect. 9.5), once the RT is constrained using O_4 observations, trace gas profile retrievals can be performed using the same OEM approach. For example, Fig. 9.33 shows how a limb-DOAS NO_2 profile reflects the elevated concentrations in the polluted boundary layer. This retrieval has a degree of freedom of 12.5, showing that nearly the entire profiles is due to information from the observations and not from the a-priori profile (black line in Fig. 9.33). These examples illustrate that a combination of limb observations during an aircraft ascent or descent with detailed RT calculations and an OEM inversion provides highly accurate vertical aerosol and trace gas profiles throughout the troposphere.

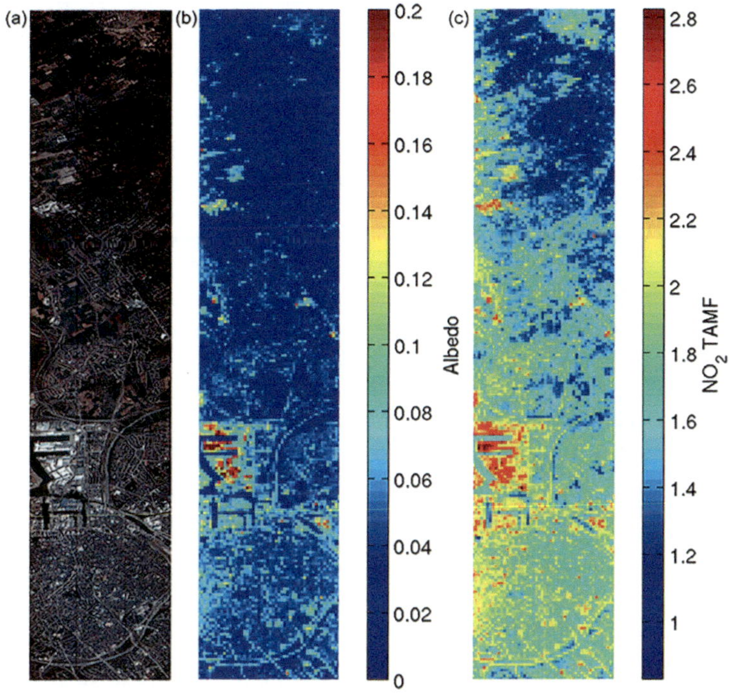

Fig. 9.30 Variation of the tropospheric AMF (TAMF) over an urban area. (**a**) true colour image of Antwerp measurement data set (15 April 2015) (**b**) measured albedo and (**c**) computed TAMFs show a strong dependene on the urban canopy and its albedo. Reprinted with permission from Tack et al. (2017),

9.7 Satellite Observed Scattered Light

Atmospheric trace gas observations from space have become increasingly popular in recent decades (see Chap. 11 for examples). Satellite DOAS instruments measure sunlight backscattered from the atmosphere and the surface (Fig. 9.34). Similar to aircraft observations, two viewing strategies have been adopted: nadir and limb. Many aspects of the RT for satellite observations are similar to those for aircraft-based observations discussed in Sect. 9.6. However, satellite observations face particular challenges that will be discussed here.

9.7.1 Radiative Transfer in Nadir Geometry

The most widely used application of satellite DOAS is mapping of trace gases across the globe using a nadir viewing strategy. It should be noted here that nadir viewing strategies typically include off-nadir viewing directions perpendicular to the satellites flight direction that allow the observation of a broad swath of the earths atmosphere. The specific viewing elevation angle is normally considered in nadir RT calculations and will not be discussed in detail below.

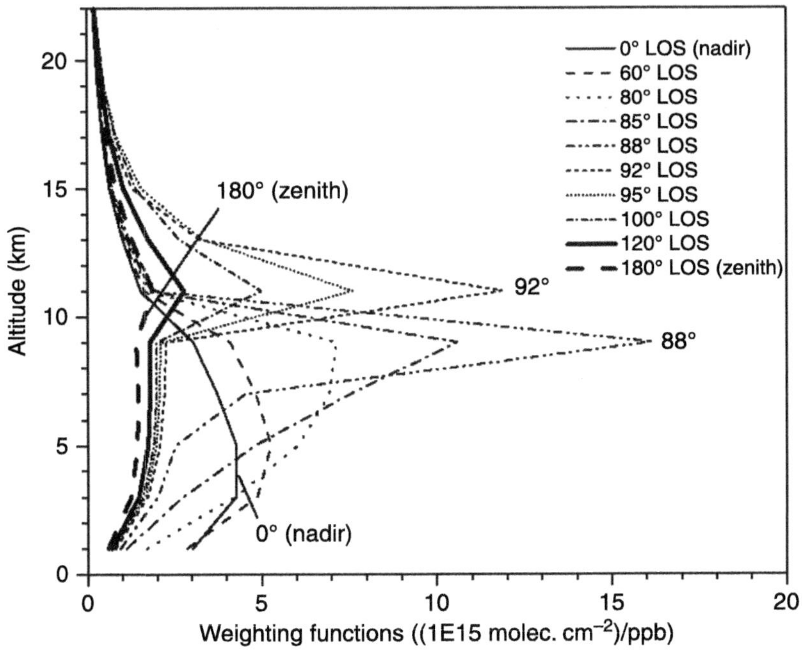

Fig. 9.31 Calculated weighting functions (Box AMFs) for a flight altitude of 10 km and NO_2 profiles anticipated for mid latitudes on the northern hemisphere spring at 51.6° SZA. Each weighting function corresponds to a different line of sight (LOS). Nadir direction is 0°. The magnitude of the weighting functions is small at the surface and above 15 km, revealing that the slant columns are not very sensitive to NO_2 in these regions. Reprinted with permission from Bruns et al. 2004, © Optical Society of America. All rights reserved

In near-nadir geometry (see Fig. 9.34), the instrument looks down to a point on earth below its current position. The radiation observed by the satellite instrument originates from solar light backscattered or reflected from the atmosphere and the earth's surface.

The AMF for the case of pure backscattering from the earth's surface (Fig. 9.34a) can be approximated by geometric considerations (Palmer et al. 2001; Martin et al. 2002). Neglecting the curvature of the earth, the (geometric) AMF for an SZA ϑ and the observation angle (angle between nadir and actual observation direction), α, is given by:

$$A_{\text{surface}} = 1/\cos\vartheta + 1/\cos\alpha \tag{9.24}$$

Since the radiation passes the atmosphere twice, A_{surface} is always greater than 2 (see Fig. 9.35).

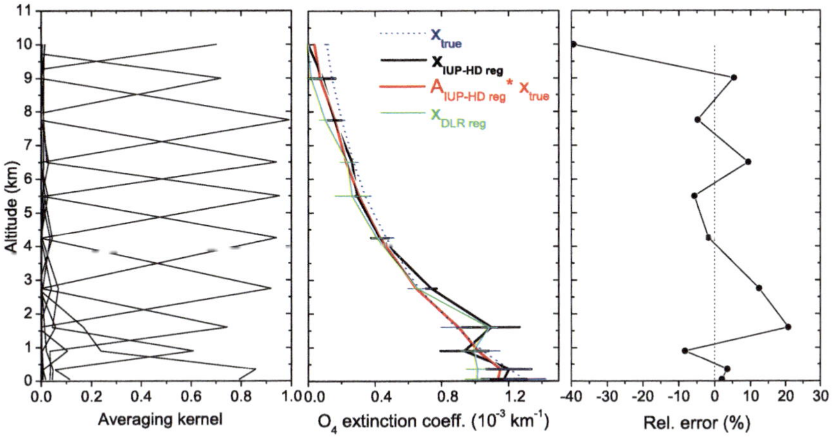

Fig. 9.32 Retrieval of the vertical distribution of the O_4 extinction coefficient from airborne limb DOAS observations. Averaging kernels (left panel) show the sensitivity to a ~2km interval around the flight altitude of the observations. The retrieved O_4 extinction coefficient (black) is compared to in-situ observations (green), showing the excellent retrieval results. The relaive error of the retrieval is shown in the right panel. Overall this retrieval had ~8 degrees of freedom, i.e. it added ~8 pieces of information to the a-priori profile. Reprinted with permission from Prados-Roman et al. (2011),

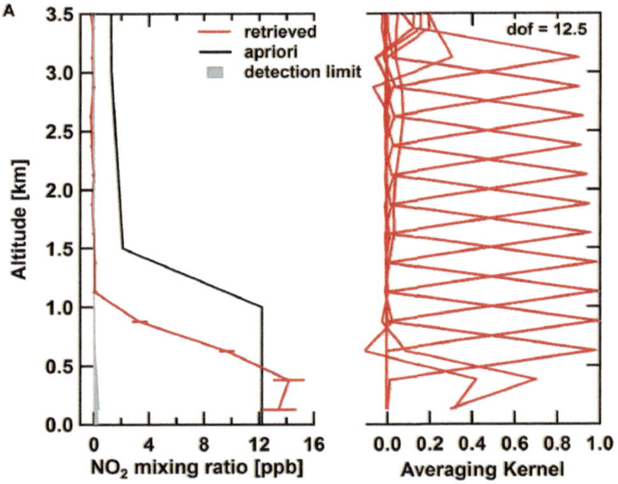

Fig. 9.33 Vertical profile retrieval and averaging kernels for NO_2 from limb DOAS observations during an approach to an airfield. As expected NO_2 is mostly located in the boundary layer below 1km altitude. The averaging kernels shows that the retrieval is equally sensitive for all altitudes. Reprinted with permission from Baidar et al. (2013),

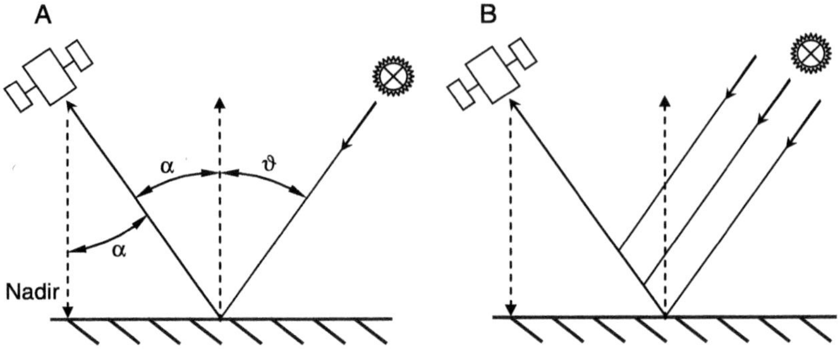

Fig. 9.34 Geometry of a (near) nadir viewing satellite instrument and a sketch of the associated radiation transport in the atmosphere: **a** Without scattering in the atmosphere the instrument would only record radiation scattered back from the surface, the AMF would always exceed 2. **b** In reality, in addition to reflection from the surface, as in the case of other Scattered-Light DOAS geometries, there is an infinite number of possible paths of scattered radiation between the light source (i.e. the sun) and the instrument

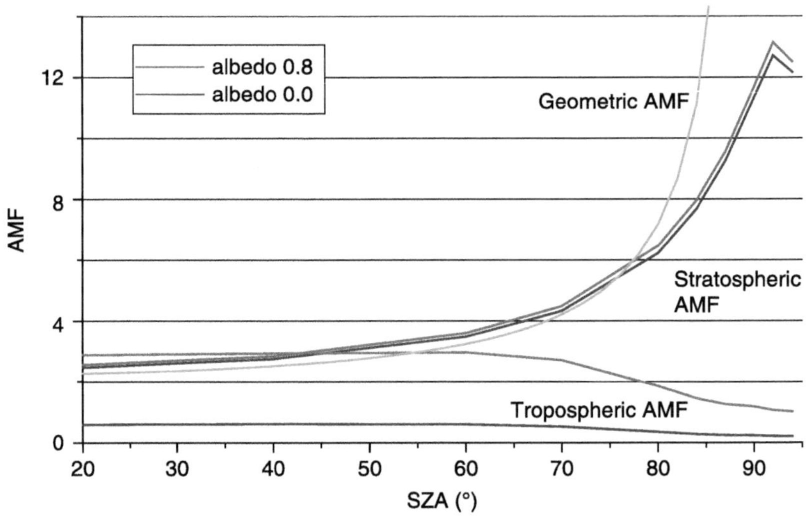

Fig. 9.35 Airmass factors for a nadir viewing satellite instrument as a function of solar zenith angle ϑ for two ground albedo values. Three cases are shown: Geometric AMF ($1/\cos\vartheta$), AMF for a tropospheric trace gas profile, AMF for a stratospheric trace gas profile

Fig. 9.36 Box airmass factors for a nadir viewing satellite – instrument as a function of box (trace gas layer) altitude for 360 nm and 630 nm and for two ground albedo values (2% and 80%), respectively (Figure courtesy of T. Wagner)

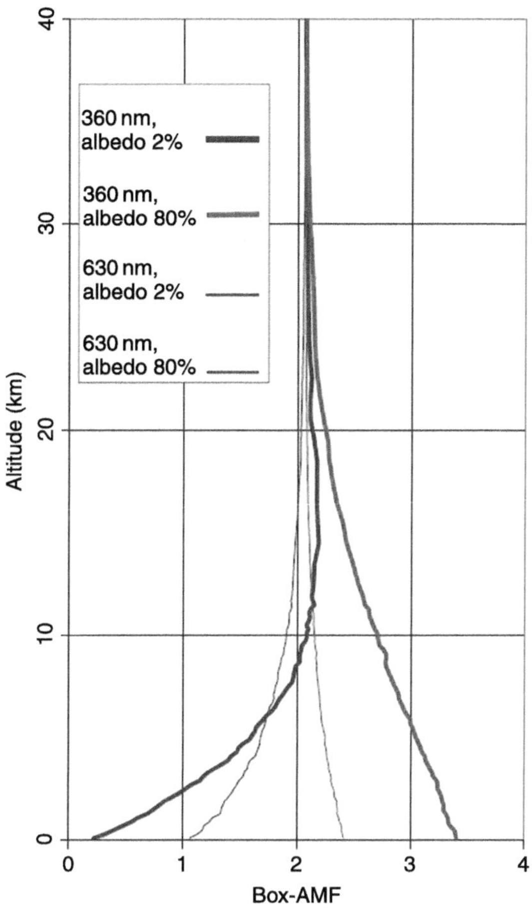

In reality, however, a fraction of the radiation will originate in the atmosphere, as a consequence of Rayleigh scattering on air molecules and Mie scattering on aerosol (Fig. 9.34b). Figure 9.35 compares the geometric AMFs and those calculated with a RT model that includes atmospheric scattering for a stratospheric and tropospheric absorber. The stratospheric AMF, i.e. the AMF for gases in the stratosphere, is close to the geometric value up to a SZA of $\sim 75°$ and shows little dependence on the surface albedo. In contrast, there is a considerable difference between the tropospheric and the geometric AMF, particularly for low albedos.

The BAMFs become altitude dependent as a consequence of atmospheric scattering (Fig. 9.35). In the UV around 360 nm the stratospheric BAMF is close to the geometric value. In the troposphere, however, the BAMF decreases to lower values for low albedos, while it increases towards the surface at high albedos (Fig. 9.36). The AMF also becomes wavelength dependent when scattering is considered, as the contribution of Rayleigh scattering increases towards shorter wavelengths. Consequently, the contribution of the lower atmosphere decreases for lower wavelengths, and satellite observations in the UV are not very sensitive to trace gases in the atmospheric boundary layer (Fig. 9.36). Mie scattering by aerosol will enhance the fraction of backscattered light above an aerosol layer while decreasing the contribution of the trace gases below or inside the layer.

Similarly, trace gas absorptions, in particular those of ozone, can change the contribution of the lower atmosphere. Below 300 nm, where the ozone layer absorbs most radiation, the troposphere cannot be observed. On the other hand, this wavelength dependence can be used for vertical trace gas profiling if a gas is measured at different wavelengths. The quantitative description of the impact of atmospheric scattering and absorption on AMFs and Box-AMFs requires RT model calculations, especially for tropospheric gases. The general approach to derive tropospheric VCDs is based on the idea of BAMFs or weighting functions (Sect. 9.4.3), see e.g. Palmer et al. (2001) and Martin et al. (2002). This approach allows the separation of the radiative transfer effects from the dependence of the VCD on the trace gas vertical profile or profile shape. The BAMFs are calculated by RT models, and are often made available in the form of look-up tables to increase computational efficiency. Trace gas profile shapes are most often extracted from chemical transport models, as further discussed below. Radiative transfer models, such as those listed in Table 9.2, are suitable for these RT calculations. Lorente et al. (2017), for example, report that four different RT models agreed within 1.5% when simulating top-of-the-atmosphere reflectances, within 3% for BAMFs, and 6% for tropospheric AMFs when identical input data are used. The differences increase considerably when the same RT models are driven by the input data from different research groups. The uncertainty of AMF calculations is thus mostly a consequence of the often poorly constrained input data chosen for specific retrievals (Lorente et al. 2017; Boersma et al. 2018). This is especially true for surface albedo, aerosol, and the a-priori NO_2 concentration profiles.

Two other factors impact the retrieval of trace gases from satellites. The presence of clouds impacts the RT in the atmosphere by blocking the underlying atmosphere as well as by changing the reflected intensity due to their large albedo. For the retrieval of tropospheric trace gas columns for gases that are present throughout the entire atmosphere, such as NO_2, the separation of the tropospheric and stratospheric component is also important. Here we will briefly discuss each of these factors.

- **Separation of tropospheric from stratospheric column**: Most current trace gas retrievals for NO_2 or other traces gases present in both troposphere and stratosphere employ a three step approach to convert surface reflected radiances to tropospheric VCDs. The first step is a DOAS retrieval to derive a total SCD. The second step separates the stratospheric and tropospheric SCD. The third step then retrieves the tropospheric VCD. In the case of NO_2, which has substantial concentrations in the stratosphere, the accuracy of the second step has a considerable impact on retrieval of the tropospheric VCD (see e.g. Bucsela et al. 2006; Boersma et al. 2007, 2011, 2018; Beirle et al. 2016a and Sect. 11.8). Several stratosphere-troposphere-separation (STS) algorithms are available. The earliest successful approach, the reference sector method (RSM), uses the latitudinal distribution of NO_2 SCDs across the remote Pacific to estimate the global stratospheric NO_2 fields (Richter and Burrows 2002; Martin et al. 2002; Beirle et al. 2003 and Sect.

11.9). The RSM method is based on the assumption of longitudinal homogeneity and negligible tropospheric NO_2 over the central Pacific "reference region". However, it has been found that this method can lead to systematic biases in the STS as the assumptions are not always fully met (Beirle et al. 2016a). A number of variations of the RSM apply filters to identify and remove potentially polluted pixel (see for example Bucsela et al. 2013). In the newest evolution of RSM, weighting functions, rather than filters or masks, are applied to derive the stratospheric NO_2 field (Beirle et al. 2016a; Liu et al. 2021). Theses functions weigh the pollution levels, cloudiness, and tropospheric residue to determine the extent that the stratospheric column contributes to the total SCD. A weighted convolution is then applied to derive global stratospheric NO_2 maps that are used in the STS correction.

Two other STS approaches have emerged over the years: (1) Limb observations from the SCIAMACHY satellite instrument were used to independently determine the stratospheric column and thus correct nadir observations (Hilboll et al. 2013). However, this approach is not available for most nadir instrument as they do not allow for simultaneous limb observations. (2) In lieu of independent measurements, stratospheric columns can be extracted from chemical transport models and used in the STS (Hilboll et al. 2013; Dirksen et al. 2011; Beirle et al. 2016a).

- **Surface reflectivity**: The accurate representation of surface reflectivity is crucial for RT calculations of satellite observations (Palmer et al. 2001; Boersma et al. 2011; Vasilkov et al. 2017). As shown in Fig. 9.37, the tropospheric AMF varies from below 0.5 at low surface albedo/reflectivity to close to 1.5 at high surface albedo/reflectivity. One of the challenges in determining the reflectivity for nadir satellite RT calculations is the high variability across the globe, i.e. from low reflectivities over the ocean to high reflectivities over snow. The reflectivity not only varies in space but also in time, such as in the case of seasonal snow. Reflectivities have been directly derived from the radiances observed by the respective satllite instrument, often assuming a lambertian behavior, i.e. without directional dependence (Martin et al. 2002; Kleipool et al. 2008; Boersma et al. 2011; Bucsela et al. 2013; Boersma et al. 2018). Improvements in the representation of surface reflectivities can be achieved by including directionality, i.e. quantifying how reflectivity depends on solar and viewing angle (Lin et al. 2014). Geometry-dependent lambertian equivalent reflectivities (LER) can be calculated using data from other satellite instruments, such as MODIS (Lin et al. 2014). The effect of the directionaly of the surface reflectivity can be substantial because surface BAMFs, and thus observation of trace gases at the surface, are highly dependent on this value. Vasilkov et al. (2017), for example, finds that geometry-dependent LERs increase NO_2 tropospheric columns by up to 50% in highly polluted areas.
- **Clouds and aerosol**: Another RT problem particular to satellite observations arises from the spatial averaging of satellite observations and the probability to encounter clouds within this ground pixel. Clouds will mask the fraction of the total trace gas column below the cloud. The total intensity received by a satellite instrument (neglecting radiation reflected from the surface below the cloud) is given by: $I_S = I_0 \cdot (f \cdot B_{Cloud} + (1 - f) \cdot B_{Surface})$, where B_{Cloud} and $B_{Surface}$ denote the albedo of cloud and surface, respectively, and f is the fraction of the pixel covered

by the cloud. Unfortunately, the effect of clouds is larger than expected at the first glance. For instance, assuming $B_{Cloud} = 0.8$, $B_{Surface} = 0.05$ and $f = 0.5$, only about 6% of the signal comes from the cloud-free part of the pixel. Accordingly, the derived column density will be dominated from the cloud-covered part of the pixel. Figure 9.37 a and b illustrate the strong dependence of the tropospheric AMF on cloud fraction as well as cloud pressure. The AMF can drop close to zero for high and dense clouds.

Correction of the cloud effect is possible, but requires a very precise determination of the cloud fraction f. While detection of clouds from satellite, e.g. through thermal IR radiometry, has been in use for a long time, DOAS-type satellite instruments are better served by cloud detection using data from the same instrument. A series of algorithms have been developed for the retrieval of cloud parameters. Two different observables are typically used for cloud retrievals: (1) The SCD of the O_2-A-Band or of the various O_4 bands are reduced in the presence of clouds. The SCD depends on cloud coverage within the pixel, cloud albedo, and cloud top height. O_2 and O_4 are used by numerous cloud detection algorithms, such as the official GOME cloud product ICFA (Initial Cloud Fitting Algorithm) (Kuze and Chance 1994; Acarreta et al. 2004), the OMI $O_2 - O_2$ operational algorithm (OMCLDO2) (Veefkind et al. 2016) and the FRESCO algorithm (Fast REtrieval Scheme for Clouds from the Oxygen-A-Band) (Koelemeijer et al. 2001; Wang et al. 2008). (2) The main idea of a second class of algorithms is that clouds can also be identified through the overall intensity of reflected light because clouds are usually brighter than the underlying surface (Grzegorski et al. 2006). The intensities are mainly independent of cloud top height, but they depend on cloud coverage and cloud albedo. This approach is applied using small spectral windows of instruments with moderate spectral resolution, onboard radiometers, or Polarisation Monitoring devices (PMD).

All of these algorithms retrieve an effective cloud fraction, a parameter that combines cloud coverage (cloud abundance) and cloud albedo in the observed ground footprint, as well as a cloud pressure, that indicates the altitude of the cloud. These parameters are used in the retrieval of tropospheric VCD.

Finally, it should be noted that cloud retrieval algorithms do not distinguish between clouds and aerosol. For example a thick aerosol layer under cloud free conditions will be represented by a small cloud fraction (Veefkind et al. 2016). The cloud algorithm thus provides an inherent aerosol correction (Boersma et al. 2018).

- **Trace gas concentration profiles**: The shape of the atmospheric concentration profile is an important component in the retrieval of VCDs. For tropospheric VCDs these profiles are typically supplied by chemical transport models, such as TM5-MP (Williams et al. 2017; Boersma et al. 2018) and GEOS-Chem (Lin et al. 2015a). Many of these models use grids that are coarser than the pixel size of current satellite instruments. It is thus often necessary to interpolate the model trace gas profiles to the satellite pixel scale. The quality of these models has an impact on the accuracy of the trace gas profiles, and the reader is referred to the extensive literature on chemical transport model performance.

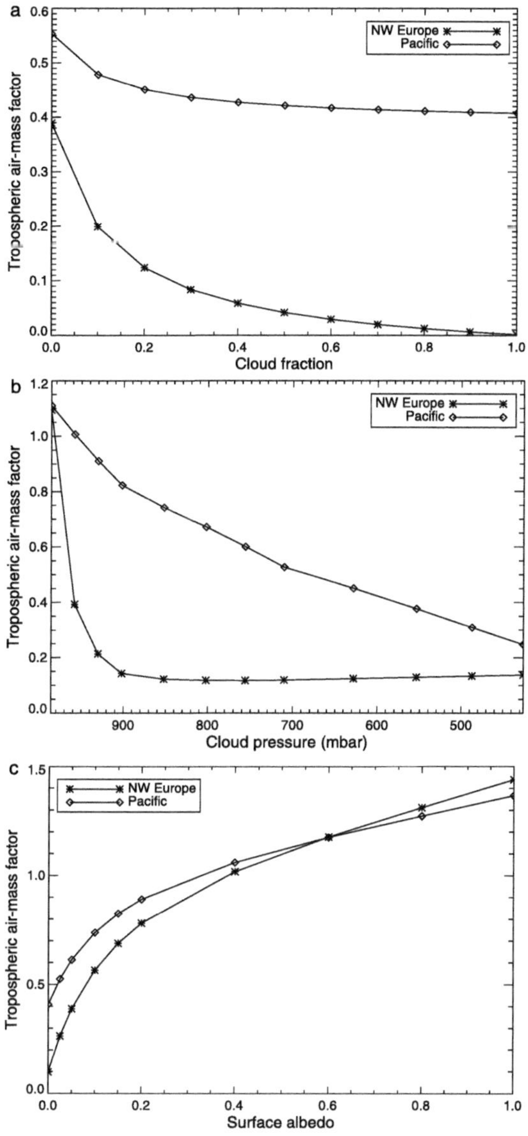

Fig. 9.37 Dependence of the tropospheric AMF on (**a**, top panel) cloud fraction in the observed ground pixel, (**b**, centre panel) cloud pressure and (**c**, bottom panel) surface albedo. The AMFs are shown for two scenarios: northwestern Europe and the Pacific. Reprinted with permission from Boersma et al. (2004), © American Geophysical Union. All rights reserved

The previous paragraphs illustrate the difficulty in retrieving trace gas VCDs from satellite nadir measurements. Despite these challenges, satellite DOAS data have provided unique insights into the composition of the global atmosphere. Examples of the various satellite DOAS applications are given in Chap. 11.

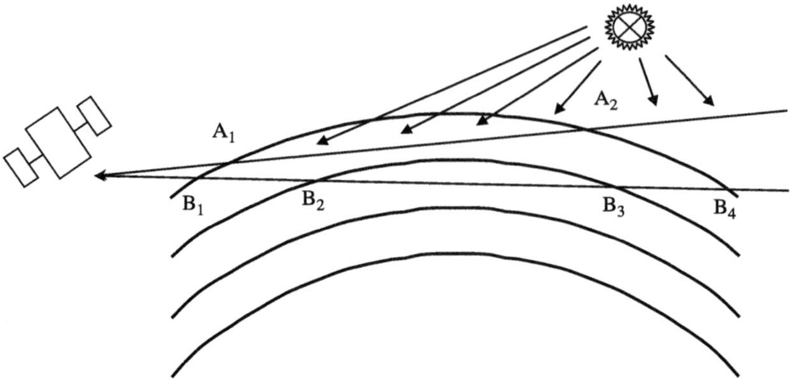

Fig. 9.38 Geometry of a limb viewing satellite instrument and a sketch of the associated radiation transport in the atmosphere. As in the case of other Scattered-light DOAS geometries there are an infinite number of possible light paths between the light source (i.e. the sun) and the instrument

9.7.2 The Analysis of Satellite-Limb Scattered Light Observations

While the goal of nadir observations is mapping pollution across the globe, limb satellite instruments focus on the vertical dimension instead (see e.g. McLinden et al. 2012; Puķīte et al. 2010; Kühl et al. 2008). In limb-view geometry, the radiation scattered from the edge of the earth's disk (the limb) is observed by a suitable telescope (Fig. 9.38). This mode of observation allows the derivation of vertical trace gas concentration profiles at relatively high accuracy and vertical resolution (von Savigny et al. 2003; Haley et al. 2004; Puķīte et al. 2010; Rozanov et al. 2011). The RT in this case is dominated by the scattering events at different altitudes within the atmosphere. Many of the RT models discussed above are able to operate in limb geometry, if sphericity of the atmosphere is included (Puķīte et al. 2010; Rozanov et al. 2011). The retrieval of vertical trace gas profiles generally follows an OEM approach, similar to that used in aircraft limb DOAS (Sect. 9.6), with the observations for different tangent heights forming the measurement vector and limb weighting functions (WFs) calculated by the RT model as the forward model. Because of the very long light path and the narrow weighting functions, satellite limb retrievals are a powerful tool to derive concentration profiles in the stratosphere (Haley et al. 2004; Puķīte et al. 2010; Rozanov et al. 2011).

9.8 Raman Scattering in the Atmosphere

Radiative transfer in the atmosphere is largely dominated by gas absorption as well as Rayleigh and Mie scattering. There are, however, other atmospheric phenomena that can also impact passive DOAS. Among those, Raman scattering is likely the most significant. Raman scattering has only a minor impact on RT calculations, such as those presented in the previous sections, but it has an impact on DOAS retrievals.

9.8.1 The Influence of Rotational Raman Scattering, the 'Ring Effect'

Named after Grainger and Ring (1962), the Ring effect manifests itself by reducing the optical density of Fraunhofer lines observed at large solar zenith angles (SZA), compared to those at small SZAs. This reduction is on the order of a few percent. However, because atmospheric trace gas absorptions can be more than an order of magnitude smaller than the Ring effect, an accurate correction is required.

Several processes, such as rotational and vibrational Raman scattering, aerosol fluorescence, etc. have been suggested as explanations of the Ring effect. However, a series of investigations (Bussemer 1993, Fish and Jones 1995, Burrows et al. 1995, Joiner et al. 1995, Aben et al. 2001) show convincingly that Rotational Raman Scattering, RRS, is the primary cause of the Ring effect.

In short, light intensity scattered into a passive DOAS instrument can be expressed as:

$$I_{meas} = I_{Rayleigh} + I_{Mie} + I_{Raman} = I_{elastic} + I_{Raman} \tag{9.25}$$

The accurate determination of $I_{elastic}$ and I_{Raman} requires detailed radiative transfer calculations for each observation. However, Schmeltekopf et al. (1987) proposed an approximation which is based on the inclusion of a "Ring Spectrum" in the spectral analysis of the observations (see Chap. 8).

Based on the logarithm of the measured spectrum, $\ln(I_{meas})$, and the equation above, the following approximation can be made.

$$\ln(I_{meas}) = \ln\left(I_{elastic} \cdot \frac{I_{elastic} + I_{Raman}}{I_{elastic}}\right) \tag{9.26}$$

$$= \ln(I_{elastic}) + \ln\left(1 + \frac{I_{Raman}}{I_{elastic}}\right) \approx \ln(I_{elastic}) + \frac{I_{Raman}}{I_{elastic}} \tag{9.27}$$

where the ratio of the Raman and the elastic part of the intensity is considered the Ring spectrum:

$$I_{Ring} = \frac{I_{Raman}}{I_{elastic}} \tag{9.28}$$

This approximation has been proven to be simple and effective. An experimentaland a numerical approach exist to determine a Ring spectrum:

(1) Experimental approach: The first approach is based on the different polarisation properties of atmospheric scattering processes. Rayleigh scattering by air molecules is highly polarised for a scattering angles near 90° (see Sect. 4.2.2). Light scattered by rotational Raman scattering, on the other hand, is only weakly polarised (see Sect. 4.2.3). By measuring the intensity of light polarised perpendicular and parallel to the scattering plane, the rotational Raman component, and therefore a "Ring spectrum", can be determined (Schmeltekopf et al. 1987; Solomon et al. 1987).

This approach faces a number of challenges. Mie scattering also contributes to the fraction of non-polarised light in the scattered solar radiation. The presence of aerosol or clouds therefore makes the determination of the Ring spectrum difficult. Also, the atmospheric light paths for different polarisations may be different, and

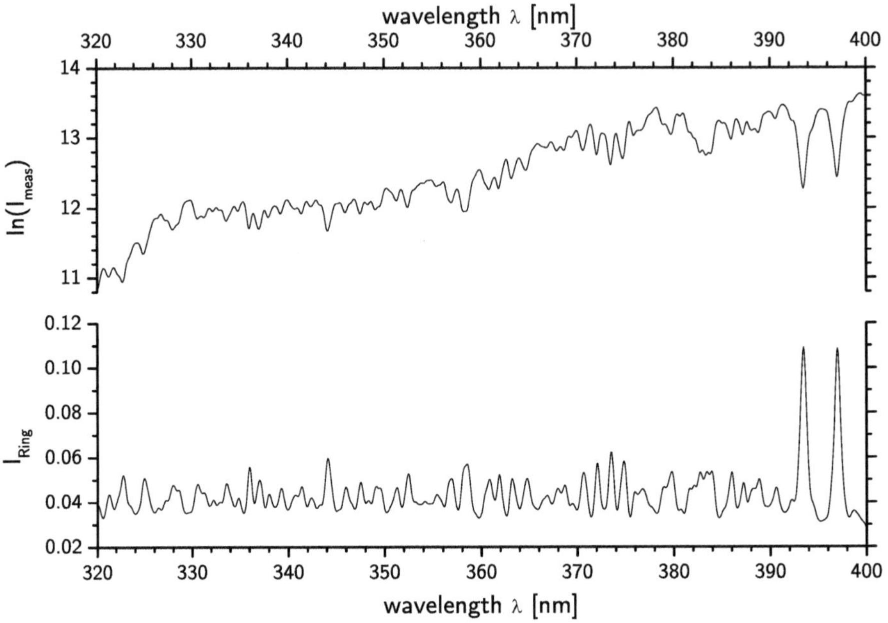

Fig. 9.39 Sample Ring spectrum (I($_{Ring}$)) calculated for the evaluation of UV spectra taken during ALERT2000. Shown is also the logarithm of the Fraunhofer reference spectrum(I_{meas}) used for the calculation. The spectrum was taken on April 22, 2000 at 15:41 UT at a local solar zenith angle of 70° and zenith observation direction. Reprinted from Hönninger, PhD Thesis, 2002, University of Heidelberg

can thus contain different trace gas absorptions, which can affect the DOAS fit of these gases.

(2) Numerical approach: The second approach uses the known energies of the rotational states of the two main constituents of the atmosphere, O_2 and N_2, to calculate the cross-section for rotational Raman scattering. This can be realised, either by including Raman scattering into radiative transfer models (e.g. Bussemer 1993; Fish and Jones 1995; Funk 2000), or by calculating the pure ratio of the cross-sections for Raman and Rayleigh scattering. In many realisations this calculation is based on measured Fraunhofer spectra (e.g. DOASIS (Kraus 2006)) leading to a Raman cross section which has all the spectral characteristics of the respective instrument. The rotational Raman spectrum is then divided by the measured Fraunhofer spectrum to determine the Ring spectrum (Fig. 9.39). Note, the Fraunhofer spectrum must be corrected for rotational Raman scattering to represent pure elastic scattering. In most cases, this approach leads to an excellent correction of the Ring effect.

9.8.2 Vibrational Raman Scattering, the 'Vibrational Ring Effect'

In addition to Rotational Raman Scattering, RRS, Vibrational Raman Scattering, VRS, also occurs in the atmosphere or back-scattered from water (e.g. over the ocean). Compared to RRS (see Sect. 9.8.1) the wavelength shift of VRS is much

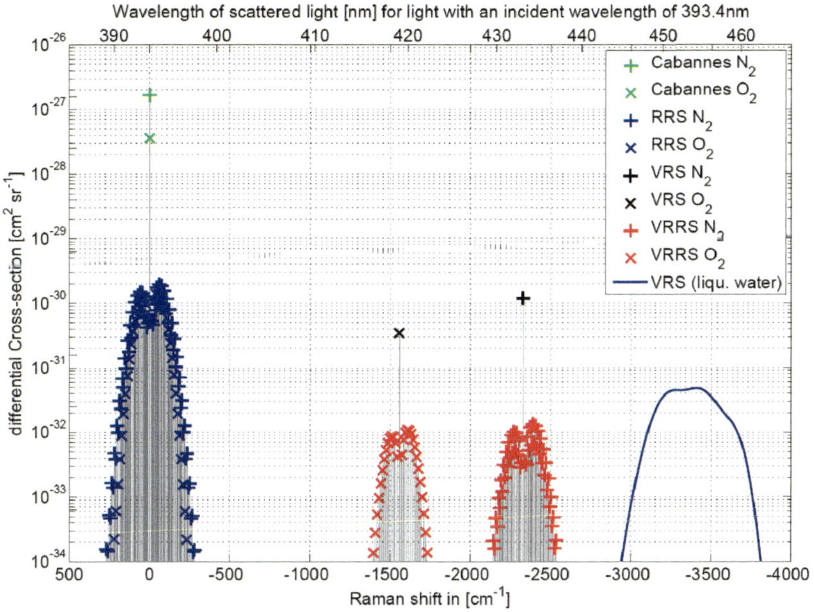

Fig. 9.40 Comparison of scattering cross-sections of rotational and vibrational Raman lines. The energy shift ΔE is given in cm^{-1}. At $\Delta E = 0$: N_2 and O_2 Cabannes lines, around $\Delta E = 0$: rotational Raman lines (RRS) of O_2 and N_2. Vibrational (VRS) and Rotational-vibrational Raman scattering (VRRS) lines, of O_2 and N_2 are located at (and around for VRRS) $\Delta E = 1556.3\,cm^{-1}$ and $2329.9\,cm^{-1}$, respectively. For a wavelength of 393 nm the wavelength shifts correspond to 25.6 nm and 39.6 nm, respectively. The RRS, VRS, and VRRS line cross-sections are scaled with 0.8 for N_2 and 0.2 for O_2, to reflect their importance in the atmosphere. The cross-sections were calculated for a scattering angle of $\Theta = 180°$. For comparison the Raman response of liquid water according to Haltrin and Kattawar (1993) (blue line around $3400cm^{-1}$) is also shown, scaled arbitrarily. Reprinted with permission from Lampel et al. (2015), © CC 3.0. All rights reserved

larger, on the order of several 10 nm in the near UV (see e.g. Lampel et al. 2015). Therefore, it manifests itself by additional lines, i.e. red-shifted (Stokes) "Fraunhofer Ghosts" of (strong) Fraunhofer lines, rather than a filling in of Fraunhofer lines. Blue shifted (anti Stokes) VRS lines are extremely weak, since only a very small fraction of the atmospheric N_2 and O_2 molecules is vibrationally excited. The intensity of VRS is about two orders of magnitude smaller than that of RRS (see Fig. 9.40), therefore VRS-lines are usually not observed. In fact the existence of VRS-lines was only experimentally verified recently by Lampel et al. (2015). They found that the relative contribution of VRS to the optical density of RRS for N_2 and O_2 are $2.1 \pm 0.3\%$ and $0.6 \pm 0.2\%$, respectively. Nevertheless, future instruments with higher sensitivity may need correction of VRS-effects. Also, as mentioned above, VRS in water can play a role in scattered sunlight measurements over the ocean (see e.g. Haltrin and Kattawar 1993 and Fig. 9.40). In particular for satellite measurements this may need to be considered.

Sample Application of 'Active' DOAS with Artificial Light Sources

10

10.1 Sample Application of 'Active' DOAS with Artificial Light Sources

Over the past four decades, active DOAS measurements have been made for a multitude of purposes. From the initial studies of the atmospheric composition in the mid 1970s, a wide variety of designs and applications in the open atmosphere, as well as in the laboratory, have been developed. Broadband lasers have been used as a light source (Rothe et al. 1974; Perner et al. 1976; Perner et al. 1991; Amerding et al. 1991; Dorn et al. 1993, 1996, 1998; Brauers et al. 1996, 1999; Brandenburger et al. 1998). However, the majority of the designs rely on "conventional" light sources such as arc-lamps or LEDs. There are a large number of reports on active Long-Path DOAS measurements. Here only a number of examples can be given: (Bonafe et al. 1976; Kuznetzov and Nigmatullina 1977; Noxon et al. 1978, 1980; Platt 1977; Evangelisti et al. 1978; Platt et al. 1979; Platt and Perner 1980, 1983; Johnston and McKenzie 1984; Dorn and Platt 1986; Edner et al. 1986, 1990, 1992, 1993a, 1994a; Axelsson et al. 1990a, 1990b, 1995; Axelsson and Lauber 1992; Amerding et al. 1991; Wagner 1990; Hallstadius et al. 1991; Galle et al. 1991; Biermann et al. 1991; Stevens and Vossler 1991; Gall et al. 1991; Fayt et al. 1992; Plane and Nien 1991; Vandaele et al. 1992; Hausmann et al. 1992, 1997; Stevens et al. 1993; Hausmann and Platt 1994; Martini et al. 1994; Plane and Smith 1995; Evangelisti et al. 1995; Heintz et al. 1996; Harder et al. 1997a; Flentje et al. 1997; Lamp et al. 1998; Reisinger et al. 1998; Russwurm 1999; Allan et al. 1999, 2000a; Hebestreit et al. 1999; Alicke, 2000; Gölz et al. 2001; Kim and Kim 2001; Geyer et al. 1999, 2001a, b, c, 2002, 2003a, b; Matveev et al. 2001; Volkamer et al. 2001; Alicke et al. 2002; Veitel et al. 2002; Lee et al. 2002; Yu et al. 2004; Saiz-Lopez et al. 2004a, 2004a; Lee et al. 2005; Zingler and Platt 2005, Kern et al. 2009; Pöhler 2010; Pöhler et al. 2010; Seitz et al. 2010; Fuchs et al. 2012; Stutz et al. 2016; Nasse et al. 2019; Saiz-Lopez et al. 2007;

U. Platt and J. Stutz, *Differential Optical Absorption Spectroscopy*, Geophysics and Environmental Physics, https://doi.org/10.1007/978-3-662-68639-3_10

Frieß et al. 2011; Wong et al. 2011, 2012; Pinto et al. 2014; Tsai et al. 2014; Holla et al. 2015; Tsai et al. 2018).

In addition there are a small number of studies using artificial 'lights of opportunity' to perform DOAS measurements (Yoshii et al. 2003; Fuqi et al. 2005). It is impossible to review the entire breadth of DOAS studies over the past four decades. However, it is instructive to consider examples for some uses of active DOAS. The intention of this chapter is to provide an overview of the abilities DOAS offers to research in atmospheric chemistry and physics. We hope that this overview will inspire the development of even more applications in the future.

10.2 Air Pollution Studies and Monitoring Applications

Motivated by the desire to understand and mitigate air quality problems in urban areas such as Los Angeles, one of the earliest uses for active DOAS was the investigation of the processes leading to the formation of urban air pollution. Today, DOAS is used as a tool to monitor air quality and also as a benchmark for other techniques. DOAS is an Environmental Protection Agency (EPA) approved method to monitor primary pollutants such as O_3, NO_2, and SO_2. In addition, the Association of German Engineers (Verein Deutscher Ingenieure, VDI) issued a guideline for the testing of DOAS instruments for air pollution monitoring (VDI 2005). A number of commercial DOAS instruments are available today. All these instruments work according to the principles described in earlier chapters. The advantages of DOAS for monitoring purposes are its high accuracy and ability to measure O_3, NO_2, and SO_2 simultaneously.

10.2.1 Measurement of Urban Pollutants

The measurement of pollutants is an important undertaking, both for air pollution-related research and the monitoring of air quality.

Ozone, *nitrogen dioxide*, and *sulfur dioxide* are some of the most important species connected to air pollution. SO_2 and NO_2 (in the form of NO which is rapidly converted to NO_2; see Chap. 2) are released from combustion sources. Ozone is formed throughout the day by photochemical processes, and is the main component of the so-called Los Angeles Smog (see Chap. 2). All three species are regulated in most industrialised countries, and considerable effort is put into their continuous measurement by air quality monitoring.

Ozone, NO_2, and SO_2 are, in most cases, the strongest absorbers in DOAS spectra acquired in polluted areas. These three trace gases are thus the most commonly measured species in active DOAS applications. Figure 10.1 shows an example of an analysis of O_3, NO_2, SO_2, and HCHO in the UV by a long-path DOAS instrument such as those described in Chap. 7. The figure illustrates again how the analysis procedure outlined in Chap. 8 is used to separate the different absorption structures

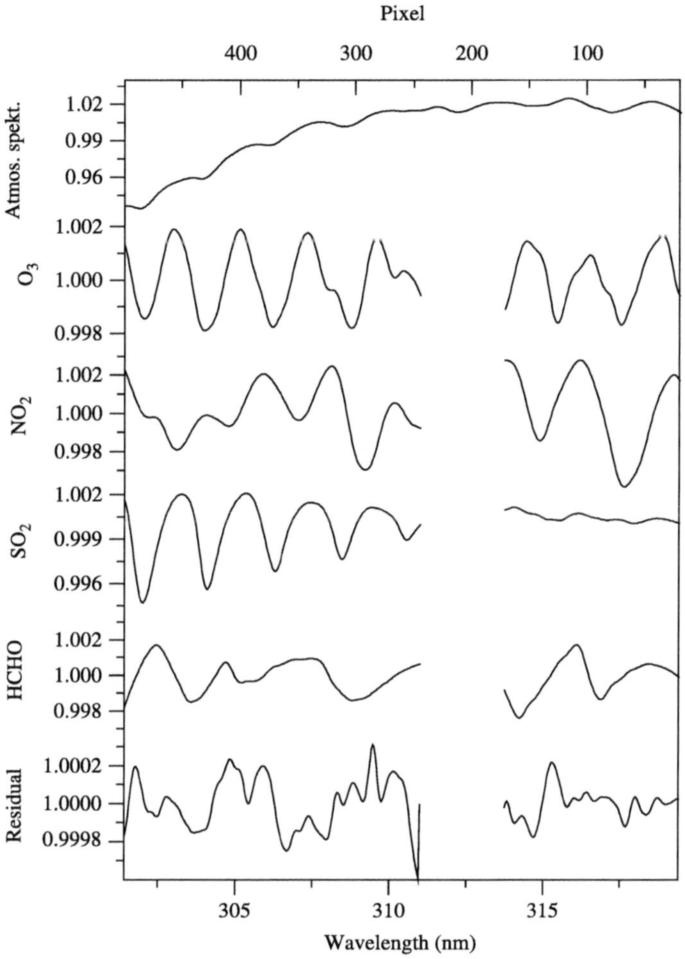

Fig. 10.1 Example of a spectral analysis of O_3, NO_2, SO_2, and HCHO in the polluted air of Heidelberg. The absorptions identified in the atmospheric spectrum (*top trace*) were: O_3, 21.1 ± 0.5 ppb; SO_2, 0.64 ± 0.01 ppb; and HCHO, 3.7 ± 0.1 ppb. The NO_2 mixing ratio was not determined since this wavelength interval is not optimal for its analysis. The missing area around 312 nm was excluded from the analysis procedure

in an atmospheric absorption spectrum. The wavelength range displayed in Fig. 10.1 is typically chosen for the detection of O_3, SO_2, and HCHO in long-path DOAS applications. An example of the measurement of NO_2 was given in Chap. 8.

Figures 10.2 and 10.3 show examples of measured time series of O_3, NO_2, SO_2, and a number of other trace gases during field experiments in Milan, Italy, and Houston, USA, respectively. The goal of both the Pianura Padania Produzione di Ozono (Po-Valley Ozone Production, PIPAPO) study in Milan and the Texas Air Quality Study (TEXAQS) in Houston was to investigate the formation of ozone and particles in the polluted environment of these cities.

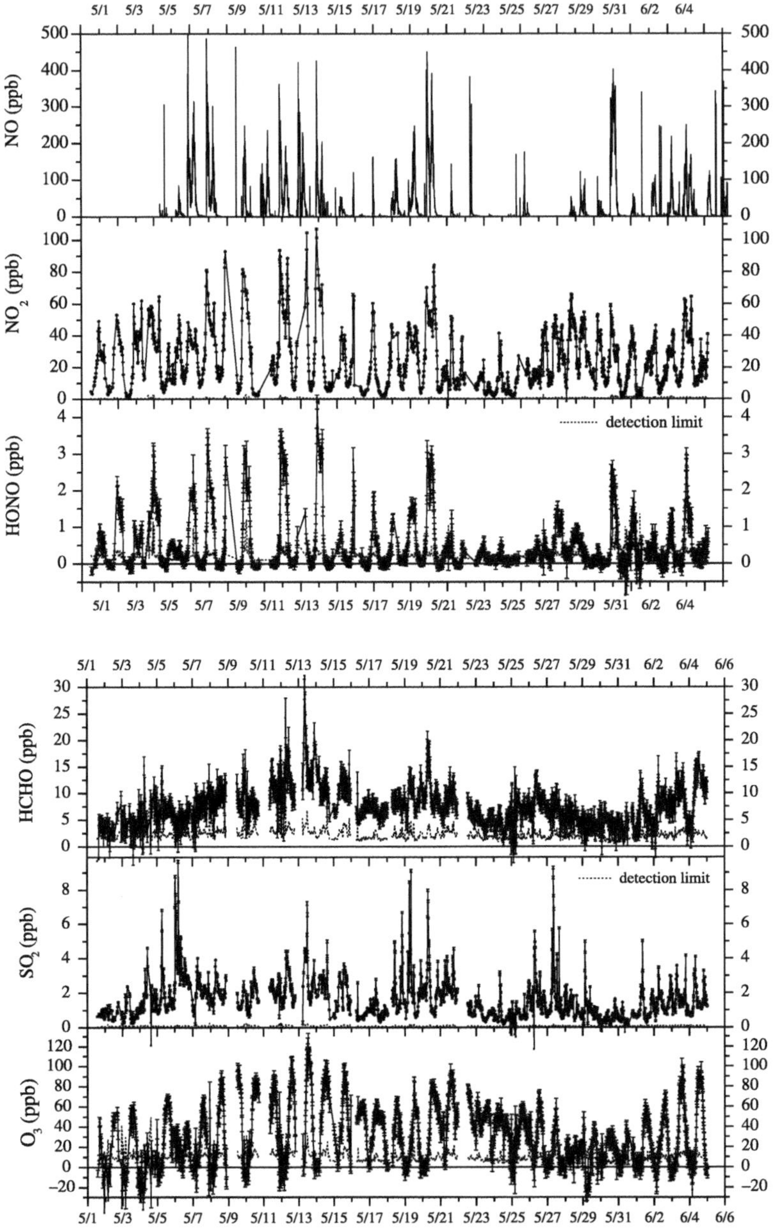

Fig. 10.2 Time series of O_3, NO_2, SO_2 HCHO, and HONO mixing ratios measured during the PIPAPO study in Milan, Italy. In addition, NO (top trace) was measured by chemoluminescence. Reprinted with permission from Alicke et al. (2000), ©American Geophysical Union. All rights reserved

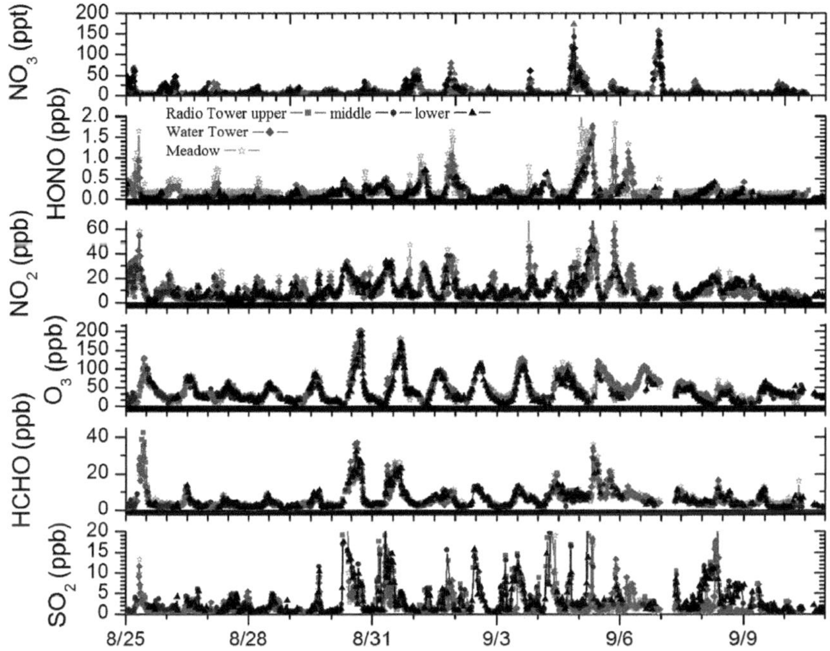

Fig. 10.3 O_3, NO_2, and SO_2, HONO, and NO_3 mixing ratios during the Texas Air Quality Study (TEXAQS) in summer 2000. The measurements in Houston, Texas (USA) were performed with two long-path DOAS instruments on five different light paths. A more detailed description of the setup is given in Stutz, et al. (2004a)

The measurements in Milan were made with a long-path DOAS instrument that employed retro-reflectors to fold the light path once. This instrument was set up ~8 km north of the city centre of Milan at a small airport. The instrument alternated on three light paths of different elevations (Fig. 10.6). Only the data from the upper light path are shown in Fig. 10.2. The mixing ratios of ozone in Milan show strong diurnal variations, with low levels at night when emission of NO converts O_3 to NO_2, and high levels after noon when ozone is photolytically formed (see Chap. 2). Mixing ratios of ozone exceeded 100 ppb on several occasions. Nitrogen dioxide shows a diurnal variation opposite to O_3, with high values at night and low values during the day. The NO_2 mixing ratios reached up to 100 ppb, showing that Milan was heavily polluted. The NO data, measured by a chemiluminescence instrument (see Chap. 5), also show extremely high mixing ratios. While the O_3 and NO_2 data are closely related to each other, the SO_2 data in Milan behaves quite differently. This can be explained by the different sources and chemistry of SO_2 which shows a higher short-term variability. Levels of SO_2 reached up to 8 ppb.

The observations in Houston (Fig. 10.3) were made using two DOAS instruments measuring on five different light paths (Fig. 10.12). A discussion of this setup will be given in Sect. 10.2.2. The instruments were located 8 km southeast of downtown Houston, close to an industrial area with a large number of refineries. In general, the behaviour of O_3 and NO_2 is similar to that in Milan, and is typical for a polluted

Table 10.1 Overview of the species measured in Milan, Italy and the observed mixing ratios (adapted from Alicke et al. 2002)

Species	Wavelength window (nm)	Absorption cross-section	Maximum mixing ratio (ppb/date)	Detection limit (ppb)
HONO	375–336	(Stutz et al. 2000)	4.4 ± 0.2, 13.5.98 21:25	0.2
NO_2	375–336	(Harder et al. 1997a)	115.2 ± 0.6, 13.5.98 21:08	0.6
NO	–		480, 7.5.98 21:30	0.25
SO_2	322–303	(Vandaele et al. 1994)	9.1 ± 0.22, 19.5.98 8:47	0.18
O_3	322–303	(Bass and Paur 1985)	123 ± 9, 13.5.98 12:37	8.5
HCHO	322–303	(Cantrell et al. 1990)	33.6 ± 1.8, 13.5.98 8:32	1.6

The distance between telescope and retro-reflectors was 1.25 km
*All times GMT (local—2 h)

urban area. The levels of NO_2 are lower in Houston. However, ozone reached mixing ratios above 200 ppb on two occasions. SO_2 levels were also higher in Houston than in Milan, most likely due to emissions from the nearby refineries.

The time series from these two field experiments (Figs. 10.2 and 10.3) illustrate the ability of modern long-path DOAS instruments to continuously measure O_3, NO_2, and SO_2 at a time resolution of 5–15 min. The detection limits reached during these two campaigns are shown in Tables 10.1 and 10.2. One will notice that the detection limits vary with the light path length. This is caused by the presence of residual structures in the spectra of both experiments (see Chap. 8). The detection limits in Tables 10.1 and 10.2 give an indication of the performance of modern DOAS instruments.

Formaldehyde plays an important role in the chemistry of the troposphere. It is formed by the oxidation of most hydrocarbons, and also directly emitted from combustion sources. The photolysis of HCHO forms HO_2 radicals which, in the presence of NO, are rapidly converted into OH. HCHO is thus an important HO_X precursor. DOAS measurements for the PIPAPO and TEXAQS field experiments are shown in Figs. 10.2 and 10.3. Typical levels in polluted urban areas are $\sim$10 ppb. Only in heavily polluted environments with a fast photochemistry, such as Houston, are higher HCHO levels observed. The HCHO levels in Houston (Fig. 10.3) roughly follow the O_3 mixing ratios. This is due to the link of ozone photochemical formation with an increased rate of hydrocarbon oxidation, and thus HCHO production.

HCHO mixing ratios are an important parameter for the study of urban air quality. The importance of HCHO as a HO_X radical precursor is shown in Fig. 10.4. The DOAS data during PIPAPO (Fig. 10.2) were used to determine the HO_X formation due to the photolysis of O_3, HCHO, and HONO, and the ozonolysis of alkenes. In

Table 10.2 Summary of data coverage, average and best detection limits for the DOAS results in Houston, Texas (USA), see details in Stutz et al. (2004a)

	Lightpath 'meadow'	Lightpath 'water tower'	Lightpaths 'radio tower'
Distance (km)	0.75	1.9	6.1
Data coverage	08/20–09/10	08/18–09/10	08/24–09/12
Detection limit NO_3 (ppt)			
Average	23	8	2.6
Best	2	1.5	0.8
Detection limit HONO (ppt)			
Average	320	90	38
Best	120	50	16
Detection limit NO_2 (ppb)			
Average	0.88	0.4	0.13
Best	0.24	0.2	0.05
Detection limit O_3 (ppb)			
Average	9	4	2.5
Best	4.7	1.5	0.9
Detection limit HCHO (ppt)			
Average	1	0.46	0.34
Best	0.5	0.15	0.11
Detection limit SO_2 (ppt)			
Average	260	130	180
Best	140	40	40

The last column refers to all three light paths ending at the radio tower (red, blue, and black colour in Fig. 10.3)

the polluted environment of Milan, HCHO photolysis is the most important HO_X precursor. The correct representation of HCHO in air quality models, as well as the accurate determination of HCHO concentrations in the field, is thus an important aspect of atmospheric chemistry research.

Finally, it should be noted here that the measurement of HCHO in the field is still a major challenge (e.g. Cardenas et al. 2000; Hak et al. 2005). The capabilities that DOAS offers for the investigation of HCHO chemistry have thus far not been entirely explored.

The photolysis of nitrous acid (HONO) can be the dominant HO_x source in the early morning, as shown in Fig. 10.4. Nitrous acid was an elusive compound until its discovery in the atmosphere by DOAS in 1979 (Perner and Platt 1979). Fig. 10.5 shows the first identification of HONO absorptions in a DOAS spectrum. Since the original measurement of HONO, DOAS has become the most reliable technique to measure HONO in the atmosphere. The high selectivity and the absence of interference with other trace gases, in this case primarily NO_2, make DOAS an excellent choice to measure HONO. Although the importance of HONO as a HO_X precursor has been known for decades, there are still major uncertainties about the mechanisms of its formation. HONO is formed through the heterogeneous conversion of NO_2 on

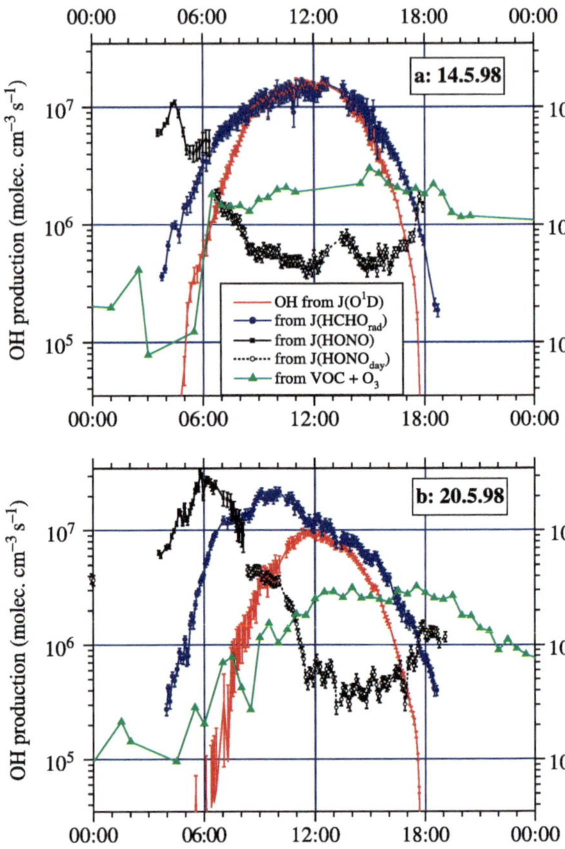

Fig. 10.4 Comparison of three different OH radical producing species for 14.5.1998 (*graph a*) and 20.5.98 (*graph b*). The data are plotted on a logarithmic scale for a better overview. The impact of HONO photolysis on the OH formation in the early morning hours can be easily seen. The estimated daytime OH production from HONO (see text) is shown with a *dotted line* Reprinted with permission from Alicke et al. (2000), ©American Geophysical Union. All rights reserved

various surfaces, as described in Chap. 2. Details about the nature of the surfaces or the mechanisms of the conversion are unclear. The investigation of HONO formation is thus an active field of research, where DOAS is playing an important role.

Modern DOAS applications have moved away from the simple measurement of HONO to more process-oriented studies of the formation of HONO. As mentioned earlier, it is suspected that the source of HONO in the atmosphere is the heterogeneous conversion of NO_2. Because DOAS measures both NO_2 and HONO in a single absorption spectrum, it offers exciting opportunities to study this chemistry.

Following the hypothesis that the deposition of NO_2 leads to the formation of HONO on the ground (e.g. Perner and Platt 1979; Stutz et al. 2004b), one would expect a positive NO_2 gradient (smaller concentration at the ground than above), and a negative HONO gradient above the ground. Based on this hypothesis, Stutz et al. 2002b set up an experiment to measure gradients and fluxes of these two species during the PIPAPO experiment. Figure 10.6 shows the setup, which was based on a modern long-path DOAS instrument (see Chap. 7) that sequentially aimed at three retro-reflectors mounted at different heights on a 4 m-high tower. The distance between telescope and retro-reflectors was 1.25 km.

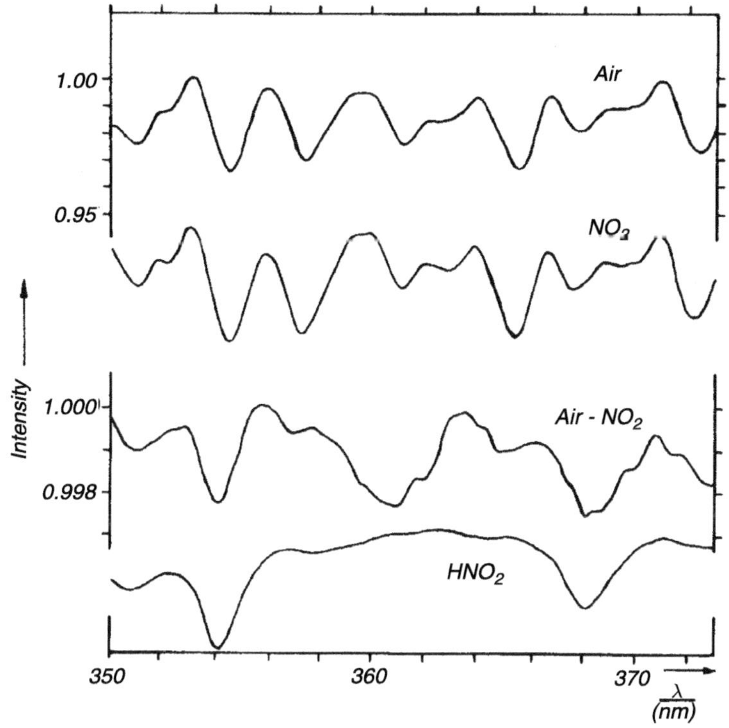

Fig. 10.5 First atmospheric absorption spectrum of nitrous acid Reprinted with permission from Perner and Platt (1979), ©American Geophysical Union. All rights reserved

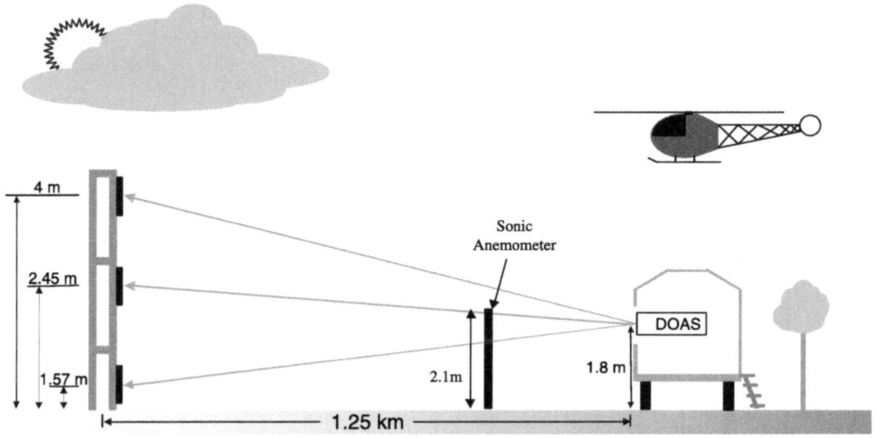

Fig. 10.6 Setup of the experiment to measure gradients and fluxes of NO_2 and HONO during the PIPAPO experiment. The DOAS instrument aimed sequentially at the three retro-reflectors mounted on the tower at 1.25 km distance. Reprinted with permission from Stutz et al. (2002b), ©American Geophysical Union. All rights reserved

The top panels of Figs. 10.7 and 10.8 show the NO_2 and HONO data acquired during one night. Figure 10.7, Panel b and Fig. 10.8, Panel b show the gradients calculated based on the original absorptions, where the uncertainty is determined from the error propagation of the original observations. The bottom panels then describe the fluxes and the deposition velocities of the two trace gases. The data show that during this night a deposition of both gases was observed. Based on this observation, Stutz et al. (2002a) suggest that the loss of HONO on surfaces is an important process in the atmosphere. They conclude, based on data from other nights, that the HONO loss outweighs its formation if the $HONO/NO_2$ concentration ratio exceeds $\sim 3\%$. This result shows that there is a maximum amount of HONO formation in the atmosphere. There is also clear indication of HONO emissions from cars, in particular from DOAS studies of HONO and $HONO/NO_X$ ratios in automobile exhaust

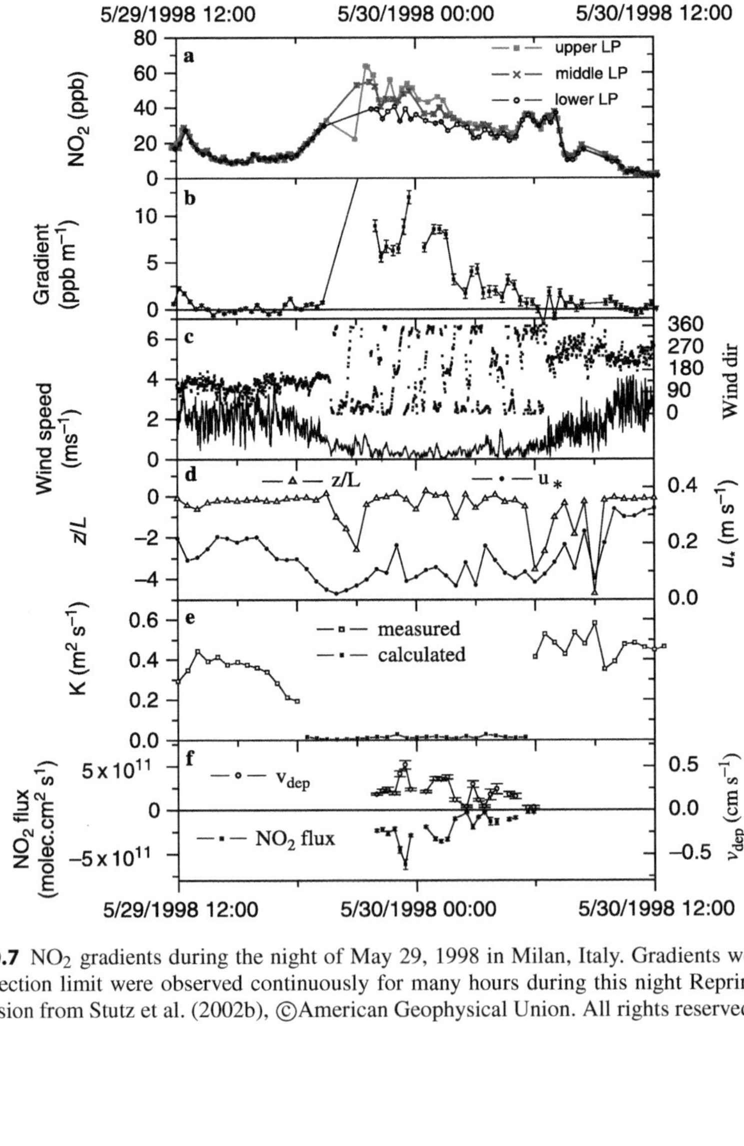

Fig. 10.7 NO_2 gradients during the night of May 29, 1998 in Milan, Italy. Gradients well above the detection limit were observed continuously for many hours during this night Reprinted with permission from Stutz et al. (2002b), ©American Geophysical Union. All rights reserved

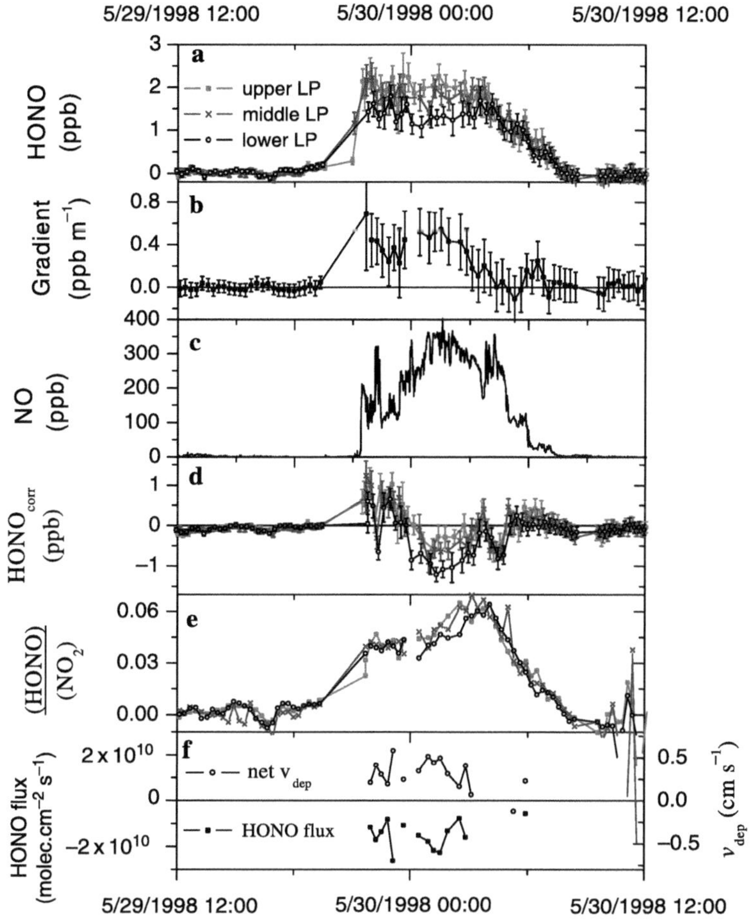

Fig. 10.8 HONO gradients during the night of May 29: (**a**) shows the mixing ratios on the individual light paths (LP). The gradient is displayed in (**b**). To remove the influence of direct HONO emissions, we calculated $HONO_{corr}$ by subtracting 0.65% of the NO_x (not shown here) from the HONO mixing ratios (**d**). The ratio of HONO to NO_2 mixing ratios is displayed in (**e**). The fluxes and net deposition velocities is shown in (**f**) Reprinted with permission from Stutz et al. (2002b), ©American Geophysical Union. All rights reserved

(Perner and Platt 1980; Kessler 1984; Kirchstetter et al. 1996; Lammel and Cape 1996).

The example shown here illustrates how modern DOAS instruments can be used in dedicated setups to answer specific questions on atmospheric chemistry.

Ammonia plays an important role as a reducing agent in the atmosphere. Through reaction with acids, such as H_2SO_4 and HNO_3, it contributes to the formation of particulate matter, and is thus an important component of smog. High numbers of particles are often found in areas where urban pollution is mixed with rural air containing elevated levels of NH_3. The most important sources of ammonia are the bacterial activity in the decomposition of animal waste, soil, and biomass burning.

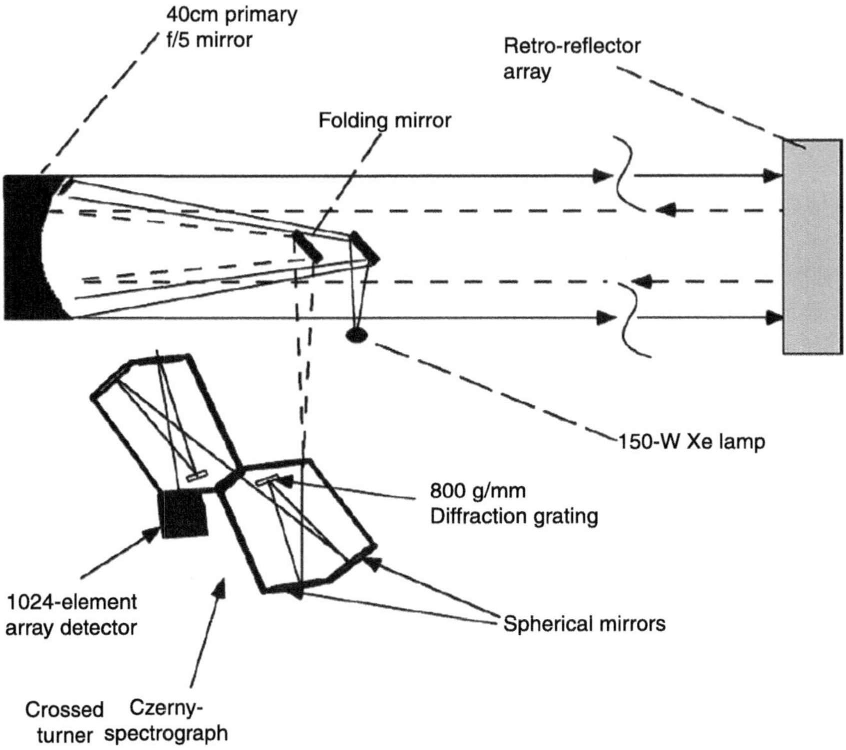

Fig. 10.9 Instrumental setup for the measurement of ammonia emission from a dairy farm. Reprinted with permission from Mount et al. (2002),

Measurements of ammonia with DOAS have, for example, been reported for flue gas analysis (Mellqvist and Rosén 1996), near roads (Burgard et al. 2006), as well as in ambient air (Neftel et al. 1990, Gall et al. 1991, Volten et al. 2012; Sintermann et al. 2016). Here, we show an example of the measurement of ammonia concentrations at a dairy farm. Mount et al. (2002) used a coaxial long-path DOAS system with an array of retro-reflectors located at a distance of up to 150 m (Fig. 10.9). The main difference from the setups described in Chap. 7 is the introduction of a double spectrograph for improved stray light suppression in the UV. Because NH_3 absorbs between 190 and 220 nm (Fig. 10.10, right panel), at even smaller wavelengths than aromatics, stray light suppression is critical for this application.

The left panel of Fig. 10.10 shows an example of a raw spectrum recorded with the instrument shown in Fig. 10.9. The intensity rapidly decreases towards shorter wavelengths. In general, the light intensity is low compared with long-path instruments operating in the visible range. However, the intensity is sufficient to allow the identification of ammonia in ambient air. The right panel of Fig. 10.10 shows a comparison of a spectrum measured with the light path passing over a NH_3 source compared with a differential absorption cross-section of NH_3. Taking into account saturation effects in the raw ratio, the agreement between the two curves is excellent.

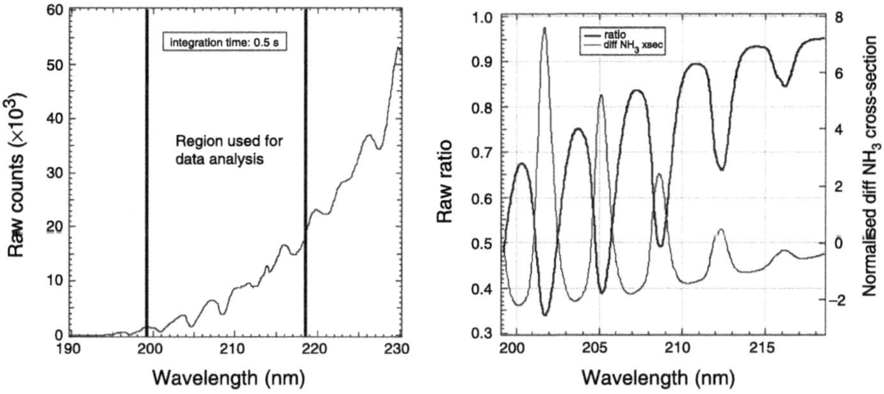

Fig. 10.10 Left panel: Raw spectrum measured on a 68 m path with an integration time of 0.5 s (Mount et al. 2002). The decrease of the intensity towards shorter wavelengths can be clearly seen. **Right panel**: Comparison of the ratio of two spectra with and without NH_3 in the light beam, to the absorption cross-section of NH_3. The difference in shape is caused by saturation effects in the raw ratio. Reprinted with permission from Mount et al. (2002), ©Elsevier. All rights reserved

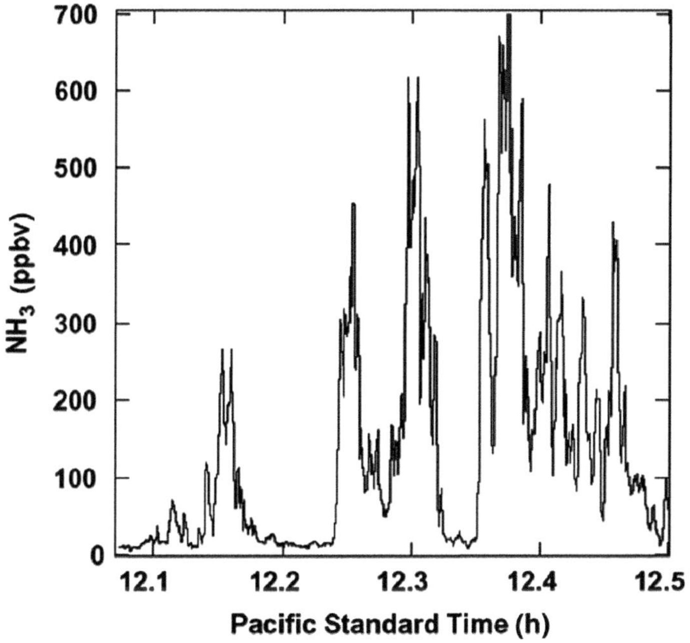

Fig. 10.11 Example of ammonia mixing ratios measured downwind of a slurry lagoon (Mount et al. 2002). The integration time for one measurement was 0.6 s. Reprinted with permission from Mount et al. (2002), ©Elsevier. All rights reserved

Mount et al. (2002) used this instrument to monitor the release of ammonia on a diary farm. Figure 10.11 shows 1 h of high frequency measurement downwind of a slurry lagoon. The rapid variations are caused by shifting winds and turbulent cells moving into the light path. The observations by Mount et al. (2002) illustrate how active DOAS instruments can be applied to fenceline monitoring, in this case of the release of ammonia from a dairy farm. Use of DOAS for the measurement of NH_3 has thus far not been extensively explored and there is great potential for further developments.

10.2.2 Vertical Profiles of Air Pollution by Multiple DOAS Light Beams

As described above, active DOAS instruments can be used to measure the vertical distribution of trace gases. This was first explored by Platt in 1977, with the goal of measuring SO_2 deposition velocities. While this technique was not used for nearly two decades, it was revived around the year 2000 to study nocturnal chemical processes. The investigation of NO_2 to HONO conversion was a first example of vertical profiling. Here, we want to discuss another example used to study the vertical variation of nocturnal chemistry.

Two long-path DOAS instruments were set up in Houston, Texas, during TEXAQS 2000 (see also Sect. 10.2.1 above). The measurements were made on five different light paths, as illustrated in Fig. 10.12. The first DOAS system measured on two short light paths, taking $\sim$ 5 min for one series of measurements. The second system took $\sim$ 15 min to measure all three long light paths. The raw data of these measurements are shown in Fig. 10.13.

A simple deconvolution process was then used to derive trace gas mixing ratios averaged in the boxes indicated in Fig. 10.12. Because the measurements were made sequentially, all data were linearly interpolated to the time of the upper long path measurement. The path integrated concentration along the ith light path is denoted as S_i, and the concentrations in a specific height interval as C_i, where the light paths are numbered from the ground upwards. The data of the two lowest paths directly represent the concentrations at 2-m altitude and in the height interval (2–44 m). The

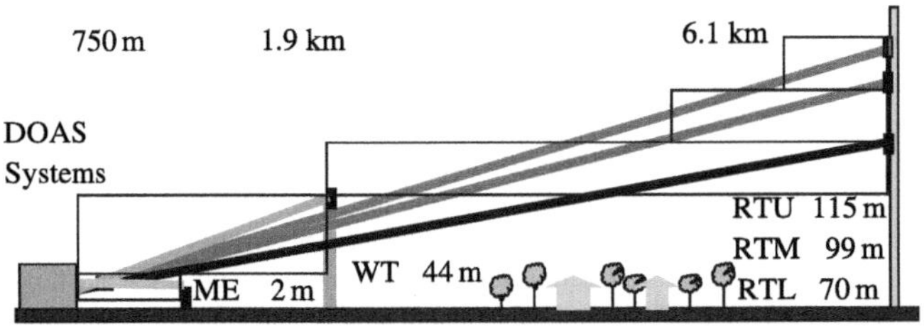

Fig. 10.12 Setup during the TEXAQS 2000 experiment. Five retro-reflector arrays were mounted at different distances and altitudes. The measurements were performed by two DOAS instruments. Reprinted with permission from Stutz et al. (2004a), ©American Geophysical Union. All rights reserved

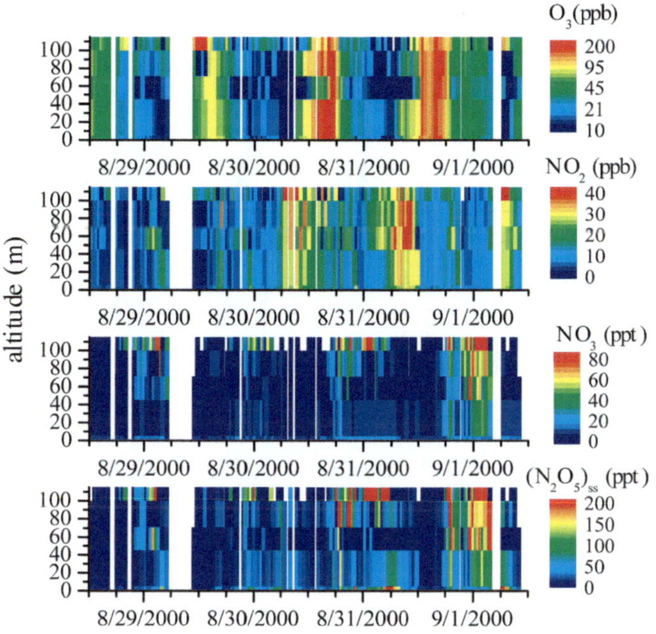

Fig. 10.13 Observations of the spatial distribution of trace gas mixing ratios during 4 d in Houston, TX. The N_2O_5 mixing ratios are calculated from the steady state of measured NO_2, NO_3, and N_2O_5. The O_3 mixing ratios are displayed in a logarithmic colour coding. Reprinted with permission from Stutz et al. (2004a), ©American Geophysical Union. All rights reserved

other concentrations C_3, C_4, and C_5 in the height intervals, 44–70 m, 70–99 m, and 99–115 m, respectively, are calculated from the ground upwards with the following equation:

$$C_i = \frac{h_i - H}{h_i - h_{i-1}} S_i - \frac{h_{i-1} - H}{h_i - h_{i-1}} S_{i-1}$$

An important assumption, which underlies the application of this equation, is that the trace gases are homogeneously distributed in the horizontal direction. This can be tested by using a number of techniques outlined by Stutz et al. (2004a). It is also essential to determine the errors of the concentrations averaged over the height intervals. Since differences in the measurements are formed and the effective absorptions path in the boxes is shorter than the initial path length, the errors of the box averaged concentrations increase. One has to be careful to assess whether the measured concentrations are statistically meaningful. Four days of box-averaged mixing ratios during times when horizontal uniformity was encountered and statistically significant trace gas levels were found are shown in Fig. 10.13. During a number of instances, vertical profiles of the various trace gases were observed. Clearest are the nocturnal maxima of NO_3 in the upper nocturnal boundary layer each night. On several occasions, vertically non-uniform distributions of O_3 and NO_2 are observed. For example, on August 31, a plume of NO_2 between 50 and 90 m was encountered.

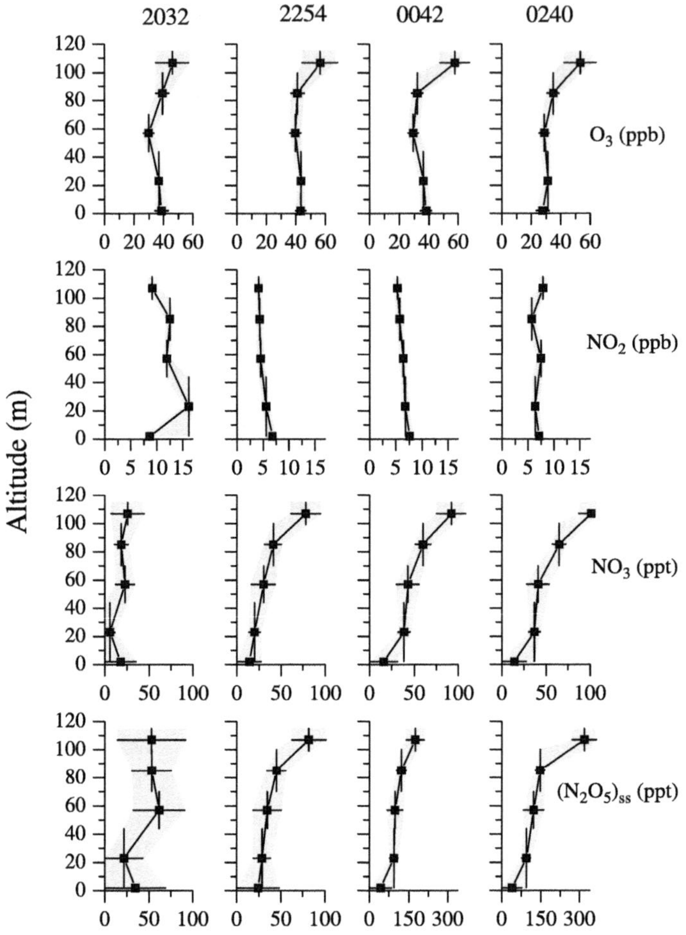

Fig. 10.14 Vertical mixing ratio profiles of O_3, NO_2, NO_3, and N_2O_5 near Houston, Texas (USA) during the night of 8/31–9/1, 2000 at four different times (noted on *top of the graphs*). The N_2O_5 mixing ratios shown are calculated from the steady state of measured NO_2, NO_3, and N_2O_5. Reprinted with permission from Stutz et al. (2004a),

The nocturnal profiles of the various gases are more clearly seen in Fig. 10.14. The profiles show that the levels of all gases vary with altitude. This variation, and in particular that of NO_3, the dominant nocturnal radical, leads to an altitude-dependent nocturnal chemistry. Similar experiments have confirmed these results. Figures 10.13 and 10.14 show examples of the abilities of modern DOAS instruments to perform process studies in the atmosphere. Active DOAS continues to contribute to the advancement of our understanding of pollution in the atmosphere.

10.3 Investigation of Free Radical Processes in the Atmosphere

As outlined in Chap. 2, free radicals play a central role in many chemical systems, such as flames, living cells, or in the atmosphere. Due to their high chemical reactivity, radicals are the driving force for most chemical processes in the atmosphere. In particular, they initiate and carry reaction chains. Thus, knowledge of the concentration of those species in the atmosphere is a key requirement for the investigation of atmospheric chemistry. For instance, degradation of most oxidisable trace gases in the atmosphere, such as hydrocarbons, carbon monoxide, or HCFC's, is initiated by free radicals.

As a consequence of their high reactivity, the steady state concentration of free radicals in the atmosphere is generally quite low, even compared with atmospheric trace gas standards. For instance, the atmospheric lifetime of the OH radical never exceeds 1 s. A consequence of this short lifetime is that transport processes can be largely neglected when considering the budgets of most free radicals. Therefore, measurement of the instantaneous concentration of free radicals, together with a comprehensive set of other trace gases linked with the radical's chemical cycles, allows the study of in situ chemical reactions in the atmosphere (Perner et al. 1976, 1987a; Platt et al. 1987, 1988, 2002; Dorn et al. 1988, 1993, 1996; Mount 1992; Mount and Eisele 1992; Eisele et al. 1994; Brauers et al. 1996, 1999; Hausmann et al. 1997; Hofzumahaus et al. 1998).

Several key free radicals of interest to atmospheric chemistry can be measured by DOAS with very high (sub—ppt) sensitivity, including hydroxyl radicals (OH), nitrate radicals (NO_3), and halogen oxide radicals (ClO, OClO, BrO, OBrO, and IO). Most of these species were detected for the first time in the atmosphere by DOAS.

In this section, we will discuss some examples of the measurement of these radical species. Again, we can only show examples of the many measurements that have been made of radical species.

10.3.1 Measurement of OH Radicals by DOAS

A series of OH measurement campaigns were performed with the DOAS technique, initially using optomechanical scanning of the spectra (Perner et al. 1976, 1987a, Platt et al. 1988), and in later instruments employing diode array detectors (Dorn et al. 1988; Hofzumahaus et al. 1991), see Sect. 7.8.2. Observed OH radicals levels ranged from the detection limit ($1–2 \times 10^6$ molec. cm^{-3}) to about 10^7 molec. cm^{-3}. Figures 10.15, 10.16, and 10.17 show the optical setup using a frequency doubled dye laser as light source, and provide examples of a spectrum and a diurnal profile of the OH concentration.

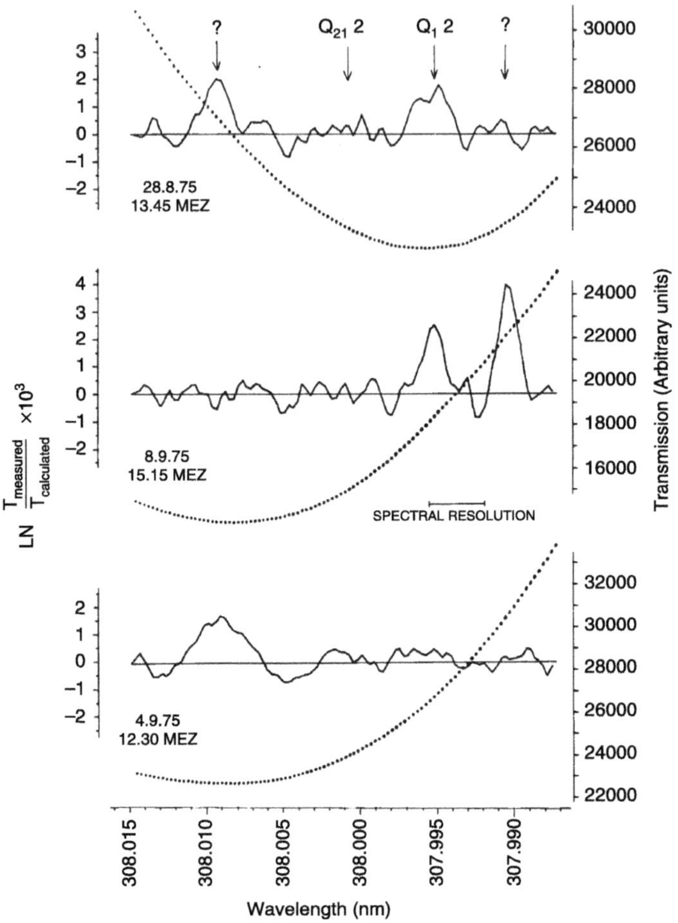

Fig. 10.15 High-resolution spectrum of OH. The DOAS instrument was used for the first reliable detection of OH in the atmosphere Reprinted with permission from Perner et al. (1976), ©American Geophysical Union. All rights reserved

Other implementations of DOAS OH instruments are based on multi-reflection cells with base path lengths of 20–40 m (Brauers et al. 1996). This instrument reaches detection limits down to 1.5×10^6 molec. cm^{-3} in an integration time of about 200 s. Figure 10.18 shows the different absorption features that can be found in the wavelength window used for the measurement of OH. In particular, interferences of SO_2, HCHO, and naphthalene are observed.

Figure 10.19 shows the identification of the OH absorptions in an atmospheric spectrum. The spectrum was measured using the multi-channel scanning technique (MCST), which is explained in detail in Sect. 7.13.2. It should be noted that the amplitude of the residual is extremely low in this measurement, with a peak-to-peak equivalent optical density of $\sim 10^{-4}$.

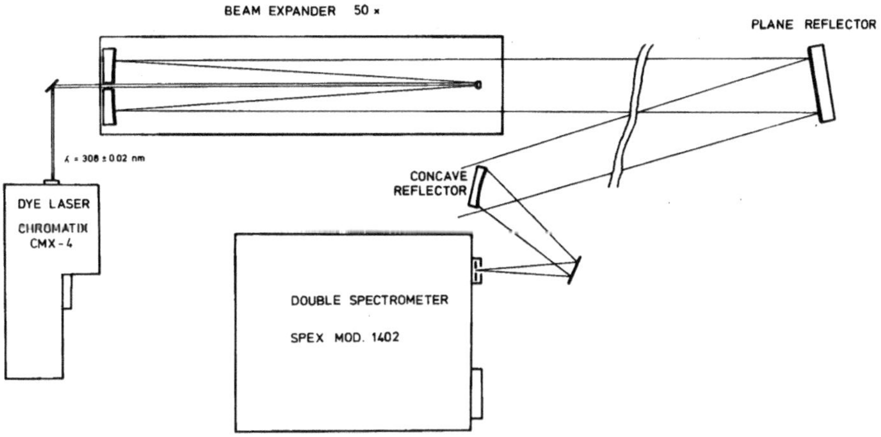

Fig. 10.16 Optical setup of the high-resolution DOAS instrument used for the first reliable detection of OH in the atmosphere (see Perner et al. 1976). A broadband, frequency-doubled dye laser serves as light source. Its beam is expanded to 50-times its original diameter, sent out to the atmosphere, reflected by a plane mirror, received by a 15cm dia. telescope, and focussed onto the entrance slit of a spectrometer. Reprinted with permission from Perner et al. (1976), ©American Geophysical Union. All rights reserved

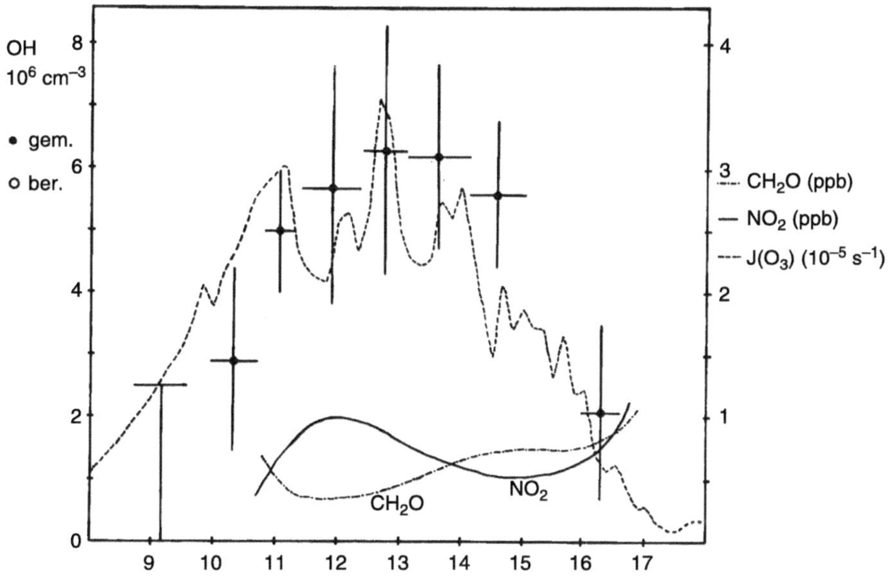

Fig. 10.17 Diurnal OH profile measured with a later version of the OH DOAS instrument (see Fig. 7.54). Reprinted with permission from Platt et al. (1987), ©2023 Informa UK Limited. All rights reserved

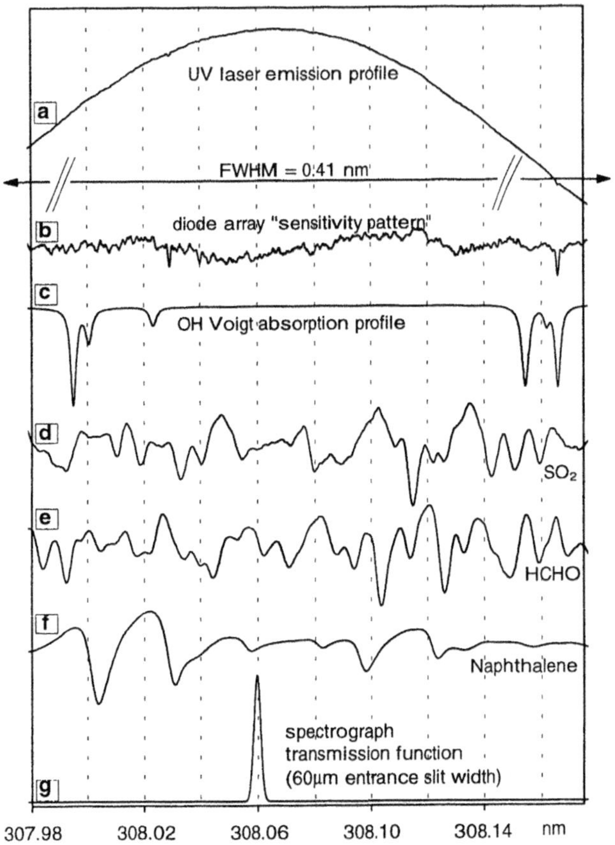

Fig. 10.18 The OH absorption spectrum (*trace c*) together with the spectra of several interfering species SO_2, CH_2O, and naphthalene (*traces d–f*). The zero points of the individual curves are suppressed. Also shown are the laser intensity profile (*trace a*) and the diode array sensitivity pattern. Reprinted with permission from Hausmann et al. (1997), ©American Geophysical Union. All rights reserved

The measurements in Figs. 10.17 and 10.20 show the diurnal profile of OH with a maximum around noon. The comparison of the OH concentration with $J(O^1D)$ in Fig. 10.20, confirms the strong influence of solar radiation on OH levels.

Fig. 10.19 Ambient OH measurement in Pennewitt, Germany, August 16, 1994. The atmospheric spectrum (*top trace*) is dominated by the absorptions of SO_2 as shown in *trace d* of Fig. 10.18. After the removal of this absorption, the structures of OH can be clearly identified in the spectrum (*bottom trace*). The absorptions correspond to an OH concentration of $8.8 \cdot \times 10^6$ molec. cm^{-3}. Reprinted with permission from Hausmann et al. (1997), ©American Geophysical Union. All rights reserved

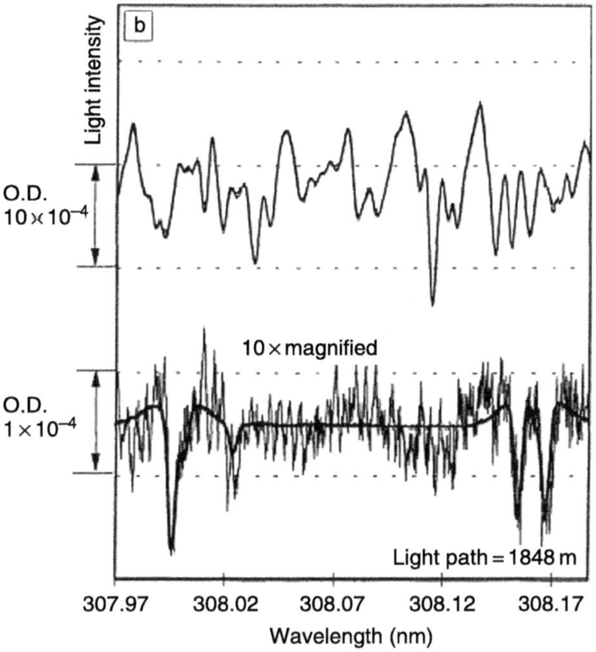

Fig. 10.20 Diurnal profiles of OH concentration and ozone photolysis frequency at a rural site in north-eastern Germany (**top panel**). Also shown are the concentrations of HCHO and SO_2 that are determined in the measurement (**bottom panel**). Reprinted with permission from Dorn et al. (1996), ©American Geophysical Union. All rights reserved

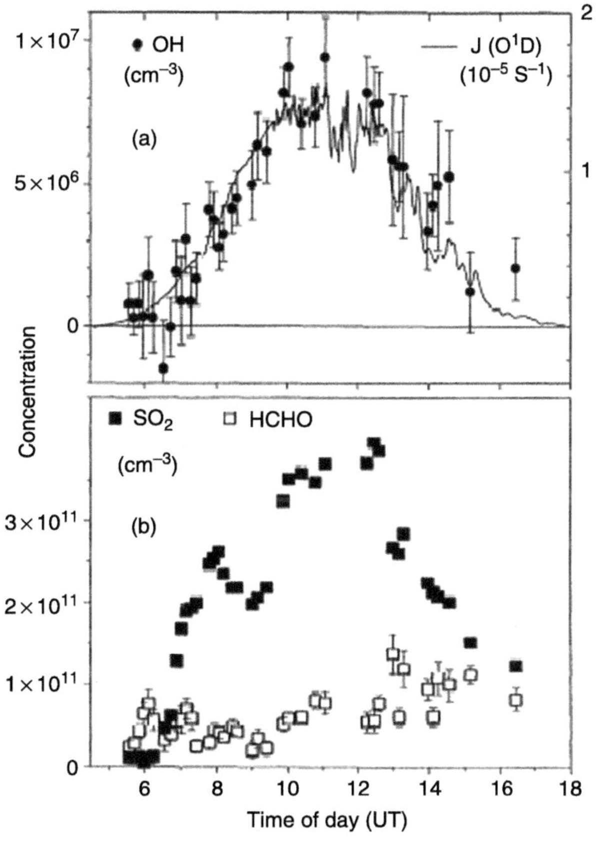

10.3.2 Measurement of NO_3 Radicals

NO_3 and OH, are in a sense complementary to each other (see Chap. 2). OH radicals are directly produced by photochemical reactions, thus their concentration is highest at noon and low at night. Conversely, nitrate radicals are effectively destroyed during the day, thus they will reach high concentrations predominately at night. Both species are responsible for the removal of hydrocarbons and NO_X from the atmosphere. The product of NO_3 and OH reactions with NO_2 is nitric acid and aerosol nitrate. We already discussed the role DOAS played in the detection of OH radicals. Here, we will concentrate on the importance of DOAS measurements for nocturnal radical chemistry, i.e. NO_3 chemistry.

NO_3 was first identified in the troposphere spectroscopically by Platt et al. (1979) and Noxon et al. (1980). Figure 10.21 shows the first reported NO_3 measurement by active DOAS. Due to the width of the NO_3 absorption bands and location in the visible part of the spectrum NO_3 can readily be measured by active DOAS using a variety of thermal light sources (incandescent lamps, Xe-arc lamps). Since then a

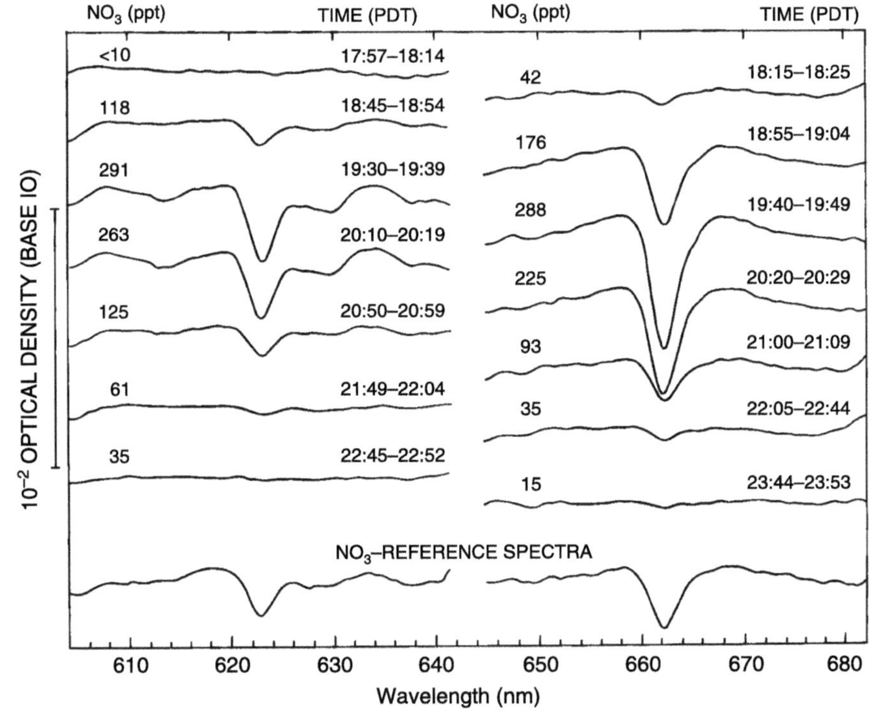

Fig. 10.21 Example of the evolution of the NO_3 radical spectra recorded in an urban area (Riverside, Los Angeles, USA on Sept. 12, 1979). The absorption structures due to H_2O and O_2 are already removed. Reprinted with permission from Platt et al. (1980a), ©American Geophysical Union. All rights reserved

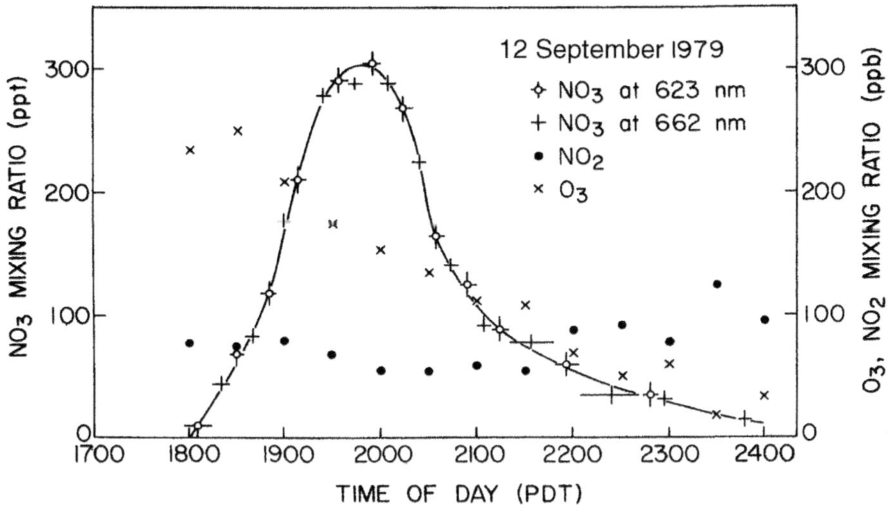

Fig. 10.22 Example of NO_3 radical time series in an urban area (Riverside, Los Angeles, USA on Sept. 12, 1979) after sunset. Data are derived from the series of spectra shown in Fig. 10.21. Reprinted with permission from Platt et al. (1980a), ©American Geophysical Union. All rights reserved

large number of NO_3 measurements in the atmosphere have been reported (Platt et al. 1980a, 1981, 1984, 1990; Harris et al. 1983; Brauers et al. 1990; Platt 1991; Plane and Nien 1991; Winer and Biermann 1991; Platt and Heintz, 1994b; Platt and Hausmann 1994a; Plane and Smith 1995; Platt and Janssen 1996a; Heintz et al. 1996; Carslaw et al. 1997; Aliwell and Jones 1998; Allan et al. 1999, 2000a; Geyer et al. 1999, 2001a, 2001b, c; Gölz et al. 2001; Ball et al. 2001; Geyer and Platt 2002; Stutz et al. 2002b; Geyer et al. 2003a, b; Kern et al. 2006). A challenge in the measurement of NO_3 is the interference with a water-absorption band system around 650 nm (see, e.g. Figure 10.23). Because the spectral structure of this band system is much narrower than that of NO_3, filtering procedures are often used to reduce the effect of water (Geyer et al. 1999). Another method to remove this water band is the use of daytime atmospheric spectra as water references. This technique has been employed in most NO_3 measurements in the past (see top panel in Fig. 10.24). However, as pointed out by Geyer et al. (2003), NO_3 levels can be elevated under conditions with high O_3 mixing ratios ($[O_3] > 100$ ppb) during the day. In this case, more care has to be taken in the choice of water reference spectra. Geyer et al. (2003) used daytime spectra taken during periods of high NO mixing ratios as water references.

Because DOAS is the most commonly used technique to measure the atmospheric concentration of NO_3, there are many examples of its application. Figures 10.21 and 10.23 show NO_3 spectra in an urban and rural environment, respectively. Figure 10.22 presents the corresponding time series of NO_3 after sunset in an urban environment. Because NO_3 also plays an important role in marine areas, we also show an example taken in the remote marine boundary layer (Fig. 10.24). This spectrum illustrates that very low NO_3 mixing ratios, close to 1 ppt, can be measured accurately.

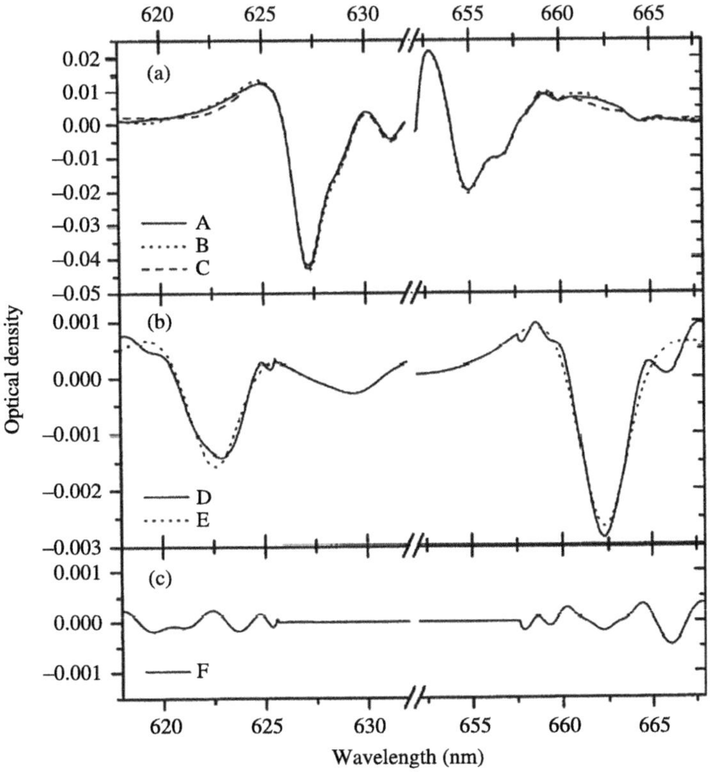

Fig. 10.23 Example of nighttime NO_3 radical absorption in a rural area (August 5, Pabsthum, Germany). (Geyer et al. 1999). The *top panel (a)* shows the atmospheric spectrum at 2150 UT (**trace A**) and two daytime references (**traces B, C**), which are dominated by water absorptions. The *middle panel (b)* shows a comparison of the scaled NO_3 reference with added residual (**D**) and the NO_3 reference (E)). The optical density of NO_3 absorption is $\sim 0.35\%$ which, in this case, was equivalent to 9.8 ± 1.1 ppt. The *bottom panel (c)* shows the residual of the analysis (**trace F**). Details about the analysis procedure can be found in Reprinted with permission from Geyer et al. (1999), ©American Geophysical Union. All rights reserved

10.3.2.1 Observations in the Marine Boundary Layer

A number of NO_3 DOAS observations in the marine boundary layer have been reported (e.g. Heintz et al. 1996; Martinez et al. 2022). Here, we will review data measured at Mace Head, Ireland and Weybourne, England (Allan et al. 2000a; Carslaw 1997).

Figure 10.25 shows an overview of a time series of NO_3 measurements. The maximum NO_3 mixing ratios observed at Mace Head are up to 40 ppt. A clear dependence on the origin of the air is observed. Clean atlantic or polar air has low NO_3 levels, while continental air shows much higher NO_3. This can be explained by

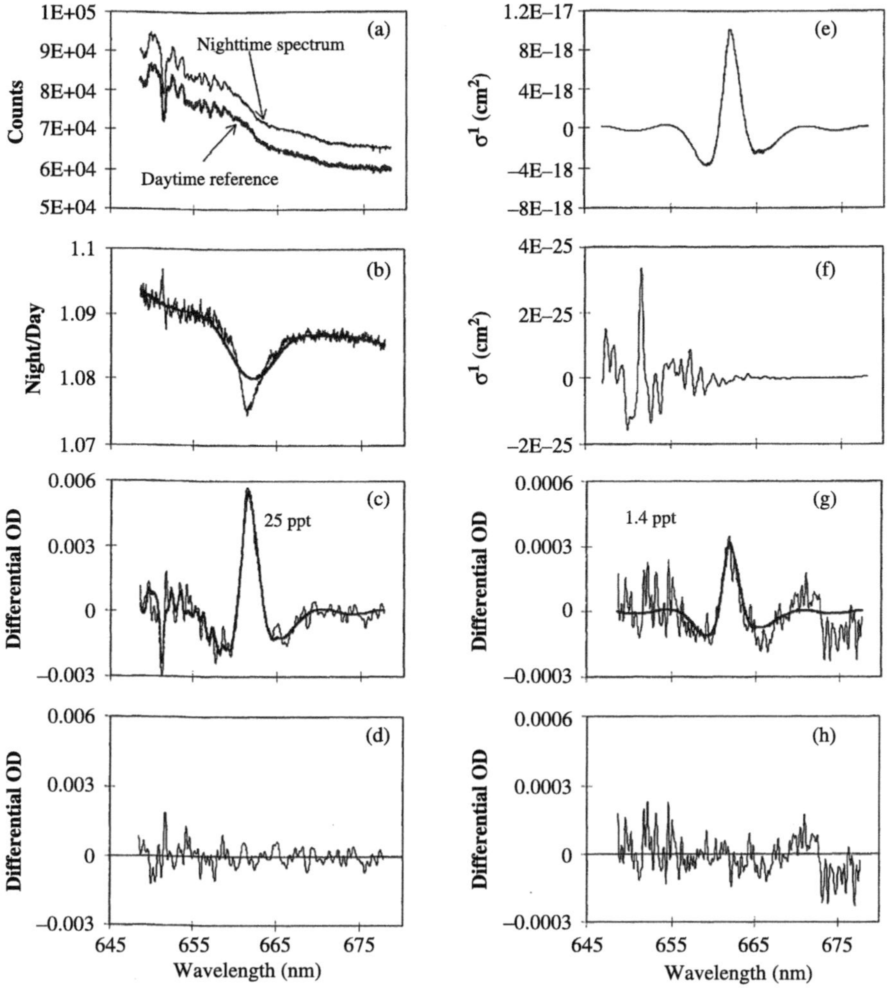

Fig. 10.24 Example of nighttime NO_3 radical absorption in the marine boundary layer (Allan et al. 2000a): (**a**) shows the daytime reference and the nocturnal spectrum; (**b**) is the ratio of the two spectra with a low-pass filter; (**c**) and (**g**) show the fit of the H_2O and NO_2, cross-sections (**e and f**) and (**d**) is the residual. The result of the fit gave a NO_3-mixing ratio of 24.7 ± 0.7 ppt. An example with 1.9 ± 0.2 ppt of NO_3 is shown in (**g**). The measurements were made with a long-path DOAS instrument on a 8.4-km-long light path at Mace Head, Ireland. Details on the instrument and the analysis procedure can be found in Reprinted with permission from Allan et al. (2000b), ©American Geophysical Union. All rights reserved

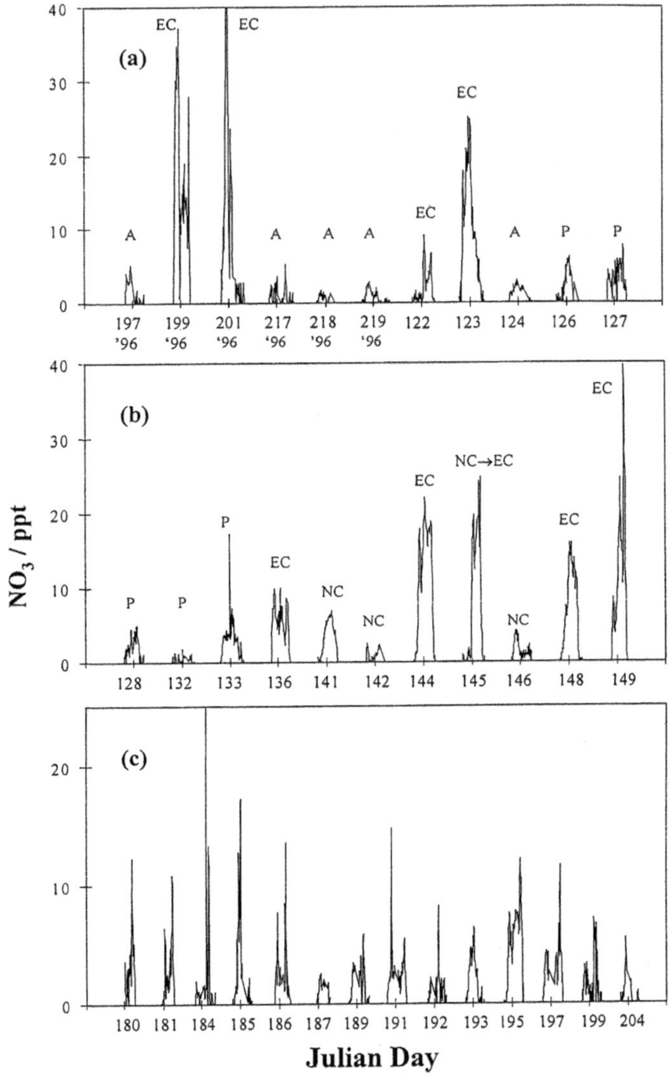

Fig. 10.25 NO_3 time series collected in the marine boundary layer of Mace Head, Ireland (**a** and **b**). The letters denote the origin of the observed air masses: A, atlantic; P, polar marine; EC, easterly continental; NC, northerly continental. Reprinted with permission from Allan et al. (2000a),

the much higher NO_x levels in these air masses, which increases the formation rate of NO_3.

The data in Fig. 10.25 can be used to derive the removal rate of NO_X at night through the calculation of N_2O_5 and its uptake on the aerosol. In addition, it can be used to determine the nocturnal oxidation of DMS. Figure 10.26 shows another application of these measurements: a comparison of NO_3 and $HO_2 + RO_2$ levels.

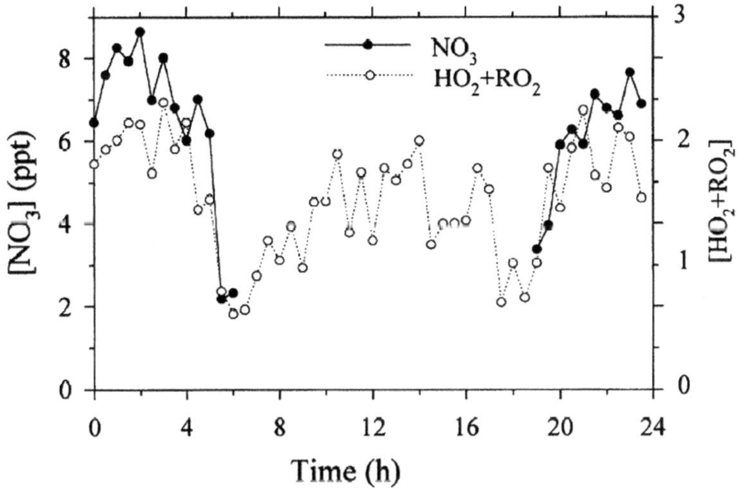

Fig. 10.26 Diurnal variation of NO_3 and $HO_2 + RO_2$ produced by averaging half-hourly data taken during 5 d. Reprinted with permission from Carslaw et al. (1997), ©American Geophysical Union. All rights reserved

This correlation is due to the oxidation of organic species by NO_3 at night (see Sect. 2.5.2).

10.3.2.2 Observations in Rural and Forested Areas

Measurements of NO_3 in rural and forested areas are of particular significance since emissions of monoterpenes, which rapidly react with NO_3, are often highest there. A continental long-term study was reported at Lindenberg near Berlin, Germany by Geyer et al. 2001a (Fig. 10.27). Observations of this type offer the possibility of statistical investigations of the nighttime NO_3 and N_2O_5 concentrations and the NO_3 production and degradation frequencies (see Sect. 2.5). In addition, the mean contribution and the seasonal variation of the different sink mechanisms of NO_3, which provides an essential tool for quantifying the average effect of NO_3 on the oxidation capacity of the atmosphere, can be studied.

10.3.2.3 Observations in Urban Areas

Figure 10.21 showed examples of urban spectra of NO_3. Surprisingly, few measurements of NO_3 have been performed in this environment. This is partially explained by the often high NO levels at the ground in cities, which efficiently destroy NO_3. As shown in Figs. 10.13 and 10.14, the levels of NO_3 are quite high in the upper nocturnal boundary layer. Mixing ratios above 100 ppt are frequently observed in cities, in which the NO_3 precursors O_3 and NO_2, are elevated. Studies by Geyer

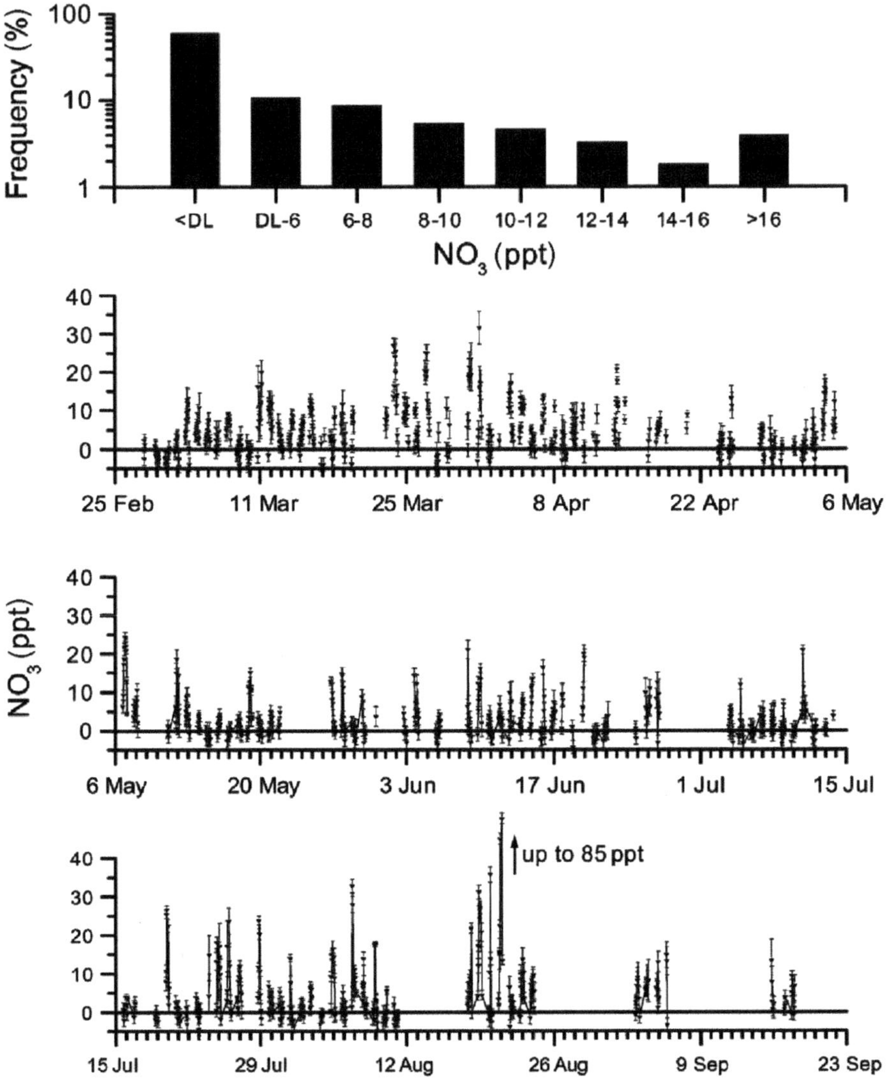

Fig. 10.27 Nighttime NO_3 radical concentration profiles observed during a long-term campaign in a semi-rural area (Lindenberg near Berlin, Germany). Time series and frequency distribution (top panel, in 2-ppt intervals centred at the number given, DL=detection limit) of NO_3 observed by DOAS measurements averaged over a 10-km light path from February 27 to September 18, 1998. During the night from August 20–21, 1998, NO_3 reached peak mixing ratios up to 85 ppt. Reprinted with permission from Geyer et al. (2000),

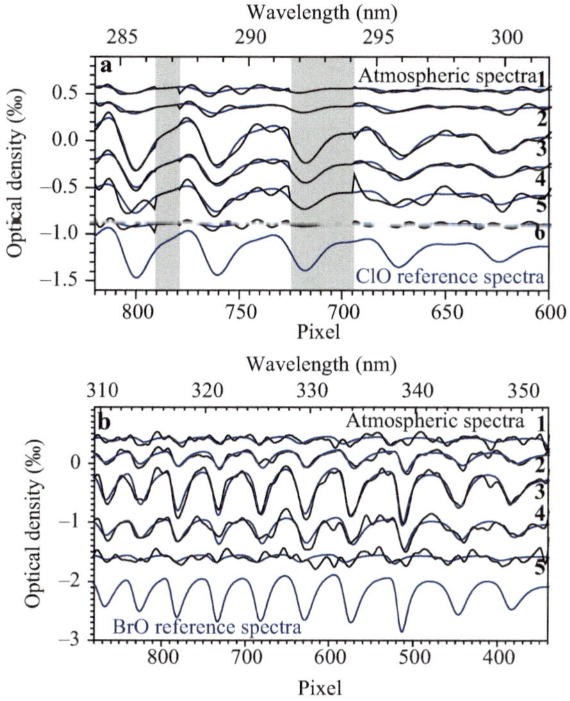

Fig. 10.28 ClO (**a**) and BrO (**b**) absorption structures recorded at the Great Salt Lake during October 14, 2000. The *black lines* are the atmospheric spectra after removing all absorptions except those by the respective halogen oxides. The *blue lines* are the respective fitted reference spectra that were calculated from a literature absorption cross-section (Orphal et al. in press; Sander and Friedl 1989). Spectra are offset to each other to simplify the graph. The *grey intervals* in (**a**) mark spectral regions that were excluded from the fit Reprinted with permission from Stutz et al. (2002b), ©American Geophysical Union. All rights reserved

and Stutz (Geyer et al. 2003a, b; Stutz et al. 2004b) showed that NO_3 can play an important role for the removal of NO_x and VOCs emitted by traffic in cities.

10.3.3 Measurement of Halogen Oxides

The discovery of boundary-layer ozone depletion events in the Arctic (Oltmans and Komhyr 1986; Barrie et al. 1988; Bottenheim et al. 1990; Niki and Becker 1993) has increased the interest in the chemistry of reactive halogens in the troposphere. DOAS measurements by Hausmann and Platt (1994) showed for the first time that these depletion events, during which ozone drops from ~ 40 ppb to unmeasurable levels (<2 ppb), are associated with high concentrations of reactive bromine, in particular BrO. Sources of gas phase Br are most likely heterogeneous reactions of

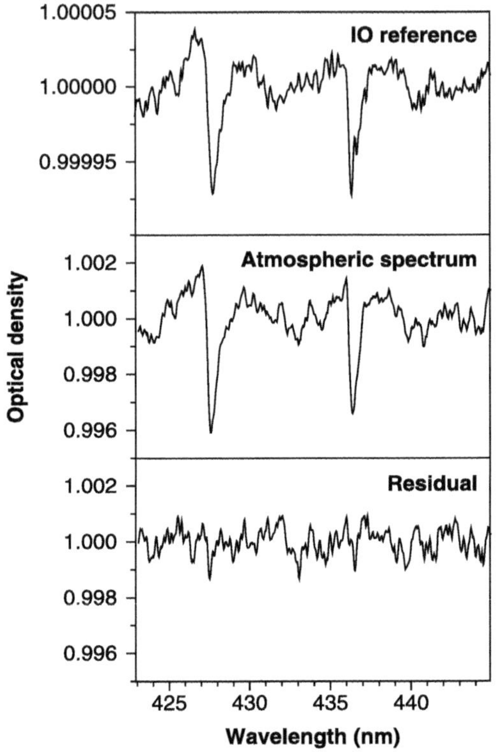

Fig. 10.29 Comparison of atmospheric spectrum after the removal of NO_2 and H_2O absorptions (centre panel) with the IO spectrum (top panel) clearly shows the presence of IO. Reprinted with permission from Alicke et al. (1999), ©1999, Macmillan Magazines Ltd.. All rights reserved

oxides of nitrogen with sea-salt surfaces (Barrie et al. 1988; Finlayson-Pitts et al. 1990; Fan and Jacob 1992). Details of this chemistry are discussed in Chap. 2.

Here, we will present a number of examples of the measurement of halogen oxides and point out how these observations have advanced our understanding of halogen chemistry.

First, we show examples of the identification of the various halogen oxides in the atmosphere. Figures 10.28 and 10.32 show the spectra of ClO and BrO at the Great Salt Lake in Utah (Stutz et al. 2002b) (see Fig. 10.33). Similar examples for the detection of BrO can be found for the Dead Sea (Hebestreit et al. 1999) and polar regions (e.g. Hausmann and Platt 1994; Tuckermann et al. 1997). The detection of ClO is more challenging due to the shorter wavelengths and the weaker absorptions.

Figures 10.29 and 10.30 show the identification of IO (Alicke et al. 1999) and OIO (Allan et al. 2001) in the atmosphere of Mace Head. Measurements of IO have been made at other locations and there is ample evidence that IO is more ubiquitous than previously assumed (Platt and Hönninger 2003; Saiz-Lopez et al. 2004a; Zingler and Platt 2005; Peters 2004; Peters et al. 2005).

The study of the chemistry of BrO in polar regions was the first scientific application for the capability of DOAS to measure halogen oxides. Figure 10.31 shows data from a field experiment in Ny Alesund, Spitsbergen, during Spring 1996. The

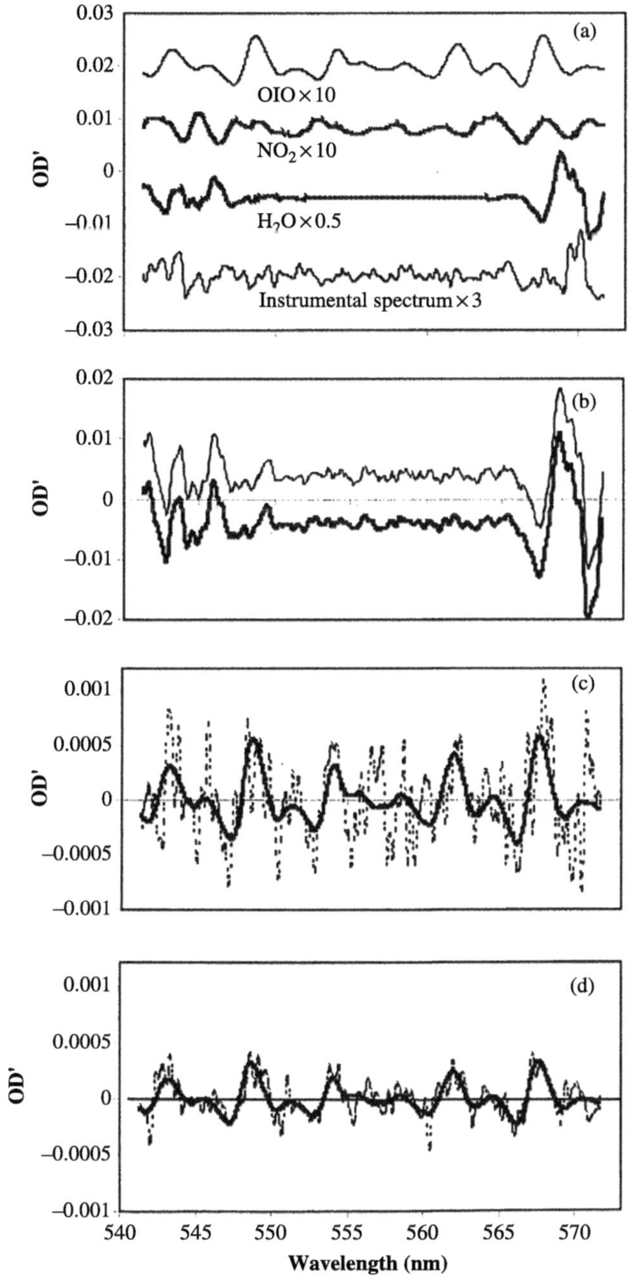

Fig. 10.30 OIO analysis from an atmospheric spectrum. *Panel* (**a**) shows the reference spectra used in the analysis; *Panel* (**b**) depicts the atmospheric spectrum containing all the absorbers. In *panel* (**c**) the absorptions of NO_2 and H2O have been removed. The OIO absorptions in this spectrum correspond to a mixing ratio of 1.8 ± 0.3 ppt. *Panel* (**d**) shows the residual of the analysis. Reprinted with permission from Allan et al. (2001),

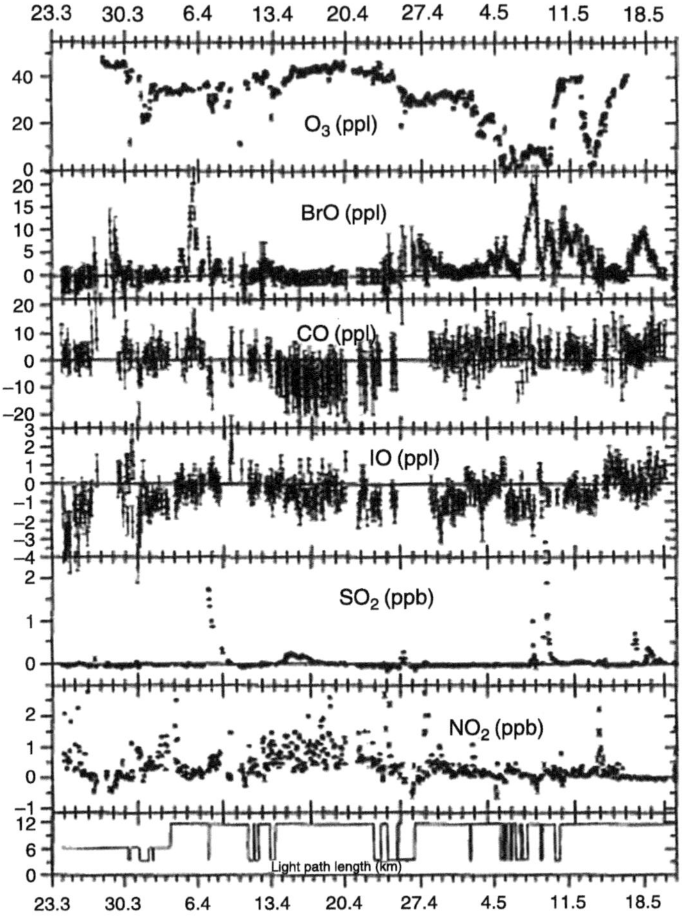

Fig. 10.31 DOAS measurements during Spring 1996 in Ny Alesung, Spitsbergen. The anti-correlation between BrO and O_3 can clearly be seen. Errors are 1σ uncertainties. Reprinted from Tuckermann, Diploma Thesis, 1, 996 University of Heidelberg

onset of ozone decay coincides with elevated BrO levels, providing evidence for the catalytic O_3 destruction cycles described in Chap. 2. The observations at the Dead Sea, Israel, (Fig. 10.32) confirm these results. The Dead Sea has the highest BrO levels thus far reported. With up to 100 ppt of BrO, the complete destruction of O_3 proceeds very fast and influences wide areas in the Dead Sea valley (Hebestreit et al. 1999; Matveev et al. 2001) (Fig. 10.32).

DOAS has provided evidence for the presence of four halogen oxides and is the most reliable method to measure these compounds today. Since tropospheric halogen chemistry is an active research field, DOAS will continue to make contributions by providing accurate measurements of these species.

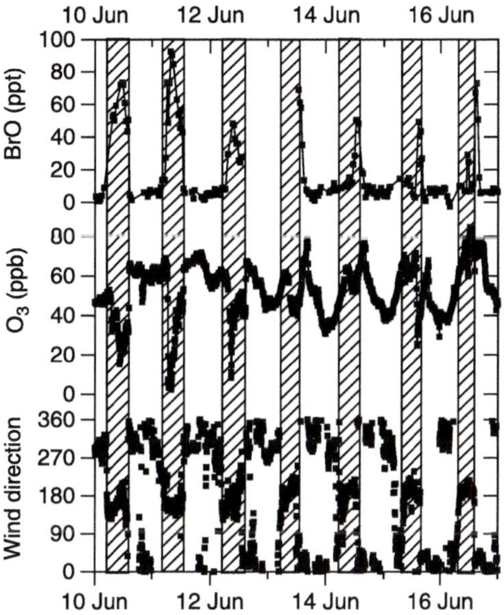

Fig. 10.32 DOAS measurements at the Dead Sea, Israel. The regular pattern in O_3 and BrO is caused by the changing wind direction and the chemical destruction of ozone. Reprinted with permission from Hebestreit et al. (1999), ©American Association for the Advancement of Science. All rights reserved

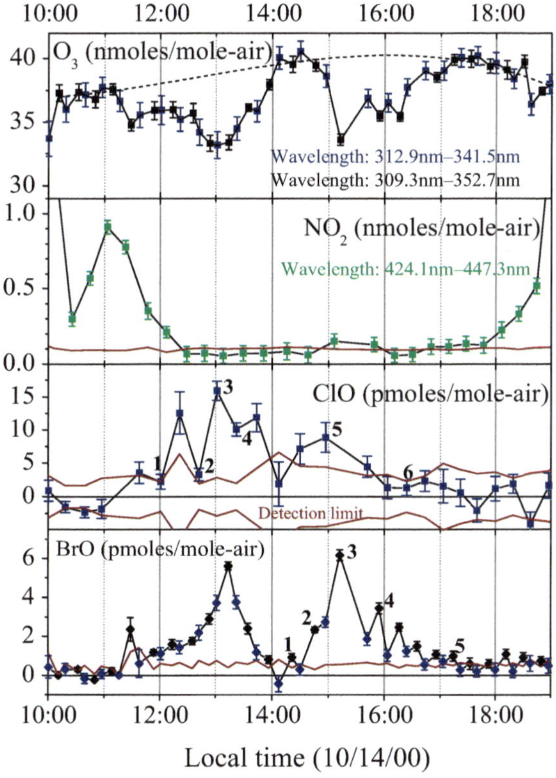

Fig. 10.33 Mixing ratios of O_3, NO_2, BrO, and ClO measured at ground level on October 14, 2000 at the Great Salt Lake, Utah, USA. The different symbols in the ozone and BrO traces indicate the different grating position used in the measurements. The *solid brown lines* show the detection limit determined for each individual spectrum. Reprinted with permission from Stutz et al. (2002a), ©American Geophysical Union. All rights reserved

10.4 Fenceline Monitoring of Industrial Facilities and Emission Measurements

A practical application for active DOAS instrument that has gained popularity in recent years is fenceline monitoring, especially around petrochemical facilities (Fig. 10.34). The target species are aromatic hydrocarbons, in particular benzene, ethyl-benzene, toluene, and xylenes, which are often abbreviated to BTEX. The motivation for monitoring BTEX are their detrimental effects on human health in neighborhoods adjacent to refineries and similar facilities. It is now well known that the incident of cancer rates in these areas is higher that those further away from the facility (see e.g. Domingo et al. 2020; Williams et al. 2020; Johnston and Cushing 2020; Smith 2010). To address this environmental justice problem several states in the United States now mandate the installation of fenceline monitors to alert neighbors and the facility of elevated BTEX levels (Bay Area AQMD 2016; South Coast AQMD 2017; EPA 2015). The South Coast Air Quality District in the Los Angeles basin, for example, has implemented Rule 1180, which requires the installation UV open-path fenceline instruments around the seven refineries in its area (South Coast AQMD 2017). Similar rules have been passed in the Bay Area of California, the State of California, and other areas in the United States (Bay Area AQMD 2016; California 2017). As a consequence, California now hosts over one hundred commercial UV DOAS instruments that are operated continuously. Here we will briefly discuss the use of DOAS-type instruments as fenceline BTEX monitors and show some examples of their usage.

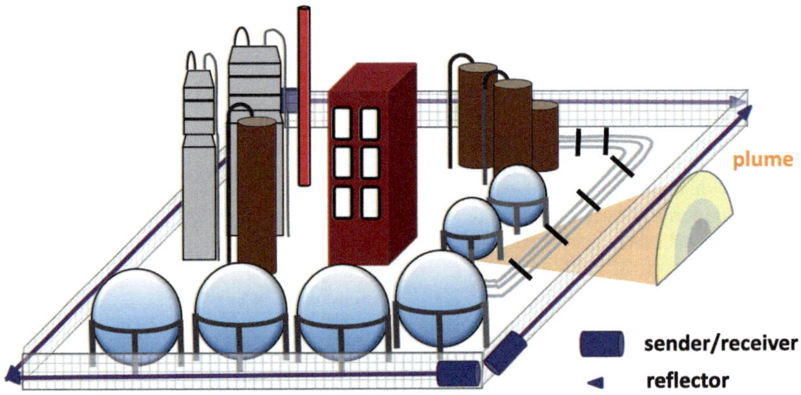

Fig. 10.34 Concept of a fenceline monitoring system for an industrial facility

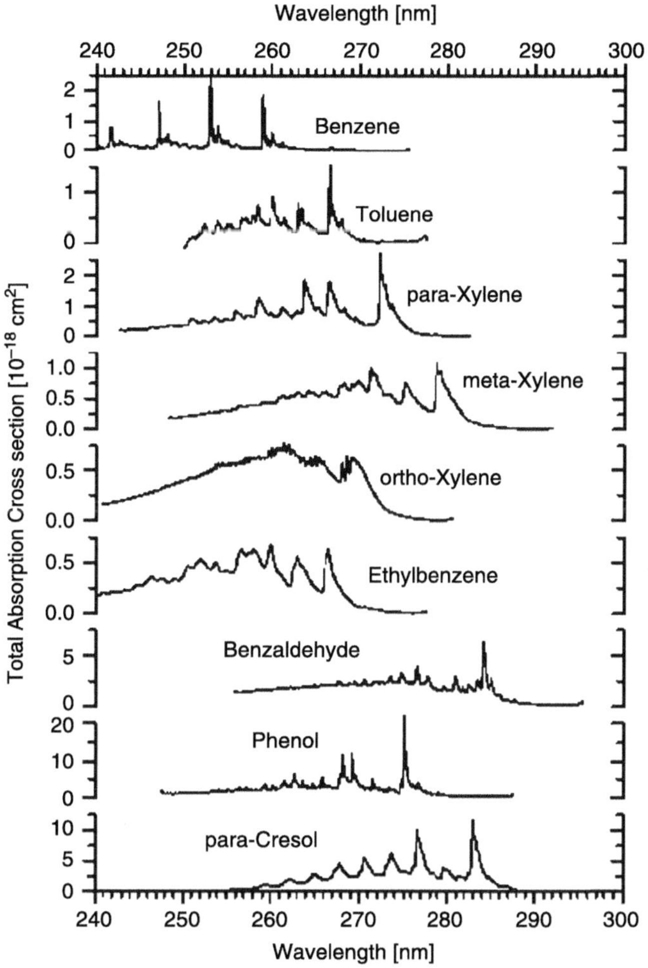

Fig. 10.35 Absorption cross-section of various aromatic hydrocarbons in the UV wavelength range data from Fally et al. (2009); Etzkorn et al. (1999)

10.4.1 Fenceline Monitoring Instruments and Realtime Spectral Analysis

The main practical application of DOAS fenceline instruments is the monitoring of BTEX. Consequently most fenceline instruments (see Fig. 10.34) focus on the spectral regions from 250–290 nm that contain the strongest absorption bands of these aromatic hydrocarbons (Fig. 10.35). Some instruments extend this range to longer wavelengths in order to include SO_2 and NO_2, which are also emitted by petrochemical facilities. One of the main constraints of BTEX observations is the increased Rayleigh and Mie extinction in the 250–290 nm range (see Chap. 3), which limits the maximum path length that can be achieved; thus absorption path lengths exceeding 1–1.5 km are not practical. Because DOAS fenceline measurements are dominated by companies contracted by the respective facilities (see e.g. Schill et al. 2022; Sanchez et al. 2019), many of the data analysis retrievals are proprietary. There are few research grade BTEX fenceline observations and we therefore will include ambient observations of BTEX in this short review (Kurtenbach et al. 2002; Jobson et al. 2010; Stutz et al. 2016). All BTEX DOAS instruments face challenges with strong absorbers in the 250–290 nm region. Oxygen absorption in the Herzberg band as well as due to $O_2 - O_2$ and $O_2 - N_2$ dimers (Fally et al. 2000) dominate the absorption spectrum (see Fig. 10.36). The O_2 concentration and consequently its atmospheric absorptions varies only by about 10% (or in the case of the dimers $\sim 20\%$) due to air temperature and pressure variations. The O_2 absorption lines are in saturation at typical absorption path lengths. At the limited resolution of most DOAS instruments, the shape of these absorptions thus depends mainly on the path length. Details of this effect are discussed in Chap. 6 (Sect. 6.6). Because the modelling of these absorptions based on literature cross-section is difficult, other solutions have to be found. Ozone is also a strong absorber in this wavelength range and, during polluted daytime conditions with high concentrations, the absolute absorptions can severely reduce the transmissivity of the atmosphere. Ozone is highly variable in urban areas and has to be included in any spectral analysis of BTEX compounds.

A challenge in DOAS BTEX measurement is spectrometer stray light. The most common light source in active DOAS, i.e. a Xe-arc lamp, has its emission maximum in the visible range, with much lower intensities in the UV. The imperfect suppression of visible light in spectrometers leads to stray light levels that can be 10% and more of the signal between 250 and 300 nm. This stray light changes optical densities of the various absorbers, making the spectral analysis of the absorption spectra difficult. Solutions to this problem are the use of filters or different light sources, such as UV LEDs (Stutz et al. 2016).

Spectral analysis procedures for BTEX vary widely in particular with respect to the description of the oxygen (and to a lesser extent ozone) bands. In many cases an atmospheric reference spectrum, recorded during times with expected low BTEX concentrations, is used in the data analysis procedure. While this simplifies the analysis by easing the requirement of highly accurate oxygen reference spectra, it cedes one of the main advantages of DOAS, i.e. the retrieval of absolute trace gas concentrations. Multireflection systems in enclosures or in reaction chambers (e.g. Volkamer et al. 1998) can also circumvent this challenge by filling their measurement volume with pure air, thus directly measuring the oxygen and oxygen dimer reference spectrum. Research grade DOAS systems, on the other hand, have evolved in recent years to retrieve BTEX solely based on literature absorption cross-sections (Volkamer et al. 1998; Jobson et al. 2010; Stutz et al. 2016). These approaches generally follow the principles outlined in Chap. 8. Special attention has to be paid to determining the instrument function accurately and to use high-resolution O_2 cross-sections (Fally et al. 2000) and the exact light path length to simulate the oxygen reference spectra. The absorption cross-sections of benzene, toluene, and xylenes are available at high spectral resolution (Fally et al. 2009) and can be directly used in the convolution process. Unfortunately, absorption cross-sections of other aromatics are not yet

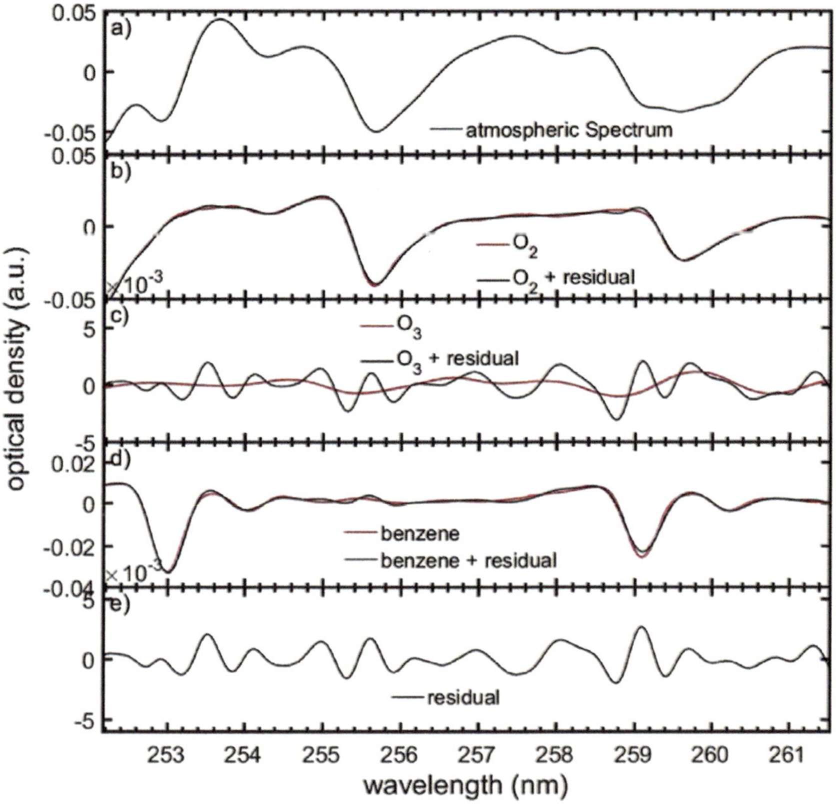

Fig. 10.36 Example of a fenceline benzene retrieval measured over a 540 m long absorption path in Houston, TX, USA Stutz et al. (2016). Aside from the dominating oxygen absorption, the spectrum contains 33 ± 0.9ppb of benzene. Ozone absorption was negligible in the early morning (2/13/2015 6:43 CST) spectrum. Reprinted with permission from Stutz et al. (2016), ©Elsevier. All rights reserved

available at high spectral resolution (see Etzkorn et al. 1999; Trost et al. 1997) and care has to be taken in the convolution process to adjust for the instrument function of the reference spectra.

10.4.2 Principle of Fenceline Monitoring

The basic idea behind open path fenceline monitoring is to surround an entire industrial facility using multiple monostatic, i.e. sender /receiver and retroreflector, or bistatic, i.e. separated sender and receiver, instruments along the entire fenceline (Fig. 10.34). With this setup a more or less continuous absorption path can be set up to measure BTEX leaving the facility. This approach has several advantages. First, the likelihood that a plume of elevated BTEX (Fig. 10.34) is captured by one of the DOAS instruments is much higher compared to in-situ observations, which are not

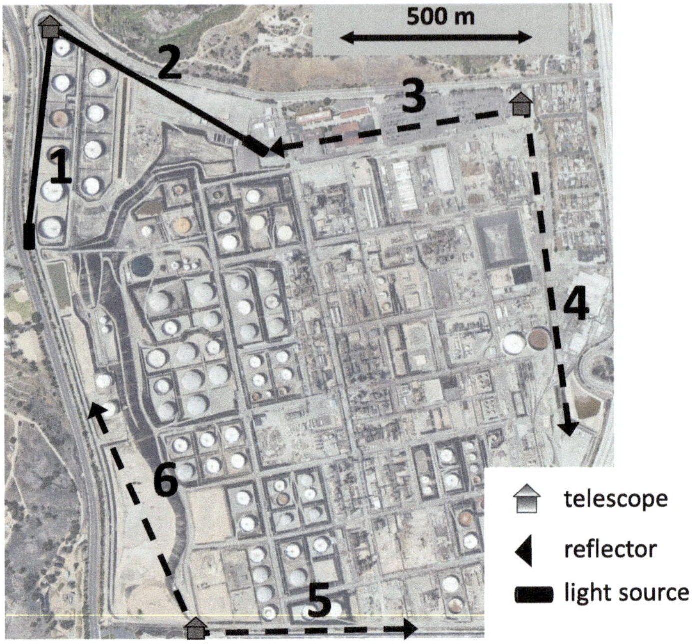

Fig. 10.37 Setup of a fenceline monitoring system at a Southern Californian refinery. Six open-path DOAS systems surround the entire refinery and provide realtime BTEX observations. Solid lines describe monostatic systems while the dahed line are bistatic DOAS instruments

able to provide gap-free sampling along the fenceline. Secondly, having upwind and downwind paths allows determination of whether the source of elevated BTEX is inside or outside the facility, i.e. the presence of a difference between upwind and downwind mixing ratios points to a source inside the fenceline paths. Thirdly, DOAS instruments offer near realtime observations, which makes fenceline monitoring useful as an alarm system for the facility (South Coast AQMD 2017).

More than 50 BTEX DOAS instruments are currently in operation around the seven refineries in the Los Angeles area, USA, alone, and there are many more around other refineries in the USA and globally. An example of a setup at a refinery in the Los Angeles area is shown in Fig. 10.37. Six absorption paths surround most of the facility. Paths 1 and 2 are based on two fixed commercial DOAS instruments, while paths 3,4 and 5,6 are scanned by a dedicated research grade longpath DOAS system (Stutz et al. 2016). This fenceline system provides realtime data on all light paths to a public website. In addition, it sends alarm messages via email and phone to residents if observed mixing ratios exceed certain threshold, for example 8 ppb for benzene.

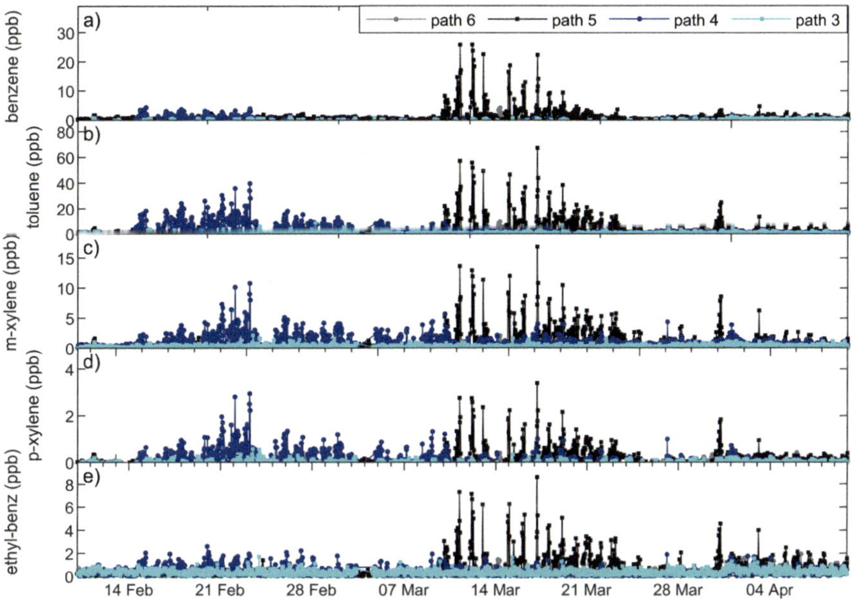

Fig. 10.38 Example of research results from the fenceline system shown in Fig. 10.37. Absorption paths 3 and 6 show low mixing ratios during the entire period. Path 4 encountered elevated toluene and xylenes, with lower mixing ratios of benzene and ethyl-benzene in February. See text for details

Published examples of fenceline BTEX data are sparse, although data are typically reported in realtime on specific websites. Figure 10.38 shows an example from the research part of the fenceline system in Fig. 10.37. Absorption paths 3 and 6 show low mixing ratios during the entire period. Path 4 encountered elevated toluene and xylenes, with lower mixing ratios of benzene and ethyl-benzene, in February. These observations were attributed to coating activities underneath this light path. The high BTEX values in March were associated with an equipment malfunction in the refinery. Benzene mixing ratios exceeded 8 ppb and triggered a public alarm, exemplifying the use of the fenceline system as a warning system for unplanned BTEX releases. It should be noted that the high variability of BTEX mixing ratios on Path 5 are caused by local meteorological effects at this location.

The previous example illustrates the value that active DOAS instruments can add to industrial emission monitoring. Neighborhoods surrounding refineries and other petrochemical facilities are highly impacted by emissions from these facilities. By providing realtime continuous observations of these emissions and warning and informing the residents, DOAS plays an important role in the improvement of the health and quality of life of these populations.

10.5 Investigation in Photoreactors (Smog Chambers) by DOAS

Since its development in the late-1970s DOAS has also been used in aerosol and smog chambers. To our knowledge the first measurements were made in Riverside in 1979 to investigate the formation of nitrous acid.

Comprehensive DOAS measurements were performed in the EUROpean PHOto REactor (EUPHORE) in Valencia, Spain (Becker 1996). The setup is shown in Fig. 10.39. In particular, kinetic and mechanistic studies of the degradation of monocyclic aromatic species were followed by DOAS (Volkamer 2001; Volkamer et al. 2001, 2002). A sample time series of photochemical p-xylene degradation and corresponding formation of the degradation products p-tolualdehyde, 2,5-dimethylphenol, and glyoxal, all measured by DOAS, is shown in Fig. 10.40.

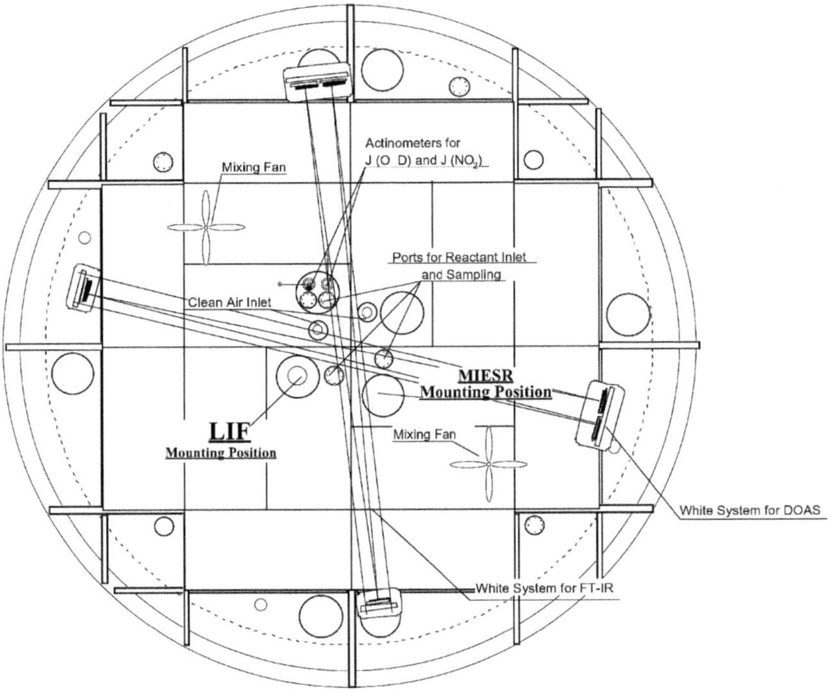

Fig. 10.39 Schematic of the DOAS multi-reflection system (White system) in the EUPHORE chamber in Valencia, Spain. Reprinted with permission from Volkamer et al. (2002), ©PCCP Owner Societies. All rights reserved

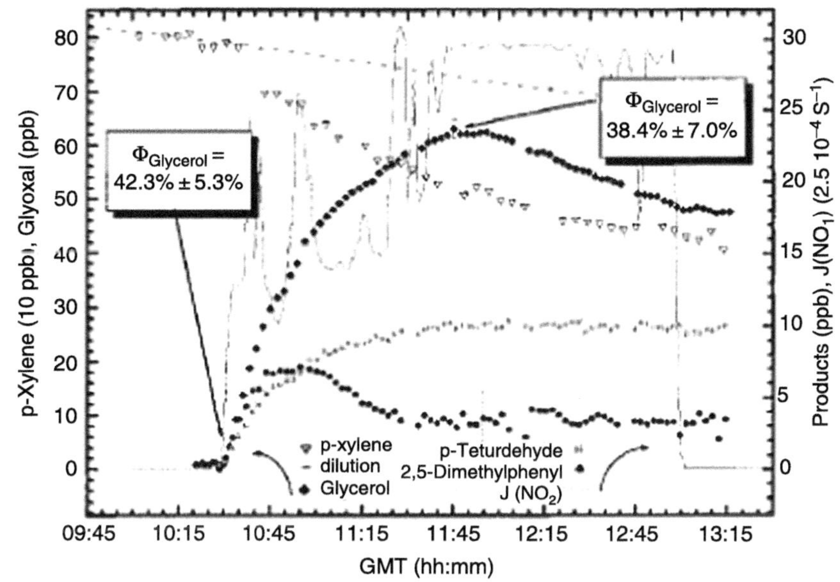

Fig. 10.40 Concentration time profile of various aromatic hydrocarbons during the oxidation of p-xylene Reprinted with permission from Volkamer et al. (2002), ©PCCP Owner Societies. All rights reserved

10.6 Validation of Active DOAS

An important aspect for all measurement techniques is the determination of the observational accuracy. For measurements in the open atmosphere, only one method is available to determine the accuracy of a measurement: the comparison of different analytical techniques. The statistical approach, i.e. repeating a measurement several times, is not available due to the ever changing conditions in the atmosphere.

As described in Chap. 6, DOAS is an absolute analytical method. Calibrations are not needed and the accuracy is determined by the accuracy of the absorption cross-sections used in the analysis of the DOAS spectra. However, as in all analytical techniques, the possibility of unknown systematic errors always has to be considered. An intercomparison between DOAS and other techniques is thus a valuable exercise. Often intercomparisons between DOAS and other techniques are used for the benefit of the other techniques which may have calibration problems, for example OH-LIF systems.

Shown here are a number of examples for intercomparisons for the most important trace gases. Measurements of O_3 and SO_2 were made by a long-path DOAS instrument, a UV ozone monitor, and a commercial SO_2 monitor at the University of East Anglia Atmospheric Observatory at Weybourne, England (Stutz 1996). Weybourne is located at the east coast of England, in an area with few pollution sources. The observed air masses were thus fairly well mixed, facilitating the comparison of

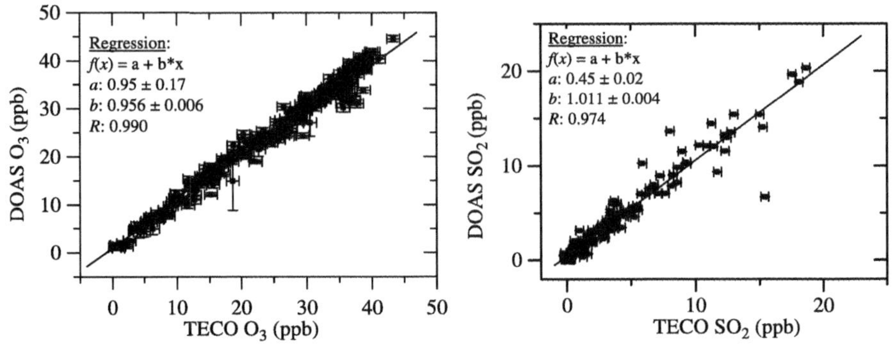

Fig. 10.41 Intercomparison between in situ O_3 and SO_2 instrument and a long-path DOAS system in Weybourne, England. Reprinted from Stutz, PhD Thesis, 1, 996University of Heidelberg

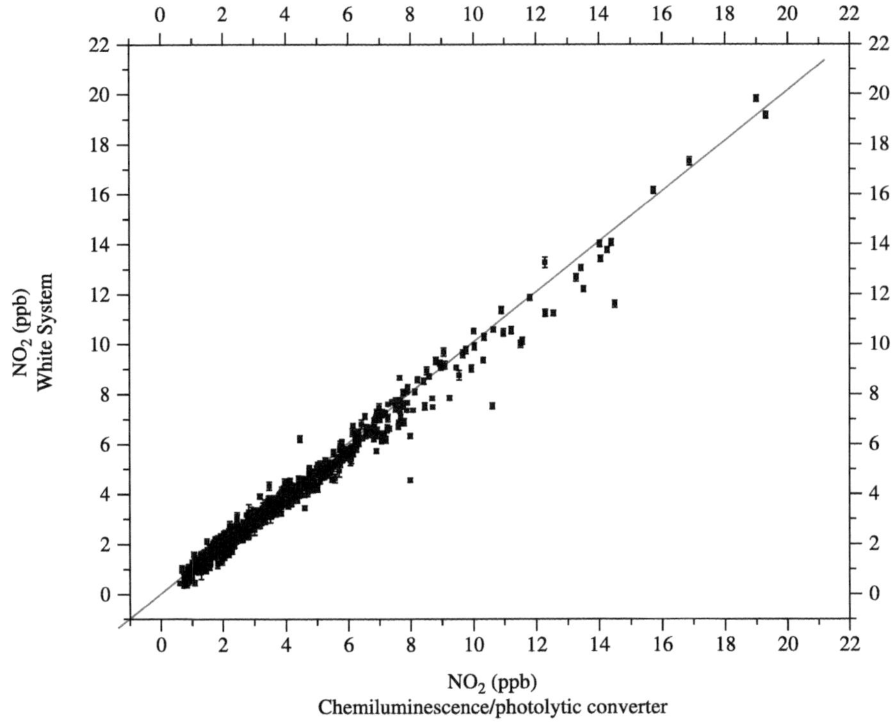

Fig. 10.42 Intercomparison of a multi-reflection DOAS instrument and an in situ NO_2 monitor. Both instruments were set up side-by-side at an altitude of 8 m above the ground on a meadow 50 km southeast of Berlin, Germany, during the BERLIOZ experiment. Reprinted with permission from Alicke et al. (2003), ©American Geophysical Union. All rights reserved

the in situ technique with the DOAS data which were averaged over a 3.5-km-long light path. Figure 10.41 shows the excellent agreement between the methods. The difference between the DOAS and the in situ O_3 measurement is ~5%, which is in the range of the accuracy of both the methods. While the slope of the SO_2 correlation is 1.01, the data is considerably more scattered than that of O_3. This is mainly caused by localised SO_2 sources close to the observatory, for example ships cruising off the coast. In both cases, small offsets were found. The reason for these offsets is unclear.

Figure 10.42 shows the intercomparison of a multi reflection DOAS system with 15 m base path length and a NO_2 in situ instrument based on the photolytic conversion of NO_2 to NO followed by the detection of NO through chemoluminescence (Alicke et al. 2003). This intercomparison was performed during the BERLIOZ campaign in 1998, ~ 50 km southeast of the city centre of Berlin, Germany. A linear regression weighted by the error of the DOAS system yields a slope of 1.006 ± 0.005 (White system against chemoluminescence system). The intercept of (36 ± 19) ppt is statistically insignificant. The excellent agreement between the two methods shows the accuracy of both methods.

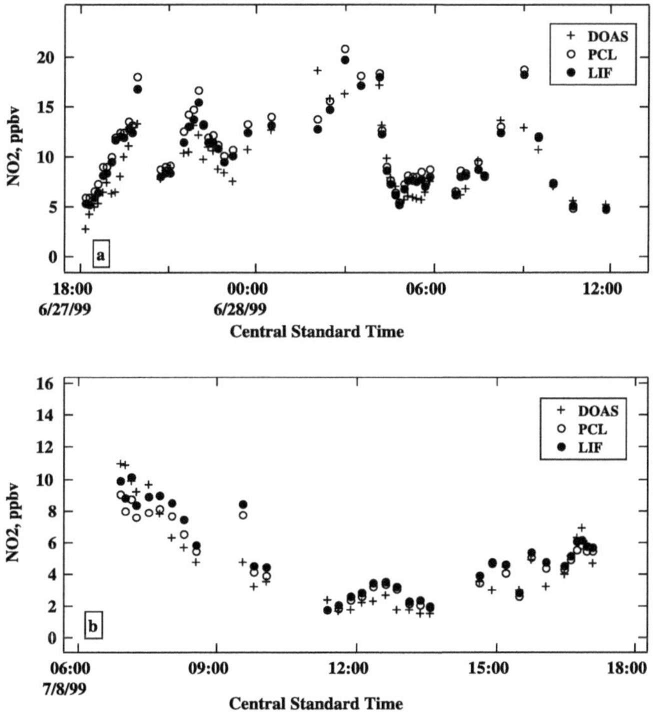

Fig. 10.43 Comparison of three NO_2 measurement methods: DOAS, Photolytic Converter/chemoluminescence, PCL, and Laser Induced Fluorescence, LIF (from Thornton et al. 2000); during the Southern Oxidant Study (SOS) 1999, in Nashville, TN, USA. The measurement was performed in a shallow river valley. The DOAS light beam crossed the entire extent (vertical and horizontal) of the valley, while the in situ instruments sampled 10 m above the ground. Reprinted with permission from Thornton et al. (2003), ©American Geophysical Union. All rights reserved

Another intercomparison between three different methods is shown in Fig. 10.43 (Thornton et al. 2003). Here, a long-path DOAS instrument (pathlength 2×1.35 km) was compared with two co-located point measurements by a Photolytic Converter/chemoluminescence instrument (PCL), and a laser induced fluorescence (LIF) system. The comparison shows that, in many situations, the agreement between the two in-situ instruments is better than with the DOAS instrument. This apparent disagreement is caused by the spatial averaging of the DOAS system. While an in situ instrument observes only a small air volume, the DOAS data represent an average over an extended air mass.

Under certain meteorological conditions, this can lead to a disagreement between DOAS and other methods. Whenever data from a long-path DOAS instrument is used, this averaging has to be considered. An example of this effect can also be observed in Fig. 10.13, where the spatial inhomogeneity of certain trace gases is illustrated.

During the same field experiment, intercomparison between a TDLS HCHO instrument (Fried A, personal communication) and the long-path DOAS was performed. While the statistical errors of both the methods were ~ 0.2–0.5 ppb, Fig. 10.44 shows an excellent agreement between the two methods. The slope of 1.07 ± 0.01 is within the range of the accuracy of both instruments.

The various intercomparisons show the accuracy of DOAS when compared with other methods. In general, the agreement is within the accuracies determined by

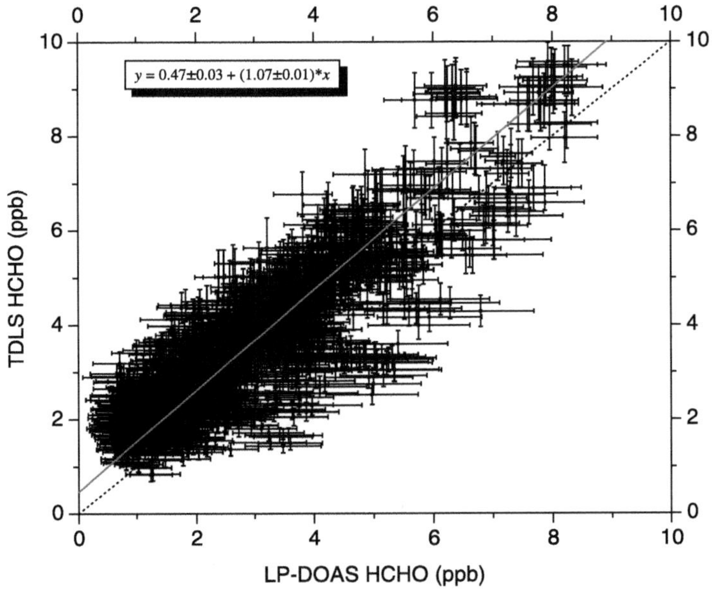

Fig. 10.44 Intercomparison of simultaneous formaldehyde measurements by a Tuneable Diode Laser Spectrometer (TDLS) in situ instrument and a long-path DOAS system during the Southern Oxidant Study (SOS), 1999 in Nashville, TN, USA. Shown is the entire dataset during 4 weeks of this experiment. (TDLS HCHO data courtesy of A. Fried, NCAR)

the different absorption cross-sections. However, spatial inhomogenities can have a considerable influence on the quality of the intercomparisons.

While many intercomparisons for pollutants have been made, few examples exist for radical species. Geyer et al. (1999), showed a comparison between a long-path DOAS instrument and a matrix isolation electron spin resonance (MIESR) method. These measurements were performed during the BERLIOZ experiment in 1998. The authors also included data from an older field experiment in Deuselbach, Germany in 1983. The comparison of the measured data (Fig. 10.45) shows an excellent agreement, in particular considering that the DOAS are measured over a 3.5-km-long path while the MIESR data were sampled at one point. In addition, the MIESR data are averaged over 1 h, while the DOAS time resolution was ~ 5 min. A correlation plot comparing the two methods during the BERLIOZ and Deuselbach experiments also illustrates the excellent agreement between the two methods (Fig. 10.46).

One of the most challenging measurements is that of the OH radical. Because of its high reactivity and low concentrations, atmospheric measurements typically have large uncertainties. It is thus essential to compare different analytical methods of this elusive radical to assess how well we can determine atmospheric OH concentration. Brauers et al. (1996) showed one of the first intercomparisons between a DOAS and a LIF instrument. The DOAS instrument was based on a multi-reflection system with

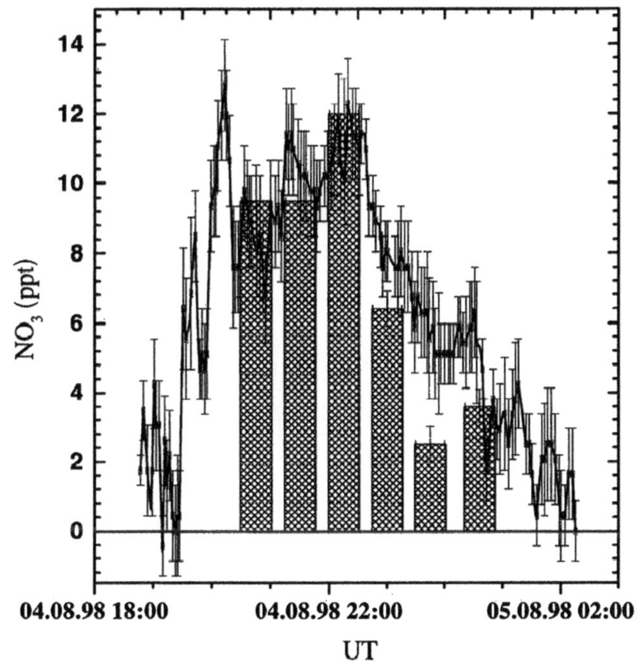

Fig. 10.45 Observation performed by a long-path DOAS instrument (*line*) and a matrix isolation electron spin resonance method (*bars*) during the BERLIOZ experiment, near Berlin, Germany. Reprinted with permission from Geyer et al. (1999), ©American Geophysical Union. All rights reserved

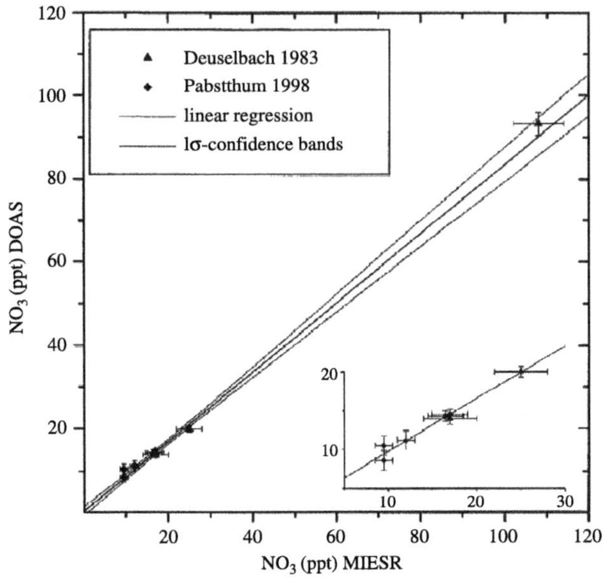

Fig. 10.46 Correlation between DOAS and MIESR measurements from two field experiments. Data marked with *triangles* were measured in Deuselbach in May/June 1983, while data denoted by *diamonds* were obtained during BERLIOZ in summer 1998. Reprinted with permission from Geyer et al. (1999), ©American Geophysical Union. All rights reserved

Fig. 10.47 Intercomparison of OH measurements by a multi-reflection DOAS system and a laser-induced fluorescence instrument. Reprinted with permission from Brauers et al. (1996), ©American Geophysical Union. All rights reserved

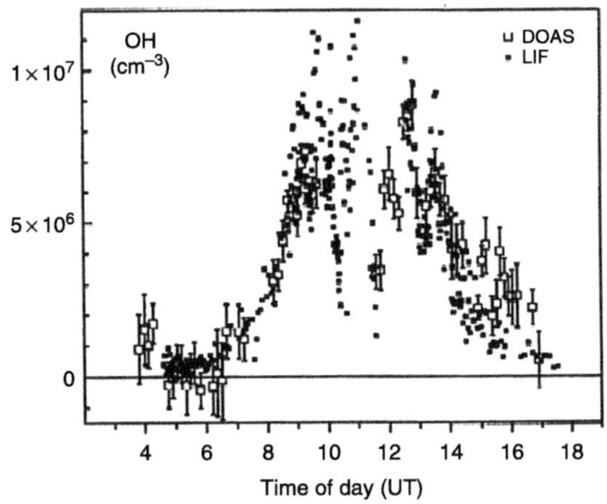

20 m base path length (see also above). Figure 10.47 shows the comparison between the data of both instruments during a diurnal cycle. The agreement is quite good, as can be seen in the correlation of data over an 11-day period (Fig. 10.48).

Mount et al. (1997), showed a similar OH comparison between a long-path DOAS instrument and a chemical ionisation mass spectrometer, confirming the accuracy of the measurements (Fig. 10.49).

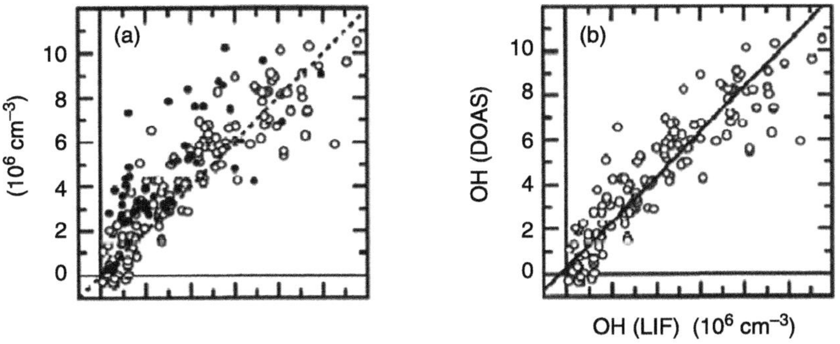

Fig. 10.48 Correlation of DOAS and laser-induced fluorescence data. *Panel* (**a**) shows the entire dataset for 11 d in August 1994, while *Panel* (**b**) presents a reduced dataset, where data from a windsector from 285–20 was excluded. The correlation coefficient was 0.9. The *solid line* represent the weighted linear fit given by $[OH]_{DOAS} = (1.01 \pm 0.04) \times [OH]_{LIF} + (0.28 \pm 0.15)$. Reprinted with permission from Brauers et al. (1996),

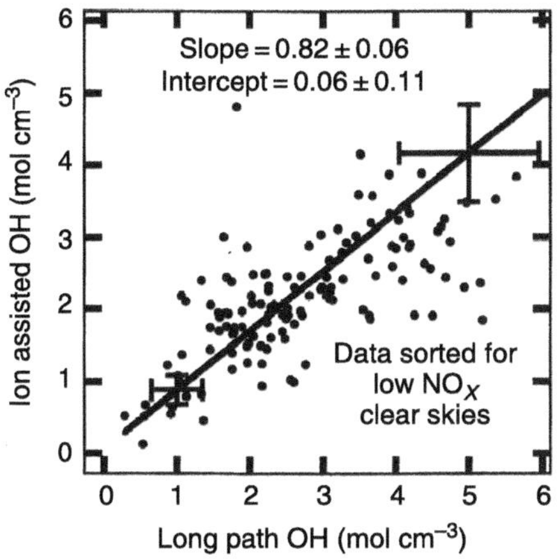

Fig. 10.49 Intercomparison of OH measurements by a long-path DOAS system and a chemical ionisation mass spectrometer in the mountains near Boulder, CO (Mount and Eisele 1992, Mount et al. 1997). Only data from clear days with NO_X levels below 500 ppt are shown. Reprinted with permission from Mount et al. (1997),

Sample Application of 'Passive' DOAS 11

For almost a century, atmospheric trace gas abundances have been measured by their absorption of sunlight. There are several possible measurement geometries, which have specific applications, advantages, and drawbacks:

- Direct sunlight (or moon and starlight)
- Scattered sunlight in zenith view geometry (ZSL-DOAS)
- Scattered sunlight in 'off-axis' view geometry
- Scattered sunlight in 'multi-axis' view geometry (MAX-DOAS).

In scattered light DOAS applications, determination of the (effective) length of the light path is more difficult than for active DOAS (see Chap. 10). The methods used to overcome this challenge through the calculation of air mass factors (AMF) were introduced in Chap. 9.

Direct sunlight is (during daylight hours) readily available and allows a relatively simple calculation of the optical path (see Chap. 9). The practical realisation of this approach requires a manual or automatic mechanism to point the instrument to the sun (or moon or star). A famous application of direct light spectroscopy is the determination of the O_3 total column density, which has been routinely performed since the technique was introduced by S. Dobson during the 1920s (Dobson and Harrison, 1926; see Chap. 6). Clearly, all sunlight reaching the surface of earth has to traverse the entire atmosphere, picking up the spectral signature of its constituents. Thus, it is straightforward to determine the total column density of a trace gas from direct light measurements. However, there is little or no information about the altitude distribution of the species. The nighttime composition of the atmosphere can be probed by using direct moonlight or starlight. Only recently have direct sunlight measurements of trace gases other than ozone been implemented, e.g. Herman et al. (2009, 2019); Park et al. (2018); Spinei et al. (2018); Fioletov et al. (2016). In many

U. Platt and J. Stutz, *Differential Optical Absorption Spectroscopy*, Geophysics and Environmental Physics, https://doi.org/10.1007/978-3-662-68639-3_11

cases the motivation of these observations is to validate satellite data. We will discuss direct sunlight measurements in Sect. 11.1.

Determination of atmospheric trace gases by **scattered** sunlight spectroscopy has been used to probe the atmosphere for three quarters of a century. Initially most of the studies concentrated on stratospheric species, in particular ozone (see Chap. 9), almost invariably using the zenith view geometry. The influence of tropospheric gases and aerosol was seen as a nuisance, merely reducing the accuracy of the measurements. However, the recognition that using off-axis viewing geometries can enhance the sensitivity to tropospheric trace gases, led to the development of the MAX-DOAS method (see Chap. 9). MAX-DOAS is now the most widely applied ground-based scattered light DOAS method and is used in air pollution monitoring as well as for studying chemistry in polluted and clean environments. The multi-axis approach also led to DOAS applications designed to quantify emissions from point or area sources, such as the mobile MAX-DOAS or imaging-DOAS techniques. Examples of MAX-DOAS applications are given in Sects. 11.3.2–11.3.3, and emission measurement schemes using MAX-DOAS and imaging DOAS are presented in Sect. 11.4.

Another area of rapid development is the observation of atmospheric composition from space. While use of this technology began in the 1960s with the mapping of the stratospheric ozone column (TOMS), recent decades have seen enormous progress. Pioneering instruments recorded, for the first time, complete spectra at sufficient spectral resolution ($\Delta\lambda < 1$nm) to allow DOAS retrievals of trace gas columns, such as the Global Ozone Monitoring Experiment (GOME) and the Scanning Imaging Absorption Spectrometer for Atmospheric Cartography (SCIAMACHY). Satellite instruments have broadened our view of atmospheric chemistry and air pollution to the global scale. With every new generation of instruments, for example the Ozone Monitoring Instrument (OMI), GOME-2, the TROPOspheric Monitoring Instrument (TropOMI), finer spatial scales could be observed, expanding the application of satellite DOAS. The latest iteration of space-based DOAS systems, for example Geostationary Environment Monitoring Spectrometer (GEMS) and Tropospheric Emissions: Monitoring Pollution (TEMPO), are now in geostationary orbit, allowing for spatially and temporally highly resolved observations. Section 11.7 provides examples of various satellite applications.

11.1 Atmospheric Measurements by Direct Light Spectroscopy

Direct light spectroscopy using radiation from celestial bodies (sun, moon, or stars) allows the direct determination of total trace gas column densities. There are several advantages of direct light spectroscopy. The light path enhancement is large at large solar zenith angles, enhancing the sensitivity of the measurements. The light path length in direct light observations is also well defined, making the conversion from slant column densities, SCD, to vertical column densities, VCD, more accurate. For satellite validations, this provides a major advantage over scattered light DOAS (see Chap. 9). Spectroscopic analysis of direct sunlight also benefits from the high intensity and the absence of the Ring effect (Chap. 9).

The use of direct moonlight or starlight allows passive DOAS observation at night, since intensities of scattered light are too low for current technology. Moonlight or starlight therefore allow the study of molecules that are destroyed by sunlight, such as chlorine dioxide (OClO) or the nitrate radical (NO_3).

One disadvantage of direct light DOAS is that Fraunhofer signatures vary when imaging different parts of the sun, as discussed in Chap. 9. Moreover, some manual or automatic mechanism to point the instrument to the sun (or moon/star) is required, which can make direct light measurements from mobile carriers difficult. Nevertheless, direct light DOAS measurements from aircraft and balloon have been reported, as outlined in Sect. 11.1.2.

11.1.1 Ground-Based Measurement of Atmospheric Species

Direct sunlight DOAS from the ground has seen increased popularity in the past decade, as the uncertainties of converting scattered light DOAS observations into VCDs have been more fully understood (Chap. 9). Driven by the desire to provide more accurate trace gas VCDs for satellite validation, several direct sun viewing instruments have been developed (Herman et al. 2009, 2015; Tzortziou et al. 2012; Park et al. 2018; Spinei et al. 2018; Fioletov et al. 2016). In particular, the Pandora instrument, which was developed at NASA, has become a staple of satellite validations (Herman et al. 2018; Pinardi et al. 2020; Judd et al. 2020; Kim et al. 2017, 2023). For example, Kim et al. (2023) used four Pandora instruments to validate the Korean geostationary GEMS instrument (Fig. 11.1). The comparison is impacted by the different footprints of the observations, i.e. the Pandora instruments observe a smaller and often more polluted area compared to the the larger satellite footprint. When filtered for cases with spatial inhomogeneities, Kim et al. (2023) found that Pandora and GEMS NO_2 VCDs are highly correlated ($R^2 = 0.69 - 0.81$).

Some of the most scientifically successful moonlight DOAS measurements were those of the OClO column density, for example, performed at the McMurdo Antarctic station in 1986 by Solomon et al. (1987b) (Figs. 11.2 and 11.3). A clear increase of OClO with increasing direct light AMF (dashed line) is observed.

Direct moonlight measurements were also made in Kiruna, Sweden, by Wagner et al. (2002c) in order to obtain improved absorption cross-sections of the oxygen dimer $(O_2)_2$ or O_4. The investigated O_4 bands were found to show an increase of the peak absorption with decreasing temperature, ranging from $\approx 13\%/100\,K$ at 477.3 and 532.2 nm, $\approx 20\%$ at 360.5 and 577.2 nm to $\approx 33\%$ at 380.2 and 630.0 nm. Moreover, with the exception of the band at 380.2 nm, the O_4 absorption cross-sections were found to be somewhat larger than previous measurements (see Appendix B).

For the separation of stratospheric and tropospheric column densities of BrO, a combination of direct sunlight and ZSL-DOAS was employed in a measurement campaign at the Arrival Heights Antarctic station during Austral spring of 2002 (Schofield et al. 2004; 2006).

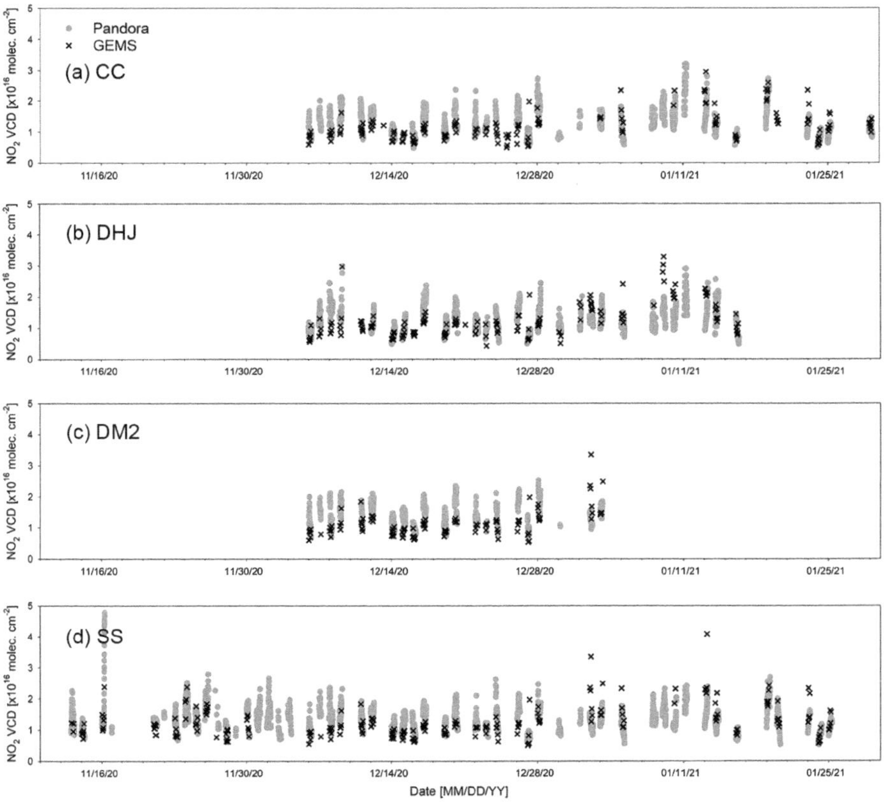

Fig. 11.1 Comparison of hourly NO_2 VCDs from Pandora instruments (full grey circles) and GEMS (black crosses) in South-Korea (SS: Seaosn, CC: Seosan, DHJ: Daeholi, DM2: Dongmoon-2dong). Reprinted with permission from Kim et al. (2023),

11.1.2 Balloon and Aircraft-Borne Measurement of Stratospheric Species

Direct light balloon-borne DOAS spectroscopy allows the determination of vertical profiles of several trace gases in the atmosphere (e.g. Rigaud et al. 1983; Pommereau and Piquard 1994a; 1994b; Renard et al. 1996; 1997a; 1997b; 1998; 2000a; 2000b Harder et al. 1998; 2000; Ferlemann et al. 1998; 2000; Fitzenberger et al. 2000; Pundt et al. 2002; Bösch et al. 2003; Weidner et al. 2005; Butz et al. 2006; Dorf et al. 2006, Kreycy et al. 2013; Berthet et al. 2017). The precise imaging of the sun from an unmanned, airborne platform usually requires sophisticated hardware that relies on active orientation control of the balloon payload and mirror steering, as described by Hawat et al. (1998). However, some simple, passive entrance optics are also in use, which rely on the radiation scattered from a diffuser plate close to the entrance optics (see below).

There are two types of observation geometries which allow the retrieval of vertical trace gas profiles. Both geometries are shown in Fig. 11.4: (1) Observation dur-

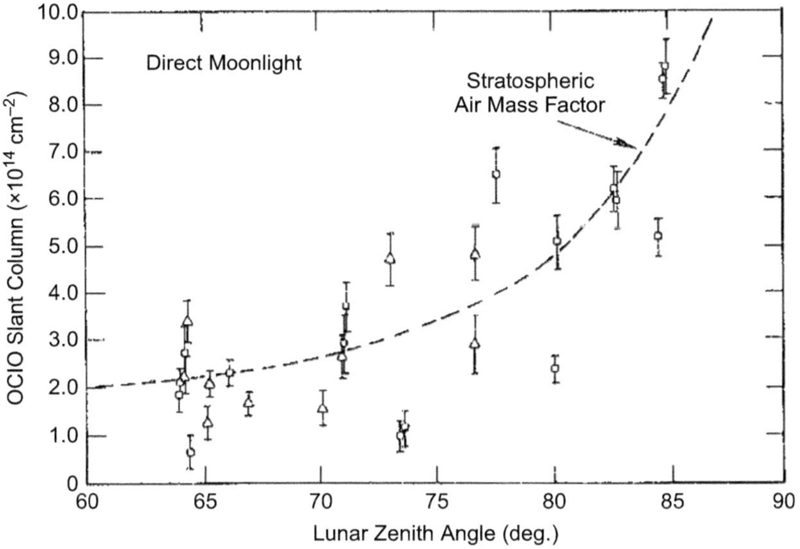

Fig. 11.2 Variation of the OClO SCD at McMurdo, Antarctica during the nights of 16–19 September 1986 as a function of lunar zenith angle. Reprinted with permission from Solomon et al. (1987b), ©American Geophysical Union. All rights reserved

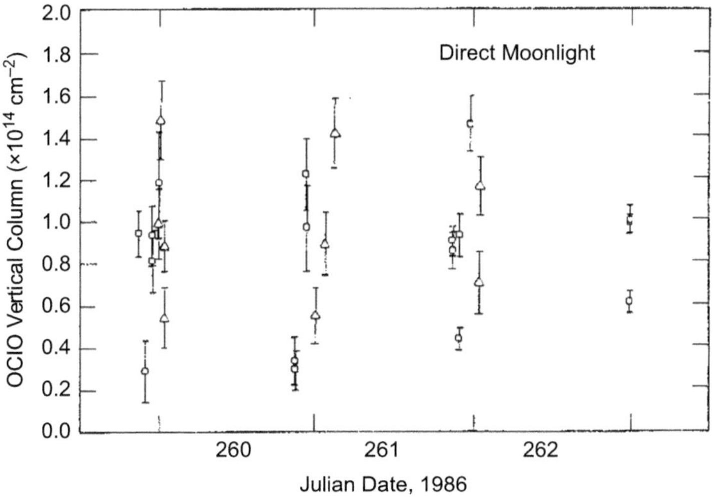

Fig. 11.3 Variation of the OClO VCDs at McMurdo, Antarctica, during the nights of 16–19 September 1986 Reprinted with permission from Solomon et al. (1987b), ©American Geophysical Union. All rights reserved

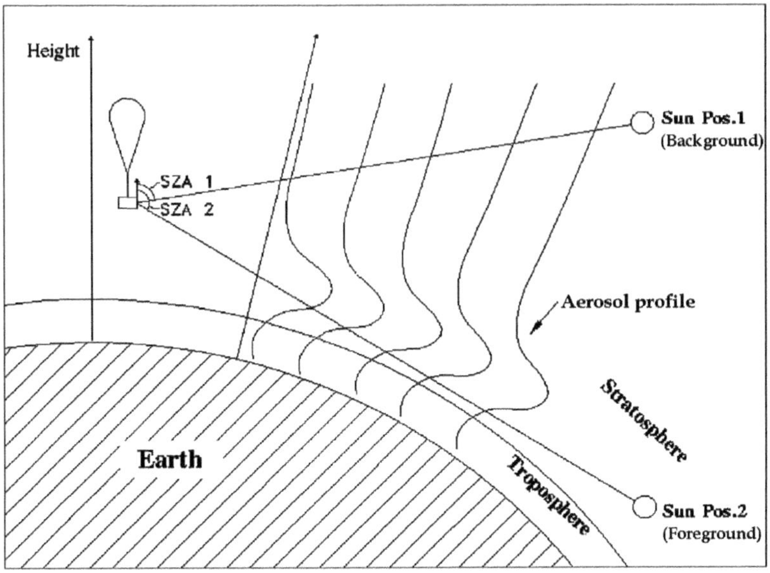

Fig. 11.4 Schematic of direct sunlight DOAS balloon measurements. During balloon ascent, typically at high solar elevations (Sun Pos. 1), differentiation of the consecutive series of trace gas column measurements yields a vertical profile. At constant balloon altitude (float level), another profile is derived during sunset or sunrise. Reprinted from Osterkam, Diploma Thesis, 1997, University of Heidelberg

ing ascent (Pos. 1) and (2) Occultation during constant payload altitude (the 'float period', Pos. 2). The analysis of the geometry during ascent is straightforward, with the air mass factor given by $A(\vartheta) \approx 1/\cos\vartheta$, (with ϑ denoting the SZA). Small corrections to the AMF due to the earth's curvature are also necessary. The average trace gas concentration between two altitude levels of the instrument z_i and z_{i+1} is then determined by taking the difference between the associated vertical column densities (VCDs) above the instrument:

$$\bar{c}\,(z_i, z_{i+1}) = V\,(z_{i+1}) - V\,(z_i) = \frac{S\,(z_{i+1}, \vartheta)}{A\,(\vartheta)} - \frac{S\,(z_i, \vartheta)}{A\,(\vartheta)}$$

where V and S denote the vertical and slant trace gas columns, respectively. Occultation experiments during the float period require more sophisticated de-convolution techniques, as described by Ferlemann et al. (1998).

Figure 11.5 shows an example of the trajectory of a balloon, and the 'line of sight' as a function of altitude (pressure) and solar zenith angle (SZA) for ascent and float period. The BrO slant column density is given by the integral of the BrO concentration (contour lines) times the path length of the beam traversing the respective atmospheric layer. Note that, in these pressure–SZA plots, the rays are not straight lines. Figure 11.6 compares several BrO profiles determined by in-situ techniques and balloon-borne direct sunlight DOAS measurements.

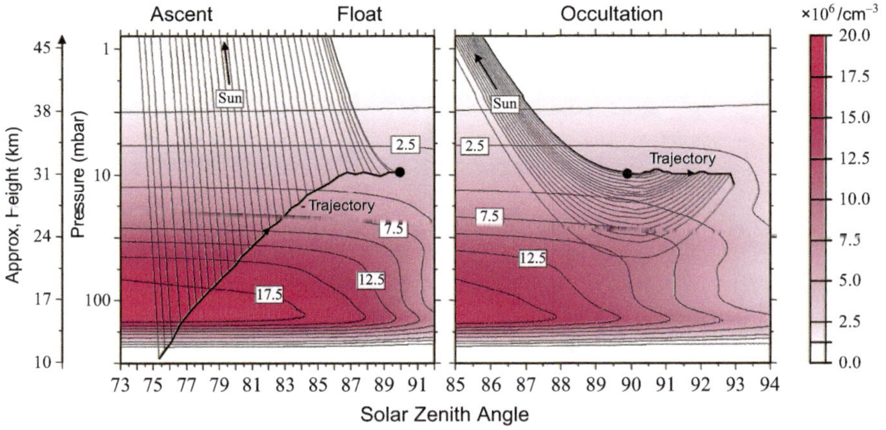

Fig. 11.5 Model simulations of stratospheric BrO as a function of SZA and pressure (altitude) at sunset for a balloon position at 40.5°N, 7.5°W on 23 November, 1996. Superimposed is the observation geometry of the balloon direct sunlight DOAS BrO measurements. **Left panel**: Balloon trajectory and 'line of sight' of each 10th spectrum. **Right panel**: Same for solar occultation (sunset). The amount of BrO along the line of sight is given by the integral of the local BrO concentration (*contour lines*) times the path element of each light beam traversing the atmospheric layer. Reprinted with permission from Harder et al. (2000),

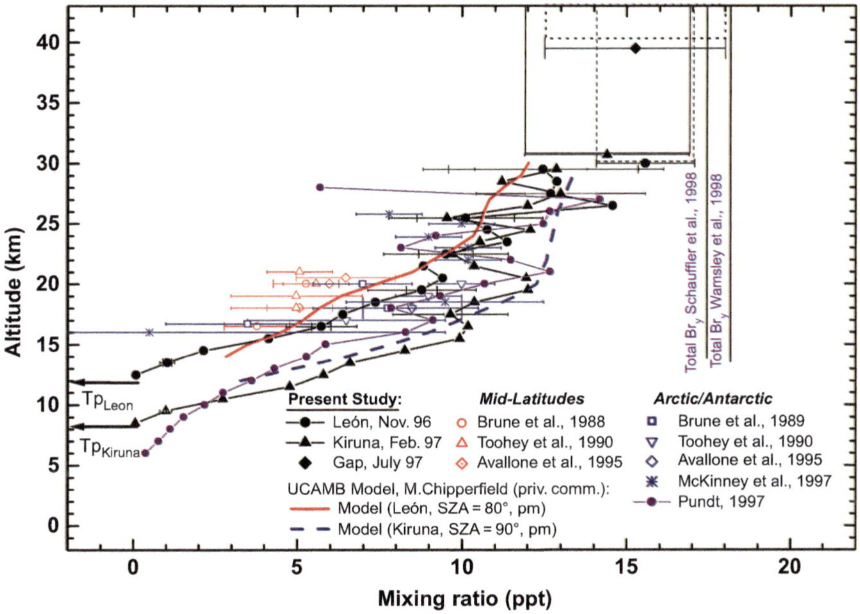

Fig. 11.6 Comparison of BrO profiles measured by several groups and techniques, including balloon-borne direct sunlight DOAS spectroscopy. Reprinted with permission from Harder et al. (1998),

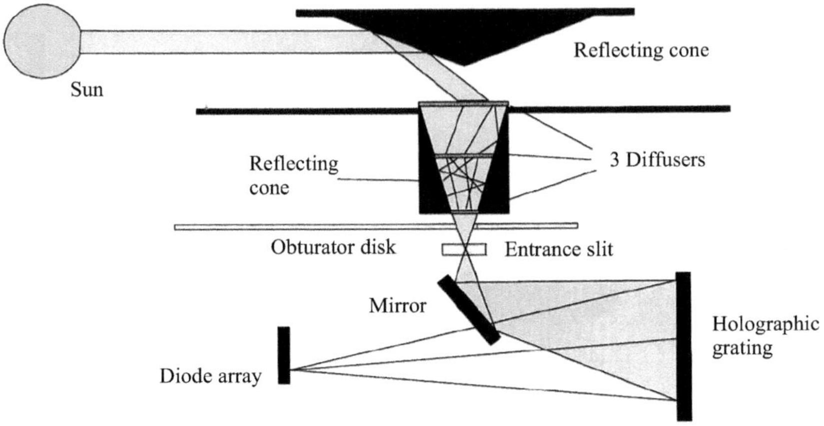

Fig. 11.7 Layout of the 'SAOZ-BrO' balloon spectrometer and entrance optics (Pundt, 1997). Radiation is accepted from a −5 to +15° elevation range and all azimuth directions by a downward pointing, reflecting cone and by a set of three diffuser plates mounted inside a second reflecting cone in front of the spectrometer. Reprinted with permission from Pundt et al. (2002), ©American Geophysical Union. All rights reserved

An interesting example of a balloon borne direct sunlight design is the Systéme D'Analyse par Observations Zénithales (SAOZ) (Pommereau and Piquard 1994a; 1994b), which consists of a small, lightweight spectrometer (spectral range 270–620 nm) capable of determining NO_2 and O_3 absorption spectra. Passive optics accept direct sunlight and reflect it through a set of diffuser plates into the spectrometer. This setup is similar to the SAOZ-BrO shown in Fig. 11.7. While the setup is simple and has the great advantage of not requiring moving parts or control electronics, its drawback is the large acceptance angle. A fraction of scattered light is also recorded along with the direct sunlight, which can complicate the evaluation of the data. On the other hand, the design makes the instrument very simple and lightweight, requiring only small balloons. The instrument has been successfully flown more than a hundred times (Pundt et al. 2002). In addition, an improved version of the instrument—the 'SAOZ-BrO' (spectral range 350–420 nm) was developed, which had more light throughput and better sampling of the spectrum. SAOZ-BrO was added to the payload in a series of 16 balloon flights which took place between late winter of 1997 and spring of 2000. Primarily from an Arctic site (Kiruna, Sweden, 68°N) but also from mid-latitude sites (southern France, 43/44°N).

From the flights in the winters of 1998, 1999, and 2000, average profiles of O_3, NO_2, OClO, and BrO are shown in Fig. 11.8. During the first two winters, relatively 'warm' stratospheric conditions prevailed, whereas in 2000 a cold polar vortex showed signs of chlorine activation. While there appears to be little difference in BrO between the different states of the stratosphere, OClO is greatly enhanced in the cold vortex.

Aircraft-based direct light measurements also require sophisticated entrance optics to image the light source into the spectrometer. Measurements have been performed from aircraft to study nighttime levels of OClO, NO_2, and NO_3 in the Arctic and Antarctic stratosphere (Solomon et al. 1988; Schiller et al. 1990; Wahner

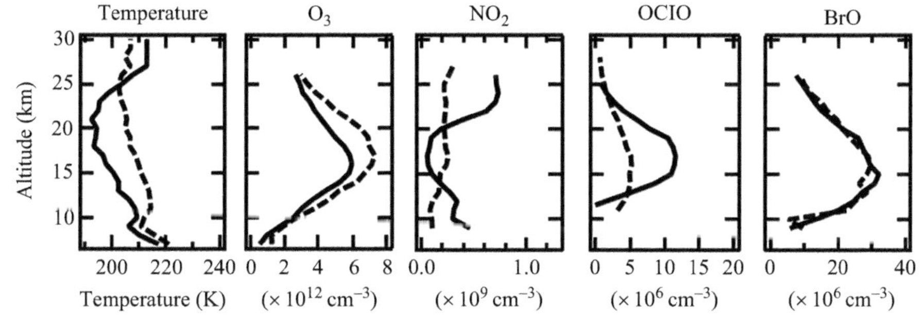

Fig. 11.8 Average vertical distribution of temperature, sunset ozone, NO_2 and OClO and daytime BrO observed by SAOZ and the SAOZ-BrO instruments flown on the same balloon in the cold, chlorine activated winter vortex of 2000 (*solid lines*) and in warmer conditions in 1998 and 1999 (*dashed lines*). Reprinted with permission from Pundt et al. (2002), ©American Geophysical Union. All rights reserved

et al. 1989a; 1990a; 1990b; Pfeilsticker et al. 1997b; Erle 1999). For instance, inside the winter polar vortices, nighttime OClO VCDs of $(1–2.5) \cdot 10^{14}\ cm^{-2}$ were found, with the higher values prevailing in Antarctica.

11.2 Stratospheric Measurements by Ground-Based Scattered Light DOAS

Since the 1930s, scattered light spectroscopy has been used to determine vertical profiles of stratospheric ozone (Götz et al. 1934). Plans to develop a fleet of supersonic aircraft in the 1970s led to an increased interest in stratospheric NO_2. However, it was only after the discovery of the ozone hole in 1985 that the list of observed stratospheric trace species expanded further to include OClO, BrO, O_4, NO_3, and HCHO. Scattered light DOAS measurements played an important role in the study of these species. Compared to direct light observations (see Sect. 11.1), scattered light observations typically use much simpler entrance optics. However, scattered light spectroscopy requires the determination of the AMF, which is much more complicated than in the case of direct light measurements (see Chap. 9). Nevertheless, scattered zenith light DOAS contributed greatly to our understanding of stratospheric processes: Noxon (1975), Noxon et al. (1979); Harrison (1979); Syed and Harrison (1980); McMahon and Simmons (1980); McKenzie and Johnston (1982; 1983; 1984); Pommereau et al. (1982, 1984); Solomon et al. (1987a, 1987b); (1989b; 1993a; 1994a; 1994b); Mount et al. (1987); Pommereau and Goutail (1988a; 1988b); Johnston and McKenzie (1989); Sanders et al. (1989; 1999); Wahner et al. (1990a); McKenzie et al. (1991); Kreher (1991); Johnston et al. (1992); Fiedler et al. (1993); Kioke et al. (1993, 1994, 1999); Elokhov and Gruzdev (1993; 1995); Pommereau and Piquard (1994b); Pfeilsticker and Platt (1994); Perner et al. (1994); Van Roozendael et al. (1994a; 1994b; 1994c; 1997); Kondo et al. (1994); Kreher et al. (1995; 1997; 1999); Hofmann et al. (1995); Gil et al. (1996; 2000); Senne et al. (1966); Vaughan et al. (1997);

Pfeilsticker et al. (1997a; 1999a; 1999b); Slusser et al. (1997); Sarkissian et al. (1997); Goutail et al. (1999); Eisinger et al. (1997); Aliwell et al. (1997; 2002); De Maziere et al. (1998); Roscoe et al. (1999, 2001); Richter et al. (1999); Liley et al. (2000); Fish et al. (2000); Wittrock et al. (2000a); Sinnhuber et al. (2002); Tørnkvist et al. (2002); Hendrick et al. (2004); and Schofield et al. (2004; 2006); Hendrick et al. (2008, 2009); Pinardi et al. (2022); Cheng et al. (2021); Hendrick et al. (2012); Bortoli et al. (2013).

The use of ground-based zenith DOAS to study the stratosphere has largely been superseded by the use of modern satellite instruments, which provide observations at a similar timescale, i.e. daily, as ground-based instruments for each location, but provide this data for the entire globe. Consequently the number of scientific publications using ground-based zenith DOAS has considerably decreased.

11.2.1 Determination of Stratospheric NO_2 and O_3 from the Ground

Ozone measurements by the Umkehr technique are not the topic of this book, so we only give a very brief account, and mention O_3 measurements only as part of other observations. The Umkehr technique inverts measurements of the intensity ratio of zenith scattered sunlight at a pair of UV wavelengths (e.g. 311.0 and 329.0 nm) recorded at a series of SZAs between about 70° and 90° (Götz et al. 1934; Mateer et al. 1996) to determine O_3 vertical profiles from the upper troposphere through the stratosphere. The measurements are usually made with Dobson spectrophotometers (Dobson and Harrison 1926) operated in zenith scattered light mode. Sample results are given in Fig. 11.9 showing differential and integral ozone profiles over Arosa, Switzerland (Götz et al. 1934).

The first species other than ozone to be measured by a scattered light DOAS-type technique was NO_2. The total column of stratospheric NO_2 was first determined from zenith scattered sunlight spectra by Brewer using the structured absorption of this molecule around 450 nm (Brewer et al. 1973). Brewer et al. (1973) also noted the increase of the NO_2 column density during the course of the day. Later, J. Noxon discovered the sharp decline of the wintertime NO_2 column at about 50°N in northern Canada (Fig. 11.10). The phenomenon, now referred to as 'Noxon Cliff' is due to conversion of most of the NO_x into nitric acid (via NO_3–N_2O_5) in the absence of sunlight (Noxon 1975, 1979; Noxon et al. 1978, 1979).

In the following years, a large number of measurements of stratospheric O_3 and NO_2 by passive DOAS were reported. These measurements were primarily made from ground-based stations, but aircrafts, balloons, and ships were also used. Here, we present a small selection of ground-based observations of stratospheric NO_2 using zenith scattered sunlight.

A time series of stratospheric NO_2 columns measured by zenith scattered sunlight DOAS during morning and evening twilight at Lauder, New Zealand between 1981 and 1999, is shown in Fig. 11.11. The NO_2 columns at morning and evening twilight show a pronounced seasonal cycle, and the increase of the NO_2 column

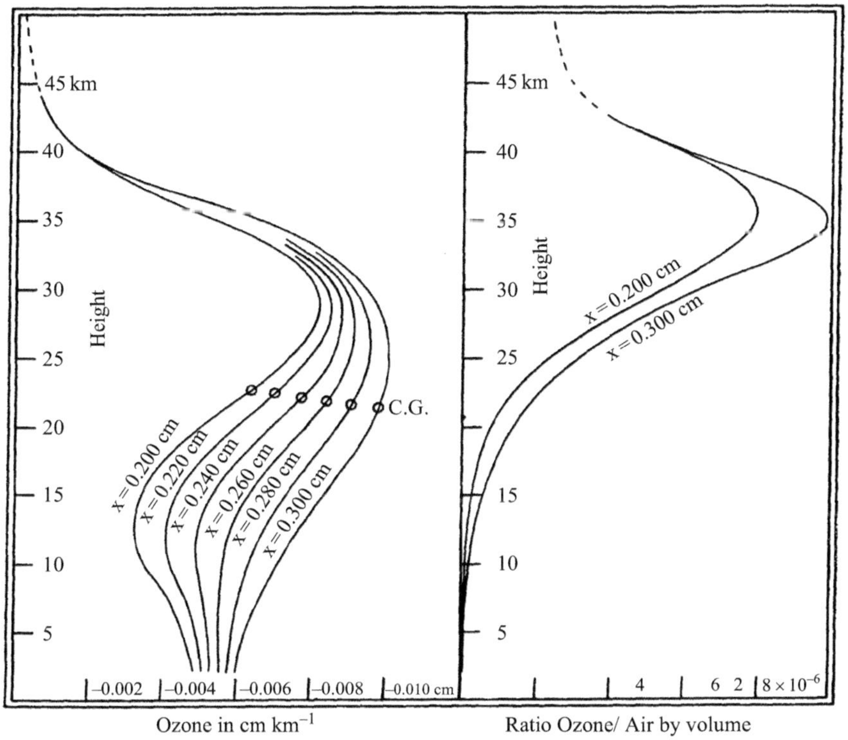

Fig. 11.9 Differential and integral ozone profiles over Arosa, Switzerland, determined in 1932 by the Umkehr technique. The data are derived from an inversion of measurements of the intensity ratio at a pair of UV wavelengths recorded in a series of SZAs from about 70° to 90°. Reprinted with permission from Götz et al. (1934), ©Royal Society. All rights reserved

during daytime is clearly seen. Distinct variations of the annual cycle are due to the combined effects of the solar cycle, the quasi-biennial oscillation QBO, and the El Nino southern oscillation ENSO. The effects of volcanic eruptions in 1982 and 1992 are also visible. In addition, a trend in the NO_2 column of +5% per decade was found. The morning/evening ratio has a clear seasonal cycle, but appears to be unaffected by the other periodic variables and volcanic events.

Figure 11.12 shows plots of the NO_2 and O_3 VCDs derived from ZSL-DOAS measurements during two cruises of the research vessel Polarstern in 1990 and 1993 from Bremerhaven, Germany to Ushuaia, Argentina (1990) (Kreher et al. 1995) and Cape Town, South Africa (1993) (Senne et al. 1966). Spectra recorded at 90° SZA were evaluated around 445 ± 7 and 493 ± 23 nm for NO_2 and O_3, respectively. The minimum around the equator in both species, as well as the increase of the NO_2 column during daylight hours, is clearly visible. In addition, the height of the NO_2 maximum is calculated (Fig. 11.12).

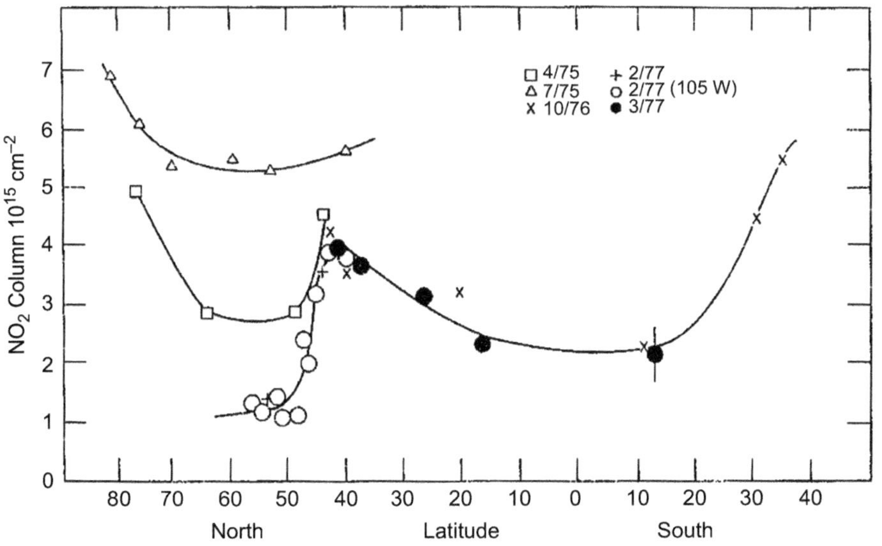

Fig. 11.10 Measurements of stratospheric NO_2 vertical column by zenith scattered sunlight during evening twilight at various latitudes. Reprinted with permission from Noxon (1979), ©American Geophysical Union. All rights reserved

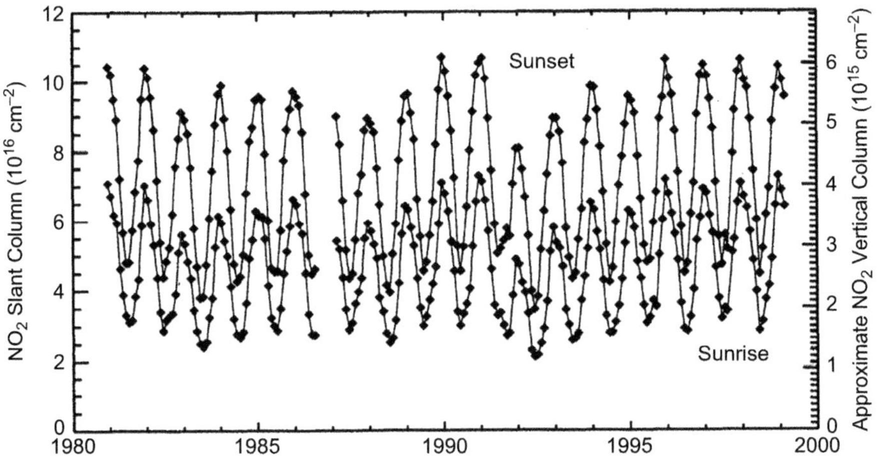

Fig. 11.11 Monthly means of stratospheric NO_2 slant columns (*left scale*) measured by zenith scattered sunlight DOAS during morning and evening twilight (labelled as 'sunrise' and 'sunset') at Lauder, New Zealand, between 1981 and 1999. Approximate vertical columns are indicated at the *right-hand scale*, assuming a fixed AMF of 17.5. Reprinted with permission from Liley et al. (2000), ©American Geophysical Union. All rights reserved

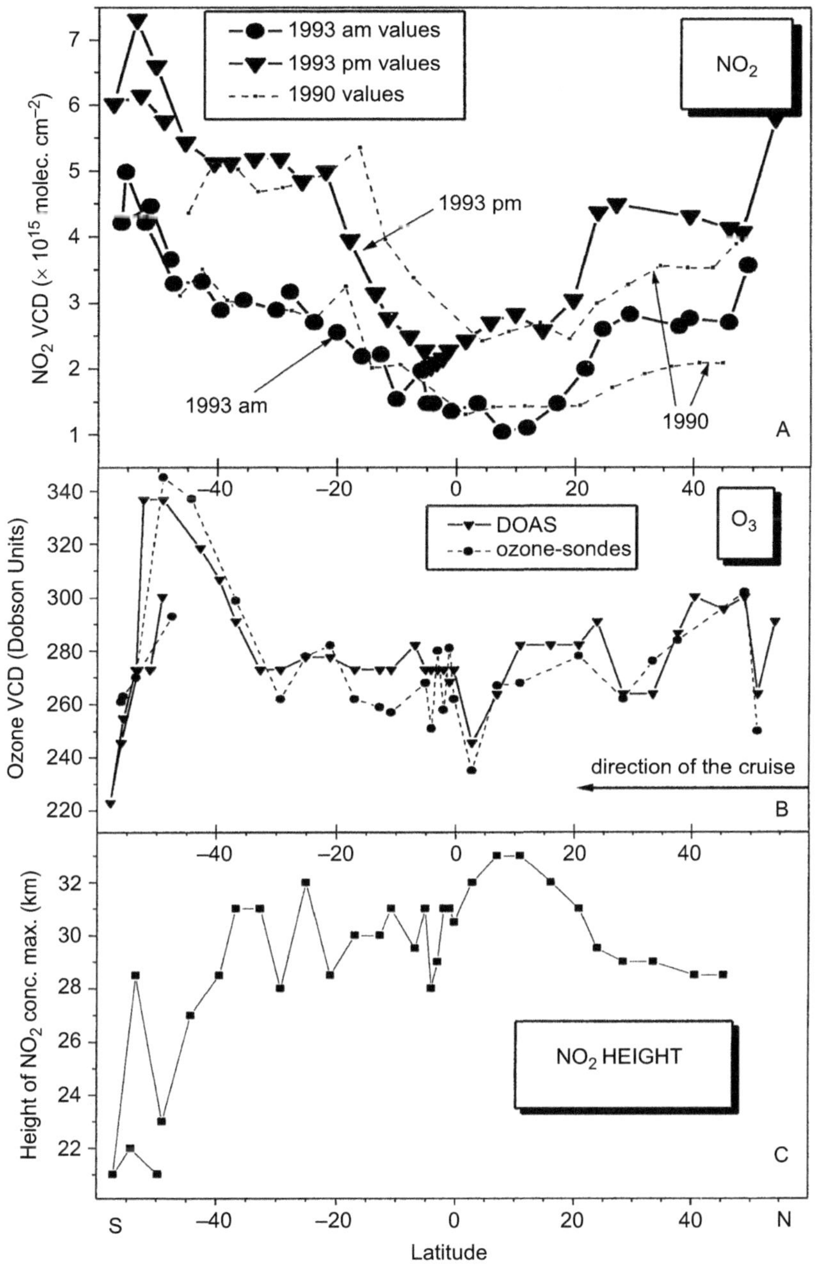

Fig. 11.12 Latitudinal cross-sections of NO_2 (**a**) and O_3 (**b**) as recorded by zenith scattered light DOAS during cruises of the research vessel Polarstern from Bremerhaven, Germany, to Ushuaia, Argentina (1990), and Cape Town, South Africa (1993). In addition, the calculated height of the NO_2 maximum is shown in (**c**). Reprinted with permission from Senne et al. (1966), ©American Geophysical Union. All rights reserved

11.2.2 Observation of Halogen Radicals in the Polar Stratosphere

During the 1990s, improved technology allowed the study of weaker atmospheric absorbers than O_3 and NO_2, in particular halogen oxide radicals. Of the halogen monoxide radicals FO, ClO, BrO, and IO, only the latter two have structured absorption features that are at sufficiently long wavelengths to be easily detected by scattered light absorption spectroscopy. The structured spectrum of ClO is centred around 290 nm, with the longest wavelength band near 308 nm. The low intensities in this wavelength range make it very difficult to detect ClO with ground-based or aircraft-based instruments, even at the relatively high levels reached under ozone hole conditions (see Sect 2.12). Stratospheric ClO might be detectable with future, improved instrumentation or in direct sunlight. However, some problems might arise from the overlapping structured absorption of ozone. Many observations of polar stratospheric (and partly tropospheric, see Sect. 11.3) IO and BrO by ground-based instruments have been reported. (Pundt et al. 1998; Wittrock et al. 2000a; Frieß et al. 2001; Solomon et al. 1989b, 1993a; Wahner et al. 1989a, 1990b; Arpag et al. 1994; Fish et al. 1995; Kreher et al. 1997; Frieß 2001; Frieß et al. 1999; 2004; 2005; Van Roozendael et al. 2002; 2003; Eisinger et al. 1997; Aliwell et al. 1997; 2002; Richter et al. 1999; Sinnhuber et al. 2002; Tørnkvist et al. 2002; Schofield et al. 2004; 2006 Hendrick et al. 2008, 2009; Pinardi et al. 2022)

As outlined in Sect. 2.12, the symmetric OClO molecule is a good indicator of stratospheric chlorine activation. It is also much easier to measure than ClO, since its prominent absorption bands are around 360 nm. The detection of OClO in the Antarctic stratosphere during the ozone hole period in 1986 by Solomon et al. (1987b) provided important evidence for perturbed chlorine chemistry as the primary cause for the dramatic ozone loss (see Chap. 2). Since then, a series of further measurements of OClO by ground and aircraft, as well as satellite-based passive DOAS, have been made (Solomon et al. 1987b; 1988; 1989b; 1990; 1993a; Schiller et al. 1990; Perner et al. 1991; 1994; Sanders et al. 1993; 1996; Fiedler et al. 1993; Pommereau and Piquard 1994a; 1994b; Kreher et al. 1996; Gil et al. 1996; Renard et al. 1997a; 1997b; Pfeilsticker et al. 1997a; Sanders et al. 1999; Tørnkvist et al. 2002; Pinardi et al. 2022). Figure 11.13 shows OClO column data from Solomon et al. (1987b) at the McMurdo Antarctic station by evening twilight measurements. Direct moonlight measurements were made at the same time (see Sect. 11.1.1, Figs. 11.2 and 11.3).

Some results encompassing simultaneous ozone, NO_2, BrO, and OClO time series recorded in Kiruna at the Arctic circle during spring of 1995 are shown in Fig. 11.14 (Otten 1997; Otten et al. 1998). The measurements were made by ZSL-DOAS during morning and evening twilight, respectively. The OClO data show a clear correlation, with periods of high potential vorticity (PV) in the stratosphere and low stratospheric temperatures.

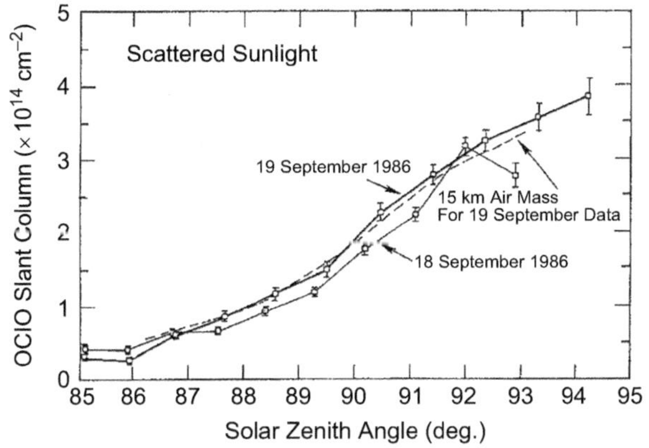

Fig. 11.13 Variation of the OClO SCDs at McMurdo, Antarctica during twilight of 18 and 19 September, 1986. *Dashed line*: calculated AMFs for 19 September. Reprinted with permission from Solomon et al. (1987b),

11.2.3 Halogen Radical Observation in the Mid-Latitude Stratosphere

Ground-based measurements of stratospheric BrO have been made at a number of mid-latitude sites (Arpag et al. 1994; Aliwell et al. 1997; 2002; Fish et al. 1997; Eisinger et al. 1997; Richter et al. 1999; Sinnhuber et al. 2002; Van Roozendael et al. 2002; Cheng et al. 2021). Figure 11.15 shows the measured and modelled difference in the morning–evening slant column densities (SCDs) of BrO at mid-latitudes in the northern and southern hemisphere (Sinnhuber et al. 2002). BrO PM–AM differences show a distinct annual cycle, with a maximum in winter.

11.2.4 Observation of Stratospheric Trace Gas Profiles

Since the 1930s, scattered light spectroscopy has been used to determine vertical profiles of stratospheric ozone by the Umkehr technique (Götz et al. 1934), rather than the total column density as described in Sect. 11.2.1. Despite significant improvements of the inversion algorithms (Mateer et al. 1996; Rodgers, 1990; 2000) surprisingly few attempts to derive vertical distributions of trace gases from ground-based zenith scattered light UV-vis spectroscopy have been reported (McKenzie et al. 1991; Preston et al. 1997; Denis et al. 2003; Schofield et al. 2004; 2006; Hendrick et al. 2004). However, the vertical resolution of all these measurements is on the order of the atmospheric scale height. Preston et al. (1997) and Denis et al. (2003) retrieved the NO_2 vertical distribution in the stratosphere from zenith sky observations using the optimal estimation method (Rodgers, 1990; 2000). In contrast, Schofield et al. (2004; 2006) applied the retrieval algorithm to a combination of ground-based zenith-sky and direct-sun measurements of BrO. Due to the light path geometry, zenith-sky and

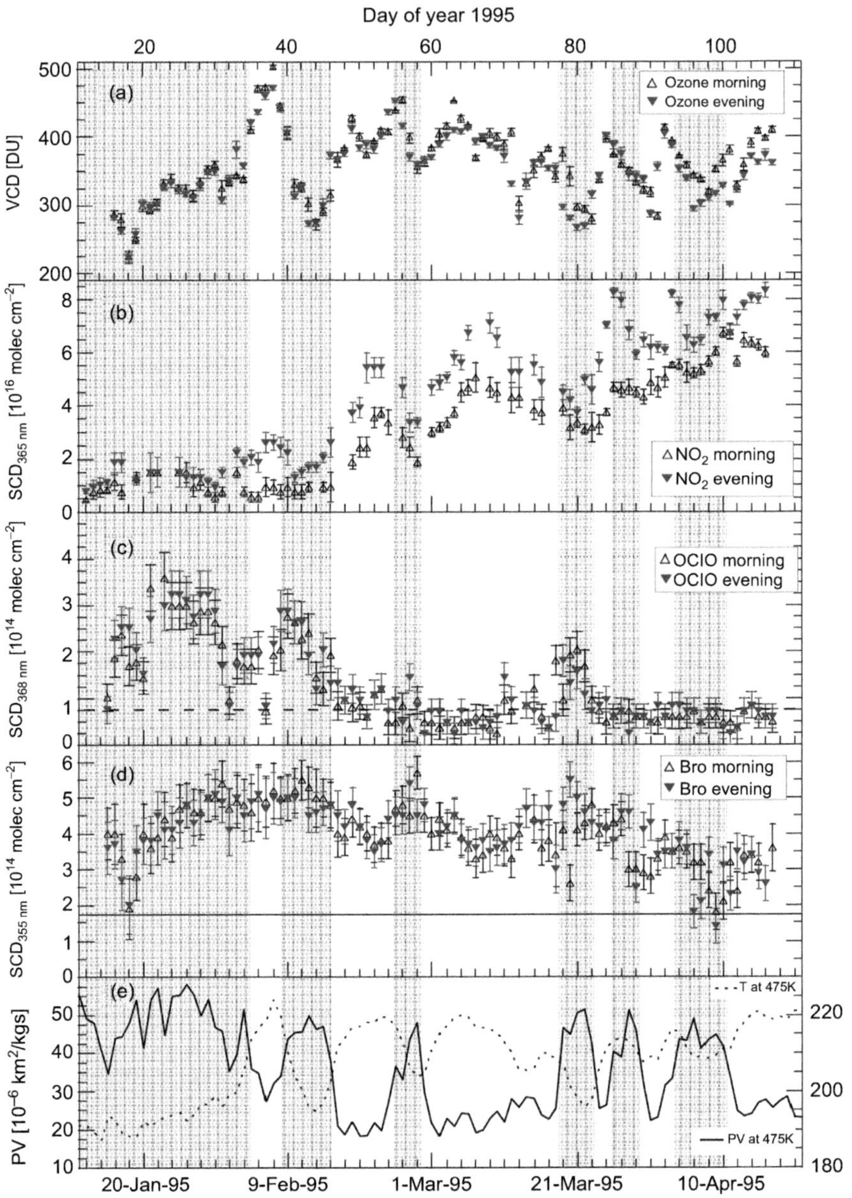

Fig. 11.14 Variation of the ozone VCD (86°–89° SZA) and NO_2, BrO, and OClO SCDs at Kiruna, Sweden during morning and evening twilight (90° SZA) January through April, 1995. *Bottom panel*: Temperature at 475 K isentropic surface (*dashed line*), potential vorticity, PV (ECMWF 1996). Reprinted from Otten, Ph.D. Thesis, (1997), University of Heidelberg

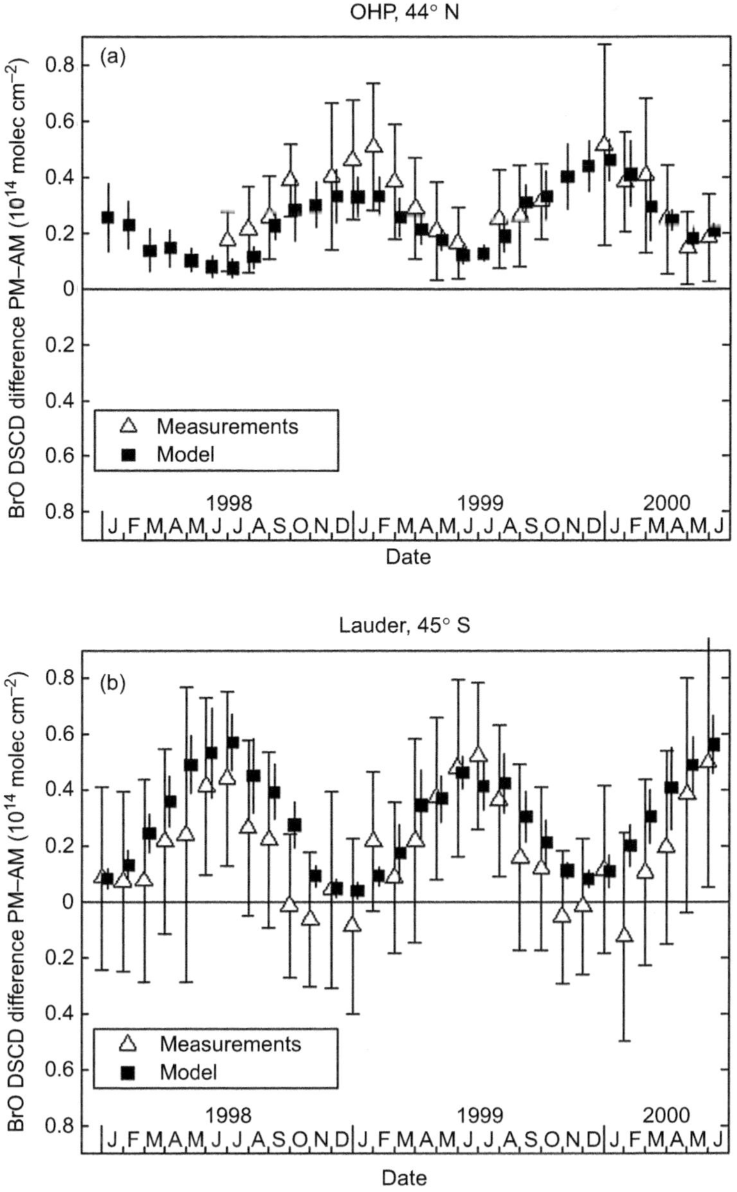

Fig. 11.15 Comparison of the measured and modelled difference between the evening and morning BrO DSCD for Observatoire de Haute-Provence (OHP, 44°N) and Lauder (45°S). The error bars represent monthly averages and their standard deviation (1σ). Reprinted with permission from Sinnhuber et al. (2002),

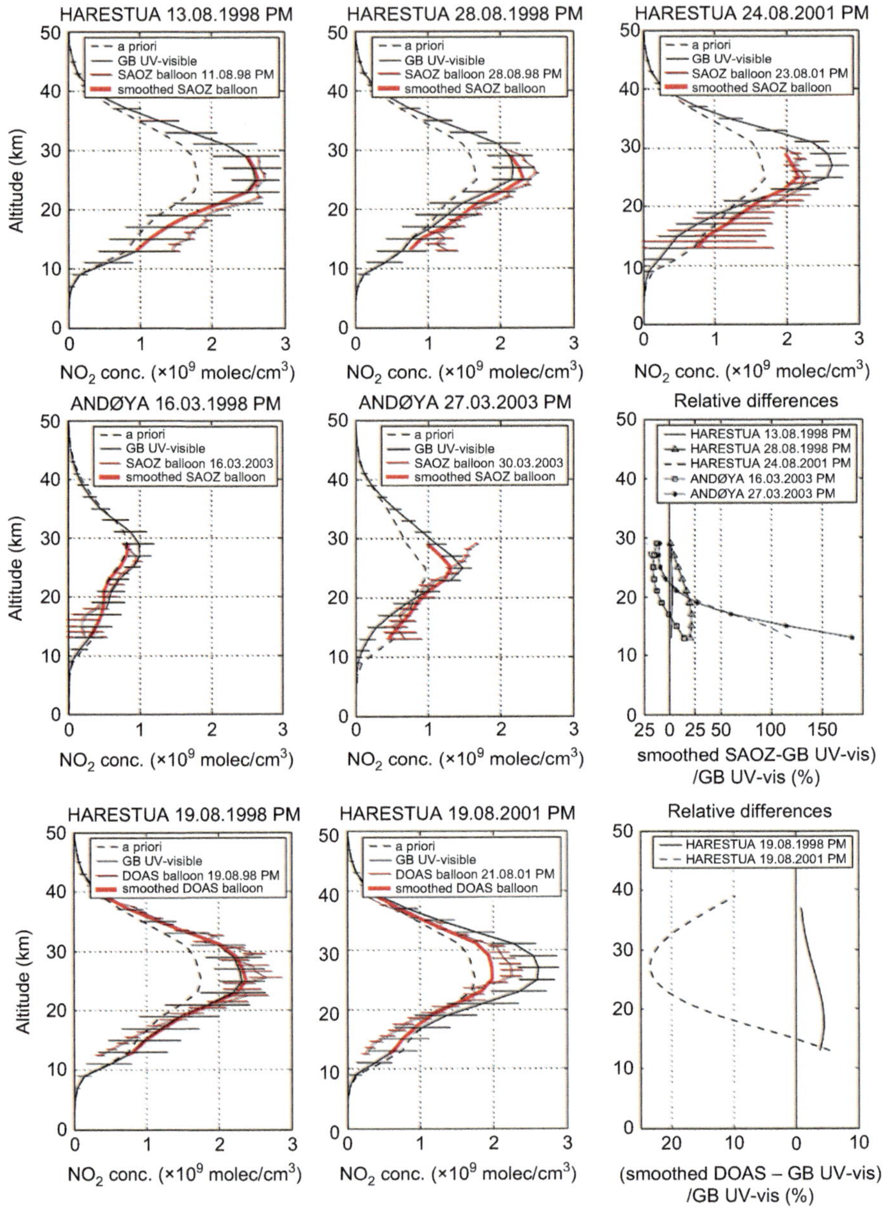

Fig. 11.16 Comparison between NO_2 profiles derived from ground-based UV-vis measurements at Harestua (sunset, summer conditions) and Andoya, Norway (sunset, late winter–early spring conditions) and SAOZ and DOAS balloon profiles (all balloons were launched from Kiruna, Sweden in 1998, 2001, and 2003). The SAOZ/DOAS balloon profiles are smoothed by convolution with the ground-based UV-vis averaging kernels. The relative differences appear in the *right lower plot*. Reprinted with permission from Hendrick et al. (2004),

direct-sun observations are dominantly sensitive to the stratosphere and the troposphere, respectively. Therefore, combining both observation geometries in a single retrieval provides information on both stratospheric and tropospheric absorbers.

The study by Hendrick et al. (2004) presents a comprehensive comparison of NO_2 profiles retrieved from zenith scattered UV-vis DOAS measurements, with balloon and satellite observations (Sects. 11.1.2 and 11.8) showing agreement at the $<25\%$ level (Fig. 11.16).

All studies stressed the impact of photochemistry on the retrieved profile information. Many relevant trace species (e.g. NO_2 and BrO) display a strong diurnal variation, which considerably complicates the retrieval since the observed variation of the measurements with the SZA depends not only on the scattering geometry, but also on photochemistry (see e.g. Roscoe and Pyle 1987). For instance, the concentration of NO_2 and BrO increases and decreases with increasing SZA, respectively.

11.3 Measurement of Tropospheric Species by Ground-Based DOAS

The earliest use of scattered sunlight to probe the troposphere was through the COSPEC technique, which was developed in the late 1960s. It was the first attempt to study tropospheric species by analysing a wider spectral range with the help of an optomechanical correlator (Millan et al. 1969; Davies 1970; Hoff and Millan 1981; Stoiber and Jepsen 1973; and Hoff et al. 1992).

Early passive DOAS applications focused on stratospheric gases. However, tropospheric observations became more widespread in recent years. In particular "off-axis" observational geometries, i.e. viewing at several elevations away from the zenith direction, have allowed the observation of tropospheric trace gases and their vertical profiles. In particular, Multi Axis DOAS (MAX-DOAS) has been shown to be a powerful technique for the study of tropospheric constituents, see e.g. Hönninger et al. (2004a); Wagner et al. (2004); Bruns et al. (2004), and the discussion of MAX-DOAS AMFs in Chap. 9. Due to its capabilities and relatively simple technical setup MAX-DOAS is now in widespread use. In fact there are so many reports on MAX-DOAS measurements, technology, and evaluation procedures that we can only list a small selection here: Ground-based measurements with MAX-DOAS have been reported by Kanaya et al. (2014); Chan et al. (2018); Javed et al. (2019); Hong et al. (2022); De Smedt et al. (2021); Yang et al. (2019); Kumar et al. (2020); Ren et al. (2021); Davis et al. (2019); Ryan et al. (2018); Chan et al. (2020); Gratsea et al. (2016); Wagner et al. (2013a); Zhang et al. (2021); Gomez et al. (2014); Li et al. (2013); Benavent et al. (2019); Garcia-Nieto et al. (2018); Frieß et al. (2011); Holla et al. (2015); Peterson et al. (2017); Simpson et al. (2017, 2018); Ma et al. (2020); Tirpitz et al. (2022); Frieß et al. (2022). More examples can be found in a series of reports from intercomparison exercises, see e.g. Pinardi et al. (2013); Frieß et al. (2019); Wang et al. (2020); Kreher et al. (2020); Tirpitz et al. (2021). A series of ship-based MAX-DOAS measurements have also been performed, see for instance Tan

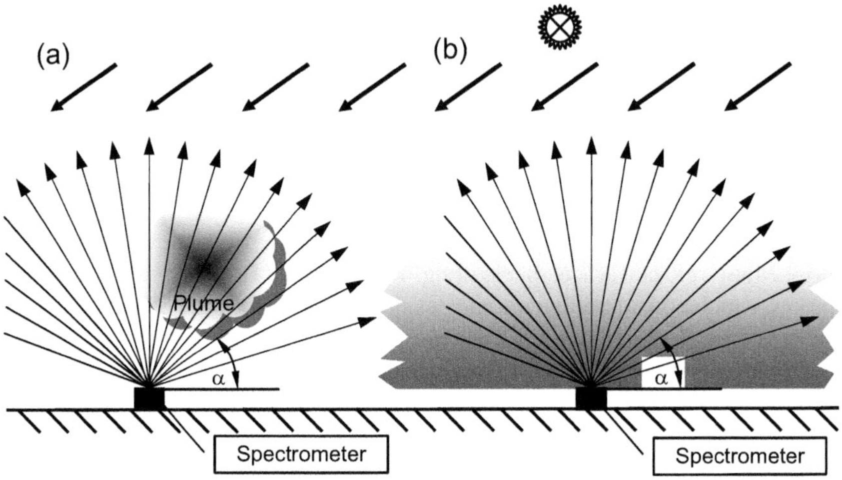

Fig. 11.17 There are two extreme cases in the distribution of trace gases. **a** A localised trace gas plume (e.g. a stack plume); **b** An (infinitely) extended trace gas layer

et al. 2018; Mahajan et al. 2019; Takashima et al. 2022. MAX-DOAS techniques are particularly useful for airborne measurements, also known as AMAX-DOAS, see e.g. Dix et al. (2009); General et al. (2014); Volkamer et al. (2015); Stutz et al. (2017).

As illustrated in Fig. 11.17, two extreme cases of the spatial distribution of trace gases have to be considered for DOAS applications:

1. A localised trace gas plume (e.g. a stack plume) with very low trace gas concentrations outside the plume. In this case, reference spectra can be taken by pointing the instrument away from the plume. A scan through the plume will reveal the column densities inside the plume.
2. An infinitely extended trace gas layer, where the trace gas concentration only depends on altitude. Some elements of the vertical trace gas distribution can be derived from observation of column densities seen at different observation angles, as described in Chap. 9.

Real conditions will be in between the two extreme cases. Nevertheless, they are a useful tool in considering the principle of the different passive DOAS techniques.

Scanning of trace gas plumes (Fig. 11.17a) emitted by various sources such as volcanoes, power plants, or cities is now systematically used as a tool for studying the emission and transformation of gases in the atmosphere. We discuss these applications in Sect. 11.4.

Here we will focus on the second case (Fig. 11.17b), where we assume that trace gases are horizontally evenly distributed, thus allowing the retrieval of trace gas profiles, as explained in detail in Chap. 9. We discuss a few examples on the scientific use of these MAX-DOAS measurements.

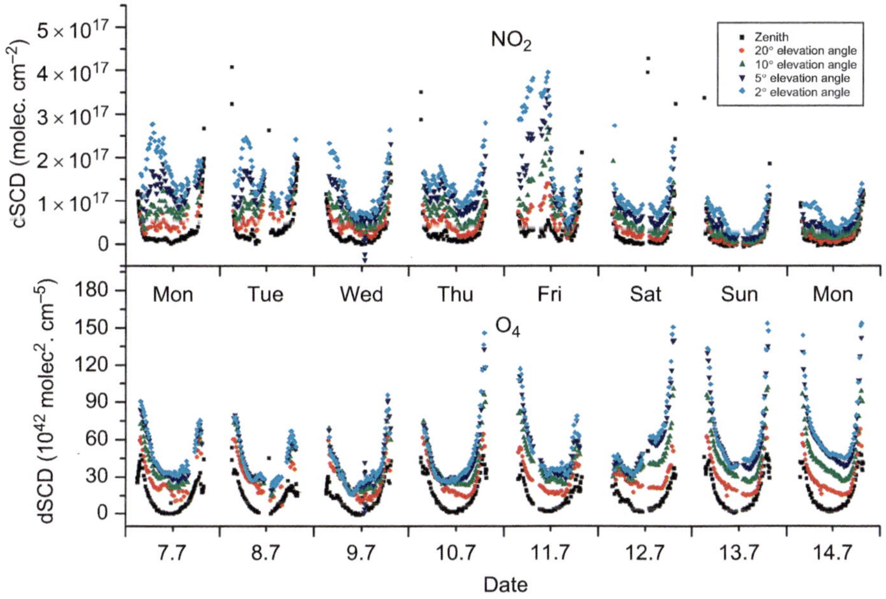

Fig. 11.18 NO_2 (*upper panel*) and O_4 (*lower panel*) SCDs recorded in Heidelberg, Germany during 8 d in July 2003. For evaluation, a reference spectrum of the 6th of July at noon was used. The elevation angle is indicated by the symbol colour as denoted in the legend. Reprinted with permission from Sinreich et al. (2005), ©The Royal Society of Chemistry (RSC). All rights reserved

11.3.1 MAX-DOAS Observations in Polluted Regions

MAX-DOAS instruments are relatively simple and easy to operate, especially over long time periods. Despite the challenges imposed by the radiative transfer and under-constrained concentration retrievals (see Chap. 9), MAX-DOAS is a unique tool for monitoring of air quality and studying atmospheric chemistry in polluted areas. One of the earliest MAX-DOAS measurements of O_4 and urban NO_2 were performed in Heidelberg, Germany during July, 2003 (Fig. 11.18). Both, O_4 and urban NO_2 DSCDs show the behavior expected from MAX-DOAS observation, with larger values for lower elevation angles. The presence of elevated surface NO_2 is illustrated by the deviation from the U-shaped behavior of O_4, in particular in the mornings of 7.7, 8.7, and 11.7, when NO_2 at low elevation angles is substantially higher compared to the zenith values.

Considerable progress in the interpretation of MAX-DOAS observations has been made since these first measurements. As described in detail in Chap. 9 new retrieval algorithms now allow to the conversion of DSCDs observed at multiple elevation angles into concentration- or mixing ratio vertical profiles. In polluted areas, where trace gas- and aerosol levels are typically high, these retrievals can provide unique information on pollutants and trace gases involved in the formation of urban smog. Figure 11.19 gives an example of such a retrieval (Wang et al. 2019d). The colour plots in Fig. 11.19 b show how pollution is concentrated near the surface, typically in the lowest 500 m. The primary pollutants NO_2 and SO_2 show highest mixing ratios in the morning, while mixing ratios of secondary species, such as HCHO and CHOCHO, increase due to photochemistry throughout the day. HONO is photolyzed in the early morning and has low, but non-negligible, daytime mixing ratios (see also Fig. 11.19c). The averaging kernels and the number of degrees of freedom of retrievals are shown in Fig. 11.19c. The retrieval adds between 2.5–4.5 pieces of information to the a-priori profiles (Fig. 11.19d), which gives a measure of the quality of the retrieved profiles. The progress in MAX-DOAS profile retrievals further strengthens its role as

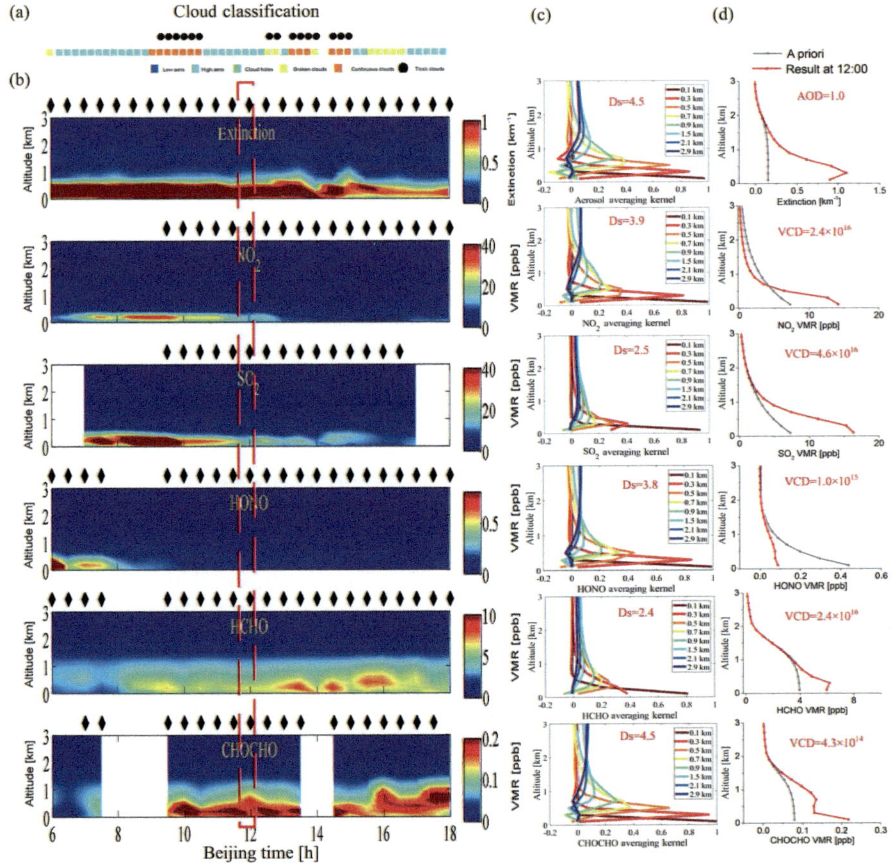

Fig. 11.19 MAX-DOAS measurements on 11 May 2016 in a highly polluted rural area in China. **a** Aerosol and cloud classification. **b** Vertical profiles of aerosol extinction (including cloud-contaminated results) and mixing ratio profiles of NO_2, SO_2, HONO, HCHO and CHOCHO retrieved using the methods presented in Chap. 9). Black diamonds indicate times of "high confidence profiles". The information content provided by the retrieval is illustrated as Averaging Kernels (AK) in (**c**), and D_s denoted the number of degrees of freedom for each retrievals. **d** comparison of a-priori and retrieved profiles at noon. Reprinted with permission from Wang et al. (2019d),

an important air pollution monitoring tool, and MAX-DOAS monitoring networks that provide this data on a daily basis have begun to emerge, see e.g. Verhoelst et al. (2021); De Smedt et al. (2021); Kanaya et al. (2014).

11.3.2 MAX-DOAS Observations of Halogen Oxides at Mid-Latitudes

MAX-DOAS plays an important role in studying the chemistry of halogen species in the troposphere. There are too many MAX-DOAS based studies of halogen oxides in the current scientific literature to be discussed here. We will therefore restrict this discussion to two examples of such studies at different geographical locations. Field measurements of reactive halogen and sulfur gas emissions were performed at the world's largest salt pan (area $\approx$ 10000 km^2), the Salar de Uyuni, Southern Altiplano, Bolivia (19.8°–20.7°S, 67.0°–68.2°W, >3600 m above sea level, see Fig. 11.20), during the dry season in October/November 2002 by Hönninger et al. (2004b). BrO and SO_2 were studied by ground-based scattered light MAX-DOAS at various locations upwind of the Salar and on the surface of the salt pan.

Significant amounts of BrO were found at all locations, with the lowest levels at the northern edge of the Salar (upwind), while BrO SCDs (differential with respect to zenith) of up to 3.7×10^{14} molec. cm^{-2} were observed inside the Salar. Using the MAX-DOAS vertical profile information, BrO mixing ratios of >20 ppt were

Fig. 11.20 Map of the Salar de Uyuni and surrounding area. Measurement sites are marked 'Salar 1' through 'Salar 4'. Reprinted with permission from Hönninger et al. (2004b), ©American Geophysical Union. All rights reserved

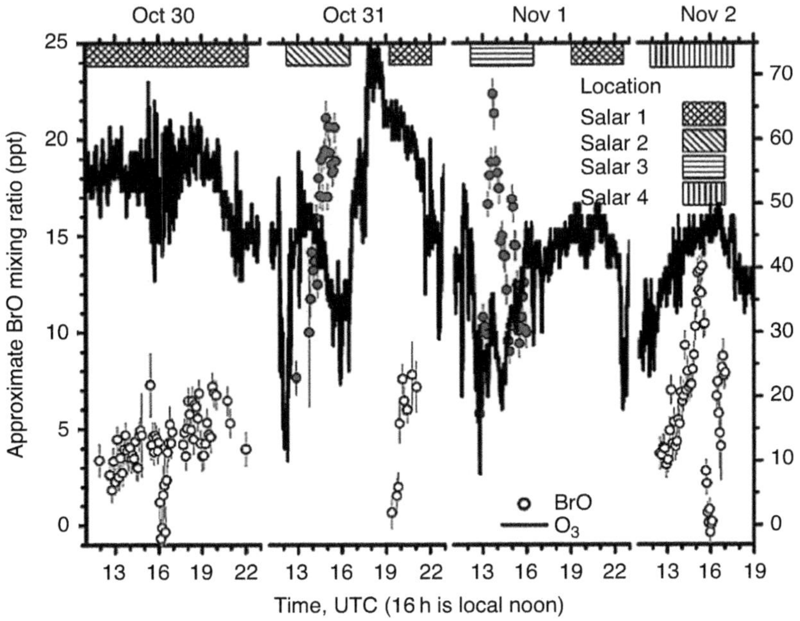

Fig. 11.21 MAX-DOAS time series of BrO recorded from 30 October to 2 November, 2002. Measurement locations are indicated at the top. Approximate BrO mixing ratios are derived from the MAX-DOAS data and in-situ ozone mixing ratios. Note that the ozone time series consists of measurements at different locations including the transects, as indicated at the top. BrO data from Salar2 and Salar3 are shown in colour/solid symbols. Reprinted with permission from Hönninger et al. (2004b), ©American Geophysical Union. All rights reserved

derived, which can have significant impact on tropospheric ozone chemistry both locally (Fig. 11.21) and regionally. In addition, SO_2 SCDs of $>5 \times 10^{16}$ molec. cm^{-2} were found around the Salar by the same technique, suggesting emissions from activity of nearby volcanoes.

11.3.3 Halogen Oxide Radicals in the Polar Troposphere

In the Arctic, passive DOAS measurements of BrO have been performed near Kiruna, Sweden (Frieß et al. 1999), as well as at Alert and the Hudson Bay, Canada (Hönninger and Platt, 2001; Hönninger et al. 2004b). Figure 11.22 shows a time series of simultaneous BrO measurements by active long-path DOAS and passive MAX-DOAS. A comparison of the data in Fig. 11.23 shows good agreement between both techniques. Tropospheric IO was measured by Wittrock et al. (2000a; 2004) in Ny Alesund on the island of Spitsbergen.

In Antarctica, several studies at Neumayer Station targeted IO and BrO abundances (Frieß 2001; Frieß et al. 2001; 2004; 2005). Tropospheric BrO was also observed at Arrival Heights, Antarctica (Kreher et al. 1997; Schofield et al. 2006). IO was also found at the Indian Antarctic bases of Bharati and Maitri, see Fig. 11.24 (Mahajan et al. 2022).

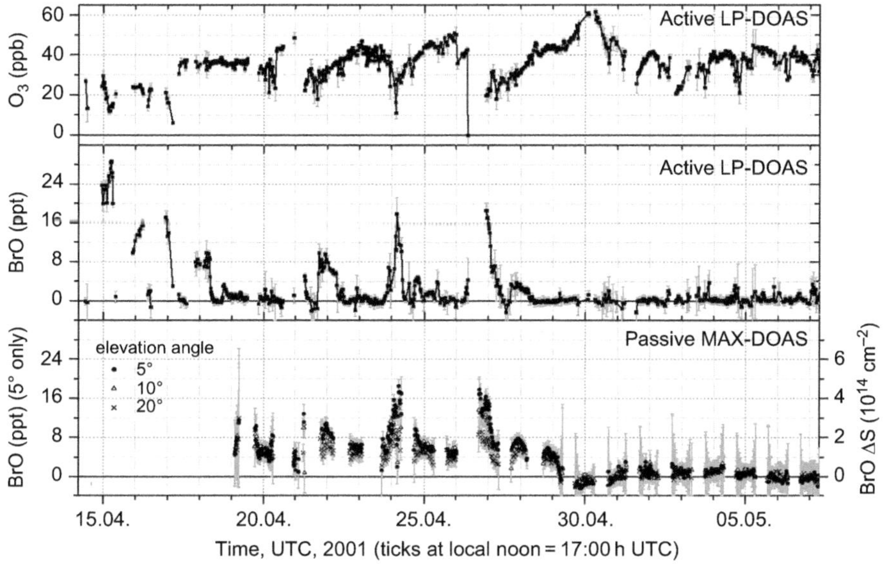

Fig. 11.22 Time series of LP-DOAS O_3 (*top*), LP-DOAS BrO (*centre*) and MAX-DOAS BrO (*bottom*) recorded at Kuujjuarapik, Nunavik, Quebec, Canada in the Hudson Bay (55°N and 75°W) during April 2001. Data for different elevation angles are shown in different colours/symbols. MAX-DOAS column densities are indicated on the right scale. The MAX-DOAS mixing ratio scale is valid for the 5° data points only. Reprinted with permission from Hönninger et al. (2004b), ©American Geophysical Union. All rights reserved

11.3.4 Trace Gases in the Marine Environment

Ship-borne MAX-DOAS observations have given access to the marine environment and led to a number of exciting research areas. Initially the focus was on the measurement of NO_2, BrO and later IO over the open ocean. For example, Leser et al. (2003) reported MAX-DOAS observations of NO_2, O_4, and BrO uring a cruise of the German research vessel Polarstern from Bremerhaven, Germany to Cape Town, South Africa in October, 2000. From the DSCDs at different elevation angles, concentrations of these absorbers in the marine boundary layer were derived. Boundary layer NO_2 was found near Europe and the Canary Islands. During a period of two days in the region north of the Canary Islands (around 35°N and 13°W), significant levels of boundary layer BrO (on the order of 1 ppt) were found (see Fig. 11.25).

Since this study, numerous other MAX-DOAS observations of BrO as well as IO have greatly contributed to our understanding of halogen chemistry in the marine boundary layer Martin et al. (2009); Großmann et al. (2013); Mahajan et al. (2012, 2021); Prados-Roman et al. (2015); Gómez Martín et al. (2013); Mahajan et al. (2019); Takashima et al. (2022); Inamdar et al. (2020).

MAX-DOAS measurements have also contributed to our understanding of organic chemistry in the marine boundary layer. For instance Sinreich et al. (2010) report the presence of glyoxal, CHOCHO, over the Pacific ocean, based on their MAX-DOAS observation onboard the Ronald H. Brown research vessel. As shown in

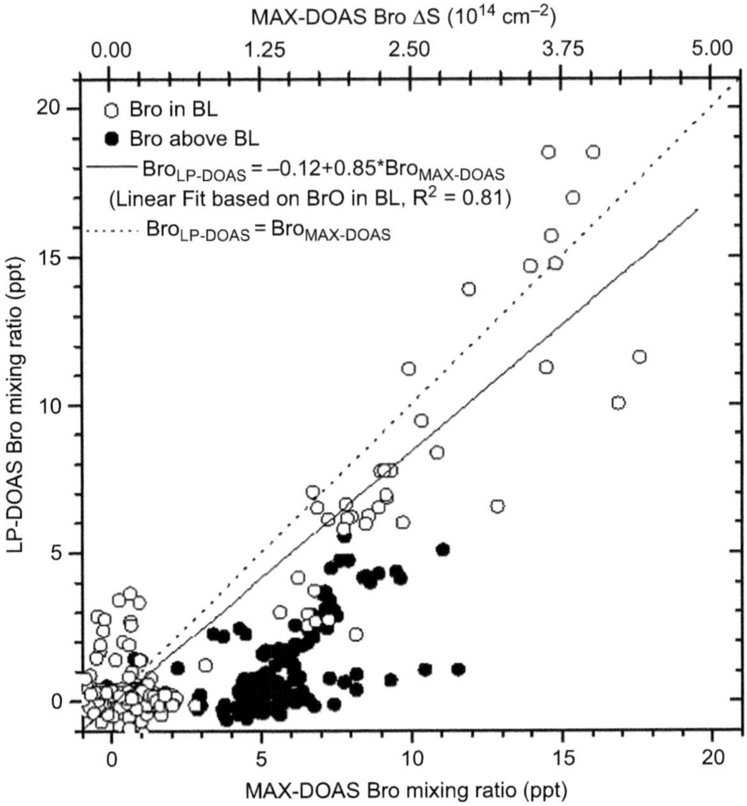

Fig. 11.23 Scatter plot of BrO mixing ratios derived by active LP-DOAS BrO versus passive MAX-DOAS. *Filled symbols* indicate MAX-DOAS signatures of BrO located above the boundary layer. The linear fit based on BrO in the boundary layer (*open symbols*) shows good agreement between both techniques. Reprinted with permission from Hönninger et al. (2004b), ©American Geophysical Union. All rights reserved

Fig. 11.26 the MAX-DOAS DSCDs are dependent on the elevation angle, a behavior that clearly identifies the presence of glyoxal. The abundance of glyoxal over the ocean is surprising due to its short lifetime and high solubility. Nevertheless the MAX-DOAS observations point to a level of up to 140 ppt of glyoxal in the marine boundary layer. The presence of glyoxal in the MBL has since been confirmed by other MAX-DOAS observations, e.g. Mahajan et al. (2014); Walker et al. (2022) as well as from various satellite measurements, Vrekoussis et al. (2009); Alvarado et al. (2014); Lerot et al. (2021).

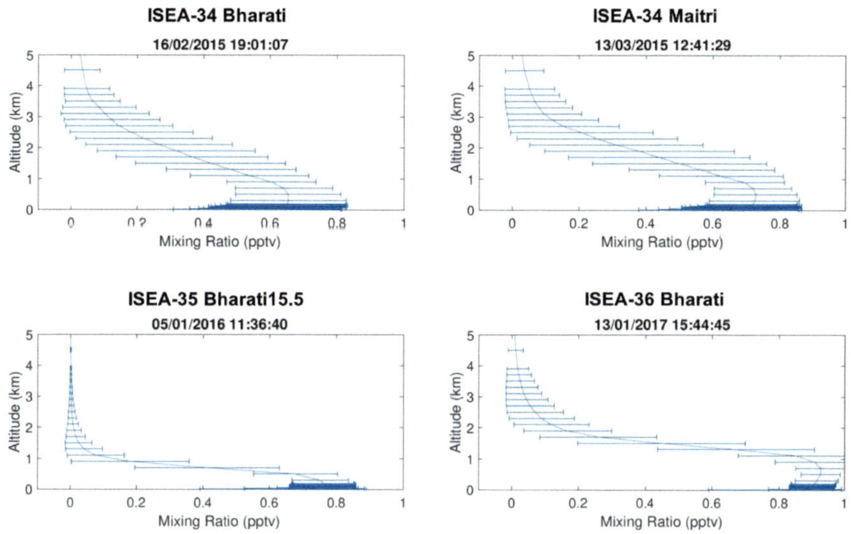

Fig. 11.24 Examples of summer IO vertical profiles at the Indian Antarctic bases of Bharati and Maitri. IO mixing ratios are highest near the surface with values between 0.6 and 0.9 ppt. Times are in UTC. Reprinted with permission from Mahajan et al. (2021), ©CC 4.0. All rights reserved

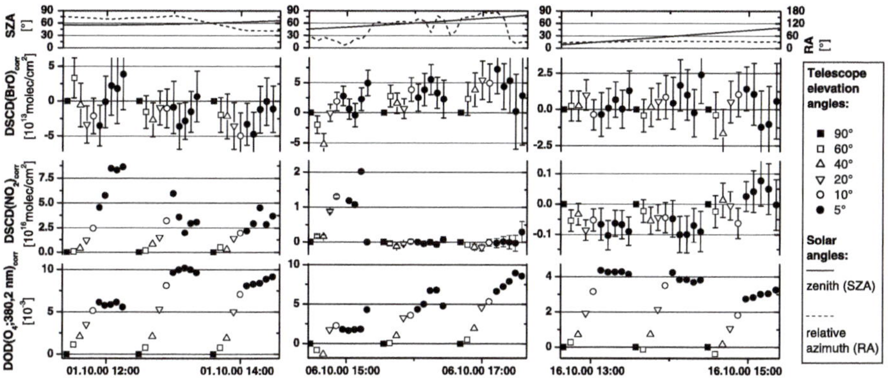

Fig. 11.25 Typical time series of (differential) MAX-DOAS SCDs of BrO (*top*), NO_2 (*centre*), and O_4 (*bottom*) at various observation elevation angles, as indicated by different *symbols*, during a trip of the research vessel Polarstern in October, 2000. Also given are SZA and relative azimuth RA (azimuth angle between viewing direction and sun). Error bars indicate 1σ fit errors. **Left panel**: On 1 October in the English Channel high NO_2 SCDs at low elevation angles indicate elevated NO_2. **Centre panel**: The pattern of the BrO series during the afternoon of 6 October, north of the Canary Islands, is typical for BrO in the marine BL. Boundary layer NO_2 levels are below the detection limit after 15:30 h. **Right panel**: On 16 October, in the remote Atlantic Southwest of Africa, both NO_2 and BrO are below the detection limits (note expanded scale). Reprinted with permission from Leser et al. (2003), ©American Geophysical Union. All rights reserved

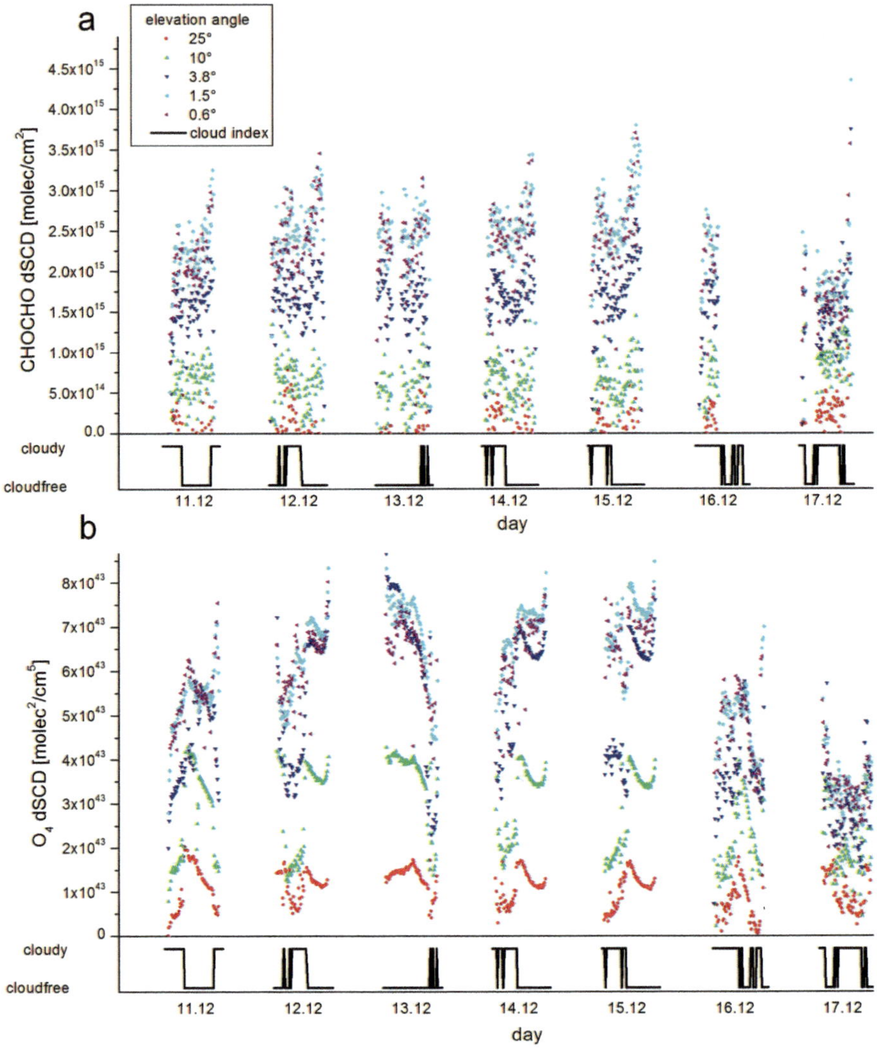

Fig. 11.26 Glyoxal (**a**) and O_4 (**b**) DSCDs measured by MAX-DOAS on board the RV Ronald H. Brown during the Tropical Atmosphere Ocean (TAO) experiment from 11 through 17 December 2008 (Sinreich et al. 2010). The DSCDs at different elevation angles are colour coded, and show the expected separation with elevation angles, i.e. higher DSCDs at smaller elevations. The cloud index is high for cloudy conditions and low value in the absence of clouds. Reprinted with permission from Sinreich et al. (2010),

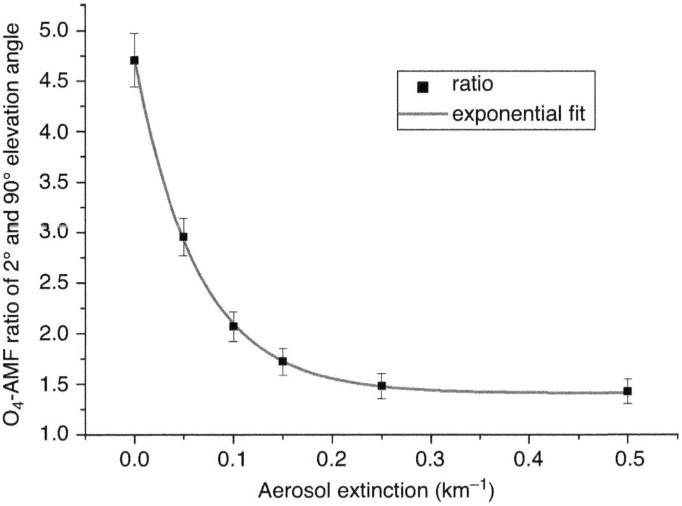

Fig. 11.27 Ratio of O_4 AMFs calculated for 2° and 90° elevation angles as a function of the aerosol extinction coefficient (assuming a constant aerosol extinction within the lowest 2 km of the atmosphere). The method is sensitive to aerosol extinctions between about 0.25 and 0.01 km^{-1} (corresponding to visibility ranges of about 40–16 km). Reprinted with permission from Sinreich et al. (2005), ©The Royal Society of Chemistry (RSC). All rights reserved

11.3.5 Determination of Aerosol Properties from MAX-DOAS Observations

DOAS was historically developed as a method to measure atmospheric trace gases. However, the quantification of the structured UV-vis absorptions of gases with well known atmospheric abundances, such as O_2 and O_4, enables MAX-DOAS to quantify aerosol through their impact on atmospheric radiative transfer. Figure 11.27 illustrates the relationship between the aerosol extinction coefficient and the ratio of the O_4 AMFs, and thus measured O_4 column densities, for 2° and 90° observation elevation angles. This capability was recognised early on in the development of MAX-DOAS (Sinreich 2004; Bruns et al. 2004; Sinreich et al. 2005; Frieß et al. 2006) and has since found wider use. By using sophisticated radiative transfer models with optimal estimation method-based inversion schemes, it is possible to derive aerosol extinction coefficients and their profiles from MAX-DOAS measurements. One of the earlier successful retrievals is illustrated in Fig. 11.28 where 476 nm aerosol extinction coefficients measured by MAX-DOAS and LIDAR in the lowest km of the atmosphere show very good agreement (Irie et al. 2008).

The retrieval of aerosol extinction profiles is also shown in the top panel of Fig. 11.19b, where, in addition to quantification of the aerosol extinction, the extent of the planetary boundary layer is illustrated (Wang et al. 2019d). In this example, the averaging kernels for the aerosol retrieval in the lowest kilometre of the atmosphere are close to 1 (Fig. 11.19c). The retrieved aerosol extinction coefficients are therefore quite accurate in the planetary boundary layer, confirming the finding from Irie

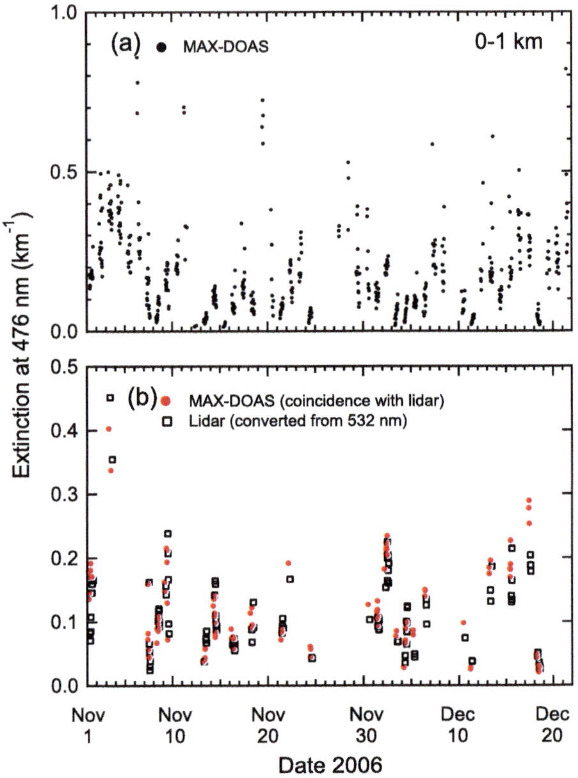

Fig. 11.28 **a** MAX-DOAS aerosol extinction coefficient at 476 nm for the layer from 0 to 1 km in Tsukuba, Japan. **b** Comparison of coincident MAX-DOAS (red) and lidar (black) observation. Lidar data was converted from 532 to 476 nm. Reprinted with permission from Irie et al. (2008), ©CC 2.5. All rights reserved

et al. (2008). Finally, it should be noted that aerosol extinction profiles are measured simultaneously with those of various trace gases, thus considerably expanding the air pollution monitoring capabilities of MAX-DOAS, as well as opening new venues in the investigation of multi-phase atmospheric chemistry.

11.3.6 Determination of NO_3 Vertical Profiles

A particular property of the nitrate radical (NO_3) molecule, its rapid photolysis in direct sunlight (photolytic lifetime of around 5 s), allows determination of NO_3 vertical concentration profiles in morning (or evening) twilight. During twilight, only higher layers of the atmosphere are illuminated, while the Earth's surface is still shaded. In the illuminated (higher) part of the atmosphere there is essentially no NO_3 due to rapid photolysis. Twilight illumination is due to radiation scattered down to the surface from the higher layers. This radiation can be observed by an instrument looking to the zenith (or away from the sun). The recorded spectra thus contain the signature of the NO_3 column remaining in the shaded (lower) part of the atmosphere. During sunrise the shaded part of the atmosphere continuously shrinks until sunrise (when it is zero), as does the remaining NO_3 column. From the resulting time series (and simple geometric considerations) the vertical NO_3 concentration profile can be calculated. The technique was originally suggested by Smith and Solomon (1990), who derived a profile from the stratosphere down to 3 km. Smith et al. (1993) repeated the experiment in Antarctica, finding negligible levels of NO_3 in the troposphere. Fish et al. (1999) modelled near-ground NO_3 concentrations of $<1.3 \times 10^9\,cm^{-3}$ ($\approx$50 ppt) and an increase to $(1.75 - 2.5) \times 10^9\,cm^{-3}$ ($\approx$65–90 ppt) within a layer between 300 and 1700 m altitude, followed by a sharp decrease above 1700 m. Povey et al. (1998) found average nocturnal NO_3 abundances of about 100–200 ppt, in boxes centred at 1.5 and 1.2 km, respectively. Similar levels and relatively long inferred free-troposphere NO_3 lifetimes were reported by Aliwell and Jones (1998). Experimental verification of these NO_3 profile structures was reported by von Friedeburg et al. (2002), Allan et al. (2002), and Coe et al. (Coe02).

11.4 Emission Flux Measurements

A practical application of passive DOAS observations is the measurements of emission fluxes of various trace gases from anthropogenic and natural sources. These emissions can originate from point sources such as smoke stacks, volcanoes, or wildfires, or can be distributed over wide areas, often called area sources, such as traffic in cities. In all cases the ultimate goal of this application is to quantify the amount of a specific trace gas that is emitted from the source in a given time interval, i.e. the emission flux. This flux provides important information on the contribution of the source to air pollution or, in the case of volcanoes, provides information on geological processes.

Here we will introduce various DOAS methodologies and their applications, such as mobile direct sun or scattered light observation and imaging DOAS. It should be noted that direct sunlight measurements with FTIR instruments have been applied for over a decade to validate reported VOC emissions from petrochemical facilities. Since our discussion is restricted to UV-vis DOAS observation, we refer the reader to Mellqvist et al. (2010); Duffell et al. (2001); Kille et al. (2017) for details on these IR Solar Occultation Flux measurements.

11.4.1 Mobile DOAS Measurements of Point Source Emission Fluxes

Mobile DOAS flux measurements are based on mounting a spectrometer looking into the zenith or directly into the sun on a vehicle and driving around the source or source area of interest. Figure 11.29 shows emission flux observation from a refinery. The spectrograph is driven around the facility, measuring all absorbing trace gases in an upwind and downwind slice. In the case of an isolated source without upwind emissions, such as volcanoes, the upwind leg can often be omitted.

Quantitatively the mass flux through a slice is determined by the overlap of the profile of the gas concentration, c(z), with altitude, z, and the profile of the wind component perpendicular to the surface, u'(z). The product of the perpendicular wind component and the concentration are integrated from the ground to a specific altitude, h, for example the mixing layer height for anthropogenic surface emissions.

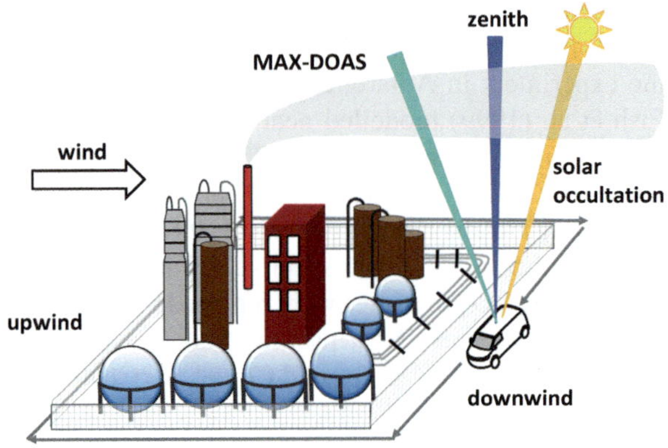

Fig. 11.29 Concept of a mobile emission flux measurement system for an industrial facility

This integral is then further integrated along the path of the mobile spectrometer along the respective leg:

$$Flux = J_c = \int_{x_1}^{x_2} \left(\int_0^h c(x, z) \cdot u'(x, z) dz \right) dx \tag{11.1}$$

When an upwind flux measurement, is available the difference between the downwind and upwind flux is the net flux of emissions between the two slices. The challenge of applying Eq. 11.1 is that many DOAS measurements, such as direct sun or zenith observations, do not provide a concentration profile, making additional assumptions, such as estimating altitude and widths of a trace gas plume, necessary for the use of Eq. 11.1 (Mellqvist et al. 2010; Johansson et al. 2014a). The measurement of the wind profile $u'(x, z)$ can be achieved by using parallel balloon sonde or wind LIDAR/SODAR observations. Methods to estimate the missing parameters for Eq. 11.1 have begun to emerge and are discussed in Sect. 11.4.4. The uncertainty of the mobile DOAS emission flux estimates is primarily determined by the unknown concentration profiles and is often in the range of 30% (Johansson et al. 2014a; Kille et al. 2017, 2022).

Mobile DOAS point source emission observations have been applied to power plants, chemical facilities, and volcanoes. The example in Fig. 11.30 shows formaldehyde, HCHO, emissions from a chemical manufacturing plant in the Houston Ship Channel. The formaldehyde flux in this plume was calculated to be approximately 33 kg/h (Johansson et al. 2014b).

11.4.2 Mobile DOAS Measurements of Area Sources

The concept of mobile DOAS emission flux measurements can also be applied to larger areas, such as cities, where is it particularly useful since few other methods exist to directly measure emissions. Some of the challenges encountered for point sources are reduced as the sources are often close to the surface and the larger scale of the measurements allows the trace gases to be more uniformly mixed in the atmosphere.

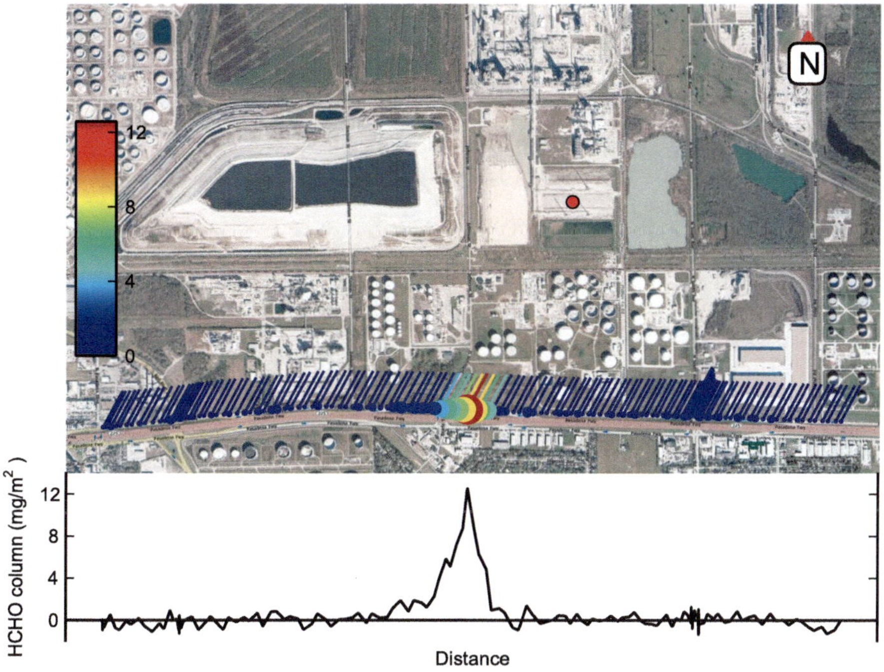

Fig. 11.30 Example of the observation of HCHO fluxes from a chemical manufacturing facility in Houston, TX, USA, at 18:33 CDT on 18 May 2009. Each circle corresponds to one Mobile DOAS observations. Size and colour of the circle indicate the magnitude of the HCHO column and the lines from each circle indicates the wind direction. Reprinted with permission from Johansson et al. (2014b), ©American Geophysical Union. All rights reserved

The determination of wind speed remains a challenge, but stationary profiling measurements are often more readily available in urban areas. An additional challenge is chemical transformations that can occur over the larger scale of the observations. It is thus often more appropriate to compare the observation with air quality model outputs to validate/update the emissions. Figure 11.31 shows an example of a mobile MAX-DOAS measurement around Paris, France, performed during the MEGAPOLI experiment in 2010. Based on these observations and an estimate of the NO_2/NO ratio a daily average NO_X emissions for Paris of 4.0×10^{25}molec. s^{-1} for summer and of 6.9×10^{25}molec. s^{-1} in winter was derived (Shaiganfar et al. 2017). These emissions were about 1.5–2 times larger than those implemented in air quality models of Paris at that time, showing the value of independent validation of emission inventories by mobile DOAS observations.

11.4.3 Emission Measurements from Air and Space

With increased spatial resolution of satellite DOAS instrument and the development of airborne DOAS instruments, the use of these observations to quantify emissions

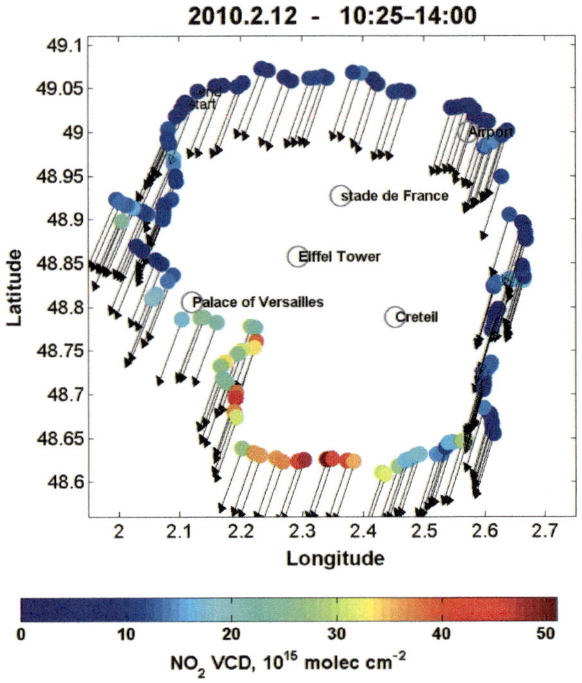

Fig. 11.31 Mobile DOAS observations of NO_2 VCDs derived around Paris on 12 February 2010. Each dot indicates an individual DOAS measurement. Arrows represent wind speed and direction from the regional CHIMERE chemical transport model. Reprinted with permission from Shaiganfar et al. (2017), ©CC 3.0. All rights reserved

has gained popularity. While the basic idea of these measurements is the same as for groundbased observations, determination of wind speeds and direction can be a challenge, especially for satellite observations. Despite the often sparse data, i.e. once a day for polar orbiting satellites, the longterm datasets as well as the ability to study remote sources, makes satellite observations a valuable tool to study emissions.

An example of the determination of the emission flux from a point source, in this case two South-African power plants, from the TROPOMI satellite instrument is shown in Fig. 11.32. Integration across the NOx plume (Fig. 11.32b) yields an emission of $E_{NO_x} = 5.5\,\mathrm{mol\,s^{-1}}$ (Lange et al. 2022). This illustrates the capability of modern high spatial resolution satellite instruments to monitor emissions from point sources anywhere on the globe.

11.4.4 Emission from Volcanic and Other Point Sources

As mentioned above, passive absorption spectroscopy has become wide-spread for the quantification of plume emissions from localised sources such as smoke stacks, industrial plants, urban centres, wildfires, and volcanoes. Some early measurements were reported by by Milan (1980) and Edner et al. (1994b).

In this application, column densities are measured in a series of angles approximately perpendicular to the propagation direction of the plume (Fig. 11.33). From a simple geometric approach, the approximate extent of the plume can be inferred,

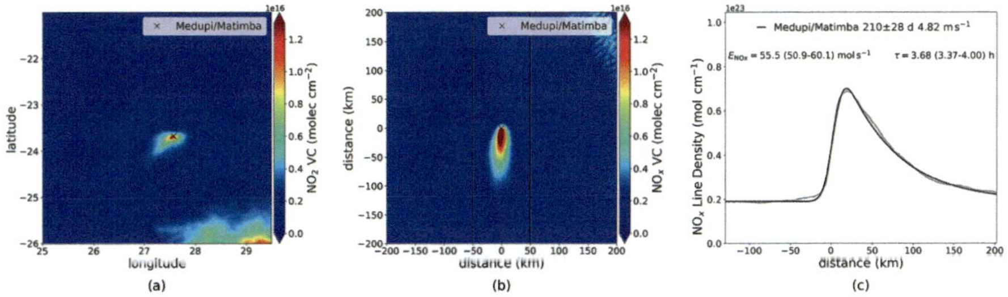

Fig. 11.32 **a** TROPOMI observations of NO_2 (**panel a**) and NO_x (**panel b**) tropospheric VCDs around the Medupi and Matimba power plants in South Africa averaged from 1 March 2018 to 29 February 2020. Only days with wind speeds > $2ms^{-1}$ were considered. Please note that the NO_X VCD is rotated with its wind direction around the centre of the source (cross). (**panel c**) The NO_X VCD integrated perpendicular to the wind direction (NO_X line density) as a function of distance to the source calculated for the ±50 km sector (grey) and fit results (black) with estimated NO_X emissions and lifetime. Reprinted with permission from Lange et al. (2022),

based solely on the DOAS measurements and wind direction. Each column density $S(\alpha)$ is an integral of the trace gas concentration along the line of sight through the plume (arrows Fig. 11.33a). Integrating over the spatial coordinate $s \approx R \cdot \alpha$ associated with the observation elevation angle α (see Fig. 11.33), where R denotes the distance from the spectrometer to the centre of the plume, the amount of trace gas Q (in molec. cm^{-1}) per unit length (along the propagation direction) of plume is derived (similar to Eq. 11.1):

$$Q = R \int S(\alpha)\, d\alpha \approx R \iint c(s) ds\, d\alpha = \int_{A} c(\mathbf{x})\, dA' \tag{11.2}$$

By using two spectrometers, the plume can be triangulated from the elevation angles α and β, under which the centre of the plume is seen. The trace gas flux J_C from the point source can be calculated if the wind speed v_P at the position of the plume is known, as $J_C = Q \cdot v_P$ (with Q from Eq. 11.2). The wind speed can either be derived from meteorological data, or it can be determined by observing the temporal variation $S(t)$ in the plume at different downwind distances from the source. From the spatial separation and the temporal correlation of the fluctuations in $S(t)$, the wind speed can be calculated (see e.g. Galle et al. 2004; McGonigle et al. 2005; Johansson et al. 2009b; Galle et al. 2010).

Many measurements of SO_2, BrO and OClO in volcanic plumes have been performed by passive DOAS techniques (Galle et al. 2003; Bobrowski et al. 2003; 2007; McGonigle et al. 2005; Bobrowski 2005; Oppenheimer et al. 2006, Salerno et al. 2009; Galle et al. 2010; Platt and Bobrowski 2015; Platt et al. 2018; Dinger et al. 2021; Gliß et al. 2015). Figure 11.34 shows measurements of the Soufrier Hills volcano plume on Montserrat. The BrO plume width (FWHM) seen from the instrument location ranged over 34°. The distance between the plume centre and the observation point was about 1 km. The corresponding plume width is ≈760 m (2σ). Combined

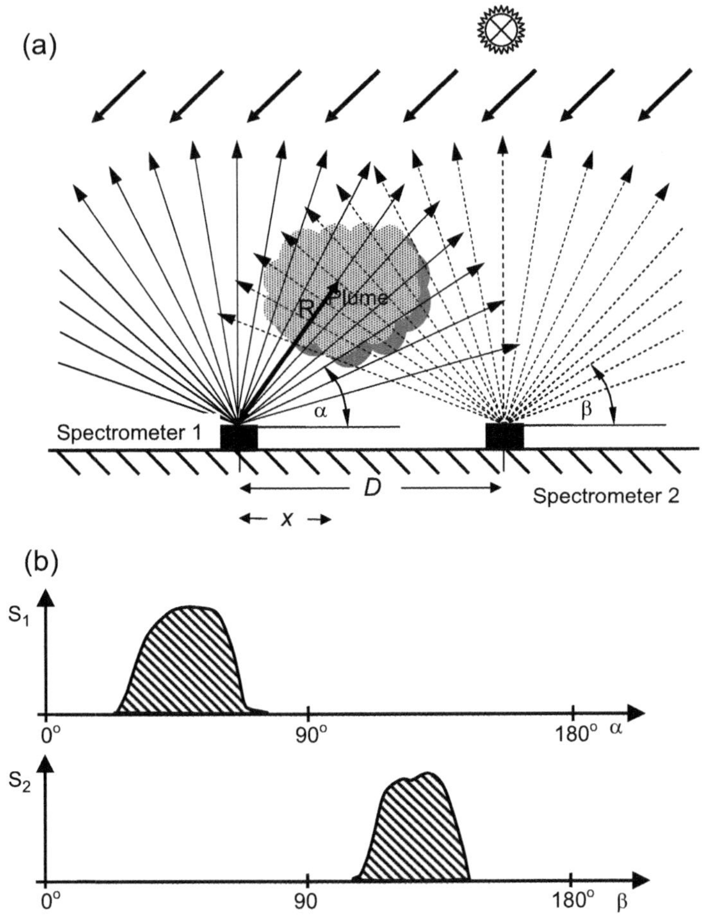

Fig. 11.33 Experimental approach to characterize the extent, trace gas content, and trace gas flux in a plume (**a**). The column densities S_1 and S_2 seen by spectrometers 1 and 2 when scanning in a plane (approximately) perpendicular to the propagation direction of the plume is sketched in (**b**). One spectrometer is enough to locate the plume. Two or more spectrometers allow triangulation of the plume

with the measured SCDs (BrO $\approx 2 \times 10^{15}$, $SO_2 \approx 1.75 \times 10^{18}$ molec. cm^{-2}), mixing ratios in the plume of $\approx$1 ppb for BrO and $\approx$1 ppm SO_2, were derived (Bobrowski et al. 2003).

The second plume-scan example in Fig. 11.35 shows repeated scans through the plume of a large power station (Alcanitz, Spain) yielding simultaneously measured SCDs of NO_2 and SO_2 (Bobrowski 2005).

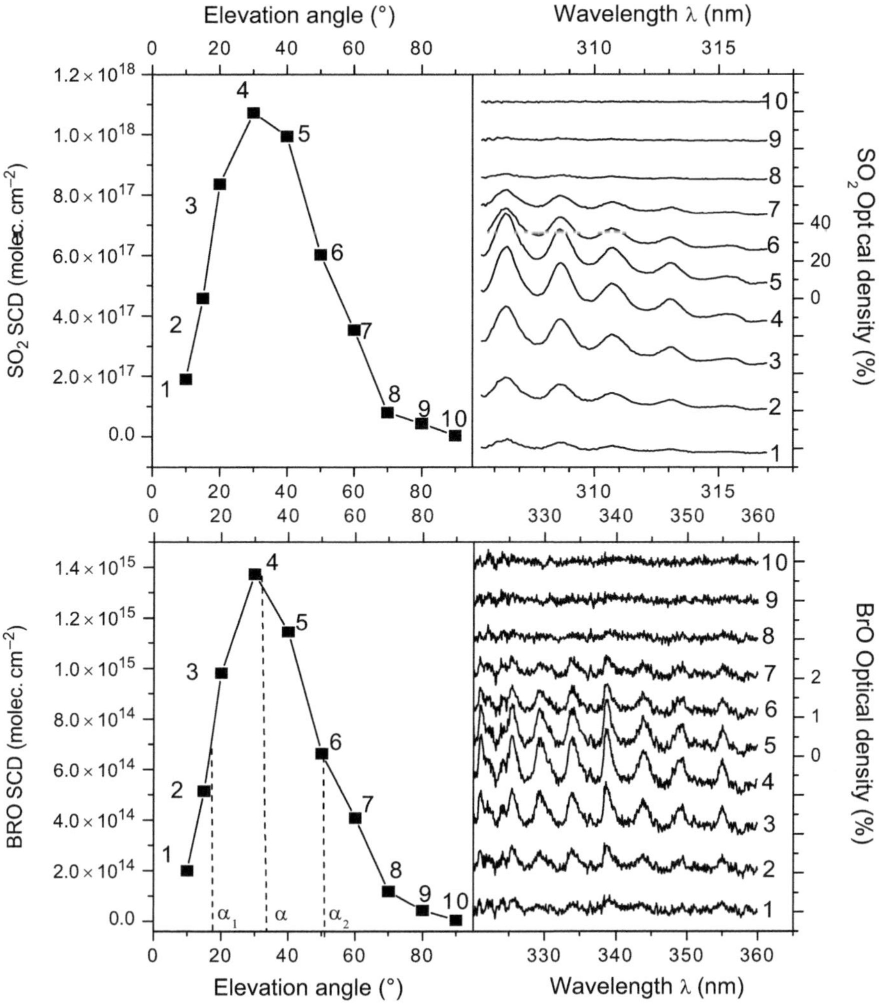

Fig. 11.34 Cross-sections of the plume from Monserat volcano. The observation point was about 4 km from the summit, on 25 May, 2002 where the plume was observed over an angular range of 34°. Measured BrO and SO_2 spectra corresponding to the individual viewing directions are shown on the *right*. (In both panels, subsequent spectra, except spectrum 5, are vertically shifted for better illustration.) Reprinted with permission from Bobrowski et al. (2003), ©Macmillan Magazines Ltd.. All rights reserved

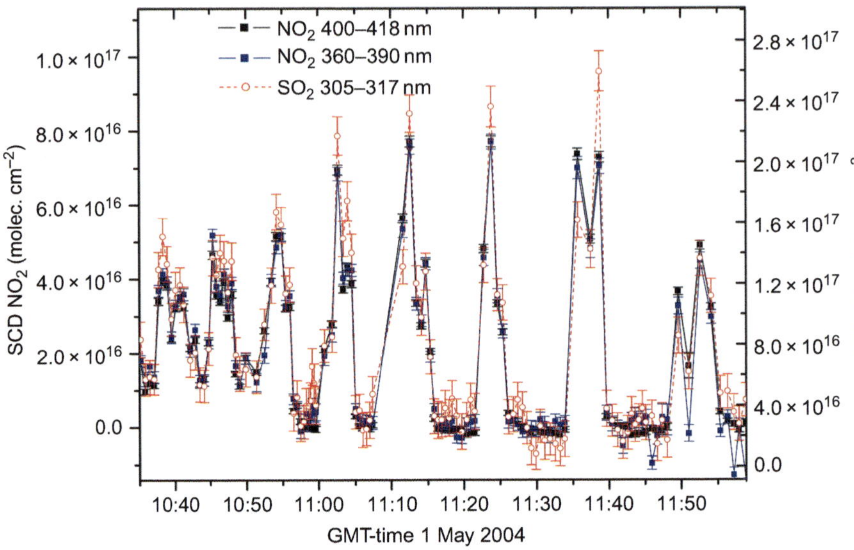

Fig. 11.35 Repeated scans through the plume of the Alcanitz power station in northern Spain. The evaluation yields simultaneous SCDs of NO_2 and SO_2. Reprinted from Bobrowski, Ph.D. Thesis, 2005, University of Heidelberg

11.4.5 Emission Measurements with Imaging DOAS (I-DOAS)

In cases when it is not possible to drive under or around an emission source, for example in the case of industrial emission within an inaccessible facility or remote volcanoes, DOAS offers another approach to measure fluxes. In the imaging-DOAS, I-DOAS, approach, a slice through a plume is measured from a stationary location by using modern imaging spectroscopic technology to simultaneously observe multiple viewing directions (Lohberger et al. 2004; Bobrowski et al. 2006; Louban et al. 2009; Pikelnaya et al. 2013). Alternatively, a fast scanner and spectrometer can be used to slice through the plume, but this approach has thus far not been realised. The simultaneous or rapid measurement in multiple viewing directions allows for a near instantaneous determination of a flux. When the ability to scan in a second direction is added, it allows visualisation of a plume and its upwind conditions. Figure 11.36 describes the I-DOAS principle (see also Fig. 7.64 in Chap. 7), where an imaging spectrometer in combination with a two-dimensional CCD is used to project a plume onto the entrance slit of a spectrometer. The imaging spectrometer then disperses the light at different locations along the slit leading to spatially separated spectra on the 2D CCD detector. A DOAS analysis is then performed on each of these spectra, basically treating each spectrum as a separate spectrometers, resulting in a column of slant column densities. With the addition of a horizontal scanner, this procedure can then be repeated along the plume, providing a slant column image of the trace gas.

There are various examples for the application of I-DOAS measurements (Heue et al. 2008; Louban et al. 2009; Lee et al. 2009; General et al. 2014; Luo et al. 2015;

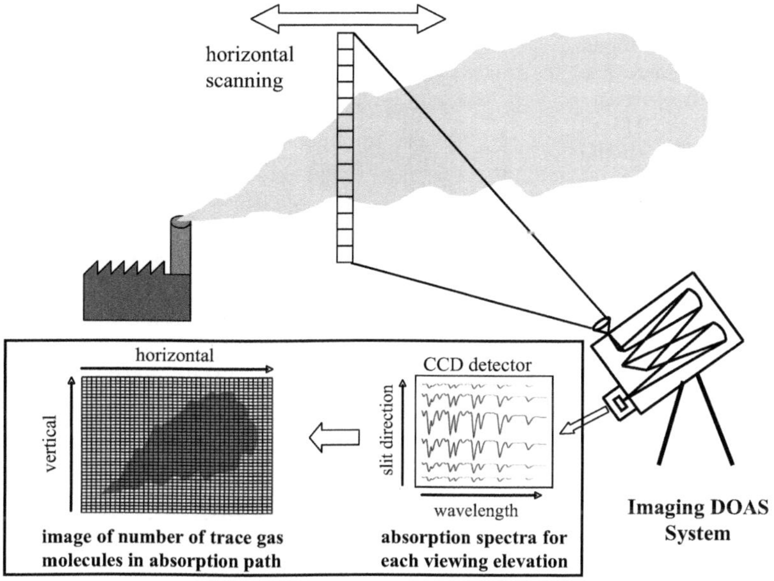

Fig. 11.36 The push-broom scanning I-DOAS measurement principle

Zhang et al. 2024; Platt et al. 2018). Figure 11.37 shows HCHO emissions from a flare in a petrochemical facility (Pikelnaya et al. 2013). Panels a and b show the HCHO VCD image and its uncertainty, with the plume starting at the stack (row 20, pixel 20) and the wind flowing from left to right. The VCD integrated over the rows, i.e. the slice through the entire plume, in panel c show the increase of HCHO downwind of the stack. The difference between the upwind (red crosses) and downwind (green crosses) together with the wind speed are used to determine the emission flux of this flare.

Imaging the SO_2 and BrO-distribution in a volcanic plume (from Mt. Etna, Italy) was reported e.g. by Louban et al. (2009). Figure 11.38 shows simultaneous images of the two gases. A good correlation of the two species is obvious, while a more detailed analysis (not shown) reveals somewhat enhanced BrO/SO_2 ratios towards the edges of the plume. While the scanning I-DOAS measurements already proved to be useful in volcanic research, their main disadvantage is the long measurement times of minutes per image. Therefore the much faster technique of FPI-Imaging (Imaging Fabry-Pérot Interferometer Correlation Spectroscopy, IFPICS) was developed (see Sect. 7.4.3), allowing image acquisition times around a second and thus 'real time' observation of plume processes (Kuhn et al. 2014). Image series of SO_2 and BrO (Fig. 11.38) were reported by e.g. Kuhn et al. (2014, 2019, 2021); Fuchs et al. (2021); Nies et al. (2023); Lv et al. (2023). It should be noted here that software packages have been developed to analyze images and images series, e.g. the Pyplis Python Software Toolbox (Gliß et al. 2017).

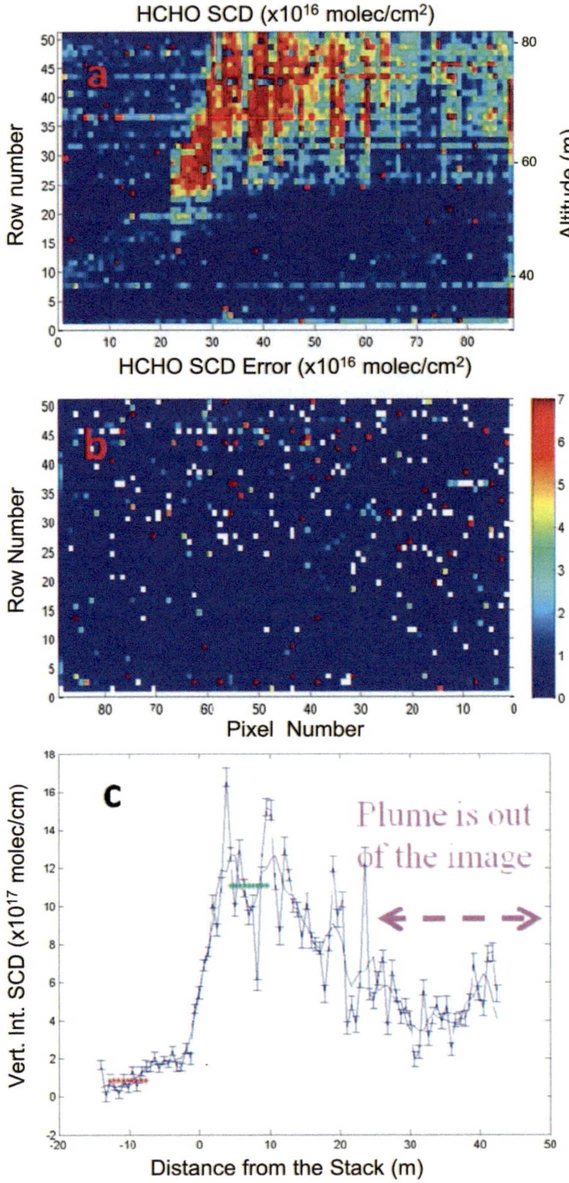

Fig. 11.37 HCHO flux calculations example for an ethylene- burning flare of a chemical facility observed by the I-DOAS on 29 May 2009 at 21:12 UTC. HCHO **a** SCD and **b** SCD error images, and **c** vertically integrated HCHO SCDs. Green crosses in panel **c** show average vertically integrated HCHO SCDs used for the flux calculation. Red crosses in panel **c** show the average background vertically integrated HCHO SCD upwind of the flare. The flux of HCHO is equal to 8.50 ± 2.75 kg/h. Reprinted with permission from Pikelnaya et al. (2013), ©American Geophysical Union. All rights reserved

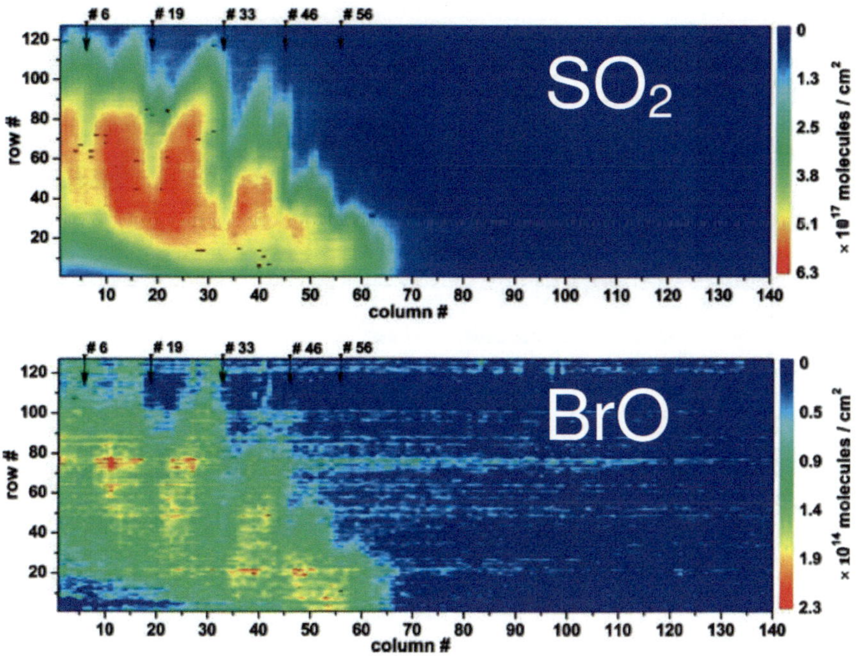

Fig. 11.38 Slant column densities of SO_2 (top) and BrO (bottom) evaluations of I-DOAS data measured on May 9th, 2005. The figure shows the lateral view of the Mt. Etna emission plume, a good correlation of the two species is obvious. Marks on the upper edge indicate its strongly structured characteristics. Reprinted with permission from Louban et al. (2009),

Figure 11.39 shows the schematics and a photo of a prototype FPI-imaging instrument (by Fuchs et al. 2021) used for the quantitative imaging of volcanic SO_2 and BrO. A sample plume image of SO_2 SCDs from Mt. Etna is shown in Fig. 11.40, where up to 10^{18}molec/cm^2 were observed. More data and image sequences can be found in Fuchs et al. (2021).

11.5 Scattered Light Aircraft Measurements of Stratospheric Species

Scattered light DOAS measurements of stratospheric species from airborne platforms offer several advantages over ground-based observations. In upward-viewing geometries, the influence of the atmosphere below the platform is usually small. Thus the fraction of the atmosphere above the platform can be probed. This can be used, for example, to selectively probe the stratosphere. In addition, during ascent or descent of the platform, vertical profiles can be recorded (see Sect. 9.6). In recent years scattered light aircraft measurements of stratospheric species have been largely replaced by measurements from satellite platforms (see Sect. 11.8).

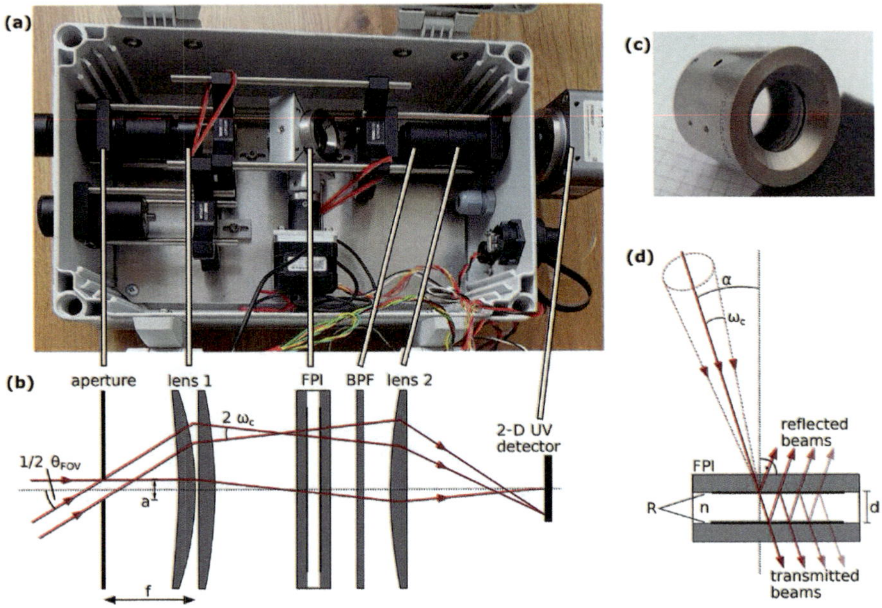

Fig. 11.39 Photograph (**a**) and schematics (**b**) of an Imaging Fabry-Pérot (IFPICS) instrument. The physical dimensions are 200mm (**w**) by 350 mm (**l**) by 130mm (**h**), weight is 4.8 kg. The imaging (**b**) uses a telecentric optical setup. Incident radiation is parallelised by an entrance aperture and lens 1 before traversing the FPI and the band pass filter (BPF). The maximum half cone opening angle ω_C is given by the aperture diameter a and the focal length f of lens 1. The camera field of view is $\theta_{FOV} = 18°$. Lens 2 maps the image onto a 2D UV-sensitive CMOS detector. **c** Photograph of the static air-spaced etalon (FPI) provided by SLS Optics Ltd. **d** Sketch of the FPI. Reprinted with permission from Fuchs et al. (2021),

Early studies of stratospheric species employed the then customary zenith scattered light geometries on aircraft platforms to map stratospheric O_3, NO_2, BrO, and OClO. Aircraft campaigns in both Antarctica and the Arctic were performed on a DC-8 (Wahner et al. 1989a; 1989b; 1990a; 1990b; Schiller et al. 1990), in the northern hemisphere on a Transall transport aircraft (Pfeilsticker and Platt 1994; Pfeilsticker et al. 1997b; 1999; Erle et al. 1998), and other, smaller aircraft (McElroy et al. 1999; Petritoli et al. 2002; Melamed et al. 2003).

An example of a comparison of measured and modelled differential SCDs of BrO and OClO is given in Fig. 11.41. The dependence of BrO and OCIO on aerosol surface area and temperature was studied using a sensitivity analysis with the model (Erle et al. 1998).

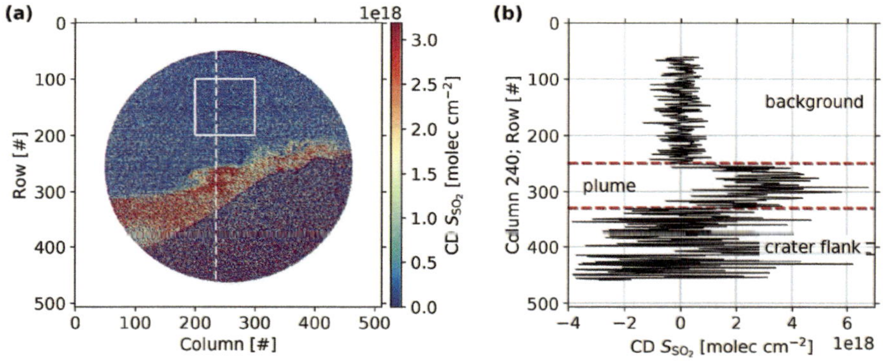

Fig. 11.40 **a** Volcanic plume SO_2 SCD distribution calculated from images acquired with the IFPICS prototype (see Fig. 11.39) using the instrument forward model conversion function. The plume-free area indicated by a white square (100 by 100 pixels) is used to correct for atmospheric background and to obtain an estimation for the detection limit. **b** Individual SO_2 CD column 240 (indicated by dashed white line in (**a**)) showing that the background, plume, and crater flank region are clearly distinguishable. High data scatter in the crater flank region (below the plume) is due to low radiances there. Reprinted with permission from Fuchs et al. (2021), ©CC 4.0. All rights reserved

11.6 Scattered Light Aircraft Measurements of Tropospheric Species

The installation of scattered light DOAS instruments on airborne platforms has opened new opportunities for mapping pollution at high spatial resolution and probing the entire extent of the troposphere as well as the lower stratosphere. As already outlined in Chap. 9, there are three approaches for airborne DOAS observations. (1) nadir viewing geometries, (2) limb geometries (3) multi-axis geometries. Of these, the first two have found the most use. Despite its advantages the latter has only been used in a few cases. For example, one of the earliest Airborne MAX-DOAS systems consisted of two groups of telescopes mounted at the top and bottom of the aircraft body (Heue et al. 2004a, b; Bruns et al. 2004; 2006). Besides a nadir, and zenith viewing telescope, telescopes collecting radiation from several angles (e.g. ±2°) above and below the aircraft were installed. The multiple viewing directions allowed the derivation of vertical profiles during a flight at one altitude, which can be displayed as two-dimensional NO_2 distribution (Fig. 11.42). In this case enhanced levels of NO_2 in the boundary layer are clearly discernible when the aircraft traverses several highways and the city of Bologna. Other results from this instrument were reported by Wang et al. (2004), Fix et al. (2005), Heue et al. (2005), and Bruns et al. (2006).

Spatial mapping of air pollution from aircraft has become popular in recent years, both as a tool to better understand the dispersion of pollutant from point and areas sources, as well as to test satellite instrument precursors. Inter-comparison campaigns in Europe have been performed to evaluated the capabilities of multiple instruments. Figure 11.43 shows NO_2 VCDs from the AROMAPEX experiment, which was held in Berlin, Germany, in April 2016 (Tack et al. 2019). Four nadir viewing AMAX-

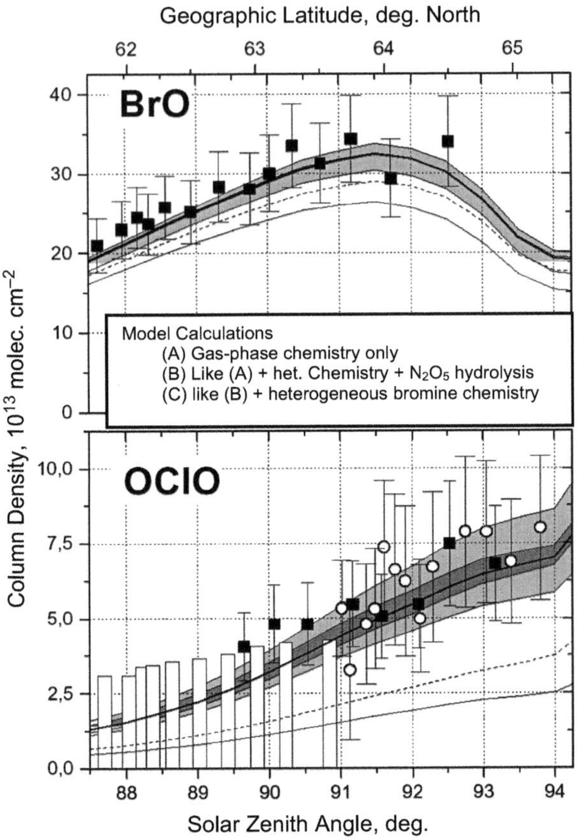

Fig. 11.41 Comparison of measured and modelled differential SCDs of BrO and OClO as a function of the SZA for 13 December, 1994. Measurements, indicated by symbols with error bars, were made by zenith scattered light DOAS on board the Transall research aircraft of the German air force. The geographical latitude of the measurements is shown at the *top* of the plot. The area shaded in *light grey* indicates the effect of multiplying or dividing the aerosol surface by a factor of 2, respectively. The area shaded in *dark grey* indicates the effect of reducing or raising the temperature of ±3 K, respectively. Reprinted with permission from Erle et al. (1998),

DOAS instruments installed in two aircraft were compared to each other, showing excellent agreement in identifying the plumes of major point sources (for example Reuter West power plant in the upper left corner of the NO_2 VCD maps), urban areas sources, as well as major roadways. The comparison also clearly shows the different mapping strategies of each instrument, i.e. their coverage and spatial resolution.

The ability to probe the atmosphere in the vertical using airborne limb DOAS observations has also given new insights into the global chemistry of halogens and other trace gases. When aircraft ascend or descend through a large altitude range, vertical concentration profiles can be retrieved using the methods explained in Chap. 9. For instance, BrO and IO profiles through the tropical and subtropical troposphere (Fig. 11.44) show that BrO increases from low ($\ll 1$ ppt) mixing ratios at the surface to 2–3 ppt near the tropopause and lower stratosphere (Wang et al. 2015). In contrast, IO is elevated near the ocean surface and rapidly decreases to a tropospheric background of around 0.1 ppt. Despite the low mixing ratios, halogen oxides have a significant impact on the global ozone budget.

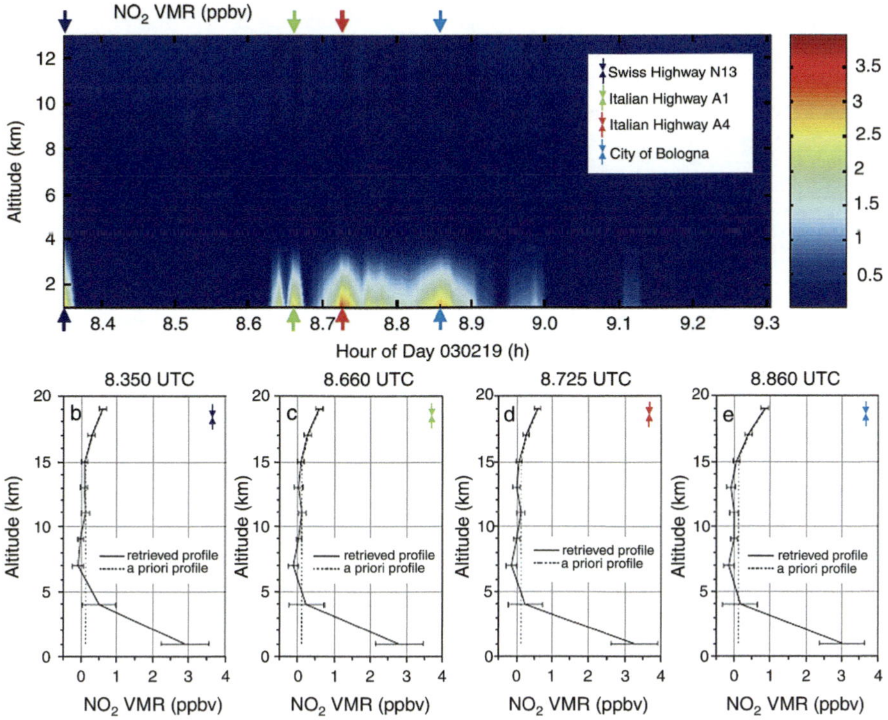

Fig. 11.42 Retrieved NO_2 profiles of a Falcon flight from Basel, Switzerland, to Tozeur, Tunesia, crossing the Alps and northern Italy on 19 February 2003. The top panel (**a**) shows a colour contour of all retrieved NO_2 profiles. Plots (**b–d**) in the lower panel show the retrieved profiles of tropospheric NO_2 mixing ratios at 8.350, 8.660, 8.725, and 8.860 UTC, respectively. The *dark blue* and *green arrows* in panel (**a**) mark the spatial positions of the profiles shown in plots (**b**) and (**c**), the *red* and *light blue arrows* mark the positions of the profiles represented by plots (**d**) and (**e**). Reprinted with permission from Bruns et al. (2006), ©CC 2.5. All rights reserved

11.7 Satellite Observations Using DOAS Techniques

Optical absorption spectroscopy, and in particular DOAS, is an important and widespread application for the remote sensing of atmospheric composition from space, for reviews see e.g. Burrows et al. (2011); Gonzalez Abad et al. (2019). In Chap. 6, several modes of observation were presented. These include:

- Nadir observations of sunlight reflected by the earth's surface or the atmosphere (earthshine)
- Limb observations of sunlight scattered at the horizon
- Solar occultation i.e. observation of sunrise/sunset through the atmosphere
- Lunar occultation
- Stellar occultation

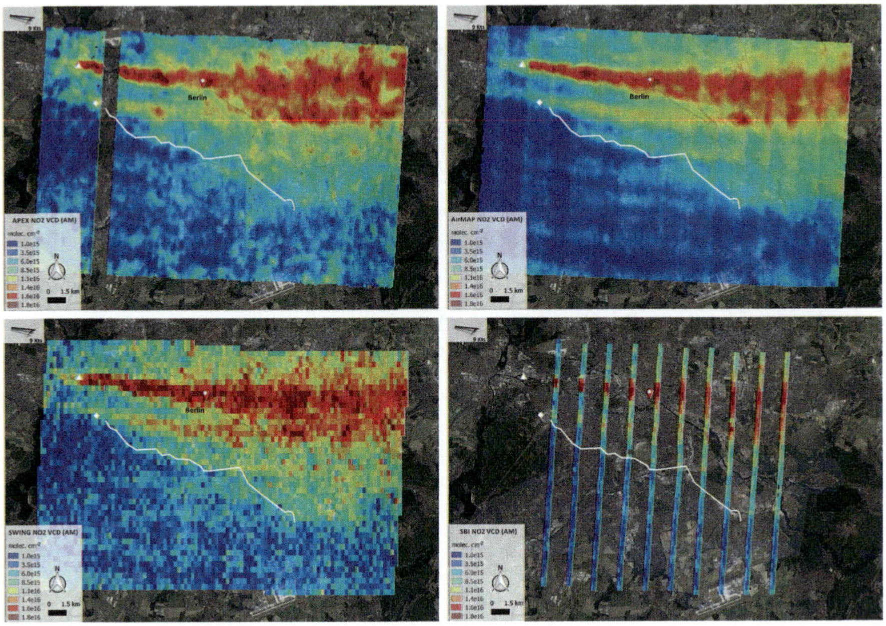

Fig. 11.43 Comparison of tropospheric NO_2 VCDs retrieved during the AROMAPEX experiment by four instruments: The APEX (Airborne Prism Experiment), AirMAP (Airborne imaging DOAS instrument for Measurement Atmospheric Pollution), SWING (Small Whiskbroom Imager for atmospheric composition monitoriNG), and SBI (Spectrolite Breadboard Instrument) for the morning flight over Berlin on 21 April 2016. NO_2 emissions are dominted by the Reuter West power plant (white triangle) and Messe Berlin (white diamond). Highways A100 and A113, south of the city of Berlin, are marked by the white line. Reprinted with permission from Tack et al. (2019), ©CC 4.0. All rights reserved

In addition to the observation mode, the orbit of the satellite largely determines its performance. The orbits can be categorised as Low Earth Orbits (LEO) or GEO-stationary Orbits (GEO). The 1st Lagrange point L1 (between Sun and Earth) is being considered as position for Earth-observing satellites and a first satellite (Deep Space Climate Observatory (DSCOVR) mission) has been deployed (Valero et al. 2021). Most satellites capable of DOAS-type measurements are in LEOs. However, DOAS-type Earth observing satellite instruments in GEO-orbits are being deployed. The "Geostationary Environment Monitoring Spectrometer" (GEMS) over Asia and the "Tropospheric Emissions: Monitoring of Pollution" (TEMPO) over North America are already in orbit, and Sentinel-4 over Europe soon will follow. Geostationary satellite instruments only observe about one third of the globe. However, these observations are continuous, allowing for monitoring of the diurnal variations of trace gas abundances.

A particular LEO is the (nearly) polar orbit, which allows observation of the entire surface of earth with the same instrument. Among the polar orbits, the sun synchronous orbit has the special advantage of observing earth always at (nearly) the same local time (the equator crossing time, ECT), typically between 11:00 and 14:00,

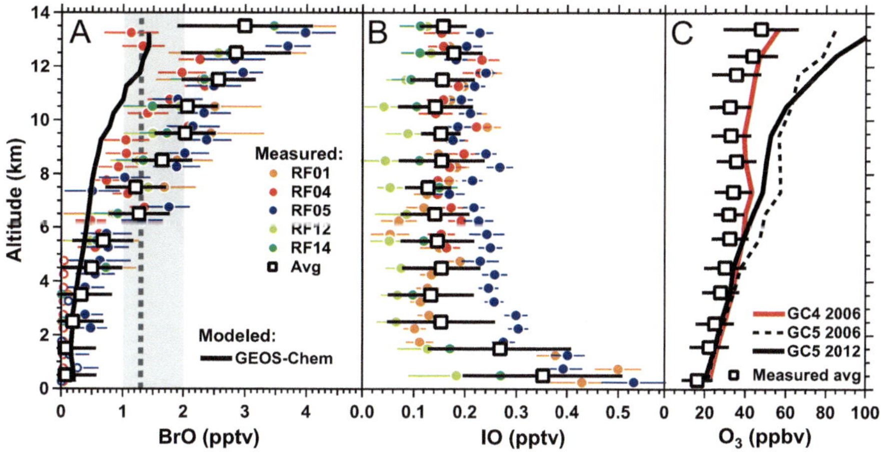

Fig. 11.44 Mixing ratio profiles of BrO (A, solid circles), IO (B, solid circles) measured by AMAX-DOAS, together with in-situ O_3 (C) during five research flights (RF01, RF04, RF05, RF12, and RF14) on board of the US National Science Foundation's HIAPER aircraft during the Tropical Ocean tRoposphere Exchange of Reactive halogen species and Oxygenated VOC (TORERO) field experiment (Wang et al. 2015). Campaign-averaged profiles are displayed as open boxes. Also shown in (A) is the campaign average GEOS-Chem (Goddard Earth Observation System—Chemistry model) base case simulation (solid black line) and the tropospheric BrO background from satellite (grey shading) and AMAX-DOAS data (dashed grayvertical line). Reprinted with permission from Wang et al. (2015)

as shown in Fig. 11.45. In this orbit, the non-sphericity of earth, in combination with a small tilt of the orbit (inclination with respect to the equator >90°), leads to a precession, keeping the plane of the orbit at a fixed angle relative to the line sun–earth. In principle, a train of LEO satellites with different ECTs could provide diurnal time resolution, as pointed out by Burrows et al. (2011).

Presently, the instruments GOME-2, OMI, OMPS, TROPOMI, and EMI (Zhang et al. 2020) (Table 11.1) are on polar, sun synchronous orbits. The first generation of polar, sun synchronous satellite instruments, such as GOME and SCHIAMACHY, which have contributed greatly to our understanding of global atmospheric composition, have been decommissioned. Fig. 11.46 shows the scanning scheme of the GOME instrument.

The GOME and GOME-2 instruments are based on four spectrometers of moderate resolution that cover the spectral range from 290 to 790 nm with 1024 channels each. Analysing the spectrum of sunlight reflected by the earth (earthshine) using the DOAS technique yields column densities of a series of trace gases in the atmosphere (e.g. NO_2, BrO, OClO, SO_2, H_2O, O_3, and O_4). The GOME instruments monitor the earth in nadir view (looking downward approximately perpendicular to the earth's surface) and record a spectrum every 1.5 s. Using a scanning mirror, the instruments record three pixels perpendicular to the direction of flight and one 'backscan' pixel at the end of each scan. The size of each 'ground pixel' of the original GOME instrument, i.e. the footprint on the ground, was normally 320 by

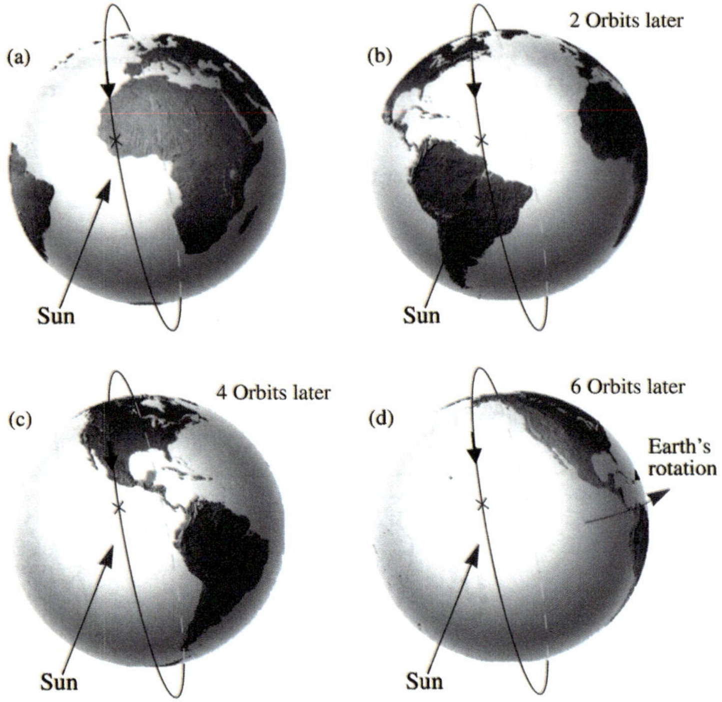

Fig. 11.45 Illustration of a sun synchronous polar satellite orbit, for example with an altitude of 800 km and an orbiting time of 100 min. The satellite crosses the equator always at the same local time (the Equator Crossing time, ECT). Examples for such satellites are: ERS-1, ERS-2 (GOME): 10:30 equator crossing time; ENVISAT (SCIAMACHY) 11:00 equator crossing time ; OMI, TROPOMI, EMI: $\approx$ 13:40 ECT

40 km (see Fig. 11.46). Additionally, a polarisation monitoring device (PMD) (Bednarz 1995) simultaneously scanned the earth with a higher spatial but lower spectral resolution, measuring integrated intensities in three spectral bands between 295 and 745 nm. Each GOME pixel is spatially resolved by 16 subpixels of 20×40 km each. This subpixel information can be used for the determination of the cloud fraction and the ground albedo, as described in Guzzi et al. (1998), Wenig (1998) and Wenig et al. (1999), and Grzegorski et al. (2006). More recent instruments, e.g. GOME-2, SCIAMACHY, OMI, and TROPOMI offer smaller ground pixels, additional infrared (IR) channels (SCIAMACHY, TROPOMI), or employ simultaneous observations of all cross-track spatial pixels, i.e. pushbroom scanning instead of whiskbroom scanning (OMI, TROPOMI, EMI).

11.8 Satellite Observations of Stratospheric Species

Mapping of stratospheric species, in particular ozone (starting in 1970 with BUV on Nimbus 4) and NO_2 (starting 1979 with SAGE I), is the oldest application of

Table 11.1 Satellite instruments, species measured, satellite platform, and orbit. (The list is not intended to be complete, but merely to illustrate past and current instrumentation.)

Name	Target species	Satellite platform	Orbit
Atmospheric trace molecule spectroscopy (ATMOS)	O_3, NO_x, N_2O_5, ClO, NO_2, HCl, HF, CH_4, CFCs, etc. (upper troposphere)	Space Shuttle Spacelab-3 (1985), ATLAS-1,2 and 3 (1992,1993, 1994)	LEO, inclined
Backscatter ultraviolet (BUV) ozone experiment	O_3 (profiles)	Nimbus-4 (1970–1974)	LEO, Polar
Environmental Trace Gases Monitoring Instrument (EMI)	O_3, NO_2, H_2O, BrO, OClO, SO_2, HCHO, CHOCHO, clouds, aerosol	GaoFen-5 (2018–present)	LEO, Polar, Sun Sync.
Geostationary Environment Monitoring Spectrometer (GEMS)	O_3, NO_2, H_2O, BrO, OClO, SO_2, HCHO, CHOCHO, clouds, aerosol	GEO-KOMPSAT-2B (2020–present)	GEO
Global ozone monitoring experiment (GOME)	O_3, NO_2, H_2O, BrO, OClO, SO_2, HCHO, clouds, aerosol	ESA-ERS-2 (1995–2011)	LEO, Polar, Sun Sync.
Global ozone monitoring experiment-2 (GOME-2)	O_3, NO_2, H_2O, BrO, OClO, SO_2, HCHO, clouds, aerosol	METOP-1 to METOP-3 (2006, 2012, 2018)	LEO, Polar, Sun Sync.
Global ozone monitoring by occultation of stars (GOMOS)	O_3, NO_2, upper troposphere	ESA ENVISAT (2002)	LEO, Polar, Sun Sync.
Imaging atmospheric sounding instrument (IASI)	O_3, CO, CH_4, N_2O, SO_2	METOP-1 (2006)	LEO, Polar, Sun Sync.
Ozone dynamics ultraviolet spectrometer (ODUS)	SO_2, NO_2, BrO, OClO	GCOM-A1 Prog, Japan (2005)	LEO, inclined
Ozone monitoring instrument (OMI)	O_3, SO_2, NO_2, BrO, HCHO	NASA-EOS-CHEM (2004)	LEO, Polar, Sun Sync.
Optical Spectrograph and Infrared Imager System (OSIRIS)	O_3, NO_2	ODIN, Sweden (2001–present)	LEO, Polar, Sun Sync.
Polar Ozone and Aerosol Measurement (POAM-II, III)	O_3, NO_2, H_2O, aerosol in the upper troposphere	SPOT-3 (1993–1996), SPOT-4 (1998–	LEO, Polar, Sun Sync.
Sentinel-4	O_3, SO_2, NO_2, BrO, HCHO, CHOCHO, aerosol	Copernicus, ESA (planned 2024)	GEO

(continued)

Table 11.1 (continued)

Name	Target species	Satellite platform	Orbit
Stratospheric Aerosol and Gas Experiment 1 (SAGE I)	O_3, NO_2, H_2O, aerosol in the upper troposphere	NASA Atm. Explorer Mission (1979–1981)	LEO, inclined
Stratospheric Aerosol and Gas Experiment 2 (SAGE II)	O_3, NO_2, H_2O, aerosol in the upper troposphere	Earth Radiation Budget Sat. (1984–2005)	LEO, inclined
Stratospheric Aerosol and Gas Experiment 3 (SAGE III)	O_3, OClO, BrO, NO_2, NO_3, aerosols	Meteor 3M (2001); Inter. Space Stat.	LEO, inclined
Solar backscatter ultraviolet (SBUV) ozone experiment	O_3 profiles	Nimbus-7 (1979–1990)	LEO, Polar
Scanning imaging absorption spectrometer for atmospheric cartography (SCIAMACHY)	O_2, O_3, O_4, NO, NO_2, N_2O, BrO, OClO H_2CO, H_2O, SO_2, HCHO, CO, CO_2, CH_4, clouds, aerosols, p, T, col. and profiles	ESA-ENVISAT (2001–2012)	LEO, Polar, Sun Sync.
Tropospheric Emissions: Monitoring of Pollution (TEMPO)	NO_2, O_3, HCHO	Intelsat 40e	GEO
Total ozone monitoring spectrometer (TOMS)	O_3	Nimbus 7 (1979–1992) ADEOS (1996–1997) Earth Probe (1996–2005) Meteor (1992–1994)	LEO, Polar
TROPOspheric Monitoring Instrument (TROPOMI)	O_3, SO_2, NO_2, BrO, HCHO	ESA, NSO (2017)	LEO, Polar, Sun Sync.

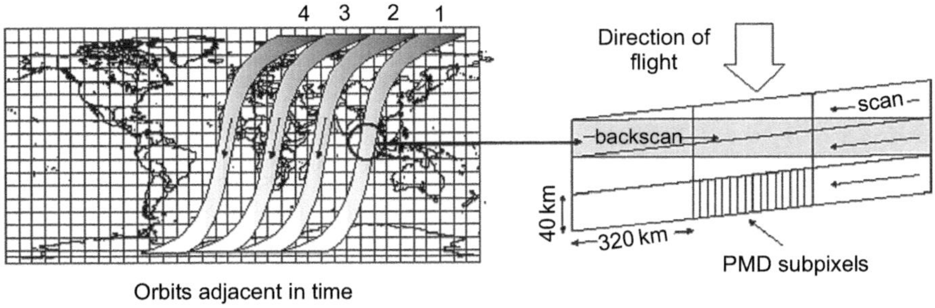

Fig. 11.46 Whiskbroom scanning scheme of the GOME 1 instrument. Reprinted with permission from Leue et al. (2001), ©American Geophysical Union. All rights reserved

satellite sensors. An example is satellite generated maps of the Antarctic ozone hole (although it was not originally discovered by satellite sensors), well known to the public from numerous presentations in the media.

A general problem in determining stratospheric species from space is that nadir viewing satellite instruments (which are the majority) determine the sum of tropospheric and stratospheric partial columns (TPC and SPC). Separation of both contributions to the total column can be achieved by several techniques that were described in Chap. 9. In some cases, like for O_3 and OClO, the SPC always dominates the total column, thus separation is straightforward. In other cases, such as NO_2, SO_2, or BrO the SPC and TPC can be of similar magnitude at certain times and locations. However, over a majority of the globe, and in particular over most parts of the oceans, the SPC dominates the total column. These spatial differences, as well as measurements in the presence of clouds (cloud slicing), can be used to separate both partial columns. Since TPCs and SPCs are determined simultaneously, the quality of their separation influences the quality of both measurements, stratospheric and tropospheric columns. In this section we will give a few selected examples of satellite DOAS measurements of stratospheric trace gases.

11.8.1 Stratospheric O_3

The DOAS technique offers a number of well-known advantages over the Dobson-type evaluation of ozone absorption. Consequently, DOAS is now the primary approach to retrieve the total O_3 column from spectra recorded by nadir-looking satellite instruments such as GOME, SCIAMACHY, GOME-2, OMI, and TROPOMI (e.g. Lambert et al. 2001; Weber et al. 2003; Garane et al. 2019; Mettig et al. 2021; Chan et al. 2023). Figure 11.47 shows as an example the daily minimum ozone vertical column found polewards from 50°S during the Antarctic winters 1995–2001 (shaded area) in contrast to the winter of 2002, when a vortex-split occurred.

11.8.2 Stratospheric NO_2

Early interest in stratospheric NO_2 was motivated by its impact on stratospheric ozone levels. Satellite NO_2 observations of the SPC, showed the variation of the stratospheric NO_2 VCD as a function of latitude and season, and elucidated several important aspects of stratospheric chemistry and dynamics, e.g. the Noxon-Cliff, zonal symmetry, or inter-hemispheric differences.

The GOME instrument made measurements over the globe for nearly 8 years. These data were used by Richter and Borrows (2002) and Wenig et al. (2004) to construct global maps of stratospheric NO_2. Figure 11.48 shows the time series of the zonally averaged stratospheric NO_2 VCD as a function of latitude and time from 1996 to 2000. The seasonal variability growing from mid- to high latitudes with maximum NO_2 columns during polar summer is clearly visible. Time series for average NO_2 VCDs over Europe, USA, and Brazil are shown in Fig. 11.49.

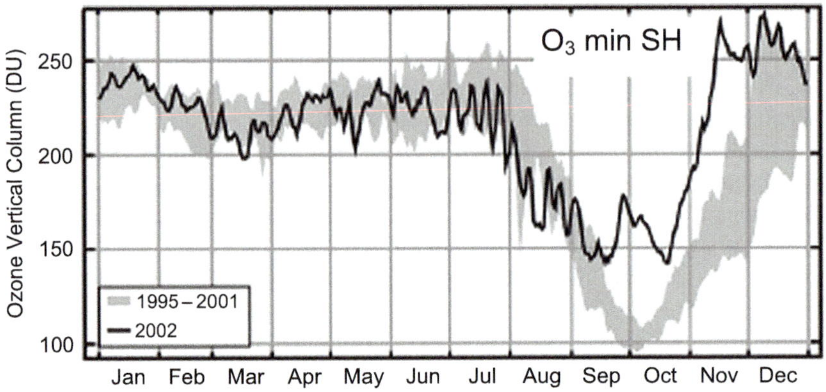

Fig. 11.47 Daily minimum GOME total ozone observed polewards of 50°S during the Austral winter of 2002 (*solid line*). The *shaded* area indicates the range observed during the winters of 1995–2001. Reprinted with permission from Richter et al. (2005b), ©American Meteorological Society. All rights reserved

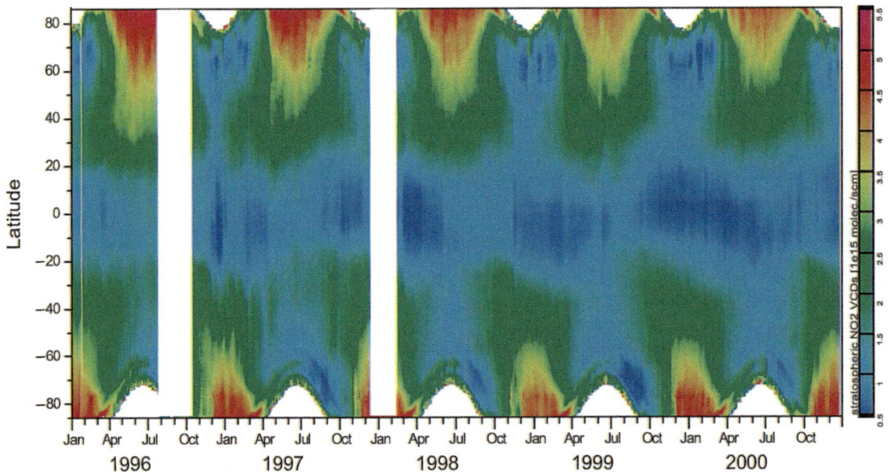

Fig. 11.48 Time series of the zonally averaged stratospheric NO_2 VCD (scale: $0\ldots5.5 \times 10^{15}$ molec. cm^{-2}) as a function of latitude and time from 1996 to 2000. Reprinted from Wenig, Ph.D. Thesis, 2001, University of Heidelberg

In recent years interest in the NO_2 SPC was mostly motivated by the need to accurately separate the the SPC and TPC to measure tropospheric NO_2 resulting from anthropogenic pollution, see e.g. Boersma et al. (2018); Beirle et al. (2016a); Bucsela et al. (2013).

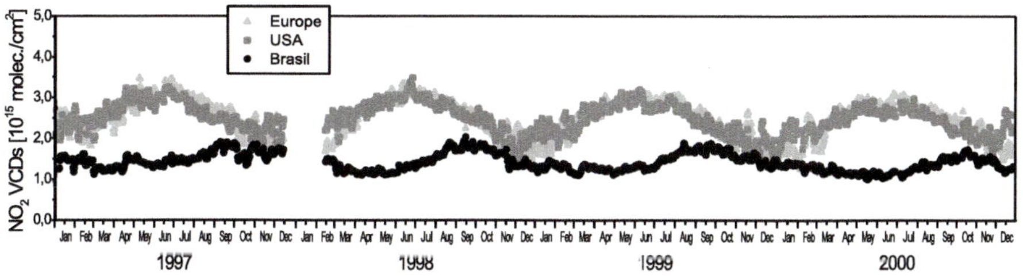

Fig. 11.49 Time series of the NO_2 VCD in units of 10^{15} molec. cm^{-2} (January 1997 to December 2000). DOAS evaluation using a fixed Fraunhofer reference spectrum (from 1 June, 1997). Reprinted with permission from Wenig et al. (2004), ©American Geophysical Union. All rights reserved

11.8.3 Stratospheric OClO

Several studies demonstrated that the observation of OClO is a good indicator for stratospheric chlorine activation. The GOME instrument allowed daily observation of this species in polar regions. The first global data set of OClO and continuous time series of its occurrence in both winter stratospheres was reported by Wagner et al. (2001a, 2002b). Figure 11.50 shows that OClO regularly occurs over Antarctica with little variation in total amount and temporal variation during the different winters, while the OClO occurrence is much more variable in the Arctic winter stratosphere. The primary reason for this is the larger dynamic activity resulting in warmer temperatures during the Arctic winter. About 40% larger OClO column amounts were

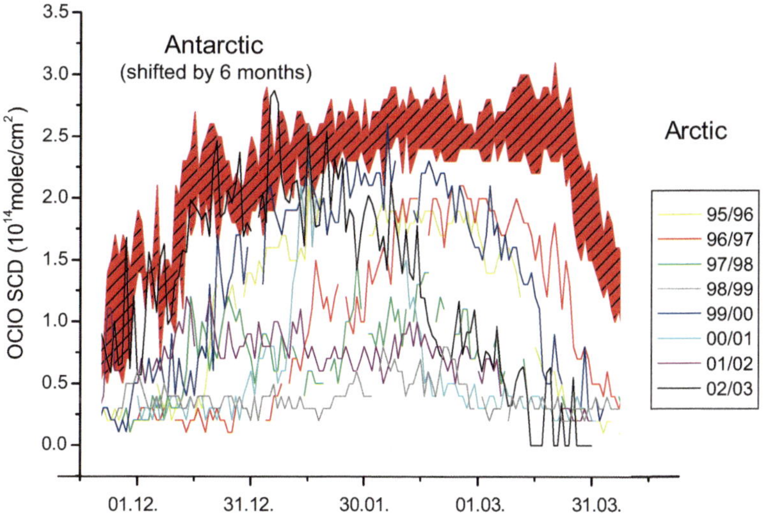

Fig. 11.50 Maximum daily OClO SCD measured by GOME over the Arctic (*thin coloured lines*) and the Antarctic (*red hatched area*) in the winters 1995 through 2003. For comparison with the measurements over the Arctic, the Antarctic OClO SCD measurements are shifted by 6 months. Note the small year-to-year variability of the chlorine activation in the Antarctic compared to the Arctic. Reprinted from Kuehl, Ph.D. Thesis, (2005), University of Heidelberg

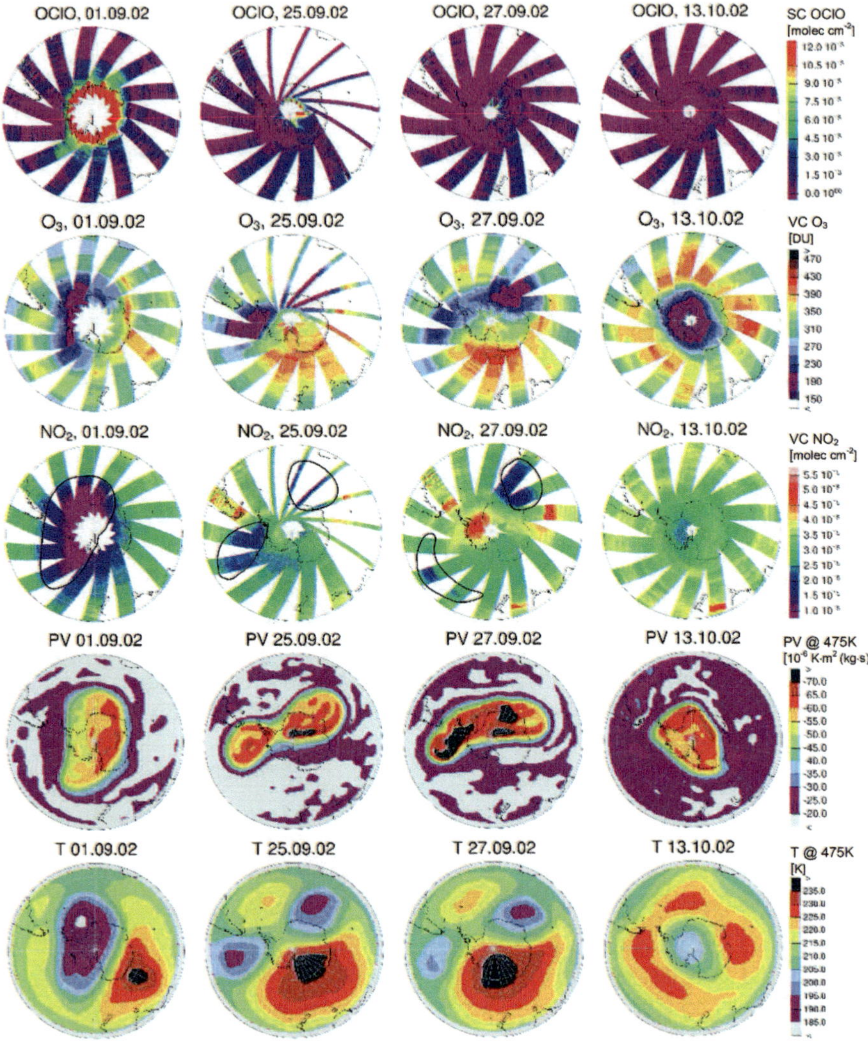

Fig. 11.51 The Antarctic vortex split in Austral spring of 2002: OClO (slant columns in molec. cm^{-2}, not corrected for light path enhancement), O_3 vertical columns (in Dobson units), and NO_2 vertical columns (in molec. cm^{-2}) in comparison to potential vorticity and temperature from ECMWF (at 475 K potential temperature surface). The data before the major warming event (9 and 25 September) on 26 September and after the warming (26 September and 31 October) are shown. Reprinted with permission from Richter et al. (2005b), ©American Meteorological Society. All rights reserved

found in the Antarctic polar stratosphere than in its northern counterpart, a further indication for a significantly more efficient chlorine activation in the Antarctic late winter and spring stratosphere.

An overview of the ability to map stratospheric species by satellite-based DOAS observations is given in Fig. 11.51, which shows OClO SCDs, O_3 vertical columns, and NO_2 vertical columns (in molec. cm^{-2}) during the Antarctic vortex split in Austral spring of 2002. The results show (see Richter et al. 2004) that chlorine activation in the vortex, as indicated by the OClO abundance, was similar to previous years until the major warming on 26 September, 2002, after which it decreased rapidly. Similarly, NO_2 columns were only slightly larger than in previous years before the warming, indicating strong denoxification and probably also denitrification. Similar dependencies of OClO on meteorological conditions were found by Puķīte et al. (2022) analyzing TROPOMI-satellite data.

11.9 Satellite Observations of Tropospheric Species

The recognition that air pollution is a global phenomenon was in large driven by the first observations of tropospheric NO_2)!troposheric from the GOME and SCHIAMACHY instruments. Since these early observations satellite DOAS observations have become a valuable tool to monitor air quality from space and to understand processes that impact atmospheric composition on a global scale, such as those controlling tropospheric ozone. The satellite data also provide a direct comparison with the output from chemical transport models (CTM) on global and regional scales and are used for validation and sensitivity assessment of these models. This leads to substantial improvement of accuracy and reliability of CTMs. Satellite measurements are also useful in providing real boundary conditions for operational models. In addition, source strengths of trace gases can be derived (Leue et al. 2001; Palmer et al. 2003).

A large number of trace gases, including, ozone, NO_2, SO_2, HCHO, Glyoxal, BrO, IO, and water vapour can be measured from satellites in the troposphere. As mentioned above, some tropospheric trace gases of interest (e.g. O_3 and NO_2) are also abundant in the stratosphere, while others (such as SO_2, HCHO, CHOCHO, or water vapour) are not. Various procedures have been developed to separate the tropospheric signal from stratospheric contributions (see Chap. 9). An additional challenge in observing tropospheric species from space is the presence of clouds (see Sect. 9.7). Consequently, cloud detection and quantification schemes have to be used. If high precision is required, as for instance in the case of long-lived greenhouse gases (CH_4 and CO_2), the effects of clouds and aerosol on radiative transfer have to be described very accurately.

Satellite measurement of tropospheric trace gas abundances has become an essential element of tropospheric chemistry and pollution research. This is also reflected in the increasing number of publications using this data, which we will not be able to fully discuss here. Instead we will present historical, and current examples of satellite DOAS data and briefly discuss their application.

11.9.1 Tropospheric O_3 from Space

Despite ozone's strong absorption features in the UV and visible spectral ranges, which led to its early spectroscopic detection in the atmosphere (see Sect. 11.2), the

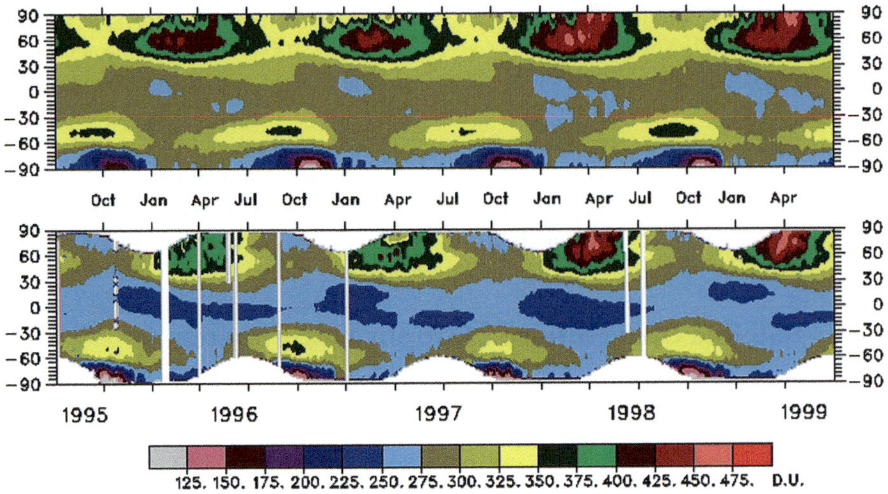

Fig. 11.52 The global total ozone column of the years 1995–1999: The lower part of the figure shows the latitude-time plot of 1 July 1995–30 June 1999 observations of the O_3 column with GOME. The upper part shows four individual years of a multi-year simulation with the chemistry-climate model ECHAM4.L39(DLR)/CHEM. (from Borrell et al. 2001)

ozone in the troposphere is among the most difficult species to derive from satellite platforms. The challenge is to separate the tropospheric partial O_3 column from the ten times larger stratospheric column (see Fig. 11.52).

Several DOAS-based retrieval algorithms for tropospheric O_3 have been developed and successfully applied (e.g. Lambert et al. 2001, Edwards et al. 2003). Results and comparison of different satellite products can for instance be found in the Tropospheric Ozone Assessment Report (TOAR) (Gaudel et al. 2018). Algorithm descriptions are found in e.g. Miles et al. (2015); Heue et al. (2022).

11.9.2 Tropospheric NO_2 From Space

Today a multitude of satellite NO_2-column measurements are available. The data include global maps, emission from shipping, analysis of weekly and seasonal variation, longterm trends, the effects of Covid-19, emission from point sources, such as biomass burning and lightning, and even the coriolis effect on plume propagation (Beirle et al. 2003a, 2004b; 2004c, Riess et al. 2022; Lange et al. 2022; Georgoulias et al. 2019; Beirle et al. 2011, 2019a, 2021; Barré et al. 2021; Griffin et al. 2021; Beirle et al. 2006, 2010; Maseko et al. 2021; Potts et al. 2023).

Figure 11.53 shows one of the earliest global maps of tropospheric NO_2 satellite measurements, the mean global NO_2 VCD for the year of 1997 (Leue et al. 1998, 2001). Like all later measurements (e.g. Beirle 2004, and see above) it shows a clear correlation between areas of high NO_2 and industrialised regions.

The satellite instruments following GOME had much higher spatial resolution, and the ground pixle size has been reduced from 40×80 km (GOME-1) down to

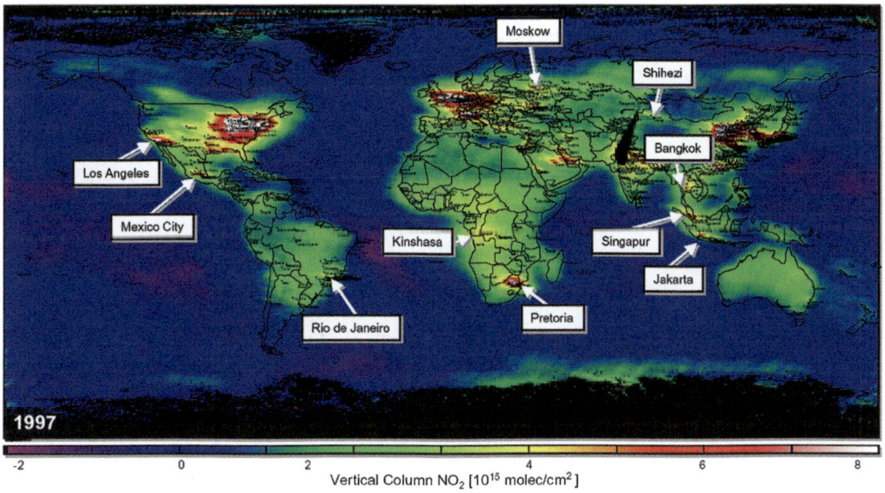

Fig. 11.53 One of the first global maps of air pollution observed from space shows the mean tropospheric NO_2 VCD for the year of 1997 measured by GOME. Reprinted with permission from Leue et al. (2001), ©American Geophysical Union. All rights reserved

5.5×7 km (TROPOMI). Figure 11.54 shows a global map of NO_2 from GOME, SCIAMACHY, and GOME-2 data for the period April 1996–September 2017.

The high NO_2 VCDs are associated with major cities across North America and Europe, along with other sites such as Mexico City in Central America, and coal-fired power plants in the eastern Highveld plateau of South Africa. The highest NO_2 levels anywhere on earth were found above north-eastern China. This is the most prominent change from previous satellite surveys, and reflects the country's rapid pace of industrialisation (Richter et al. 2005; van der A et al. 2006). Across Southeast Asia and much of Africa, NO_2 produced by biomass burning is seen. Ship tracks (NO_2 from ship exhaust stacks; see also Beirle et al. 2004b; Beirle 2004) are also visible in some locations, for instance at the Red Sea and the Indian Ocean between the southern tip of India and Indonesia.

GOME satellite data was also used to derive the weekly variation of the NO_2 column densities of populated or industrial areas of the world (Beirle et al. 2003a). Figure 11.55 shows the normalised amplitude of the NO_2 variation as a function of the day of week averaged over the period from 1996 to 2001. There is a pronounced weekly cycle in Western Europe, North America, and parts of Asia (Korea and Japan), while the cycle is absent in Northeast China. More comprehensive data (also from other satellite instruments) can be found e.g. in Stavrakou et al. (2020); Lange et al. (2022).

Biomass burning events have been studied using multiple satellite sensors by Edwards et al. 2003; Griffin et al. (2021); van der Velde et al. (2021); Castellanos et al. (2014). Another area of research where satellite measurements are an important tool is studies of the intercontinental transport of air pollution (Spichtinger et al. 2001; Wenig et al. 2003; Ghude et al. 2013; Park et al. 2016; Johnson et al. 2021). The new instruments such as OMI, TROPOMI, GEMS, or TEMPO also allow the study of

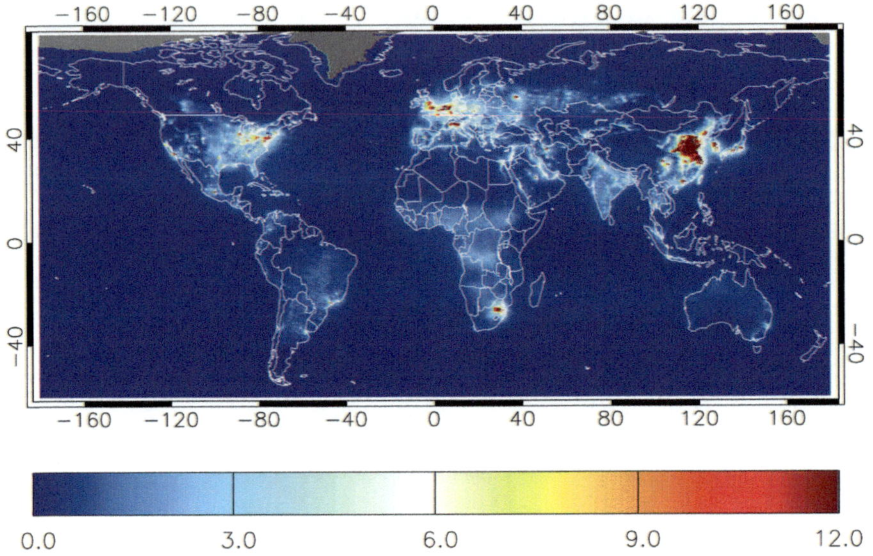

Fig. 11.54 Global map of the tropospheric NO_2 VCD (in 10^{15} molecules cm^{-2}) from GOME, SCIAMACHY, and GOME-2 data for the period April 1996-September 2017. Reprinted with permission from Georgoulias et al. (2019),

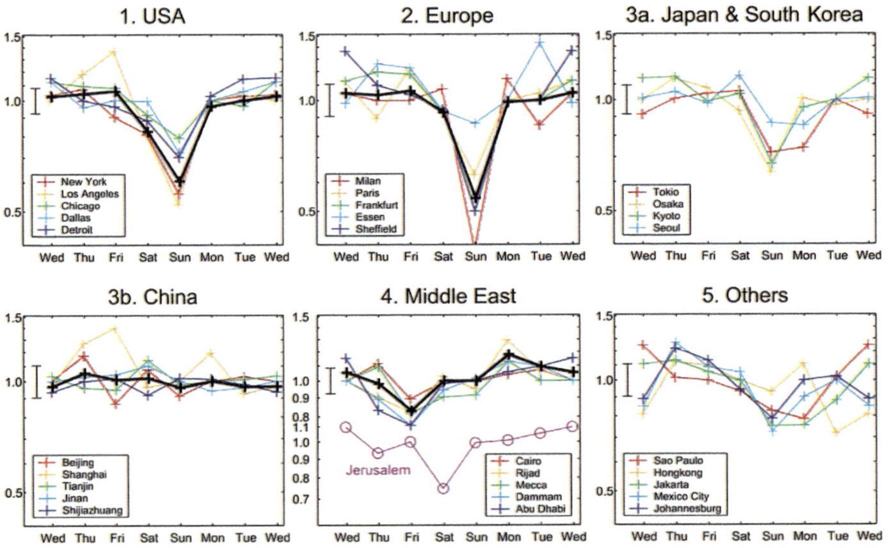

Fig. 11.55 The weekly cycle of air pollution: Six-year mean (1996–2001) of GOME tropospheric NO_2 VCDs as a function of day of week. Reprinted with permission from Beirle et al. (2003a),

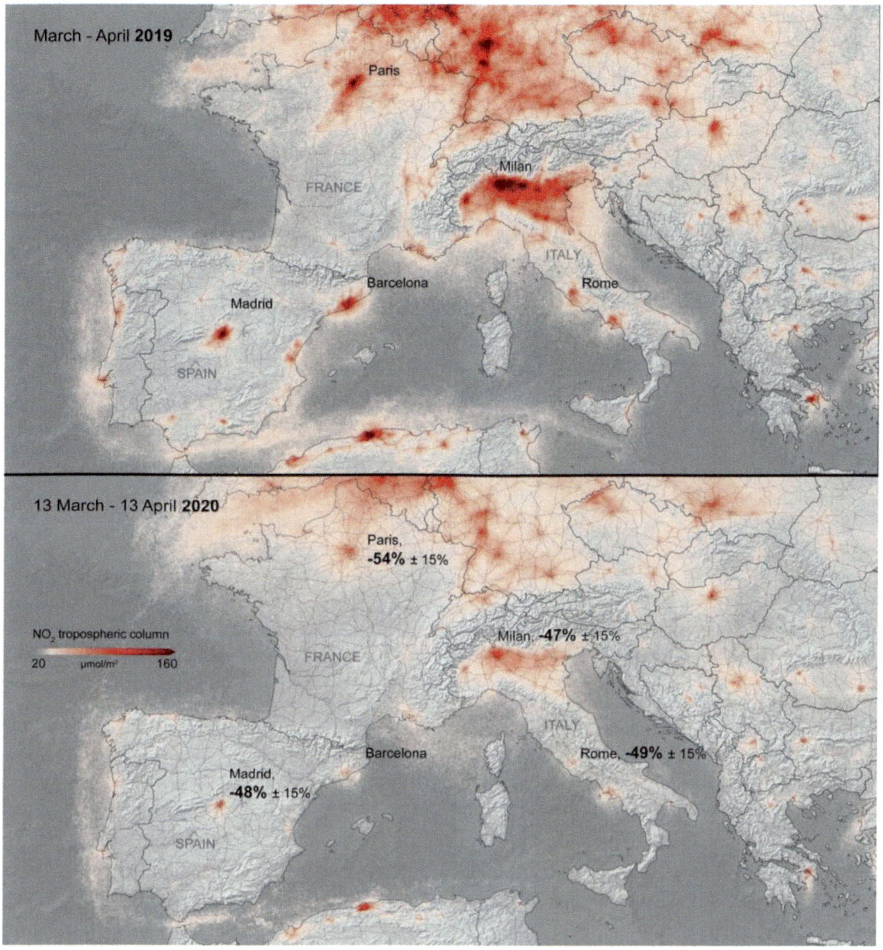

Fig. 11.56 The NO_2 columns over Europe before (March-April 2019) and during the Covid-19 pandemic (March-April 2021). Units are μmole/m^2, 160 μmole/m^2 equal 9.6×10^{15} molec./cm^2. Up to 54% reduction in NO_2 is observed. (from ESA 2020)

regional air quality and effects of global events such as the COVID-19 pandemic, as can be seen from Fig. 11.56 which shows the reduction of the NO_2 column during spring of 2020 compared to spring of 2019.

11.9.3 Tropospheric Formaldehyde from Space

Formaldehyde (CH_2O or HCHO) can also be measured by satellite instruments. It is a valuable indicator of anthropogenic or natural combustion (biomass burning), degradation of methane, and the oxidation of non-methane volatile organic compounds, such as isoprene. HCHO has been quantified in many satellite studies.

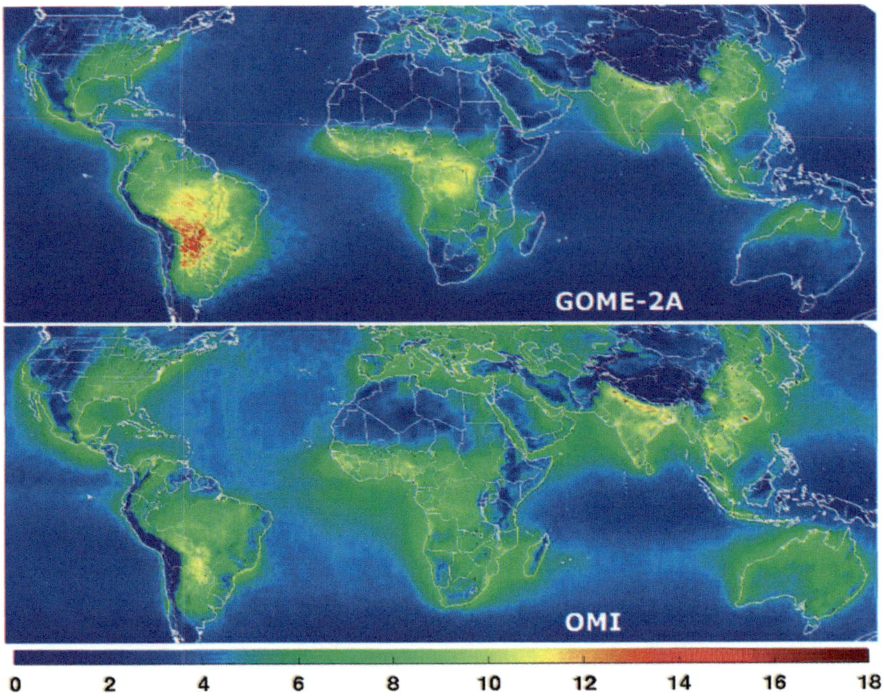

Fig. 11.57 Global average distribution of the formaldehyde total vertical column densities (VCDs) averaged 2007 to 2013 in units of 10^{15}molec.cm^{-2}. Highest VCDs are seen over the tropical regions of South America and Africa as well as over India, south-eastern China, and the south-eastern United States. Reprinted with permission from De Smedt et al. (2015), ©CC 4.0. All rights reserved

Elevated HCHO levels are found in forested areas and in regions with biomass burning, as well as over regions with anthropogenic pollution (e.g. Wittrock et al. 2000b; Palmer et al. 2003; Vrekoussis et al. 2010; Alvarado et al. 2020; De Smedt et al. 2015, 2021). Figure 11.57 shows the global average distribution of formaldehyde for the years 2007 to 2013 derived from the GOME-2a and OMI satellite instruments (De Smedt et al. 2015). The highest columns are seen over the tropical regions of South America and Africa as well as over India, south-eastern China and the south-eastern United States.

11.9.4 Tropospheric Glyoxal from Space

Glyoxal (CHOCHO) is a short lived degradation product of non-methane volatile hydrocarbons (NMHC), in particular isoprene. Globally, its sources are predominantly biogenic with some contribution from human activities and forest fires (e.g. Alvarado et al. 2020). In contrast to HCHO, glyoxal is not produced as a product of methane degradation. Therefore the combined observation of both species offers the potential to quantify NMHC emission and degradation (see e.g. Wittrock et al. 2006;

Stavrakou et al. 2009; Vrekoussis et al. 2010; Chan Miller et al. 2014; Mahajan et al. 2014; Cao et al. 2018; Lerot et al. 2021). A comparison of global vertical glyoxal and formaldehyde columns is shown in Fig. 11.58. The patterns of the two species show similarities, which is not surprising given that they originate from the same type of source (biogenic VOCs and biomass burning). Enhanced values of both species are seen over the tropical and subtropical land areas, in particular those covered with dense vegetation (rain forest) and affected by frequent biomass burning events, i.e. in South America, Africa, and Indonesia (Vrekoussis et al. 2010).

11.9.5 Tropospheric SO_2 from Space

In satellite spectra, the characteristic absorption structures of SO_2 can be readily identified in the spectral range of about 315–327 nm. Using DOAS, integrated columns of SO_2 can be determined. However, due to the increased Rayleigh scat-

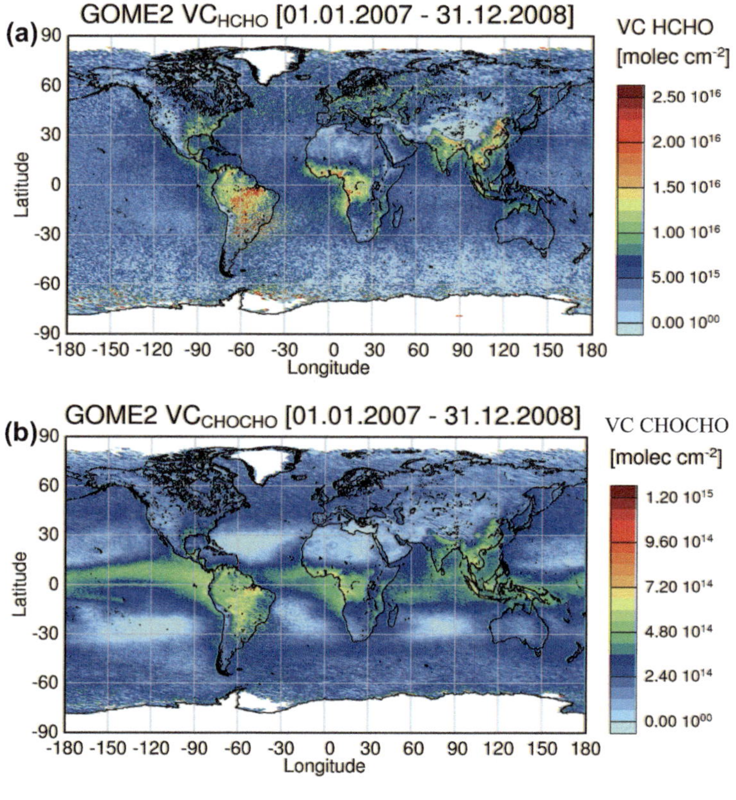

Fig. 11.58 Global view of formaldehyde (HCHO, a) and glyoxal (CHOCHO, b) total vertical columns in the years 2007–2008. Highest levels are seen at low latitudes over areas with tropical vegetation. Reprinted with permission from Vrekoussis et al. (2010), ©CC 3.0. All rights reserved

tering at these short wavelengths, a large fraction of light observed from space is already scattered back in the mid-troposphere. Therefore, the sensitivity to the lowest part of the troposphere is strongly reduced compared to the upper troposphere. This effect can be accounted for by making a priori assumptions on the shape of the vertical SO_2 distribution. Details of the assumed SO_2 profile have a significant impact on the final result. Some details of SO_2-retrieval algorithm development can, for example, be found in Li et al. (2017) and Theys et al. (2017).

While the global background concentration of SO_2 is difficult to quantify by spaceborne observations, hot-spots of elevated column densities caused by industrial activities or volcanic eruptions can readily be observed. Figure 11.59 shows a global map of SO_2 vertical columns averaged over the period of 2004 to 2009 (Theys et al. 2015). Hot-spots are seen in some industrial areas like in eastern China, north-east Sibiria (Norilsk smelter), and in South Africa (Highveld industrial complex). Further hot-spots in central Africa and at the west coast of South America are due to volcanic activities. Volcanic emission plumes can be monitored over several days (e.g. Eisinger and Burrows 1998, Beirle et al. 2014; Theys et al. 2015; Warnach et al. 2019; Theys et al. 2019; Burton et al. 2021; Lamotte et al. 2023). Satellites thus also allow monitoring of global (chemical) evolution of volcanic SO_2 emissions (see e.g. McLinden et al. 2016, Carn et al. 2016, 2017). Figure 11.60 shows the evolution of the the volcanic plume originating from an eruption of Mount Etna, Italy, on 25 December 2018 using TROPOMI maps of SO_2 VCDs, BrO VCDs, and the ratios of both VCDs. The maps span a longitude range of about 20° (or about 1800 km), and a transport time of about 24 h. Lower BrO/SO_2 ratios are detected in the centre of the plume, where SO_2 VCDs are highest (Warnach 2022). Observations of this type

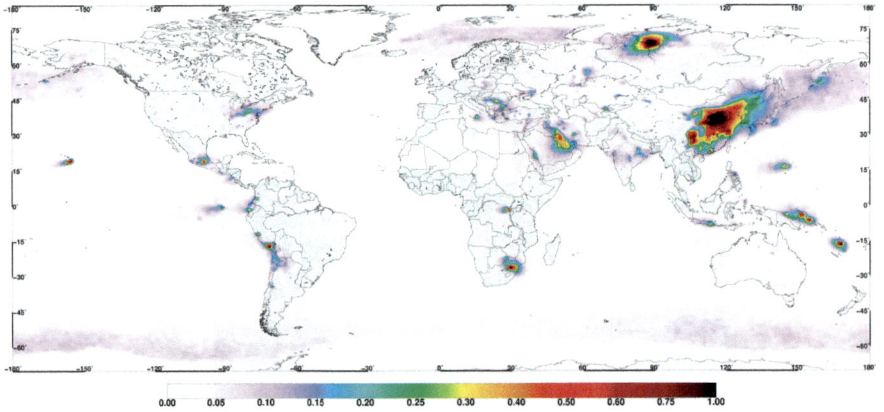

Fig. 11.59 Global SO_2 total vertical columns averaged over 2004 to 2009 from OMI spectra (in Dobson Units, DU, 1 DU = 2.69×10^{16} molec. cm^{-2}). Some of the hot-spots visible on the map are due to industrial areas (e.g. eastern China, North east Sibiria, South Africa) or to volcanic activities (e.g. in central Africa, at the west coast of South America), see text. (Some elevated SO_2 values over clean regions at high latitudes could be caused by residual interference with ozone or the Ring effect.) Reprinted with permission from Theys et al. (2015), ©American Geophysical Union. All rights reserved

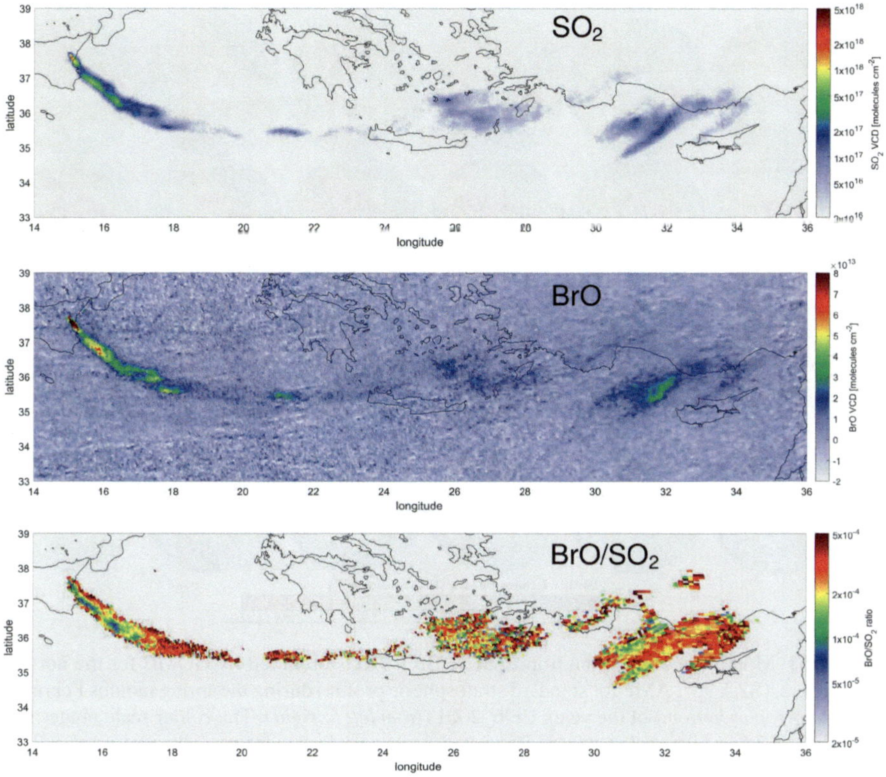

Fig. 11.60 TROPOMI satellite images of the plume originating from an eruption of Mount Etna, Italy, on 25 December 2018. Top panel: SO_2 VCD (logarithmic scale), Centre panel: BrO VCD, Bottom panel: BrO/SO_2 ratio calculated for each pixel (logarithmic scale) Warnach 2022.Reprinted from Warnach, Ph.D. Thesis, (2022), University of Heidelberg

can give information about the volcanic emission rate during the last 24 h although the satellite overflight takes place only once per day. Moreover, insight into chemical processes in the plume can be gained from the BrO/SO_2 ratio. This also holds for minor explosions and continuous outgassing processes, which are difficult to monitor from the ground.

11.9.6 Tropospheric BrO from Space

Polar BrO

Spatial and temporal evolution of enhanced BrO concentrations in the boundary layer on a global scale were first monitored by satellites (see Fig. 11.61; Wagner and Platt, 1998; Hegels et al. 1998; Wagner et al. 2001b; Richter et al. 1998, 2002; Hollwedel et al. 2004).

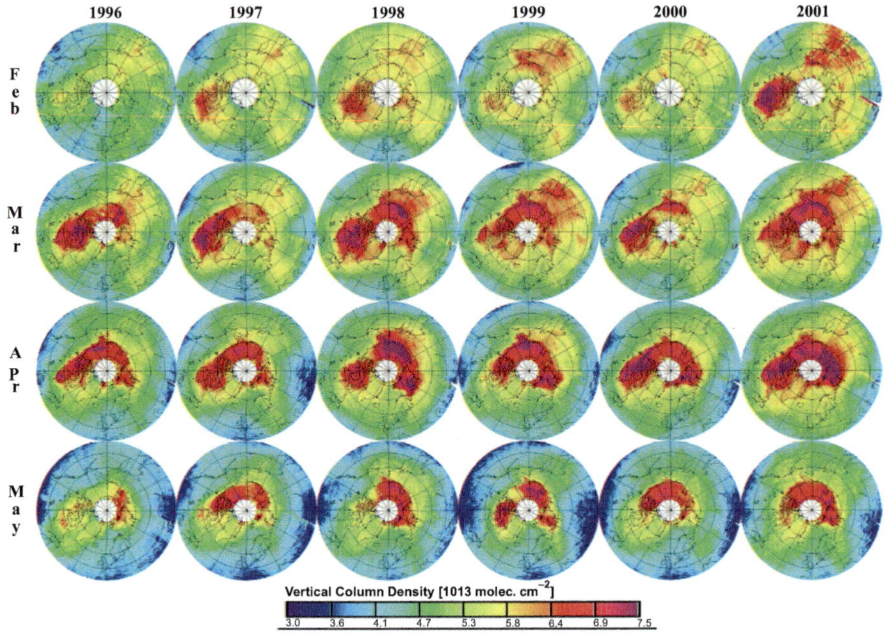

Fig. 11.61 Monthly means of the tropospheric BrO VCDs observed by GOME for the northern hemisphere. (SZA 90°, AMF for standard stratospheric profile) during the spring months February–May (*from top to bottom*) of the years 1996–2001 (*from left to right*). The colour scale ranges from 3×10^{13} to 7.5×10^{13} molec. cm^{-2}. Reprinted with permission from Hollwedel et al. (2004),

Similar to the cases of O_3 and NO_2, the fact that BrO is a normal constituent of stratospheric air complicates the tropospheric BrO retrievals. Over most of the globe the SPC dominates the BrO total VCD, only in 'hot spot' areas (e.g. volcanic plumes, salt lakes, and polar regions) can the TPC be comparable to the SPC (in fact maximum TPCs can be several times larger than average SPCs). For small hot spots the SPC can be interpolated from the area surrounding the hot spot (see e.g. Warnach 2022; Warnach et al. 2023). For large hot spots (like polar regions), however, the spatial variability of the BrO SPC cannot be neglected (see e.g. Salawitch et al. 2010).

Two techniques for separating SPC and TPC have been developed: The SPC can be modelled (e.g. Salawitch et al. 2010; Choi et al. 2012; Seo et al. 2019, 2020), which leads to a 'hybrid data product' composed of model results and satellite data (which, is close to the true BrO TPC). Alternatively, a BrO TPC can be constructed using only data from the same satellite instrument (O_3, NO_2, and BrO), where the O_3 column is a measure of the tropopause height and NO_2 indicates the sequestration of BrO as $BrONO_2$ (see e.g. Sihler et al. 2012; Hörmann et al. 2016; Schöne 2023).

Wagner et al. (2001b) were the first to conclude that the BrO concentrations in the boundary layer measured by GOME are in agreement with those from previous ground-based observations, and that elevated boundary layer BrO coincides with ozone depletion events observed by ground-based instruments. As shown in

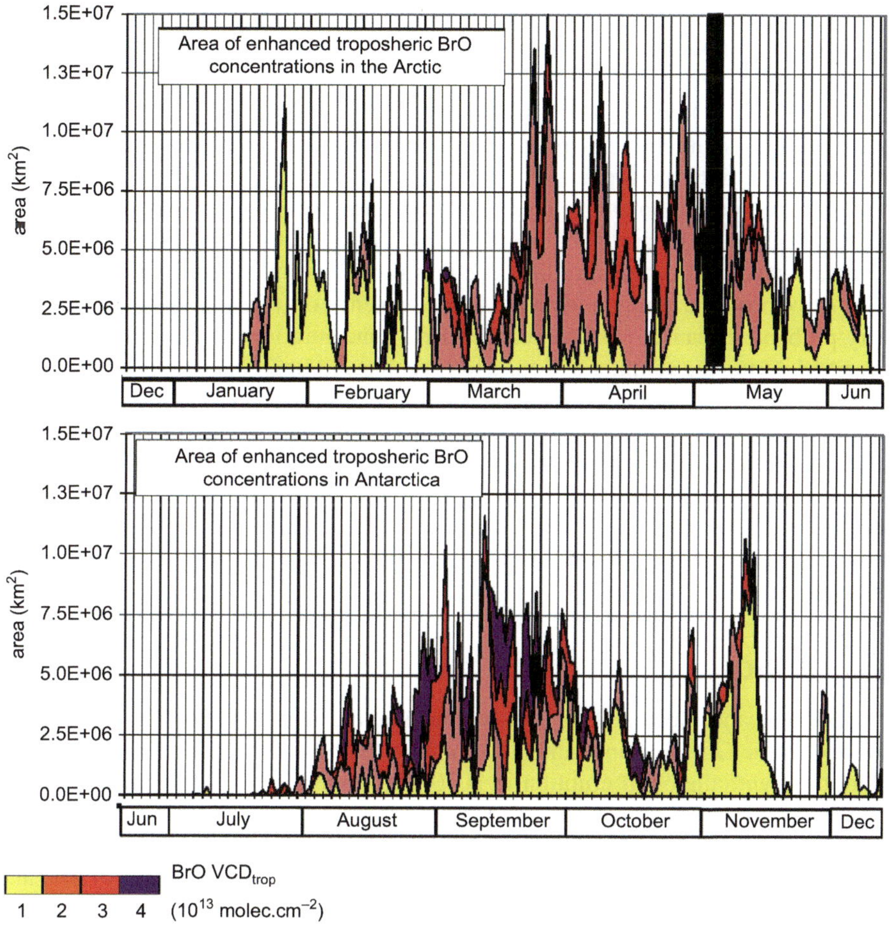

Fig. 11.62 Temporal evolution of the total area where enhanced tropospheric BrO VCDs are observed by GOME for the northern (*upper panel*) and the southern hemisphere (*bottom panel*). The magnitude of the tropospheric BrO VCD is indicated by different colours. It was determined using the stratospheric AMF and can thus serve only as a rough indicator. The *black box* in the *upper panel* indicates a period with no data. Reprinted with permission from Wagner et al. (2001b), ©American Geophysical Union. All rights reserved

Fig. 11.62, the events start to appear in January and July for the northern and southern hemisphere, respectively. The onset of the events is most probably determined by the end of polar night over regions where sea ice surfaces are present. The last events of the year were observed during June and December, probably determined by the melting of the sea ice surfaces. The highest boundary layer BrO concentrations and the largest spatial extension of the 'bromine clouds' occur during March/April and August/September in the northern and southern hemisphere, respectively. During this time, about 8×10^{30}–10×10^{30} BrO molecules ($\approx$1000–1300 t of Br) are present in the polar boundary layer. The total area reaches 1.5×10^7 km^2, about

3% of the surface of earth. The average BrO load (from the beginning of August to the middle of November in the Antarctic, and from the middle of January to the beginning of June in the Arctic) is about 3×10^{30}–4×10^{30} molecules (equivalent to 400–550 t of Br). In the northern hemisphere, the areas are up to 20% larger compared with the southern hemisphere; however, in the southern hemisphere, the peak BrO concentrations are higher.

Enhanced marine boundary layer BrO concentrations may be undetected due to the small ground albedo of the open oceans. Thus, the total area where enhanced boundary layer BrO concentrations occur is probably underestimated. The release of reactive bromine compounds without the presence of (one year old) sea ice is not very probable because: (1) no enhanced BrO concentrations in coastal regions are observed during the summer periods and (2) enhanced BrO concentrations above the Antarctic continent or above Greenland are rarely found and are probably related to transport processes. This, in particular, confirms the assumptions of Lehrer (1999). In summary, it appears that whenever the conditions for the autocatalytic amplification mechanisms are fulfilled, high concentrations of reactive bromine compounds are present. This might indicate that small concentrations of reactive bromine (possibly not measurable with current instrumentation) are nearly always present in the troposphere.

Volcanic BrO

The first volcanic BrO satellite measurements of the Kasatochi volcano plume, performed by the GOME-2 instrument, were reported by Theys et al. (2009). BrO data from the Eyjafjallajökull eruption in 2010 were reported by Heue et al. (2011). A comprehensive study of volcanic BrO by Hörmann et al. 2013, found clear enhancement of BrO in $\approx$ 25% of the examined volcanoes plumes and large spatial variations in the observed BrO/SO_2 ratios, ranging from 10^{-5} to 10^{-4}. An example of BrO and BrO/SO_2 measurements in a volcanic plume is given in Fig. 11.60. Further volcanic BrO data measured from space have been reported by (see e.g. Warnach et al. 2019, 2020 and Fig. 11.60).

BrO over Salt Lakes

Events of enhanced boundary layer BrO concentrations can also be seen from space at mid-latitudes, for example around the Caspian Sea, the Rann of Kutch salt lake, and the Great Salt Lake (USA) (Hörmann et al. 2016; Suleiman et al. 2019; Seo et al. 2019). Since high BrO levels were found by ground based measurements (see Sect. 11.3) over salt lakes it is to be expected that more measurements of enhanced BrO column densities in these environments will be reported in the future.

11.9.7 Tropospheric Carbon Monoxide from Space

Carbon monoxide (CO), like carbon dioxide (CO_2), and methane (CH_4) has relatively strong absorption bands in the infrared region, in particular in the near IR (also known as short-wave IR, SW-IR). Some satellite DOAS instruments, such as SCIAMACHY on Envisat (no longer in operation) and TROPOMI on Sentinel-5 Precursor, cover the near IR spectra range. Other satellite instruments (e.g. MIPAS, GOSAT, OCO-2, TES) have been developed to study the mid-IR spectral region. The evaluation of their spectra is based on particular algorithms relying on thermal emissions. Here we restrict the discussion to near IR instruments, which can be evaluated by DOAS techniques (for a comparison of the approaches, see e.g. Platt et al. 2018).

Two problems affecting the retrieval in the near IR have to be considered. First, the measurement of the VCD of a well-mixed gas scales linearly with the total column of air, and thus with surface pressure. Hence, a proxy for surface pressure is required in order to convert column density data into mixing ratios. Second, clouds and aerosols can substantially alter the light path of the recorded photons. Stray light from adjacent reflective surfaces can also bias retrievals (e.g. Butz et al. 2010, 2012). While these problems affect all satellite measurements of tropospheric species, they are particularly severe for CO, CO_2 and CH_4, since the variations ΔS in their column densities are rather small with $\Delta S_{CO_2} < \Delta S_{CH_4} < \Delta S_{CO}$. The above problems can be overcome by two approaches (see e.g. Butz et al. 2010): (1) Use a less variable gas (e.g. CO_2 for CH_4) as proxy for the light path or (2) Apply a "physics-based" method, simultaneously retrieving trace gas concentrations and scattering properties of the atmosphere.

For DOAS CO retrieval, a fit window has to be selected as a compromise between the higher instrument sensitivity at shorter wavelengths and a better signal-to-noise ratio at longer wavelengths. One also has to consider that the CO absorption lines are overlaid by strong absorptions by CH_4 and H_2O. Frankenberg et al. (2005), for example chose a window between 2324 and 2335 nm.

In contrast to the UV/vis spectral region, the near IR also exhibits peculiarities that make it somewhat more difficult to apply than, 'classical' DOAS algorithms. This is mostly due to the strong sensitivity of the line shape to temperature and pressure in the IR (see Sect. 3.6.4), and the presence of unresolved absorption lines (see e.g. Frankenberg et al. 2008). To overcome these problems, Buchwitz et al. (2000) introduced the concept of Weighting Function Modified DOAS (WFM-DOAS), which was further developed to WFMD (Schneising et al. 2023). Other research groups, such as Schrijver (2004), also developed modified algorithms for the near-IR spectral region; some preliminary results are shown by Buchwitz et al. (2004) and Gloudemans et al. (2004). Frankenberg et al. (2004) developed the 'Iterative Maximum A Posteriori'-DOAS (IMAP-DOAS), which was applied to obtain the global CO and CH_4 column density maps shown in Figs. 11.63 and 11.64. In addition to these modifications of the classical DOAS algorithm, instrumental problems of higher dark currents of the near-IR detectors are an important issue and have to be analysed and corrected in detail. Since these early studies the retrievals of CO in the near-IR have been further developed, and we refer the reader to recent literature on this topic.

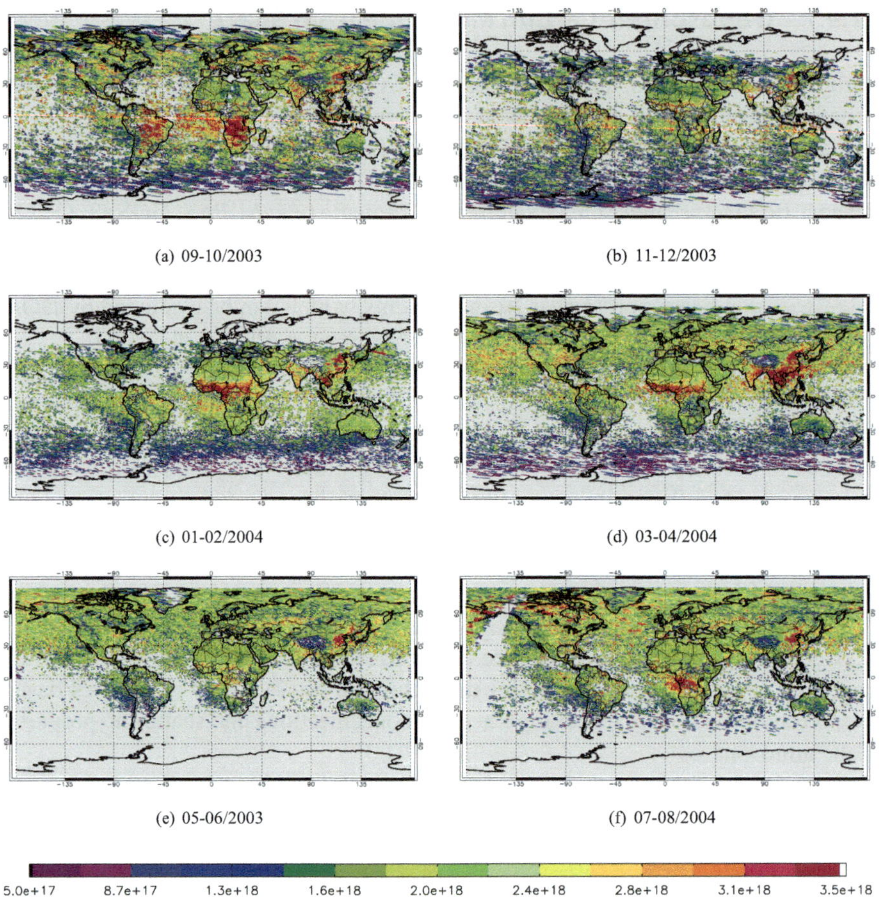

Fig. 11.63 Maps of maximum CO VCDs (molec. cm^{-2}) within a two-month period from September 2003 to August 2004, as derived from SCIAMACHY near-IR spectra. The most striking enhancements are seen in equatorial America and Africa, where the high CO levels are probably due to biomass burning. Reprinted with permission from Frankenberg et al. (2005), ©CC 2.5. All rights reserved

11.9.8 Tropospheric Carbon Dioxide from Space

For carbon dioxide, a suitable retrieval window is located in the range of 1562–1585 nm. The challenge is the small variation in its total columns (see above) across the globe, see e.g. Wagner et al. (2008). Global CO_2 hot spots, seasonal variations and trends have been identified in a number of studies, such as Schneising et al. (2012); Hou et al. (2013); Zhang et al. (2014); Buchwitz et al. (2015, 2017).

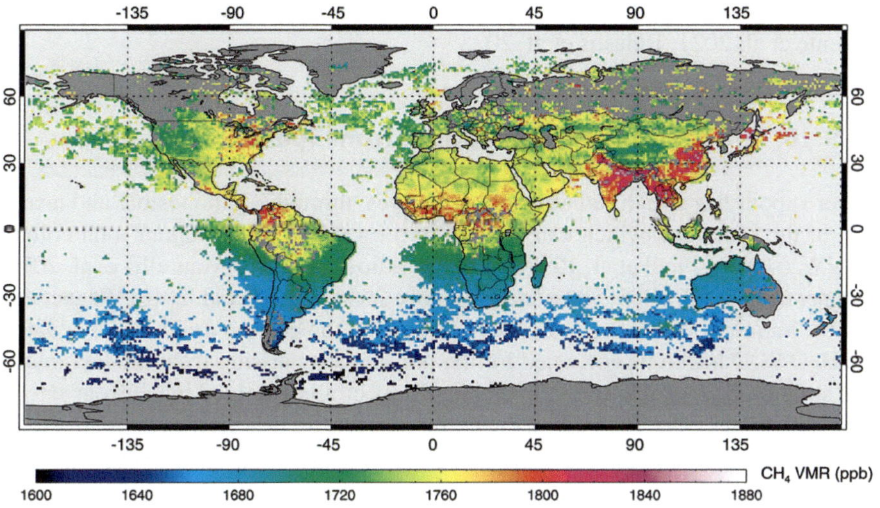

Fig. 11.64 Global map of column-averaged methane mixing-ratios (in ppb). Data are averaged over the time period of August through November 2003 on a 1° ×° 1 horizontal grid. At least five measurements (and up to 150) are taken for each grid cell. Few observations are available over the ocean due to the low ocean reflectivity substantially reducing the signal-to-noise ratio there. Occasionally, sun glint or clouds at low altitudes allow measurement over the ocean. Note that the scale only spans a concentration range of 1740 ± 140 ppb. Reprinted with permission from Frankenberg et al. (2005), ©American Association for the Advancement of Science. All rights reserved

11.9.9 Tropospheric Methane

Like CO, the greenhouse gas methane (CH_4) has relatively strong absorption bands in the near IR. For its retrieval, similar problems have to be overcome as in the case of CO (see Sect. 11.9.7). An additional difficulty arises from the fact that the CH_4 lifetime in the atmosphere is around 10 years (compared to a few months in the case of CO), and therefore the CH_4 mixing ratio is quite uniform throughout the global troposphere. In order to resolve structures in the CH_4 distribution due to sources and sinks, the column density has to be measured with a precision of about 1%. Frankenberg et al. (2005) solved the problem by ratioing the CH_4 columns derived from satellite near-IR measurements (SCIAMACHY on ENVISAT; see Table 11.1) by CO_2 columns derived from IR data from the same instrument. As discussed above, CO_2 exhibits much smaller variations in its total columns and the retrieval window (1562–1585 nm) is spectrally close to the CH_4 retrieval window (1630–1670 nm). Thus, it is well suited as a proxy for both surface pressure (or, in the presence of clouds, cloud top pressure) and changes in the light path due to aerosols and partial cloud cover. Consequently, it is possible to derive the column-averaged dry (no water vapour included) volume mixing ratio (VMR) of CH_4 in the atmosphere from the ratio of the CH_4 VCD and the CO_2 VCD scaled to the averaged atmospheric CO_2 mixing ratio (Frankenberg et al. 2005, 2006). More recent CH_4 measurements using SCIAMACHY spectra have been reported by Wagner et al. (2008); Schneising et al.

(2012); Buchwitz et al. (2015, 2017). TROPOMI CH_4-data were derived by e.g. Lorente et al. 2021; Balasus et al. 2023.

11.9.10 Tropospheric Water Vapour from Space

Water vapour has structured absorption features throughout the visible and near IR part of the spectrum, which can be used to retrieve the water vapour total column, TCVW, (see e.g. Noël et al. 1999; 2002; Casadio et al. 2000; Maurellis et al. 2000b; Wagner et al. 2003, 2005; Lang and Lawrence 2004, Grossi et al. 2015; Wagner et al. 2021b; Vaquero-Martinez et al. 2022). Following the customary DOAS evaluation, the H_2O SCD has to be corrected for the non-linearity arising from the fact that the fine structured H_2O absorption lines are not spectrally resolved by current satellite instruments (see Sect. 6.7). Then, the corrected H_2O SCD is divided by a 'measured' AMF, which is derived from the simultaneously evaluated O_2 or O_4 absorption. Thus, the effects of radiation transport and clouds can largely be eliminated. Compared to 'traditional' water vapour measurements from satellites, the DOAS data have essentially uniform sensitivity to H_2O throughout the atmosphere, thus they see the bulk of the H_2O close to the surface. Initial measurements of the water vapour total column mostly used spectral ranges at the long-wavelength end of the visible (i.e. "red" between about 500–700nm), because the H_2O absorption is strongest there. However, a disadvantage of evaluating absorption spectra in the "red" (or near IR) region is the relatively low albedo of the ocean surface at these wavelengths adversely affecting the signal-to-noise ratio. Wagner et al. (2013b) suggested applying retrievals in the "blue" spectral range (around 442 nm). Evaluation in the "blue" range has the additional advantage that the the H_2O absorption is much weaker (but still sufficiently large) than in the "red", making the retrieval quasi-linear. Results of this approach have been reported by Wang et al. (2019a); Borger et al. (2020).

An example, where the DOAS fit was performed in the wavelength interval 611–673 nm is shown in Fig. 11.65 (see also Wagner et al. 2005 and Wagner et al. 2021b). A band of high H_2O VCDs always surrounds the tropics according to the high surface temperatures. In most months, the local maxima are located over the continents, e.g. Brazil or Indonesia. Local minima at the equator are found at the east coast of Africa. As expected, the belt of maximum H_2O VCD follows the latitudinal variation of the Intertropical Convergence Zone (ITCZ) location with season; this variation is especially pronounced above the continents. During El-Niño periods, the global pattern of the H_2O VCD changes significantly. In addition, a very strong increase of H_2O in the equatorial Pacific, and a shift southwards can be recognised. This shift is most obvious at the end of 1997 and the beginning of 1998.

11.10 Solar Induced Fluorescence (SIF)

A more recent application of a DOAS-type measurement is the study of photosynthesis and carbon uptake by the earth's biosphere via remote sensing of Solar-Induced

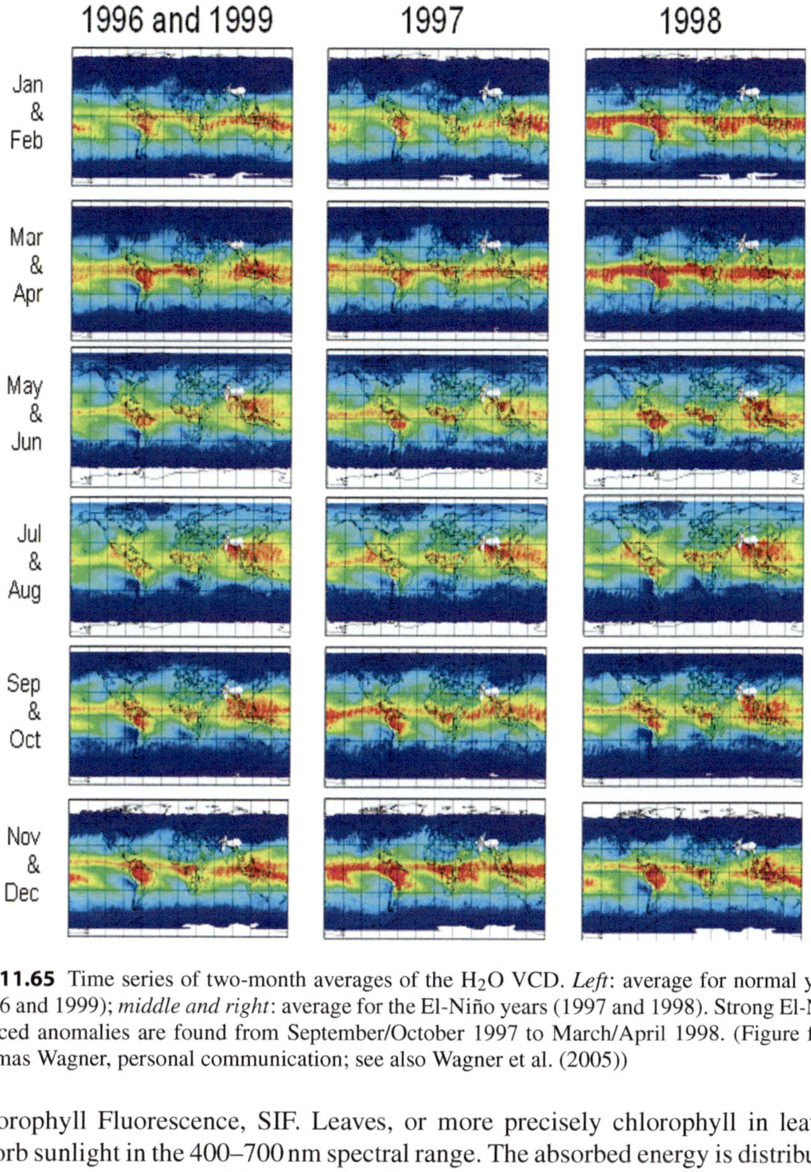

Fig. 11.65 Time series of two-month averages of the H_2O VCD. *Left*: average for normal years (1996 and 1999); *middle and right*: average for the El-Niño years (1997 and 1998). Strong El-Niño induced anomalies are found from September/October 1997 to March/April 1998. (Figure from Thomas Wagner, personal communication; see also Wagner et al. (2005))

Chlorophyll Fluorescence, SIF. Leaves, or more precisely chlorophyll in leaves, absorb sunlight in the 400–700 nm spectral range. The absorbed energy is distributed into various pathways inside the leaves. The most important pathway is photosynthesis, where CO_2, taken up from the atmosphere, and water are converted into organic compounds, such as sugars and starch. This pathway is associated with the carbon uptake plants use to grow, which is often quantified as gross primary productivity (GPP). GPP is an important parameter to understand the impact the biosphere has on the global CO_2 budget and thus directly related to the role of the biosphere in climate change. Another part of the energy absorbed by leaves is diverted into thermal energy dissipation, a mechanism plants regulate to counteract potential damage from excess energy under environmental stress conditions (low temperatures, high light levels, drought, etc.). Finally, 1–3% of the absorbed energy is re-emitted as a fluorescence signal in the 650–850nm spectral range. On large spatial and temporal scales, such as those observed from satellites, this SIF signal correlates linearily with GPP, making it a powerful proxy for monitoring plant photosynthesis and carbon uptake on regional and global scales. It is beyond the scope of this book to discuss the intricacies of the GPP-SIF relationship, and the reader is encouraged to consult recent literature on this topic (e.g., Porcar-Castell et al. 2014, 2021; Rascher et al. 2015; Frankenberg and Berry 2018; Meroni et al. 2009; Mohammed et al. 2019; Sun et al. 2023a, b). The details of the DOAS-type retrieval of the SIF-signal is described in detail below. SIF is observable on a global scale from space from various satellites, such as the Greenhouse Gas Observing Satellite (GOSAT) (Frankenberg et al. 2011a, b, 2012; Joiner et al. 2011, 2012b; Guanter et al. 2012, 2013), Global Ozone Monitoring Experiment (GOME-2) (Joiner et al. 2013) as well as NASA's Orbiting Carbon Observatory-2 (OCO-2) satellite (Joiner et al. 2012b; Wolanin et al. 2015), OMI and TropOMI (Köhler et al. 2018).

Recently, spectrometer systems have been developed to retrieve SIF from above-canopy towers, unmanned aerial vehicles (UAVs), or aircraft to link leaf to global scale data (e.g., Moya et al. 1998, 2004; Meroni et al. 2009; Rascher et al. 2009; Guanter et al. 2013; Rascher et al. 2015; Cogliati et al. 2015; Middleton et al. 2017; Grossmann et al. 2018). For example, measurements of canopy SIF at 760 nm were performed in temperate deciduous forests by using the spectrometer system FluoSpec (Yang et al. 2015, 2017). The ground-based MRI and SFLUOR box instruments were used to measure SIF and vegetation indices of different canopies such as sugar beet, grassland, or a lawn carpet (Cogliati et al. 2015).

In the following sections we discuss the general setup of SIF instruments and the principles and application of SIF remote sensing, especially as it relates to the DOAS technique.

11.10.1 SIF Instruments

SIF instruments are similar to passive DOAS systems. They generally consist of a spectrograph-detector combination that is coupled to a telescope-scanner unit. For ground-based instruments the coupling is usually achieved through fibre optics. The main difference is that SIF measurements require the comparison of skylight (direct solar + scattered) to light reflected from a canopy. Figure 11.66 illustrates such as ground-based system, where two separate spectrometers, one measuring from 670 to 732 nm and the other from 729 to 784 nm, measure red and far-red SIF, respectively. A long fibre connects the spectrometers to a telescope-scanner unit that is typically mounted on a tower above the canopy. This PhotSpec instrument also includes a broadband spectrometer to measure canopy reflectance.

Groundbased SIF systems typically aim at an upward looking diffuser for the solar spectrum and then aim downward at a canopy.

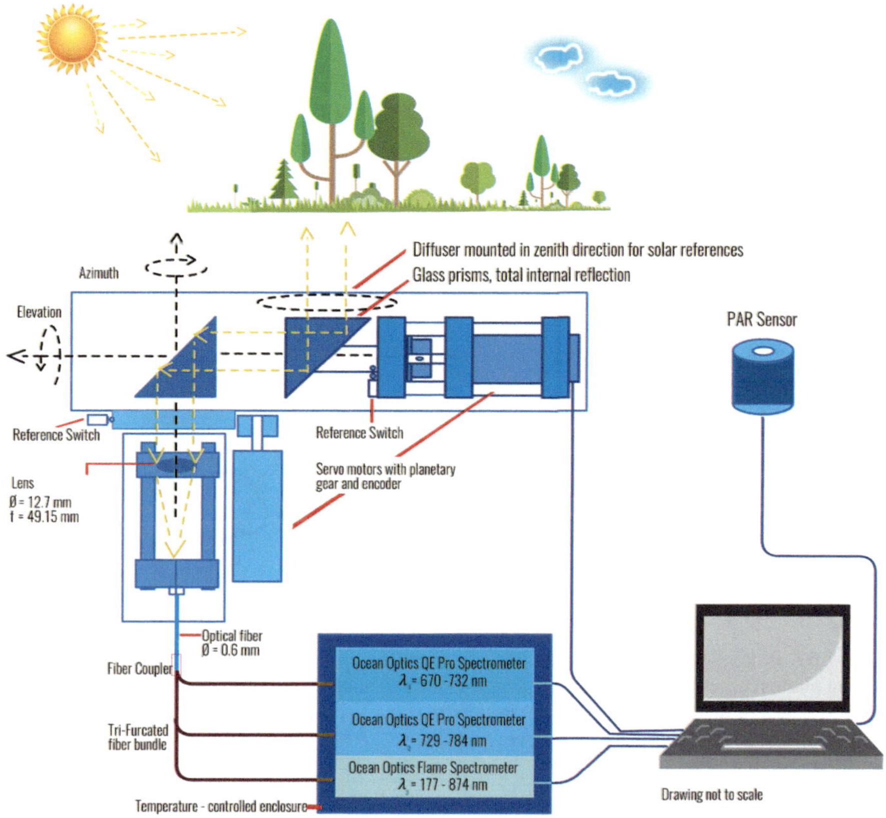

Fig. 11.66 The PhotoSpec instrument described by Grossmann et al. 2018 includes a thermally stabilised triple spectrograph unit that is connected to a telescope-scanner unit through an optical fibre. Reprinted with permission from Grossmann et al. (2018),

Most satellite SIF observations have been performed with satellites that were not specifically designed for this application. For example, TropOMI, which is predominately a trace gas remote sensing instrument, includes the wavelengths needed for the retrieval of SIF. Because the references needed for SIF retrievals have to include the atmospheric absorptions of the reflected light, the solar references for the SIF retrieval are usually determined from reflective vegetation free surfaces. The observations thus do not require any specific accommodations compared to trace gas retrievals.

11.10.2 Physical Basis of SIF Retrievals

To develop the theoretical framework of SIF remote sensing we first consider how sunlight interacts with canopies (Fig. 11.67). Direct and indirect solar radiation fall onto a canopy where it is absorbed by leaves as well as other material (soil, branches,

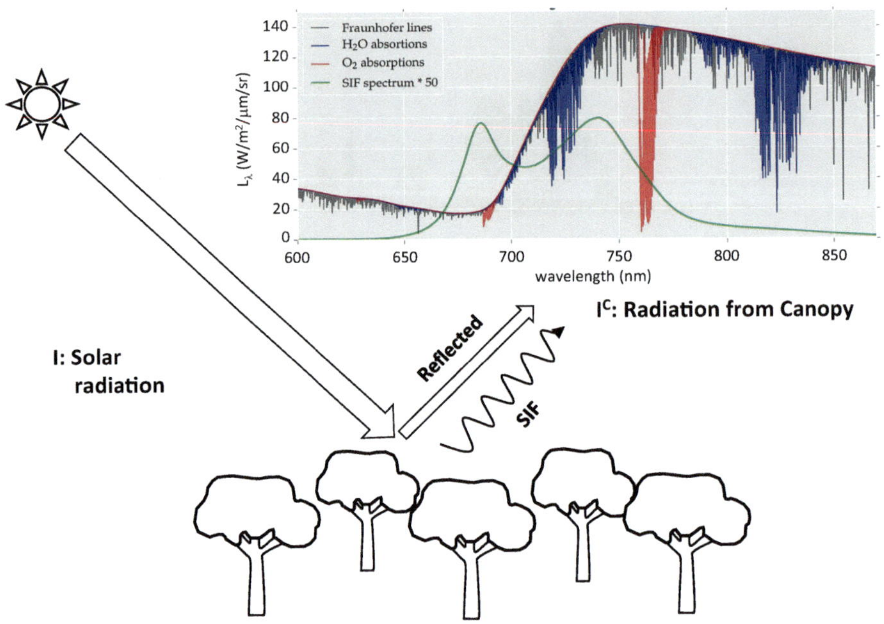

Fig. 11.67 Principle of SIF observations. Light from the sun is reflected by a vegetation canopy. During photosynthesis the SIF signal is added to the reflected signal. The insert shows the spectral characteristics of the solar spectrum, absorption by water vapour and oxygen, and the fluorescence spectrum. Insert adapted from Frankenberg and Berry (2018)

trunks). The absorption by green leaves or needles is dominated by chlorophyll and primarily occurs in the 400–700 nm range. The solar irradiance integrated from 400–700 nm is therefore often called Photosynthetic Active Radiation, PAR. Chlorophyll imposes a strong absorption in the 400–700nm range that is very prominent in spectrum of reflected solar light (Fig. 11.67). It should be noted that the reflected spectrum contains Fraunhofer lines as well as signatures of all atmospheric absorbers.

While much of this absorbed energy is used for photosynthesis, a small percentage is re-emitted by chlorophyll fluorescence in two broad bands centered at 685 nm (red) and 740 nm (far-red) (e.g., Genty et al. 1989; Krause and Weis 1991; Baker 2008; Porcar-Castell et al. 2014).

It is this small unstructured component of the reflected light that a SIF retrieval aims to quantify. Because the SIF signal is overlayed by the reflected signal a direct determintion of the SIF signal is not possible and all current implementations use spectroscopic methods to determine SIF. The idea is to use a (narrow) spectral interval where the solar intensity is lower or (ideally) zero to measure the fluorescence signal. (This is possible since the fluorescence at a given wavelength is excited by a broad range of incoming solar radiation.) Most studies use either the oxygen absorption bands at 760 nm ($O_2 - A$) or at 690 nm ($O_2 - B$) (e.g., Meroni et al. 2009; Rascher et al. 2009, 2015; Rossini et al. 2015, and references therein) or solar Fraunhofer lines to determine SIF (e.g., Plascyk and Gabriel 1975). The latter is used for most satellite and many ground-based SIF retrievals (e.g., Joiner et al. 2011, 2012b, 2013;

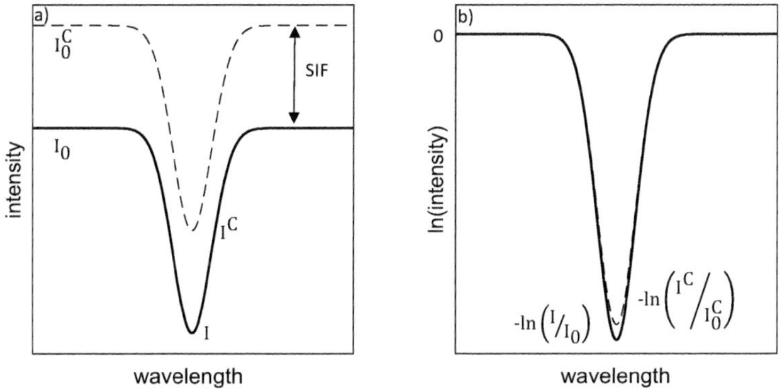

Fig. 11.68 Fraunhofer line in-filling due to the presence of an additive SIF signal. The intensities at the shoulder and centre of a Fraunhofer line, I_0 and I, respectively, are changed by the addition of the SIF signal to I_0^C and I^C. As a consequence the optical densities of the canopy intensities, $-\ln(I^C/I_0^C)$, are lower than those from direct sunlight $-\ln(I/I_0)$. Reprinted with permission from Grossmann et al. (2018),

Frankenberg et al. 2011b, 2014; Guanter et al. 2013; Köhler et al. 2015; Aas et al. 2019), because the Fraunhofer line approach is less sensitive to atmospheric scattering and thus more suited for observations in the presence of clouds (e.g., Yang et al. 2017). While we will focus on Fraunhofer line retrievals, the theoretical considerations are the same for the oxygen band based retrievals.

SIF retrievals are based on the influence the additive SIF signal has on the depth of an absorption line and presented within the framework of the correction of additive signals in Section 8.6. This is often referred to as the Fraunhofer in-filling method (Guanter et al. 2013). We start by discussing how SIF impacts the optical densities of an absorption band $-\ln(I/I_0)$, such as a Fraunhofer line. Here I is the intensity in the band centre and I_0 is the intensity at the band shoulder (Fig. 11.68). The respective intensities leaving a sun illuminated canopy are, I_0^{C} and I^{C}, which can be written as:

$$\begin{aligned} I_0^{\mathrm{C}} &= a^{\mathrm{C}} \cdot I_0 + I_{\mathrm{SIF}} \\ I^{\mathrm{C}} &= a^{\mathrm{C}} \cdot I + I_{\mathrm{SIF}} \end{aligned} \tag{11.3}$$

where a^{C} is the canopy reflectance and I_{SIF} is the SIF signal emitted by the canopy. $\ln(\frac{I^{\mathrm{C}}}{I_0^{\mathrm{C}}})$ can now be approximated by (see Sect. 8.6 for the derivation):

$$\ln\left(\frac{I^{\mathrm{C}}}{I_0^{\mathrm{C}}}\right) \approx \ln\left(\frac{I}{I_0}\right) \cdot \left(1 - \frac{I_{\mathrm{SIF}}}{a^{\mathrm{C}} \cdot I_0}\right) \quad (11.4)$$

The optical density of a spectral band is reduced by the ratio of the SIF radiance, I_{SIF}, and the intensity reflected by the canopy $a^{\mathrm{C}} \cdot I_0$. The reduction is typically between 0 and 0.03 in the far-red. The effect in the red wavelength range is larger, between 0 and 0.3, because the smaller a^{C} increases the effect of the additive SIF intensity.

11.10.3 SIF Retrieval Approach

Many tower based applications use the reduction of the oxygen A-band optical density in a canopy spectrum compared to a temporally close recorded sky spectrum to derive SIF (Meroni et al. 2009 and references therein). For satellite retrievals this approach is very challenging, therefore for this application solar Fraunhofer lines are used as the spectral structure. In both cases the retrieval idea is the same in that a skylight/atmospheric reference spectrum, determined by pointing the SIF instrument onto an upward facing diffuser or onto a non-fluorescent diffusive surface, $I^{\mathrm{D}}(\lambda)$, is compared to a spectrum of the radiance reflected from a plant/canopy $I^{\mathrm{C}}(\lambda)$.

$$I^{\mathrm{D}}(\lambda) = T(\lambda) \cdot (a^{\mathrm{D}}(\lambda) \cdot I(\lambda) + I_{\mathrm{stray}}(\lambda)) + I_{\mathrm{DC}} + I_{\mathrm{offset}} \quad (11.5)$$

$$I^{C}(\lambda) = T(\lambda) \cdot (a^{\mathrm{C}}(\lambda) \cdot I(\lambda) + I_{\mathrm{stray}}(\lambda) + I_{\mathrm{SIF}}(\lambda)) + I_{\mathrm{DC}} + I_{\mathrm{offset}} \quad (11.6)$$

Here $I(\lambda)$ is the solar intensity (we ignore the radiance-irradiance conversion for simplicity and just use the term intensity). $I_{\mathrm{stray}}(\lambda)$ is the spectrometer stray light. $a^{\mathrm{D}}(\lambda)$ is the reflectivity/transmissivity of the optical element used to measure the solar reference. $T(\lambda)$ is the instrument sensitivity. We also consider the detector dark current, I_{DC} and electronic offset I_{offset}.

We assume that spectrometer stray light is negligible and that dark current and electronic offset are accurately corrected. In addition, the detector sensitivity, $T(\lambda)$, cancels out when comparing the two spectra and will also be omitted. We can thus simplify Eqs. 11.5 and 11.6 to:

$$\begin{aligned} I^{\mathrm{D}}(\lambda) &= a^{\mathrm{D}}(\lambda) \cdot I_0(\lambda) \\ I^{\mathrm{C}}(\lambda) &= a^{\mathrm{C}}(\lambda) \cdot I_0(\lambda) + I_{\mathrm{SIF}}(\lambda) \end{aligned} \quad (11.7)$$

Using a similar mathematical approximation as described in detail in Section 8.6, we can now relate the canopy signal, $I^{\mathrm{C}}(\lambda)$, to the solar signal, $I^{\mathrm{D}}(\lambda)$.

$$\ln(I^{\mathrm{C}}(\lambda)) \approx \ln(I^{\mathrm{D}}(\lambda)) + \ln\left(\frac{a^{\mathrm{C}}(\lambda)}{a^{\mathrm{D}}(\lambda)}\right) + \frac{I_{\mathrm{SIF}}(\lambda)}{I^{\mathrm{C}}(\lambda)} \quad (11.8)$$

This equation is of a similar form than the one already derived for an additive signal, and we can thus use a DOAS fit with the model function $f(\lambda)$ (Eq. 11.9) to determine the additive SIF signal:

$$f(\lambda) = \ln(I^{\mathrm{D}}(\lambda)) + P_{\alpha}(\lambda) + C_{\mathrm{SIF}} \frac{R_{\mathrm{SIF}}(\lambda)}{I^{\mathrm{C}}(\lambda)} \tag{11.9}$$

The polynomial of order α $P_{\alpha}(\lambda)$ describe the smoothly varying ratio $\ln(a^{\mathrm{C}}(\lambda)/a^{\mathrm{D}}(\lambda))$. $R_{\mathrm{SIF}}(\lambda)$ is a reference spectrum for the shape of the SIF emission signal. The ratio of $R_{\mathrm{SIF}}(\lambda)/I^{\mathrm{C}}(\lambda)$ is scaled in the fit with C_{SIF}, in order to dervie the magnitude of the SIF signal: $\mathrm{C_{SIF}} \cdot \mathrm{R_{SIF}}(\lambda)$.

The error of the linearisation in Eq. 11.9 is on the order of $-\frac{1}{2}\left(\frac{I_{\mathrm{SIF}}(\lambda)}{I^{\mathrm{C}}(\lambda)}\right)^2$ and introduces a relative negative error(bias) on $\frac{I_{\mathrm{SIF}}(\lambda)}{I^{\mathrm{D}}(\lambda)}$ of 0 - 1.5% in the far-red wavelength range and 0 - 15% in the red wavelength range. Grossmann et al. 2018 describes a two-step linearised SIF retrieval that overcomes this issue by first retrieving the SIF signal using Eq. 11.9, subtracting it from $I^{\mathrm{C}}(\lambda)$, and then again applying another fit with Eq. 11.9 to determine the second order of the approximation in Eq. 11.8.

Figure 11.69 shows an example of a SIF retrieval in a 6 nm wide window around 683 nm. This window was chosen by Grossmann et al. (2018) to minimise the effects of atmospheric absorbers, while still allowing for the presence of strong Fraunhofer lines. The main spectral structures in this wavelength range are solar Fraunhofer lines overlaid by the broad band reflectivities of the diffuser and canopy (top panel in Fig. 11.69). The retrieved SIF reference spectrum $C_{\mathrm{SIF}} \frac{R_{\mathrm{SIF}}(j)}{I^{C}(j)}$ is illustrated by the red curve in the middle panel panel of Fig. 11.69. The comparison between the retrieved spectrum and the retrieved spectrum with added fit residual (black line, middle panel in Fig. 11.69) illustrates the quality of the fit. The fit residual is mostly noise above 681.5 nm, with a small unidentified structure below 681.5 nm. After applying a radiometric calibration, the resulting SIF signal is found to be $2.26 \pm 0.08\,\mathrm{mW\,m^{-2}sr^{-1}nm^{-1}}$.

It should be noted that SIF retrievals can also be performed using a fully non-linear retrieval, as described for GOSAT and OCO-2 (Frankenberg et al. 2011a). The linear retrievals, however, have the advantage of faster computing times and a better convergence to the optimal solution of the fit.

11.10.4 Instrumental Effects on SIF Retrievals

In contrast to normal DOAS retrievals, where instrumental effects often have small impacts on the retrieved quantities, SIF retrievals are highly sensitive to non-ideal instrument behavior. It is thus useful to briefly review these impacts, following the description in Grossmann et al. (2018).

Dark current and offset correction: Correcting electronic offset and dark current is typically achieved by measuring the respective effects with a darkened detector and very short and long exposure times, respectively. It is typically assumed that

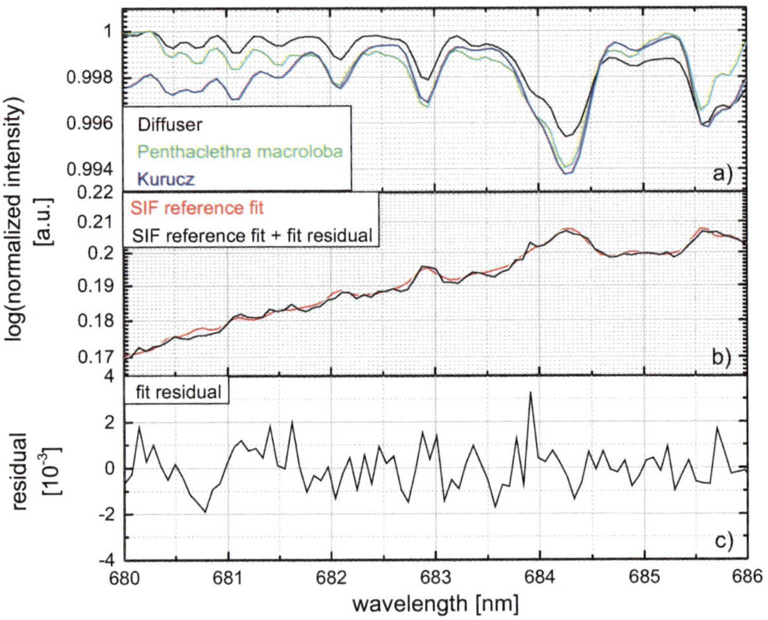

Fig. 11.69 Example for a SIF spectral retrieval in the red wavelength range for a *Penthaclethra macroloba* spectrum recorded on April 21, 2017 at 13:09 (LT) at an SZA of 23°, and a fit RMS of $9.11 \cdot 10^{-4}$. Panel **a** compares diffuser and canopy spectra with those from a solar reference (Kurucz). Panel **b** shows the fit results (red) compared to the fit result with the added retrieval residual (black). The residual, i.e., the unexplained spectral structure, of the retrieval is shown in panel **c**. The calibrated SIF signal of this spectrum is 2.26±0.08 mW $m^{-2}sr^{-1}nm^{-1}$. Reprinted with permission from Grossmann et al. (2018), ©Elsevier. All rights reserved

uncertainties of this correction result in very small errors in a DOAS retrieval. However, in SIF retrievals any errors in correcting dark current and electronic offset lead to a direct absolute error in the SIF signal. This error can be reduced to insignificant levels by cooling the detector to reduce its dark signal and by regularly measuring both effects.

Detector non-linearity Detector non-linearity on the order of up to 1% is a common problem in DOAS applications (see Sect. 7.5.1). However, in most cases its effect is fairly minor and many application overcome this issue by ensuring that the observations are always performed at the same detector saturation level, for example by varying the exposure time. For SIF retrievals the effect of non-linearity is more pronounced (Grossmann et al. 2018). We can illustrate this by investigating how the depth of an absorption band defined by I and I_0 changes for a given nonlinearity. Here, nonlinearity is defined as the deviation from the proportionality between the incoming photon flux L and L_0, and the measured detector signal I and I_0, in the band centre and shoulder, respectively:

$$\begin{aligned} I_0 &= d_1 \cdot L_0 + d_2 \cdot L_0^2 \\ I &= d_1 \cdot L + d_2 \cdot L^2 \\ &= d_1 \cdot F \cdot L_0 + d_2 \cdot F^2 \cdot L_0^2 \quad \text{where} \quad F = \frac{L}{L_0} \end{aligned} \tag{11.10}$$

The first term is the linear/proportional response and the second quadratic term denotes the nonlinearity. For simplicity we ignore higher order nonlinear terms. We can now define the optical depth of the Fraunhofer band as:

$$\begin{aligned} \ln\left(\frac{I}{I_0}\right) &= \ln\left(\frac{d_1 \cdot F \cdot L_0 + d_2 \cdot F^2 \cdot L_0^2}{d_1 \cdot L_0 + d_2 \cdot L_0^2}\right) \quad \text{with} \quad F = \frac{L}{L_0} \quad \text{and} \quad \frac{d_2}{d_1} \text{ small} \\ &\approx \ln(F) + \frac{d_2}{d_1} \cdot F \cdot L_0 - \frac{d_2}{d_1} \cdot L_0 \\ &= \ln(F) + \frac{d_2}{d_1} \cdot L_0 \cdot (F - 1) \quad \text{with NL} = \frac{d_2 \cdot L_0^2}{d_1 \cdot L_0} = \frac{d_2}{d_1} \cdot L_0 \\ &= \ln(F) - \text{NL} \cdot (1 - \text{F}) \end{aligned} \tag{11.11}$$

The relative nonlinearity is determined as the ratio of the quadratic and linear terms in Eq. 11.10 and L_0. $\ln(F)$ is the optical density of the absorption line, and $(1 - F) \approx \ln(F)$ for a small F. The relative change of the optical density is thus directly proportional to NL. We can now apply this approximation to the general SIF linearisation (Eq. 11.4):

$$\begin{aligned} \ln\left(\frac{I + I_{\text{SIF}}}{I_0 + I_{\text{SIF}}}\right) &= \ln\left(\frac{I}{I_0}\right)\left(1 - \frac{I_{\text{SIF}}}{I_0}\right) \\ &= \ln(F) - \ln(F) \cdot \frac{I_{\text{SIF}}}{I_0} \\ &\approx \ln(F) - (1 - F) \cdot \frac{I_{\text{SIF}}}{I_0} \end{aligned} \tag{11.12}$$

Comparing Eq. 11.12 to Eq. 11.11 illustrates that the nonlinearity NL influences the optical density of the absorption line in the same way as I_{SIF}/I_0. Consequently the error on the retrieved SIF signal is:

$$\Delta I_{\text{SIF}} = I_0 \cdot \text{NL} \tag{11.13}$$

Because far-red I_{SIF}/I_0 is between 1 and 3%, the requirements for detector linearity are high. A 0.1% nonlinearity over the detector saturation range translates to an error in the SIF signal of 3–10 %. Choosing a highly linear detector or performing accurate nonlinearity corrections are thus essential when performing SIF observations.

Spectrometer stray light:
Incomplete suppression of light originating from wavelengths outside the measurement window, i.e. stray light, poses another challenge in SIF retrievals. Grossmann et al. (2018) investigated the effect of stray light on SIF retrievals using the linearisation Eq. 11.7. They assumed a constant stray light signal I_{stray} across a narrow Fraunhofer band and derived a linear approximation for the joint impact of I_{SIF} and I^{C}_{stray}:

$$\begin{aligned} I^{\text{C}} &= a^{C} \cdot I + I_{\text{SIF}} + I^{\text{C}}_{\text{stray}} \\ I^{\text{C}}_{0} &= a^{\text{C}} \cdot I_0 + I_{\text{SIF}} + I^{\text{C}}_{\text{stray}} \\ \ln\left(\frac{I^{\text{C}}}{I^{\text{C}}_{0}}\right) &\approx \ln\left(\frac{I}{I_0}\right) \cdot \left(1 - \frac{I^{C}_{\text{stray}}}{a^{\text{C}} \cdot I_0} - \frac{I_{\text{SIF}}}{a^{\text{C}} \cdot I_0}\right) \end{aligned} \tag{11.14}$$

Stray light in the spectrometer has the same impact on the optical density of an absorption line as the SIF signal. Using the same formalism as above, the solar measurement with spectrometer stray light can be written as follows:

$$\begin{aligned} I^{\text{D}} &= a^{\text{D}} \cdot I + I^{\text{D}}_{\text{stray}} \\ I^{\text{D}}_{0} &= a^{\text{D}} \cdot I_0 + I^{\text{D}}_{\text{stray}} \\ \ln\left(\frac{I^{\text{D}}}{I^{\text{D}}_{0}}\right) &\approx \ln\left(\frac{I}{I_0}\right) \cdot \left(1 - \frac{I^{\text{D}}_{\text{stray}}}{a^{\text{D}} \cdot I_0}\right) \end{aligned} \tag{11.15}$$

In a SIF retrieval $\ln\left(\frac{I^{\text{D}}}{I^{\text{D}}_{0}}\right)$ is typically used as $\ln\left(\frac{I}{I_0}\right)$ in Eq. 11.14. If the relative stray light signal $\frac{I^{\text{C}}_{\text{stray}}}{I_0}$ and $\frac{I^{\text{D}}_{\text{stray}}}{I_0}$ in both observations is the same, the stray light effect would cancel out. However, in reality the relative stray light levels are different, as the overall spectrum outside the measurement range, i.e. the main source of the stray light is quite different due to the strong chlorophyll absorption between 400–700 nm. The error on relative SIF, $\ln(\frac{I_{\text{SIF}}}{I_0})$, caused by spectrograph stray light is thus:

$$\Delta I^{\text{rel}}_{\text{SIF,stray}} = \frac{I^{\text{D}}_{\text{stray}}}{a^{\text{D}} \cdot I_0} - \frac{I^{\text{C}}_{\text{stray}}}{a^{\text{C}} \cdot I_0} \tag{11.16}$$

The effect of spectrograph stray light on the SIF retrieval is caused by the difference in the stray light in the solar reference and canopy spectrum. It is very challenging to accurately measure spectrograph stray light, in particular under circumstances where the overall spectral shape of the light entering the spectrograph varies. The most effective approach to reduce this error is thus to reduce spectrograph stray light, for example through the use of filters.

Grossmann et al. (2018) present a back-of-the-envelop estimate of the effect of spectograph stray light. They developed a general description of how each wavelength of incoming radiation $I_{in}(\lambda_{in})$, contributes to stray light through a translation function $\delta(\lambda, \lambda_{in})$:

$$\begin{aligned} I_{stray}(\lambda) &= \sum \delta(\lambda, \lambda_{in}) \cdot I_{in}(\lambda_{in}) \\ &= I_{tot} \cdot \sum \delta(\lambda, \lambda_{in}) \cdot f(\lambda_{in}) \qquad \text{where} \quad I_{in}(\lambda_{in}) = I_{tot} \cdot f(\lambda_{in}) \end{aligned} \tag{11.17}$$

Because $\delta(\lambda, \lambda_{in})$ is usually not known they make the simplified assumption that the relative stray light is $\frac{I^D_{stray}}{a^D I_0} \approx 0.005$ at 682 and 745 nm, which is in the range typically found in DOAS applications. They further assume that all incoming wavelengths contribute equally to stray light, i.e. $\delta(\lambda, \lambda_{in}) \approx 1.8 \cdot 10^{-5}$. Using observations of the spectral characteristics of the reflectance of a basil leaf and transmission of a ground glass diffuser they estimate I_{stray} for the diffuser and canopy case for an instrument measuring in the red and far-red SIF wavelength range.

Without any additional measure to reduce stray light they estimate $\frac{I^D_{stray}}{I^D_0} \approx 5 \cdot 10^{-3}$ and $\frac{I^C_{stray}}{I^C_0} \approx 8 \cdot 10^{-4}$ at 745nm. Consequently the bias on the relative SIF signal is $\Delta I^{rel}_{SIF,stray} \approx -0.0042$ which, compared to a typical relative SIF value of $\frac{I_{SIF}}{I_0} \approx 0.01$, results in a -42% bias on the SIF signal. This large bias is due to absorptions by chlorophyll-a on the canopy spectrum between 400–700 nm, i.e., the relative stray light contribution at 745 nm is reduced in the canopy spectrum due to the lower radiances between 400 - 700nm. The same calculation at 682 nm yields $\Delta I^{rel}_{SIF,stray} \approx -0.0037$. However, because this wavelength is impacted by chlorophyll-a absorption of $\frac{I_{SIF}}{I_0} \approx 0.1$, the stray light bias on SIF is only $\approx -3.7\%$.

Both calculations show that spectrograph stray light can introduce a considerable uncertainty in SIF observations if no additional measures are taken. Inclusion of high-pass glass filters that cut out light below 590 nm in the red or 695 nm in the far-red leads to much lower stray light biases of $\Delta I^{rel}_{SIF,stray} \approx -0.0026$ or 2.5% at 682 nm and $\Delta I^{rel}_{SIF,stray} \approx -10^{-4}$ or 1% at 745 nm. The use of these filters thus considerably reduces the effect of stray light in SIF applications.

11.10.5 Examples of SIF Applications

SIF has become a valuable tool to study the biosphere in recent years. Accordingly there is now a large number of publications on the technique as well as the results of these observations. Because a review of these publications goes beyond the scope of this book we will only show two example of SIF applications here.

The first example shows how SIF instruments mounted on towers above a vegetation canopy can be used to track net carbon uptake, here expressed as GPP. As shown in Fig. 11.70, SIF tracks GPP over one year of observations in an apline evergreen forest. Other commonly used reflectance based metrics, such as the normalised difference vegetation index (NDVI), which is based on "greenness" fails in this environment. The reason for the good SIF-GPP agreement is due to the fact that both

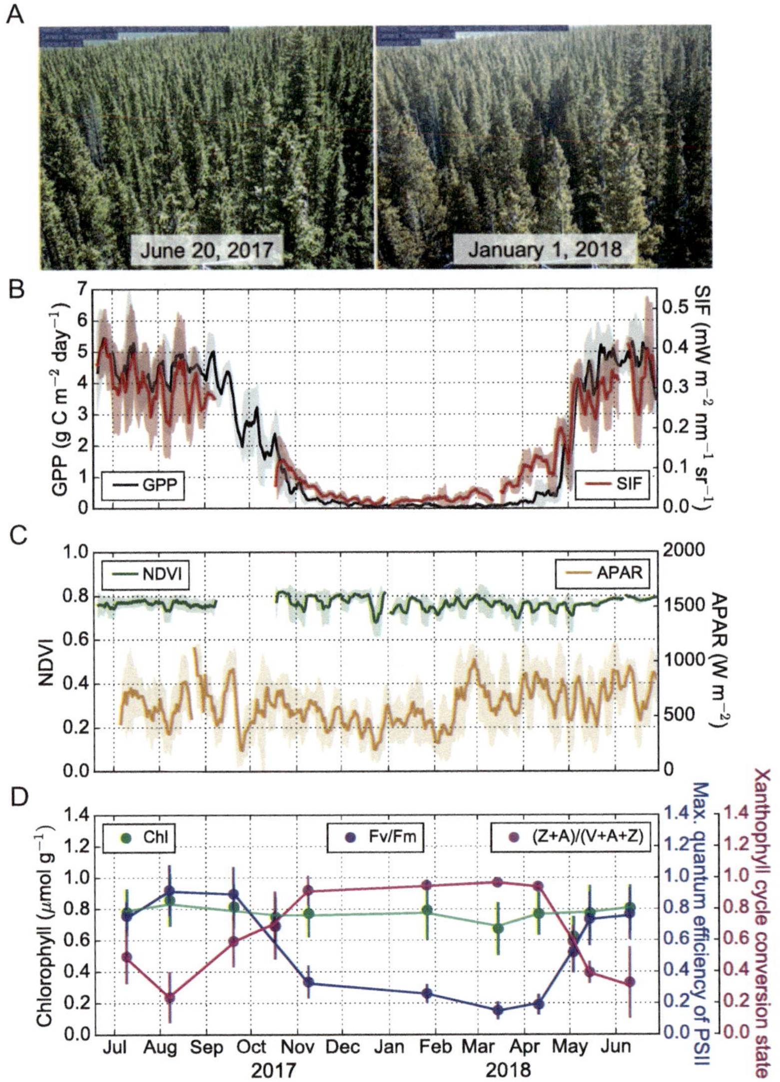

Fig. 11.70 Seasonal dynamics of GPP, far-red SIF, NDVI, APAR, needle pigments, and fluorescence at Niwot Ridge, CO. Reprinted with permission from Magney et al. (2019)

are influenced by the same biochemical mechanism, i.e. non-photochemical quenching by the Xanthophyll cycle (Panel D in Fig. 11.70), in the needles that regulated photosynthesis down in winter (Magney et al. 2019).

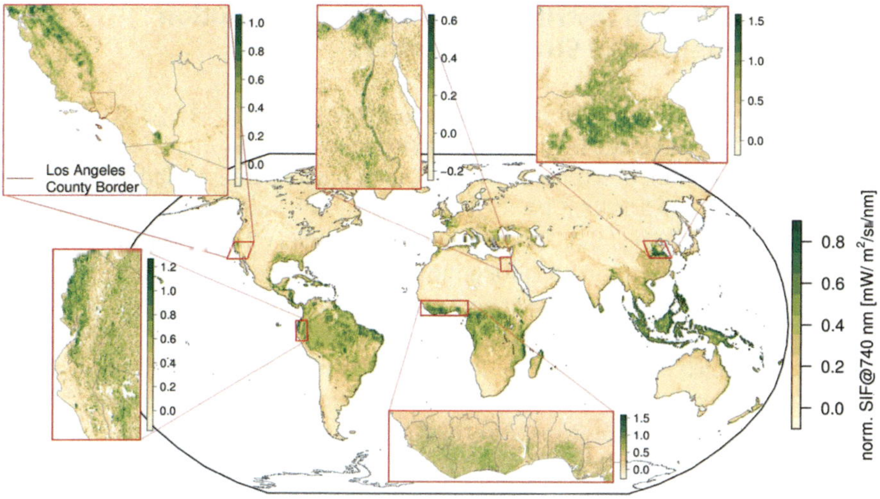

Fig. 11.71 The global solar-induced chlorophyll fluorescence (SIF) map is based on data acquired between 8 and 15 April 2018 (114 orbits, 16M measurements over land). Retrievals are gridded to a spatial resolution of 0.2° × 0.2°. The zoom-ins are gridded to 0.05° × 0.05°. Reprinted with permission from Köhler et al. (2018), ©American Geophysical Union. All rights reserved

The ability of SIF to track GPP makes it a promising tool to monitor the biosphere on a global scale from satellites. Figure 11.71 shows an example of one week of SIF data from the TROPOMI instrument (Köhler et al. 2018). Future satellite SIF instruments will be able to provide daily maps of SIF and thus better constrain the carbon uptake by the biosphere and its response to climate change.

11.11 Determination of Photon Path Lengths by 'Reversed DOAS'

An interesting application of DOAS 'reverses' the customary DOAS approach, which relies on determining an unknown concentration with a known light path, from the measured optical density, which is a product of both quantities. The lengths of photon paths e.g. in clouds can be measured by making use of the known concentrations of atmospheric constituents, such as molecular oxygen (O_2), tropospheric ozone, or the oxygen dimer $(O_2)_2$, O_4.

11.11.1 Average Path Lengths from Low Resolution Measurement of Weak Absorbers

Oxygen dimers have very wide spectral features (>1 nm), and they can therefore be monitored by simple, low-resolution instruments (Erle et al. 1995). Time series indicating average photon path lengths in clouds of almost 30 km at Kiruna (Sweden) were reported by Wagner et al. (1998).

11.11.2 Path Length Distributions from High Resolution Measurement of Strong Absorbers

By analysing the absorption of individual rotational lines (e.g. of O_2), it is possible to infer not only the average light path length in clouds, but also elements of the path-length distribution function (photon PDF) of solar photons transmitted to the ground (Pfeilsticker et al. 1995; 1998a; 1998b; Pfeilsticker 1999; Veitel et al. 1998; Wagner et al. 1998; Min et al. 2001; Funk 2000; Funk and Pfeilsticker 2003). The knowledge of photon PDFs is of primary interest for atmospheric radiative transfer, and thus for the climate. The method relies on the analysis of highly resolved oxygen A-band (762–775 nm) spectra (Fig. 11.72) observed with a small field of view ($\approx 1°$) telescope in zenith scattered skylight. It uses the fact that solar photons are randomly scattered by molecules, aerosols, and cloud droplets while traversing the atmosphere. Since a fraction of the solar photons is absorbed in the atmosphere, they travel different distances, depending on opacity. Spectral intervals with a changing atmospheric opacity but known absorber concentration and spectroscopic constants, such as in the spectral region of the oxygen A-band, thus contain information on the photon path probability density function, PDF. Studies have shown that the photon path PDF is largely correlated with the total amount of liquid water in the atmosphere, the spatial arrangement of clouds, cloud inhomogeneity, and solar illumination (Pfeilsticker et al. 1998; Pfeilsticker 1999; Veitel et al. 1998; Min et al. 2001; Funk and Pfeilsticker, 2003). For example, for thunderstorm cumulus nimbus (Cb) clouds, average photon path lengths of up to 100 km have been observed; whereas for stratus (Sc) clouds, photon paths are only 50–100% larger than for the direct and slant path of the sun's rays. Moreover, it appears that even though the spatial distribution of cloud droplet is inhomogeneous, with the moments of the density behaving like multi-fractals, photon PDFs tend to be mono-fractal, mostly due to the so-called radiative smoothing for optically thick clouds (Savigny et al. 1999).

11.11.3 Measurement of Trace Gases Inside Clouds

The phenomenon of light path extension inside clouds described in Sect. 11.11 allows determination of trace gas concentrations inside clouds. By analysing the radiation originating from the bottom of a cloud, the light path extension inside the cloud can be determined (e.g. from O_4), while the column density of trace gases (e.g. NO_2) can be simultaneously determined. By comparing to a situation without cloud at a different time or point at the sky in the case of patchy cloudiness, the excess trace gas column due to the cloud can be determined. By dividing the excess trace gas column density by the extra light path in the cloud, the average trace gas concentration inside the cloud can be derived. This technique was used by Noxon (1976), Franzblau and Popp (1989), Winterrath et al. (1999), and Langford et al. (2004) to derive the

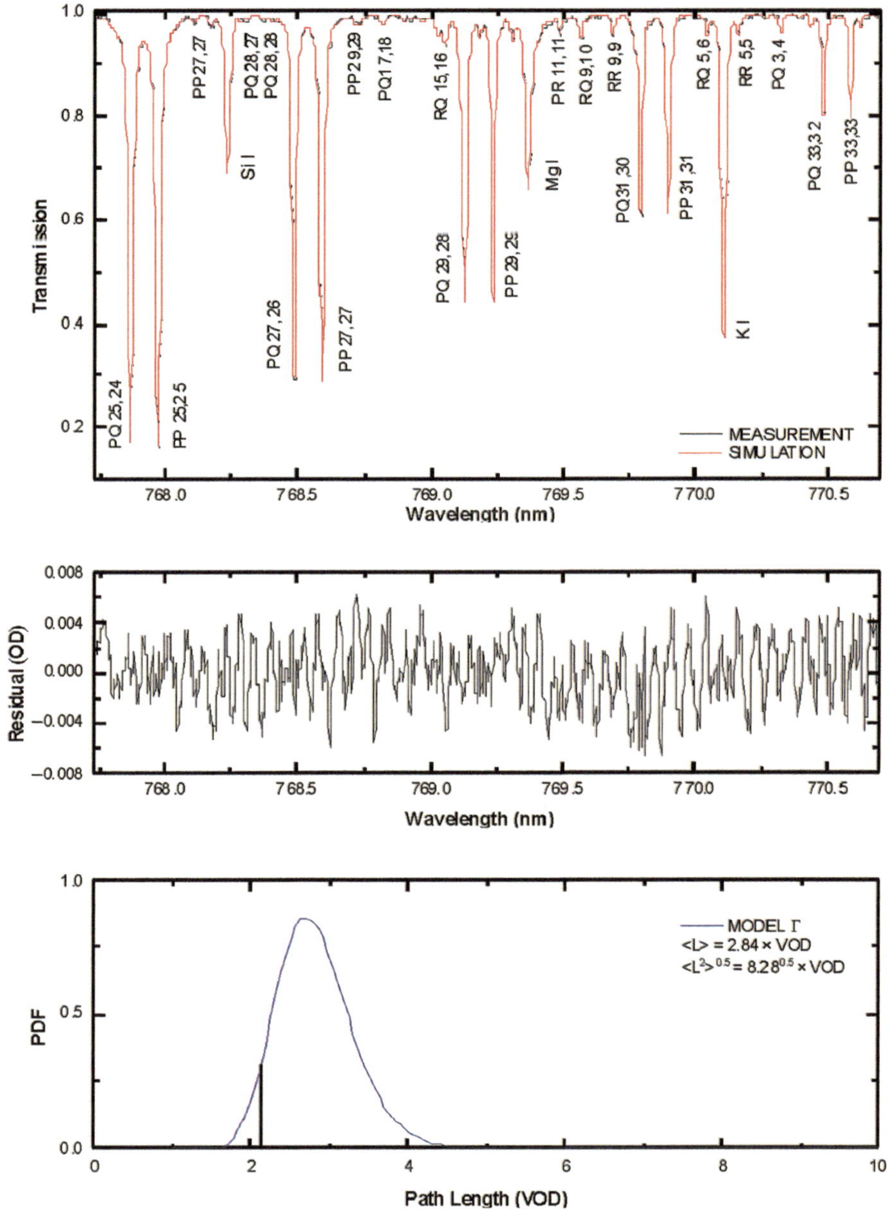

Fig. 11.72 Measured and modelled oxygen A-band spectrum (*upper panel*), inferred residual spectrum (*middle panel*) and inferred photon PDF for the observation between 12:32 and 12:33 UT over Cabauw/NL on 23 September, 2001. The *black vertical line* indicates the optical path for the direct sunlight. Photon path lengths are given in units of vertical atmosphere (VOD). Reprinted from Scholl, Ph.D. Thesis, 2006, University of Heidelberg

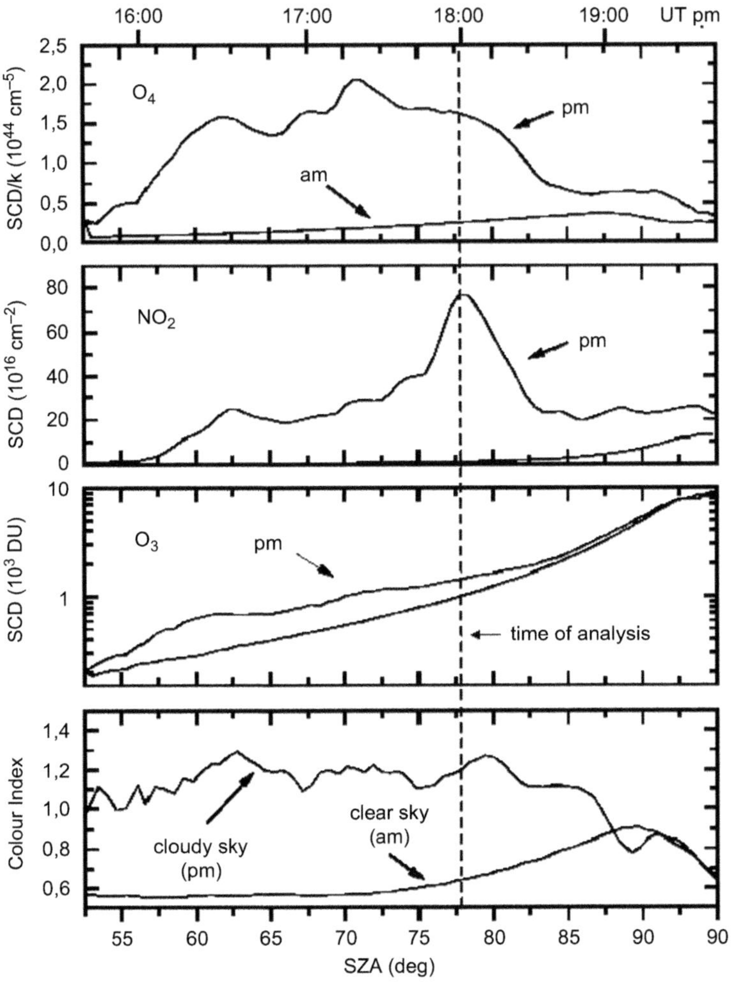

Fig. 11.73 The time development of the observed SCDs of various trace gases and the cloud index. The evening values during the cloudy period are clearly enhanced. Note the logarithmic scale for O_3. The *dashed line* indicates the time of analysis. Reprinted with permission from Winterrath et al. (1999), ©American Geophysical Union. All rights reserved

evolution of NO_2 during thunderstorm events and thus to derive the NO_x production by lightning. Figure 11.73 shows the evolution of the absorption due to O_4, NO_2, and O_3 in a thunderstorm cloud.

12 DOAS: Yesterday, Today, and Tomorrow

The use of DOAS to monitor and study the atmosphere has seen substantial growth in the past two decades. This trend will likely continue in the future. This final chapter will provide a historical perspective in order to provide an outlook on the future of DOAS, with respect to both, its further technological development, and upcoming scientific and monitoring applications. Following the approach taken throughout this book, we distinguish between active DOAS, which employs artificial light sources, and passive DOAS, which uses sunlight.

12.1 Passive DOAS Applications

Beginning with the direct sunlight measurements of total ozone in 1930s by Dobson, the passive DOAS method has branched out in a multitude of different applications (Fig. 12.1). Direct sunlight or moonlight applications have seen a slow resurgence, driven by the need to accurately validate satellite observations. Balloon borne DOAS measurements using direct sunlight (Ferlemann et al. 1998; Harder et al. 1998; Kreycy et al. 2013; Berthet et al. 2017) have given important insights into the vertical distribution of trace gases such as NO_2 and BrO. A few direct moonlight measurements of photolabile species, such as NO_3, have also been reported. Ground-based solar occultation observations are now regularly performed by Pandora instruments (Herman et al. 2009; Spinei et al. 2018), which have been deployed throughout the world (https://pandora.gsfc.nasa.gov/). Satellite solar occultation measurements were also performed by the now defunct SCIAMACHY (Scanning Imaging Spectrometer for Atmospheric Chartography) satellite instrument that was launched in 2002 (Burrows et al. 1991).

U. Platt and J. Stutz, *Differential Optical Absorption Spectroscopy*, Geophysics and Environmental Physics, https://doi.org/10.1007/978-3-662-68639-3_12

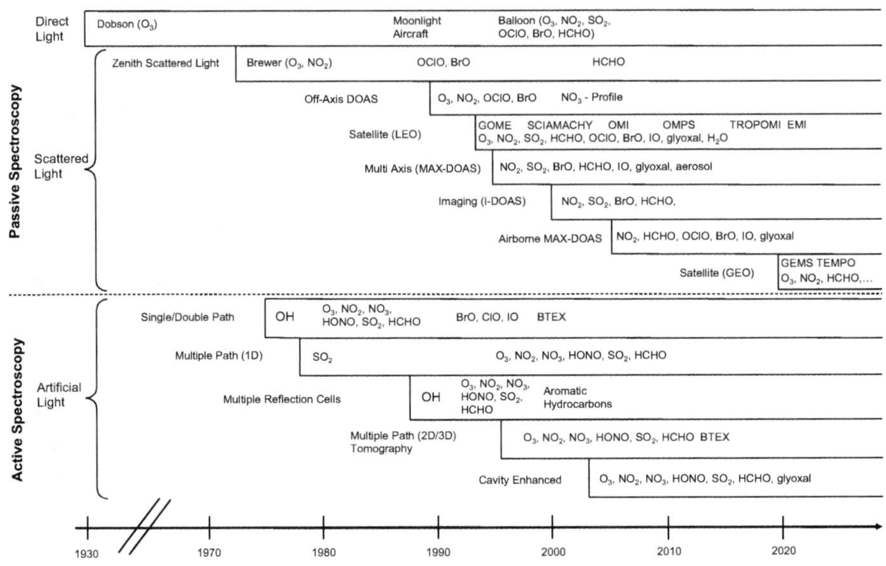

Fig. 12.1 The evolution of passive (scattered light) and active (artificial light) Differential Optical Absorption Spectroscopy since 1930

The measurement of scattered sunlight, which was initiated by Noxon in 1975 by aiming his instrument at the zenith, led to a better understanding of both, ozone and NO_2 in the stratosphere. Driven by the discovery of the Antarctic ozone hole in 1985, DOAS rapidly found its place as a key tool to study the halogen-catalysed ozone destruction mechanisms and the budget of nitrogen species. The measurements of BrO and OClO by Solomon et al. (1987b, 1989b) revealed that chlorine and bromine chemistry play a crucial role in the formation of the ozone hole. While still in use in monitoring networks, such as the Network for the Detection of Atmospheric Composition Change (NDACC), zenith DOAS has primarily been superseded by modern satellite observations, which provide similar information but across the entire globe. Followed by this success, variations of the zenith-light technique were developed by aiming the telescope to lower elevations and increasing the sensitivity, for example for NO_3 (e.g. Sanders et al. 1987). This advance culminated in the development of Multi-Axis DOAS, which uses multiple viewing angles to obtain trace gas and aerosol vertical distributions. MAX-DOAS also expands the use of passive DOAS systems from mostly stratospheric applications to the measurements of tropospheric trace gases. Due to its capabilities and ease of use, MAX-DOAS has become the most wide-spread DOAS-technique. The fact that MAX-DOAS instruments can be built small and inexpensively makes them ideal for observations in remote and difficult to access areas, such as polar regions, volcanoes, and the ocean. MAX-DOAS instruments have also been deployed on mobile platforms, such as ships, cars, and aircraft. The purpose of these measurements is predominately the study and monitoring of trace gases such as NO_2, SO_2, HCHO, BrO, IO, etc.

One of the most exciting developments is the expansion of DOAS to satellite-based measurements. With the launch of GOME (Global Ozone Monitoring Instrument) on board ERS-2 (Earth Research Satellite 2 by ESA) in 1995, a new era of DOAS measurements began. GOME's unique ability to measure multiple trace gases, such as O_3, NO_2, BrO, HCHO, and SO_2, on a global scale has expanded the scale of DOAS applications from local and regional applications, to global coverage. For the first time, species in the troposphere and stratosphere could be routinely monitored from space. The next generation of satellite instruments, OMI, TROPOMI, GEMS, TEMPO, etc., are now operational, providing a unique view of the global atmosphere.

The following sections will discuss established and emerging passive techniques that we expect to play a key role for the continued use of DOAS in the future.

12.1.1 MAX-DOAS

The rapid development of MAX-DOAS over the past decade has led to a large number of practical applications. Other uses of MAX-DOAS are in the development phase, and some are currently just ideas that may or may not be developed.

MAX-DOAS is an inexpensive and easy to use technique to monitor levels of air pollutants. MAX-DOAS offers two decisive advantages over traditional air pollution monitoring instruments: (1) It provides altitude averaged measurement of trace gas abundances in the lower atmosphere (typically a few hundred metres). (2) MAX-DOAS provides vertical trace gas profiles. Due to vertical and horizontal averaging, MAX-DOAS measurements are only weakly influenced by local effects and/or small-scale transport phenomena and thus offer substantial benefits for monitoring of total amounts of gases such as NO_2 or HCHO. In recent years, networks of MAX-DOAS instruments have been set up to provide long-term observations of air pollutants (Verhoelst et al. 2021; De Smedt et al. 2021; Kanaya et al. 2014). One can expect the further expansion of these networks for air quality monitoring.

Quantification of trace gas emissions is another expanding application. MAX-DOAS provides a powerful tool to directly measure fluxes of trace gases from point and area sources. An example of the emission monitoring capabilities is the remote measurement of SO_2 emitted from volcanoes. Small spectrometers have made it possible to develop miniaturised instruments that can easily be carried to volcano craters. Continuous measurements of SO_2 and other gases are often used in the monitoring of the state of a volcano, with the goal of predicting its next eruption. Due to the advantages of these instruments this technique has become widespread, for instance the NOVAC-network (see e.g. Galle et al. 2010) has more than 100 instruments installed at more than 40 volcanoes world-wide. Further growth of these monitoring networks can be expected.

Monitoring of pollutant emissions from industrial point sources and area sources has been demonstrated in various research applications. For example, ground-based and mobile MAX-DOAS has been used to monitor the composition of smoke stack plumes and NO_x emission from cities. These studies show that it is possible to use the path-averaging capabilities of MAX-DOAS to validate industrial and ship emissions in an operational way. First steps in this direction have been taken and it is likely that this application will find wider use in the future.

Finally, a potential extension of MAX-DOAS to the near IR (SWIR), which will allow measurements of CH_4, HCl, CO, and other species, is made possible by the introduction of Fabry-Pérot spectrographs with superior étendue and thus light throughput. Since MAX-DOAS relies on light scattering in the atmosphere, which decreases with λ^{-4} one would think that at longer wavelengths there is insufficient light (e.g. at $\lambda = 1500$ nm there will be $\approx$3 orders of magnitude less intensity compared to 300 nm). However, as pointed out by Kuhn et al. (2023) the better performance of Fabry-Pérot spectrographs could compensate for at least two orders of magnitude, thus potentially allowing MAX-DOAS observations in the near infrared.

12.1.2 Aerosol and Cloud Monitoring

While DOAS has primarily been used for the measurement of atmospheric trace gases, the ability to measure gases with constant mixing ratios in the atmosphere, i.e. O_2 and O_4, allows for new applications in cloud and aerosol monitoring. Several studies have demonstrated the ability of MAX-DOAS to measure aerosol optical depth, AOD, a quantity that is typically measured by dedicated spectroradiometers. The ability to determine aerosol extinction profiles in the lower troposphere has also been demonstrated, but the reliability of these observations still needs to be improved to become a standard MAX-DOAS product. Further progress is expected from adding information on polarisation to the measured intensity spectra. As shown by Tirpitz et al. (2022), more detailed information on the aerosol can be derived from measuring this additional parameter. The continued development of retrievals of aerosol extinction profiles, and potential aerosol optical properties, offers exciting opportunities to expand the capabilities of DOAS to monitor particulate matter in the troposphere.

12.1.3 Imaging DOAS

Imaging DOAS, also known as I-DOAS, is an extension of MAX-DOAS where the use of a two-dimensional array detector together with a horizontal scanner allows the measurement of trace gas slant column density images. Several studies have shown the potential to monitor volcano emissions and industrial emissions from smoke stacks and flares with I-DOAS. However, the full potential of I-DOAS for emission flux monitoring remains to be explored.

Recent developments make use of Fabry-Pérot interferometers as comb-filters matched to the periodic absorption structure of the target gas (e.g. SO_2, HCHO, or BrO). This approach provides acceptable specificity but at a much faster data acquisition rate compared to I-DOAS (Kuhn et al. 2014; Fuchs et al. 2021; Kuhn et al. 2019).

12.1.4 Passive DOAS Tomography

Another promising application of MAX-DOAS that remains to be explored is the tomographic observation of 2D or 3D trace gas distributions by passive instruments (tomography with active DOAS will be discussed in Sect. 12.2.6). The basic idea is to deploy many simple MAX-DOAS systems throughout an urban area or around an emission source. The instruments are then consecutively aimed in many different directions, so that viewing geometries cross at multiple locations (Frins et al. 2006). The tomographic analysis (Hartl et al. 2006; Johansson et al. 2009a; Casaballe et al. 2020) of these observations results in 2D or even 3D concentration fields that can be used to quantify emissions from pollution sources or initialise regional air quality models.

12.1.5 Satellite Instruments

One of the most exciting implementations of DOAS during the past three decades has been the success of the space-borne instruments. Starting with the GOME instrument, DOAS observations from space have changed our view of the global atmosphere. The next step in this development was taken in 2002 with the launch of SCIAMACHY. Its smaller ground-pixel size, the expansion of the usable wavelength range into the IR, and the ability to make limb and occultation measurements added new capabilities to space-borne DOAS. Since then a series of further Earth-observing satellites with sensors based on DOAS principles were launched. Examples are the instruments GOME-2a, GOME-2b, GOME-2c, OMI, TROPOMI (precursor of the forthcoming Sentinel-5 instrument), and others on Low Earth Orbit (LEO) satellites with sun-synchronous orbits. With these new instruments the trend of decreasing ground-pixel sizes is continuing (for instance, TROPOMI has a ground pixle size of 7 x 3.5 km^2 for most spectral bands). In addition, satellites on Geostationary Orbits (GEO) are coming on-line. These instruments record the diurnal variation of the atmospheric composition. Examples include the Geostationary Environment Monitoring Spectrometer, (GEMS, launched 2020) and Tropospheric Emissions: Monitoring of Pollution, (TEMPO, launched 2023), as well as the forthcoming Sentinel-4 instruments.

The most useful application for DOAS from space is the continuous monitoring of tropospheric pollution levels which, like today's meteorological satellites, provide maps of the 'chemical weather'. Geostationary satellites, with their 1–2 h time resolution, are a major improvement towards this application. As the data of these instrument becomes more available, new uses of these observations, such as emission monitoring, will become more widespread. The higher spatial resolution of modern satellite instruments also allows air quality monitoring at the neighborhood scale, which will help highlight societal inequities in air pollution exposure across the world.

There are many more targets of research and observation to be explored as new satellite capabilities are introduced. Future hardware developments include swarms of small satellites, which could be based on the results of ongoing spectrograph miniaturisation (Platt et al. 2021).

12.2 Active DOAS Applications

The first active DOAS instruments were developed in the 1970s as research tools for the measurement of various radicals, in particular OH. To develop systems capable of measuring O_3, NO_2, and SO_2, as well as NO_3 and HONO, the 'broadband' laser used in these initial systems was soon replaced by thermal light sources (e.g. Xe arc lamps) or LEDs, which give a much broader spectrum and are technologically simpler. These early instruments were based on opto-mechanical 'slotted disk' detectors, which have now been replaced by solid state detector arrays. These new detectors increased the measurement frequency and allowed the use of longer light paths, thus improving the detection limits. The introduction of combined transmitter/receiver assemblies with retroreflectors like the coaxial telescope or fibre DOAS setups greatly simplified the instruments and their deployment in the field. The desire to fold the long light path in a small volume led to the development of multi-reflection DOAS systems.

The number of trace gases that are measured with DOAS has expanded over the past decade, and now includes air pollutants (O_3, NO_2, HCHO, HONO, SO_2, NO), atmospheric radicals (OH, NO_3), all tropospheric halogen oxides, many atmospheric hydrocarbons, glyoxal, and ammonia.

Advances in active DOAS over the past two decades have been made through technological developments such as the use of cavities and new light sources. In addition, classical longpath DOAS has found a new practical application in fenceline monitoring, which has led to its more widespread use. In the following sections we will discuss many of these advances and attempt to identify future directions for active DOAS.

12.2.1 New Trace Gases

The number of species measured by active DOAS has been increasing over the past few decades, in large part due to the enhancement of detection limits and new applications for this technique. However, there is still about an order of magnitude in sensitivity to be gained in active DOAS applications and it is likely that new compounds will be available for observation. For example, very recent studies show that isoprene and 1,3-butadiene can be measured between 200–230 nm (Gao et al. 2021; Zhang et al. 2023). Other potential candidates include polycyclic aromatics and oxygenated hydrocarbons such as acrolein or ketones.

12.2.2 Near-Infrared Measurements

A number of trace gases that play an important role in atmospheric chemistry and climate research, such as CH_4 and CO, absorb in the near-infrared wavelength region. While the observation of these compounds by DOAS has been implemented from satellites, the expansion of active DOAS to the near infrared has not yet been realised. It is possible that more trace gases can be observed in this wavelength region, including hydrocarbons, which can be detected by overtone spectra of the C–H stretch between 1 and 2 μm.

12.2.3 Aliphatic and Aromatic Hydrocarbon (BTEX) Monitoring

The measurement of the ambient concentrations of aromatic hydrocarbons, such as benzene, ethylbenzene, toluene, and xylenes (known as BTEX), has been demonstrated in research applications. New longpath DOAS instruments based on UV LEDs have overcome challenges in the accurate description of strong absorptions of oxygen and oxygen dimers that overlay absorptions by aromatic hydrocarbons. These systems have been used to study BTEX in neighborhoods adjacent to refineries (Stutz et al. 2016). Operational BTEX monitoring has been implemented by the South Coast Air Quality Management District (SCAQMD) which operates several extractive DOAS instruments based on multi-reflection White cells to monitor aromatic hydrocarbons in environmental justice neighborhoods in southern California.

12.2.4 Air Pollution Monitoring

A number of commercial active DOAS instruments are available on the market to monitor pollutants such as O_3, SO_2, and NO_2. While some instruments have been employed by air quality agencies, the potential of DOAS for monitoring air pollutants has not been fully exploited. DOAS offers a number of advantages over other methods that are often not recognised. Due to the simultaneous measurement capability, one DOAS instrument can replace a whole set of instruments, which are individually dedicated to the measurement of O_3, SO_2, NO_2, and HCHO. DOAS is an absolute method that does not need to be calibrated, therefore there are no additional costs associated with calibrations. DOAS instruments are also easy to automate, thus remote control can be implemented. Finally, DOAS measurements provide a 'photograph' of the chemical composition of the atmosphere which can be used as incontestable evidence in court.

An important step towards air pollution monitoring is the development of certification and verification standards by the German VDI for active (Kommission Reinhaltung der Luft KRDL) VDI-Normenausschuss (2013) and passive (Kommission Reinhaltung der Luft KRDL) VDI-Normenausschuss (2019) DOAS—based instruments. Certification efforts outside of Germany would be highly advantageous and could further expand the use of DOAS. Ultimately, networks of active and passive DOAS instruments could supplement existing air quality monitoring networks.

12.2.5 Fenceline Monitoring

Fenceline monitoring, i.e. measuring trace gases along the fence of an industrial facility, has become an important application of active DOAS. The State of California, USA and its local air quality management districts, such as the SCAQMD, introduced legislation in 2017 and 2018 that requires the use of fenceline monitoring of BTEX and other air toxics around refineries. Refineries in California are now surrounded by multiple active longpath DOAS instruments that provide realtime BTEX data. These systems act as early detection systems of BTEX releases from refineries. The use of this many active DOAS instruments has also triggered new interest in developing methodologies to test and validate DOAS instruments to ensure their accuracy in detecting BTEX. Fenceline BTEX monitoring by DOAS can also be found at refineries in other countries, but to our knowledge the efforts in California are the first example of DOAS being used to fulfill regulatory air monitoring requirements. There are currently discussions on the expansion of fenceline monitoring to refineries in other states in the USA, and use of this approach for other industrial facilities. There is thus great potential to further expand the use of DOAS to provide crucial information to neighborhoods adjacent to large petrochemical facilities.

Efforts are underway to develop other DOAS based approaches to monitor emissions inside industrial facilities before they reach the fenceline. Such an Optical Tent, which consists of multiple LP-DOAS light paths covering the area of a refinery, is being tested in Los Angeles, USA. The Optical Tent identifies elevated BTEX levels inside the refinery, allowing the operator to quickly mitigate equipment malfunctions and thus reduce emissions.

12.2.6 Active DOAS Tomography

Active DOAS can be used to make tomographic measurements. Setting up multiple active DOAS instruments with multiple intersecting light beams, together with established tomographic methods, allows the determination of 2D or even 3D concentration fields. For example, Stutz et al. (2016); Olaguer et al. (2017); Olaguer (2020) showed how a setup of two long-path DOAS instruments with four reflectors each can be used to monitor the spatial distribution of BTEX in a neighborhood. The challenge of setting up and operating such systems is outweighed by the benefits of providing spatially resolved observations over an extended area and the ability to identify pollution hot-spots and sources. Further development of this technology, for example as leak detectors in industrial facilities, would be highly beneficial.

12.2.7 Range Resolved Technology/Broadband LIDAR

Another application that has not been extensively explored is the use of the DOAS approach in range resolved techniques, i.e. LIDAR. In contrast to classical LIDARs, which are based on one or multiple lasers, a DOAS LIDAR would require a high-intensity broadband pulsed light source, that allows the detection of an entire spectrum. While the detection capabilities for time resolved measurements of the returning light are available with modern CCD detectors, a suitable light source has not yet been found. An approach, which circumvents the light-source problem, relies on the "Scheimpflug technique" (see Sect. 7.8.5). In this technique the beam of a continuous light source is directed to the zenith (or another direction), while a narrow-field of view detector system observes the beam under a small angle. A DOAS LIDAR could give unique insights into the vertical distribution of trace gases and, by using horizontal viewing direction, could also be used to observe emissions.

12.2.8 Optical Resonator Techniques (CE-DOAS)

Optical resonator techniques effectively allow point measurements with the advantages of the long-path DOAS approach. Compared to multi-reflection optics, optical resonators allow much longer light paths and can routinely pack a several kilometer long light path in a 50 cm base-path resonator that fits into a 19-inch box, such as those used for most in-situ instruments (Sect. 7.7.3). Because of these advantages CE-DOAS instruments are in widespread use for the measurement of all species accessible to DOAS. The small size of these instruments also allows the use of CE-DOAS in laboratory settings. With the availability of commercial instruments one can expect that CE-DOAS will find wider use in atmospheric field and laboratory research, as well as air quality monitoring.

12.3 Development of the Underlying Technology

In the past, the advancement of DOAS has often gone hand in hand with the availability of new technology. For example, the multiplex advantage of new solid-state detectors, such as photodiode arrays, has reduced the noise in the spectroscopic detection, thus allowing for longer light paths in active DOAS applications, and shorter exposure times in passive DOAS measurements. The availability of inexpensive miniaturised spectrometers (Sect. 7.12.5 and e.g. Yang et al. 2021) has made it possible to build small and light MAX-DOAS instruments that can be carried to a volcano crater to routinely monitor SO_2, BrO, OClO, and other gases.

Improvements could also come from better databases, e.g. for trace gas reference spectra and Fraunhofer catalog spectra. The latter would allow absolute SCD-determinations in passive DOAS. Many such examples can be found in the history of DOAS. It is thus valuable to speculate how new technology will influence the development of DOAS in the future.

12.3.1 New Light-Sources

Illumination technology has seen an increased use of light emitting diodes (LEDs). The high brightness of modern LEDs makes them a natural choice for DOAS applications. Advantages include the low power consumption, longevity, and high stability of these sources. LEDs are now used in most CE-DOAS instruments and the availability of UV LEDs in the 250–300nm range has allowed the development of new LP-DOAS instruments to measure BTEX. As LED technology further advances, one can expect improvements in DOAS instrument performance, and new applications, such as expanding the use of LEDs to longer (IR) and shorter (UV) wavelengths.

In addition, the traditional Xe-arc lamp has been superseded by the Laser Driven Light Source (LDLS) technology, which offers higher brightness, more stability, and a longer lifetime.

Tunable diode lasers (TDLs) have also seen a rapid development in recent years. The introduction of the quantum cascade lasers, which can be operated at ambient temperatures, makes these systems easier to use and also less expensive. The scanning ability results in absorption spectra that can be analysed using DOAS techniques. The combination of TDLS and DOAS would open novel applications in the infrared wavelength region.

The development of a broadband pulsed LIDAR that is based on DOAS depends on the availability of suitable light sources. Further development of lasers, flash lamps, and other sources could make this instrument possible.

12.3.2 New Detectors

The development of photodiode arrays, and more recently CCD and CMOS detectors, has considerably impacted DOAS. Both expensive high-quality and low-cost CCD detectors are available today. The development of this technology is, however, still advancing, and new detectors become available every year. For active DOAS, high capacity (large full well capacity) detectors are not yet available, imposing limitations on the achievable noise levels. Faster readout speed would also improve the performance of active DOAS instruments.

DOAS has mostly been restricted to the UV and visible wavelength range. The upper wavelength limits were initially determined by the availability of suitable photomultiplier tubes, and later by the bandgap limit of silicon used in photodiode arrays and CCDs. New semiconductor technology has led to the development of solid-state detector arrays based on Indium-Gallium-Arsenide (InGaAs) that also work above the silicon-limit at wavelength up to 2.4 μm. These detectors open a new spectral window for DOAS. As the development of these detectors advances, and their cost decreases, DOAS can expand its IR capabilities.

12.3.3 New (and Old) Spectrographs

It is likely that the future will see entirely new spectrograph designs (Aas et al. 2019). Grundmann (2019a, b) proposes a special type of diode array as the only element of a spectroscopic instrument. The semiconductor band-gap of the detector pixel gradually increases with the pixel number. Each consecutive pixel thus observes a slightly smaller spectral range. The difference of the signal between adjacent pixel is equivalent to the signal measured in a small wavelength interval, and a spectral resolution of to $\approx$1 nm could be achieved. This device would function as a ultra-compact spectrograph.

Very traditional spectrograph designs, such as the Fabry Pérot interferometer (Perot and Fabry 1899) could come into more wide-spread use (Sect. 7.4.3), since new manufacturing methods of super-polished surfaces and advanced assembly allow high-performance instruments which could give superior light throughput in standard DOAS applications. This could lead to improved imaging spectrometers (e.g. Kuhn et al. 2014, 2019), and provide very compact picometer resolution instruments (Kuhn et al. 2021).

12.3.4 New Software

A goal for future DOAS applications is to further advance the automation of DOAS instruments. While advances have been made in the control of instruments and the long-term operation, improvements in the analysis software with respect to the fully automated extraction of trace gas concentrations from the absorption spectra is needed. To reduce the need for initial manual spectral evaluation, and to simplify the process of data reduction, methods must be developed to automatically recognise changes in the atmosphere and the instrument, and correct for these effects. In addition, algorithms for automatic quality control and improved error calculations are needed. These improvements will benefit the use of DOAS as a monitoring technique, where ease of operation and reliability is a key component.

Further advances in software development are also expected for MAX-DOAS applications. Radiative transfer models required for the interpretation of MAX-DOAS data are continuously being improved. Several algorithms to invert the observations in order to derive vertical profiles of trace gases and aerosol are in use. Considerable advancement in these retrieval methods, for instance using better descriptions of aerosol scattering and the inclusion of polarisation information, can be expected.

12.3.5 Improved System Design

The design of DOAS instruments can proceed in several directions, including (1) The improvement of current detection limits and the expansion to new wavelength regions. (2) The development of smaller, more automated, and less expensive DOAS systems. (3) The development of entirely new applications such as the observation of chlorophyll fluorescence.

The detection limits of current active DOAS instruments are about one order of magnitude above the physical limits imposed by photon noise. There is thus great potential in improving these instruments to measure lower trace gas levels, improve the accuracy of the measurements, and expand the pool of observable species.

The development of miniaturised spectrometers has allowed the construction of smaller DOAS instruments and it is expected that this trend will continue. The lower power consumption of the systems also allows for operation in remote locations. The low price of miniature spectrometers also makes it very likely that the number of DOAS instruments will increase, and that more observation networks will be established. Ultimately, one can imagine that further miniaturised ‘personal’ DOAS instruments may become a tool for trace gas monitoring for the interested scientist and personnel working in air quality control.

The use of DOAS on mobile platforms has grown in recent years and, with the decrease of instrument size, we expect an increased use of this approach. In particular, the use of miniature DOAS instruments on small drones offers exciting opportunities.

12.4 Conclusions

The DOAS technique plays an important role in observing the earth’s atmosphere and in providing insights into the mechanism controlling its composition. Technological development and introduction of new applications of DOAS will strengthen its contribution to atmospheric research and monitoring and ensure its continued success.

Spectral Positions of Emission Lines from Calibration Lamps and Lasers

A

Emission lines are taken from the following internet site:
https://lweb.cfa.harvard.edu/amp/ampdata/kurucz23/sekur.html

A.1 Cadmium Lines

Vaccum wavelength/nm	Air wavelength /nm	Log_gf	Element (name)	E_lower_lev. /cm^{-1})	J upper
214.5068	214.4393	−0.108	Cd II	0.000	1.5
226.5720	226.5019	−0.340	Cd II	0.000	0.5
228.8724	228.8018	0.336	Cd I	0.000	1.0
230.7334	230.6624	−0.980	Cd I	30656.130	0.0
231.3464	231.2753	−0.468	Cd II	46618.550	2.5
233.0003	232.9288	−0.960	Cd I	31826.996	1.0
257.3699	257.2928	−0.194	Cd II	44136.080	0.5
283.7726	283.6892	−0.980	Cd I	30114.017	1.0
288.1581	288.0736	−0.760	Cd I	30656.130	2.0
298.1492	298.0622	−0.410	Cd I	31826.996	3.0
340.4629	340.3652	−0.350	Cd I	30114.017	1.0
346.7195	346.6202	0.140	Cd I	30656.130	2.0
346.8650	346.7657	−0.360	Cd I	30656.130	1.0
361.1540	361.0510	0.410	Cd I	31826.996	3.0
361.3905	361.2875	−0.300	Cd I	31826.996	2.0
467.9458	467.8149	−0.970	Cd I	30114.017	1.0
480.1256	479.9914	−0.400	Cd I	30656.130	1.0
508.7242	508.5824	−0.230	Cd I	31826.996	1.0
644.0249	643.8470	−0.010	Cd I	43692.474	2.0

U. Platt and J. Stutz, *Differential Optical Absorption Spectroscopy*, Geophysics and Environmental Physics, https://doi.org/10.1007/978-3-662-68639-3

A.2 Mercury Lines

Wl_vac/nm	Wl_air/nm	log_gf	Element (name)	E_lower_lev ./cm^{-1}
226.0960	226.0260	-0.532	Hg II	51485.000
253.7279	253.6517	−1.382	Hg I	0.000
257.7057	257.6285	−1.714	Hg I	39412.300
275.3591	275.2777	−1.712	Hg I	37645.080
284.8516	284.7679	−0.101	Hg II	60608.000
289.4440	289.3592	−1.267	Hg I	39412.300
292.6262	292.5406	−1.481	Hg I	44042.977
296.8145	296.7278	0.470	Hg I	37645.080
302.2375	302.1495	−0.270	Hg I	44042.977
312.6571	312.5665	0.060	Hg I	39412.300
313.2453	313.1545	−0.040	Hg I	39412.300
313.2746	313.1838	−0.040	Hg I	39412.300
334.2434	334.1473	−1.100	Hg I	44042.977
365.5873	365.4832	0.140	Hg I	44042.977
366.3918	366.2875	−0.430	Hg I	44042.977
366.4318	366.3275	0.040	Hg I	44042.977
398.5017	398.3890	−1.730	Hg II	35514.000
404.7702	404.6559	−0.863	Hg I	37645.080
407.8978	407.7827	−1.470	Hg I	39412.300
435.9548	435.8323	−0.383	Hg I	39412.300
491.7435	491.6062	−1.506	Hg I	54068.781
546.2249	546.0731	−0.137	Hg I	44042.977
577.1193	576.9593	0.630	Hg I	54068.781
579.2266	579.0660	0.710	Hg I	54068.781

A.3 Hydrogen Lines

Wl_vac/nm	Wl_air /nm	Log_gf	Element (name)	E_lower_lev. ref./cm^{-1}
102.5722		−0.801	H I	0.000
121.5671		−0.080	H I	0.000
397.1195	397.0072	−0.993	H I	82259.105
410.2892	410.1734	−0.753	H I	82259.105
434.1682	434.0462	−0.447	H I	82259.105
486.2681	486.1323	−0.020	H I	82259.105
656.2822	656.1010	0.710	H I	82281.662
656.4610	656.2797	0.710	H I	82259.105
901.7388	901.4913	−0.901	H I	97492.302
923.1550	922.9017	−0.735	H I	97492.302
954.8592	954.5973	−0.540	H I	97492.302
1005.2128	1004.9373	−0.303	H I	97492.302
1094.1089	1093.8093	0.002	H I	97492.302

A.4 Neon I Lines

Wl_vac/nm	Wl_air/nm	Log_gf	Element (name)	E_lower_lev ./cm^(-1)
576.6018	576.4419	−0.370	NE I	149657.042
585.4110	585.2488	−0.460	NE I	135888.717
590.4098	590.2462	−0.423	NE I	150858.508
594.6481	594.4834	−0.120	NE I	134041.840
607.6019	607.4337	−0.470	NE I	134459.287
609.7851	609.6163	−0.270	NE I	134459.287
614.4763	614.3063	−0.350	NE I	134041.840
633.6180	633.4428	−0.310	NE I	134041.840
638.4756	638.2991	−0.260	NE I	134459.287
640.4016	640.2246	0.360	NE I	134041.840
650.8326	650.6528	0.030	NE I	134459.287
653.4687	653.2882	−0.440	NE I	134818.640
660.0776	659.8953	−0.360	NE I	135888.717
668.0120	667.8276	−0.400	NE I	135888.717
671.8897	671.7043	−0.310	NE I	135888.717
693.1379	692.9467	0.030	NE I	135888.717

A.5 Zinc Lines

Wl_vac/nm	Wl_air/nm	Log_gf	Element (name)	E_lower_lev./ cm^{-1}
202.6136	202.5483	−0.086	Zn II	0.000
206.2663	206.2004	−0.377	Zn II	0.000
210.0604	209.9937	−0.032	Zn II	49355.040
213.9247	213.8573	0.161	Zn I	0.000
250.2743	250.1989	−0.495	Zn II	48481.000
255.8715	255.7948	−0.194	Zn II	49355.040
277.1671	277.0853	−0.576	Zn I	32501.421
280.1688	280.0862	−0.303	Zn I	32890.352
328.3274	328.2328	−0.377	Zn I	32311.350
330.3535	330.2584	−0.057	Zn I	32501.421
330.3892	330.2941	−0.534	Zn I	32501.421
334.5977	334.5015	0.246	Zn I	32890.352
334.6532	334.5570	−0.502	Zn I	32890.352
468.1444	468.0134	−0.815	Zn I	32311.350
472.3474	472.2153	−0.338	Zn I	32501.421
481.1872	481.0528	−0.137	Zn I	32890.352
636.4097	636.2338	0.150	Zn I	46745.413

Absorption Spectra of Molecules Measurable by DOAS

B

In Appendix B, we summarise the spectral features and approximate wavelengths of strong absorption structures of the most important species that are observed by DOAS (see Table B.1 for an overview). Absorption cross-sections were measured during several studies; note that we give no recommendation as to which one to use. Comprehensive lists of absorption cross-section data can be found in compilations such as the Mainz spectral atlas (www.uv-vis-spectral-atlas-mainz.org, Keller-Rudek et al. 2013).

B.1 Nitric Oxide, NO

There are four UV transitions of the NO molecule:

γ-band: $A^2\Sigma^+ \leftarrow X^2\Pi$
β-band: $B^2\Pi^+ \leftarrow X^2\Pi$
δ-band: $C^2\Sigma^+ \leftarrow X^2\Pi$
ϵ-band: $D^2\Pi^+ \leftarrow X^2\Pi$

The energy difference of the γ-band corresponds to wavelengths near 214 and 226 nm (e.g. Tajime et al. 1978; Mellqvist and Rosén 1996); the wavelengths of the other bands are even shorter, and they are in a region where the O_2 absorption does not permit absorption paths exceeding a few metres in the atmosphere. The γ-band has been used successfully to measure the abundance of NO in the atmosphere (Gall et al. 1991; Edner et al. 1992; Edner et al. 1990; Mount et al. 2002), in smoke stacks (Gall et al. 1991; Mellqvist et al. 1992; Mellqvist and Rosén, 1996), and in car exhaust (Perner et al. 1987a). It should be noted that strong atmospheric absorption and scattering does not allow light paths longer than about 1 km in this wavelength range. Moreover, the absorption spectrum of NO consists of many narrow rotational lines, which are not resolved at spectral resolutions of typical DOAS instruments

U. Platt and J. Stutz, *Differential Optical Absorption Spectroscopy*, Geophysics and Environmental Physics, https://doi.org/10.1007/978-3-662-68639-3

Table B.1 Substances detectable by active UV/vis absorption spectroscopy. Detection limits were calculated for a minimum detectable optical density of 5×10^{-4}

Species	Wavelength interval (nm)	Approx. diff. absorption cross-section 10^{-19} cm^2/molec.	Transition	Detection limit ($L = 5$ km, 20°C) ppb
SO_2	200–230	65	$c^1B_2 \leftarrow X^1A_1$	0.03[a]
	290–310	5.7	$A^1B_1 \leftarrow X^1A_1$	0.06
	340–390	0.005	$a^3B_2 \leftarrow X^1A_1$	80
CS_2	320–340	0.4		0.9
NO	200–230	24	$A^2\Sigma^+ \leftarrow X^2\Pi,$	0.08[a]
NO_2	330–500	2.5	$A^2B_1 \leftarrow X^2A_1$	0.15
NO_3	600–670	200	$\tilde{B}^2E' \leftarrow \tilde{X}^2A_2$	0.002
NH_3	200–230	180	$A^1A_2 \leftarrow X^1A_1$	0.009[a]
HNO_2	330–380	5.1	$A^1A \leftarrow X^1A$	0.07
O_3	300–330	0.1	$C \leftarrow X^1A_1$	3,7
CH_2O	300–360	0.48	$A^1A_2 \leftarrow X^1A_1$	0.8
CHOCHO	400–480	10	$A^1A_u - X^1A_g$	0.04
ClO	260–300	35	$X^2\Pi \leftarrow A$	0.01
BrO	300–360	104	$X^2\Pi \leftarrow A$	0.004
IO	400–470	170	$X^2\Pi \leftarrow A$	0.002
OClO	325–435	116	$A^2A_2 \leftarrow X^2B_1$	0.004
OBrO	390–610	150	$C^2A_2 \leftarrow X^2B_1$	0.003
OIO	470–670	270	$^2B_2 \leftarrow X^2B_1$	0.002
[a]Benzene	240–270	21.9	$A^1B_{2u} \leftarrow X^1A_{1g}$	0.17
[a]Toluene	250–280	12.8		0.3
[a]Xylene (o/m/p)	250–280	2.1/6.6/20.3		1.8/0.6/0.2
[a]Phenol	260–290	198		0.02
[a]Cresol (o/m/p)	250–280	20.1/31.8/87.2		0.18/0.11/0.04
[a]Benzaldehyde	280–290	44		0.08

[a] Due to strong absorption/scattering only L = 1 km assumed

(≈0.5 nm), thus saturation effects can already occur at NO levels found in polluted areas (e.g. reducing the apparent NO SCD by 20% at S = $10^{17} cm^{-2}$, see Lutz 2019). Literature data also appear to suffer from saturation effects (see e.g. Trad et al. 2005).

B.2 Nitrogen Dioxide, NO_2

The NO_2 molecule has a continuous absorption below 200 nm with a maximum at 170 nm, a short-wave UV system with band-heads at 235.0 and 249.1 nm ($B^2B_2 \leftarrow X^2A_1$ transition), and an extended system with absorptions from the UV (about 320 nm) through the green spectral region ($A^2B_1 \leftarrow X^2A_1$ transition). This latter system is responsible for the brown colour of gaseous NO_2; it is almost exclusively used for the DOAS measurement of NO_2 in the atmosphere and allows detection limits on the order of ~50 ppt at 10 km light path length.

There are numerous measurements of the NO_2 absorption cross-section (e.g. Hall and Blacet 1952; Schneider et al. 1987; Davidson et al. 1988; Harwood and Jones 1994; Coquart et al. 1995; Merienne et al. 1995; Vandaele et al. 1996, 1998 2003; Harder et al. 1997a; 1997b; Burrows et al. 1998; Voigt et al. 2002; Vandaele et al. 2003). The temperature dependence of the cross-section was studied by Davidson et al. (1988), Harder et al. (1997a; 1997b), Burrows et al. (1998), Voigt et al. (2002), Vandaele et al. (2003), and Nizkorodov et al. (2004). A review can be found at Orphal (2003).

Points to watch when determining NO_2 absorption cross-sections are: (1) the formation of N_2O_4 in equilibrium with NO_2 (e.g. Harwood and Jones 1994, Platt and Kuhn 2019):

$$NO_2 + NO_2 \Leftrightarrow N_2O_4$$

N_2O_4 has a smooth absorption spectrum (e.g. Hall and Blacet 1952). While it is unimportant in the atmosphere, at the high concentrations typically used in laboratory determinations of the NO_2 cross-sections, it might form in appreciable amounts. It will thus lower the apparent NO_2 cross-section. (2) There might be heterogeneous formation of HONO (see Sect. 2.6.3) in NO_2 containing vessels (e.g. absorption cuvettes, see Platt and Kuhn 2019) unless they are kept very dry. Besides lowering the amount of NO_2, the highly structured HONO spectrum (see Sect. B.11) may contaminate the NO_2 spectrum.

B.3 Ammonia, NH_3

While the ground state of NH_3 is tetrahedral with C_{3v} symmetry, the electronically excited states are planar with D_{3h} symmetry. There are extended vibrational progressions of the $A^1A_2 \leftarrow X^1A_1$ and $A^3A_2 \leftarrow X^1A_1$ transitions. The $A^1A_2 \leftarrow X^1A_1$ transition gives rise to a band system from 216.8 nm to about 170 nm, the longer wavelength bands of which are used for observation of ammonia by DOAS (e.g. Edner et al. 1990; Mellqvist et al. 1992; Mellqvist and Rosén 1996; Mount et al. 2002). Cross section data from room temperature to 590K were reported by Weng et al. (2021). Note that the strong atmospheric absorption and scattering does not allow light paths longer than about 1 km in this short UV wavelength range, as in the case of NO (Sect. B.1).

B.4 Formaldehyde, HCHO

The HCHO molecule undergoes a change in configuration upon transition from the electronic ground state (X^1A_1 plane molecule with C_{2v} symmetry) to the exited state ($X^2A_1C_s$ symmetry). As a molecule with $n = 4$ atoms, formaldehyde has $3n - 6 =$

6 different vibrational modes. Absorptions due to formaldehyde ($A^1A_2 \leftarrow X^1A_1$ transition) are found between 260–360 nm (Cantrell et al. 1990; Meller and Moortgart 2000; Pope et al. 2005a; 2005b). High resolution UV absorption cross-sections were reported by Pope et al. (2005b), Smith et al. (2006); Brauers et al. (2007).

This structure has been used by numerous atmospheric studies to determine the atmospheric formaldehyde concentration by active and passive DOAS techniques (Chaps. 10 and 11).

B.5 Glyoxal, CHOCHO

The glyoxal (CHO-CHO) molecule exhibits a structured $A^1A_u - X^1A_g$ and $a^3A_u - X^1A_g$ band system in the blue spectral region (about 400–480 nm) and the UV region (250–350 nm), respectively, and a diffuse B–X transition. The absorption spectrum of the molecule was studied by several authors (e.g. Plum et al. 1983; Zhu et al. 1996; Orlando and Tyndall, 2001; Horowitz et al. 2001; Volkamer 2001; Chen and Zhu 2003). High-resolution absorption cross-sections were reported by Volkamer et al. (2005a). Methylglyoxal (CH_3CO-CHO) also shows structured (though weaker) absorption in this spectral range and can be measured. A number of measurements using the bands of glyoxal and in some cases also of methylglyoxal in the blue spectral region have been reported (Volkamer et al. 2001 2005b; Wittrock et al. 2006; Sinreich et al. 2007; Volkamer et al. 2015; Gratsea et al. 2016; Cao et al. 2018; Benavent et al. 2019; Liu et al. 2019; Lerot et al. 2021). The spectrum exhibits very narrow band spectral structures, which are not normally resolved by DOAS instruments; thus, apparent deviations from Lambert–Beer's law can occur, as demonstrated in Fig. B.1. Many of the previous determinations (e.g. Plum et al. 1983; Zhu et al. 1996; Orlando and Tyndall 2001) did not fully resolve the spectral structures (Fig. B.1).

B.6 Sulfur Dioxide, SO_2

The sulfur dioxide molecule exhibits three band systems with narrow, discrete vibrational bands.

1. The strongest system, in the range of 180–235 nm, is due to transitions from the ground state to several electronic states, in particular the $c^1B_2 \leftarrow X^1A_1$ transition.
2. A relatively strong system from 260 to 340 nm is due to the $A^1B_1 \leftarrow X^1A_1$ transition.
3. A weak system in the spectral range from 340 to 390 nm corresponds to the spin-forbidden $a^3B_2 \leftarrow X^1A_1$ transition.

DOAS measurements typically make use of the spectral range from about 290 to 310, where the strongest bands of the $A^1B_2 \leftarrow X^1A_1$ transition are located (Brand and Srikameswaran 1972; Brand et al. 1972; Hamada and Merer 1974; Brand and Nanes

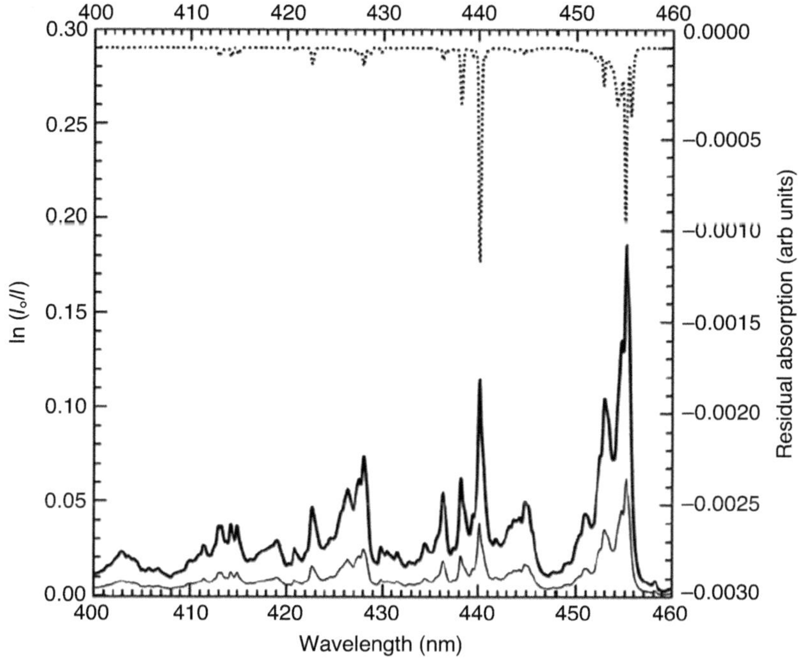

Fig. B.1 Apparent deviation from Lambert–Beer's law for glyoxal. Two high-resolution spectra corresponding to column densities of 10^{17} molec./cm^2 (*thin line*) and 3×10^{17} molec./cm^2 (*thick line*) were convoluted to 0.25 nm spectral resolution (FWHM); residual structure (*dotted line*) after the two spectra were scaled to match column density and subtracted. This distortion of the spectrum can be understood from the considerable ro-vibrational structure, which, when observed at low spectral resolution, causes the apparent absorption cross-section to become lower at higher column densities. Reprinted with permission from Volkamer et al. (2005a), ©The Royal Society of Chemistry (RSC) on behalf of the European Society for Photobiology and the European Photochemistry Association. All rights reserved

1973; Marx et al. 1980; Wu and Judge 1981; Brassington 1981; Brassington et al. 1984; Thomsen 1990; Vandaele et al. 2009; Blackie et al. 2011). In polluted air, SO_2 dominates the absorption structure in this spectral range. Note that the structures of the O_3 Huggins bands (see Sect. B.8) are similar to those of SO_2 in spectral width (though about two orders of magnitude weaker), and thus may cause interferences in clean air if not taken into account, particularly at low spectral resolution. The SO_2 absorption in the 360 to 390 nm range is, despite being rather weak (see e.g. Hermans et al. 2009), interesting for situations where high SO_2 concentrations prevail, e.g. in volcanic plumes (Bobrowski et al. 2010). The SO_2 absorption structures around 230 nm and 300 nm have been used in a large number of active and passive DOAS studies (Chaps. 10 and 11).

B.7 Carbon Disulfide, CS_2

The CS_2 molecule exhibits two strongly structured absorption systems:

1. Between 190 and 210 nm,
2. In the near UV between 290 and about 360 nm.

Determinations of the cross-sections in both systems were reported by e.g. Ahmed and Kumar (1992) and Grosch et al. (2015).

B.8 Ozone, O_3

Although ozone is similar in structure to SO_2, it exhibits only diffuse bands. There are four band systems in the UV-visible spectral range:

1. The Hartley system from 220 to 300 nm.
2. The Huggins system from 300 to 374 nm.
3. The Chappuis system from 550 to 610 nm.
4. In addition, there is a further system of transitions in the 400–1000 nm region (Bass and Paur 1985; Anderson and Mauersberger 1992; Burkholder and Talukdar 1994; Burrows et al. 1999b; Voigt et al. 2001; Orphal, 2003).

Table B.2 summarises the relevant states and some transitions of the O_3 molecule. DOAS measurements of O_3 with artificial light sources and from satellites largely make use of the structured absorption of the Hartley and Huggins bands (e.g. Axelsson et al. 1990b; Edner et al. 1994b; Hausmann and Platt 1994), while stratospheric measurements with ground-based ZSL-DOAS instruments frequently use the structures of the Chappuis band (e.g. Hoiskar et al. 1997; Sarkissian et al. 1997; Vaughan et al. 1997; van Roozendael et al. 1998; Roscoe et al. 2001a; 2001b). A summary and critical review of the UV and visible absorption cross-section of the O_3 molecule

Table B.2 The relevant states and some transitions of the O_3 molecule

Electronic state	Energy of electronic state, T_0 cm^{-1}	Vibrational energies, cm^{-1}	Denomination of transition from X^1A_1 ground state		
		ν_1	ν_2	ν_3	
D	33000		300		Hartley
C	28447	636	351		Huggins
B^1B_1	16625	1099			Chappuis
A^3B_1	10000		567		
X^1A_1	0	1110	705	1042	

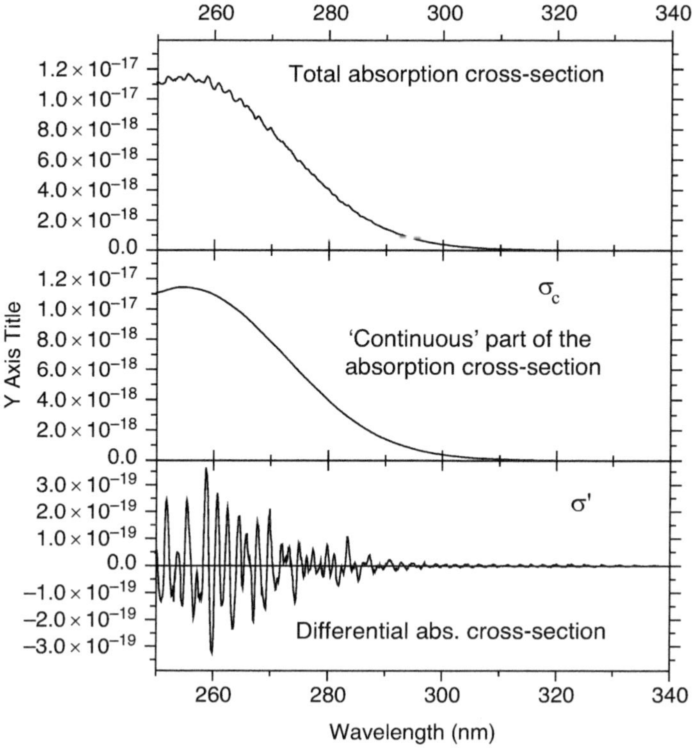

Fig. B.2 The total- and differential absorption cross-section spectra of ozone in the Huggins band (Geyer, 2000). Note that this is one of the few cases where there is a large difference between σ and σ'. Reprinted from Geyer, PhD Thesis, 2000, University of Heidelberg

can be found in Orphal et al. (2016). The evaluation is reported by the activity "Absorption Cross-Sections of Ozone" (ACSO), which started in 2008 as a joint initiative of the International Ozone Commission (IO3C), the World Meteorological Organization (WMO) and the IGACO ("Integrated Global Atmospheric Chemistry Observations") O_3/UV subgroup to study, evaluate, and recommend the most suitable ozone absorption cross-section laboratory data to be used in atmospheric ozone measurements.

It must be kept in mind that the structured absorption of all bands are very weak as compared to the total absorption; this is illustrated in Fig. B.2 showing the total- and differential absorption cross-section spectra of ozone in the Huggins band.

B.9 Monocyclic Aromatic Hydrocarbons

Benzene and its derivatives exhibit a strong band system in the vacuum–UV, and narrow, distinct bands between 240 and 280 nm. In addition, there is a very weak system of benzene in the 300 to 340 nm range. Since the vacuum–UV transition is in

Table B.3 Electronic terms and transitions of the benzene molecule (see e.g. Hollas)

Electronic term of C_6H_6	Spectral range of the transition from X^1A_{1g} ground state (nm)	Electric dipole transition probability
A^1B_{2u}	227–267	Orbital-forbidden, spin-allowed
B^1B_{1u}	185–205	Orbital-forbidden, spin-allowed
C^1E_{1u}	170–180	Allowed
a^3B_{1u}	300–340	Orbital- and spin-forbidden

a range where the atmosphere is not sufficiently transparent and the $a^3B_{1u} \leftarrow X^1A_{1g}$ transition is too weak for DOAS purposes, the $A^1B_{2u} \leftarrow X^1A_{1g}$ transition is used (see Table B.3). Table B.4 gives an overview of absorption cross-sections and DOAS detection limits of a number of aromatics.

As discussed by Volkamer et al. (1997), care has to be taken that spectrometers with sufficient spectral resolution are used, the overlap of the Hertzberg bands of molecular oxygen is taken into account, and correct absorption cross-sections for the aromatic compounds are taken. Unfortunately, in most of the literature on atmospheric DOAS measurements of aromatic compounds no information is given about the absorption cross-sections used. Only Axelsson et al. (1995), Trost et al. (1997), Etzkorn et al. (1999), and Fally et al. (2009), Stutz et al. (2016) presented their measured absorption cross-sections or gave references. When comparing determinations of absorption cross-sections of aromatic compounds [Pantos et al. (1978): benzene; Milton et al. (1992): benzene and toluene; Suto et al. (1992): benzene and the isomers of xylene; Axelsson et al. (1995): benzene, toluene, the isomers of xylene, ethylbenzene, 1, 2, 4-trimethylbenzene, and 1, 3, 5-trimethylbenzene; Trost et al. (1997), Etzkorn et al. (1999): benzene, toluene, the isomers of xylene, phenol, benzaldehyde, and the isomers of cresol], differences of up to a factor of two occur between reported values (see Table B.5).

B.10 Polycyclic Aromatic Hydrocarbons

Polycyclic aromatic hydrocarbons also show structured absorption spectra in the near UV, which allow sensitive determination by DOAS: an important example is naphthalene and its derivatives. Besides the medium resolution structures (on the order of 0.5 nm width) usually used by DOAS, there are also narrow-band (0.001 nm) spectral structures, which allow detection of these species by high-resolution DOAS (George and Morris 1968; Majewski and Meerts 1984; Neuroth et al. 1991).

B.11 Nitrous Acid, HONO

Nitrous acid, HONO (HNO_2), has strong absorption bands in the 300–400 nm spectral range. It has been extensively studied by DOAS, as described in Sect. 2.6.3. There have been many determinations of the HONO absorption cross-section; a summary

Table B.4 Summary of absorption cross-sections and typical DOAS detection limits for a series of monocyclic aromatic hydrocarbons (adapted from Etzkorn et al. 1999). The second column shows the differential cross-section (σ' in units of 10^{-18} cm^2 at a spectral resolution (FWHM) of 0.146 nm) after high pass filtering for the strongest band together with the corresponding wavelength. The fourth column gives typical DOAS detection limits (assuming a light path of 480 m and a minimal detectable optical density of 0.001 at 25 ± 3°C)

Aromatic species (stated purity)	Differential cross-section of the strongest band (wavelength) 10^{-18} cm^2 (nm)	Uncertainty of the differential cross section %	DOAS detection limit ppb	Uncertainty of the absolute cross-section 10^{-18} cm^2
Benzene (>99.8%)	2.19 (252.9)	7.2	0.39	0.21
Toluene (99.8%)	1.28 (266.6)	2.6	0.66	0.24
p-xylene (99%)	2.03 (272.2)	5.6	0.42	0.17
m-xylene (99%)	0.66 (270.6)	3.5	1.28	0.08
o-xylene (>97%)	0.21 (269.0)	5.1	3.99	0.05
1,3,5-trimethylbenzene (99%)	0.34 (270.7)	2.9	2.51	0.11
1,2,4-trimethylbenzene (98%)	0.50 (274.3)	4.9	1.71	0.08
Styrene (>99%)	1.55 (287.7)	8.3	0.55	0.17
Ethylbenzene[1] (99%)	0.38 (266.5)	3.2	2.25	0.10
Benzaldehyde (>99%)	4.40 (284.1)	4.7	0.19	1.82
o-tolulaldehyde (97%)	1.31 (292.2)	3.9	0.65	0.34
m-tolulaldehyde (97%)	1.26 (292.4)	3.5	0.67	0.44
p-tolulaldehyde (97%)	1.84 (284.5)	5.8	0.46	0.26
2,4,6-trimethylphenol (97%)	0.74 (283.0)	4.7	1.14	0.30
Phenol(>99.5%)	19.84 (275.1)	2.6	0.04	1.63
p-cresol (99%)	8.72 (283.0)	5.1	0.10	0.84
o-cresol (>99%)	2.01 (275.8)	2.8	0.42	0.60
m-cresol (99%)	3.18 (277.9)	4.1	0.27	0.49
2,3-dimethylphenol (99%)	2.16 (277.3)	3.7	0.39	0.41
2,4-dimethylphenol (98%)	1.27 (283.3)	9.2	0.66	0.32
2,5-dimethylphenol (>99%)	1.13 (263.6)	7.9	0.75	0.32
2,6-dimethylphenol (>99.5%)	0.94 (275.8)	7.6	0.90	0.30
3,4-dimethylphenol (99%)	1.24 (283.3)	4.6	0.68	0.34
3,5-dimethylphenol (>99%)	2.85 (279.1)	8.0	0.30	0.80

Table B.5 Comparison of literature reports of differential UV absorption cross-sections of a series of monocyclic aromatic compounds. For each compound, the differential cross-section of the strongest band was divided by the differential cross-section obtained by Etzkorn et al. (1999). The different spectral resolutions have been taken into account (from Etzkorn et al., 1999)

Compound	Trost et al. 1997	Suto et al. 1992	Axelsson 1995	Milton et al. 1992	Pantos et al. 1978
Benzene	0.89	1.05	0.56	0.59	0.90
Toluene	0.85		0.98	0.46	
o-xylene	1.12	1.06	1.18		
m-xylene	1.06	1.13	1.22		
p-xylene	1.12	1.02	2.80		
Ethylbenzene			1.93		
1, 2, 4-trimethylbenzene			0.37		
1, 3, 5-trimethylbenzene			0.93		
Phenol	1.13				
Benzaldehyde	1.12				
o-cresol	0.44				
m-cresol	0.63				
p-cresol	0.60				

is given in Table B.6. It can be seen that most of the data are in good agreement when the differences in resolution are considered (see also intercomparison by Gratien et al. 2009). In addition, Table B.7 gives the spectral positions of the different HONO bands and the assignment to the cis and trans isomers of the molecule.

B.12 Halogen Monoxides

The halogen monoxide radicals have strongly banded spectra of the $A \leftarrow X^2\Pi$ transition in the near UV (ClO, BrO) and blue (IO) spectral regions. The peak cross-sections increase from just below 10^{-17} cm^2 for ClO to several times 10^{-17} cm^2 for IO, thus allowing ppt sensitivities for DOAS measurements of these species. The halogen monoxide molecules partly exhibit high spectral resolution structures (e.g. Neuroth et al. 1991), which can give rise to apparent non-linearity effects if not resolved (see also Appendix B.5). Moreover, in passive DOAS measurements, interference of the high resolution absorption structures with solar Fraunhofer lines can lead to interferences, the so called 'solar I_0 Effect' (see e.g. Mohammed-Tahrin, 2001).

B.12.1 Chlorine Monoxide, ClO

The ClO molecule has a strong, structured absorption system ($A^2\Pi - X^2\Pi$) in the spectral range 270–310 nm. It has been subject to numerous studies (e.g. Johnston

Table B.6 Comparison of the differential and absolute absorption cross-sections of nitrous acid (HONO) with literature data (data from Stutz et al. 2000)

Author	Spectral resolution	Total absorption cross-section σ at 354 nm	σ at 354 nm with a resolution of 1 nm	Difference to Stutz et al. (2000)	Difference to the σ' of Stutz et al. (2000)
	nm	10^{-20} cm^2	10^{-20} cm^2	%	%
Johnston and Graham (1974)	0.87	13.6 ± 2.7	13.0 ± 2.7	-70	-70 ± 0.1
Cox and Derwent, (1976, 1977)	< 0.1	67 ± 7	58.8 ± 6	26	9.7 ± 1.2
Stockwell and Calvert (1978)	< 1	49.6 ± 4.9	48.9 ± 4.9	13	30 ± 1
Perner and Platt (1979)	0.6	–	–	–	13
Vasudev, 1990	–	49.6 ± 4.9	43.6 ± 4.3	0.8	0.4 ± 0.8
Bongartz et al. (1994)	0.1	54.9 ± 2	46.44 ± 1.7	7.4	4.8 ± 0.1
Febo et al. (1996)		–	–	–	?
Pagsberg et al. (1997)	–	50.2 ± 5	44.29 ± 5	2.4	-12 ± 0.5
Brust et al. (1999)	0.5	38.9 ± 6	34.7 ± 5	-25	-26.8 ± 0.2
Stutz et al. (2000)	0.08	51.9 ± 0.1	43.24 ± 0.1	–	–

et al. 1966; Rigaud et al. 1977; Mandelman and Nicholls 1977; Jourdain et al. 1978; Sander and Friedl 1990; Simon et al. 1990; Trolier et al. 1990; Howie et al. 1999). There are few measurements of ClO in the atmosphere (Chaps. 10 and 11).

B.12.2 Bromine Monoxide, BrO

Strong bands of the $A(^2\Pi_{3/2}) \leftarrow X(^2\Pi_{3/2})$ transition of BrO are in the spectral region 310–350 nm. Quantitative determination of spectra was reported by several authors (e.g. Cox et al. 1982; Wahner et al. 1988; Wilmouth et al. 1999; Fleischmann et al. 2004). These spectra were employed by numerous researchers to study the abundance of BrO, both in the stratosphere and troposhere (Chaps. 10 and 11).

Table B.7 Comparison of the spectral positions (in cm^{-1}) of the different HONO bands of the cis and trans isomers of the molecule table, but (data from Stutz et al. 2000)

Band	Bongartz et al. (1994)	King and Moule (1962)	Stutz et al. (2000) ($\pm 2\sigma$)
Trans HONO			
0	26035	26034 ± 10	26036.3 ± 1.4
1	27153	27150 ± 2	27153.7 ± 1.5
2	28220	28225 ± 2	28227.8 ± 1.6
3	29249	29518 ± 2	29257.2 ± 3.4
4	30227	30231 ± 2	30230.3 ± 3.7
5	31158	31175 ± 10	31170.0 ± 3.9
6	32076	32075 ± 30	32078.2 ± 4.1
Cis HONO			
0	26312		
1	27430	27420 ± 50	27441.4 ± 3.0
2	28496	28495 ± 2	28495.0 ± 3.2
3	29513	29518 ± 2	29523.3 ± 3.5
4	30502	30505 ± 2	30510.8 ± 3.7
5	31456	31461 ± 10	31453.5 ± 4.0
6	32344	32350 ± 30	32359.6 ± 4.2

B.12.3 Iodine Monoxide, IO

The standard IO spectral retrieval uses the wavelength interval 425–465 nm, where five strong bands of the 'cold' (4-0, 3-0, 2-0, 1-0, 0-0) and two weaker bands of the 'hot' (3-1, 2-1) vibrational absorption of the IO $A^2\Pi \rightarrow X^2\Pi$ electronic transitions are located. The total absorption cross-section of the IO molecule was measured by a number of authors (Lazlo et al. 1995; Himmelmann et al. 1996; Harwood et al. 1997; Hönninger 1999; Cox et al. 1999; Dillon et al. 2005). These spectra were employed by several researchers to study the abundance of IO in the troposphere (Chaps. 10 and 11).

B.13 Halogen Dioxides

The symmetric halogen dioxide molecules exhibit strongly structured absorption spectra in the near UV (OClO) and visible (OBrO, OIO) spectral ranges, which have been used to make sensitive measurements of OClO and OIO in the atmosphere.

B.13.1 Chlorine Dioxide, OClO

Studies of the UV absorption cross-section of the OClO $A^2A_2 \leftarrow X^2B_1$ transition in the wavelength range 300–440 nm were performed by Wahner et al. (1987) and Kromminga et al. (2003) at spectral resolutions better than 0.1 nm. These bands have been used in numerous studies of stratospheric OClO as described in Chap. 11.

B.13.2 Bromine Dioxide, OBrO

Studies of the UV absorption cross-section of the OBrO $C^2A_2 \leftarrow X^2B_1$ transition in the wavelength range 390–620 nm were reported by Rattigan et al. (1996), Knight et al. (2000), and Fleischmann and Burrows (2002).

B.13.3 Iodine Dioxide, OIO

There are few absolute determinations of the structured absorption cross-section of OIO in the spectral range between 467–667 nm (Himmelmann et al. 1996; Cox et al. 1999; Ashworth et al. 2002; Joseph et al. 2005; Tucceri et al. 2006). The peak absorption cross-section at 548.6 nm is estimated at 2.7×10^{-17} cm^2/molec.

B.14 Molecular Iodine (I_2)

The electronic transition of the $B^3\Pi\left(O_u^+\right) \leftarrow X^1\Sigma_g^X$ electronic transition exhibits a distinct, regular vibrational structure in the range 500–630 nm, which can be used for DOAS evaluations. Absolute determinations of the cross-section were reported by Goy and Pritchard (1964), Ogryzlo and Thomas (1965), Tellinghuisen (1973), Saiz-Lopez et al. (2004c), and Spietz et al. (2006). The dissociation energy of I_2 corresponds to a threshold wavelength of 533 nm (Saiz-Lopez et al. 2004c). Atmospheric determinations of I_2 using the above transition were reported by Saiz-Lopez et al. (2004a) and Peters et al. (2005).

B.15 Water Vapour, H_2O

Water vapour has a series of bands in the visible and near IR spectral regions, which are suitable for DOAS measurements. Data on line strengths are found in Rothman et al. (1992, 1998), with several recent corrections (e.g. Belmiloud et al. 2000; Coheur et al. 2002; Schermaul et al. 2002; Mérienne et al. 2003). The bands correspond to the excitations 4ν, $3\nu + \delta$, 3ν, and $2\nu + \delta$, (located around 725, 820, 943, and 1136 nm, respectively), where ν denotes a quantum number of the stretching modes and δ of the bending mode; details of the line assignment can be found at Tolchenov et al. (2005).

The water molecule partly exhibits high spectral resolution structures, which can give rise to apparent non-linearity effects if not resolved (see also B.5) or corrected for. At high water vapour column densities ($S \approx 4 \times 10^{23}$ cm^{-2}), the individual H_2O lines have optical densities in excess of unity. Considering the variation of the water vapour column amount, a model study was carried out to investigate the effects of this non-linear dependence of H_2O absorption bands on the results of the NO_3 evaluation. Different optical densities of water vapour can be simulated by applying Lambert–Beer's law to the high-resolution water absorption cross-section (Mandin et al. 1986) and then smoothing the calculated spectrum to simulate the instrumental resolution. This simulation demonstrates that the column amount-dependent absorption structure of H_2O can be interpolated by simultaneously fitting two water reference spectra, which bracket the optical density of the atmospheric spectrum (see Geyer et al. 2000).

B.16 Nitrate Radical, NO_3

The absorption cross-section $\sigma(\lambda)$ of the nitrate radical shows strong features in the red spectral region, with prominent bands at wavelengths of 662 and 623 nm (Fig. 10.21). Various studies of its wavelength and temperature dependency have been performed in recent years (see Table B.8). Most of the studies report similar relative shapes of the spectrum; however, there are discrepancies concerning the absolute value of $\sigma(\lambda)$. The value of the cross-section at 298 K and 662 nm published by different authors varies from as low as 1.21×10^{-17} cm^2 (Mitchell et al. 1984) up to 2.49×10^{-17} cm^2 (Magnotta and Johnston, 1980). Recent publications reported $\sigma(622\,\text{nm}) = (2.23 \pm 0.22)10^{-17}$ cm^2 (Yokelson et al. 1994). In addition, the temperature dependence of the NO_3 absorption cross-section is not clear. While Sander (1986), Ravishankara and Mauldin (1986), and Yokelson et al. (1994) determined an increase in $\sigma(\lambda)$ towards lower temperatures (Table B.8), Cantrell et al. (1987) did not find a temperature effect. In many DOAS observations the temperature-independent cross-section $\sigma(662\,\text{nm}) = (2.1 \pm 0.2) \times 10^{-17}$ cm^2 recommended by Wayne et al. (1991) is used as reference spectrum for the NO_3 evaluation. This cross-section was derived by averaging the data of four studies of Ravishankara and Mauldin (1986), Sander (1986), Cantrell et al. (1987), and Canosa-Mas et al. (1987).

B.17 OH Radicals

Several active DOAS studies of atmospheric OH have been performed using the rotational lines of the $A\,^2\Sigma^+(v' = 0) \leftarrow X^2\Pi_{3/2}(v''=0)$ and $A\,^2\Sigma^+(v'=0) \leftarrow X^2\Pi_{1/2}(v''=0)$ transitions near 308 nm. Due to the relatively small reduced mass and corresponding small momentum of inertia of the OH radical, the spacing of the individual rotational lines is relatively large (of the order of 0.15 nm). Under ambient conditions (1 atmosphere, 300K), the strongest rotational lines are $Q_1(2)$ (307.9951 nm), $Q_1(3)$ (308.1541 nm), and $P_1(1)$ (308.1665 nm), with line widths (FWHM) around 0.0025 nm (Dorn et al. 1995); Doppler broadening and pressure

Table B.8 Results of studies of the NO_3 absorption cross-section since 1978. Reported absolute values of the peak absorption cross-section at 662 nm vary from (1.21 to 2.48) × 10^{-17} cm^2. The temperature dependence of the cross-section is not understood at present (adapted from Geyer et al. 2000)

Studies	Technique	Temp. [K]	σ (662 nm) [10^{-17} cm^2]	Temperature dependence of σ (662 nm) [10^{-17} cm^2]
Graham and Johnston, (1978)	Modulated photolysis	298	1.71	not studied
Magnotta and Johnston, (1980)	Calibration of quantum yield	298	2.48	not studied
Mitchell et al. (1980)	Differential photomultiplier	294	1.21	not studied
Marinelli et al. (1982)	Tuneable dye laser	298	1.90	not studied
Ravishankara and Wine, (1983)	Discharge flow, tuneable dye laser	298	1.78	not studied
Cox et al. (1984)	Modulated photolysis, diode array spectrometer	298	1.63 ± 0.15	not studied
Burrows et al. (1985)	Modulated photolysis	298	1.85 ± 0.56	not studied
Ravishankara and Mauldin, (1986)	Discharge flow, tuneable dye laser	298	1.90 ± 0.22	4.65–0.00932 T
Sander, (1986)	Flash photolysis, diode array spectrometer	298	2.28 ± 0.34	4.19–0.0064 T
Cantrell et al. (1987)	Fourier transform spectrometer	298	2.02 ± 0.2	not found
Canosa-Mas et al. (1987)	Differential photomultiplier	296	2.23 ± 0.35	not studied
Wayne et al. (1991)	Averaged value	–	2.10 ± 0.20	–
Yokelson et al. (1994)	Diode array spectrometer	298	2.23 ± 0.22	4.56–0.00788 T
Orphal et al. (2016)	Fourier transform spectroscopy	294		

broadening contribute approximately equally to the line width. Spectroscopic parameters for the OH radical have been measured by many authors (e.g. Dieke and Crosswhite 1961; German 1975; Dimpfl and Kinsey 1979; Goldman and Gillis 1981). Recent measurements of the UV absorption cross-section of the OH radical were reported by Dorn et al. (1995). In contrast to most other DOAS applications, the specific and sensitive detection of atmospheric OH requires instruments that are able resolve the individual rotational lines. The narrow absorption lines dictate the use of a light source with high spectral intensity. For example, OH can be detected using a laser with an emission bandwidth broader than the OH lines in combination with a high-resolution spectrometer (Hübler et al. 1984; Perner et al. 1987; Platt et al. 1987 1988; Dorn et al. 1988; Hofzumahaus et al. 1991; Mount 1992), or a narrow-band laser scanning the OH line (Zellner and Hägele 1985; Amerding et al. 1990 1992). Figure 10.21 shows a portion of the OH absorption spectrum. Several species, e.g. SO_2 and CH_2O, absorb at similar wavelengths to OH (Hübler et al. 1984; Neuroth et al. 1991).

B.18 Oxygen, O_2

The oxygen molecule exhibits several absorption systems (see Table B.9), which are relevant for DOAS applications:

1. The $B^3\Sigma_g^- \leftarrow X^3\Sigma_g^-$ transition known as Schuman–Runge bands in the short-wavelength UV makes the atmosphere essentially opaque below about 200 nm, preventing DOAS measurements at shorter wavelengths.
2. The $A^3\Sigma_g^+ \leftarrow X^3\Sigma_g^-$ transition of the Herzberg I system in the spectral range 242–286 nm is overlapping the absorption features of most monocyclic aromatic species (see Sect. B.9 and Volkamer et al. 1998).
3. The near IR and visible spectral range has been used to determine optical path lengths in the atmosphere as described in Sect. 11.11.2 and for cloud detection by satellite sensors.
4. An additional forbidden system gives rise to absorption in the near IR (Smith and Newnham, 1999; Newman et al.; 1999, 2000; Smith et al. 2001).

B.19 Oxygen Dimer, O_4 or $(O_2)_2$

The broadband absorptions in gaseous O_2 (Fig. B.3) were first described by Janssen et al. (1885, 1886), who also showed that the intensity of these bands varies with the square of the oxygen pressure (i.e. O_2 concentration). The absorption of oxygen molecules in the UV, vis, and near IR spectral ranges is due to different types of chemical bonds and transitions. Besides the discrete, structured ro-vibrational bands of the electronic transition of the O_2 monomer (see Sect. B.18) and the structured bands of the bound van der Waal's molecule O_4, oxygen also shows broad unstructured

Table B.9 The absorption bands of oxygen. The atmospheric bands are in the VIS and NIR spectral regions. The band strength is the sum of all line strengths of the band. The spectroscopic data is taken from the HITRAN96 database (from Gamache et al. 1998)

Electronic transition	Vibrational transition	Band centre (nm)	Band strength (cm/molec.)	Designation
$B^3\Sigma_g^- \leftarrow X^3\Sigma_g^-$	–	UV 175.9	–	Schumann–Runge
$A^3\Sigma_g^+ \leftarrow X^3\Sigma_g^-$	–	UV 242–286	–	Herzberg I
$b^1\Sigma_g^+ \leftarrow X^3\Sigma_g^-$	$(0 \leftarrow 0)$	762.19	2.24×10^{-22}	A band
$b^1\Sigma_g^+ \leftarrow X^3\Sigma_g^-$	$(1 \leftarrow 0)$	688.47	1.49×10^{-23}	B band
$b^1\Sigma_g^+ \leftarrow X^3\Sigma_g^-$	$(2 \leftarrow 0)$	628.85	4.63×10^{-25}	γ band
$b^1\Sigma_g^+ \leftarrow X^3\Sigma_g^-$	$(1 \leftarrow 1)$	771.07	9.53×10^{-26}	
$b^1\Sigma_g^+ \leftarrow X^3\Sigma_g^-$	$(0 \leftarrow 1)$	864.75	7.88×10^{-27}	
$a^1\Delta_g \leftarrow X^3\Sigma_g^-$	$(0 \leftarrow 0)$	1268.6	3.68×10^{-24}	
$a^1\Delta_g \leftarrow X^3\Sigma_g^-$	$(1 \leftarrow 0)$	1067.7	9.53×10^{-27}	
$a^1\Delta_g \leftarrow X^3\Sigma_g^-$	$(0 \leftarrow 1)$	1580.8	2.75×10^{-28}	

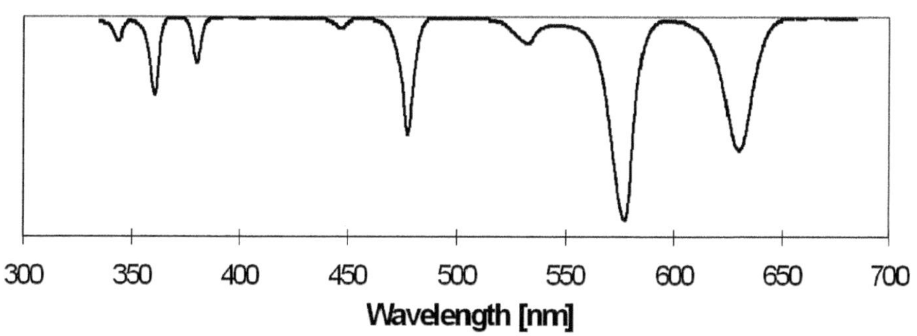

Fig. B.3 Absorption spectrum of oxygen dimers (O_4). Reprinted with permission from Wagner et al. 2007a,

absorptions (Fig. B.3) due to the collision induced oxygen dimer $(O_2)_2$ (see Greenblatt et al. 1990; Solomon et al. 1998, and references therein). Since the original work of Janssen et al. (1885, 1886), several studies investigated the $(O_2)_2$ absorptions demonstrating that, under atmospheric conditions, these bands contain no fine structure (see Solomon et al. 1998; Naus and Ubachs, 1999, Thalman and Volkamer 2013; Lampel et al. 2018, and references therein). Greenblatt et al. (1990) concluded from the weak temperature dependence of the $(O_2)_2$ absorption that, under atmospheric conditions, they are most probably related to an oxygen collision complex rather than a bound dimer. The O_4 absorption bands in the UV-vis spectral regions (360–630 nm) belong to simultaneous transitions between two ground state oxygen molecules and electronically excited states of both oxygen molecules, as summarised in Table B.10. At longer wavelengths, transitions of only one excited oxygen molecule occur and, under atmospheric conditions, N_2 can serve as a collision partner (see Solomon et al. 1998; Smith and Newnham, 2000, and references therein). Direct moonlight mea-

Table B.10 Wavelengths and transition assignments of the O_4 absorption bands in the UV-vis spectral region (adapted from Wagner et al. 2002c and Lampel et al. 2018)

Wavelength (nm)	Upper state of transition (from $^3\Sigma_g^- + ^3\Sigma_g^-$ ground state)
328.0$^+$	$^1\Sigma_g^+ + ^1\Sigma_g^+$ ($\nu = 3$)
343.4*	$^1\Sigma_g^+ + ^1\Sigma_g^+$ ($\nu = 2$)
360.5	$^1\Sigma_g^+ + ^1\Sigma_g^+$ ($\nu = 1$)
380.2	$^1\Sigma_g^+ + ^1\Sigma_g^+$ ($\nu = 0$)
419.0$^+$	$^1\Sigma_g^+ + ^1\Delta_g$ ($\nu = 2$)
446.7*	$^1\Sigma_g^+ + ^1\Delta_g$ ($\nu = 1$)
477.3	$^1\Sigma_g^+ + ^1\Delta_g$ ($\nu = 0$)
532.2	$^1\Delta_g + ^1\Delta_g$ ($\nu = 2$)
577.2	$^1\Delta_g + ^1\Delta_g$ ($\nu = 1$)
630.0	$^1\Delta_g + ^1\Delta_g$ ($\nu = 0$)

Table B.11 O_4 absorption cross-sections, as determined in the laboratory and the atmosphere from several studies (adapted from Wagner et al. 2002c)

O_4 band	Perner and Platt (1980)	Greenblatt et al. (1990)	Greenblatt et al. (1990)	Volkamer, 1996	Osterkamp et al. (1998)	Newnham and Ballard (1998)	Naus and Ubachs (1999)	Wagner et al. (2002c)*		
Temp. nm	279 K	296 K	296 K	296 K	204 K	256 K	223 K	283 K	294 K	242 K
360	5.4 ± 1.5	4.1 ± 0.4	5.7 ± 0.6	5.42 ± 0.7						5.70 ± 0.5
380	< 1.4	2.4 ± 0.2	3.7 ± 0.4	2.4 ± 0.2						2.44 ± 0.4
477	5.9 ± 1.8	6.3 ± 0.6	7.6 ± 1.3	6.1 ± 0.3	7.9 ± 0.3	7.0 ± 0.3	6.99 ± 0.35	8.34 ± 0.83		7.80 ± 0.2
532				1.3 ± 0.3	1.4 ± 0.2	1.2 ± 0.2	1.31 ± 0.2	1.23 ± 0.38		1.74 ± 0.5
577	16 ± 6	11 ± 1	–	10.3 ± 0.3	12.2 ± 0.4	13.6 ± 0.4	12.61 ± 0.11	11.75 ± 0.2	11.41 ± 0.5	13.50 ± 0.4
630	–	7.2 ± 0.7	–	6.2 ± 0.6			8.8 ± 0.13	7.9 ± 0.15	7.55 ± 0.5	9.61 ± 0.3

* Direct moonlight measurements in the atmosphere

surements by Wagner et al. (2002c) yielded improved absorption cross-sections of O_4. The investigated O_4 bands were found to show an increase of the peak absorption with decreasing temperature ranging from ≈13%/100 K at 477.3 and 532.2 nm, ≈20% at 360.5 and 577.2 nm to ≈33% at 380.2 and 630.0 nm. Moreover (except for the band at 380.2 nm), the O_4 absorption cross-sections were found to be somewhat larger than previous measurements (see Table B.11).

Atmospheric absorptions of O_4 were first reported by Perner and Platt (1980), and are important in atmospheric radiative transfer modelling.

Literature

C

Internet Addresses

Keller-Rudek, H. and Moortgat, G.K. (2005), MPI-Mainz-UV-VIS Spectral Atlas of Gaseous Molecules. URL: www.atmosphere.mpg.de/spectral-atlas-mainz

Smith, P.L., Heise, C., Esmond, J.R., Kurucz, R.L. (1995), Atomic spectral line database from CD-ROM 23 of R.L. Kurucz

Kurucz, R.L. and Bell, B. (1995), Atomic Line Data Kurucz CD-ROM No. 23. Cambridge, Mass.: Smithsonian Astrophysical Observatory. URL: https://lweb.cfa.harvard.edu/amp/ampdata/kurucz23/sekur.html, last accessed on Dec. 20, 2023

Science-softCon UV/Vis Spectra Data Base URL: http://www.science-softcon.de/

Rothman, L.S., Barbe, A., Chris Benner, D., Brown, L.R., Camy-Peyret, C., Carleer, M.R., Chance, K., Clerbaux, C., Dana, V., Devi, V.M., Fayt, A., Flaud, J.-M., Gamache, R.R., Goldman, A., Jacquemart, D., Jucks, K.W., Lafferty, W.J., Mandin, J.-Y., Massie, S.T., Nemtchinov, V., Newnham, D.A., Perrin, A., Rinsland, C.P., Schroeder, J., Smith, K.M., Smith, M.A.H., Tang, K., Toth, R.A., Vander Auwera, J., Varanasi, P., Yoshino, K. (2003), The HITRAN molecular spectroscopic database: edition of 2000 including updates through 2001, Journal of Quantitative Spectroscopy & Radiative Transfer **82**, 5–44. http://www.hitran.org

U. Platt and J. Stutz, *Differential Optical Absorption Spectroscopy*, Geophysics and Environmental Physics, https://doi.org/10.1007/978-3-662-68639-3

References

A. Aalto, G. Genty, T. Laurila, J. Toivonen, Incoherent broadband cavity enhanced absorption spectroscopy using supercontinuum and superluminescent diode sources. Opt. Express **23**(19), 25225–25234 (2015). https://doi.org/10.1364/OE.23.025225

W. Aas, A. Mortier, V. Bowersox, R. Cherian, G. Faluvegi, H. Fagerli, J. Hand, Z. Klimont, C. Galy-Lacaux, C.M.B. Lehmann, C.L. Myhre, G. Myhre, D. Olivié, K. Sato, J. Quaas, P.S.P. Rao, M. Schulz, D. Shindell, R.B. Skeie, A. Stein, T. Takemura, S. Tsyro, R. Vet, X. Xu, Global and regional trends of atmospheric sulfur. Sci. Rep. **9**(1), 953 (2019). https://doi.org/10.1038/s41598-018-37304-0

I. Aben, D.M. Stam, F. Helderman, The ring effect in skylight polarisation. Geophys. Res. Lett. **28**(3), 519–522 (2001)

J.R. Acarreta, J.F. de Haan, P. Stammes, Cloud pressure retrieval using the O_2-O_2 absorption band at 477 nm. J. Geophys. Res.: Atmos. **109**(D5) (2004). https://doi.org/10.1029/2003JD003915

S.M. Ahmed, V. Kumar, Measurement of photoabsorption and fluorescence cross sections for carbon disulfide at 188–213 and 287.5–339.5 nm. J. Phys. **39**, 367–380 (1992)

H. Akimoto, *Introduction to Atmospheric Chemistry* (Springer Japan, Tokyo, 2016), pp. 1–9. https://doi.org/10.1007/978-4-431-55870-5_1

H. Akimoto, H. Tanimoto, Review of comprehensive measurements of speciated NOy and its chemistry: need for quantifying the role of heterogeneous processes of HNO_3 and HONO. Aerosol Air Qual. Res. **21**(2), 200395 (2021). https://doi.org/10.4209/aaqr.2020.07.0395

D.L. Albritton, A.L. Schmeltekopf, R.N. Zare, An introduction to the least-squares fitting of spectroscopic data, in *Molecular Spectroscopy: Modern Research*. ed. by R.K. Narahar, M.W. Weldon (Academic, Orlando, 1976)

M. Aldener, S.S. Brown, H. Stark, J.S. Daniel, A.R. Ravishankara, Near-IR absorption of water vapor: pressure dependence of line strengths and an upper limit for continuum absorption. J. Mol. Spectrosc. **232**, 223–230 (2005)

B. Alexander, L.J. Mickley, Paleo-perspectives on potential future changes in the oxidative capacity of the atmosphere due to climate change and anthropogenic emissions. Curr. Pollut. Rep. **1**(2), 57–69 (2015). https://doi.org/10.1007/s40726-015-0006-0

B. Alicke, *Messung von troposphärischen Halogenoxidradikalen in mittleren Breiten* (Diplomarbeit, Ruprecht-Karls-Universität Heidelberg, Germany, 1997)

B. Alicke, K. Hebestreit, J. Stutz, U. Platt, Detection of iodine oxide in the marine boundary layer. Nature **397**, 572–573 (1999)

B. Alicke, The role of nitrous acid in the boundary layer. Ph.D. thesis, University of Heidelberg (2000)

U. Platt and J. Stutz, *Differential Optical Absorption Spectroscopy*, Geophysics and Environmental Physics, https://doi.org/10.1007/978-3-662-68639-3

B. Alicke, U. Platt, J. Stutz, Impact of nitrous acid photolysis on the total hydroxyl radical budget during the limitation of oxidant production/Pianura Padana Produzione di Ozono study in Milan. J. Geophys. Res. **107**(D22), 8196 (2002). https://doi.org/10.1029/2000JD000075

B. Alicke, A. Geyer, A. Hofzumahaus, F. Holland, S. Konrad, H.W. Pätz, J. Schäfer, J. Stutz, A. Volz-Thomas, U. Platt, OH formation by HONO photolysis during the BERLIOZ experiment. J. Geophys. Res. **108**(D4), 8247 (2003). https://doi.org/10.1029/2001JD000579

S.R. Aliwell, R.L. Jones, D.J. Fish, Mid-latitude observations of the seasonal variation of BrO 1. Zenith-sky measurements. Geophys. Res. Lett. **24**, 1195–1198 (1997)

S.R. Aliwell, R.L. Jones, Measurements of tropospheric NO_3 at midlatitude. J. Geophys. Res.: Atmos. **103**(D5), 5719–5727 (1998). https://doi.org/10.1029/97JD03119

S.R. Aliwell, M. van Roozendael, P.V. Johnston, A. Richter, T. Wagner, D.W. Arlander, J.P. Burrows, D.J. Fish, R.L. Jones, K.K. Tørnkvist, J.-C. Lambert, K. Pfeilsticker, I. Pundt, Analysis for BrO in zenith-sky spectra: an intercomparison exercise for analysis improvement. J. Geophys. Res.: Atmos. **107**(D14), ACH 10–1–ACH 10–20 (2002). https://doi.org/10.1029/2001JD000329

D. Allan, Statistics of atomic frequency standards. Proc. IEEE **54**(2), 221–230 (1966). https://doi.org/10.1109/PROC.1966.4634

B.J. Allan, N. Carlslaw, H. Coe, R.A. Burgess, J.M.C. Plane, Observation of the nitrate radical in the marine boundary layer. J. Atmos. Chem. **33**, 129–154 (1999)

B.J. Allan, G. McFiggans, J.M.C. Plane, H. Coe, Observation of iodine monoxide in the remote marine boundary layer. J. Geophys. Res. **105**, 14363–14369 (2000a)

B.J. Allan, G. McFiggans, H. Coe, J.M.C. Plane, G.G. McFayden, The nitrate radical in the remote marine boundary layer. J. Geophys. Res. **105**(D19), 24191–24204 (2000b)

B.J. Allan, J.M.C. Plane, G. McFiggans, Observations of OIO in the remote marine boundary layer. Geophys. Res. Lett. **28**, 1945–1948 (2001)

B.J. Allan, J.M.C. Plane, H. Coe, J. Shillito, Observation of NO_3 concentration profiles in the troposphere. J. Geophys. Res. **107**(D21), 4588 (2002). https://doi.org/10.1029/2002JD002112

J.D. Allan, P.I. Williams, J. Najera, J.D. Whitehead, M.J. Flynn, J.W. Taylor, D. Liu, E. Darbyshire, L.J. Carpenter, R. Chance, S.J. Andrews, S.C. Hackenberg, G. McFiggans, Iodine observed in new particle formation events in the Arctic atmosphere during ACCACIA. Atmos. Chem. Phys. **15**(10), 5599–5609 (2015). https://doi.org/10.5194/acp-15-5599-2015

L.M.A. Alvarado, A. Richter, M. Vrekoussis, F. Wittrock, A. Hilboll, S.F. Schreier, J.P. Burrows, An improved glyoxal retrieval from OMI measurements. Atmos. Meas. Tech. **7**(12), 4133–4150 (2014). https://doi.org/10.5194/amt-7-4133-2014

L.M.A. Alvarado, A. Richter, M. Vrekoussis, A. Hilboll, A.B. Kalisz Hedegaard, O. Schneising, J.P. Burrows, Unexpected long-range transport of glyoxal and formaldehyde observed from the Copernicus Sentinel-5 precursor satellite during the 2018 canadian wildfires. Atmos. Chem. Phys. **20**(4), 2057–2072 (2020). https://doi.org/10.5194/acp-20-2057-2020

S. Amann, Q. Duong-Ederer, T. Haist, B. Sierk, B. Guldimann, W. Osten, Characterization of fiber-based slit homogenizer devices in the NIR and SWIR, in *International Conference on Space Optics-ICSO 2018*, vol. 11180 (SPIE, 2019), pp. 2276–2286

W. Amerding, A. Herbert, T. Schindler, M. Spiekermann, F.J. Comes, In situ measurements of tropospheric OH radicals-a challenge for the experimentalist. Ber. Bunsenges. Phys. Chem. **94**, 776–781 (1990)

W. Amerding, A. Herbert, M. Spiekermann, J. Walter, F.J. Comes, Fast scanning DOAS-A very promising technique for monitoring OH and other tropospheric trace gases. Fresenius J. Anal. Chem. **340**, 654–660 (1991)

W. Amerding, A. Herbert, M. Spiekermann, J. Walter, F.J. Comes, Ein schnell durchstimmbares Laserspektrometer. Ber. Bunsenges. Phys. Chem. **96**, 314 (1992)

W. Amerding, M. Spiekermann, R. Grigonis, J. Walter, A. Herbert, F.J. Comes, Fast scanning laser DOAS for local monitoring of trace gases, in particular tropospheric OH radicals. Ber. Bunsenges. Phys. Chem. **96**, 314–318 (1992)

M. Ammann, M. Kalberer, D.T. Jost, L. Tobler, E. Rössler, D. Piguet, H.W. Gäggeler, U. Baltensperger, Heterogeneous production of nitrous acid on soot in polluted air masses. Nature **395**, 157–160 (1998)

S.T. Andersen, L.J. Carpenter, C. Reed, J.D. Lee, R. Chance, T. Sherwen, A.R. Vaughan, J. Stewart, P.M. Edwards, W.J. Bloss, R. Sommariva, L.R. Crilley, G.J. Nott, L. Neves, K. Read, D.E. Heard, P.W. Seakins, L.K. Whalley, G.A. Boustead, L.T. Fleming, D. Stone, K.W. Fomba, Extensive field evidence for the release of HONO from the photolysis of nitrate aerosols. Sci. Adv. **9**(3), eadd6266 (2023). https://doi.org/10.1126/sciadv.add6266

D.E. Anderson, S.A. Lloyd, Polar twilight UV-visible radiation field: perturbations due to multiple scattering, ozone depletion, stratospheric clouds, and surface albedo. J. Geophys. Res. **95**, 7429–7434 (1990)

S.M. Anderson, K. Mauersberger, Laser measurements of ozone absorption cross sections in the Chappuis band. Geophys. Res. Lett. **19**(9), 933–936 (1992)

M.O. Andreae, R.J. Ferek, F. Bermond, K.P. Byrd, R.B. Chatfield, R.T. Engstrom, S. Hardin, P.D. Houmere, F. LeMarrec, H. Raemdonck, Dimethyl sulfide in the marine atmosphere. J. Geophys. Res. **90**, 12891–12900 (1985)

M.D. Andres-Hernandez, J. Notholt, J. Hjorth, O. Schrems, A DOAS study on the origin of nitrous acid at urban and non-urban sites. Atmos. Environ. **30**(2), 175–180 (1996)

L.C. Andrews, *Field Guide to Atmospheric Optics, SPIE Field Guides*, vol. FG02 (SPIE Press, Bellingham, 2004)

A. Angström, On the atmospheric transmission of sun radiation and on dust in the air. Geogr. Ann. Stockholm **11**, 156–166 (1929)

A. Angström, On the atmospheric transmission of sun radiation. II. Geogr. Ann. Stockholn **12**, 130–159 (1930)

B.R. Appel, A.M. Winer, Y. Tokiwa, H.W. Biermann, Comparison of atmospheric nitrous acid measurements by annular denuder and differential optical absorption systems. Atmos. Environ. **24A**, 611–616 (1990)

K.A. Arpag, P.V. Johnston, H.L. Miller, R.W. Sanders, S. Solomon, Observations of the stratospheric BrO column over Colorado, 40N. J. Geophys. Res. **99**, 8175–8181 (1994)

S. Arrhenius, On the influence of carbonic acid in the air upon the temperature of the ground. Lond., Edinb., Dublin Philos. Mag. J. Sci. **41**, 237–276 (1896)

D. Asaf, D. Pedersen, V. Matveev, M. Peleg, C. Kern, J. Zingler, U. Platt, M. Luria, Long-term measurements of NO_3 radical at a semiarid urban site: 1. Extreme concentration events and their oxidation capacity. Environ. Sci. Technol. **43**(24), 9117–9123 (2009). https://doi.org/10.1021/es900798b

S.H. Ashworth, B.J. Allan, J.M.C. Plane, High resolution spectroscopy of the OIO radical: implications for the ozone depleting potential of iodine. Geophys. Res. Lett. **29**(10), 1456–1459 (2002)

R. Atkinson, W.P.L. Carter, J.N. Pitts Jr., A.M. Winer, Measurements of nitrous acid in an urban area. Atmos. Environ. **20**, 408–409 (1986)

R. Atkinson, D.L. Baulch, R.A. Cox, R.F. Hampson, J.A. Kerr, J. Troe, Evaluated kinetic and photochemical data for atmospheric chemistry: supplement III. J. Phys. Chem. Ref. Data **18**, 881–1095 (1989)

R. Atkinson, Gas-phase tropospheric chemistry of organic compounds: a review. Atmos. Environ. **24A**, 1–41 (1990)

R. Atkinson, D.L. Baulch, R.A. Cox, J.N. Crowley, R.F. Hampson, R.G. Hynes, M.E. Jenkin, M.J. Rossi, J. Troe, Evaluated kinetic and photochemical data for atmospheric chemistry: vol. I—gas phase reactions of O_x, HO_x, NO_x and SO_x species. Atmos. Chem. Phys. **4**(6), 1461–1738 (2004). https://doi.org/10.5194/acp-4-1461-2004

R. Atkinson, D.L. Baulch, R.A. Cox, J.N. Crowley, R.F. Hampson, R.G. Hynes, M.E. Jenkin, M.J. Rossi, J. Troe, I. Subcommittee, Evaluated kinetic and photochemical data for atmospheric chemistry: vol. II—gas phase reactions of organic species. Atmos. Chem. Phys. **6**(11), 3625–4055 (2006). https://doi.org/10.5194/acp-6-3625-2006

B. Aumont, F. Chervier, S. Laval, Contribution of HONO sources to the $NO_X/HO_X/O_3$ chemistry in the polluted boundary layer. Atmos. Environ. **37**, 487–498 (2003)

I. Avrutsky, K. Chaganti, I. Salakhutdinov, G. Auner, Concept of a miniature optical spectrometer using integrated optical and micro-optical components. Appl. Opt. **45**(30), 7811–7817 (2006). https://doi.org/10.1364/AO.45.007811

H. Axelsson, B. Galle, K. Gustavsson, P. Regnarsson, M. Rudin, A transmitting/receiving telescope for DOAS-measurements using retroreflector technique, in *Digest of Topical Meeting on Optical Remote Sensing of the Atmosphere*, vol. 4 (OSA, 1990a), pp. 641–644

H. Axelsson, H. Edner, B. Galle, P. Ragnarson, M. Rudin, Differential optical absorption spectroscopy (DOAS) measurements of ozone in the 280–290 nm wavelength region. Appl. Spectrosc. **44**, 1654–1658 (1990b)

H. Axelsson, A. Eilard, A. Emanuelsson, B. Galle, H. Edner, P. Ragnarson, H. Kloo, Measurement of aromatic hydrocarbons with the DOAS technique. Appl. Spectrosc. **49**, 1254–1260 (1995)

L. Axelsson, A. Lauber, Measurement of sulfur dioxide with the differential optical absorption technique combined with Fourier transformation. Appl. Spectrosc. **46**, 1832–1836 (1992)

A. Baccarini, L. Karlsson, J. Dommen, P. Duplessis, J. Vüllers, I.M. Brooks, A. Saiz-Lopez, M. Salter, M. Tjernström, U. Baltensperger, P. Zieger, J. Schmale, Frequent new particle formation over the high Arctic pack ice by enhanced iodine emissions. Nat. Commun. **11**(1), 4924 (2020). https://doi.org/10.1038/s41467-020-18551-0

S. Baidar, H. Oetjen, S. Coburn, B. Dix, I. Ortega, R. Sinreich, R. Volkamer, The CU Airborne MAX-DOAS instrument: vertical profiling of aerosol extinction and trace gases. Atmos. Meas. Tech. **6**(3), 719–739 (2013). https://doi.org/10.5194/amt-6-719-2013

N.R. Baker, Chlorophyll fluorescence: a probe of photosynthesis in vivo. Annu. Rev. Plant Biol. **59**(1), 89–113 (2008). https://doi.org/10.1146/annurev.arplant.59.032607.092759

N. Balasus, D.J. Jacob, A. Lorente, J.D. Maasakkers, R.J. Parker, H. Boesch, S. Chen, M.M. Kelp, H. Nesser, D.J. Varon, A blended TROPOMI+GOSAT satellite data product for atmospheric methane using machine learning to correct retrieval biases. Atmos. Meas. Tech. **16**(16), 3787–3807 (2023). https://doi.org/10.5194/amt-16-3787-2023

S.M. Ball, I.M. Povey, E.G. Norton, R.L. Jones, Broadband cavity ringdown spectroscopy of the NO_3 radical. Chem. Phys. Lett. **342**, 113–120 (2001)

S.M. Ball, R.L. Jones, Broadband cavity ringdown spectroscopy. Chem. Rev. **103**, 5239 (2003)

S.M. Ball, J.M. Langridge, R.L. Jones, Broadband cavity enhanced absorption spectroscopy using light emitting diodes. Chem. Phys. Lett. **398**, 68–74 (2004)

J.E. Barnes, S. Bronner, R. Beck, N. Parikh, Boundary layer scattering measurements with a charge-coupled device camera lidar. Appl. Opt. **42**(15), 2647–2652 (2003)

J.E. Barnes, N.P. Sharma, T.B. Kaplan, Atmospheric aerosol profiling with a bistatic imaging lidar system. Appl. Opt. **46**(15), 2922–2929 (2007)

J. Barré, H. Petetin, A. Colette, M. Guevara, V.-H. Peuch, L. Rouil, R. Engelen, A. Inness, J. Flemming, C. Pérez García-Pando, D. Bowdalo, F. Meleux, C. Geels, J.H. Christensen, M. Gauss, A. Benedictow, S. Tsyro, E. Friese, J. Struzewska, J.W. Kaminski, J. Douros, R. Timmermans, L. Robertson, M. Adani, O. Jorba, M. Joly, R. Kouznetsov, Estimating lockdown-induced european NO_2 changes using satellite and surface observations and air quality models. Atmos. Chem. Phys. **21**(9), 7373–7394 (2021). https://doi.org/10.5194/acp-21-7373-2021

G. Barrefors, Monitoring of benzene, toluene and p-xylene in urban air with differential optical absorption spectroscopy technique. Sci. Total Environ. **189**(190), 287–292 (1996)

J.J. Barrett, S.A. Myers, New interferometric method for studying periodic spectra using a fabry-perot interferometer. J. Opt. Soc. Am. **61**(9), 1246–1251 (1971). https://doi.org/10.1364/JOSA.61.001246

L.A. Barrie, J.W. Bottenheim, R.C. Shnell, P.J. Crutzen, R.A. Rasmussen, Ozone destruction and photochemical reactions at polar sunrise in the lower Arctic atmosphere. Nature **334**, 138–141 (1988)

L. Barrie, U. Platt, Arctic tropospheric chemistry: an overview. Tellus B **49**, 449–454 (1997)

A.R. Barringer, J.H. Davies, A.J. Moffat, The problems and potential in monitoring pollution from satellites, AIAA paper, presented at the AIAA earth resources observations and information systems meeting, Annapolis, MD, 2–4 March 1970, A1AA Paper No. 70–305 (1970), pp. 62–75

A.M. Bass, R.J. Paur, The ultraviolet cross section of ozone. I. The measurements, in *Proceedings of the Quadrennial Ozone Symposium, Chalkidiki*, ed. by C. Zerefos, A. Ghazy (1985), pp. 606–616

D.R. Bates, M. Nicolet, The photochemistry of the atmospheric water vapor. J. Geophys. Res. **55**, 301 (1950)

D.R. Bates, A.E. Witherspoon, The photochemistry of some minor constituents of the earth's atmosphere (CO_2, CO, CH_4, N_2O). Mon. Not. R. Astron. Soc. **112**, 101 (1952)

T.S. Bates, R.J. Charlson, R.H. Gammon, Evidence for the climatic role of marine biogenic sulfur. Nature **329**, 319–321 (1987)

Bay Area AQMD, Air District Regulation 12, Rule 15: Petroleum Refining Emissions Tracking (2016). https://www.baaqmd.gov/rules-and-compliance/rules/regulation-12-rule-15--petroleum-refining-emissions-tracking

K.H. Becker, U. Schurath, T. Tatarczyk, Fluorescence determination of low formaldehyde concentrations in air by dye laser excitation. Appl. Opt. **14**, 310–313 (1975a)

K.H. Becker, A. Inocennio, U. Schurath, The reaction of ozone with hydrogen sulfide and its organic derivatives. Int. J. Chem. Kinet. **7**, 205–220 (1975b)

K.H. Becker (ed.), The European Photoreactor EUPHORE, Final Report of the EC-Project, Contract EV5V-CT92-0059. Department of Physical Chemistry, Bergische Universität Wuppertal (1996)

W.S. Beckett, M.B. Russi, A.D. Haber, R.M. Rivkin, J.R. Sullivan, Z. Tameroglu, V. Mohsenin, B.P. Leaderer, Effect of nitrous acid on lung function in asthmatics: a chamber study. Environ. Health Perspect. **103**, 372–375 (1995)

F. Bednarz (ed.), *GOME* (Global Ozone Monitoring Experiment. Users manual, European Space Research and Technology Centre (ESTEC) (Frascati, Italy, 1995)

M. Begoin, A. Richter, M. Weber, L. Kaleschke, X. Tian-Kunze, A. Stohl, N. Theys, J.P. Burrows, Satellite observations of long range transport of a large BrO plume in the arctic. Atmos. Chem. Phys. **10**(14), 6515–6526 (2010). https://doi.org/10.5194/acp-10-6515-2010

S. Beilke, H. Markusch, D. Jost, Measurements of NO-oxidation in power plant plumes by correlation spectroscopy, in ed. by B. Versino, H. Ott, *Proceedings of 2nd European Symposium on the Physico-Chemical Behaviour of Atmospheric Pollutants*, Varese, 29 September to 1 October 1981, D. Reidel Publishing Company (Dordrecht, Holland, 1981), pp. 448–459

S. Beirle, U. Platt, M. Wenig, T. Wagner, Weekly cycle of NO_2 by GOME measurements: a signature of anthropogenic sources. Atmos. Chem. Phys. **3**, 2225–2232 (2003)

S. Beirle, Estimating source strengths and lifetime of nitrogen oxides from satellite data. Ph.D. thesis, University of Heidelberg, Heidelberg (2004)

S. Beirle, U. Platt, M. Wenig, T. Wagner, Highly resolved global distribution of tropospheric NO_2 using GOME narrow swath mode data. Atmos. Chem. Phys. **4**, 1913–1924 (2004a)

S. Beirle, U. Platt, R. von Glasow, M. Wenig, T. Wagner, Estimate of nitrogen oxide emission from shipping by satellite remote sensing. Geophys. Res. Lett. **31**, L18102 (2004b). https://doi.org/10.1029/2004GL020312

S. Beirle, U. Platt, M. Wenig, T. Wagner, NO_X production by lightning estimated with GOME. Adv. Space Res. **34**, 793–797 (2004c)

S. Beirle, N. Spichtinger, A. Stohl, K.L. Cummins, T. Turner, D. Boccippio, O.R. Cooper, M. Wenig, M. Grzegorski, U. Platt, T. Wagner, Estimating the NO_x produced by lightning from GOME and NLDN data: a case study in the gulf of Mexico. Atmos. Chem. Phys. **6**(4), 1075–1089 (2006). https://doi.org/10.5194/acp-6-1075-2006

S. Beirle, H. Huntrieser, T. Wagner, Direct satellite observation of lightning-produced NO_x. Atmos. Chem. Phys. **10**(22), 10965–10986 (2010). https://acp.copernicus.org/articles/10/10965/2010/

S. Beirle, K.F. Boersma, U. Platt, M.G. Lawrence, T. Wagner, Megacity emissions and lifetimes of nitrogen oxides probed from space. Science **333**(6050), 1737–1739 (2011). https://www.science.org/doi/abs/10.1126/science.1207824

S. Beirle, H. Sihler, T. Wagner, Linearisation of the effects of spectral shift and stretch in DOAS analysis. Atmos. Meas. Tech. **6**(3), 661–675 (2013). https://amt.copernicus.org/articles/6/661/2013/

S. Beirle, C. Hörmann, M. Penning de Vries, S. Dörner, C. Kern, T. Wagner, Estimating the volcanic emission rate and atmospheric lifetime of SO_2 from space: a case study for Kīlauea volcano,

Hawai'i. Atmos. Chem. Phys. **14**(16), 8309–8322 (2014). https://acp.copernicus.org/articles/14/8309/2014/

S. Beirle, C. Hörmann, P. Jöckel, S. Liu, M. Penning de Vries, A. Pozzer, H. Sihler, P. Valks, T. Wagner, The STRatospheric estimation algorithm from Mainz (STREAM): estimating stratospheric NO_2 from nadir-viewing satellites by weighted convolution. Atmos. Meas. Tech. **9**(7), 2753–2779 (2016a). https://doi.org/10.5194/amt-9-2753-2016

S. Beirle, J. Lampel, C. Lerot, H. Sihler, T. Wagner, Parameterizing the instrumental spectral response function and its changes by a super-gaussian and its derivatives. Atmos. Meas. Tech. **10**(2), 581–598 (2017). https://doi.org/10.5194/amt-10-581-2017

S. Beirle, C. Borger, S. Dörner, A. Li, Z. Hu, F. Liu, Y. Wang, T. Wagner, Pinpointing nitrogen oxide emissions from space. Sci. Adv. **5**(11) (2019a). https://doi.org/10.1126/sciadv.aax9800

S. Beirle, S. Dörner, S. Donner, J. Remmers, Y. Wang, T. Wagner, The Mainz profile algorithm (MAPA). Atmos. Meas. Tech. **12**(3), 1785–1806 (2019b). https://doi.org/10.5194/amt-12-1785-2019

S. Beirle, C. Borger, S. Dörner, H. Eskes, V. Kumar, A. de Laat, T. Wagner, Catalog of NO_x emissions from point sources as derived from the divergence of the NO_2 flux for TROPOMI. Earth Syst. Sci. Data **13**(6), 2995–3012 (2021). https://essd.copernicus.org/articles/13/2995/2021/

S. Beirle, C. Borger, S. Dörner, V. Kumar, T. Wagner, Calculating the vertical column density of O_4 during daytime from surface values of pressure, temperature, and relative humidity. Atmos. Meas. Tech. **15**(4), 987–1006 (2022). https://doi.org/10.5194/amt-15-987-2022

M.L. Bell D.L. Davis, Reassessment of the lethal london fog of 1952: novel indicators of acute and chronic consequences of acute exposure to air pollution. Environ. Health Perspect. **109**, 389–394 (2001). https://doi.org/10.2307/3434786

D. Belmiloud, R. Schermaul, K.M. Smith, N.F. Zobov, J.W. Brault, R.C.M. Learner, D.A. Newnham, J. Tennyson, New studies of the visible and near infra-red absorption by water vapour and some problems with the database. Geophys. Res. Lett. **27**(22), 3703–3706 (2000)

N. Benavent, D. Garcia-Nieto, S. Wang, A. Saiz-Lopez, MAX-DOAS measurements and vertical profiles of glyoxal and formaldehyde in Madrid. Spain. Atmos. Environ. **199**, 357–367 (2019). https://doi.org/10.1016/j.atmosenv.2018.11.047

R.E. Benedick, *Ozone Diplomacy* (Harvard University Press, Cambridge, 1991)

S. Bergman, Lehrbuch der Experimentalphysik, Bd. III, Optik, F. Matossi (Walter de Gruyter & Co., Berlin, 1966)

H. Berresheim, T. Elste, C. Plass-Dülmer, F.L. Eisele, D.J. Tanner, Chemical ionization mass spectrometer for long-term measurements of atmospheric OH and H_2SO_4. Int. J. Mass Spectrom. **202**, 91–109 (2000)

G. Berthet, F. Jégou, V. Catoire, G. Krysztofiak, J.-B. Renard, A.E. Bourassa, D.A. Degenstein, C. Brogniez, M. Dorf, S. Kreycy, K. Pfeilsticker, B. Werner, F. Lefèvre, T.J. Roberts, T. Lurton, D. Vignelles, N. Bègue, Q. Bourgeois, D. Daugeron, M. Chartier, C. Robert, B. Gaubicher, C. Guimbaud, Impact of a moderate volcanic eruption on chemistry in the lower stratosphere: balloon-borne observations and model calculations. Atmos. Chem. Phys. **17**(3), 2229–2253 (2017). https://doi.org/10.5194/acp-17-2229-2017

P.R. Bevington, *Data Reduction and Error Analysis for the Physical Sciences* (McGraw-Hill, New York, 1969)

H.W. Biermann, E.C. Tuazon, A.M. Winer, T.J. Wallington, J.N. Pitts, Simultaneous absolute measurements of gaseous nitrogen species in urban ambient air by long pathlength infrared and ultraviolet-visible spectroscopy. Atmos. Environ. **22**, 1545–1554 (1988)

H.W. Biermann, M. Green, J.N. Neider, Long-pathlength DOAS (differential optical absorption spectrometer) system for the in situ measurement of xylene in indoor air, in *Measurements of Atmospheric Gases*, vol. 1433, ed. by H.I. Schiff (SPIE-The International Society of Optical Engineering, Bellingham, 1991), pp.2–7

R.G. Bingham, Grating spectrometers and spectrographs re-examined. R. Astron. Soc., Q. J. **20**, 395–421 (1979). https://ui.adsabs.harvard.edu/abs/1979QJRAS..20..395B. Provided by the SAO/NASA Astrophysics Data System

G.A. Bishop, D.H. Stedman, Reactive nitrogen species emission trends in three light-/medium-duty united states fleets. Environ. Sci. Technol. **49**(18), 11234–11240 (2015)

D. Blackie, R. Blackwell-Whitehead, G. Stark, J.C. Pickering, P.L. Smith, J. Rufus, A.P. Thorne, High-resolution photoabsorption cross-section measurements of SO_2 at 198 K from 213 to 325 nm. J. Geophys. Res.: Planets **116**(E3) (2011). https://doi.org/10.1029/2010JE003707

W. Bloss, T. Gravestock, D.E. Heard, T. Ingham, G.P. Johnson, J.D.L. Lee, Application of a compact all solid-state laser system to the in situ detection of atmospheric OH, HO_2, NO and IO by laser-induced fluorescence. J. Environ. Monit. **5**, 21–28 (2003)

N. Bobrowski, G. Hönninger, B. Galle, U. Platt, Detection of bromine monoxide in a volcanic plume. Nature **423**, 273–276 (2003)

N. Bobrowski, Volcanic gas studies by multi axis differential optical absorption spectroscopy. Ph.D. thesis, University of Heidelberg, Heidelberg (2005)

N. Bobrowski, G. Honninger, F. Lohberger, U. Platt, IDOAS: a new monitoring technique to study the 2D distribution of volcanic gas emissions. J. Volcanol. Geotherm. Res. **150**(4), 329–338 (2006). https://doi.org/10.1016/j.jvolgeores.2005.05.004

N. Bobrowski, R.V. Glasow, A. Aiuppa, S. Inguaggiato, I. Louban, O.W. Ibrahim, U. Platt, Reactive halogen chemistry in volcanic plumes. J. Geophys. Res. **112**, DO6311 (2007). https://doi.org/10.1029/2006JD007206

N. Bobrowski, U. Platt, SO_2/BrO ratios studied in five volcanic plumes. J. Volcanol. Geotherm. Res. **166**(3), 147–160 (2007). https://doi.org/10.1016/j.jvolgeores.2007.07.003

N. Bobrowski, C. Kern, U. Platt, C. Hörmann, T. Wagner, Novel SO_2 spectral evaluation scheme using the 360–390 nm wavelength range. Atmos. Meas. Tech. **3**(4), 879–891 (2010). https://amt.copernicus.org/articles/3/879/2010/

G.E. Bodeker, H. Shiona, H. Eskes, Indicators of Antarctic ozone depletion. Atmos. Chem. Phys. **5**, 2603–2615 (2005)

G.E. Bodeker, S. Kremser, Indicators of Antarctic ozone depletion: 1979–2019. Atmos. Chem. Phys. **21**(7), 5289–5300 (2021). https://doi.org/10.5194/acp-21-5289-2021

K.F. Boersma, H.J. Eskes, E.J. Brinksma, Error analysis for tropospheric NO_2 retrieval from space: error analysis for tropospheric NO_2. J. Geophys. Res.: Atmos. **109**(D4), D04311 (2004). https://doi.org/10.1029/2003JD003962

K.F. Boersma, H.J. Eskes, J.P. Veefkind, E.J. Brinksma, R.J. van der A, M. Sneep, G.H.J. van den Oord, P.F. Levelt, P. Stammes, J.F. Gleason, E.J. Bucsela, Near-real time retrieval of tropospheric NO_2 from OMI. Atmos. Chem. Phys. **7**(8), 2103–2118 (2007). https://doi.org/10.5194/acp-7-2103-2007

K.F. Boersma, H.J. Eskes, R.J. Dirksen, R.J. van der A, J.P. Veefkind, P. Stammes, V. Huijnen, Q.L. Kleipool, M. Sneep, J. Claas, J. Leitão, A. Richter, Y. Zhou, D. Brunner, An improved tropospheric NO_2 column retrieval algorithm for the ozone monitoring instrument. Atmos. Meas. Tech. **4**(9), 1905–1928 (2011). https://doi.org/10.5194/amt-4-1905-2011

K.F. Boersma, H.J. Eskes, A. Richter, I. De Smedt, A. Lorente, S. Beirle, J.H.G.M. van Geffen, M. Zara, E. Peters, M. van Roozendael, T. Wagner, J.D. Maasakkers, R.J. van der A, J. Nightingale, A. de Rudder, H. Irie, G. Pinardi, J.-C. Lambert, S.C. Compernolle, Improving algorithms and uncertainty estimates for satellite NO_2 retrievals: results from the quality assurance for the essential climate variables (qa4ecv) project. Atmos. Meas. Tech. **11**(12), 6651–6678 (2018). https://doi.org/10.5194/amt-11-6651-2018

K. Bognar, X. Zhao, K. Strong, R.Y.-W. Chang, U. Frieß, P.L. Hayes, A. McClure-Begley, S. Morris, S. Tremblay, A. Vicente-Luis, Measurements of tropospheric bromine monoxide over four halogen activation seasons in the Canadian high arctic. J. Geophys. Res.: Atmos. **125**(18), e2020JD033015 (2020). https://doi.org/10.1029/2020JD033015

K. Bogumil, J. Orphal, T. Homann, S. Voigt, P. Spietz, O. Fleischmann, A. Vogel, M. Hartmann, H. Kromminga, H. Bovensmann, J. Frerick, J. Burrows, Measurements of molecular absorption spectra with the SCIAMACHY pre-flight model: instrument characterization and reference data for atmospheric remote-sensing in the 230–2380 nm region. J. Photochem. Photobiol., A: Chem. **157**(2), 167–184 (2003). https://doi.org/10.1016/S1010-6030(03)00062-5

U. Bonafe, G. Cesari, G. Giovanelli, T. Tirabassi, O. Vittori, Mask correlation spectrophotometry advanced methodology for atmospheric measurements. Atmos. Environ. **10**, 469–474 (1976)
A. Bongartz, J. Kamens, F. Welter, U. Schurath, Near-UV absorption cross sections and trans/cis equilibrium of nitrous acid. J. Phys. Chem. **95**, 1076–1082 (1991)
C. Borger, S. Beirle, S. Dörner, H. Sihler, T. Wagner, Total column water vapour retrieval from S-5P/TROPOMI in the visible blue spectral range. Atmos. Meas. Tech. 13(5), 2751–2783 (2020). https://amt.copernicus.org/articles/13/2751/2020/
P. Borrell, J.P. Burrows, U. Platt, C. Zehner, Determining tropospheric concentrations of trace gases from space. ESA Bull. **107**, 72–81 (2001)
P. Borrell, J.P. Burrows, A. Richter, U. Platt, T. Wagner, New directions: new developments in satellite capabilities for probing the chemistry of the troposphere. Atmos. Environ. **37**, 2567–2570 (2003)
D. Bortoli, F. Ravegnani, G. Giovanelli, P.S. Kulkarni, M. Anton, M.J. Costa, A.M. Silva, Fifteen years of stratospheric nitrogen dioxide and ozone measurements in Antarctica. AIP Conf. Proc. **1531**(1), 300–303 (2013). https://doi.org/10.1063/1.4804766
H. Bösch, Studies of the stratospheric nitrogen and iodine chemistry by balloon-borne DOAS measurements and model calculations. Ph.D. thesis, University of Heidelberg (2002)
H. Bösch, C. Camy-Peyret, M.P. Chipperfield, R. Fitzenberger, H. Harder, U. Platt, K. Pfeilsticker, Upper limits of stratospheric IO and OIO inferred from center-to-limb-darkening-corrected balloon-borne solar occultation visible spectra: implications for total gaseous iodine and stratospheric ozone. J. Geophys. Res. **108**, 445 (2003). https://doi.org/10.1029/2002JD003078
T. Bösch, V. Rozanov, A. Richter, E. Peters, A. Rozanov, F. Wittrock, A. Merlaud, J. Lampel, S. Schmitt, M. de Haij, S. Berkhout, B. Henzing, A. Apituley, M. den Hoed, J. Vonk, M. Tiefengraber, M. Müller, J.P. Burrows, BOREAS–a new MAX-DOAS profile retrieval algorithm for aerosols and trace gases. Atmos. Meas. Tech. **11**(12), 6833–6859 (2018). https://doi.org/10.5194/amt-11-6833-2018
J. Bossmeyer, Ship-based multi-axis differential optical absorption spectroscopy measurements of tropospheric trace gases over the Atlantic Ocean: new measurement concepts. Diploma Thesis, University of Heidelberg, Germany (2002)
J.W. Bottenheim, A.G. Gallant, K.A. Brice, Measurements of NO_y species and O_3 at 82°N latitude. Geophys. Res. Lett. **13**, 113–116 (1986)
J.W. Bottenheim, L.W. Barrie, E. Atlas, L.E. Heidt, H. Niki, R.A. Rasmussen, P.B. Shepson, Depletion of lower tropospheric ozone during Arctic spring: the polar sunrise experiment 1988. J. Geophys. Res. **95**, 18555–18568 (1990)
P. Braesicke, J.L. Neu, V.E. Fioletov, S. Godin-Beekmann, D. Hubert, I. Petropavlovskikh, M. Shiotani, B.-M. Sinnhuber, W. Ball, K.-L. Chang, R. Damadeo, S. Dhomse, S. Frith, A. Gaudel, B. Hassler, R. Hossaini, S. Kremser, S. Misios, O. Morgenstern, R.J. Salawitch, V. Sofieva, K. Tourpali, O. Tweedy, D. Zawada, Update on global ozone: past, present, and future, Chap. 3 in WMO scientific assessment of ozone depletion (2018). Technical Report. https://elib.dlr.de/135132/
J.C.D. Brand, K. Srikameswaran, The Π^*-Π (2350 A) band system of sulphur dioxide. Chem. Phys. Lett. **15**, 130–132 (1972)
J.C.D. Brand, R. Nanes, The 3400–3000 A absorption of sulphur dioxide. J. Mol. Spectrosc. **46**, 194–199 (1973)
J.C.D. Brand, V.T. Jones, C. DiLauro, The $^3B_1$-$^1A_1$ band system of sulphur dioxide: rotational analysis of the (010), (100), and (110) bands. J. Mol. Spectrosc. **45**, 404–411 (1973)
U. Brandenburger, T. Brauers, H.-P. Dorn, M. Hausmann, D.H. Ehhalt, In-situ measurement of tropospheric hydroxyl radicals by folded long-path laser absorption during the field campaign POPCORN in 1994. J. Atmos. Chem. **31**, 181–204 (1998)
R. Brandtjen, T. Klüpfel, D. Perner, Airborne measurements during the European Arctic stratospheric ozone experiment: observation of OClO. Geophys. Res. Lett. **21**, 1363–1366 (1994)
G. Brasseur, S. Solomon, *Aeronomy of the Middle Atmosphere*, 2nd edn. (D. Reidel Publishing Company, Dordrecht, 1986)
G.P. Brasseur D.J. Jacob, *Modeling of Atmospheric Chemistry* (Cambridge University Press, 2017)

D.J. Brassington, Sulphur dioxide absorption cross-section measurements from 290 to 317 nm. Appl. Opt. **20**, 3774–3779 (1981)

D.J. Brassington, R.C. Felton, B.W. Jolliffe, B.R. Marx, J.T.M. Moncrieff, Errors in spectroscopic measurements of SO_2 due to nonexponential absorption of laser radiation, with application to the remote monitoring of atmospheric pollution. Appl. Opt. **23**, 469–475 (1984)

T. Brauers, H.-P. Dorn, U. Platt, Spectroscopic measurements of NO_2, O_3, SO_2, IO and NO_3 in maritime air, in ed. by G. Restelli, G. Angeletti, *Physico-Chemical Behaviour of Atmospheric Pollutants, Proceedings of the 5th European Symposium, Varese, Italia* (Kluwer Academic Publishers, Dordrecht, 1990), pp. 237–242

T. Brauers, *FZ-Jülich* (Unpublished results, Germany, 1992)

T. Brauers, M. Hausmann, U. Brandenburger, H.-P. Dorn, Improvement of differential optical absorption spectroscopy using multichannel scanning technique. Appl. Opt. **34**(21), 4472–4479 (1995)

T. Brauers, U. Aschmutat, U. Brandenburger, H.-P. Dorn, M. Hausmann, M. Heßling, A. Hofzumahaus, F. Holland, C. Plass-Dülmer, D.H. Ehhalt, Intercomparison of tropospheric OH radical measurements by multiple folded long-path laser absorption and laser induced fluorescence. Geophys. Res. Lett. **23**, 2545–2548 (1996)

T. Brauers, M. Hausmann, A. Bister, A. Kraus, H.-P. Dorn, OH radicals in the boundary layer of the Atlantic Ocean: 1. Measurements by long-path laser absorption spectroscopy. J. Geophys. Res. **106**, 7399–7414 (1999)

T. Brauers, J. Bossmeyer, H.-P. Dorn, E. Schlosser, R. Tillmann, R. Wegener, A. Wahner, Investigation of the formaldehyde differential absorption cross section at high and low spectral resolution in the simulation chamber SAPHIR. Atmos. Chem. Phys. **7**(13), 3579–3586 (2007). https://acp.copernicus.org/articles/7/3579/2007/

T. Breider, M. Chipperfield, N. Richards, K. Carslaw, G. Mann, D. Spracklen, Impact of BrO on DMS in the remote marine boundary layer. Geophys. Res. Lett. **37**, 01 (2010). https://doi.org/10.1029/2009GL040868

C.A.M. Brenninkmeijer, M.R. Manning, D.C. Lowe, G.A. Wallace, R.J. Sparks, A. Volz-Thomas, Interhemispheric asymmetry in OH abundance inferred from measurements of atmospheric ^{14}CO. Nature **356**, 50–52 (1992)

A.W. Brewer, C.T. McElroy, J.B. Kerr, Nitrogen dioxide concentrations in the atmosphere. Nature **246**, 129–133 (1973)

P. Brimblecombe, M. Heymann, *TRAP-45-analysis of tropospheric air pollution problems and air pollution abatement strategies in Europe since 1945, A Subproject in EUROTRAC-2* (International Scientific Secretariat, GSF-Forschungszentrum für Umwelt and Gesundheit GmbH, Munich, Germany, 1998)

E.J. Brinksma, G. Pinardi, H. Volten, R. Braak, A. Richter, A. Schönhardt, M. van Roozendael, C. Fayt, C. Hermans, R.J. Dirksen, T. Vlemmix, A.J.C. Berkhout, D.P.J. Swart, H. Oetjen, F. Wittrock, T. Wagner, O.W. Ibrahim, G. de Leeuw, M. Moerman, R.L. Curier, E.A. Celarier, A. Cede, W.H. Knap, J.P. Veefkind, H.J. Eskes, M. Allaart, R. Rothe, A.J.M. Piters, P.F. Levelt, The 2005 and 2006 DANDELIONS NO_2 and aerosol intercomparison campaigns. J. Geophys. Res.: Atmos. **113**(D16) (2008). https://doi.org/10.1029/2007JD008808

R. Bröske, J. Kleffmann, P. Wiesen, Heterogeneous conversion of NO_2 on secondary organic aerosol surfaces: a possible source of nitrous acid (HONO) in the atmosphere? J. Atmos. Chem. Phys. **3**, 469–474 (2003)

S.S. Brown, H. Stark, S.J. Ciciora, A.R. Ravishankara, In-situ measurement of atmospheric NO_3 and N_2O_5 via cavity ring-down spectroscopy. Geophys. Res. Lett. **28**(17), 3227–3230 (2001)

S.S. Brown, H. Stark, S.J. Ciciora, R.J. McLaughlin, A.R. Ravishankara, Simultaneous in situ detection of atmospheric NO_3 and N_2O_5 via cavity ring-down spectroscopy. Rev. Sci. Instrum. **73**(9), 3291–3301 (2002)

S.S. Brown, Absorption spectroscopy in high-finesse cavities for atmospheric studies. Chem. Rev. **103**(12), 5219–5238 (2003)

S.S. Brown, H. Stark, T.B. Ryerson, E.J. Williams, D.K. Nicks Jr., M. Trainer, F.C. Fehsenfeld, A.R. Ravishankara, Nitrogen oxides in the nocturnal boundary layer: simultaneous in situ measurements of NO_3, N_2O_5, NO_2, NO, and O_3. J. Geophys. Res. **108**(D9), 4299 (2003). https://doi.org/10.1029/2002JD002917

S.S. Brown, J. Stutz, Nighttime radical observations and chemistry. Chem. Soc. Rev. **41**(19), 6405–6447 (2012)

W.H. Brune, P.S. Stevens, J.H. Mather, Measuring OH and HO_2 in the troposphere by laser induced fluorescence at low pressure. J. Geophys. Res. **99**, 3543–3557 (1995). https://doi.org/10.1029/93JD03342

W.H. Brune, D.O. Miller, A.B. Thames, H.M. Allen, E.C. Apel, D.R. Blake, T.P. Bui, R. Commane, J.D. Crounse, B.C. Daube, G.S. Diskin, J.P. DiGangi, J.W. Elkins, S.R. Hall, T.F. Hanisco, R.A. Hannun, E.J. Hintsa, R.S. Hornbrook, M.J. Kim, K. McKain, F.L. Moore, J.A. Neuman, J.M. Nicely, J. Peischl, T.B. Ryerson, J.M. St. Clair, C. Sweeney, A.P. Teng, C. Thompson, K. Ullmann, P.R. Veres, P.O. Wennberg, G.M. Wolfe, Exploring oxidation in the remote free troposphere: insights from atmospheric tomography (Atom). J. Geophys. Res.: Atmos. **125**(1), e2019JD031685 (2020). https://doi.org/10.1029/2019JD031685

B. Brunekreef, S.T. Holgate, Air pollution and health. Lancet **360**, 1233–1242 (2002)

M. Bruns, S.A. Buehler, J.P. Burrows, K.-P. Heue, U. Platt, I. Pundt, A. Richter, A. Rozanov, T. Wagner, P. Wang, Retrieval of profile information from airborne multi axis UV/visible skylight absorption measurements. Appl. Opt. **43**(22), 4415–4426 (2004)

M. Bruns, S.A. Buehler, J.P. Burrows, A. Richter, A. Rozanov, P. Wang, K.P. Heue, U. Platt, I. Pundt, T. Wagner, NO_2 profile retrieval using airborne multi axis UV-visible skylight absorption measurements over central Europe. Atmos. Chem. Phys. **6**, 3049–3058 (2006)

M. Buchwitz, V. Rozanov, J. Burrows, A near-infrared optimised DOAS method for the fast global retrieval of atmospheric CH_4, CO, CO_2, H_2O, and N_2O total column amounts from SCIAMACHY Envisat-1 nadir radiances. J. Geophys. Res. **105**, 15231–15245 (2000)

M. Buchwitz, M. Reuter, O. Schneising, H. Boesch, S. Guerlet, B. Dils, I. Aben, R. Armante, P. Bergamaschi, T. Blumenstock, H. Bovensmann, D. Brunner, B. Buchmann, J. Burrows, A. Butz, A. Chédin, F. Chevallier, C. Crevoisier, N. Deutscher, C. Frankenberg, F. Hase, O. Hasekamp, J. Heymann, T. Kaminski, A. Laeng, G. Lichtenberg, M. de Mazière, S. Noël, J. Notholt, J. Orphal, C. Popp, R. Parker, M. Scholze, R. Sussmann, G. Stiller, T. Warneke, C. Zehner, A. Bril, D. Crisp, D. Griffith, A. Kuze, C. O'Dell, S. Oshchepkov, V. Sherlock, H. Suto, P. Wennberg, D. Wunch, T. Yokota, Y. Yoshida, The greenhouse gas climate change initiative (GHG-CCI): comparison and quality assessment of near-surface-sensitive satellite-derived CO_2 and CH_4 global data sets. Remote Sens. Environ. **162**, 344–362 (2015). https://doi.org/10.1016/j.rse.2013.04.024

M. Buchwitz, M. Reuter, O. Schneising, W. Hewson, R. Detmers, H. Boesch, O. Hasekamp, I. Aben, H. Bovensmann, J. Burrows, A. Butz, F. Chevallier, B. Dils, C. Frankenberg, J. Heymann, G. Lichtenberg, M. De Mazière, J. Notholt, R. Parker, T. Warneke, C. Zehner, D. Griffith, N. Deutscher, A. Kuze, H. Suto, D. Wunch, Global satellite observations of column-averaged carbon dioxide and methane: the GHG-CCI XCO_2 and XCH_4 CRDP3 data set. Remote Sens. Environ. **203**, 276–295 (2017). https://doi.org/10.1016/j.rse.2016.12.027

E. Bucsela, E. Celarier, M. Wenig, J. Gleason, J. Veefkind, K. Boersma, E. Brinksma, Algorithm for NO_2 vertical column retrieval from the ozone monitoring instrument. IEEE Trans. Geosci. Remote Sens. **44**(5), 1245–1258 (2006). https://doi.org/10.1109/TGRS.2005.863715

E.J. Bucsela, N.A. Krotkov, E.A. Celarier, L.N. Lamsal, W.H. Swartz, P.K. Bhartia, K.F. Boersma, J.P. Veefkind, J.F. Gleason, K.E. Pickering, A new stratospheric and tropospheric NO_2 retrieval algorithm for nadir-viewing satellite instruments: applications to OMI. Atmos. Meas. Tech. **6**(10), 2607–2626 (2013). https://doi.org/10.5194/amt-6-2607-2013

D.A. Burgard, T.R. Dalton, G.A. Bishop, J.R. Starkey, D.H. Stedman, Nitrogen dioxide, sulfur dioxide, and ammonia detector for remote sensing of vehicle emissions. Rev. Sci. Instrum. **77**(1), 014101 (2006). https://doi.org/10.1063/1.2162432

J.B. Burkholder, R.K. Talukdar, Temperature dependence of the ozone absorption cross section over the wavelength range 410 to 760 nm. Geophys. Res. Lett. **21**, 581–584 (1994)

J.B. Burkholder, J. Curtius, A.R. Ravishankara, E.R. Lovejoy, Laboratory studies of the homogeneous nucleation of iodine oxides. Atmos. Chem. Phys. **4**, 19–34 (2004)

J. Burkholder, S. Sander, J. Abbatt, J. Barker, C. Cappa, J. Crounse, T.S. Dibble, R. Huie, C. Kolb, M. Kurylo, V. Orkin, C. Percival, D. Wilmouth, P.H. Wine, Chemical Kinetics and Photochemical Data for Use in Atmospheric Studies, Evaluation No. 19. Publication 19-5, Jet Propulsion Laboratory, Pasadena (2016). http://jpldataeval.jpl.nasa.gov

J.P. Burrows, G.S. Tyndall, G.K. Moortgat, Absorption spectrum of NO_3 and kinetics of the reactions of NO_3 with NO_2, Cl, and several stable atmospheric species at 298 K. J. Phys. Chem. **89**, 4848–4856 (1985)

J.P. Burrows, K.V. Chance, P.J. Crutzen, J. Fishman, J.E. Fredericks, J.C. Geary, T.J. Johnson, G.W. Harris, I.S.A. Isaksen, H. Kelder, G.K. Moortgat, C. Muller, D. Perner, U. Platt, J.-P. Pommereau, H. Rodhe, E. Roeckner, W. Schneider, P. Simon, H. Sundqvist, J. Vercheval, SCIAMACHY phase A study—scientific requirements specification, report to European space agency (1991)

J. Burrows, M. Vountas, H. Haug, K. Chance, L. Marquard, K. Muirhead, U. Platt, A. Richter, V. Rozanov, Study of the ring effect, final report for ESA contract 109996/94/NL/CN, ESA ITT No. AO/1-2778/94/NL/CN (1995)

J. Burrows, U. Platt, K. Chance, M. Vountas, V. Rozanov, A. Richter, H. Haug, L. Marquard, *Study of the ring effect* (European Space Agency, Noordiwijk, 1996)

J.P. Burrows, A. Dehn, B. Deters, S. Himmelmann, A. Richter, S. Voigt, J. Orphal, Atmospheric remote-sensing reference data from GOME: part I. temperature-dependent absorption cross-sections of NO_2 in the 231–794 nm range. J. Quant. Spectrosc. Radiat. Transf. **60**, 1025–1031 (1998)

J.P. Burrows, A. Richter, A. Dehn, B. Deters, S. Himmelmann, Atmospheric remote-sensing reference data from GOME: part 2. temperature-dependent absorption cross sections of O_3 in the 231–794 nm range. J. Quant. Spectrosc. Radiat. Transf. **61**, 509–517 (1999a)

J.P. Burrows, M. Weber, M. Buchwitz, V. Rozanov, A. Ladstätter-Weißenmayer, A. Richter, R. DeBeek, R. Hoogen, K. Bramstedt, K.-U. Eichmann, M. Eisinger, D. Perner, The global ozone monitoring experiment (GOME): mission concept and first scientific results. J. Atmos. Sci. **56**, 151–171 (1999b)

J.P. Burrows, U. Platt, P. Borrell, The remote sensing of tropospheric composition from space. Springer Sci. Bus. Media (2011). https://doi.org/10.1007/978-3-64214791-3

M. Burton, C. Hayer, C. Miller, B. Christenson, Insights into the 9 December 2019 eruption of Whakaari/White Island from analysis of TROPOMI SO_2 imagery. Sci. Adv. **7**(25), eabg1218 (2021)

M. Bussemer, Der Ring-Effekt: Ursachen und Einfluß auf die spektroskopische Messung stratosphärischer Spurenstoffe. Diploma thesis in Physics, University Heidelberg (1993)

N.I. Butkovskaya, G. Le Bras, Mechanism of the $NO_3 + DMS$ reaction by discharge flow mass spectrometry. J. Phys. Chem. **98**, 2582–2591 (1994)

A. Butz, H. Bösch, C. Camy-Peyret, M. Chipperfield, M. Dorf, G. Dufour, K. Grunow, P. Jeseck, S. Kühl, S. Payan, I. Pepin, J. Pukite, A. Rozanov, C. von Savigny, C. Sioris, T. Wagner, F. Weidner, K. Pfeilsticker, Inter-comparison of stratospheric O_3 and NO_2 abundances retrieved from balloon borne direct sun observations and Envisat/SCIAMACHY limb measurements. J. Atmos. Chem. Phys. **6**, 1293–1314 (2006)

A. Butz, H. Bösch, C. Camy-Peyret, M.P. Chipperfield, M. Dorf, S. Kreycy, L. Kritten, C. Prados-Román, J. Schwärzle, K. Pfeilsticker, Constraints on inorganic gaseous iodine in the tropical upper troposphere and stratosphere inferred from balloon-borne solar occultation observations. Atmos. Chem. Phys. **9**(18), 7229–7242 (2009). https://doi.org/10.5194/acp-9-7229-2009

A. Butz, O.P. Hasekamp, C. Frankenberg, J. Vidot, I. Aben, CH_4 retrievals from space-based solar backscatter measurements: performance evaluation against simulated aerosol and cirrus loaded scenes. J. Geophys. Res.: Atmos. **115**(D24) (2010). https://doi.org/10.1029/2010JD014514

A. Butz, A. Galli, O. Hasekamp, J. Landgraf, P. Tol, I. Aben, TROPOMI aboard Sentinel-5 precursor: prospective performance of CH_4 retrievals for aerosol and cirrus loaded atmospheres. Remote Sens. Environ. **120**, 267–276 (2012). https://doi.org/10.1016/j.rse.2011.05.030

California. California Assembly Bill No. 1647 Muratsuchi. Petroleum refineries: air monitoring systems (2017). https://leginfo.legislature.ca.gov/faces/billTextClient.xhtml?bill_id=201720180AB1647

J.G. Calvert, Hydrocarbon involvement in photochemical smog formation in Los Angeles atmosphere. Environ. Sci. Technol. **10**, 256–262 (1976)

J.G. Calvert, A. Lazrus, G.L. Kok, B.G. Heikes, J.G. Walega, J. Lind, C.A. Cantrell, Chemical mechanisms of acid generation in the troposphere. Nature **317**, 27–35 (1985)

J.G. Calvert, G. Yarwood, A.M. Dunker, An evaluation of the mechanism of nitrous acid formation in the urban atmosphere. Res. Chem. Intermed. **20**(3–5), 463–502 (1994)

K. Calvin, D. Dasgupta, G. Krinner, A. Mukherji, P.W. Thorne, C. Trisos, J. Romero, P. Aldunce, K. Barrett, G. Blanco, et al., IPCC, 2023: Climate Change 2023: Synthesis Report, in *Contribution of Working Groups I, II and III to the Sixth Assessment Report of the Intergovernmental Panel on Climate Change*, ed. by Core Writing Team, H. Lee, J. Romero (IPCC, Geneva, Switzerland, 2023). Technical report, Intergovernmental Panel on Climate Change (IPCC). https://www.ipcc.ch/report/ar6/syr/

M.J. Campbell, J.C. Sheppard, B.F. Au, Measurement of hydroxyl concentration in boundary-layer air by monitoring CO oxidation. Geophys. Res. Lett. **6**, 175–178 (1979)

M.J. Campbell, J.C. Farmer, C.A. Fitzner, M.N. Henry, J.C. Sheppard, R.J. Hardy, J.F. Hopper, V. Muralidhar, Radiocarbon tracer measurements of atmospheric hydroxyl radical concentrations. J. Atmos. Chem. **4**, 413–427 (1986)

C. Camy-Peyret, P. Jeseck, T. Hawat, G. Durry, E. Berubeé, L. Rochette, D. Huguenin, The LPMA balloon borne FTIR spectrometer for remote sensing of the atmospheric constituents, in ed. by T.A. Blix *Proceedings of 12th ESA Symposium on Rocket and Balloon Programmes and Related Research, Lillehammer, Norway, 29 May–1 June 1995*, ESA SP #370, Paris: European Space Agency (ESA) (1995), pp. 323–328

C. Camy-Peyret, B. Bergqvist, B. Galle, M. Carleer, C. Clerbaux, R. Colin, C. Fayt, F. Goutail, M. Nunes-Pinharanda, J.P. Pommereau, H. Hausmann, U. Platt, I. Pundt, T. Rudolph, C. Hermans, P.C. Simon, A.C. Vandaele, J.M.C. Plane, N. Smith, Intercomparison of instruments for tropospheric measurements using differential optical absorption spectroscopy. J. Atmos. Chem. **23**, 51–80 (1996)

C. Camy-Peyret, T. Hawat, P. Jeseck, S. Payan, Recent results obtained by the LPMA balloon-borne experiment during SESAME and the ILAS validation campaign European Rocket and Balloon Programmes and Related Research, Proceedings of the 13th ESA Symposium held 26–29 May, 1997 in Öland, Sweden. Edited by Brigitte Kaldeich-Schürmann. ESA SP-397. Paris: European Space Agency, 1997., p.223

C. Camy-Peyret, S. Payan, P. Jeseck, Y. Té, T. Hawat, K. Pfeilsticker, H. Harder, R. Fitzenberger, H. Bösch, Recent results obtained with the LPMA and DOAS balloon-borne instruments during the ILAS, SABINE and THESEO campaigns, in *Proceedings of 14th ESA Symposium on Rocket and Balloon Programmes and Related Research* (1999)

Canrad-Hanovia, Compact arc lamps, data sheet. Canrad-Hanovia Inc., Newark (1986)

C.A. Cantrell, D.H. Stedman, G.J. Wendel, Measurement of atmospheric peroxy radicals by chemical amplification. Anal. Chem. **56**, 1496–1502 (1984)

C.A. Cantrell, J.A. Davidson, A.H. McDaniel, R.E. Shetter, J.G. Calvert, Temperature-dependent formaldehyde cross section in the near-ultraviolet spectral region. J. Phys. Chem. **94**, 3902–3908 (1990)

C.A. Cantrell, R.E. Shetter, J.A. Lind, A.H. Mcdaniel, J.G. Calvert, D.D. Parrish, F.C. Fehsenfeld, M.P. Buhr, M. Trainer, An improved chemical amplifier technique for peroxy radical measurements. J. Geophys. Res. **98**, 2897–2909 (1993)

H. Cao, T.-M. Fu, L. Zhang, D.K. Henze, C.C. Miller, C. Lerot, G.G. Abad, I. De Smedt, Zhang Q, van Roozendael M, Hendrick F, Chance K, Li J, Zheng J, Zhao Y, Adjoint inversion of Chinese non-methane volatile organic compound emissions using space-based observations of formaldehyde and glyoxal. Atmos. Chem. Phys. **18**(20), 15017–15046 (2018). https://acp.copernicus.org/articles/18/15017/2018/

L.M. Cardenas, D.J. Brassington, B.J. Allan, H. Coe, B. Alicke, U. Platt, K.M. Wilson, J.M.C. Plane, S.A. Penkett, Intercomparison of formaldehyde measurements in clean and polluted atmospheres. J. Atmos. Chem. **37**, 53–80 (2000)

S. Carn, L. Clarisse, A. Prata, Multi-decadal satellite measurements of global volcanic degassing. J. Volcanol. Geotherm. Res. **311**, 99–134 (2016). https://doi.org/10.1016/j.jvolgeores.2016.01.002

S.A. Carn, V.E. Fioletov, C.A. McLinden, C. Li, N.A. Krotkov, A decade of global volcanic SO_2 emissions measured from space. Sci. Rep. **7**(1), 44095 (2017). https://doi.org/10.1038/srep44095

J. Caron, B. Kruizinga, R. Vink, Slit homogenizers for earth observation spectrometers: overview on performance, present and future designs, in *International Conference on Space Optics-ICSO 2018*, vol. 11180 (SPIE, 2019), pp. 402–417

L.J. Carpenter, W.T. Sturges, S.A. Penkett, P.S. Liss, B. Alicke, K. Hebestreit, U. Platt, Short-lived alkyl iodides and bromides at Mace Head, Ireland: links to biogenic sources and halogen oxide production. J. Geophys. Res. **104**, 1679–1689 (1999)

L.J. Carpenter, K. Hebestreit, U. Platt, P.S. Liss, Coastal zone production of IO precursors: a 2-dimensional study. Atmos. Chem. Phys. **1**, 9–18 (2001)

L.J. Carpenter, S.M. MacDonald, M.D. Shaw, R. Kumar, R.W. Saunders, R. Parthipan, J. Wilson, J.M.C. Plane, Atmospheric iodine levels influenced by sea surface emissions of inorganic iodine. Nat. Geosci. **6**(2), 108–111 (2013). https://doi.org/10.1038/ngeo1687

M.A. Carroll, R.W. Sanders, S. Solomon, A.L. Schmeltekopf, Visible and near-ultraviolet spectroscopy at McMurdo Station, Antarctica, 6. Observations of BrO. J. Geophys. Res. **94**(D14), 16633–16638 (1989)

N. Carslaw, L. Carpenter, J.M.C. Plane, B.J. Allan, R.A. Burgess, K. Clemitshaw, C.H. Coe, S.A. Penkett, Simultaneous observations of nitrate and peroxy radicals in the marine boundary layer. J. Geophys. Res. **102**, 18917–18933 (1997)

N. Carslaw, D.J. Creasey, D.E. Heard, A.C. Lewis, J.B. McQuaid, M.J. Pilling, P.S. Monks, B.J. Bandy, S.A. Penkett, Modeling OH, HO_2, and RO_2 radicals in the marine boundary layer, 1, model construction and comparison with field measurements. J. Geophys. Res. **104**, 30241–30255 (1999)

N. Casaballe, M.D. Martino, M. Osorio, J.A. Ferrari, T. Wagner, E. Frins E. Improved algorithm with adaptive regularization for tomographic reconstruction of gas distributions using DOAS measurements. Appl. Opt. **59**(13), D179–D188 (2020). https://opg.optica.org/ao/abstract.cfm?URI=ao-59-13-D179

S. Casadio, C. Zehner, G. Pisacane, E. Putz, Empirical retrieval of the atmospheric air mass factor (ERA) for the measurement of water vapour vertical content using GOME data. Geophys. Res. Lett. **27**(10), 1483–1486 (2000)

P. Castellanos, K.F. Boersma, G.R. van der Werf, Satellite observations indicate substantial spatiotemporal variability in biomass burning NO_x emission factors for South America. Atmos. Chem. Phys. **14**(8), 3929–3943 (2014). https://doi.org/10.5194/acp-14-3929-2014

H. Cauer, Schwankungen der Jodmenge der Luft in Mitteleuropa, deren Ursachen und deren Bedeutung für den Jodgehalt unserer Nahrung (Auszug). Angew. Chem. **52**(11), 625–628 (1939)

C. Chan Miller, G. Gonzalez Abad, H. Wang, X. Liu, T. Kurosu, D.J. Jacob, K. Chance, Glyoxal retrieval from the ozone monitoring instrument. Atmos. Meas. Tech. **7**(11), 3891–3907 (2014). https://doi.org/10.5194/amt-7-3891-2014

J.K. Chan, S. Parasurama, R. Atlas, R. Xu, U.A. Jongebloed, B. Alexander, J.M. Langenhan, J.A. Thornton, J.A. Riffell, Olfaction in the anthropocene: NO_3 negatively affects floral scent and nocturnal pollination. Science **383**(6683), 607–611 (2024). https://www.science.org/doi/abs/10.1126/science.adi0858

K.L. Chan, M. Wiegner, M. Wenig, D. Pöhler, Observations of tropospheric aerosols and NO_2 in Hong Kong over 5 years using ground based MAX-DOAS. Sci. Total Environ. **619–620**, 1545–1556 (2018). https://doi.org/10.1016/j.scitotenv.2017.10.153

K.L. Chan, M. Wiegner, J. van Geffen, I. De Smedt, C. Alberti, Z. Cheng, S. Ye, M. Wenig, Max-DOAS measurements of tropospheric NO_2 and HCHO in Munich and the comparison to OMI and TROPOMI satellite observations. Atmos. Meas. Tech. **13**(8), 4499–4520 (2020). https://amt.copernicus.org/articles/13/4499/2020/

K.L. Chan, P. Valks, K.-P. Heue, R. Lutz, P. Hedelt, D. Loyola, G. Pinardi, M. van Roozendael, F. Hendrick, T. Wagner, V. Kumar, A. Bais, A. Piters, H. Irie, H. Takashima, Y. Kanaya, Y. Choi, K. Park, J. Chong, A. Cede, U. Frieß, A. Richter, J. Ma, N. Benavent, R. Holla, O. Postylyakov, C. Rivera Cárdenas, M. Wenig, Global Ozone Monitoring Experiment-2 (GOME-2) daily and monthly level-3 products of atmospheric trace gas columns. Earth Syst. Sci. Data **15**(4), 1831–1870 (2023). https://doi.org/10.5194/essd-15-1831-2023

K. Chance, P.I. Palmer, R.J.D. Spurr, R.V. Martin, T.P. Kurosu, D.J. Jacob et al., Satellite observations of formaldehyde over North America from GOME. Geophy. Res. Lett. **27**(21), 3461–3464 (2000)

K. Chance, T.P. Kurosu, C.E. Sioris, Undersampling correction for arraydetector-based satellite spectrometers. Appl. Opt. **44**(7), 1296–1304 (2005)

K. Chance, R.L. Kurucz, An improved high-resolution solar reference spectrum for earth's atmosphere measurements in the ultraviolet, visible, and near infrared. J. Quant. Spectrosc. Radiat. Transf. **111**(9), 1289–1295 (2010)

S. Chapman, On ozone and atomic oxygen in the upper atmosphere. Philos. Mag. **7**, 369–383 (1930)

J. Chappuis, Sur le spectre d'absorption de l'ozone. C.R. Acad. Sci. **91**, 985 (1880)

R.J. Charlson, H. Rhode, Factors controlling the acidity of natural rainwater. Nature **295**, 683–685 (1982)

R.J. Charlson, J.E. Lovelock, M.O. Andreae, S.G. Warren, Oceanic phytoplankton, atmospheric sulphur, cloud albedo and climate. Nature **326**, 655–661 (1987)

Y.Q. Chen, L. Zhu, Wavelength-dependent photolysis of glyoxal in the 290–420 nm region. J. Phys. Chem. A **107**(23), 4643–4651 (2003)

Q. Chen, T. Sherwen, M. Evans, B. Alexander. DMS oxidation and sulfur aerosol formation in the marine troposphere: a focus on reactive halogen and multiphase chemistry. Atmos. Chem. Phys. **18**(18), 13617–13637 (2018). https://acp.copernicus.org/articles/18/13617/2018/

W. Chen, D.S. Venables, 3—broadband optical cavity methods, in *Advances in Spectroscopic Monitoring of the Atmosphere*, ed. by W. Chen, D.S. Venables, M.W. Sigrist (Elsevier, 2021), pp. 95–158. https://doi.org/10.1016/B978-0-12-815014-6.00009-9

S. Cheng, J. Ma, X. Zheng, M. Gu, S. Donner, S. Dörner, W. Zhang, J. Du, X. Li, Z. Liang, J. Lv, T. Wagner, Retrieval of O_3, NO_2, BrO and OClO columns from ground-based zenith scattered light DOAS measurements in summer and autumn over the northern Tibetan plateau. Remote Sens. **13**(21), 4242 (2021). https://doi.org/10.3390/rs13214242

S.M. Chernin, E.G. Barskaya, Optical multipass matrix systems. Appl. Opt. **30**, 51–57 (1991)

S.M. Chernin, New generation of multipass systems in high resolution spectroscopy. Spectrochim. Acta A **52**, 1009–1022 (1996)

M. Chin, D.D. Davis, Global sources and sinks of OCS and CS_2 and their distribution. Global Biogeochem. Cycles **7**, 321–337 (1993)

M. Chin, D.J. Jacob, Anthropogenic and natural contributions to tropospheric sulphate: a global model analysis. J. Geophys. Res. **101**, 18691–18699 (1996)

M.P. Chipperfield, S. Bekki, S. Dhomse, N.R.P. Harris, B. Hassler, R. Hossaini, W. Steinbrecht, R. Thiéblemont, M. Weber, Detecting recovery of the stratospheric ozone layer. Nature **549**(7671), 211–218 (2017). https://doi.org/10.1038/nature23681

S. Choi, Y. Wang, R.J. Salawitch, T. Canty, J. Joiner, T. Zeng, T.P. Kurosu, K. Chance, A. Richter, L.G. Huey, J. Liao, J.A. Neuman, J.B. Nowak, J.E. Dibb, A.J. Weinheimer, G. Diskin, T.B. Ryerson, A. da Silva, J. Curry, D. Kinnison, S. Tilmes, P.F. Levelt, Analysis of satellite-derived arctic tropospheric BrO columns in conjunction with aircraft measurements during arctas and arcpac. Atmos. Chem. Phys. **12**(3), 1255–1285 (2012). https://acp.copernicus.org/articles/12/1255/2012/

D. Clarke, H. Basurah, Polarization measurements of the ring effect in the daytime sky. Planet. Space Sci. **37**, 627–630 (1989)

K. Clémer, M. Van Roozendael, C. Fayt, F. Hendrick, C. Hermans, G. Pinardi, R. Spurr, P. Wang, M. De Mazière, Multiple wavelength retrieval of tropospheric aerosol optical properties from MAXDOAS measurements in Beijing. Atmos. Meas. Tech. **3**(4), 863–878 (2010). https://doi.org/10.5194/amt-3-863-2010

K.C. Clemitshaw, L.J. Carpenter, S.A. Penkett, M.E. Jenkin, A calibrated peroxy radical chemical amplifier for ground-based tropospheric measurements. J. Geophys. Res. **102**, 25405–25416 (1997)

K.C. Clemitshaw, A review of instrumentation and measurement techniques for ground-based and airborne field studies of gas-phase tropospheric chemistry. Crit. Rev. Environ. Sci. Technol. **34**, 1–108 (2004). https://doi.org/10.1080/10643380490265117

S. Coburn, I. Ortega, R. Thalman, B. Blomquist, C.W. Fairall, R. Volkamer, Measurements of diurnal variations and eddy covariance (EC) fluxes of glyoxal in the tropical marine boundary layer: description of the fast LED-CE-DOAS instrument. Atmos. Meas. Tech. **7**(10), 3579–3595 (2014). https://amt.copernicus.org/articles/7/3579/2014/

H. Coe, R.L. Jones, R. Colin, M. Carleer, R.M. Harrison, J. Peak, J.M.C. Plane, N. Smith, B. Allan, K.C. Clemitshaw, R.A. Burgess, U. Platt, T. Etzkorn, J. Stutz, J.P. Pommereau, F. Goutail, M. Nunes-Pinharanda, P. Simon, C. Hermans, A.-C. Vandaele, A comparison of differential optical absorption spectrometers for measurement of NO_2, O_3, SO_2 and HONO, in ed. by P.M. Borrell, P. Borrell, T. Cvitaš, K. Kelly, W. Seiler, *Proceedings of EUROTRAC Symposium 1996: Transport and Transformation of Pollutants* (Computational Mechanics Publications, Southampton, 1997), pp. 757–762

H. Coe, B.J. Allan, J.M.C. Plane, Retrievals of vertical profiles of NO_3 from zenith sky measurements using an optimal estimation method. J. Geophys. Res. **107**(D21), 4587 (2002). https://doi.org/10.1029/2002JD002111

S. Cogliati, M. Rossini, T. Julitta, M. Meroni, A. Schickling, A. Burkart, F. Pinto, U. Rascher, R. Colombo, Continuous and long-term measurements of reflectance and sun-induced chlorophyll fluorescence by using novel automated field spectroscopy systems. Remote Sens. Environ. **164**, 270–281 (2015). https://doi.org/10.1016/j.rse.2015.03.027

P.-F. Coheur, S. Fally, M. Carleer, C. Clerbaux, R. Colin, A. Jenouvrier, M.-F. Mérienne, C. Hermans, A.C. Vandaele, New water vapour line parameters in the 26000–13000 cm^{-1} region. J. Quant. Spectrosc. Radiat. Transf. **74**(4), 493–510 (2002)

M. Coldewey-Egbers, M. Weber, M. Buchwitz, J.P. Burrows, Application of a modified DOAS method for total ozone retrieval from GOME data at high polar latitudes. Adv. Space Res. **34**(4), 749–753 (2004). https://doi.org/10.1016/j.asr.2003.05.051

M. Coldewey-Egbers, M. Weber, L.N. Lamsal, R. de Beek, M. Buchwitz, J.P. Burrows, Total ozone retrieval from GOME UV spectral data using the weighting function DOAS approach. Atmos. Chem. Phys. **5**(4), 1015–1025 (2005). https://doi.org/10.5194/acp-5-1015-2005

R. Colin, M. Carleer, P.C. Simon, A.C. Vandaele, P. Dufour, C. Fayt, Atmospheric absorption measurement by Fourier transform DOAS, EUROTRAC annual report 1991 (TOPAS subproject) (1991), pp. 14–46

B. Coquart, A. Jenouvrier, M.F. Merienne, The NO_2 absorption spectrum. II. Absorption cross section at low temperatures in the 400–500 nm region. J. Atmos. Chem. **21**, 251–261 (1995)

A. Cornu, Observation de la limite ultraviolette du spectre solaire a diverses altitudes. C.R. Acad. Sci. **89**, 808 (1879)

R.A. Cox, D.W. Sheppard, M.P. Stevens, Absorption coefficients and kinetics of the BrO radical using molecular modulation. J. Photochem. **19**, 189–207 (1982)

R.A. Cox, W.J. Bloss, R.L. Jones, D.M. Rowley, OIO and the atmsopheric cycle of iodine. Geophys. Res. Lett. **26**, 1857–1860 (1999)

S. Crowell, T. Haist, M. Tscherpel, J. Caron, E. Burgh, B. Moore III., Performance and polarization response of slit homogenizers for the GeoCarb mission. Atmos. Meas. Tech. **16**(1), 195–208 (2023). https://doi.org/10.5194/amt-16-195-2023

P.J. Crutzen, The influence of nitrogen oxides on the atmospheric ozone content. Qart. J. Roy. Met. Soc. **96**, 320–325 (1970)

P.J. Crutzen, A discussion of the chemistry of some minor constituents in the stratosphere and troposphere. Pure Appl. Geophys. **106–108**, 1385–1399 (1973)

P.J. Crutzen, Photochemical reactions initiated by and influencing ozone in unpolluted tropospheric air. Tellus **26**, 47–57 (1974)

P.J. Crutzen, R. Müller, Ch. Brühl, T. Peter, On the potential importance of the gas phase reaction $CH_3O_2 + ClO \longrightarrow ClOO + CH_3O$ and the heterogeneous reaction $HOCl + HCl \longrightarrow H_2O + Cl_2$ in "ozone hole" chemistry. Geophys. Res. Lett. **19**, 1113–1116 (1992)

R.W. Cunningham, Comparison of three methods for determining fit parameter uncertainties for the Marquardt compromise. Comput. Phys. **7**(5), 570 (1993)

K.D. Custard, K.A. Pratt, S. Wang, P.B. Shepson, Constraints on arctic atmospheric chlorine production through measurements and simulations of Cl_2 and ClO. Environ. Sci. Technol. **50**(22), 12394–12400 (2016). https://doi.org/10.1021/acs.est.6b03909

A. Dahlback, K. Stamnes, A New spherical model for computing the radiation-field available for photolysis and heating at twilight. Plan. Spa. Sci. **39**(5), 671–683 (1991)

A. Dahlback, P. Rairoux, B. Stein, M. Del Guasta, E. Kyrö, L. Stefanutti, N. Larsen, G. Braathen, Effects of stratospheric aerosols from the Mt. Pinatubo eruption on ozone measuremnts at Sodankylá Finland in 1991/92. Geophys. Res. Lett. **21**(13), 1399–1402 (1994)

J.V. Dave, C.L. Mateer, A preliminary study on the possibility of estimating total atmospheric ozone from satellite measurements. J. Atmos. Sci. **24**, 414–427 (1967)

J.A. Davidson, C.A. Cantrell, A.H. McDaniel, R.E. Shetters, S. Madronich, J.G. Calvert, Visible-ultraviolet absorption cross sections for NO_2 as a function of temperature. J. Geophys. Res. **93**, 7105–7112 (1988)

J.H. Davies, Correlation spectroscopy. Anal. Chem. **42**, 101A-112A (1970)

J.H. Davies, N.D. van Egmond, R. Wiens, H. Zwick, Recent developments in environmental sensing with the Barringer correlation spectrometer. Can. J. Remote. Sens. **1**, 85–94 (1975)

W.E. Davies, G. Vaughan, F.M. O'Connor, Observations of near-zero ozone concentrations in the upper troposphere at mid-latitudes. Geophys. Res. Lett. **25**, 1173–1176 (1998)

D.D. Davis, M.O. Rodgers, S.D. Fischer, K. Asai, An experimental assessment of the O_3/H_2O interference problem the detection of natural levels of OH via laser induced fluorescence. Geophys. Res. Lett. **8**, 69–72 (1981)

D. Davis, J. Crawford, S. Liu, S. McKeen, A. Bandy, D. Thornton, F. Rowland, D. Blake, Potential impact of iodine on tropospheric levels of ozone and other critical oxidants. J. Geophys. Res. **101**, 2135–2147 (1996)

Z.Y.W. Davis, S. Baray, C.A. McLinden, A. Khanbabakhani, W. Fujs, C. Csukat, J. Debosz, R. McLaren, Estimation of NO_X and SO_2 emissions from Sarnia, Ontario, using a mobile MAX-DOAS (multi-axis differential optical absorption spectroscopy) and a NO_X analyzer. Atmos. Chem. Phys. **19**(22), 13871–13889 (2019). https://acp.copernicus.org/articles/19/13871/2019/

Z.Y.W. Davis, R. McLaren, Recommendations for spectral fitting of SO_2 from miniature multi-axis differential optical absorption spectroscopy (MAX-DOAS) measurements. Atmos. Meas. Tech. **13**(7), 3993–4008 (2020). https://amt.copernicus.org/articles/13/3993/2020/

R. de Beek, M. Vountas, V.V. Rozanov, A. Richter, J.P. Burrows, The ring effect in the cloudy atmosphere. Geophys. Res. Lett. **28**(4), 721–724 (2001)

M. De Maziere, M. Van Roozendael, C. Hermans, P.C. Simon, P. Demoulin, G. Roland, R. Zander, Quantitative evaluation of the post-Pinatubo NO_2 reduction and recovery, based on 10 years of FTIR and UV-Visible. J. Geophys. Res. **103**, 10849–10858 (1998)

I. De Smedt, T. Stavrakou, F. Hendrick, T. Danckaert, T. Vlemmix, G. Pinardi, N. Theys, C. Lerot, C. Gielen, C. Vigouroux, Hermans C, Fayt C, Veefkind P, Müller J-F, Van Roozendael M, Diurnal, seasonal and long-term variations of global formaldehyde columns inferred from combined OMI and GOME-2 observations. Atmos. Chem. Phys. **15**(21), 12519–12545 (2015). https://acp.copernicus.org/articles/15/12519/2015/

I. De Smedt, G. Pinardi, C. Vigouroux, S. Compernolle, A. Bais, N. Benavent, F. Boersma, K.-L. Chan, S. Donner, K.-U. Eichmann, P. Hedelt, F. Hendrick, H. Irie, V. Kumar, J.-C. Lambert, B. Langerock, C. Lerot, C. Liu, D. Loyola, A. Piters, A. Richter, C. Rivera Cárdenas, F. Romahn, R.G. Ryan, V. Sinha, N. Theys, J. Vlietinck, T. Wagner, T. Wang, H. Yu, M. van Roozendael, Comparative assessment of TROPOMI and OMI formaldehyde observations and validation against MAX-DOAS network column measurements. Atmos. Chem. Phys. **21**(16), 12561–12593 (2021). https://doi.org/10.5194/acp-21-12561-2021

D.A. de Wolf, Optical propagation through turbulent air. Opt. Laser Technol. **11**(1), 29–36 (1979)

E. Dekemper, J. Vanhamel, B. van Opstal, D. Fussen, The AOTF-based NO_2 camera. Atmos. Meas. Tech. **9**(12), 6025–6034 (2016). https://amt.copernicus.org/articles/9/6025/2016/

V.-K.R. der Luft (KRdL) VDI-Normenausschuss, Air quality—atmospheric measurements near ground with active differential optical absorption spectroscopy (DOAS)—ambient air and diffuse emission measurements; German version EN 16253:2013. Technical report, VDI (2013)

V.-K.R. der Luft (KRdL) VDI-Normenausschuss, Remote sensing—atmospheric measurements using passive DOAS—gaseous emissions and ambient air measurements VDI 4212:2019. Technical report, VDI (2019)

T. Deutschmann, S. Beirle, U. Frieß, M. Grzegorski, C. Kern, L. Kritten, U. Platt, C. Prados-Román, J. Pukīte, T. Wagner, B. Werner, K. Pfeilsticker, The Monte Carlo atmospheric radiative transfer model McArtim: introduction and validation of Jacobians and 3D features. J. Quant. Spectrosc. Radiat. Transf. **112**(6), 1119–1137 (2011). https://doi.org/10.1016/j.jqsrt.2010.12.009

D.R. Dieke, H.M. Crosswhite, The ultraviolet bands of OH, fundamental data. J. Quant. Spectrosc. Radiat. Transf. **2**, 97–199 (1961)

T.J. Dillon, M.E. Tucceri, D. Hölscher, J.N. Crowley, Absorption cross-section of IO at 427.2 nm and 298 K. J. Photochem. Photobiol. A. Chem. **176**, 3–14 (2005)

W.L. Dimpfl, J.L. Kinsey, Radiative lifetimes of OH ($A\ ^2\Sigma$) and Einstein coefficients of the A-X system of OH and OD. J. Quant. Spectrosc. Radiat. Transf. **21**, 233–241 (1979)

B.M. Dinelli, P. Raspollini, M. Gai, L. Sgheri, M. Ridolfi, S. Ceccherini, F. Barbara, N. Zoppetti, E. Castelli, E. Papandrea, P. Pettinari, A. Dehn, A. Dudhia, M. Kiefer, A. Piro, J.-M. Flaud, M. López-Puertas, D. Moore, J. Remedios, M. Bianchini, The ESA MIPAS/ENVISAT level2-v8 dataset: 10 years of measurements retrieved with ORM v8.22. Atmos. Meas. Tech. **14**(12), 7975–7998 (2021). https://doi.org/10.5194/amt-14-7975-2021

F. Dinger, T. Kleinbek, S. Dörner, N. Bobrowski, U. Platt, T. Wagner, M. Ibarra, E. Espinoza, SO_2 and BrO emissions of Masaya volcano from 2014 to 2020. Atmos. Chem. Phys. **21**(12), 9367–9404 (2021). https://acp.copernicus.org/articles/21/9367/2021/

R.J. Dirksen, K.F. Boersma, H.J. Eskes, D.V. Ionov, E.J. Bucsela, P.F. Levelt, H.M. Kelder, Evaluation of stratospheric NO_2 retrieved from the Ozone Monitoring Instrument: intercomparison, diurnal cycle, and trending. J. Geophys. Res. **116**(D8), D08305 (2011). https://doi.org/10.1029/2010JD014943

B. Dix, C.A.M. Brenninkmeijer, U. Frieß, T. Wagner, U. Platt, Airborne multi-axis DOAS measurements of atmospheric trace gases on Caribic long-distance flights. Atmos. Meas. Tech. **2**(2), 639–652 (2009)

B. Dix, T.K. Koenig, R. Volkamer, Parameterization retrieval of trace gas volume mixing ratios from Airborne MAX-DOAS. Atmos. Meas. Tech. **9**(11), 5655–5675 (2016). https://doi.org/10.5194/amt-9-5655-2016

G.M.B. Dobson, D.N. Harrison, Measurements of the amount of ozone in the earth's atmosphere and its relation to other geophysical conditions, Part 1. Proc. R. Soc. Lond. A **110**, 660–693 (1926)

G.M.B. Dobson, A photoelectric spectrophotometer for measuring the amount of atmospheric ozone. Proc. Phys. Soc. **43**, 324–339 (1931)

G.M.B. Dobson, Forty years' research on atmospheric ozone at Oxford: a history. Appl. Opt. **7**, 387–406 (1968)

J.L. Domingo, M. Marquès, M. Nadal, M. Schuhmacher, Health risks for the population living near petrochemical industrial complexes. 1. Cancer risks: a review of the scientific literature. Environ. Res. **186**, 109495 (2020). https://doi.org/10.1016/j.envres.2020.109495

M. Dorf, H. Bösch, A. Butz, C. Camy-Peyret, M.P. Chipperfield, A. Engel, F. Goutail, K. Grunow, F. Hendrick, S. Hrechanyy, B. Naujokat, J.-P. Pommereau, M. Van Roozendael, C. Sioris, F. Stroh, F. Weidner, K. Pfeilsticker, Balloon-borne stratospheric BrO measurements: comparison with Envisat/SCIAMACHY BrO limb profiles. Atmos. Chem. Phys. **6**, 2483–2501 (2006)

M. Dorf, A. Butz, C. Camy-Peyret, M.P. Chipperfield, L. Kritten, K. Pfeilsticker, Bromine in the tropical troposphere and stratosphere as derived from balloon-borne BrO observations. Atmos. Chem. Phys. **8**(23), 7265–7271 (2008). https://doi.org/10.5194/acp-8-7265-2008

H.-P. Dorn, U. Platt, Eine empfindliche optische Nachweismethode für Spurenstoffe in der Atmosphäre. Elektrizitätswirtschaft **24**, 967–970 (1986)

H.-P. Dorn, J. Callies, U. Platt, D.H. Ehhalt, Measurement of tropospheric OH concentrations by laser long-path absorption spectroscopy. Tellus B **40**, 437–445 (1988)

H.-P. Dorn, R. Neuroth, T. Brauers, U. Brandenburger, D.H. Ehhalt, Measurement of tropospheric OH radical concentrations by differential UV laser long-path absorption, in H.I. Schiff, U. Platt, *Proceedings of SPIE Symposium on Optical Methods in Atmospheric chemistry* (1993), pp. 361–366

H.-P. Dorn, R. Neuroth, A. Hofzumahaus, Investigation of OH absorption cross sections of rotational transitions in the $A^2\Sigma^+$, $v' = 0 < - X^2\Pi$, $v'' = 0$ band under atmospheric conditions: implications for tropospheric long-path absorption measurements. J. Geophys. Res. **100**, 7397–7409 (1995)

H.P. Dorn, U. Brandenburger, T. Brauers, M. Hausmann, D.H. Ehhalt, In- situ detection of tropospheric OH radicals by folded long- path laser absorption: results from the POPCORN field campaign in August 1994. Geophys. Res. Lett. **23**, 2537–2540 (1996)

H.-P. Dorn, R.L. Apodaca, S.M. Ball, T. Brauers, S.S. Brown, J.N. Crowley, Dubé WP, Fuchs H, Häseler R, Heitmann U, Jones RL, Kiendler-Scharr A, Labazan I, Langridge JM, Meinen J, Mentel TF, Platt U, Pöhler D, Rohrer F, Ruth AA, Schlosser E, Schuster G, Shillings AJL, Simpson WR, Thieser J, Tillmann R, Varma R, Venables DS, Wahner A, Intercomparison of NO_3 radical detection instruments in the atmosphere simulation chamber SAPHIR. Atmos. Meas. Tech. **6**(5), 1111–1140 (2013). https://amt.copernicus.org/articles/6/1111/2013/

O. Dubovik, B. Holben, T.F. Eck, A. Smirnov, Y.J. Kaufman, M.D. King, D. Tanré, I. Slutsker, Variability of absorption and optical properties of key aerosol types observed in worldwide locations. J. Atmos. Sci. **59**(3), 590–608 (2002). https://journals.ametsoc.org/view/journals/atsc/59/3/1520-0469_2002_059_0590_voaaop_2.0.co_2.xml

H. Duffell, C. Oppenheimer, M. Burton, Volcanic gas emission rates measured by solar occultation spectroscopy. Geophys. Res. Lett. **28**(16), 3131–3134 (2001). https://doi.org/10.1029/2000GL012425

M. Duncianu, A. Lahib, A. Tomas, P.S. Stevens, S. Dusanter, Characterization of a chemical amplifier for peroxy radical measurements in the atmosphere. Atmos. Environ. **222**, 117106 (2020). https://doi.org/10.1016/j.atmosenv.2019.117106

G. Durry, G. Megie, Atmospheric CH_4 and H_2O monitoring with near-infrared InGaAs laser diodes by the SDLA, a balloonborne spectrometer for tropospheric and stratospheric in situ measurements. Appl. Opt. **38**(36), 7342–7354 (1999)

V.L. Dvortsov, M.A. Geller, S. Solomon, S.M. Schauffler, E.L. Atlas, D.R. Blake, Rethinking reactive halogen budgets in the midlatitude lower stratosphere. Geophys. Res. Lett. **26**, 1699–1702 (1999)

H.D. Eckhardt, Simple model of corner reflector phenomena. Appl. Opt. **10**, 1559–1566 (1971)

H. Edner, A. Sunesson, S. Svanberg, L. Uneus, S. Wallin, Differential optical absorption spectroscopy system used for atmospheric mercury monitoring. Appl. Opt. **25**, 403–409 (1986)

H. Edner, R. Amer, P. Ragnarson, M. Rudin, S. Svanberg, Atmospheric NH_3 monitoring by long-path UV absorption spectroscopy, in *Proceedings of International Congress on Optical Science and Engineering. Environment and Pollution Measurement Sensors and Systems* (Den Haag, 1990)

H. Edner, P. Ragnarson, S. Svanberg, A multi-path DOAS system for large area pollution monitoring, in P.M. Borrell, et al., *Proceedings of EUROTRAC Symposium 1992* (SPB Academic Publishing BV, Den Haag, 1992), pp. 220–223

H. Edner, P. Ragnarson, S. Spännare, S. Svanberg, Differential optical absorption spectroscopy (DOAS) system for urban atmospheric pollution monitoring. Appl. Opt. **32**, 327–333 (1993a)

H. Edner, P. Ragnarsson, S. Svanberg, E. Wallinder, Simultaneous tropospheric ozone monitoring using lidar and DOAS systems. Lund Rep. Atomic Phys. LRAP **155** (1994a)

H. Edner, P. Ragnarson, S. Svanberg, E. Wallinder, E. Ferrara, R. Cioni, B. Raco, G. Taddeucci, Total fluxes of sulfur-dioxide from the Italian volcanoes Etna, Stromboli and Vulcano measured

by differential absorption LIDAR and passive differential optical-absorption spectroscopy. J. Geophys. Res. **99**(D9), 18827–18838 (1994b)
D.P. Edwards, J.-F. Lamarque, J.-L. Attié, L.K. Emmons, A. Richter, J.-P. Cammas, J.C. Gille, G.L. Francis, M.N. Deeter, J. Warner, D.C. Ziskin, L.V. Lyjak, J.R. Drummond, J.P. Burrows, Tropospheric ozone over the tropical Atlantic: a satellite perspective. J. Geophys. Res. **108**(D8), 4237 (2003). https://doi.org/10.1029/2002JD002927
D.H. Ehhalt, J.W. Drummond, The tropospheric cycle of NO_X, in *Chemistry of the Unpolluted and Polluted Troposphere* (D. Reidel Publishing, 1982), p. 219
D.H. Ehhalt, H.-P. Dorn, D. Poppe, The chemistry of the hydroxyl radical in surface air. Proc. Roy. Soc. Edinburgh **97B**, 17–34 (1991)
D.H. Ehhalt, Gas phase chemistry of the troposphere, in *Global Aspects of Atmospheric Chemistry. Deutsch Bunsen-Gesellschaft für Physikalische Chemie*, ed. by R. Zellner, (Steinkopf Verlag, Darmstadt, Springer, New York, 1999)
R. Eisberg R. Resnick, *Quantum Physics of Atoms, Molecules, Solids, Nuclei, and Particles*, 2nd edn. (Wiley, 1985). https://vdocument.in/eisberg-resnick-quantum-mechanics.html?page=1
F.L. Eisele, D.J. Tanner, Ion-assisted tropospheric OH measurements. J. Geophys. Res. **D5**, 9295–9308 (1991)
F.L. Eisele, G.H. Mount, F.C. Fehsenfeld, J. Harder, E. Madronicj, D.D. Parrish, J. Roberts, M. Trainer, D. Tanner, Intercomparison of tropospheric OH and ancillary trace gas measurements at Fritz Peak observatory. Colorado. J. Geophys. Res. **D99**, 18605–18626 (1994)
F.L. Eisele, D.J. Tanner, C.A. Cantrell, J.G. Calvert, Measurements and steady state calculations of OH concentrations at Mauna Loa observatory. J. Geophys. Res. **101**, 14665–14679 (1996)
M. Eisinger, A. Richter, A. Ladstätter-Weißmayer, J.P. Burrows, DOAS zenith sky observations: 1. BrO measurements over Bremen (53°N) 1993–1994. J. Atmos. Chem. **26**, 93–108 (1997)
M. Eisinger, J.P. Burrows, Tropospheric sulfur dioxide observed by the ERS-2 GOME instrument. Geophys. Res. Lett. **25**, 4177–4180 (1998)
A.S. Elokhov, A.N. Gruzdev, Spectrometric measurements of total NO_2 in different regions of the globe. SPIE **2107**, 111–121 (1993)
A.S. Elokhov, A.N. Gruzdev, Estimation of tropospheric and stratospheric NO_2 from spectrometric measurements of column NO_2 abundances. SPIE **2506**, 444–455 (1995)
C. Emde, R. Buras-Schnell, A. Kylling, B. Mayer, J. Gasteiger, U. Hamann, J. Kylling, B. Richter, C. Pause, T. Dowling, L. Bugliaro, The libRadtran software package for radiative transfer calculations (version 2.0.1). Geosci. Model. Dev. **9**(5), 1647–1672 (2016). https://doi.org/10.5194/gmd-9-1647-2016
C. Emde, H. Yu, A. Kylling, M. van Roozendael, K. Stebel, B. Veihelmann, B. Mayer, Impact of 3D cloud structures on the atmospheric trace gas products from UV-Vis sounders - Part 1: synthetic dataset for validation of trace gas retrieval algorithms. Atmos. Meas. Tech. **15**(5), 1587–1608 (2022). https://amt.copernicus.org/articles/15/1587/2022/
R. Engeln, G. Berden, R. Peeters, G. Meijer, Cavity enhanced absorption and cavity enhanced magnetic rotation spectroscopy. Rev. Sci. Instrum. **69**, 3763–3769 (1998)
U.S. EPA, 40 CFR CH 1. / 63.658: Fenceline monitoring provisions (2015). https://www.ecfr.gov/current/title-40/chapter-I/subchapter-C/part-63/subpart-CC/section-63.658
F. Erle, K. Pfeilsticker, U. Platt, On the influence of tropospheric clouds on zenith-scattered-light measurements of stratospheric species. Geophys. Res. Lett. **22**, 2725–2728 (1995)
F. Erle, A. Grendel, D. Perner, U. Platt, K. Pfeilsticker, Evidence of heterogeneous bromine chemistry on cold stratospheric sulphate aerosols. Geophys. Res. Lett. **25**, 4329–4332 (1998)
F. Erle, Untersuchungen zur Halogenaktivierung der winterlichen arktischen Stratosphäre anhand flugzeuggestützter spektroskopischer Messungen. Ph.D. thesis, University of Heidelberg (1999)
ESA, Air pollution remains low as europeans stay at home (2020). https://www.esa.int/Applications/Observing_the_Earth/Copernicus/Sentinel-5P/Air_pollution_remains_low_as_Europeans_stay_at_home
ESA, The GOME users manual, in ed. by F. Bednarz (ESA Publication Division, ESTEC, Noordwijk, The Netherlands, 1995)

T. Etzkorn, B. Klotz, S. Sørensen, I. Patroescu, I. Barnes, K. Becker, U. Platt, Gas-phase absorption cross sections of 24 monocyclic aromatic hydrocarbons in the UV and IR spectral ranges. Atmos. Environ. **33**(4), 525–540 (1999). http://linkinghub.elsevier.com/retrieve/pii/S1352231098002891

EUROTRAC Final Report, *Instrument Development for Atmospheric Research and Monitoring*, in J. Bösenberg, D. Brassington, P.C. Simon, Chapter 11, vol. 8 (Springer, Berlin, Heidelberg, New York, 1997)

F. Evangelisti, G. Giovanelli, G. Orsi, T. Tirabassi, O. Vittori, Application features of mask correlation spectrophotometry to long horizontal paths. Atmos. Environ. **12**, 1125–1131 (1978)

F. Evangelisti, A. Baroncelli, P. Bonasoni, G. Giovanelli, F. Ravegnani, Differential optical absorption spectrometer for measurement of tropospheric pollutants. Appl. Opt. **34**, 2737–2744 (1995)

C. Fabry, L'absorption de l'ultraviolet par l'ozone et la limite du spectre solaire. J. Phys. Radium **3**, 196–206 (1913)

S. Fally, A. Vandaele, M. Carleer, C. Hermans, A. Jenouvrier, M. Mérienne, B. Coquart, R. Colin, Fourier transform spectroscopy of the O_2 Herzberg Bands. III. Absorption cross sections of the collision-induced bands and of the Herzberg continuum. J. Mol. Spectrosc. **204**(1), 10–20 (2000). https://doi.org/10.1006/jmsp.2000.8204

S. Fally, M. Carleer, A.C. Vandaele, UV Fourier transform absorption cross sections of benzene, toluene, meta-, ortho-, and para-xylene. J. Quant. Spectrosc. Radiat. Transf. **110**(9–10), 766–782 (2009). https://doi.org/10.1016/j.jqsrt.2008.11.014

O.G. Famy, M.J. Famy, Mutagenicity of N-α-Acetoxyethyl-N-ethylnitrosamine and N, N-Diethylnitrosamine in relation to the mechanism of metabolic activation of Dialkylnitrosamines. Cancer Res. **36**, 4504–4512 (1976)

S.-M. Fan, D.J. Jacob, Surface ozone depletion in the Arctic spring sustained by bromine reactions on aerosols. Nature **359**, 522–524 (1992)

X. Fang, J.A. Pyle, M.P. Chipperfield, J.S. Daniel, S. Park, R.G. Prinn, Challenges for the recovery of the ozone layer. Nat. Geosci. **12**(8), 592–596 (2019). https://doi.org/10.1038/s41561-019-0422-7

J.C. Farman, B.G. Gardiner, J.D. Shanklin, Large losses of total ozone in Antarctica reveal seasonal ClO_X/NO_X interaction. Nature **315**, 207–210 (1985)

W.G. Fastie, Ebert spectrometer reflections. Phys. Today **4**(1), 37–43 (1991)

C. Fayt, P. Dufour, C. Hermans, M. van Roozendael, P.C. Simon, Instrument and software development for DOAS measurements of atmospheric constituents, in *Proceedings of EUROTRAC Symposium 1992*, ed. by P.M. Borrell, et al. (SPB Academic Publishing BV, Den Haag, 1992), pp. 231–233

A. Febo, C. Perrino, I. Allegrini, Measurement of nitrous acid in Milan, Italy, by DOAS and diffusion denuders. Atmos. Environ. **30**(21), 3599–3609 (1996)

M.E. Feigelson, *Radiation in a Cloudy Atmosphere* (D. Reidel Publishing Company, Dordrecht, 1981)

C.C. Felton, J.C. Sheppard, M.J. Campbell, The radiochemical hydroxyl radical measurement method. Environ. Sci. Technol. **24**, 1841–1847 (1990)

J. Fenger, Urban air quality. Atmos. Environ. **33**(29), 4877–4900 (1999)

F. Ferlemann, *Ballongestützte Messung stratosphärischer Spurengase mit differentieller optischer Absorptionsspektroskopie (in German)*. Ph.D. Thesis, University of Heidelberg (1998). http://www.ub.uni-heidelberg.de/archiv/468

F. Ferlemann, C. Camy-Peyret, R. Fitzenberger, H. Harder, T. Hawat, H. Osterkamp, M. Schneider, D. Perner, U. Platt, P. Vradelis, K. Pfeilsticker, Stratospheric BrO profiles measured at different latitudes and seasons: instrument description, spectral analysis and profile retrieval. Geophys. Res. Lett. **25**(20), 3847–3850 (1998). https://doi.org/10.1029/1998GL900020

F. Ferlemann, N. Bauer, R. Fitzenberger, H. Harder, H. Osterkamp, D. Perner, U. Platt, M. Schneider, P. Vradelis, K. Pfeilsticker, Differential optical absorption spectroscopy instrument for stratospheric balloon-borne trace gas studies. Appl. Opt. **39**, 2377–2386 (2000)

M. Fickel, H. Delgado Granados, On the use of different spectral windows in DOAS evaluations: effects on the estimation of SO_2 emission rate and mixing ratios during strong emission of

Popocatépetl volcano. Chem. Geol. **462**, 67–73 (2017). https://doi.org/10.1016/j.chemgeo.2017.05.001

S. Fickert, J.W. Adams, J.N. Crowley, Activation of Br_2 and BrCl via uptake of HOBr onto aqueous salt solutions. J. Geophys. Res. **104**, 23719–23728 (1999)

M. Fiedler, H. Frank, T. Gomer, M. Hausmann, K. Pfeilsticker, U. Platt, Groundbased spectroscopic measurements of stratospheric NO_2 and OClO in Arctic winter 1989/90. Geophys. Res. Lett. **20**, 963–966 (1993)

S.E. Fiedler, A. Hese, A.A. Ruth, Incoherent broad-band cavity-enhanced absorption spectroscopy. Chem. Phys. Lett. **371**(3), 284–294 (2003). https://doi.org/10.1016/S0009-2614(03)00263-X

S.E. Fiedler, Incoherent broad-band cavity-enhanced absorption spectroscopy. Ph.D. thesis, D83, Faculty II—Mathematics and Sciences, Technical University of Berlin (TU) (2005)

F. Filsinger, MAX-DOAS measurements of tropospheric BrO at the Hudson Bay. Diploma thesis, University of Heidelberg (2004)

H. Finkenzeller, D. Pöhler, M. Horbanski, J. Lampel, U. Platt, New methods for the calibration of optical resonators: integrated calibration by means of optical modulation (ICOM) and narrow-band cavity ring-down (NB-CRD). Atmos. Meas. Tech. **16**(5), 1343–1356 (2023). https://amt.copernicus.org/articles/16/1343/2023/

B.J. Finlayson-Pitts, S.N. Johnson, The reaction of NO_2 with NaBr: possible source of BrNO in polluted marine atmospheres. Atmos. Environ. **22**, 1107–1112 (1988)

B.J. Finlayson-Pitts, M.J. Ezell, J.N. Pitts, Formation of chemically active chlorine compounds by reactions of atmospheric NaCl particles with gaseous N_2O_5 and $ClONO_2$. Nature **337**, 241–244 (1989)

B.J. Finlayson-Pitts, F.E. Livingston, H.N. Berko, Ozone destruction and bromine photo chemistry in the Arctic spring. Nature **343**, 622–625 (1990)

B.J. Finlayson-Pitts, J.N. Pitts, *Chemistry of the Upper and Lower Atmosphere: Theory, Experiments and Applications*, vol. xxii (Academic, San Diego, 2000), p.969

B.J. Finlayson-Pitts, L.M. Wingen, A.L. Sumner, D. Syomin, K.A. Ramazan, The heterogeneous hydrolysis of NO_2 in laboratory systems and in outdoor and indoor atmospheres: an integrated mechanism. Phys. Chem. Chem. Phys. **5**, 223–242 (2003)

V.E. Fioletov, C.A. McLinden, A. Cede, J. Davies, C. Mihele, S. Netcheva, S.-M. Li, J. O'Brien, Sulfur dioxide (SO_2) vertical column density measurements by Pandora spectrometer over the Canadian oil sands. Atmos. Meas. Tech. **9**(7), 2961–2976 (2016). https://doi.org/10.5194/amt-9-2961-2016

H. Fischer, Remote sensing of atmospheric trace gases. Interdisc. Sci. Rev. **18**(3), 185–191 (1993)

H. Fischer, H. Oelhaf, Remote sensing of vertical profiles of atmospheric trace constituents with MIPAS limb emission spectrometers. Appl. Opt. **35**(16), 2787–2796 (1996)

H. Fischer, M. Birk, C. Blom, B. Carli, M. Carlotti, T. von Clarmann, L. Delbouille, A. Dudhia, D. Ehhalt, Endemann M, Flaud JM, Gessner R, Kleinert A, Koopman R, Langen J, López-Puertas M, Mosner P, Nett H, Oelhaf H, Perron G, Remedios J, Ridolfi M, Stiller G, Zander R, MIPAS: an instrument for atmospheric and climate research. Atmos. Chem. Phys. **8**(8), 2151–2188 (2008). https://acp.copernicus.org/articles/8/2151/2008/

D.J. Fish, S.R. Aliwell, R.L. Jones, Mid-latitude observations of the seasonal variation of BrO: 2. interpretation and modelling study. Geophys. Res. Lett. **24**, 1199–1202 (1997)

D.J. Fish, H.K. Roscoe, P.V. Johnston, Possible causes of stratospheric NO_2 trends observed at Lauder New Zealand. Geophys. Res. Lett. **27**, 3313–3316 (2000)

R. Fitzenberger, H. Bösch, C. Camy-Peyret, M.P. Chipperfield, H. Harder, U. Platt, B.M. Sinnhuber, T. Wagner, K. Pfeilsticker, First profile measurements of tropospheric BrO. Geophys. Res. Lett. **27**(18), 2921–2924 (2000). https://doi.org/10.1029/2000GL011531

A. Fix, G. Ehret, H. Flentje, G. Poberaj, M. Gottwald, H. Finkenzeller, H. Bremer, M. Bruns, J.P. Burrows, A. Kleinböhl, H. Küllmann, J. Kuttippurath, A. Richter, P. Wang, K.-P. Heue, U. Platt, I. Pundt, T. Wagner, SCIAMACHY validation by aircraft remote measurements: design, execution, and first results of the SCIA-VALUE mission. Atmos. Chem. Phys. **5**, 1273–1289 (2005)

O.C. Fleischmann, J.P. Burrows, University of Bremen, Germany, unpublished results (2002)

O.C. Fleischmann, J. Orphal, J.P. Burrows, New ultraviolet absorption cross-sections of BrO at atmospheric temperatures measured by time-windowing Fourier transform spectroscopy. J. Photochem. Photobiol., A: Chem. **168**, 117–132 (2004)

Z.L. Fleming, R.M. Doherty, E. von Schneidemesser, C.S. Malley, O.R. Cooper, J.P. Pinto, A. Colette, X. Xu, D. Simpson, M.G. Schultz, A.S. Lefohn, S. Hamad, R. Moolla, S. Solberg, Z. Feng, Tropospheric ozone assessment report: present-day ozone distribution and trends relevant to human health. Elementa: Sci. Anthropocene **6**, 12 (2018). https://doi.org/10.1525/elementa.273

H. Flentje, R. Dubois, J. Heintzenberg, H.J. Karbach, Retrieval of aerosol properties from boundary layer extinction measurements with a DOAS system. Geophys. Res. Lett. **24**, 2019–2022 (1997)

S. Flock, LED-Lidar—theoretische und experimentelle Machbarkeitsstudie zur Realisierung eines bistatischen, LED-basierten Lidarsystems (in German). Master's Thesis, University of Heidelberg (2012)

K.L. Foster, R.A. Plastridge, J.W. Bottenheim, P.B. Shepson, B.J. Finlayson-Pitts, C.W. Spicer, The role of Br_2 and BrCl in surface ozone destruction at polar sunrise. Science **291**, 471–474 (2001)

P. Francis, M.R. Burton, C. Oppenheimer, Remote measurements of volcanic gas compositions by solar occultation spectroscopy. Nature **396**, 567–569 (1998)

H. Frank, U. Platt, Advanced calculation procedures for the interpretation of skylight measurements, in *Proceedings of First European Ozone Workshop* (Schliersee, 1990), pp. 65–68

H. Frank, Ein Strahlungstransportmodell zur Interpretation von spektroskopischen Spurenstoffmessungen in der Erdatmosphäre. Diploma thesis, University of Heidelberg (1991)

C. Frankenberg, U. Platt, T. Wagner, Iterative maximum a posteriori (IMAP)-DOAS for trace gas retrieval of strong absorbers: model studies for CH_4 and CO_2 retrieval from near infrared spectra of SCIAMACHY onboard ENVISAT. Atmos. Chem. Phys. **4**, 6067–6106 (2004)

C. Frankenberg, U. Platt, T. Wagner, Retrieval of CO from SCIAMACHY onboard ENVISAT detection of strongly polluted areas and seasonal patterns in global CO abundances. Atmos. Chem. Phys. **4**, 8425–8438 (2004)

C. Frankenberg, J.F. Meirink, M. van Weele, U. Platt, T. Wagner, Assessing methane emissions from global space-borne observations. Science **308**, 1010–1014 (2005)

C. Frankenberg, J.F. Meirink, P. Bergamaschi, A. Goede, M. Heimann, S. Körner, U. Platt, M. van Weele, T. Wagner, Satellite chartography of atmospheric methane from SCIAMACHY onboard ENVISAT: (I) Analysis of the years 2003 and 2004. J. Geophys. Res. **111**, D07303 (2006). https://doi.org/10.1029/2005JD006235

C. Frankenberg, T. Warneke, A. Butz, I. Aben, F. Hase, P. Spietz, L.R. Brown, Pressure broadening in the $2\nu_3$ band of methane and its implication on atmospheric retrievals. Atmos. Chem. Phys. **8**(17), 5061–5075 (2008). https://acp.copernicus.org/articles/8/5061/2008/

C. Frankenberg, A. Butz, G.C. Toon, Disentangling chlorophyll fluorescence from atmospheric scattering effects in O_2 A-band spectra of reflected sun-light. Geophys. Res. Lett. **38**(3), L03801 (2011). https://doi.org/10.1029/2010GL045896

C. Frankenberg, J.B. Fisher, J. Worden, G. Badgley, S.S. Saatchi, J.-E. Lee, G.C. Toon, A. Butz, M. Jung, A. Kuze, T. Yokota, New global observations of the terrestrial carbon cycle from GOSAT: patterns of plant fluorescence with gross primary productivity. Geophys. Res. Lett. **38**(17), L17706 (2011). https://doi.org/10.1029/2011GL048738

C. Frankenberg, C. O'Dell, L. Guanter, J. McDuffie, Remote sensing of near-infrared chlorophyll fluorescence from space in scattering atmospheres: implications for its retrieval and interferences with atmospheric CO_2 retrievals. Atmos. Meas. Tech. **5**(8), 2081–2094 (2012). https://doi.org/10.5194/amt-5-2081-2012

C. Frankenberg, C. O'Dell, J. Berry, L. Guanter, J. Joiner, P. Köhler, R. Pollock, T.E. Taylor, Prospects for chlorophyll fluorescence remote sensing from the Orbiting Carbon Observatory-2. Remote Sens. Environ. **147**, 1–12 (2014). https://doi.org/10.1016/j.rse.2014.02.007

C. Frankenberg, J. Berry, 3.10—Solar induced chlorophyll fluorescence: origins, relation to photosynthesis and retrieval, in *Comprehensive Remote Sensing*, ed. by S. Liang (Elsevier, Oxford, 2018), pp. 143–162. https://doi.org/10.1016/B978-0-12-409548-9.10632-3

E. Franzblau, C.J. Popp, Nitrogen oxides produced from lightning. J. Geophys. Res. **84**(D8), 11089–11104 (1989)

W. Fricke, S. Beilke, Changing concentrations and deposition of sulfur and nitrogen compounds in Central Europe between 1980 and 1992, in *Air Pollution Research Report 47, CEC, Proceedings of Joint Workshop CEC/BIATEX of Eurotrac, 4–7 May 1993, Aveiro, Portugal*, ed. by J. Slanina, G. Angeletti, S. Beilke (1993) pp. 9–30

W. Fricke, K. Uhse, Anteile von Witterung und Emissionsminderung am SO_2-Rückgang in Deutschland. Staub-Reinhaltung der Luft **54**, 289–296 (1994)

U. Frieß, M. Chipperfield, C. Otten, U. Platt, J. Pyle, T. Wagner, K. Pfeilsticker, Intercomparison of measured and modelled BrO slant column amounts for the Arctic winter and spring 1994/95. Geophys. Res. Lett. **26**, 1861–1864 (1999)

U. Frieß, Spectroscopic measurements of atmospheric trace gases at Neumayer-station, Antarctica. Ph.D. thesis, University of Heidelberg, Heidelberg (2001)

U. Frieß, T. Wagner, I. Pundt, K. Pfeilsticker, U. Platt, Spectroscopic measurements of tropospheric iodine oxide at Neumayer station. Antarctica. Geophys. Res. Lett. **28**, 1941–1944 (2001)

U. Frieß, J. Hollwedel, G. König-Langlo, T. Wagner, U. Platt, Dynamics and chemistry of tropospheric bromine explosion events in the Antarctic coastal region. J. Geophys. Res. **109**, D06305 (2004). https://doi.org/10.1029/2003JD004133

U. Frieß, K. Kreher, P.V. Johnston, U. Platt, Ground-based DOAS measurements of stratospheric trace gases at two Antarctic stations during the 2002 ozone hole period. J. Atmos. Sci. **62**(3), 765–777 (2005). https://doi.org/10.1175/JAS-3319.1

U. Frieß, P.S. Monks, J.J. Remedios, A. Rozanov, R. Sinreich, T. Wagner, U. Platt, MAX-DOAS O_4 measurements: a new technique to derive information on atmospheric aerosols. (II) Modelling studies. J. Geophys. Res. **111**, D14203 (2006). https://doi.org/10.1029/2005JD006618

U. Frieß, H. Sihler, R. Sander, D. Pöhler, S. Yilmaz, U. Platt, The vertical distribution of BrO and aerosols in the arctic: measurements by active and passive differential optical absorption spectroscopy. J. Geophys. Res.: Atmos. **116**(D14) (2011). https://doi.org/10.1029/2011JD015938

U. Frieß, H. Klein Baltink, S. Beirle, K. Clémer, F. Hendrick, B. Henzing, H. Irie, G. de Leeuw, A. Li, M.M. Moerman, M. van Roozendael, R. Shaiganfar, T. Wagner, Y. Wang, P. Xie, S. Yilmaz, P. Zieger, Intercomparison of aerosol extinction profiles retrieved from MAX-DOAS measurements. Atmos. Meas. Tech. **9**(7), 3205–3222 (2016). https://doi.org/10.5194/amt-9-3205-2016

U. Frieß, S. Beirle, L. Alvarado Bonilla, T. Bösch, M.M. Friedrich, F. Hendrick, A. Piters, A. Richter, M. van Roozendael, V.V. Rozanov, E. Spinei, J.-L. Tirpitz, T. Vlemmix, T. Wagner, Y. Wang, Intercomparison of MAX-DOAS vertical profile retrieval algorithms: studies using synthetic data. Atmos. Meas. Tech. **12**(4), 2155–2181 (2019). https://doi.org/10.5194/amt-12-2155-2019

U. Frieß, K. Kreher, R. Querel, H. Schmithüsen, D. Smale, R. Weller, U. Platt, Source mechanisms and transport patterns of tropospheric BrO: findings from long-term MAX-DOAS measurements at two antarctic stations. EGUsphere 1–40, 2022 (2022). https://egusphere.copernicus.org/preprints/2022/egusphere-2022-1074/

E. Frins, N. Bobrowski, U. Platt, T. Wagner, Tomographic MAX-DOAS observations of sun illuminated targets: a new technique providing well defined absorption paths in the boundary layer. Appl. Opt. **45**(24), 6227–6240 (2006)

E. Frins, U. Platt, T. Wagner, High spatial resolution measurements of NO_2 applying topographic target light scattering-differential optical absorption spectroscopy (total-DOAS). Atmos. Chem. Phys. **8**(24), 7595–7601 (2008). https://acp.copernicus.org/articles/8/7595/2008/

C. Fuchs, J. Kuhn, N. Bobrowski, U. Platt, Quantitative imaging of volcanic SO_2 plumes using Fabry-Pérot interferometer correlation spectroscopy. Atmos. Meas. Tech. **14**(1), 295–307 (2021). https://amt.copernicus.org/articles/14/295/2021/

H. Fuchs, H.-P. Dorn, M. Bachner, B. Bohn, T. Brauers, S. Gomm, A. Hofzumahaus, F. Holland, S. Nehr, F. Rohrer, R. Tillmann, A. Wahner, Comparison of OH concentration measurements by DOAS and LIF during Saphir chamber experiments at high OH reactivity and low NO concentration. Atmos. Meas. Tech. **5**(7), 1611–1626 (2012). https://doi.org/10.5194/amt-5-1611-2012

R. Fuller, P.J. Landrigan, K. Balakrishnan, G. Bathan, S. Bose-O'Reilly, M. Brauer, J. Caravanos, T. Chiles, A. Cohen, L. Corra, M. Cropper, G. Ferraro, J. Hanna, D. Hanrahan, H. Hu, D. Hunter, G. Janata, R. Kupka, B. Lanphear, M. Lichtveld, K. Martin, A. Mustapha, E. Sanchez-Triana, K. Sandilya, L. Schaefli, J. Shaw, J. Seddon, W. Suk, M.M. Téllez-Rojo, C. Yan, Pollution and health: a progress update. Lancet Planet. Health **6**(6), e535–e547 (2022). https://doi.org/10.1016/S2542-5196(22)00090-0

K. Fung, D. Grosjean, Determination of nanogram amounts of carbonyls as 2,4-dinitrophenylhydrazine by high performance liquid chromatography. Aerosol Sci. Technol. **53**, 168–171 (1981)

O. Funk, Photon path length distributions for cloudy skies; oxygen a-band measurements and radiative transfer calculations. Ph.D. thesis, University of Heidelberg (2000)

O. Funk, K. Pfeilsticker, Photon path lengths distributions for cloudy skies: oxygen A-Band measurements and model calculations. Annal. Geophys. **21**, 615–626 (2003)

S. Fuqi, H. Kuze, Y. Yoshii, M. Nemoto, N. Takeuchi, T. Kimura, T. Umekawa, T. Yoshida, T. Hioki, T. Tsutsui, M. Kawasaki, Measurement of regional distribution of atmospheric NO_2 and aerosol particles with flashlight long-path optical monitoring. Atmos. Environ. **39**, 4959–4968 (2005)

GAW-Report-278, Scientific assessment of ozone depletion: 2022. Technical report, GAW (2023). https://library.wmo.int/idurl/4/58360

R. Gall, D. Perner, A. Ladstätter-Weissenmayer, Simultaneous determination of NH_3, SO_2, NO and NO_2 by direct UV-absorption in ambient air. Fresenius J. Anal. Chem. **340**, 646–649 (1991)

B. Galle, H. Axelsson, H. Edner, A. Eilard, J. Mellqvist, P. Ragnarson, S. Svanberg, L. Zetterberg, Development of DOAS for atmospheric trace species monitoring, EUROTRAC Annual Report (1991)

B. Galle, L. Klemedtsson, D.W. Griffith, Application of an FTIR for measurements of N_2O fluxes using micrometeorological methods, an ultralarge chamber system and conventional field chambers. J. Geophys. Res. **99**, 16575–16583 (1994)

B. Galle, C. Oppenheimer, A. Geyer, A. McGonigle, M. Edmonds, L. Horrocks, A miniaturised ultraviolet spectrometer for remote sensing of SO_2 fluxes: a new tool for volcano surveillance. J. Volcanol. Geotherm. Res. **119**, 214–254 (2003)

B. Galle, U. Platt, M. Van Roozendael, C. Oppenheimer, T. Hansteen, G. Boudon, M. Burton, H. Delgado, W. Strauch, E. Malavassi, G. Garzon, C. Pullinger, M. Kasereka, M. Molina, L. Molina, S. Carn, NOVAC, network for observation of volcanic and atmospheric change. Project proposal to the European Union (2004)

B. Galle, M. Johansson, C. Rivera, Y. Zhang, M. Kihlman, C. Kern, T. Lehmann, U. Platt, S. Arellano, S. Hidalgo, Network for observation of volcanic and atmospheric change (NOVAC)-a global network for volcanic gas monitoring: network layout and instrument description. J. Geophys. Res.: Atmos. **115**(D5) (2010). https://doi.org/10.1029/2009JD011823

R.R. Gamache, A. Goldman, L.S. Rothman, Improved spectral parameters for the three most abundant isotopomers of the oxygen molecule. J. Quant. Spectrosc. Radiat. Transf. **59**(3–5), 495–509 (1998)

S. Gao, S. Wang, C. Gu, J. Zhu, R. Zhang, Y. Guo, Y. Yan, B. Zhou, Study on the measurement of isoprene by differential optical absorption spectroscopy. Atmos. Meas. Tech. **14**(4), 2649–2657 (2021). https://doi.org/10.5194/amt-14-2649-2021

K. Garane, M.-E. Koukouli, T. Verhoelst, C. Lerot, K-P. Heue, V. Fioletov, D. Balis, A. Bais, Bazureau A, Dehn A, Goutail F, Granville J, Griffin D, Hubert D, Keppens A, Lambert J-C, Loyola D, McLinden C, Pazmino A, Pommereau J-P, Redondas A, Romahn F, Valks P, Van Roozendael M, Xu J, Zehner C, Zerefos C, Zimmer W, TROPOMI/S5P total ozone column data: global ground-based validation and consistency with other satellite missions. Atmos. Meas. Tech. **12**(10), 5263–5287 (2019). https://amt.copernicus.org/articles/12/5263/2019/

D. Garcia-Nieto, N. Benavent, A. Saiz-Lopez, Measurements of atmospheric HONO vertical distribution and temporal evolution in MADRID (Spain) using the MAX-DOAS technique. Sci. Total Environ. **643**, 957–966 (2018). https://doi.org/10.1016/j.scitotenv.2018.06.180

A. Gaudel, O.R. Cooper, G. Ancellet, B. Barret, A. Boynard, J.P. Burrows, C. Clerbaux, P.-F. Coheur, J. Cuesta, E. Cuevas, S. Doniki, G. Dufour, F. Ebojie, G. Foret, O. Garcia, M.J. Granados-Muñoz, J.W. Hannigan, F. Hase, B. Hassler, G. Huang, D. Hurtmans, D. Jaffe, N. Jones, P. Kalabokas, B. Kerridge, S. Kulawik, B. Latter, T. Leblanc, E. Le Flochmoën, W. Lin, J. Liu, X. Liu, E. Mahieu, A. McClure-Begley, J.L. Neu, M. Osman, M. Palm, H. Petetin, I. Petropavlovskikh, R. Querel, N. Rahpoe, A. Rozanov, M.G. Schultz, J. Schwab, R. Siddans, D. Smale, M. Steinbacher, H. Tanimoto, D.W. Tarasick, V. Thouret, A.M. Thompson, T. Trickl, E. Weatherhead, C. Wespes, H.M. Worden, C. Vigouroux, X. Xu, G. Zeng, J. Ziemke, Tropospheric ozone assessment report: present-day distribution and trends of tropospheric ozone relevant to climate and global atmospheric chemistry model evaluation. Elementa: Sci. Anthropocene **6**, 39 (2018). https://doi.org/10.1525/elementa.291

S. General, D. Pöhler, H. Sihler, N. Bobrowski, U. Frieß, J. Zielcke, M. Horbanski, P.B. Shepson, B.H. Stirm, W.R. Simpson, K. Weber, C. Fischer, U. Platt, The Heidelberg airborne imaging DOAS instrument (HAIDI)—a novel imaging DOAS device for 2-D and 3-D imaging of trace gases and aerosols. Atmos. Meas. Tech. **7**(10), 3459–3485 (2014). ISSN 1867-1381. https://doi.org/10.5194/amt-7-3459-2014

A. Genner, P. Martín-Mateos, H. Moser, B. Lendl, A quantum cascade laser-based multi-gas sensor for ambient air monitoring. Sensors **20**(7) (2020). https://doi.org/10.3390/s20071850

B. Genty, J.-M. Briantais, N.R. Baker, The relationship between the quantum yield of photosynthetic electron transport and quenching of chlorophyll fluorescence. Biochim. Biophys. Acta **990**(1), 87–92 (1989). https://doi.org/10.1016/S0304-4165(89)80016-9

G.A. George, G.C. Morris, The intensity of absorption of naphthalene from 30000-53000 cm^{-1}. J. Mol. Spectrosc. **26**, 67–71 (1968)

L.A. George, T.M. Hard, R.J. O'Brien, Measurement of free radicals OH and HO_2 in Los Angeles smog. J. Geophys. Res. **104**, 11643–11655 (1999)

A.K. Georgoulias, R.J. van der A, P. Stammes, K.F. Boersma, H.J. Eskes, Trends and trend reversal detection in 2 decades of tropospheric NO_2 satellite observations. Atmos. Chem. Phys. **19**(9), 6269–6294 (2019). https://doi.org/10.5194/acp-19-6269-2019

K.R. German, Direct measurement of the of the radiative lifetime of the A $^2\Sigma^+$(v'=0) states of OH and OD. J. Chem. Phys. **62**, 2584–2587 (1975)

M.Y. Gershenzon, S. Il'in, N.G. Fedetov, Y.M. Gershenzon, The mechanism of reactive NO_3 uptake on dry NaX (X=Cl, Br). J. Atmos. Chem. **34**, 119–135 (1999)

A. Geyer, B. Alicke, D. Mihelcic, J. Stutz, U. Platt, Comparison of tropospheric NO_3 radical measurements by differential optical absorption spectroscopy and matrix isolation electron spin resonance. J. Geophys. Res. **104**, 26097–26105 (1999)

A. Geyer, *The Role of the Nitrate Radical in the Boundary Layer-Observations and Modeling Studies*. Doctoral Thesis, University of Heidelberg (2000). https://de.book-info.com/isbn/3-89825-085-7.htm

A. Geyer, R. Ackermann, R. Dubois, B. Lohrmann, T. Müller, U. Platt, Long-term observation of nitrate radicals in the continental boundary layer near Berlin. Atmos. Environ. **35**, 3619–3631 (2001a)

A. Geyer, B. Alicke, S. Konrad, T. Schmitz, J. Stutz, U. Platt, Chemistry and oxidation capacity of the nitrate radical in the continental boundary layer near Berlin. J. Geophys. Res. **106**, 8013–8025 (2001b)

A. Geyer, U. Platt, The temperature dependence of the NO_3 degradation frequency-a new indicator for the contribution of NO_3 to VOC oxidation and NO_X removal in the atmosphere. J. Geophys. Res. **107**, 4431–4442 (2002). https://doi.org/10.1029/2001JD001215

A. Geyer, A. Hofzumahaus, F. Holland, S. Konrad, T. Klüpfel, H.-W. Pätz, D. Perner, H.-J. Schäfer, A. Volz-Thomas, U. Platt, Nighttime production of peroxy and hydroxyl radicals during the BERLIOZ campaign: observations and modeling studies. J. Geophys. Res. **108**(D4), 8249 (2003a). https://doi.org/10.1029/2001JD000656

A. Geyer, B. Alicke, R. Ackermann, M. Martinez, H. Harder, W. Brune, P. di Carlo, E. Williams, T. Jobson, S. Hall, R. Shetter, J. Stutz, Direct observations of daytime NO_3: implications for urban boundary layer chemistry. J. Geophys. Res. **108**(D12), 4368 (2003b). https://doi.org/10.1029/2002JD002967

A. Geyer, J. Stutz, The vertical structure of $OH-HO_2-RO_2$ Chemistry in the nocturnal boundary layer: a one-dimensional study. J. Geophys. Res. **109**, D16301 (2004). https://doi.org/10.1029/2003JD004425

S.D. Ghude, G.G. Pfister, C. Jena, R. van der A, L.K. Emmons, R. Kumar, Satellite constraints of nitrogen oxide (NO_x) emissions from India based on OMI observations and WRF-Chem simulations. Geophys. Res. Lett. **40**(2), 423–428 (2013). https://doi.org/10.1002/grl.50065

M. Gil, O. Puentedura, M. Yela, C. Parrondo, D. Jadhav, B. Thorkelsson, OClO, NO_2 and O_3 total columns observations over Iceland during the winter 1993/94. Geophys. Res. Lett. **23**, 3337–3340 (1996)

M. Gil, O. Puentedura, E. Yela, M. Cuevas, Behavior of NO_2 and O_3 columns during the eclipse of February 26, 1998, as measured by visible spectroscopy. J. Geophys. Res. **105**, 3583 (2000)

G. Giovanelli, T. Tirabassi, S. Sandroni, Sulphur dioxide plume structure by mask correlation spectroscopy. Atmos. Environ. **13**, 1311–1318 (1979)

W.A. Gleason, A.M. Dunker, Investigation of background radical sources in a teflon-film irradiated chamber. Environ. Sci. Technol. **23**, 970–978 (1989)

J. Gliß, N. Bobrowski, L. Vogel, D. Pöhler, U. Platt, OClO and BrO observations in the volcanic plume of Mt. Etna—implications on the chemistry of chlorine and bromine species in volcanic plumes. Atmos. Chem. Phys. **15**(10), 5659–5681 (2015). https://doi.org/10.5194/acp-15-5659-2015

J. Gliß, K. Stebel, A. Kylling, A.S. Dinger, H. Sihler, A. Sudbø, Pyplis-a python software toolbox for the analysis of SO_2 camera images for emission rate retrievals from point sources. Geosciences **7**(4) (2017). https://doi.org/10.3390/geosciences7040134

A. Goldman, J.R. Gillis, Spectral line parameters of the A $^2\Sigma^+ \leftarrow X\ ^2\Pi$ (0, 0) band of OH for atmospheric and high temperatures. J. Quant. Spectrosc. Radiat. Transf. **25**, 111–135 (1981)

A.H. Goldstein, I.E. Galbally, Known and unexplored organic constituents in the earth's atmosphere. Environ. Sci. Technol. **41**(5), 1514–1521 (2007). https://doi.org/10.1021/es072476p

C. Gölz, J. Senzig, U. Platt, NO_3 initiated oxidation of biogenic hydrocarbons. CHEMOSPHERE. Glob. Change Sci. **3**, 339–352 (2001)

T. Gomer, J. Stutz, F. Heintz, U. Platt, *MFC Handbook* (University of Heidelberg, 1995)

J.C. Gómez Martín, A.S. Mahajan, T.D. Hay, C. Prados-Román, C. Ordóñez, S.M. MacDonald, J.M. Plane, M. Sorribas, M. Gil, J.F. Paredes Mora, M.V. Agama Reyes, D.E. Oram, E. Leedham, A. Saiz-Lopez, Iodine chemistry in the eastern Pacific marine boundary layer. J. Geophys. Res.: Atmos. **118**(2), 887–904 (2013). https://doi.org/10.1002/jgrd.50132

L. Gomez, M. Navarro-Comas, O. Puentedura, Y. Gonzalez, M. Cuevas E, Gil-Ojeda, Long-path averaged mixing ratios of O_3 and NO_2 in the free troposphere from mountain MAX-DOAS. Atmos. Meas. Tech. **7**(10), 3373–3386 (2014). https://amt.copernicus.org/articles/7/3373/2014/

G. Gonzalez Abad, A.H. Souri, J. Bak, K. Chance, L.E. Flynn, N.A. Krotkov, L. Lamsal, C. Li, X. Liu, C.C. Miller, C.R. Nowlan, R. Suleiman, H. Wang, Five decades observing earth's atmospheric trace gases using ultraviolet and visible backscatter solar radiation from space. J. Quant. Spectrosc. Radiat. Transf. **238**, 106478 (2019). https://doi.org/10.1016/j.jqsrt.2019.04.030

A.L. Goodman, G.M. Underwood, V.H. Grassian, Heterogeneous reaction of NO_2: characterization of gas-phase and adsorbed products from the reaction, $2\ NO_2(g) + H_2O(a) \longrightarrow HONO(g) + NO_3(a)$ on hydrated silica particles. J. Phys. Chem. A **103**, 7217–7223 (1999)

I.E. Gordon, L.S. Rothman, C. Hill, R.V. Kochanov, Y. Tan, P.F. Bernath, M. Birk, V. Boudon, A. Campargue, K.V. Chance, B.J. Drouin, J.M. Flaud, R.R. Gamache, J.T. Hodges, D. Jacquemart, V.I. Perevalov, A. Perrin, K.P. Shine, M.A.H. Smith, J. Tennyson, G.C. Toon, H. Tran, V.G. Tyuterev, A. Barbe, A.G. Császár, V.M. Devi, T. Furtenbacher, J.J. Harrison, J.M. Hartmann, A. Jolly, T.J. Johnson, T. Karman, I. Kleiner, A.A. Kyuberis, J. Loos, O.M. Lyulin, S.T. Massie, S.N. Mikhailenko, N. Moazzen-Ahmadi, H.S.P. Müller, O.V. Naumenko, A.V. Nikitin, O.L. Polyansky, M. Rey, M. Rotger, S.W. Sharpe, K. Sung, E. Starikova, S.A. Tashkun, J.V. Auwera, G. Wagner, J. Wilzewski, P. Wciso, S. Yu, E.J. Zak, The HITRAN2016 molecular spectroscopic database. J. Quant. Spectrosc. Radiat. Transf. **203**, 3–69 (2017). https://doi.org/10.1016/j.jqsrt.2017.06.038

P.F.W. Götz, A.R. Meetham, G.M.B. Dobson, The vertical distribution of ozone in the atmosphere. Proc. R. Soc. Lond. A **145**, 416–446 (1934)

F. Goutail, J.-P. Pommereau, C. Phillips, C. Deniel, A. Sarkissian, F. Lefévre, E. Kyrö, M. Rummukainen, P. Eriksen, S.B. Andersen, B.-A. Kaastad-Hoiskar, G. Braathen, V. Dorokhov, V.U. Khatatov, Depletion of column ozone in the Arctic during the winters of 1993–94 and 1994–95. J. Atmos. Chem. **32**, 1–34 (1999)

C.A. Goy, C.A. Pritchard, Pressure dependence of the visible iodine bands. J. Mol. Spectrosc. **12**, 38–44 (1964)

J.F. Grainger, J. Ring, Anomalous Fraunhofer line profiles. Nature **193**, 762 (1962)

W.B. Grant, R.T. Menzies, A survey of laser and selected optical systems for remote measurement of pollutant gas concentrations. J. Air Pollut. Control Assoc. **33**, 187–194 (1983)

L. Grassi, R. Guzzi, Theoretical and practical consideration on the construction of a zero geometrical loss multi-pass cell based on the use of monolithic multiple-face retro-reflectors. Appl. Opt. **40**(33), 6062–6071 (2001)

A. Gratien, M. Lefort, B. Picquet-Varrault, J. Orphal, J.-F. Doussin, J.-M. Flaud, Experimental intercomparison of the absorption cross-sections of nitrous acid (HONO) in the ultraviolet and mid-infrared spectral regions. J. Quant. Spectrosc. Radiat. Transf. **110**(4), 256–263 (2009). https://doi.org/10.1016/j.jqsrt.2008.11.003

M. Gratsea, M. Vrekoussis, A. Richter, F. Wittrock, A. Schönhardt, J. Burrows, S. Kazadzis, N. Mihalopoulos, E. Gerasopoulos, Slant column MAX-DOAS measurements of nitrogen dioxide, formaldehyde, glyoxal and oxygen dimer in the urban environment of athens. Atmos. Environ. **135**, 118–131 (2016). https://doi.org/10.1016/j.atmosenv.2016.03.048

T. Green, A. Hatton, *The Claw Hypothesis: A New Perspective on the Role of Biogenic Sulphur in the Regulation of Global Climate, Chapter in: Oceanography and Marine Biology, An annual review*, vol. 52 (CRC Press, 2014). https://doi.org/10.1201/b17143

G.D. Greenblatt, J.J. Orlando, J.B. Burkholder, A.R. Ravishankara, Absorption measurements of oxygen between 330 and 1140 nm. J. Geophys. Res. **95**, 18577–18582 (1990)

D. Griffin, C.A. McLinden, E. Dammers, C. Adams, C.E. Stockwell, C. Warneke, I. Bourgeois, J. Peischl, T.B. Ryerson, K.J. Zarzana, J.P. Rowe, R. Volkamer, C. Knote, N. Kille, T.K. Koenig, C.F. Lee, D. Rollins, P.S. Rickly, J. Chen, L. Fehr, A. Bourassa, D. Degenstein, K. Hayden, C. Mihele, S.N. Wren, J. Liggio, A. Akingunola, P. Makar, Biomass burning nitrogen dioxide emissions derived from space with TROPOMI: methodology and validation. Atmos. Meas. Tech. **14**(12), 7929–7957 (2021). https://doi.org/10.5194/amt-14-7929-2021

S.M. Griffith, R.F. Hansen, S. Dusanter, V. Michoud, J.B. Gilman, W.C. Kuster, P.R. Veres, M. Graus, J.A. de Gouw, J. Roberts, C. Young, R. Washenfelder, S.S. Brown, R. Thalman, E. Waxman, R. Volkamer, C. Tsai, J. Stutz, J.H. Flynn, N. Grossberg, B. Lefer, S.L. Alvarez, B. Rappenglueck, L.H. Mielke, H.D. Osthoff, P.S. Stevens, Measurements of hydroxyl and hydroperoxy radicals during CalNex-LA: Model comparisons and radical budgets. J. Geophys. Res.: Atmos. **121**(8), 4211–4232 (2016). https://doi.org/10.1002/2015JD024358

D.W.T. Griffith, D. Pöhler, S. Schmitt, S. Hammer, S.N. Vardag, U. Platt, Long open-path measurements of greenhouse gases in air using near-infrared Fourier transform spectroscopy. Atmos. Meas. Tech. **11**(3), 1549–1563 (2018). https://amt.copernicus.org/articles/11/1549/2018/

H. Grosch, A. Fateev, S. Clausen, UV absorption cross-sections of selected sulfur-containing compounds at temperatures up to 500°C. J. Quant. Spectrosc. Radiat. Transf. **154**, 28–34 (2015). https://doi.org/10.1016/j.jqsrt.2014.11.020

M. Grossi, P. Valks , D. Loyola, B. Aberle, S. Slijkhuis, T. Wagner, S. Beirle, R. Lang, Total column water vapour measurements from GOME-2 MetOp-A and MetOp-B. Atmos. Meas. Tech. **8**(3), 1111–1133 (2015). https://amt.copernicus.org/articles/8/1111/2015/

K. Großmann, U. Frieß, E. Peters, F. Wittrock, J. Lampel, S. Yilmaz, J. Tschritter, R. Sommariva, R. von Glasow, B. Quack, K. Krüger, K. Pfeilsticker, U. Platt, Iodine monoxide in the Western Pacific marine boundary layer. Atmos. Chem. Phys. **13**(6), 3363–3378 (2013). https://doi.org/10.5194/acp-13-3363-2013

K. Grossmann, C. Frankenberg, T.S. Magney, S.C. Hurlock, U. Seibt, J. Stutz, PhotoSpec: a new instrument to measure spatially distributed red and far-red solar-induced chlorophyll fluorescence. Remote. Sens. Environ. **216**, 311–327 (2018). https://doi.org/10.1016/j.rse.2018.07.002

M. Grundmann, Modeling of a waveguide-based UV-VIS-IR spectrometer based on a lateral (IN, GA)N alloy gradient. Phys. Status Solidi (A) **216**(14), 1900170 (2019). https://onlinelibrary.wiley.com/doi/abs/10.1002/pssa.201900170

M. Grundmann, Monolithic waveguide-based linear photodetector array for use as ultracompact spectrometer. IEEE Trans. Electron Devices **66**(1), 470–477 (2019). https://doi.org/10.1109/TED.2018.2877407

M. Grzegorski, M. Wenig, U. Platt, P. Stammes, N. Fournier, T. Wagner, The Heidelberg iterative cloud retrieval utilities (HICRU) and its application to GOME data. Atmos. Chem. Phys. **6**, 4461–4476 (2006)

L. Guanter, C. Frankenberg, A. Dudhia, P.E. Lewis, J. Gómez-Dans, A. Kuze, H. Suto, R.G. Grainger, Retrieval and global assessment of terrestrial chlorophyll fluorescence from GOSAT space measurements. Remote Sens. Environ. **121**, 236–251 (2012). https://doi.org/10.1016/j.rse.2012.02.006

L. Guanter, M. Rossini, R. Colombo, M. Meroni, C. Frankenberg, J.-E. Lee, J. Joiner, Using field spectroscopy to assess the potential of statistical approaches for the retrieval of sun-induced chlorophyll fluorescence from ground and space. Remote Sens. Environ. **133**, 52–61 (2013). https://doi.org/10.1016/j.rse.2013.01.017

A. Guenther, C.N. Hewitt, D. Erickson, R. Fall, C. Geron, T. Graedel, P. Harley, L. Klinger, M. Lerdau, W.A. McKay, T. Pierce, B. Scholes, R. Steinbrecher, R. Tallamraju, J. Taylor, P. Zimmerman, A global model of natural volatile organic compound emissions. J. Geophys. Res. **100**(D5), 8873–8892 (1995). https://doi.org/10.1029/94JD02950

W. Gurlit, C. Giesemann, V. Ebert, R. Zimmermann, U. Platt, J. Wolfrum, J.P. Burrows, Lightweight diode laser spectrometer "CHILD" for balloon-borne measurements of water vapor and methane. Appl. Opt. **44**(1), 91–101 (2005)

L. Gutzwiller, F. Arens, U. Baltensperger, H.W. Gäggeler, M. Ammann, Significance of semivolatile diesel exhaust organics for secondary HONO formation. Environ. Sci. Technol. **36**, 677–682 (2002)

R. Guzzi, J. Burrows, M. Cervino, C. Levoni, E. Cattani, T. Kurosu, T. Torricella, GOME cloud and aerosol data products algorithms development. Report, ESA Contract 11572/95/NL/CN (1998)

R. Guzzi, S. Zoffoli, S. Corradini, M. Chiarini, *Information Content of the Radiative Transfer Theory* (Agenzia Spaziale Italiana, Rome Report, 2003)

A.J. Haagen-Smit, Chemistry and physiology of Los Angeles smog. Ind. Eng. Chem. **44**, 1342–1346 (1952)

A.J. Haagen-Smit, M.M. Fox, Photochemical ozone formation with hydrocarbons and automobile exhaust. J. Air Pollut. Control Assoc. **4**, 105–109 (1954)

C. Hak, I. Pundt, S. Trick, C. Kern, U. Platt, J. Dommen, C. Ordóñez, A.S.H. Prévôt, W. Junkermann, C. Astorga-Lloréns, B.R. Larsen, J. Mellqvist, A. Strandberg, Y. Yu, B. Galle, J. Kleffmann, J.C. Lörzer, G.O. Braathen, R. Volkamer, Intercomparison of four different in-situ techniques for ambient formaldehyde measurements in urban air. J. Atmos. Chem. Phys. **5**, 2881–2900 (2005)

J.B.S. Haldane, *Possible Worlds and Other Essays* (Chatto and Windus, London, 1927)

C.S. Haley, S.M. Brohede, C.E. Sioris, E. Griffioen, D.P. Murtagh, I.C. McDade, P. Eriksson, E.J. Llewellyn, A. Bazureau, F. Goutail, Retrieval of stratospheric O_3 and NO_2 profiles from Odin Optical Spectrograph and Infrared Imager System (OSIRIS) limb-scattered sunlight measurements. J. Geophys. Res.: Atmos. **109**(D16) (2004). https://doi.org/10.1029/2004JD004588

C.T. Hall, F.E. Blacet, Separation of the absorption spectra of NO_2 and N_2O_4 in the range of 2400–5000 A. J. Chem. Phys. **20**, 1745–1749 (1952)

M. Hallquist, D.J. Stewart, S.K. Stephenson, R.A. Cox, Hydrolysis of N_2O_5 on sub-micron sulfate aerosols. Phys. Chem. Chem. Phys. **5**(16), 3453–3463 (2003)

H. Hallstadius, L. Unéus, S. Wallin, System for evaluation of trace gas concentration in the atmosphere based on the differential optical absorption spectroscopy technique. Proc. Soc. Photo. Opt. Instrum. Eng. **1433**, 36–43 (1991)

V.I. Haltrin, G.W. Kattawar, Self-consistent solutions to the equation of transfer with elastic and inelastic scattering in oceanic optics: I. model. Appl. Opt. **32**(27), 5356–5367 (1993). https://doi.org/10.1364/AO.32.005356

Y. Hamada, A.J. Merer, Rotational structure at the long wavelength end of the 2900 A system of SO_2. Can. J. Phys. **52**, 1443–1457 (1974)

P.L. Hanst, A.S. Lefohn, B.W. Gay, Detection of atmospheric pollutants at parts-per-billion levels by infrared spectroscopy. Appl. Spectrosc. **27**, 188–198 (1973)

P.L. Hanst, Air pollution measurement by Fourier transform spectroscopy. Appl. Opt. **17**, 1360–1366 (1978)

P.L. Hanst, S.T. Hanst, Gas measurement in the fundamental infrared region, in *Air Monitoring by Spectroscopic Techniques, Chemical Analysis Series*, vol. 127, ed. by M.W. Sigrist (Wiley, New York, 1994), pp.335–470

T.M. Hard, C.Y. Chan, A.A. Mehrabzadeh, R.J. Obrien, Diurnal HO_2 cycles at clean air and urban sites in the troposphere. J. Geophys. Res. **97**, 9785–9794 (1992)

T.M. Hard, A.A. Mehrabzadeh, C.Y. Chan, R.J. O'Brien, FAGE measurements of tropospheric HO with measurements and model interferences. J. Geophys. Res. **97**, 9795–9817 (1992)

T.M. Hard, L.A. George, R.J. O'Brien, FAGE determination of tropospheric HO and HO_2. J. Atmos. Sci. **52**, 3354–3372 (1995)

T.M. Hard, L.A. George, R.J. O'Brien, An absolute calibration for gas-phase hydroxyl measurements. Environ. Sci. Technol. **36**, 1783–1790 (2002)

H. Harder, C. Camy-Peyret, F. Ferlemann, R. Fitzenberger, T. Hawat, H. Osterkamp, M. Schneider, D. Perner, U. Platt, P. Vradelis, K. Pfeilsticker, Stratospheric BrO profiles measured at different latitudes and seasons: atmospheric observations. Geophys. Res. Lett. **25**(20), 3843–3846 (1998). https://doi.org/10.1029/1998GL900026

H. Harder, H. Bösch, C. Camy-Peyret, M. Chipperfield, R. Fitzenberger, S. Payan, D. Perner, U. Platt, B.-M. Sinnhuber, K. Pfeilsticker, Comparison of measured and modelled stratospheric BrO: implications for the total amount of stratospheric bromine. Geophys. Res. Lett. **27**, 3695–3698 (2000)

J.W. Harder, J.W. Brault, P.V. Johnston, G.H. Mount, Temperature dependent NO_2 cross section at high spectral resolution. J. Geophys. Res. **102**(D3), 3861–3879 (1997a)

J.W. Harder, R.O. Jakoubek, G.H. Mount, Measurement of tropospheric trace gases by long-path differential absorption spectroscopy during the 1993 OH photochemistry experiment. J. Geophys. Res. **102**, 6215–6226 (1997b)

G.W. Harris, W.P.L. Carter, A.M. Winer, J.N. Pitts, U. Platt, D. Perner, Observations of nitrous acid in the Los Angeles atmosphere and implications for the predictions of ozone-precursor relationships. Environ. Sci. Technol. **16**, 414–419 (1982)

G.W. Harris, A.M. Winer, J.N. Pitts, U. Platt, D. Perner, Measurement of HONO, NO_3 and NO_2 by long-path differential optical absorption spectroscopy in the Los Angeles basin, in *Optical and Laser Remote Sensing*, vol. 39, ed. by D.K. Killinger, A. Mooradian (Springer, New York, 1983), pp.106–113

G.W. Harris, G.I. Mackay, T. Iguchi, L.K. Mayne, H.I. Schiff, Measurements of formaldehyde in the troposphere by tunable diode laser absorption spectroscopy. J. Atmos. Chem. **8**, 119–137 (1989)

A.W. Harrison, Midsummer stratospheric NO_2 at latitude 45°S. Can. J. Phys. **57**, 1110–1117 (1979)

R.M. Harrison, J.D. Peak, G.M. Collins, Tropospheric cycle of nitrous acid. J. Geophys. Res. **101**, 14429–14439 (1996)

A. Hartl, B.C. Song, I. Pundt, 2-D reconstruction of atmospheric concentration peaks from horizontal long path DOAS tomographic measurements: parametrisation and geometry within a discrete approach. Atmos. Chem. Phys. **6**(3), 847–861 (2006). https://acp.copernicus.org/articles/6/847/2006/

W.N. Hartley, On the probable absorption of solar radiation by atmospheric ozone. Chem. News **42**, 268 (1880)

W.N. Hartley, On the absorption spectrum of ozone. J. Chem. Soc. **39**, 57–60 (1881)

M.H. Harwood, R.L. Jones, Temperature dependent ultraviolet-visible absorption cross sections of NO_2 and N_2O_4: low-temperature measurements of the equilibrium constant for 2 NO_2 to N_2O_4. J. Geophys. Res. **99**, 22.955–22.964 (1994)
M. Harwood, J. Burkholder, M. Hunter, R. Fox, A. Ravishankara, Absorption cross sections and self-reaction kinetics of the IO radical. J. Phys. Chem. A **101**, 853–863 (1997)
R.A. Hashmonay, M.G. Yost, C.-F. Wu, Computed tomography of air pollutants using radial scanning path-integrated optical remote sensing. Atmos. Environ. **33**, 267–274 (1999)
D.R. Hastie, M. Weißenmayer, J.P. Burrows, G.W. Harris, Calibrated chemical amplifier for atmospheric RO_X measurements. Anal. Chem. **63**, 2048–2057 (1991)
H. Haug, Raman-Streuung von Sonnenlicht in der Erdatmosphäre, Diploma Thesis, University of Heidelberg (1996)
H. Haug, K. Pfeilsticker, U. Platt, Vibrational Raman scattering in the atmosphere, University of Heidelberg, Unpublished Manuscript (1996)
Haugen (ed.), Workshop on micrometeorology (American Meteorological Society Science Press, Ephrata, 1973)
M. Hausmann, D. Ritz, U. Platt, New coaxial "long-path-DOAS" system: first application to BrO measurement in the Arctic troposphere, in *Proceedings Europto Series. Optical Methods in Atmospheric Chemistry*, vol. 1715, ed. by H.I. Schiff, U. Platt (1992), pp. 341–352
M. Hausmann, T. Rudolf, U. Platt, *Spectroscopic measurement of bromine oxide, ozone, and nitrous acid in alert, NATO-ASI Series Subseries I "Global Environmental Change"*, vol. 7 (Springer, Berlin, 1993), pp.189–203
M. Hausmann, U. Platt, Spectroscopic measurement of bromine oxide and ozone in the high Arctic during polar sunrise experiment 1992. J. Geophys. Res. **99**, 25399–25413 (1994)
M. Hausmann, U. Brandenburger, T. Brauers, H.-P. Dorn, Detection of tropospheric OH radicals by long-path differential-optical-absorption spectroscopy: experimental setup, accuracy, and precision. J. Geophys. Res. **102**, 16011–16022 (1997)
M. Hausmann, U. Brandenburger, T. Brauers, H.-P. Dorn, Simple Monte Carlo methods to estimate the spectra evaluation error in differential-optical-absorption spectroscopy. Appl. Opt. **38**(3), 462–475 (1999)
T.M. Hawat, C. Camy-Peyret, R.J. Torguet, Suntracker for atmospheric remote sensing. Opt. Eng. **37**(05), 1633–1642 (1998)
X.-C. He, Y.J. Tham, L. Dada, M. Wang, H. Finkenzeller, D. Stolzenburg, S. Iyer, M. Simon, A. Kürten, J. Shen et al., Role of iodine oxoacids in atmospheric aerosol nucleation. Science **371**(6529), 589–595 (2021). https://doi.org/10.1126/science.abe0298
D.E. Heard, M.J. Pilling, Measurement of OH and HO_2 in the troposphere. Chem. Rev. **103**(12), 5163–5198 (2003)
D. Heard, (ed.), *Analytical Techniques for Atmospheric Measurement* (Wiley, 2006). https://doi.org/10.1002/9780470988510
D. Heard, L.K. Whalley, S.S. Brown, *Gas Analysers and Laser Techniques* (Springer International Publishing, Cham, 2021), pp. 475–508. https://doi.org/10.1007/978-3-030-52171-4_16
K. Hebestreit, J. Stutz, D. Rosen, V. Matveev, M. Peleg, M. Luria, U. Platt, First DOAS measurements of tropospheric BrO in mid latitudes. Science **283**, 55–57 (1999)
E. Hecht, *Optics*, 4th edn. (Adison Wesley, New York, 2002)
A. Heckel, Messungen troposphärischer Spurengase mit einem MAXDOAS-Instrument Nachweis von troposphärischem Formaldehyd in Norditalien während der Format Kampagne. Diploma Thesis, University of Bremen (2003)
A. Heckel, A. Richter, T. Tarsu, F. Wittrock, C. Hak, I. Pundt, W. Junkermann, J.P. Burrows, MAX-DOAS measurements of formaldehyde in the Po-Valley. Atmos. Chem. Phys. **5**(4), 909–918 (2005). https://doi.org/10.5194/acp-5-909-2005
E. Hegels, P.J. Crutzen, T. Klüpfel, D. Perner, P.J. Burrows, Globale distribution of atmospheric bromine Monoxide from GOME on Earth-observing satellite ERS 2. Geophys. Res. Lett. **25**, 3127–3130 (1998)
F. Heintz, H. Flentje, R. Dubois, U. Platt, Long-term observation of nitrate radicals at the TOR-Station Kap Arkona (Rügen). J. Geophys. Res. **101**, 22891–22910 (1996)

B. Heismann, Eine CCD-Kamera zur Messung atmosphärischer Spurenstoffe. Diploma Thesis, University of Heidelberg (1996)

F. Helleis, J. Crowley, G. Moortgat, Temperature dependent rate constants and production branching ratios for the gas phase reaction between CH_3O_2 and ClO. J. Phys. Chem. **97**, 11464–11473 (1993)

F. Helleis, J. Crowley, G. Moortgat, Temperature dependent CH_3OCl formation in the reaction between CH_3O_2 and ClO. Geophys. Res. Lett. **21**(17), 1795–1798 (1994)

F. Hendrick, B. Barret, M. Van Roozendael, H. Boesch, A. Butz, M. De Mazière, F. Goutail, C. Hermans, J.-C. Lambert, K. Pfeilsticker, J.-P., Pommereau, Retrieval of nitrogen dioxide stratospheric profiles from ground-based zenith-sky UV-visible observations: validation of the technique through correlative comparisons. Atmos. Chem. Phys. **4**, 2091–2106 (2004). https://doi.org/10.5194/acp-4-2091-2004

F. Hendrick, M. Van Roozendael, A. Kylling, A. Petritoli, A. Rozanov, S. Sanghavi, R. Schofield, C. von Friedeburg, T. Wagner, F. Wittrock, D. Fonteyn, M. De Mazière, Intercomparison exercise between different radiative transfer models used for the interpretation of ground-based zenith-sky and multi-axis DOAS observations. Atmos. Chem. Phys. **6**, 93–108 (2006)

F. Hendrick, P.V. Johnston, M. De Mazière, C. Fayt, C. Hermans, K. Kreher, N. Theys, A. Thomas, M. van Roozendael, One-decade trend analysis of stratospheric BrO over Harestua (60°N) and Lauder (45°S) reveals a decline. Geophys. Res. Lett. **35**(14) (2008). https://doi.org/10.1029/2008GL034154

F. Hendrick, A. Rozanov, P.V. Johnston, H. Bovensmann, M. De Mazière, C. Fayt, C. Hermans, K. Kreher, W. Lotz, B.-M. Sinnhuber, N. Theys, A. Thomas, J.P. Burrows, M. van Roozendael, Multi-year comparison of stratospheric BrO vertical profiles retrieved from SCIAMACHY limb and ground-based UV-visible measurements. Atmos. Meas. Tech. **2**(1), 273–285 (2009). https://doi.org/10.5194/amt-2-273-2009

F. Hendrick, E. Mahieu, G.E. Bodeker, K.F. Boersma, M.P. Chipperfield, M. De Mazière, I. de Smedt, P. Demoulin, C. Fayt, C. Hermans, K. Kreher, B. Lejeune, G. Pinardi, C. Servais, R. Stübi, R. van der A, J.-P. Vernier, M. van Roozendael, Analysis of stratospheric NO_2 trends above Jungfraujoch using ground-based UV-visible, FTIR, and satellite nadir observations. Atmos. Chem. Phys. **12**(18), 8851–8864 (2012). https://doi.org/10.5194/acp-12-8851-2012

F. Hendrick, J.-F. Müller, K. Clémer, P. Wang, M. De Mazière, C. Fayt, C. Gielen, C. Hermans, J.Z. Ma, G. Pinardi, T. Stavrakou, T. Vlemmix, M. van Roozendael, Four years of ground-based MAX-DOAS observations of HONO and NO_2 in the Beijing area. Atmos. Chem. Phys. **14**(2), 765–781 (2014). https://doi.org/10.5194/acp-14-765-2014

L. Henyey J. Greenstein, Diffuse radiation in the galaxy. Astrophys. J. **93**, 70–83 (1941). http://adsabs.harvard.edu/full/1941ApJ....93...70H

J. Herman, A. Cede, E. Spinei, G. Mount, M. Tzortziou, N. Abuhassan, NO_2 column amounts from ground-based Pandora and MFDOAS spectrometers using the direct-sun DOAS technique: Intercomparisons and application to OMI validation. J. Geophys. Res.: Atmos. **114**(D13) (2009). https://doi.org/10.1029/2009JD011848

J. Herman, R. Evans, A. Cede, N. Abuhassan, I. Petropavlovskikh, G. McConville, Comparison of ozone retrievals from the Pandora spectrometer system and Dobson spectrophotometer in Boulder. Colorado. Atmos. Meas. Tech. **8**(8), 3407–3418 (2015). https://doi.org/10.5194/amt-8-3407-2015

J. Herman, E. Spinei, A. Fried, J. Kim, J. Kim, W. Kim, A. Cede, N. Abuhassan, M. Segal-Rozenhaimer, NO_2 and HCHO measurements in Korea from 2012 to 2016 from Pandora spectrometer instruments compared with OMI retrievals and with aircraft measurements during the KORUS-AQ campaign. Atmos. Meas. Tech. **11**(8), 4583–4603 (2018). https://doi.org/10.5194/amt-11-4583-2018

J. Herman, N. Abuhassan, J. Kim, J. Kim, M. Dubey, M. Raponi, M. Tzortziou, Underestimation of column NO_2 amounts from the OMI satellite compared to diurnally varying ground-based retrievals from multiple PANDORA spectrometer instruments. Atmos. Meas. Tech. **12**(10), 5593–5612 (2019). https://doi.org/10.5194/amt-12-5593-2019

C. Hermans, A. Vandaele, S. Fally, Fourier transform measurements of SO_2 absorption cross sections: I. temperature dependence in the 24000-29000 cm-1 (345-420 nm) region. J. Quant. Spectrosc. Radiat. Transf. **110**(9), 756–765 (2009). https://doi.org/10.1016/j.jqsrt.2009.01.031

T. Hermes, Lichtquellen und Optik für die Differentielle Optische Absorptionsspektroskopie. Diploma Thesis in Physics, University of Heidelberg (1999)

D. Herriott, H. Kogelnik, R. Kompfner, Off-axis paths in spherical mirror interferometers. Appl. Opt. **3**, 523–526 (1964)

D.R. Herriott, H.J. Schulte, Folded optical delay lines. Appl. Opt. **4**(8), 883–889 (1965). https://doi.org/10.1364/AO.4.000883

K.-P. Heue, M. Bruns, J.P. Burrows, W.-D. Lee, U. Platt, I. Pundt, A. Richter, T. Wagner, P. Wang, NO_2 over the tropics and the arctic measured by the AMAXDOAS in September 2002, in *Proceedings of the 16th ESA Symposium on Rocket and Balloon Program and Related Research*, St. Gallen, 02–05 June 2003, ESA SP-530 (2003)

K.-P. Heue, M. Bruns, J.P. Burrows, C. von Friedeburg, W.-D. Lee, U. Platt, I. Pundt, A. Richter, T. Wagner, P. Wang, Validation of scientific NO_2-SCIAMACHY data using the AMAXDOAS instrument. Atmos. Chem. Phys. **5**, 1039–1051 (2005)

K.-P. Heue, T. Wagner, S.P. Broccardo, D. Walter, S.J. Piketh, K.E. Ross, S. Beirle, U. Platt, Direct observation of two dimensional trace gas distributions with an airborne Imaging DOAS instrument. Atmos. Chem. Phys. **8**(22), 6707–6717 (2008). https://doi.org/10.5194/acp-8-6707-2008

K -P. Heue, C.A.M. Brenninkmeijer, A.K. Baker, A. Rauthe-Schöch, D. Walter, T. Wagner, C. Hörmann, H. Sihler, Dix B, Frieß U, Platt U, Martinsson BG, van Velthoven PFJ, Zahn A, Ebinghaus R, SO_2 and BrO observation in the plume of the Eyjafjallajökull volcano 2010: CARIBIC and GOME-2 retrievals. Atmos. Chem. Phys. **11**(6), 2973–2989 (2011). https://acp.copernicus.org/articles/11/2973/2011/

K-P. Heue, D. Loyola, F. Romahn, W. Zimmer, S. Chabrillat, Q. Errera, J. Ziemke, N. Kramarova, Tropospheric ozone retrieval by a combination of TROPOMI/S5P measurements with BASCOE assimilated data. Atmos. Meas. Tech. **15**(19), 5563–5579 (2022). https://amt.copernicus.org/articles/15/5563/2022/

G.M. Hidy, P.K. Mueller, S.L. Altshuler, J.C. Chow, J.G. Watson, Air quality measurements-from rubber bands to tapping the rainbow. J. Air Waste Manag. Assoc. **67**(6), 637–668 (2017). https://doi.org/10.1080/10962247.2017.1308890

A. Hilboll, A. Richter, A. Rozanov, O. Hodnebrog, A. Heckel, S. Solberg, F. Stordal, J.P. Burrows, Improvements to the retrieval of tropospheric NO_2 from satellite-stratospheric correction using SCIAMACHY limb/nadir matching and comparison to Oslo CTM2 simulations. Atmos. Meas. Tech. **6**(3), 565–584 (2013). https://doi.org/10.5194/amt-6-565-2013

S. Himmelmann, J. Orphal, H. Bovensmann, A. Richter, A. Ladstätter-Weißenmayer, J.P. Burrows, First observation of the OIO molecule by time-resolved flash photolysis absorption spectroscopy. Chem. Phys. Lett. **251**, 330–334 (1996)

E.D. Hinkley (ed.), *Laser Monitoring of the Atmosphere, Topics in Applied Physics*, vol. 14 (Springer, Berlin, 1976)

J. Hirokawa, K. Onaka, Y. Kajii, H. Akimoto, Heterogeneous processes involving sodium halide particles and ozone: molecular bromine release in the marine boundary layer in the absence of nitrogen oxides. Geophys. Res. Lett. **25**, 2449–2452 (1998)

R.M. Hoff, M.M. Millan, Remote SO_2 mass flux measurements using Cospec. J. Air Poll. Cont. Assoc. **31**(4), 381–384 (1981)

R.M. Hoff, Differential SO_2 column measurements of the Mt. Pinatubo volcanic plume. Geophys. Res. Lett. **19**(2), 175–178 (1992)

T. Hoffmann, C.D. O'Dowd, J.H. Seinfeld, IO homogeneous nucleation: an explanation for coastal new particle formation. Geophys. Res. Lett. **28**(10), 1949–1952 (2001)

D. Hofmann, P. Bonasoni, M. De Maziere, F. Evangelisti, G. Giovanelli, A. Goldman, F. Goutail, J. Harder, R. Jakoubek, P. Johnston, J. Kerr, T. McElroy, R. McKenzie, G. Mount, U. Platt, J.-P. Pommereau, A. Sarkissian, P. Simon, S. Solomon, J. Stutz, A. Thomas, M. Van Roosendael, E.

Wu, Intercomparison of UV/visible spectrometers for measurement of stratospheric NO_2 for the network for the detection of stratospheric change. J. Geophys. Res. **100**, 16765–16791 (1995)
A. Hofzumahaus, H.-P. Dorn, J. Callies, U. Platt, D.H. Ehhalt, Tropospheric OH concentration measurements by laser long-path absorption spectroscopy. Atmos. Environ. **25A**, 2017–2022 (1991)
A. Hofzumahaus, U. Aschmutat, M. Heßling, F. Ehhalt, D.H. Holland, The measurement of tropospheric OH radicals by laser- induced fluorescence spectroscopy during the POPCORN field campaign. Geophys. Res. Lett. **23**, 2541–2544 (1996)
A. Hofzumahaus, T. Brauers, U. Aschmutat, U. Brandenburger, H.-P. Dorn, M. Hausmann, M. Heßling, F. Holland, C. Plass-Dülmer, M. Sedlacek, M. Weber, D.H. Ehhalt, Reply to comment by Lanzendorf, et al., Geophys. Res. Lett. **24**, 3039–3040 (1997)
A. Hofzumahaus, U. Aschmutat, U. Brandenburger, T. Brauers, H.-P. Dorn, M. Hausmann, M. Hessling, F. Holland, C. Plass-Dulmer, D.H. Ehhalt, Intercomparison of tropospheric OH measurements by different laser techniques during the POPCORN campaign 1994. J. Atmos. Chem. **31**(1–2), 227–246 (1998)
B.A.K. Hoiskar, A. Dahlbak, G. Vaughan, G.O. Braathen, F. Goutail, P. Pommereau, R. Kivi, Interpretation of ozone measurements by ground-based visible spectroscopy-a study of seasonal dependence of airmass factors for ozone on climatology data. J. Quant. Spectrosc. Radiat. Transf. **57**, 569–579 (1997)
R. Holla, S. Schmitt, U. Frieß, D. Pöhler, J. Zingler, U. Corsmeier, U. Platt, Vertical distribution of BrO in the boundary layer at the dead sea. Environ. Chem. **12**(4), 438–460 (2015). https://doi.org/10.1071/EN14224
F. Holland, M. Hessling, A. Hofzumahaus, In-situ measurement of tropospheric OH radicals by laser-induced fluorescence: a description of the KFA instrument. J. Atmos. Sci. **52**, 3393–3401 (1995)
F. Holland, U. Aschmutat, M. Heßling, A. Hofzumahaus, D.H. Ehhalt, Highly time resolved measurements of OH during POPCORN using laser-induced fluorescence spectroscopy. J. Atmos. Chem. **31**, 205–225 (1998)
F. Holland, A. Hofzumahaus, J. Schäfer, A. Kraus, H. Pätz, Measurements of OH and HO_2 radical concentrations and photolysis frequencies during BERLIOZ. J. Geophys. Res. **108**(D4), 8246 (2003). https://doi.org/10.1029/2001JD001393
A.M. Holloway, R.P. Wayne, *Atmospheric Chemistry*, vol. 3 (The Royal Society of Chemistry, 2010). https://doi.org/10.1039/9781839168628
J. Hollwedel, M. Wenig, S. Beirle, S. Kraus, S. Kühl, W. Wilms-Grabe, U. Platt, T. Wagner, Year-to- year variability of polar tropospheric BrO as seen by GOME, (Proc. COSPAR 2002). Adv. Space Res. 804–808 (2004)
Q. Hong, C. Liu, Q. Hu, Y. Zhang, C. Xing, J. Ou, W. Tan, H. Liu, X. Huang, Z. Wu, Vertical distribution and temporal evolution of formaldehyde and glyoxal derived from MAX-DOAS observations: the indicative role of VOC sources. J. Environ. Sci. **122**, 92–104 (2022). https://doi.org/10.1016/j.jes.2021.09.025
G. Hönninger, Referenzspektren reaktiver Halogenverbindungen für DOAS Messungen. Diploma Thesis, Institut für Umweltphysik, University of Heidelberg (1999)
G. Hönninger, Halogen oxide studies in the boundary layer by multi axis differential optical absorption spectroscopy and active longpath-DOAS. Ph.D. Thesis, University of Heidelberg (2002)
G. Hönninger, U. Platt, Observations of BrO and its vertical distribution during surface ozone depletion at Alert. Atmos. Environ. **36**(15), 2481–2489 (2002). https://doi.org/10.1016/S1352-2310(02)00104-8
G. Hönninger, C. von Friedeburg, U. Platt, Multi axis differential absorption spectroscopy (MAX-DOAS). Atmos. Chem. Phys. **4**, 231–254 (2004a)
G. Hönninger, N. Bobrowski, E.R. Palenque, R. Torrez, U. Platt, Reactive bromine and sulfur emissions at Salar de Uyuni, Bolivia. J. Geophys. Res. **31**, L04101 (2004b). https://doi.org/10.1029/2003GL018818

G. Hönninger, H. Leser, O. Sebastian, U. Platt, Ground-based measurements of halogen oxides at the Hudson Bay by active long path DOAS and passive MAX-DOAS. Geophys. Res. Lett. **31**, L04111 (2004). https://doi.org/10.1029/2003GL018982

R. Hoogen, V.V. Rozanov, J.P. Burrows, Ozone profiles from GOME satellite data: algorithm description and first validation. J. Geophys. Res. **104**(D7), 8263–8280 (1999)

M. Horbanski, Emissions and distribution of reactive iodine from seaweed in coastal regions - Investigations using new mobile and in-situ DOAS techniques. Ph.D. Thesis, University of Heidelberg (2015). http://www.ub.uni-heidelberg.de/archiv/20106

M. Horbanski, D. Pöhler, J. Lampel, U. Platt, The ICAD (iterative cavity-enhanced DOAS) method. Atmos. Meas. Tech. **12**(6), 3365–3381 (2019). https://amt.copernicus.org/articles/12/3365/2019/

C. Hörmann, H. Sihler, N. Bobrowski, S. Beirle, M. Penning de Vries, U. Platt, T. Wagner, Systematic investigation of bromine monoxide in volcanic plumes from space by using the GOME-2 instrument. Atmos. Chem. Phys. **13**(9), 4749–4781 (2013). https://acp.copernicus.org/articles/13/4749/2013/

C. Hörmann, H. Sihler, S. Beirle, M. Penning de Vries, U. Platt, T. Wagner, Seasonal variation of tropospheric bromine monoxide over the Rann of Kutch salt marsh seen from space. Atmos. Chem. Phys. **16**(20), 13015–13034 (2016). https://acp.copernicus.org/articles/16/13015/2016/

D. Horn, G.C. Pimentel, 2.5 km low-temperature multiple-reflection cell. Appl. Opt. **10**, 1892–1898 (1971)

A. Horowitz, R. Meller, G.K. Moortgat, The UV/Visible absorption cross section of the a-dicarbonyl compounds: pyruvic acid, biacetyl and glyoxal. J. Photochem. Photobiol., A: Chem. **146**, 19–27 (2001)

Y. Hou, S. Wang, Y. Zhou, F. Yan, J. Zhu, Analysis of the carbon dioxide concentration in the lowest atmospheric layers and the factors affecting China based on satellite observations. Int. J. Remote Sens. **34**(6), 1981–1994 (2013). https://doi.org/10.1080/01431161.2012.730159

C.J. Howard, Kinetics of the reaction of HO_2 with NO_2. J. Chem. Phys. **67**, 5258–5263 (1977)

W.H. Howie, I.C. Lane, S.M. Newman, D.A. Johnson, A.J. Orr-Ewing, The UV absorption of ClO. Phys. Chem. Chem. Phys. **1**, 3079–3085 (1999)

R.J. Huang, K. Seitz, T. Neary, C.D. O'Dowd, U. Platt, T. Hoffmann, Observations of high concentrations of I_2 and IO in coastal air supporting iodine-oxide driven coastal new particle formation. Geophys. Res. Lett. **37**(3) (2010). https://doi.org/10.1029/2009GL041467

G. Hübler, D. Perner, U. Platt, A. Tönnissen, D.H. Ehhalt, Groundlevel OH radical concentration: new measurements by optical absorption. J. Geophys. Res. **89**, 1309–1319

T. Hummel, C. Meister, C. Keim, J. Krauser, M. Wenig, Slit homogenizer introduced performance gain analysis based on the Sentinel-5/UVNS spectrometer. Atmos. Meas. Tech. **14**(8), 5459–5472 (2021). https://amt.copernicus.org/articles/14/5459/2021/

T. Hummel, C. Coatantiec, X. Gnata, T. Lamour, R. Rivière, C. Meister, A. Stute, J. Krauser, D. Weise, M. Wenig, A fibre-based 2S-slit homogenizer concept for high-precision space-based spectrometer missions. CEAS Space J. **14**(2), 239–252 (2022)

T. Hüneke, O.-A. Aderhold, J. Bounin, M. Dorf, E. Gentry, K. Grossmann, J.-U. Grooß, P. Hoor, P. Jöckel, M. Kenntner, M. Knapp, M. Knecht, D. Lörks, S. Ludmann, S. Matthes, R. Raecke, M. Reichert, J. Weimar, B. Werner, A. Zahn, H. Ziereis, K. Pfeilsticker, The novel HALO mini-DOAS instrument: inferring trace gas concentrations from airborne UV/visible limb spectroscopy under all skies using the scaling method. Atmos. Meas. Tech. **10**(11), 4209–4234 (2017). https://doi.org/10.5194/amt-10-4209-2017

M.C. Hutley, *Diffraction Gratings* (Academic, London, 1982)

G.A. Impey, P.B. Shepson, D.R. Hastie, L.A. Barrie, K.G. Anlauf, Measurements of photolyzable chlorine and bromine during the polar sunrise experiment 1995. J. Geophys. Res. **102**(D13), 16005–16010 (1997)

S. Inamdar, L. Tinel, R. Chance, L.J. Carpenter, P. Sabu, R. Chacko, S.C. Tripathy, A.U. Kerkar, A.K. Sinha, P.V. Bhaskar, A. Sarkar, R. Roy, T. Sherwen, C. Cuevas, A. Saiz-Lopez, K. Ram, A.S. Mahajan, Estimation of reactive inorganic iodine fluxes in the Indian and Southern Ocean

marine boundary layer. Atmos. Chem. Phys. **20**(20), 12093–12114 (2020). https://doi.org/10.5194/acp-20-12093-2020
IPCC, Climate Change, *The IPCC Scientific Assessment, Intergovernmental Panel on Climate Change* (Cambridge University Press, Cambridge, 1992)
IPCC: Climate change 2001. Third assessment report of the intergovernmental panel on climate change (Cambridge University Press, Cambridge, 2002)
IPCC, *Climate Change 2021—The Physical Science Basis: Working Group I Contribution to the Sixth Assessment Report of the Intergovernmental Panel on Climate Change*, 1st edn. (Cambridge University Press, 2023). https://doi.org/10.1017/9781009157896
H. Irie, Y. Kanaya, H. Akimoto, H. Iwabuchi, A. Shimizu, K. Aoki, First retrieval of tropospheric aerosol profiles using MAX-DOAS and comparison with lidar and sky radiometer measurements. Atmos. Chem. Phys. **8**(2), 341–350 (2008). https://doi.org/10.5194/acp-8-341-2008
H. Irie, H. Takashima, Y. Kanaya, K.F. Boersma, L. Gast, F. Wittrock, D. Brunner, Y. Zhou, M. Van Roozendael, Eight-component retrievals from ground-based MAX-DOAS observations. Atmos. Meas. Tech. **4**(6), 1027–1044 (2011). https://doi.org/10.5194/amt-4-1027-2011
R.G. Isaacs, W.-C. Wang, R.D. Worsham, S. Goldenberg, Multiple scattering LOWTRAN and FASCODE models. Appl. Opt. **26**, 1272–1281 (1987)
M.Z. Jacobson, *Fundamentals of Atmospheric Modeling*, 2nd edn. (Cambridge University Press, 2005)
L. Jaeglé, D.J. Jacob, W.H. Brune, I. Faloona, D. Tan, B.G. Heikes, Y. Kondo, G.W. Sachse, B. Anderson, G.L. Gregory, H.B. Singh, R. Pueschel, G. Ferry, D.R. Blake, R.E. Shetter, Photochemistry of HO_X in the upper troposphere at northern midlatitudes. J. Geophys. Res. **105**, 3877–3892 (2000)
J. Janssen, Analyse spectrale des e'le'ments de l'atmosphe're terrestre. C. R. Hebd. Seances Acad. Sci. **101**, 649–651 (1885)
J. Janssen, Sur les spectres d'absorption de l'oxyge'ne. C. R. Hebd. Seances Acad. Sci. **102**, 1352–1353 (1886)
M.A. Janssen, *An Introduction to the Passive Remote Atmospheric Remote Sensing by Microwave Radiometry* (Wiley, New York, 1993), pp.1–36
Z. Javed, C. Liu, K. Ullah, W. Tan, C. Xing, H. Liu, Investigating the effect of different meteorological conditions on MAX-DOAS observations of NO_2 and CHOCHO in Hefei, China. Atmosphere **10**(7) (2019). https://doi.org/10.3390/atmos10070353
A. Jbaily, X. Zhou, J. Liu, T.-H. Lee, L. Kamareddine, S. Verguet, F. Dominici, Air pollution exposure disparities across US population and income groups. Nature **601**(7892), 228–233 (2022). https://doi.org/10.1038/s41586-021-04190-y
M.I. Jenkin, R.A. Cox, D.J. Williams, Laboratory studies of the kinetics of formation of nitrous acid from the thermal reaction of nitrogen dioxide and water vapour. Atmos. Environ. **22**, 487–498 (1988)
J.L. Jimenez, R. Bahreini, D.R. Cocker III., H. Zhuang, V. Varutbangkul, R.C. Flagan, J.H. Seinfeld, C.D. O'Dowd, T. Hoffmann, New particle formation from photooxidation of diiodomethane (CH_2I_2). J. Geophys. Res. **108**(D10), 4318 (2003). https://doi.org/10.1029/2002JD002452
R. Jimenez, M. Taslakov, V. Simeonov, B. Calpini, F. Jeanneret, D. Hofstetter, M. Beck, J. Faist, H. van den Bergh, Ozone detection by differential absorption spectroscopy at ambient pressure with a 9.6 μm pulsed quantum-cascade laser. Appl. Phys. B **78**, 249 (2004)
B.T. Jobson, H. Niki, Y. Yokouchi, J. Bottenheim, F. Hopper, R. Leaitch, Measurements of C_2-C_6 hydrocarbons during polar sunrise experiment 1992. J. Geophys. Res. **99**, 25355–25368 (1994)
B.T. Jobson, R.A. Volkamer, E. Velasco, G. Allwine, H. Westberg, B.K. Lamb, M.L. Alexander, C.M. Berkowitz, L.T. Molina, Comparison of aromatic hydrocarbon measurements made by PTR-MS, DOAS and GC-FID during the MCMA 2003 field experiment. Atmos. Chem. Phys. **10**(4), 1989–2005 (2010). https://doi.org/10.5194/acp-10-1989-2010
M. Johansson, B. Galle, C. Rivera, Y. Zhang, Tomographic reconstruction of gas plumes using scanning DOAS. Bull. Volcanol. **71**(10), 1169–1178 (2009a). https://doi.org/10.1007/s00445-009-0292-8

M. Johansson, B. Galle, Y. Zhang, C. Rivera, D. Chen, K. Wyser, The dual-beam mini-DOAS technique–measurements of volcanic gas emission, plume height and plume speed with a single instrument. Bull. Volcanol. **71**(7), 747–751 (2009b). https://doi.org/10.1007/s00445-008-0260-8

J.K.E. Johansson, J. Mellqvist, J. Samuelsson, B. Offerle, B. Lefer, B. Rappenglück, J. Flynn, G. Yarwood, Emission measurements of alkenes, alkanes, SO_2, and NO_2 from stationary sources in Southeast Texas over a 5 year period using SOF and mobile DOAS. J. Geophys. Res.: Atmos. **119**(4), 1973–1991 (2014a). https://doi.org/10.1002/2013JD020485

J.K.E. Johansson, J. Mellqvist, J. Samuelsson, B. Offerle, J. Moldanova, B. Rappenglück, B. Lefer, J. Flynn, Quantitative measurements and modeling of industrial formaldehyde emissions in the Greater Houston area during campaigns in 2009 and 2011. J. Geophys. Res.: Atmos. **119**(7), 4303–4322 (2014b)

M.S. Johnson, K. Strawbridge, K.E. Knowland, C. Keller, M. Travis, Long-range transport of Siberian biomass burning emissions to North America during FIREX-AQ. Atmos. Environ. **252**, 118241 (2021). https://doi.org/10.1016/j.atmosenv.2021.118241

H.S. Johnston, E.D. Morris, J. Van den Bogaerde, Molecular modulation kinetic spectrometry. ClOO and ClO_2 radicals in the photolysis of clorine in oxygen. J. Am. Chem. Soc. **91**, 7712–7727 (1969)

H.S. Johnston, Reduction of stratospheric ozone by nitrogen oxide catalyst from supersonic transport exhaust. Science **173**, 517–522 (1971)

H.S. Johnston, R. Graham, Photochemistry of NO_x and HNO_x Compounds. Can. J. Chem. **52**(8), 1415–1423 (1974). https://doi.org/10.1139/cjc-52-8

H.S. Johnston, C.A. Cantrell, J.G. Calvert, Unimolecular decomposition of NO_3 to form NO and O_2 and a review of N_2O_5/NO_3 kinetics. J. Geophys. Res. **91**, 5159–5172 (1986)

P.V. Johnston, R.L. McKenzie, Long-path absorption measurements of NO_2 in rural New Zealand. Geophys. Res. Lett. **11**, 69–72 (1984)

P.V. Johnston, R.L. McKenzie, NO_2 observations at 45S during the decreasing phase of solar cycle 21, from 1980 to 1987. J. Geophys. Res. **94**, 3473–3486 (1989)

P.V. Johnston, R.L. McKenzie, J.G. Keys, W.A. Matthews, Observations of depleted stratospheric NO_2 following the Pinatubo volcanic eruption. Geophys. Res. Lett. **19**, 211–213 (1992)

P.V. Johnston, Making UV/Vis cross sections, reference Fraunhofer and synthetic spectra, Unpublished Manuscript, NIWA, Lauder (1996)

J. Johnston, L. Cushing, Chemical exposures, health, and environmental justice in communities living on the fenceline of industry. Curr. Environ. Health Rpt. **7**(1), 48–57 (2020). https://doi.org/10.1007/s40572-020-00263-8

J. Joiner, P.K. Bhartia, The determination of cloud pressures from rotational Raman scattering in satellite backscatter ultraviolet measurements. J. Geophys. Res. **100**, 23019–23026 (1995)

J. Joiner, P.K. Bhartia, R.P. Cebula, E. Hilsenrath, R.D. McPeters, Rotational-Raman scattering (ring effect) in satellite backscatter ultraviolet measurements. Appl. Opt. **34**(21), 4513–4525 (1995)

J. Joiner, Y. Yoshida, A.P. Vasilkov, Y. Yoshida, L.A. Corp, E.M. Middleton, First observations of global and seasonal terrestrial chlorophyll fluorescence from space. Biogeosciences **8**(3), 637–651 (2011). https://doi.org/10.5194/bg-8-637-2011

J. Joiner, Y. Yoshida, A.P. Vasilkov, E.M. Middleton, P.K.E. Campbell, Y. Yoshida, A. Kuze, L.A. Corp, Filling-in of near-infrared solar lines by terrestrial fluorescence and other geophysical effects: simulations and space-based observations from SCIAMACHY and GOSAT. Atmos. Meas. Tech. **5**(4), 809–829 (2012). https://doi.org/10.5194/amt-5-809-2012

J. Joiner, L. Guanter, R. Lindstrot, M. Voigt, A.P. Vasilkov, E.M. Middleton, K.F. Huemmrich, Y. Yoshida, C. Frankenberg, Global monitoring of terrestrial chlorophyll fluorescence from moderate-spectral-resolution near-infrared satellite measurements: methodology, simulations, and application to GOME-2. Atmos. Meas. Tech. **6**(10), 2803–2823 (2013). https://doi.org/10.5194/amt-6-2803-2013

D.G. Jones, Photodiode array detectors in UV-VIS spectroscopy: part I. Anal. Chem. **57**, 1057–1073 (1985a)

D.G. Jones, Photodiode array detectors in UV-VIS spectroscopy: part II. Anal. Chem. **57**, 1207–1214 (1985b)

N. Jordan, H.D. Osthoff, Quantification of nitrous acid (HONO) and nitrogen dioxide (NO_2) in ambient air by broadband cavity-enhanced absorption spectroscopy (IBBCEAS) between 361 and 388 nm. Atmos. Meas. Tech. **13**(1), 273–285 (2020). https://amt.copernicus.org/articles/13/273/2020/

D.M. Joseph, S.H. Ashworth, J.M.C. Plane, The absorption cross-section and photochemistry of OIO. J. Photochem. Photobiol., A: Chem. **176**, 68–77 (2005)

J.L. Jourdain, G. LeBras, G. Poulet, J. Combourieu, R. Rigaud, B. Leroy, UV absorption spectrum of ClO($A^2\Pi$-$X^2\Pi$) up to the (1,0) band. Chem. Phys. Lett. **57**, 109–112 (1978)

L.M. Judd, J.A. Al-Saadi, J.J. Szykman, L.C. Valin, S.J. Janz, M.G. Kowalewski, H.J. Eskes, J.P. Veefkind, A. Cede, M. Mueller, M. Gebetsberger, R. Swap, R.B. Pierce, C.R. Nowlan, G.G. Abad, A. Nehrir, D. Williams, Evaluating Sentinel-5P TROPOMI tropospheric NO_2 column densities with airborne and Pandora spectrometers near New York City and Long Island sound. Atmos. Meas. Tech. **13**(11), 6113–6140 (2020). https://doi.org/10.5194/amt-13-6113-2020

D. Jung, R. Soler, D. de la Paz, A. Notario, A. Muñoz, M. Ródenas, T. Vera, E. Borrás, R. Borge, Oxidation capacity changes in the atmosphere of large urban areas in Europe: modelling and experimental campaigns in atmospheric simulation chambers. Chemosphere **341**, 139919 (2023). https://doi.org/10.1016/j.chemosphere.2023.139919

C. Junge, C.W. Chagnon, J.E. Manson, A world-wide stratospheric aerosol layer. Science **133**(3463), 1478–1479 (1961)

C.E. Junge, *Air Chemistry and Radioactivity, International Geophysics*, vol. 4 (Academic, New York, 1963)

W. Junkermann, U. Platt, A. Volz-Thomas, A photoelectric detector for the measurement of photolysis frequencies of ozone and other atmospheric molecules. J. Atmos. Chem. **8**, 203–227 (1989)

W. Junkermann, T. Ibusuki, FTIR spectroscopic measurements of surface bond products of nitrogen oxides on aerosol surfaces: implications for heterogeneous HNO_2 production. Atmos. Environ. **26**, 3099–3103 (1992)

N. Kaiser, Off-axis-Messungen von troposphärischem NO_3. Diploma Thesis, University of Heidelberg (1997)

J.W. Kaiser, J.P. Burrows, Fast weighting functions for retrievals from limb scattering measurements. J. Quant. Spectrosc. Radiat. Transf. **77**(3), 273–283 (2003). https://doi.org/10.1016/S0022-4073(02)00125-5

M. Kalberer, M. Ammann, F. Arens, H.W. Gäggeler, U. Baltensperger, Heterogeneous formation of nitrous acid (HONO) on soot aerosol particles. J. Geophys. Res. **104**, 13825–13832 (1999)

Y. Kanaya, Y. Sadanaga, J. Matsumoto, U.K. Sharma, J. Hirokawa, Y. Kajii, H. Akimoto, Daytime HO_2 concentrations at Oki Island, Japan, in summer 1998: comparison between measurement and theory. J. Geophys. Res. **105**, 24205–24222 (2000)

Y. Kanaya, Y. Sadanaga, J. Hirokawa, Y. Kajii, H. Akimoto, Development of a ground-based LIF instrument for measuring HOx radicals: instrumentation and calibrations. J. Atmos. Chem. **38**, 73–110 (2001)

Y. Kanaya, A. Hofzumahaus, H-P. Dorn, T. Brauers, H. Fuchs, F. Holland, F. Rohrer, B. Bohn, R. Tillmann, R. Wegener, Wahner A, Kajii Y, Miyamoto K, Nishida S, Watanabe K, Yoshino A, Kubistin D, Martinez M, Rudolf M, Harder H, Berresheim H, Elste T, Plass-Dülmer C, Stange G, Kleffmann J, Elshorbany Y, Schurath U, Comparisons of observed and modeled OH and HO_2 concentrations during the ambient measurement period of the HO_XComp field campaign. Atmos. Chem. Phys. **12**(5), 2567–2585 (2012). https://acp.copernicus.org/articles/12/2567/2012/

Y. Kanaya, H. Irie, H. Takashima, H. Iwabuchi, H. Akimoto, K. Sudo, M. Gu, J. Chong, Y.J. Kim, H. Lee, A. Li, F. Si, J. Xu, P.-H. Xie, W.-Q. Liu, A. Dzhola, O. Postylyakov, V. Ivanov, E. Grechko, S. Terpugova, M. Panchenko, Long-term MAX-DOAS network observations of NO_2 in Russia and Asia (MADRAS) during the period 2007–2012: instrumentation, elucidation of climatology, and comparisons with OMI satellite observations and global model simulations. Atmos. Chem. Phys. **14**(15), 7909–7927 (2014)

J. Kasparian, M. Rodriguez, G. Méjean, J. Yu, E. Salmon, H. Wille, R. Bourayou, S. Frey, Y.-B. André, A. Mysyrowicz, R. Sauerbrey, J.-P. Wolf, L. Wöste, White-light filaments for atmospheric analysis. Science **301**, 61–64 (2003)

C.D. Keeling, R.B. Barcastow, A.E. Bainbridge, C.A. Ekdahl, P.R. Guenther, L.S. Waterman, Atmospheric carbon dioxide variations at Mauna Loa observatory. Hawaii. Tellus **28**, 538–551 (1976)

H. Keller-Rudek, G.K. Moortgat, MPI-Mainz-UV-VIS spectral atlas of gaseous molecules (2005). www.atmosphere.mpg.de/spectral-atlas-mainz

H. Keller-Rudek, G.K. Moortgat, R. Sander, R. Sörensen, The MPI-MAINZ UV/VIS spectral atlas of gaseous molecules of atmospheric interest. Earth Syst. Sci. Data **5**(2), 365–373 (2013). https://doi.org/10.5194/essd-5-365-2013

C. Kern, Applicability of light-emitting diodes as light sources for active long path DOAS measurements: a feasibility study. Diploma Thesis in Physics, Institut für Umweltphysik, University of Heidelberg (2004)

C. Kern, S. Trick, B. Rippel, U. Platt, Applicability of light-emitting diodes as light sources for active DOAS measurements. Appl. Opt. **45**, 2077–2088 (2006)

C. Kern, H. Sihler, L. Vogel, C. Rivera, M. Herrera, U. Platt, Halogen oxide measurements at Masaya Volcano, Nicaragua using active long path differential optical absorption spectroscopy. Bull. Volcanol. **71**, 659–670 (2009). https://doi.org/10.1007/s00445-008-0252-8

C. Kern, F. Kick, P. Lübcke, L. Vogel, M. Wöhrbach, U. Platt, Theoretical description of functionality, applications, and limitations of SO_2 cameras for the remote sensing of volcanic plumes. Atmos. Meas. Tech. **3**(3), 733–749 (2010). https://amt.copernicus.org/articles/3/733/2010/

C. Kern, P.J. Kelly, Weak degassing from remote Alaska volcanoes characterized with a new airborne imaging DOAS instrument and a suite of in situ sensors. Front. Earth Sci. **11** (2023). https://doi.org/10.3389/feart.2023.1088056

J.B. Kerr, C.T. McElroy, W.F. Evans, Mid-latitude summertime measurements of stratospheric NO_2. Can. J. Phys. **60**, 196–200 (1982)

H.E. Kessenich, A. Seppälä, C.J. Rodger, Potential drivers of the recent large Antarctic ozone holes. Nat. Commun. **14**(1), 7259 (2023). https://doi.org/10.1038/s41467-023-42637-0

C. Kessler, D. Perner, U. Platt, Spectroscopic measurements of nitrous acid and formaldehyde—implications for urban photochemistry, in *Proceedings of the 2nd European Symposium on Physico-Chemical Behavior of Atmospheric Pollutants*, ed. by B. Versino, H. Ott, September 29 to October 1 (Varese, 1981), pp. 393–400

C. Kessler, Gasförmige Salpetrige Säure (HNO_2) in der belasteten Atmosphäre. Ph.D. Thesis, University of Cologne (1984)

C. Kessler, U. Platt, Nitrous acid, in *polluted air masses: sources and formation pathways, in Proceeding on the 3rd European Symposium on PhysioChemical Behaviour of Atmospheric Pollutants, Varese, Italia, 10–12 Apr.* ed. by D. Reidel (Mass, Norwell, 1984), pp.412–422

M.A.K. Khalil, R.A. Rasmussen, A. Gundwardena, Atmospheric methyl bromide: trends and global mass balance. J. Geophys. Res. **98**, 2887–2896 (1993)

M.F. Khokhar, C. Frankenberg, M. Van Roozendael, S. Beirle, S. Kühl, A. Richter, U. Platt, T. Wagner, Satellite observations of atmospheric SO_2 from volcanic eruptions during the time period of 1996–2002. Adv. Space Res. **36**(5), 879–887 (2005)

N. Kille, S. Baidar, P. Handley, I. Ortega, R. Sinreich, O.R. Cooper, F. Hase, J.W. Hannigan, G. Pfister, R. Volkamer, The CU mobile Solar Occultation Flux instrument: structure functions and emission rates of NH_3, NO_2 and C_2H_6. Atmos. Meas. Tech. **10**(1), 373–392 (2017). https://doi.org/10.5194/amt-10-373-2017

N. Kille, K.J. Zarzana, J. Romero Alvarez, C.F. Lee, J.P. Rowe, B. Howard, T. Campos, A. Hills, R.S. Hornbrook, I. Ortega, W. Permar, I.T. Ku, J. Lindaas, I.B. Pollack, A.P. Sullivan, Y. Zhou, C.D. Fredrickson, B.B. Palm, Q. Peng, E.C. Apel, L. Hu, J.L.J. Collett, E.V. Fischer, F. Flocke, J.W. Hannigan, J. Thornton, R. Volkamer, The CU airborne solar occultation flux instrument: performance evaluation during BB-FLUX. ACS Earth Space Chem. **6**(3), 582–596 (2022). https://doi.org/10.1021/acsearthspacechem.1c00281

J. Kim, J. Kim, H.-K. Cho, J. Herman, S.S. Park, H.K. Lim, J.-H. Kim, K. Miyagawa, Y.G. Lee, Intercomparison of total column ozone data from the Pandora spectrophotometer with Dobson, Brewer, and OMI measurements over Seoul. Korea. Atmos. Meas. Tech. **10**(10), 3661–3676 (2017). https://doi.org/10.5194/amt-10-3661-2017

J. Kim, U. Jeong, M.-H. Ahn, J.H. Kim, R.J. Park, H. Lee, C.H. Song, Y.-S. Choi, K.-H. Lee, J.-M. Yoo et al., New era of air quality monitoring from space: Geostationary Environment Monitoring Spectrometer (GEMS). Bull. Am. Meteorol. Soc. **101**(1), E1 – E22 (2020). https://doi.org/10.1175/BAMS-D-18-0013.1

K. H. Kim, M.-Y. Kim, Comparison of an open path differential optical absorption spectroscopy system and a conventional in situ monitoring system on the basis of long term measurements of SO_2, NO_2, and O_3. Atmos. Environ. **35**, 4059–4072 (2001)

S. Kim, D. Kim, H. Hong, L.-S. Chang, H. Lee, D.-R. Kim, D. Kim, J.-A. Yu, D. Lee, U. Jeong, C.-K. Song, S.-W. Kim, S.S. Park, J. Kim, T.F. Hanisco, J. Park, W. Choi, K. Lee, First-time comparison between NO_2 vertical columns from Geostationary Environmental Monitoring Spectrometer (GEMS) and Pandora measurements. Atmos. Meas. Tech. **16**(16), 3959–3972 (2023). https://doi.org/10.5194/amt-16-3959-2023

M.D. King, E.M. Dick, W.R. Simpson, A new method for the atmospheric detection of the nitrate radical (NO_3). Atmos. Environ. **34**, 685–688 (2000)

T.W. Kirchstetter, R.A. Harley, D. Littlejohn, Measurement of nitrous acid in motor vehicle exhaust. Environ. Sci. Technol. **30**(9), 2843–2849 (1996)

J. Kleffmann, K.H. Becker, P. Wiesen, Heterogeneous NO_2 conversion processes on acid surfaces: possible atmospheric implications. Atmos. Environ. **32**, 2721–2729 (1998)

J. Kleffmann, T. Gavriloaiei, A. Hofzumahaus, F. Holland, R. Koppmann, L. Rupp, E. Schlosser, M. Siese, A. Wahner, Daytime formation of nitrous acid: a major source of OH radicals in a forest. Geophy. Res. Lett. **32**, L05818 (2005). https://doi.org/10.1029/2005GL022524

J. Kleffmann, Daytime sources of nitrous acid (HONO) in the atmospheric boundary layer. ChemPhysChem **8**(8), 1137–1144 (2007). https://doi.org/10.1002/cphc.200700016

U. Klein, I. Wohltmann, K. Lindner, K.F. Künzi, Ozone depletion and chlorine activation in the Arctic winter 1999/2000 observed in Ny-Ålesund. J. Geophys. Res. **107**(D20), 8288 (2002). https://doi.org/10.1029/2001JD000543

M.T. Kleinman, S.J. Head, E.D. Stevenson, J.G. Watson, J.C. Chow, G.M. Hidy, S.L. Altshuler, P.K. Mueller, Air quality measurements-from rubber bands to tapping the rainbow. J. Air Waste Manag. Assoc. **67**(11), 1159–1168 (2017). https://doi.org/10.1080/10962247.2017.1370309

Q.L. Kleipool, M.R. Dobber, J.F. de Haan, P.F. Levelt, Earth surface reflectance climatology from 3 years of OMI data. J. Geophys. Res.: Atmos. **113**(D18) (2008). https://doi.org/10.1029/2008JD010290

D. Kley, Tropospheric chemistry and transport. Science **276**, 1043–1045 (1997)

G. Knight, A.R. Ravishankara, J.B. Burkholder, Laboratory studies of OBrO. J. Phys. Chem. A **104**, 11121–11125 (2000)

P. Knoll, R. Singer, W. Kiefer, Improving spectroscopic techniques by a scanning multichannel method. Appl. Spectrosc. **44**, 776–782 (1990)

R.B.A. Koelemeijer, P. Stammes, J.W. Hovenier, J.F. de Haan, A fast method for retrieval of cloud parameters using oxygen A band measurements from the Global Ozone Monitoring Experiment. J. Geophys. Res. **106D**, 3475–3490 (2001)

P. Köhler, L. Guanter, J. Joiner, A linear method for the retrieval of sun-induced chlorophyll fluorescence from GOME-2 and SCIAMACHY data. Atmos. Meas. Tech. **8**(6), 2589–2608 (2015). https://doi.org/10.5194/amt-8-2589-2015

P. Köhler, C. Frankenberg, T.S. Magney, L. Guanter, J. Joiner, J. Landgraf, Global retrievals of solar-induced chlorophyll fluorescence with TROPOMI: first results and intersensor comparison to OCO-2. Geophys. Res. Lett. **45**(19), 10456–10463 (2018). https://agupubs.onlinelibrary.wiley.com/doi/abs/10.1029/2018GL079031

N. Kohno, J. Zhou, J. Li, M. Takemura, N. Ono, Y. Sadanaga, Y. Nakashima, K. Sato, S. Kato, Y. Sakamoto, Y. Kajii, Impacts of missing OH reactivity and aerosol uptake of HO_2 radicals on tropospheric O_3 production during the AQUAS-Kyoto summer campaign in 2018. Atmos. Environ. **281**, 119130 (2022). https://www.sciencedirect.com/science/article/pii/S1352231022001959

M. Koike, Y. Kondo, W.A. Matthews, P.V. Johnston, K. Yamazaki, Decrease of stratospheric NO_2 at 44° N caused by Pinatubo volcanic aerosols. Geophys. Res. Lett. **20**, 1975–1978 (1993)

M. Koike, N.B. Jones, W.A. Matthews, P.V. Johnston, R.L. McKenzie, R.L. Kinnison, J. Rodriguez, Impact of Pinatubo aerosols on the partitioning between NO_2 and HNO_3. Geophys. Res. Lett. **21**, 597–600 (1994)

M. Koike, Y. Kondo, W.A. Matthews, P.V. Johnston, P.V. Nakajima, A. Kawaguchi, H. Nakane, I. Murata, A. Budiyono et al., Assessment of the uncertainties in the NO_2 and O_3 measurements by visible spectrometers. J. Atmos. Chem. **32**, 121–145 (1999)

A.N. Kolmogorov, The local structure of turbulence in incompressible viscous fluid for very large Reynolds number. Dokl. Akad. Nauk SSSR **30**, 229–303 (1941). Reprinted in Proc. R. Soc Lond. A **434**, 15–17 (1991)

Y. Kondo, W.A. Matthews, S. Solomon, M. Koike, M. Hayashi, K. Yamazaki, H. Nakajima, K. Tsukui, Ground based measurements of column amounts of NO_2 over Syowa station. Antarctica. J. Geophys. Res. **99**, 14535–14548 (1994)

A.A. Kosterev, F.K. Tittel, Chemical sensors based on quantum cascade lasers. IEEE J. Quantum Electron. **38**, 582 (2002)

S. Kraus, *DOASIS A Framework Design for DOAS* (Dissertation, University of Mannheim, Germany, 2005)

G. Krause, E. Weis, Chlorophyll fluorescence and photosynthesis: The basics. Ann. Rev. Plant Phys. Plant Mol. Biol. **42**(1), 313–349 (1991). https://doi.org/10.1146/annurev.pp.42.060191.001525

K. Kreher, Messung der Breitenverteilung (50° N–70° S) von stratosphärischem Ozon und Stickstoffdioxid mittels optischer Absorptionsspektroskopie. Diploma Thesis, University of Heidelberg (1991)

K. Kreher, M. Fiedler, T. Gomer, J. Stutz, U. Platt, The latitudinal distribution (50°N-50°S) of NO_2 and O_3 in October/November 1990. Geophys. Res. Lett. **22**, 1217–1220 (1995)

K. Kreher, J.G. Keys, P.V. Johnston, U. Platt, X. Liu, Ground-based measurements of OClO and HCl in austral spring 1993 at arrival heights. Antarctica. Geophys. Res. Lett. **23**, 1545–1548 (1996)

K. Kreher, P.V. Johnston, S.W. Wood, U. Platt, Ground-based measurements of tropospheric and stratospheric BrO at arrival heights (78°S). Antarctica. Geophys. Res. Lett. **24**, 3021–3024 (1997)

K. Kreher, G.E. Bodeker, H. Kanzawa, H. Nakane, H. Sasano, Ozone and temperature profiles measured above Kiruna inside, at the edge of, and outside the Arctic polar vortex in February and March 1997. Geophys. Res. Lett. **26**, 715–718 (1999)

K. Kreher, M. van Roozendael, F. Hendrick, A. Apituley, E. Dimitropoulou, U. Frieß, A. Richter, T. Wagner, J. Lampel, N. Abuhassan, L. Ang, M. Anguas, A. Bais, N. Benavent, T. Bösch, K. Bognar, A. Borovski, I. Bruchkouski, A. Cede, K.L. Chan, S. Donner, T. Drosoglou, C. Fayt, H. Finkenzeller, D. Garcia-Nieto, C. Gielen, L. Gómez-Martín, N. Hao, B. Henzing, J.R. Herman, C. Hermans, S. Hoque, H. Irie, J. Jin, P. Johnston, J. Khayyam Butt, F. Khokhar, T.K. Koenig, J. Kuhn, V. Kumar, C. Liu, J. Ma, A. Merlaud, A.K. Mishra, M. Müller, M. Navarro-Comas, M. Ostendorf, A. Pazmino, E. Peters, G. Pinardi, M. Pinharanda, A. Piters, U. Platt, O. Postylyakov, C. Prados-Roman, O. Puentedura, R. Querel, A. Saiz-Lopez, A. Schönhardt, S.F. Schreier, A. Seyler, V. Sinha, E. Spinei, K. Strong, F. Tack, X. Tian, M. Tiefengraber, J.-L. Tirpitz, J. van Gent, R. Volkamer, M. Vrekoussis, S. Wang, Z. Wang, M. Wenig, F. Wittrock, P.H. Xie, J. Xu, M. Yela, C. Zhang, and X. Zhao. Intercomparison of NO_2, O_4, O_3 and HCHO slant column measurements by MAX-DOAS and zenith-sky UV-visible spectrometers during CINDI-2. Atmos. Meas. Tech. **13**(5), 2169–2208 (2020). https://doi.org/10.5194/amt-13-2169-2020

S. Kreycy, C. Camy-Peyret, M.P. Chipperfield, M. Dorf, W. Feng, R. Hossaini, L. Kritten, B. Werner, K. Pfeilsticker, Atmospheric test of the $j(BrONO_2)/k_{BrO+NO_2}$ ratio: implications for total stratospheric Br_y and bromine-mediated ozone loss. Atmos. Chem. Phys. **13**(13), 6263–6274 (2013). https://acp.copernicus.org/articles/13/6263/2013/

H. Kromminga, J. Orphal, P. Spietz, S. Voigt, J.P. Burrows, The temperature dependence (213–293 K) of the absorption cross-sections of OClO in the 340–450 nm region measured by Fourier-transform spectroscopy. J. Photochem. Photobiol., A: Chem. **157**, 149–160 (2003)

S.Kühl, J. Pukite, T. Deutschmann, U. Platt, T. Wagner, SCIAMACHY limb measurements of NO_2, BrO and OClO. Retrieval of vertical profiles: algorithm, first results, sensitivity and comparison studies. Adv. Space Res. **42**(10), 1747–1764 (2008). https://doi.org/10.1016/j.asr.2007.10.022

J.G. Kuhn, N. Bobrowski, P. Lübcke, L. Vogel, U. Platt, A Fabry-Pérot interferometer-based camera for two-dimensional mapping of SO_2 distributions. Atmos. Meas. Tech. **7**, 3705–3715 (2014)

J. Kuhn, U. Platt, N. Bobrowski, T. Wagner, Towards imaging of atmospheric trace gases using Fabry-Pérot interferometer correlation spectroscopy in the UV and visible spectral range. Atmos. Meas. Tech. **12**(1), 735–747 (2019). https://amt.copernicus.org/articles/12/735/2019/

J. Kuhn, N. Bobrowski, T. Wagner, U. Platt, Mobile and high-spectral-resolution Fabry-Pérot interferometer spectrographs for atmospheric remote sensing. Atmos. Meas. Tech. **14**(12), 7873–7892 (2021). https://amt.copernicus.org/articles/14/7873/2021/

J. Kuhn, N. Bobrowski, G. Boudoire, S. Calabrese, G. Giuffrida, M. Liuzzo, K. Karume, D. Tedesco, T. Wagner, U. Platt, High-spectral-resolution Fabry-Pérot interferometers overcome fundamental limitations of present volcanic gas remote sensing techniques. Front. Earth Sci. **11** (2023). https://doi.org/10.3389/feart.2023.1039093

L. Kuhn, J. Kuhn, T. Wagner, U. Platt, The NO_2 camera based on gas correlation spectroscopy. Atmos. Meas. Tech. **15**(5), 1395–1414 (2022). https://amt.copernicus.org/articles/15/1395/2022/

V. Kumar, S. Beirle, S. Dörner, A.K. Mishra, S. Donner, Y. Wang, V. Sinha, T. Wagner, Long-term MAX-DOAS measurements of NO_2, HCHO, and aerosols and evaluation of corresponding satellite data products over Mohali in the Indo-Gangetic plain. Atmos. Chem. Phys. **20**(22), 14183–14235 (2020). https://acp.copernicus.org/articles/20/14183/2020/

N. Künzli, R. Kaiser, S. Medina et al., Public-health impact of outdoor and traffic-related air pollution: a European assessment. Lancet **356**, 795–801 (2000)

T. Kurosu, V.V. Rozanov, J.P. Burrows, Parameterization schemes for terrestrial water clouds in the radiative transfer model GOMETRAN. J. Geophys. Res. **102**(D18), 21809–21823 (1997)

R. Kurtenbach, K.H. Becker, J.A.G. Gomes, J. Kleffmann, J.C. Lörzer, M. Spittler, P. Wiesen, R. Ackermann, A. Geyer, U. Platt, Investigations of emissions and heterogeneous formation of HONO in a road traffic tunnel. Atmos. Environ. **35**, 3385–3394 (2001)

R. Kurtenbach, R. Ackermann, K. Becker, A. Geyer, J. Gomes, J.C. Lörzer, U. Platt, P. Wiesen, Verification of the contribution of vehicular traffic to the total NMVOC emissions in Germany and the importance of the NO_3 chemistry in the city air. J. Atmos. Chem. **42**(1), 395–411 (2002). https://doi.org/10.1023/A:1015778616796

R.L. Kurucz, B. Bell, Atomic Line Data, Kurucz CD-ROM No. 23. Smithsonian Astrophysical Observatory, Cambridge, Mass. Available in the internet (1995). https://lweb.cfa.harvard.edu/amp/ampdata/kurucz23/sekur.html. Accessed 20 Dec 2023

A. Kuze, K.V. Chance, Analysis of cloud-top height and cloud coverage from satellites using the O_2 A and B bands. J. Geophys. Res. **99**, 14482–14491 (1994)

B.I. Kuznetzov, K.S. Nigmatullina, Optical determination of the nitrogen dioxide content in the atmosphere. Izv. Atmos. Oceanic Phys. **13**, 614–617 (1977)

E. Laan, J. de Vries, B. Kruizinga, H. Visser, P. Levelt, G.H.J. van den Oord, A. Maelkki, G. Leppelmeier, E. Hilsenrath, Ozone monitoring with the OMI instrument, in *Proceedings of SPIE 45th Annual Meeting (Imaging Spectrometry VI: Sensor Applications). The International Symposium on Optical Science and Technology* (San Diego, 2000), pp. 334–343

J.C. Lambert, M. Van Roozendael, P.C. Simon, J.P. Pommereau, F. Goutail, J.F. Gleason, S.B. Andersen, D.W. Arlander, N.A. Buivan, H. Claude, J. De La Noe, M. De Maziere, V. Dorokhov, P. Eriksen, A. Green, K. Karlsen Tornqvist, B.A. Kastadt Hoiskar, E. Kyro, J. Leveau, M.F. Merienne, G. Milinevsky, H.K. Roscoe, A. Sarkissian, J.D. Shanklin, J. Staehelin, C. Wahlstrom Tellefsen, G. Vaughan, Combined characterization of GOME and TOMS total ozone measurements from space using ground-based observations from the NDSC. Adv. Space Res. **26**, 1931–1940 (2001)

G. Lammel, D. Perner, The atmospheric aerosol as a source of nitrous acid in the polluted atmosphere. J. Aerosol Sci. **19**, 1199–1202 (1988)
G. Lammel, J.N. Cape, Nitrous acid and nitrite in the atmosphere. Chem. Soc. Rev. **25**, 361–369 (1996)
C. Lamotte, V. Marécal, J. Guth, G. Salerno, S. Corradini, N. Theys, S. Warnach, L. Guerrieri, H. Brenot, T. Wagner, M. Bacles, Impact of SO_2 flux estimation in the modeling of the plume of mount ETNA Christmas 2018 eruption and comparison against multiple satellite sensors. Remote Sens. **15**(3) (2023). https://doi.org/10.3390/rs15030758
T. Lamp, A. Ropertz, K. Weber, G. van Haaren, First results of ambient air measurements with different remote sensing systems over a lake in Germany. Proc. Soc. Photo. Opt. Instrum. Eng. **3534**, 162–172 (1998)
J. Lampel, U. Frieß, U. Platt, The impact of vibrational Raman scattering of air on DOAS measurements of atmospheric trace gases. Atmos. Meas. Tech. **8**(9), 3767–3787 (2015). https://amt.copernicus.org/articles/8/3767/2015/
J. Lampel, Y. Wang, A. Hilboll, S. Beirle, H. Sihler, J. Puķīte, U. Platt, T. Wagner, The tilt effect in DOAS observations. Atmos. Meas. Tech. **10**(12), 4819–4831 (2017). https://doi.org/10.5194/amt-10-4819-2017
J. Lampel, J. Zielcke, S. Schmitt, D. Pöhler, U. Frieß, U. Platt, T. Wagner, Detection of O_4 absorption around 328 and 419 nm in measured atmospheric absorption spectra. Atmos. Chem. Phys. **18**(3), 1671–1683 (2018). https://acp.copernicus.org/articles/18/1671/2018/
L.N. Lamsal, S.J. Janz, N.A. Krotkov, K.E. Pickering, R.J.D. Spurr, M.G. Kowalewski, C.P. Loughner, J.H. Crawford, W.H. Swartz, J.R. Herman, High-resolution NO_2 observations from the airborne compact atmospheric mapper: retrieval and validation. J. Geophys. Res.: Atmos. **122**(3), 1953–1970 (2017). https://doi.org/10.1002/2016JD025483
R. Lang, M.G. Lawrence, Evaluation of the hydrological cycle of MATCH driven by NCEP reanalysis data: comparison with GOME water vapor field measurements. Atmos. Chem. Phys. Discuss. **4**, 7917–7984 (2004)
K. Lange, A. Richter, J.P. Burrows, Variability of nitrogen oxide emission fluxes and lifetimes estimated from Sentinel-5P TROPOMI observations. Atmos. Chem. Phys. **22**(4), 2745–2767 (2022). https://doi.org/10.5194/acp-22-2745-2022
A.O. Langford, R.W. Portmann, J.S. Daniel, H.L. Miller, S. Solomon, Spectroscopic measurement of NO_2 in a Colorado thunderstorm: determination of the mean production by cloud-to-ground lightning flashes. J. Geophys. Res. **109**, D11304 (2004). https://doi.org/10.1029/203JD004158
K. Larche, Die Strahlung des Xenon-Hochdruckbogens hoher Leistungsaufnahme. Z. Phys. **136**, 74–86 (1953)
T. Larsson, H. Neuhaus, Notizen: the immersion grating: spectroscopic advantages and resemblance to the echelon grating. Zeitschrift für Naturforschung A **23**(12), 2130–2132 (1968). https://doi.org/10.1515/zna-1968-1245
D.J. Lary, M.P. Chipperfield, R. Toumi, T. Lenton, Heterogeneous atmospheric bromine chemistry. J. Geophys. Res. **101**, 1489–1504 (1996)
D.J. Lary, Halogens and the chemistry of the free troposphere. Atmos. Chem. Phys. **5**, 227–237 (2005)
I. Laszlo, K. Stamnes, W.J. Wiscombe, S.-C. Tsay, The discrete ordinate algorithm, DISORT for radiative transfer, in *Light Scattering Reviews: Light Scattering and Radiative Transfer*, ed. by A. Kokhanovsky, vol. 11 (Springer Praxis Books, Springer, Berlin, Heidelberg, 2016), pp. 3–65. https://doi.org/10.1007/978-3-662-49538-4_1
A. Lauer, M. Dameris, A. Richter, J.P. Burrows, Tropospheric NO_2 columns: a comparison between model and retrieved data from GOME measurements. Atmos. Chem. Phys. **2**, 67–78 (2002)
V. Lauther, B. Vogel, J. Wintel, A. Rau, P. Hoor, V. Bense, R. Müller, C.M. Volk, In situ observations of CH_2Cl_2 and $CHCl_3$ show efficient transport pathways for very short-lived species into the lower stratosphere via the Asian and the North American summer monsoon. Atmos. Chem. Phys. **22**(3), 2049–2077 (2022). https://acp.copernicus.org/articles/22/2049/2022/

M.J. Lawler, R. Sander, L.J. Carpenter, J.D. Lee, R. von Glasow, R. Sommariva, E.S. Saltzman, HOCl and Cl_2 observations in marine air. Atmos. Chem. Phys. **11**(15), 7617–7628 (2011). https://doi.org/10.5194/acp-11-7617-2011

J.P. Lawrence, J.S. Anand, J.D. Vande Hey, J. White, R.R. Leigh, P.S. Monks, R.J. Leigh, High-resolution measurements from the airborne Atmospheric Nitrogen Dioxide Imager (ANDI). Atmos. Meas. Tech. **8**(11), 4735–4754 (2015). https://doi.org/10.5194/amt-8-4735-2015

B. Lazlo, M.J. Kurylo, R.E. Huie, Absorption cross section, kinetics of formation, and self-reaction of the IO radical via laser photolysis of $N_2O/I_2/N_2$ mixtures. J. Phys. Chem. **99**, 11701–11707 (1995)

G. Le Bras, U. Platt, A possible mechanism for combined chlorine and bromine catalysed destruction of tropospheric ozone in the Arctic. Geophys. Res. Lett. **22**, 599–602 (1995)

G. Le Bras, C. Gölz, U. Platt, Production of peroxy-radicals in the DMS oxidation during night-time, in *Dimethylsulphide: Oceans, Atmosphere and Climate: Proceedings of the International Symposium, Belgirate, Italy, 13–15 October 1992*, ed. by G. Restelli, G. Angeletti (Kluwer Academic Publishers, 1993), pp. 251–260

B. Lee, K. Lehmann, J. Taylor, A. Yalin, A high-finesse broadband optical cavity using calcium fluoride prism retroreflectors. Opt. Express **22**(10), 11583–11591 (2014). https://doi.org/10.1364/OE.22.011583

C. Lee, Y.K. Kim, H. Tanimoto, N. Bobrowski, U. Platt, T. Mori, K. Yamamoto, Remote measurement of volcanic halogen oxides and observation of surface ozone depletion. Geophys. Res. Lett. **32**, L21809 (2005). https://doi.org/10.1029/2005GL023785

D.S. Lee, I. Köhler, E. Grobler, F. Rohrer, R. Sausen, L. Gallardo-Klenner, J.G.J. Olivier, F.J. Dentener, A.F. Bouwman, Estimates of global NO_X emissions and their uncertainties. Atmos. Environ. **31**, 1735–1749 (1997)

H. Lee, Y.J. Kim, C. Lee, Estimation of the rate of increase in nitrogen dioxide concentrations from power plant stacks using an imaging-DOAS. Environ. Monit. Assess. **152**(1–4), 61–70 (2009). https://doi.org/10.1007/s10661-008-0296-4

J.S. Lee, B.J. Kuk, Y.J. Kim, Development of a differential optical absorption spectroscopy (DOAS) system for the detection of atmospheric trace gas species; NO_2, SO_2, and O_3. J. Korean Phys. Soc. **41**, 693–698 (2002)

K.K. Lehmann, P.S. Johnston, P. Rabinowitz, Brewster angle prism retroreflectors for cavity enhanced spectroscopy. Appl. Opt. **48**(16), 2966–2978 (2009). https://opg.optica.org/ao/abstract.cfm?URI=ao-48-16-2966

P.A. Leighton, *Photochemistry of Air Pollution* (Academic, New York, 1961)

J. Lelieveld, P.J. Crutzen, Influence of cloud and photochemical processes on tropospheric ozone. Nature **343**, 227–233 (1990)

J. Lelieveld, P.J. Crutzen, The role of clouds in tropospheric photochemistry. J. Atmos. Chem. **12**, 229–267 (1991)

J. Lenoble, *Radiative Transfer in Scattering and Absorbing Atmospheres: Standard Computational Procedures* (A. Deepak Publishing, Hampton, 1985)

J. M. Lerner, A. Thevenon, The optics of spectroscopy, Jobin-Yvon Optical Systems/Instrumentss SA (1988)

C. Lerot, F. Hendrick, M. Van Roozendael, L.M.A Alvarado, A. Richter, I. De Smedt, N. Theys, Vlietinck J, Yu H, Van Gent J, Stavrakou T, Müller J-F, Valks P, Loyola D, Irie H, Kumar V, Wagner T, Schreier SF, Sinha V, Wang T, Wang P, Retscher C, Glyoxal tropospheric column retrievals from TROPOMI–multi-satellite intercomparison and ground-based validation. Atmos. Meas. Tech. **14**(12), 7775–7807 (2021). https://amt.copernicus.org/articles/14/7775/2021/

H. Leser, G. Hönninger, U. Platt, MAX-DOAS measurements of BrO and NO_2 in the marine boundary layer. Geophys. Res. Lett. **30**(10), 1537 (2003). https://doi.org/10.1029/2002GL015811

C. Leue, M. Wenig, U. Platt, Retrieval of atmospheric trace gas concentrations, in *Handbook of Computer Vision and Applications: Systems and Applications*, ed. by B. Jähne, H. Haußecker, P. Geißler, vol. III (Academic Press, San Diego, 1999)

C. Leue, M. Wenig, T. Wagner, U. Platt, B. Jähne, Quantitative analysis of NO_X emission from Global Ozone Monitoring Experiment satellite image sequences. J. Geophys. Res. **106**, 5493–5505 (2001)

P.F. Levelt, B. van den Oord, E. Hilsenrath, G.W. Leppelmeier, P.K. Bhartia, A. Malkki, H. Kelder, A.R.J. van der, E.J. Brinksma, R. van Oss, P. Veefkind, M. van Weele, R. Noordhoek, Science Objectives of EOS-Aura's Ozone Monitoring Instrument (OMI), in *Proceedings of Quadrennial Ozone Symposium, Sapporo, Japan* (2000), pp. 127–128

P.F. Levelt, J. Joiner, J. Tamminen, J.P. Veefkind, P.K. Bhartia, D.C. Stein Zweers, B.N. Duncan, D.G. Streets, H. Eskes, R. van der A, C. McLinden, V. Fioletov, S. Carn, J. de Laat, M. DeLand, S. Marchenko, R. McPeters, J. Ziemke, D. Fu, X. Liu, K. Pickering, A. Apituley, G. González Abad, A. Arola, F. Boersma, C. Chan Miller, K. Chance, M. de Graaf, J. Hakkarainen, S. Hassinen, I. Ialongo, Q. Kleipool, N. Krotkov, C. Li, L. Lamsal, P. Newman, C. Nowlan, R. Suleiman, L.G. Tilstra, O. Torres, H. Wang, K. Wargan, The Ozone Monitoring Instrument: overview of 14 years in space. Atmos. Chem. Phys. **18**(8), 5699–5745 (2018). https://doi.org/10.5194/acp-18-5699-2018

K. Levenberg, A method for the solution of certain non-linear problems in least squares. Quant. Appl. Math. **2**, 164–168 (1944)

C. Levoni, M. Cervino, R. Guzzi, F. Torricella, Atmospheric aerosol optical properties: a database of radiative characteristics for different components and classes. Appl. Opt. **36**(30), 8031–8041 (1997). http://www.ncbi.nlm.nih.gov/pubmed/18264333

H. Levy, Normal atmosphere: large radical and formaldehyde concentrations predicted. Science **173**, 141–143 (1971)

R.C. Levy, L.A. Remer, O. Dubovik, Global aerosol optical properties and application to moderate resolution imaging spectroradiometer aerosol retrieval over land. J. Geophys. Res.: Atmos. **112**(D13) (2007). https://doi.org/10.1029/2006JD007815

C. Li, J. Joiner, N.A. Krotkov, P.K. Bhartia, A fast and sensitive new satellite SO_2 retrieval algorithm based on principal component analysis: application to the ozone monitoring instrument. Geophys. Res. Lett. **40**(23), 6314–6318 (2013). https://doi.org/10.1002/2013GL058134

C. Li, N.A. Krotkov, S. Carn, Y. Zhang, R.J.D. Spurr, J. Joiner, New-generation NASA Aura Ozone Monitoring Instrument (OMI) volcanic SO_2 dataset: algorithm description, initial results, and continuation with the SUOMI-NPP ozone mapping and profiler suite (OMPS). Atmos. Meas. Tech. **10**(2), 445–458 (2017). https://doi.org/10.5194/amt-10-445-2017

J. Li, B.E. Carlson, O. Dubovik, A.A. Lacis, Recent trends in aerosol optical properties derived from AERONET measurements. Atmos. Chem. Phys. **14**(22), 12271–12289 (2014). https://doi.org/10.5194/acp-14-12271-2014

J. Li, H. Deng, J. Sun, B. Yu, H. Fischer, Simultaneous atmospheric CO, N_2O and H_2O detection using a single quantum cascade laser sensor based on dual-spectroscopy techniques. Sens. Actuators, B Chem. **231**, 723–732 (2016). https://doi.org/10.1016/j.snb.2016.03.089

J. Li, X. An, M. Cui, Z. Sun, C. Wang, Y. Li, Simulation study on regional atmospheric oxidation capacity and precursor sensitivity. Atmos. Environ. **263**, 118657 (2021). https://doi.org/10.1016/j.atmosenv.2021.118657

S.-M. Li, Equilibrium of particle nitrite with gas phase HONO: tropospheric measurements in the high Arctic during polar sunrise. J. Geophys. Res. **99**, 25469–25478 (1994)

J. Liao, L.G. Huey, D.J. Tanner, F.M. Flocke, J.J. Orlando, J.A. Neuman, J.B. Nowak, A.J. Weinheimer, S.R. Hall, J.N. Smith, A. Fried, R.M. Staebler, Y. Wang, J.-H. Koo, C.A. Cantrell, P. Weibring, J. Walega, D.J. Knapp, P.B. Shepson, C.R. Stephens, Observations of inorganic bromine (HOBr, BrO, and Br_2) speciation at Barrow, Alaska, in spring 2009. J. Geophys. Res.: Atmos. **117**(D14) (2012)

J. Liao, H. Sihler, L.G. Huey, J.A. Neuman, D.J. Tanner, U. Friess, U. Platt, F.M. Flocke, J.J. Orlando, P.B. Shepson, H.J. Beine, A.J. Weinheimer, S.J. Sjostedt, J.B. Nowak, D.J. Knapp, R.M. Staebler, W. Zheng, R. Sander, S.R. Hall, K. Ullmann, A comparison of Arctic BrO measurements by chemical ionization mass spectrometry and long path-differential optical absorption spectroscopy. J. Geophys. Res.: Atmos. **116**(D14) (2011). https://doi.org/10.1029/2010JD014788

J. Liao, L.G. Huey, Z. Liu, D.J. Tanner, C.A. Cantrell, J.J. Orlando, F.M. Flocke, P.B. Shepson, A.J. Weinheimer, S.R. Hall, K. Ullmann, H.J. Beine, Y. Wang, E.D. Ingall, C.R. Stephens, R.S. Hornbrook, E.C. Apel, D. Riemer, A. Fried, R.L. Mauldin, J.N. Smith, R.M. Staebler, J.A. Neuman, J.B. Nowak, High levels of molecular chlorine in the Arctic atmosphere. Nat. Geosci. **7**(2), 91–94 (2014). https://doi.org/10.1038/ngeo2046

D.R. Lide, CRC handbook of chemistry and physics, 88th ed editor-in-chief: (National institute of standards and technology) CRC Press/Taylor & Francis group: Boca Raton, fl. 2007. 2640. J. Am. Chem. Soc. **130**(1), 382 (2008). https://doi.org/10.1021/ja077011d

J.B. Liley, P.V. Johnston, R.L. McKenzie, A.J. Thomas, I.S. Boyd, Stratospheric NO_2 variations from a long time series at Lauder. New Zealand. J. Geophys. Res. **105**(D9), 11633–11640 (2000)

J.-T. Lin, R.V. Martin, K.F. Boersma, M. Sneep, P. Stammes, R. Spurr, P. Wang, M. Van Roozendael, K. Clémer, H. Irie, Retrieving tropospheric nitrogen dioxide from the Ozone Monitoring Instrument: effects of aerosols, surface reflectance anisotropy, and vertical profile of nitrogen dioxide. Atmos. Chem. Phys. **14**(3), 1441–1461 (2014). https://doi.org/10.5194/acp-14-1441-2014

J.-T. Lin, M.-Y. Liu, J.-Y. Xin, K.F. Boersma, R. Spurr, R. Martin, Q. Zhang, Influence of aerosols and surface reflectance on satellite NO_2 retrieval: seasonal and spatial characteristics and implications for NO_x emission constraints. Atmos. Chem. Phys. **15**(19), 11217–11241 (2015). https://doi.org/10.5194/acp-15-11217-2015

Z. Lin, S. Stamnes, Z. Jin, I. Laszlo, S.C. Tsay, W.J. Wiscombe, K. Stamnes, Improved discrete ordinate solutions in the presence of an anisotropically reflecting lower boundary: upgrades of the DISORT computational tool. J. Quant. Spectrosc. Radiat. Transf. **157**, 119–134 (2015). https://doi.org/10.1016/j.jqsrt.2015.02.014

S. Lindberg, S. Brooks, C.-J. Lin, K.J. Scott, M.S. Landis, R.K. Stevens, M. Goodsite, A. Richter, Dynamic oxidation of gaseous mercury in the Arctic troposphere at polar sunrise. Environ. Sci. Technol. **36**, 1245–1256 (2002)

K. Liou, *An Introduction to Atmospheric Radiation*, 2nd edn. (Academic Press, 2002)

Y. Liu, Z. Wang, Q. Sun, A.M. Erb, Z. Li, C.B. Schaaf, X. Zhang, M.O. Román, R.L. Scott, Q. Zhang, K.A. Novick, M. Syndonia Bret-Harte, S. Petroy, M. SanClements, Evaluation of the VIIRS BRDF, Albedo and NBAR products suite and an assessment of continuity with the long term MODIS record. Remote Sens. Environ. **201**, 256–274 (2017). https://doi.org/10.1016/j.rse.2017.09.020

J. Liu, X. Li, Y. Yang, H. Wang, Y. Wu, X. Lu, M. Chen, J. Hu, X. Fan, L. Zeng, Y. Zhang, An IBBCEAS system for atmospheric measurements of glyoxal and methylglyoxal in the presence of high NO_2 concentrations. Atmos. Meas. Tech. **12**(8), 4439–4453 (2019). https://doi.org/10.5194/amt-12-4439-2019

S. Liu, P. Valks, G. Pinardi, J. Xu, K.L. Chan, A. Argyrouli, R. Lutz, S. Beirle, E. Khorsandi, F. Baier, V. Huijnen, A. Bais, S. Donner, S. Dörner, M. Gratsea, F. Hendrick, D. Karagkiozidis, K. Lange, A.J.M. Piters, J. Remmers, A. Richter, M. Van Roozendael, T. Wagner, M. Wenig, D.G. Loyola, An improved TROPOMI tropospheric NO_2 research product over Europe. Atmos. Meas. Tech. **14**(11), 7297–7327 (2021). https://doi.org/10.5194/amt-14-7297-2021

N.J. Livesey, W.G. Read, L. Froidevaux, J.W. Waters, M.L. Santee, H.C. Pumphrey, D.L. Wu, Z. Shippony, R.F. Jarnot, The UARS microwave limb sounder version 5 data set: theory, characterization, and validation. J. Geophys. Res. **108**(D13), 4378 (2003). https://doi.org/10.1029/2002JD002273

L. Löfgren, Determination of benzene and toluene in urban air with differential optical absorption spectroscopy. Int. J. Environ. Anal. Chem. **47**, 69–74 (1992)

J.A. Logan, M.J. Prather, S.C. Wofsy, M.B. McElroy, Tropospheric chemistry: a global perspective. J. Geophys. Res. **86**, 7210–7254 (1981)

F. Lohberger, G. Hönninger, U. Platt, Ground-based imaging differential optical absorption spectroscopy of atmospheric gases. Appl. Opt. **43**(24), 4711–4717 (2004) http://www.ncbi.nlm.nih.gov/pubmed/15352396

W.A. Long, *Raman Spectroscopy* (McGraw-Hill, New York, 1977)

C.A. Longfellow, T. Imamura, A.R. Ravishankara, D.R. Hanson, HONO solubility and heterogeneous reactivity on sulfuric acid surfaces. J. Phys. Chem. A **102**, 3323–3332 (1998)

A. Lorente, K. Folkert Boersma, H. Yu, S. Dörner, A. Hilboll, A. Richter, M. Liu, L. N. Lamsal, M. Barkley, I. De Smedt, M. Van Roozendael, Y. Wang, T. Wagner, S. Beirle, J.-T. Lin, N. Krotkov, P. Stammes, P. Wang, H.J. Eskes, M. Krol, Structural uncertainty in air mass factor calculation for NO_2 and HCHO satellite retrievals. Atmos. Meas. Tech. **10**(3), 759–782 (2017). https://doi.org/10.5194/amt-10-759-2017

A. Lorente, T. Borsdorff, A. Butz, O. Hasekamp, J. aan de Brugh, A. Schneider, L. Wu, F. Hase, R. Kivi, D. Wunch, D.F. Pollard, K. Shiomi, N.M. Deutscher, V.A. Velazco, C.M. Roehl, P.O. Wennberg, T. Warneke, J. Landgraf, Methane retrieved from TROPOMI: improvement of the data product and validation of the first 2 years of measurements. Atmos. Meas. Tech. **14**(1), 665–684 (2021). https://doi.org/10.5194/amt-14-665-2021

I. Louban, N. Bobrowski, D. Rouwet, S. Inguaggiato, U. Platt, Imaging DOAS for volcanological applications. Bull. Volcanol. **71**(7), 753–765 (2009). https://doi.org/10.1007/s00445-008-0262-6

J.E. Lovelock, *Gaia: A New Look at Life on Earth* (Oxford University Press, Oxford, 1979)

A.G. Löwe, D. Adukpo, S. Fietkau, A. Heckel, A. Ladstätter-Weißenmayer, T. Medeke, H. Oetjen, A. Richter, F. Wittrock, J.P. Burrows, Multi-axis-DOAS observations of atmospheric trace gases at different latitudes by the global instrument network BREDOM, in *Proceedings of 10th Science Conference of IAMAS, CACGP and 7th Science Conference of IGAC, September 2002, Crete* (2002)

X. Lu, M. Qin, P. Xie, J. Duan, W. Fang, W. Liu, Observation of ambient NO_3 radicals by LP-DOAS at a rural site in North China Plain. Sci. Total Environ. **804**, 149680 (2022). https://doi.org/10.1016/j.scitotenv.2021.149680

P. Lübcke, J. Lampel, S. Arellano, N. Bobrowski, F. Dinger, B. Galle, G. Garzón, S. Hidalgo, Z. Chacón Ortiz, L. Vogel, S. Warnach, U. Platt, Retrieval of absolute SO_2 column amounts from scattered-light spectra: implications for the evaluation of data from automated DOAS networks. Atmos. Meas. Tech. **9**(12), 5677–5698 (2016). https://doi.org/10.5194/amt-9-5677-2016

Y. Luo, J. Liu, F. Si, K. Dou, K. Zhan, Estimation of the power plant plume emissions using high resolution imaging DOAS technique, in *Light, Energy and the Environment 2015, paper JTu5A.21* (Optica Publishing Group, 2015), pp. JTu5A.21. https://doi.org/10.1364/PV.2015.JTu5A.21

E. Lutz, NO cross section determination for DOAS concentration retrieval of NO_x in the UV. Master's Thesis, University of Heidelberg (2019)

Y. Lv, P. Xie, J. Xu, A. Li, Z. Hu, outao Li, H. Zhang, Z. Zhang, X. Tian, F. Hu, J. Zheng, Y. Ye, Study of SO_2 measurement based on a dual optical path Fabry-Pérot correlation spectroscopy. Opt. Express **31**(2), 2602–2620 (2023). https://doi.org/10.1364/OE.480738

J. Ma, S. Dörner, S. Donner, J. Jin, S. Cheng, J. Guo, Z. Zhang, J. Wang, P. Liu, G. Zhang, J. Pukite, J. Lampel, T. Wagner, MAX-DOAS measurements of NO_2, SO_2 HCHO, and BrO at the Mt. Waliguan WMO GAW global baseline station in the Tibetan plateau. Atmos. Chem. Phys. **20**(11), 6973–6990 (2020). https://doi.org/10.5194/acp-20-6973-2020

S.M. MacDonald, J.C. Gómez Martín, R. Chance, S. Warriner, A. Saiz-Lopez, L.J. Carpenter, J.M.C. Plane, A laboratory characterisation of inorganic iodine emissions from the sea surface: Dependence on oceanic variables and parameterisation for global modelling. Atmos. Chem. Phys. **14**(11), 5841–5852 (2014). https://doi.org/10.5194/acp-14-5841-2014

J.B. MacManus, P.L. Kebabian, M.S. Zahniser, Astigmatic mirror multipass absorption cells for long-path-length spectroscopy. Appl. Opt. **34**, 3336–3348 (1995)

T.S. Magney, D.R. Bowling, B.A. Logan, K. Grossmann, J. Stutz, P.D. Blanken, S.P. Burns, R. Cheng, M.A. Garcia, P. Köhler, S. Lopez, N.C. Parazoo, B. Raczka, D. Schimel, C. Frankenberg, Mechanistic evidence for tracking the seasonality of photosynthesis with solar-induced fluorescence. Proc. Natl. Acad. Sci. U. S. A. **116**(24), 11640–11645 (2019). https://doi.org/10.1073/pnas.1900278116

A.S. Mahajan, H. Oetjen, A. Saiz-Lopez, J.D. Lee, G.B. McFiggans, J.M.C. Plane, Reactive iodine species in a semi-polluted environment. Geophys. Res. Lett. **36**(16) (2009). https://doi.org/10.1029/2009GL038018

A.S. Mahajan, J.C. Gómez Martín, T.D. Hay, S.-J. Royer, S. Yvon-Lewis, Y. Liu, L. Hu, C. Prados-Roman, C. Ordóñez, J.M.C. Plane, A. Saiz-Lopez, Latitudinal distribution of reactive iodine in the Eastern Pacific and its link to open ocean sources. Atmos. Chem. Phys. **12**(23), 11609–11617 (2012). https://doi.org/10.5194/acp-12-11609-2012

A.S. Mahajan, C. Prados-Roman, T.D. Hay, J. Lampel, D. Pöhler, K. Großmann, J. Tschritter, U. Frieß, U. Platt, P. Johnston, K. Kreher, F. Wittrock, J.P. Burrows, J.M. Plane, A. Saiz-Lopez, Glyoxal observations in the global marine boundary layer. J. Geophys. Res.: Atmos. **119**(10), 6160–6169 (2014). https://doi.org/10.1002/2013JD021388

A.S. Mahajan, L. Tinel, S. Hulswar, C.A. Cuevas, S. Wang, S. Ghude, R.K. Naik, R.K. Mishra, P. Sabu, A. Sarkar, N. Anilkumar, A. Saiz Lopez, Observations of iodine oxide in the Indian ocean marine boundary layer: a transect from the tropics to the high latitudes. Atmos. Environ.: X **1**, 100016 (2019). https://doi.org/10.1016/j.aeaoa.2019.100016

A.S. Mahajan, M.S. Biswas, S. Beirle, T. Wagner, A. Schönhardt, N. Benavent, A. Saiz-Lopez, Observations of iodine monoxide over three summers at the Indian Antarctic bases of Bharati and Maitri. Atmos. Chem. Phys. **21**(15), 11829–11842 (2021). https://doi.org/10.5194/acp-21-11829-2021

W. Majewski, W.L. Meerts, Near-UV spectra with fully resolved rotational structure of naphthalene and perdeuterated naphthalene. J. Mol. Spectrosc. **104**, 271–281 (1984)

M. Mandelman, R.W. Nicholls, The absorption cross section and f-values for the $v'' = 0$ progression of bands and associated continuum for the ClO (A^2Pi-X^2Pi) system. J. Quant. Spectrosc. Radiat. Transf. **17**, 483–491 (1977)

I. Manisalidis, E. Stavropoulou, A. Stavropoulos, E. Bezirtzoglou, Environmental and health impacts of air pollution: a review. Front. Public Health **8** (2020). https://www.frontiersin.org/articles/10.3389/fpubh.2020.00014

G.I. Marchuk, G.A. Mikhailov, M.A. Nazaraliev, R.A. Darbinjan, B.A. Kargin, B.S. Elepov, *The Monte Carlo Methods in Atmospheric Optics*. Springer Series in Optical Sciences, vol. 12 (Springer, Berlin, Heidelberg, 1980). https://doi.org/10.1007/978-3-540-35237-2

D.W. Marquard, An algorithm for least-squares estimation of nonlinear parameters. J. Soc. Indust. Appl. Math. **11**, 431–441 (1963)

L.C. Marquard, Modellierung des Strahlungstransports in der Erdatmosphäre für absorptionsspektroskopische Messungen im ultravioletten und sichtbaren Spektralbereich, Doktoral Thesis, University of Heidelberg (1998)

L.C. Marquard, T. Wagner, U. Platt, Improved approaches for the calculation of air mass factors required for scattered light differential optical absorption spectroscopy. J. Geophys. Res. **105**, 1315–1327 (2000)

R.V. Martin, K. Chance, D.J. Jacob, T.P. Kurosu, R.J.D. Spurr, E. Bucsela, J.F. Gleason, P.I. Palmer, I. Bey, A.M. Fiore, Q. Li, R.M. Yantosca, R.B.A. Koelemeijer, An improved retrieval of tropospheric nitrogen dioxide from GOME. J. Geophys. Res. **107**(D20), 4437 (2002). https://doi.org/10.1029/2001JD001027

R.V. Martin, D.D. Parrish, T.B. Ryerson, D.K. Nicks Jr., K. Chance, T.P. Kurosu, D.J. Jacob, E.D. Sturges, A. Fried, B.P. Wert, Evaluation of GOME satellite measurements of tropospheric NO_2 and HCHO using regional data from aircraft campaigns in the southeastern United States. J. Geophys. Res. **109**, D24307 (2004). https://doi.org/10.1029/2004JD004869

M. Martin, D. Pöhler, K. Seitz, R. Sinreich, U. Platt, BrO measurements over the Eastern North-Atlantic. Atmos. Chem. Phys. **9**(24), 9545–9554 (2009). https://doi.org/10.5194/acp-9-9545-2009

M. Martinez, T. Arnold, D. Perner, The role of bromine and chlorine chemistry for arctic ozone depletion events in Ny-Ålesund and comparison with model calculations. Ann. Geophys. **7**, 941–956 (1999)

M. Martinez, D. Perner, E. Hackenthal, S. Kultzer, L. Schultz, NO_3 at Helgoland during the NORDEX campaign in October 1996. J. Geophys. Res. **105**(D18), 22685–22695 (2000)

L. Martini, R. Sladkovic, F. Slemr, P. Werle, Monitoring of air pollutants: long term intercomparison of DOAS with conventional techniques, in *Proceedings of 87th Annual Meeting & Exhibition of Air & Waste Management Association, Cincinnati, Ohio, 19–24 June 1994*

B.R. Marx, K.P. Birch, R.C. Felton, B.W. Jolliffe, W.R.C. Rowley, P.T. Woods, High-resolution spectroscopy of SO_2 using a frequency-doubled continuous-wave dye laser. Opt. Comm. **33**, 287–291 (1980)

B. Maseko, G. Feig, R. Burger, Estimating lightning NO_x production over South Africa. South African J. Sci. **117**(9/10) (2021). https://doi.org/10.17159/sajs.2021/8035

S.T. Massie, M. Hervig, HITRAN 2012 refractive indices. J. Quant. Spectrosc. Radiat. Transf. **130**, 373–380 (2013). https://doi.org/10.1016/j.jqsrt.2013.06.022

C.L. Mateer, H.U. Dutsch, J. Staehelin, Influence of a priori profiles on trend calculations from Umkehr data. J. Geophys. Res. **101**(D11), 16779–16787 (1996)

J. Matsumoto, H. Imai, N. Kosugi, Y. Kajii, Development of a measurement system of nitrate radical and dinitrogen pentoxide using a thermal conversion/laser-induced fluorescence. Rev. Sci. Instrum. **76**, 064101 (2005). https://doi.org/10.1063/1.1927098

V. Matveev, M. Peleg, D. Rosen, D.S. Tov-Alper, J. Stutz, K. Hebestreit, U. Platt, D. Blake, M. Luria, Bromine oxide-ozone interaction over the Dead Sea. J. Geophys. Res. **106**, 10375–10378 (2001)

A.N. Maurellis, R. Lang, W.J. van der Zande, A new DOAS parametrization for retrieval of trace gases with highly-structured absorption spectra. Geophys. Res. Lett. **27**, 4069–4072 (2000a)

A.N. Maurellis, R. Lang, W.J. van der Zande, I. Aben, W. Ubachs, Precipitable water column retrieval from GOME data. Geophys. Res. Lett. **27**, 903–906 (2000b)

B. Mayer, A. Kylling, Technical note: The libRadtran software package for radiative transfer calculations - description and examples of use. Atmos. Chem. Phys. **5**(7), 1855–1877 (2005). https://doi.org/10.5194/acp-5-1855-2005

B. Mayer, Radiative transfer in the cloudy atmosphere. EPJ Web Conf. **1**, 75–99 (2009). https://doi.org/10.1140/epjconf/e2009-00912-1

J.C. McConnell, G.S. Henderson, L. Barrie, J. Bottenheim, H. Niki, C.H. Langford, E.M.J. Templeton, Photochemical bromine production implicated in Arctic boundary-layer ozone depletion. Nature **355**, 150–152 (1992)

C.T. McElroy, C.A. McLinden, J.C. McConnell, Evidence for bromine monoxide in the free troposphere during Arctic polar sunrise. Nature **397**, 338–340 (1999)

A.J.S. McGonigle, D.R. Hilton, T.P. Fischer, C. Oppenheimer, Plume velocity determination for volcanic SO_2 flux measurements. Geophys. Res. Lett. **32**, L11302 (2005). https://doi.org/10.1029/2005GL022470

S.A. McKeen, M. Trainer, E.Y. Hsie, R.K. Tallamraju, S.C. Liu, On the indirect determination of atmospheric OH radical concentrations from reactive hydrocarbon measurements. J. Geophys. Res. **95**, 7493–7500 (1990)

R.L. McKenzie, P.V. Johnston, Seasonal variations in stratospheric NO_2 at 45° S. Geophys. Res. Lett. **9**, 1255–1258 (1982)

R.L. McKenzie, P.V. Johnston, Stratospheric nitrogen dioxide measurements at Arrival Heights, Antarctica, N.Z. Antarct. Rec. **5**, 12 (1983)

R.L. McKenzie, P.V. Johnston, Springtime stratospheric NO_2 in Antarctica. Geophys. Res. Lett. **11**, 73–75 (1984)

R.L. McKenzie, P.V. Johnston, C.T. McElroy, J.B. Kerr, S. Solomon, Altitude distributions of stratospheric constituents from ground-based measurements at twilight. J. Geophys. Res. **96**, 15499–15511 (1991)

C.A. McLinden, A.E. Bourassa, S. Brohede, M. Cooper, D.A. Degenstein, W.J.F. Evans, R.L. Gattinger, C.S. Haley, E.J. Llewellyn, N.D. Lloyd, P. Loewen, R.V. Martin, J.C. McConnell, I.C. McDade, D. Murtagh, L. Rieger, C.V. Savigny, P.E. Sheese, C.E. Sioris, B. Solheim, K. Strong, OSIRIS: a decade of scattered light. Bull. Am. Meteorol. Soc. **93**(12), 1845–1863 (2012). https://doi.org/10.1175/BAMS-D-11-00135.1

C.A. McLinden, V. Fioletov, M.W. Shephard, N. Krotkov, C. Li, R.V. Martin, M.D. Moran, J. Joiner, Space-based detection of missing sulfur dioxide sources of global air pollution. Nat. Geosci. **9**(7), 496–500 (2016). https://doi.org/10.1038/ngeo2724

B.B. McMahon, E.L. Simmons, Ground based measurements of atmospheric NO_2 by differential optical absorption. Nature **287**, 710–711 (1980)

J.B. McManus, P.L. Kebabian, Narrow optical interference fringes for certain setup conditions in multipass absorption cells of the Herriott type. Appl. Opt. **29**(7), 898–900 (1990)

J.B. McManus, P.L. Kebabian, M.S. Zahniser, Astigmatic mirror multipass absorption cells for long path length spectroscopy. Appl. Opt.-LP **34**(18), 3336–3348 (1995)

L. Mei, M. Brydegaard, Continuous-wave differential absorption lidar. Laser Photonics Rev. **9**(10), 629–636 (2015). https://doi.org/10.1002/lpor.201400419

A.C. Meier, A. Schönhardt, T. Bösch, A. Richter, A. Seyler, T. Ruhtz, D.-E. Constantin, R. Shaiganfar, T. Wagner, A. Merlaud, M. van Roozendael, L. Belegante, D. Nicolae, L. Georgescu, J.P. Burrows, High-resolution airborne imaging DOAS measurements of NO_2 above Bucharest during AROMAT. Atmos. Meas. Tech. **10**(5), 1831–1857 (2017). https://doi.org/10.5194/amt-10-1831-2017

J. Meinen, J. Thieser, U. Platt, T. Leisner, Technical note: Using a high finesse optical resonator to provide a long light path for differential optical absorption spectroscopy: CE-DOAS. Atmos. Chem. Phys. **10**(8), 3901–3914 (2010). https://doi.org/10.5194/acp-10-3901-2010

M.L. Melamed, S. Solomon, J.S. Daniel, A.O. Langford, R.W. Portmann, T.B. Ryerson, D.K. Nicks Jr., S.A. McKeeen, Measuring reactive nitrogen emissions from point sources using visible spectroscopy from aircracft. J. Environ. Monit. **5**, 29–34 (2003)

R. Meller, G.K. Moortgart, Temperature dependence of the absorption cross sections of formaldehyde between 223 and 323 K in the wavelength range 225–375 nm. J. Geophys. Res. **201**, 7089–7101 (2000)

A. Mellouki, M. Ammann, R.A. Cox, J.N. Crowley, H. Herrmann, M.E. Jenkin, V.F. McNeill, J. Troe, T.J. Wallington, Evaluated kinetic and photochemical data for atmospheric chemistry: volume viii - gas-phase reactions of organic species with four, or more, carbon atoms ($\geq C_4$). Atmos. Chem. Phys. **21**(6), 4797–4808 (2021). https://doi.org/10.5194/acp-21-4797-2021

J. Mellqvist, A. Rosén, H. Axelsson, Temperature dependence of the absorption spectra of nitrogen oxide, nitrogen dioxide and sulfur dioxide in the application of differential optical spectroscopy. Analyst **117**, 417–418 (1992)

J. Mellqvist, A. Rosén, DOAS for flue gas monitoring – I: temperature in the U.V./visible absorption spectra of NO, NO_2, SO_2 and NH_3. J. Quant. Spectrosc. Radiat. Transf. **56**, 187–208 (1996)

J. Mellqvist, J. Samuelsson, J. Johansson, C. Rivera, B. Lefer, S. Alvarez, J. Jolly, Measurements of industrial emissions of alkenes in Texas using the solar occultation flux method. J. Geophys. Res. **115**, 1–13 (2010). https://doi.org/10.1029/2008JD011682

T.F. Mentel, D. Bleilebens, A. Wahner, A study of nighttime nitrogen oxide oxidation in a large reaction chamber-the fate of NO_2, N_2O_5, HNO_3, and O_3 at different humidities. Atmos. Environ. **30**, 4007–4020 (1996)

M.F. Mérienne, A. Jenouvrier, B. Coquart, The NO_2 absorption spectrum. I: absorption cross-sections at ambient temperature in the 300–500 nm region. J. Atmos. Chem. **20**, 281–297 (1995)

M.F. Mérienne, A. Jenouvrier, C. Hermans, A.C. Vandaele, M. Carleer, C. Clerbaux, P.F. Coheur, R. Colin, S. Fally, M. Bach, Water vapor line parameters in the 13,000-9250 cm^{-1} region. JQSRT **82**, 99–117 (2003)

A. Merlaud, M. van Roozendael, N. Theys, C. Fayt, C. Hermans, B. Quennehen, A. Schwarzenboeck, G. Ancellet, M. Pommier, J. Pelon, J. Burkhart, A. Stohl, M. De Mazière, Airborne DOAS measurements in Arctic: vertical distributions of aerosol extinction coefficient and NO_2 concentration. Atmos. Chem. Phys. **11**(17), 9219–9236 (2011). https://doi.org/10.5194/acp-11-9219-2011

M. Meroni, M. Rossini, L. Guanter, L. Alonso, U. Rascher, R. Colombo, J. Moreno, Remote sensing of solar-induced chlorophyll fluorescence: review of methods and applications. Remote Sens. Environ. **113**(10), 2037–2051 (2009). https://doi.org/10.1016/j.rse.2009.05.003

A. Merten, J. Tschritter, U. Platt, Design of differential optical absorption spectroscopy long-path telescopes based on fiber optics. Appl. Opt. **50**(5), 738–754 (2011). https://doi.org/10.1364/AO.50.000738

N. Metropolis, S. Ulam, The Monte Carlo method. J. Am. Stat. Assoc. **44**(247), 335–341 (1949). https://doi.org/10.1080/01621459.1949.10483310

N. Mettig, M. Weber, A. Rozanov, C. Arosio, J.P. Burrows, P. Veefkind, A.M. Thompson, R. Querel, T. Leblanc, S. Godin-Beekmann, R. Kivi, M.B. Tully, Ozone profile retrieval from nadir TROPOMI measurements in the UV range. Atmos. Meas. Tech. **14**(9), 6057–6082 (2021). https://doi.org/10.5194/amt-14-6057-2021

A.A. Michelson, Xxviii. interference phenomena in a new form of refractometer. Lond., Edinb., Dublin Philos. Mag. J. Sci. **13**(81), 236–242 (1882). https://doi.org/10.1080/14786448208627176

E. Middleton, U. Rascher, L. Corp, K. Huemmrich, B. Cook, A. Noormets, A. Schickling, F. Pinto, L. Alonso, A. Damm, L. Guanter, R. Colombo, P. Campbell, D. Landis, Q. Zhang, M. Rossini, D. Schuettemeyer, R. Bianchi, The 2013 FLEX - US airborne campaign at the parker tract loblolly pine plantation in North Carolina, USA. Remote Sens. **9**, 612 (2017). https://doi.org/10.3390/rs9060612

G. Mie, Beiträge zur Optik trüber Medien, speziell kolloidaler Metallösungen. Annalen der Physik, Vierte Folge **25**(3), 377–445 (1908)

M.V. Migeotte, Lines of methane at 7.7μ in the solar spectrum. Phys. Rev. **74**, 112–113 (1948)

M. Migeotte, The fundamental band of carbon monoxide at 4.7 μm in the solar spectrum. Phys. Rev. **75**, 1108–1109 (1949)

D. Mihelcic, P. Muesgen, D.H. Ehhalt, An improved method of measuring tropospheric NO_2 and RO_2 by matrix isolation and electron spin resonance. J. Atmos. Chem. **3**, 341–361 (1985)

D. Mihelcic, D. Klemp, P. Müsgen, H.W. Pätz, A. Volz-Thomas, Simultaneous measurements of peroxy and nitrate radicals at Schauinsland. J. Atmos. Chem. **16**, 313–335 (1993)

D. Mihelcic, F. Holland, A. Hofzumahaus, L. Hoppe, S. Konrad, P. Müsgen, H.-W. Pätz, T. Schmitz, H.-J. Schäfer, T. Schmitz, A. Volz-Thomas, K. Bächmann, S. Schlomski, U. Platt, A. Geyer, B. Alicke, G. Moortgat, Peroxy radicals during BERLIOZ at Pabstthum: measurements, radical budgets, and ozone production. J. Geophys. Res. **108**(D4), 8254 (2003). https://doi.org/10.1029/2001JD001014

G.M. Miles, R. Siddans, B.J. Kerridge, B.G. Latter, N.A.D. Richards, Tropospheric ozone and ozone profiles retrieved from GOME-2 and their validation. Atmos. Meas. Tech. **8**(1), 385–398 (2015). https://doi.org/10.5194/amt-8-385-2015

M. Millan, S. Townsend, J. Davies, Study of the Barringer refractor plate correlation spectrometer as a remote sensing instrument. Utias rpt. 146, m.a.sc. thesis, University of Toronto, Toronto, Ontario, Canada (1969)

M.M. Millán, A study of the operational characteristics and optimization procedures of dispersive correlation spectrometers for the detection of trace gases in the atmosphere. Ph.D. Thesis, Universidad de Toronto, Ontario (1972)

M.M. Millan, R.M. Hoff, Dispersive correlation spectroscopy: a study of mask optimization procedures. Appl. Opt. **16**, 1609–1618 (1977)

M.M. Millán, R.M. Hoff, Remote sensing of air pollutants by correlation spectroscopy—instrumental response characteristics. Atmos. Environ. (1967) **12**(4), 853–864 (1978). https://doi.org/10.1016/0004-6981(78)90023-9

M. Millan, Recent advances in correlation spectroscopy for the remote sensing of SO_2, in *Proceedings of 4th Joint Conference on Sensing of Environmental Pollutants* (1978), pp. 40–43

M.M. Millan, Remote sensing of air pollutants: a study of some atmospheric scattering effects. Atmos. Environ. **14**, 1241–1253 (1980)

H.L. Miller, A. Weaver, R.W. Sanders, K. Arpag, S. Solomon, Measurements of arctic sunrise surface ozone depletion events at Kangerlussuaq. Greenland (67°N to 51°W), Tellus B, **49**(5), 496–509 (1997)

Q.-L. Min, L.C. Harrison, E. Clothiaux, Joint statistics of photon path length and cloud optical depth: case studies. J. Geophys. Res. **106**, 7375–7386 (2001)

K.-E. Min, R.A. Washenfelder, W.P. Dubé, A.O. Langford, P.M. Edwards, K.J. Zarzana, J. Stutz, K. Lu, F. Rohrer, Y. Zhang, S.S. Brown, A broadband cavity enhanced absorption spectrometer for aircraft measurements of glyoxal, methylglyoxal, nitrous acid, nitrogen dioxide, and water vapor. Atmos. Meas. Tech. **9**(2), 423–440 (2016). https://doi.org/10.5194/amt-9-423-2016

A. Minato, N. Sugimoto, Y. Sasano, Optical design of cube-corner retroreflectors having curved mirror surfaces. Appl. Opt. **31**, 6015–6020 (1992)

A. Minato, N. Sugimoto, Design of a four-element, hollow-cube corner retroreflector for satellites by use of a genetic algorithm. Appl. Opt. **37**, 438–442 (1998)

Modelling of low-power high-pressure discharge lamps, Fischer. Philips J. Res. **42**, 58–85 (1987)

G.H. Mohammed, R. Colombo, E.M. Middleton, U. Rascher, C. van der Tol, L. Nedbal, Y. Goulas, O. Pérez-Priego, A. Damm, M. Meroni, J. Joiner, S. Cogliati, W. Verhoef, Z. Malenovský, J.-P. Gastellu-Etchegorry, J.R. Miller, L. Guanter, J. Moreno, I. Moya, J.A. Berry, C. Frankenberg, P.J. Zarco-Tejada, Remote sensing of solar-induced chlorophyll fluorescence (SIF) in vegetation: 50 years of progress. Remote Sens. Environ. **231**, 111177 (2019). https://doi.org/10.1016/j.rse.2019.04.030

N. Mohammed-Tahrin, A.M. South, D.A. Newnham, R.L. Jones, An accurate wavelength calibration for the ozone absorption cross-section in the near-UV spectral region, and its effect on the retrieval of BrO from measurements of zenith-scattered sunlight. J. Geophys. Res. **106**(D9), 9897–9907 (2001)

M.J. Molina, F.S. Rowland, Stratospheric sink for chlorofluoromethans: chlorine atom catalyzed destruction of ozone. Nature **249**, 810 (1974)

L.T. Molina, M.J. Molina, Production of the Cl_2O_2 from the Self-Reaction of the ClO Radical. J. Phys. Chem. **91**(2), 433–436 (1987)

P.S. Monks, L.J. Carpenter, S.A. Penkett, G.P. Ayers, Night-time peroxy radical chemistry in the remote marine boundary layer over the southern ocean. Geophys. Res. Lett. **23**, 535–538 (1996)

S. Montzka, J. Butler, R.M.T. Myers, T. Swanson, A. Clarke, L. Lock, J. Elkins, Decline in the tropospheric abundance of halogen from halocarbons: implications for stratospheric ozone depletion. Science **272**, 1318–1320 (1996)

T. Mori, M. Burton. The SO_2 camera: a simple, fast and cheap method for ground-based imaging of SO_2 in volcanic plumes. Geophys. Res. Lett. **33**(24) (2006). https://doi.org/10.1029/2006GL027916

G.H. Mount, R.W. Sanders, A.L. Schmeltekopf, S. Solomon, Visible spectroscopy at McMurdo Station, Antarctica, 1. Overview and daily variations of NO_2 and O_3, austral spring, 1986. J. Geophys. Res. **92**, 8320–8328 (1987)

G.H. Mount, The measurement of tropospheric OH by long path absorption. 1. Instrumentation. J. Geophys. Res. **97**, 2427–2444 (1992)

G.H. Mount, F.L. Eisele, An intercomparison of tropospheric OH: measurements at Fritz Peak observatory. Colorado. Science **256**, 1187–1190 (1992)

G.H. Mount, R.W. Sanders, J.W. Brault, Interference effects in Reticon photodiode array detectors. Appl. Opt. **31**, 851 (1992)

G.H. Mount, F.L. Eisele, D.J. Tanner, J.W. Brault, P.V. Johnston, J.W. Harder, E.J. Williams, A. Fried, R.E. Shetter, An intercomparison of spectroscopic laser long-path and ion-assisted in situ measurements of hydroxyl concentrations during the tropospheric OH photochemistry experiment, fall 1993. J. Geophys. Res. **102**, 6437–6455 (1997). https://doi.org/10.1029/96JD02342

G.H. Mount, B. Rumburg, J. Havig, B. Lamb, H. Westberg, D. Yonge, K. Johnson, R. Kincaid, Measurement of atmospheric ammonia at a dairy using differential optical absorption spectroscopy in the mid-ultraviolet. Atmos. Environ. **36**, 1799–1810 (2002)

I. Moya, L. Camenen, G. Latouche, C. Mauxion, S. Evain, Z. Cerovic, An Instrument for the Measurement of Sunlight Excited Plant Fluorescence (Springer. Dordrecht (1998). https://doi.org/10.1007/978-94-011-3953-3_986

I. Moya, L. Camenen, S. Evain, Y. Goulas, Z. Cerovic, G. Latouche, J. Flexas, A. Ounis, A new instrument for passive remote sensing: 1. measurements of sunlight-induced chlorophyll fluorescence. Remote Sens. Environ. **91**(2), 186–197 (2004).https://doi.org/10.1016/j.rse.2004.02.012

M. Mozurkewich, J.G. Calvert, Reaction probability of N_2O_5 on aqueous aerosol. J. Geophys. Res. **93**, 15889–15896 (1988)

D. Murtagh, U. Frisk, F. Merino, M. Ridal, A. Jonsson, J. Stegman, G. Witt, P. Eriksson, C. Jiménez, G. Megie, J. de la Noë, P. Ricaud, P. Baron, J. Ramon Pardo, A. Hauchcorne, E.J. Llewellyn, J. Edward, E.J. Degenstein, R.L. Gattinger, N.D. Lloyd, W.E.J. Evans, I.C. McDade, C.S. Haley, C. Sioris, C.V. Savigny, B.H. Solheim, J.C. McConnell, K. Strong, E.H. Richardson, G.W. Leppelmeier, E. Kyrölä, H. Auvinen, L. Oikarinen, An overview of the Odin atmospheric mission. Can. J. Phys. **80**, 309–319 (2002)

Y. Nakashima, Y. Sadanaga, S. Saito, J. Hoshi, H. Ueno, Contributions of vehicular emissions and secondary formation to nitrous acid concentrations in ambient urban air in Tokyo in the winter. Sci. Total Environ. **592**, 178–186 (2017). https://doi.org/10.1016/j.scitotenv.2017.03.122

B. Nardi, W. Bellon, L.D. Oolman, T. Deshler, Spring 1996 and 1997 ozonesonde measurements over McMurdo station. Antarctica. Geophys. Res. Lett. **26**, 723–726 (1999)

T. Nash, The colorimetric estimation of formaldehyde by means of the Hantzsch reaction. Biochem. J. **55**, 416–421 (1953)

J.-M. Nasse, P.G. Eger, D. Pöhler, S. Schmitt, U. Frieß, U. Platt, Recent improvements of long-path DOAS measurements: impact on accuracy and stability of short-term and automated long-term observations. Atmos. Meas. Tech. **12**(8), 4149–4169 (2019). https://doi.org/10.5194/amt-12-4149-2019

H. Naus, W. Ubachs, Visible absorption bands of the $(O_2)_2$-collision complex at pressures below 760 Torr. Appl. Opt. **38**, 3423–3428 (1999)

A. Neftel, A. Blatter, T. Staffelbach, Gas phase measurements of NH_3 and NH_4^+ with differential optical absorption spectroscopy and gas stripping scrubber in combination with flow injection analysis, in *Proceedings of Symposium on Physico-Chemical Behaviour of Atmospheric Pollutants*, ed. by G. Restelli, G. Angeletti (Kluwer, Dordrecht, 1990)

A. Neftel, A. Blatter, R. Hesterberg, T. Staffelbach, Measurements of concentration gradients of HNO_2 and HNO_3 over a semi-natural ecosystem. Atmos. Environ. **30**, 3017–3025 (1996)

M. Nestlen, U. Platt, D. Flothmann, A new instrument for measuring dry deposition velocities of SO_2 using the eddy-correlation method, in *Proceedings of Workshop "Trockene Deposition", Oberursel* (1981)

R. Neuroth, H.-P. Dorn, U. Platt, High resolution spectral features of a series of aromatic hydrocarbons and BrO: potential interferences with OH measurements. J. Atmos. Chem. **12**, 12287–12298 (1991)

M.J. Newchurch, E.-S. Yang, D.M. Cunnold, G.C. Reinsel, J.M. Zawodny, J.M. Russell III., Evidence for slowdown in stratospheric ozone loss: first stage of ozone recovery. J. Geophys. Res. **108**(D16), 4507 (2003). https://doi.org/10.1029/2003JD003471

D.M. Newitt, L.E. Outridge, The ultraviolet absorption bands ascribed to HNO_2. J. Chem. Phys. **6**, 752–754 (1938)

S.M. Newman, I.C. Lane, A.J. Orr-Ewing, D.A. Newnham, J. Ballard, Integrated absorption intensity and Einstein coefficients for the O_2 $a^1\Delta_g \leftarrow X^3\Sigma_g^-$ transition: a comparison of cavity ringdown and high resolution Fourier transform spectroscopy with a long-path absorption cell. J. Chem. Phys. **110**(22), 10749–10757 (1999)

S.M. Newman, A.J. Orr-Ewing, D.A. Newnham, J. Ballard, Temperature and pressure dependence of linewidths and integrated absorption intensities for the $O_2^{a^1\Delta_g \leftarrow X^3\Sigma_g^-}$ (0,0) transition. J. Phys. Chem. A **104**(42), 9467–9480 (2000)

D.A. Newnham, J. Ballard, Visible absorption cross section and integrated absorption intensities of molecular oxygen (O_2 and O_4). J. Geophys. Res. **103**, 28801–28816 (1998)

N.L. Ng, S.S. Brown, A.T. Archibald, E. Atlas, R.C. Cohen, J.N. Crowley, D.A. Day, N.M. Donahue, J.L. Fry, H. Fuchs, R.J. Griffin, M.I. Guzman, H. Herrmann, A. Hodzic, Y. Iinuma, J.L. Jimenez, A. Kiendler-Scharr, B.H. Lee, D.J. Luecken, J. Mao, R. McLaren, A. Mutzel, H.D. Osthoff, B. Ouyang, B. Picquet-Varrault, U. Platt, H.O.T. Pye, Y. Rudich, R.H. Schwantes, M. Shiraiwa, J. Stutz, J.A. Thornton, A. Tilgner, B.J. Williams, R.A. Zaveri, Nitrate radicals and biogenic

volatile organic compounds: oxidation, mechanisms, and organic aerosol. Atmos. Chem. Phys. **17**(3), 2103–2162 (2017). https://doi.org/10.5194/acp-17-2103-2017
M.T. Nguyen, R. Sumathi, D. Sengupta, J. Peeters, Theoretical analysis of reactions related to the HNO_2 energy surface: OH + NO and H + NO_2. Chem. Phys. **230**(1), 1–11 (1998)
M. Nicolet, On the molecular scattering in the terrestrial atmosphere: an empirical formula for its calculation in the homosphere. Planet. Space Sci. **32**(11), 1467–1468 (1984)
A. Nies, J. Heimann, C. Fuchs, J. Kuhn, N. Bobrowski, U. Platt, Quantifying BrO and SO_2 distributions in volcanic plumes-recent advances in imaging Fabry-Pérot interferometer correlation spectroscopy. Front. Earth Sci. **11** (2023). https://doi.org/10.3389/feart.2023.1063922
H. Niki, K.H. Becker (eds.), *NATO Advanced Research Workshop: "The Tropospheric Chemistry of Ozone in the Polar Regions", NATO ASI Series, Subseries I "Global Environmental Change"* (Springer, Heidelberg, 1993)
E.G. Nisbet, M.R. Manning, E.J. Dlugokencky, S.E. Michel, X. Lan, T. Röckmann, H.A.C. Denier van der Gon, J. Schmitt, P.I. Palmer, M.N. Dyonisius, Y. Oh, R.E. Fisher, D. Lowry, J.L. France, J.W.C. White, G. Brailsford, T. Bromley, Atmospheric methane: comparison between methane's record in 2006–2022 and during glacial terminations. Glob. Biogeochem. Cycles **37**(8), e2023GB007875–2023GB007875 (2023). https://doi.org/10.1029/2023GB007875
S.A. Nizkorodov, S.P. Sander, L.R. Brown, Temperature and pressure dependence of high-resolution air-broadened absorption cross sections of NO_2 (415–525 nm). J. Phys. Chem. A **108**(22), 4864–4872 (2004). https://doi.org/10.1021/jp049461n
S. Noël, H. Bovensmann, J.P. Burrows, J. Frerick, K.V. Chance, A.H.P. Goede, Global atmospheric monitoring with SCIAMACHY. Phys. Chem. Earth (C) **24**(5), 427–434 (1999)
S. Noel, M. Buchwitz, H. Bovensmann, J.P. Burrows, Retrieval of total water vapour column amounts from GOME/ERS-2 data. Adv. Space Res. **29**(11), 1697–1702 (2002)
GOME Users Manual, SP-1182, European Space Agency, Publications Division, ESTEC, Noordwijk, The Netherlands, ed. by F. Bednarz (1995)
J. Notholt, J. Hjorth, F. Raes, Long path field measurements of aerosol parameters and trace gas concentrations-formation of nitrous acid during foggy periods. J. Aerosol Sci. **22**(S1), S411–S414 (1991)
J. Notholt, J. Hjorth, F. Raes, O. Schrems, Simultaneous long path field measurements of HNO_2, CH_2O and aerosol. Ber. Bunsenges. Phys. Chem. **96**, 290–293 (1992)
J. Notholt, G.C. Toon, R. Lehmann, B. Sen, J.-F. Blavier, Comparison of Arctic and Antarctic trace gas column abundances from ground-based FTIR spectrometry. J. Geophys. Res. **102**, 12863–12869 (1997)
C.R. Nowlan, X. Liu, J.W. Leitch, K. Chance, G. González Abad, C. Liu, P. Zoogman, J. Cole, T. Delker, W. Good, F. Murcray, L. Ruppert, D. Soo, M.B. Follette-Cook, S.J. Janz, M.G. Kowalewski, C.P. Loughner, K.E. Pickering, J.R. Herman, M.R. Beaver, R.W. Long, J.J. Szykman, L.M. Judd, P. Kelley, W.T. Luke, X. Ren, J.A. Al-Saadi, Nitrogen dioxide observations from the Geostationary Trace gas and Aerosol Sensor Optimization (GeoTASO) airborne instrument: retrieval algorithm and measurements during DISCOVER-AQ Texas 2013. Atmos. Meas. Tech. **9**(6), 2647–2668 (2016). https://doi.org/10.5194/amt-9-2647-2016
J.F. Noxon, R. Goody, Noncoherent scattering of skylight. Atmos. Ocean. Phys. **1**, 257–281 (1965)
J.F. Noxon, Nitrogen dioxide in the stratosphere and troposphere measured by ground-based absorption spectroscopy. Science **189**, 547–549 (1975)
J.F. Noxon, R.B. Norton, W.R. Henderson, Observation of atmospheric NO_3. Geophys. Res. Lett. **5**, 675–678 (1978)
J.F. Noxon, Stratospheric NO_2. 2. Global behavior, J. Geophys. Res. **84**(C8), 5067–5076 (1979)
J.F. Noxon, E.C. Whipple, R.S. Hyde, Stratospheric NO_2. 1. Observational method and behavior at midlatitudes. J. Geophys. Res. **84**(C8), 5047–5065 (1979)
J.F. Noxon, R.B. Norton, E. Marovich, NO_3 in the troposphere. Geophys. Res. Lett. **7**, 125–128 (1980)
J.F. Noxon, NO_3 and NO_2 in the mid-Pacific troposphere. J. Geophys. Res. **88**, 11017–11021 (1983)
NRC: Review of the NARSTO Draft report: an assessment of tropospheric ozone pollution—A North American perspective. The National Academy of Sciences (2000)

C.D. O'Dowd, Biogenic coastal aerosol production and its influence on aerosole radiative properties. J. Geophys. Res. **106**(D2), 1545–1549 (2001)

C.D. O'Dowd, J.L. Jimenez, R. Bahreini, R.C. Flagan, J.H. Seinfeld, K. Hämeri, L. Pirjola, M. Kulmala, S.G. Jennings, T. Hoffmann, Marine aerosol formation from biogenic iodine emissions. Nature **417**, 632–636 (2002)

H. Oetjen, Messung atmosphärischer Spurengase in Ny-Alesund, Aufbau und Inbetriebnahme eines neuen DOAS-Meßsystems, Diploma Thesis, University of Bremen (2002)

H. Oetjen, S. Baidar, N.A. Krotkov, L.N. Lamsal, M. Lechner, R. Volkamer, Airborne MAX-DOAS measurements over California: testing the NASA OMI tropospheric NO_2 product. J. Geophys. Res.: Atmos. **118**(13), 7400–7413 (2013). https://doi.org/10.1002/jgrd.50550

E.A. Ogryzlo, G.E. Thomas, Pressure dependence of the visible absorption bands of molecular iodine. J. Mol. Spectrosc. **17**, 198–202 (1965)

E.P. Olaguer, J. Stutz, M.H. Erickson, S.C. Hurlock, R. Cheung, C. Tsai, S.F. Colosimo, J. Festa, A. Wijesinghe, B.S. Neish, Real time measurement of transient event emissions of air toxics by tomographic remote sensing in tandem with mobile monitoring. Atmos. Environ. **150**, 220–228 (2017). https://doi.org/10.1016/j.atmosenv.2016.11.058

E.P. Olaguer, Overview of the benzene and other toxics exposure (BEE-TEX) field study. Environ. Health Insights **9**(s4) (2020). https://doi.org/10.1177/EHI.S15654

S.J. Oltmans, W.D. Komhyr, Surface ozone distributions and variations from 1973 to 1984 measurements at the NOAA geophysical monitoring for climate change baseline observatories. J. Geophys. Res. **91**, 5229–5236 (1986)

C. Oppenheimer, V.I. Tsanev, C.F. Braban, R.A. Cox, J.W. Adams, A. Aiuppa, N. Bobrowski, P. Delmelle, J. Barclay, A.J. McGonigle, BrO formation in volcanic plumes. Geochim. Cosmochim. Acta **70**, 2935–2941 (2006)

J.J. Orlando, G.S. Tyndall, The atmospheric chemistry of the HC(O)CO. Radical. Int. J. Chem. Kinet. **33**, 149–156 (2001)

J. Orphal, K. Bogumil, A. Dehn, B. Deters, S. Dreher, O.C. Fleischmann, M. Hartmann, S. Himmelmann, T. Homann, H. Kromminga, P. Spietz, A. Türk, A. Vogel, S. Voigt, J.P. Burrows, Laboratory spectroscopy in support of UV-visible remote-sensing of the atmosphere, in *Recent Research Developments in Physical Chemistry*. ed. by S.G. Pandalai (Trivandrum, Transworld Research, 2002), pp.15–34

J. Orphal, A critical review of the absorption cross-sections of O_3 and NO_2 in the ultraviolet and visible. J. Photochem. Photobiol., A: Chem. **157**(2), 185–209 (2003). https://doi.org/10.1016/S1010-6030(03)00061-3

J. Orphal, C.E. Fellows, P.-M. Flaud, The visible absorption spectrum of NO_3 measured by high-resolution Fourier transform spectroscopy. J. Geophys. Res. **108**(D3), 4077 (2003). https://doi.org/10.1029/2002JD002489

J. Orphal, J. Staehelin, J. Tamminen, G. Braathen, M.-R. de Backer, A. Bais, D. Balis, A. Barbe, P.K. Bhartia, M. Birk, J.B. Burkholder, K. Chance, T. von Clarmann, A. Cox, D. Degenstein, R. Evans, J.-M. Flaud, D. Flittner, S. Godin-Beekmann, V. Gorshelev, A. Gratien, E. Hare, C. Janssen, E. Kyrölä, T. McElroy, R. McPeters, M. Pastel, M. Petersen, I. Petropavlovskikh, B. Picquet-Varrault, M. Pitts, G. Labow, M. Rotger-Languereau, T. Leblanc, C. Lerot, X. Liu, P. Moussay, A. Redondas, M. Van Roozendael, S.P. Sander, M. Schneider, A. Serdyuchenko, P. Veefkind, J. Viallon, C. Viatte, G. Wagner, M. Weber, R.I. Wielgosz, C. Zehner, Absorption cross-sections of ozone in the ultraviolet and visible spectral regions: status report 2015. J. Mol. Spectrosc. **327**, 105–121 (2016)

G. Ortgies, K.H. Gericke, F.J. Comes, Is UV laser induced fluorescence a method to monitor tropospheric OH? Geophys. Res. Lett. **7**, 905–908 (1980)

Osram Brochure 'XBO—theater lamps'—technology and application. https://web.archive.org/web/20130718003525/http://assets.sylvania.com/assets/documents/ENGR_BLTN11.161355cc-1d94-4996-b6cd-a3001fea6f1a.pdf

H. Osterkamp, Messung von atmosphärischen O_4-Profilen. Diploma Thesis, University of Heidelberg, Germany (1997)

H. Osterkamp, F. Ferlemann, H. Harder, D. Perner, U. Platt, M. Schneider, K. Pfeilsticker, First measurement of the atmospheric O_4 profile, in *Proceedings of the Fourth European Symposium, 22 to 26 September, Schliersee, Germany, Air Pollution Report 66* (European Commission, Brussels, 1998), pp. 478–481

H.D. Osthoff, J.M. Roberts, A.R. Ravishankara, E.J. Williams, B.M. Lerner, R. Sommariva, T.S. Bates, D. Coffman, P.K. Quinn, J.E. Dibb, H. Stark, J.B. Burkholder, R.K. Talukdar, J. Meagher, F.C. Fehsenfeld, S.S. Brown, High levels of nitryl chloride in the polluted subtropical marine boundary layer. Nat. Geosci. **1**(5), 324–328 (2008). https://doi.org/10.1038/ngeo177

R. Oswald, T. Behrendt, M. Ermel, D. Wu, H. Su, Y. Cheng, C. Breuninger, A. Moravek, E. Mougin, C. Delon, B. Loubet, A. Pommerening-Röser, M. Sörgel, U. Pöschl, T. Hoffmann, M.O. Andreae, F.X. Meixner, I. Trebs, HONO emissions from soil bacteria as a major source of atmospheric reactive nitrogen. Science **341**(6151), 1233–1235 (2013). https://doi.org/10.1126/science.1242266

C. Otten, Messung stratosphärischer Spurenstoffe in den Wintern 1992/93 bis 1994/95 über Kiruna in Nordschweden, Ph.D. Thesis, University of Heidelberg (1997)

C. Otten, F. Ferlemann, U. Platt, T. Wagner, K. Pfeilsticker, Groundbased DOAS UV/visible measurements at Kiruna (Sweden) during the SESAME winters 1993/94 and 1994/95. J. Atmos. Chem. **30**, 141–162 (1998)

K.W. Oum, M.J. Lakin, D.O. DeHaan, T. Brauers, B.J. Finlayson-Pitts, Formation of molecular chlorine from the photolysis of ozone and aqueous sea-salt particles. Science **279**, 74–77 (1998a)

K.W. Oum, M.J. Lakin, B.J. Finlayson-Pitts, Bromine activation in the troposphere by the dark reaction of O_3 with seawater ice. Geophys. Res. Lett. **25**, 3923–3926 (1998b)

P. Pagsberg, E. Bjergbakke, E. Ratajczak, A. Sillesen, Kinetics of the gas phase reaction OH+NO(+M) $\longrightarrow$ HONO(+M) and the determination of the UV absorption cross sections of HONO. Chem. Phys. Lett. **272**(5–6), 383–390 (1997)

B.A. Paldus, R.N. Zare, CRDS an historical perspective and introduction, in *Cavity-Ringdown Spectroscopy: An Ultralow-Absorption Measurement Technique*. ed. by K.W. Busch, M.A. Busch (ACS, Washington, DC, 1999)

P.I. Palmer, D.J. Jacob, K. Chance, R.V. Martin, R.J.D. Spurr, T.P. Kurosu, I. Bey, R. Yantosca, A. Fiore, Q. Li, Air mass factor formulation for spectroscopic measurements from satellites: application to formaldehyde retrievals from the Global Ozone Monitoring Experiment. J. Geophys. Res.: Atmos. **106**(D13), 14539–14550 (2001). https://doi.org/10.1029/2000JD900772

P.I. Palmer, D.J. Jacob, A.M. Fiore, R.V. Martin, K. Chance, T.P. Kurosu, Mapping isoprene emissions over North America using formaldehyde column observations from space. J. Geophys. Res. **108**(D6), 4180 (2003). https://doi.org/10.1029/2002JD002153

J.-H. Park, A.H. Goldstein, J. Timkovsky, S. Fares, R. Weber, J. Karlik, R. Holzinger, Active atmosphere-ecosystem exchange of the vast majority of detected volatile organic compounds. Science **341**(6146), 643–647 (2013). https://doi.org/10.1126/science.1235053

J. Park, J. Ryu, D. Kim, J. Yeo, H. Lee, Long-range transport of SO_2 from Continental Asia to Northeast Asia and the Northwest Pacific Ocean: flow rate estimation using OMI data, surface in situ data, and the HYSPLIT model. Atmosphere **7**(4), 53 (2016). https://doi.org/10.3390/atmos7040053

J. Park, H. Lee, J. Kim, J. Herman, W. Kim, H. Hong, W. Choi, J. Yang, D. Kim, Retrieval accuracy of HCHO vertical column density from ground-based direct-sun measurement and first HCHO column measurement using Pandora. Remote Sens. **10**(2), 173 (2018). https://doi.org/10.3390/rs10020173

D.D. Parrish, M. Trainer, E.J. Williams, D.W. Fahey, G. Huebler, C.S. Eubank, S.C. Liu, P.C. Murphy, D.L. Albritton, F.C. Fehsenfeld, Measurements of the NO_x-O_3 photostationary state at Niwot Ridge. Colorado. J. Geophys. Res. **91**, 5361–5370 (1986)

S.E. Paulson, A.D. Sen, L. Ping, J.D. Fenske, M.J. Fox, Evidence for formation of OH radicals from the reaction of O_3 with alkenes in the gas phase. Geophys. Res. Lett. **24**(24), 3193–3196 (1997)

S.E. Paulson, M. Chung, A. Hasson, OH radical formation from the gas-phase reaction of ozone with terminal alkenes, and the relationship between structure and mechanism. J. Phys. Chem. **103**, 8125–8138 (1999)

K.F. Pearson, LIII. On lines and planes of closest fit to systems of points in space. Lond., Edinb., Dublin Philos. Mag. J. Sci. **2**(11), 559–572 (1901). https://doi.org/10.1080/14786440109462720

S.A. Penkett, N.J. Blake, P. Lightman, A.R.W. Marsh, P. Anwyl, G. Butcher, The seasonal variation of nonmethane hydrocarbons in the free troposphere over the North Atlantic Ocean: possible evidence for extensive reaction of hydrocarbons with the nitrate radical. J. Geophys. Res. **98**, 2865–2885 (1993)

R. Penndorf, Tables of the refractive index for standard air and the Rayleigh scattering coefficient for the spectral region between 0.2 and 20.0 μ and their application to atmospheric optics. J. Opt. Soc. Am. **47**, 176–182 (1957)

C.M. Penney, R.L. St. Peters, M. Lapp, Absolute rotational Raman cross sections for N_2, O_2, and CO_2. J. Opt. Soc. Am. **64**, 712–716 (1974)

L.M. Perliski, S. Solomon, Radiative influences of Pinatubo volcanic aerosols on twilight observations of NO_2 column abundances. Geophys. Res. Lett. **19**, 1923–1926 (1992)

L.M. Perliski, S. Solomon, On the evaluation of air mass factors for atmospheric near-ultraviolet and visible absorption spectroscopy. J. Geophys. Res. **98**, 10363–10374 (1993)

D. Perner, D.H. Ehhalt, H.W. Pätz, U. Platt, E.P. Röth, A. Volz, OH-radicals in the lower troposphere. Geophys. Res. Lett. **3**, 466–468 (1976)

D. Perner, U. Platt, Detection of nitrous acid in the atmosphere by differential optical absorption. Geophys. Res. Lett. **6**, 917–920 (1979)

D. Perner, U. Platt, Absorption of light in the atmosphere by collision pairs of oxygen $(O_2)_2$. Geophys. Res. Lett. **7**, 1053–1056 (1980)

D. Perner, C. Kessler, U. Platt, HNO_2, NO_2, and NO measurements in automobile engine exhaust by optical absorption, in *Monitoring of Gaseous Pollutants by Tunable Diode Laser*. ed. by R. Grisar, H. Preier, G. Schmidke, G. Restelli (Reidel, Dordrecht, 1987a), pp.116–119

D. Perner, U. Platt, M. Trainer, G. Hübler, J.W. Drummond, W. Junkermann, J. Rudolph, B. Schubert, A. Volz, D.H. Ehhalt, K.J. Rumpel, G. Helas, Measurement of tropospheric OH concentrations: a comparison of field data with model predictions. J. Atmos. Chem. **5**, 185–216 (1987b)

D. Perner, T. Klüpfel, U. Parchatka, A. Roth, T. Jörgensen, Ground-based UV-vis spectroscopy: diurnal OClO-profiles during January 1990 above Söndre Strømfjord. Greenland. Geophys. Res. Lett. **18**, 787–790 (1991)

D. Perner, A. Roth, D. Klüpfel, Groundbased measurements of stratospheric OClO, NO_2, and O_3 at Söndre Strömfjord in winter 1991/92. Geophys. Res. Lett. **21**, 1367–1370 (1994)

D. Perner, T. Arnold, J. Crowley, T. Klüpfel, M. Martinez, R. Seuwen, The measurements of active chlorine in the atmosphere by chemical amplification. J. Atmos. Chem. **34**, 9–20 (1999)

A. Perot, C. Fabry, On the application of interference phenomena to the solution of various problems of spectroscopy and metrology. Astrophys. J. **9**(2), 87–115 (1899). https://doi.org/10.1086/140557

C. Peters, Studies of reactive halogen species (RHS) in the marine and mid-latitudinal boundary layer by active longpath differential optical absorption spectroscopy, Ph.D. Thesis, University of Heidelberg (2004)

C. Peters, S. Pechtl, J. Stutz, K. Hebestreit, G. Hönninger, K.G. Heumann, A. Schwarz, J. Winterlik, U. Platt, Reactive and organic halogen species in three different European coastal environments. Atmos. Chem. Phys. **5**, 3357–3375 (2005)

P.K. Peterson, D. Pöhler, H. Sihler, J. Zielcke, S. General, U. Frieß, U. Platt, W.R. Simpson, S.V. Nghiem, P.B. Shepson, B.H. Stirm, S. Dhaniyala, T. Wagner, D.R. Caulton, J.D. Fuentes, K.A. Pratt, Observations of bromine monoxide transport in the Arctic sustained on aerosol particles. Atmos. Chem. Phys. **17**(12), 7567–7579 (2017). https://doi.org/10.5194/acp-17-7567-2017

A. Petritoli, F. Ravegnani, G. Giovanelli, D. Bortoli, U. Bonafe, I. Kostadinov, A. Oulanovsky, Off-Axis measurements of atmospheric trace gases by use of an airborne ultraviolet-visible spectrometer. Appl. Opt. **27**, 5593–5599 (2002)

I. Petropavlovskikh, P.K. Bhartia, J. DeLuisi, New Umkehr ozone profile retrieval algorithm optimized for climatological studies. Geophys. Res. Lett. **32**, L16808 (2005). https://doi.org/10.1029/2005GL023323

K. Pfeilsticker, U. Platt, Airborne measurements during the Arctic stratospheric experiment: observation of O_3 and NO_2. Geophys. Res. Lett. **21**, 1375–1378 (1994)

K. Pfeilsticker, F. Erle, O. Funk, Senne, T. Wagner, U. Platt, Can enhanced tropospheric photon path lengths explain the anomalous cloud absorption phenomenon? in *Poster, Third European Symposium on Polar Stratospheric Ozone, Schliersee, 18–22 September* (1995)

K. Pfeilsticker, C.E. Blom, R. Brandtjen, H. Fischer, N. Glatthor, A. Grendel, T. Gulde, M. Höpfner, D. Perner, C. Piesch, U. Platt, W. Renger, J. Sessler, M. Wirth, Aircraft-borne detection of stratospheric column amounts of O_3, NO_2, OClO, $ClNO_3$, HNO_3, and aerosols around the arctic vortex (79°N to 39°N) during spring 1993. 1. Observational data. J. Geophys. Res. **102**(D9), 10801–10814 (1997a)

K. Pfeilsticker, F. Erle, U. Platt, Absorption of solar radiation by atmospheric O_4. J. Atm. Sci. **54**, 933–939 (1997b)

K. Pfeilsticker, F. Erle, O. Funk, L. Marquard, T. Wagner, U. Platt, Optical path modifications due to tropospheric clouds: implications for zenith sky measurements of stratospheric gases. J. Geophys,. Res. **103**, 25323–25335 (1998a)

K. Pfeilsticker, F. Erle, O. Funk, H. Veitel, U. Platt, First geometrical path lengths probability density function derivation of the skylight from spectroscopically highly resolved oxygen A-band observations. 1. Measurement technique, atmospheric observations and model calculations. J. Geophys. Res. **103**, 11483–11504 (1998b)

K. Pfeilsticker, First geometrical path lengths probability density function derivation of the skylight from high resolution oxygen A-band spectroscopy. 2. Derivation of the Lévy index for the skylight transmitted by mid-latitude clouds. J. Geophys. Res. **104**, 4101–4116 (1999)

K. Pfeilsticker, D.W. Arlander, J.P. Burrows, F. Erle, M. Gil, F. Goutail, C. Hermans, J.-C. Lambert, U. Platt, J.-P. Pommereau, A. Richter, A. Sarkissian, M. Van Roozendael, T. Wagner, T. Winterrath, Intercomparison of the influence of tropospheric clouds on UV-visible absorption detected during the NDSC intercomparison campaign at OHP in June 1996. Geophys. Res. Lett. **26**, 1169–1173 (1999a)

K. Pfeilsticker, F. Erle, U. Platt, Observation of the stratospheric NO_2 latitudinal distribution in the northern winter hemisphere. J. Atmos. Chem. **32**, 101–120 (1999b)

A. Pierson, J. Goldstein, Stray light in spectrometers: causes and cures. Lasers. Optron. 67–74 (1989)

O. Pikelnaya, S. Hurlock, S. Trick, J. Stutz, Measurements of reactive iodine species on the Isles of Shoals. Gulf of Maine, in AGU Fall Meeting Abstracts **86**, A14A-06 (2005)

O. Pikelnaya, S.H. Hurlock, S. Trick, J. Stutz, Intercomparison of multiaxis and long-path differential optical absorption spectroscopy measurements in the marine boundary layer. J. Geophys. Res. **112**, D10S01 (2007). https://doi.org/10.1029/2006JD007727

O. Pikelnaya, J.H. Flynn, C. Tsai, J. Stutz, Imaging DOAS detection of primary formaldehyde and sulfur dioxide emissions from petrochemical flares. J. Geophys. Res. **118**, 1–13 (2013). https://doi.org/10.1002/jgrd.50643

G. Pinardi, M. Van Roozendael, N. Abuhassan, C. Adams, A. Cede, K. Clémer, C. Fayt, U. Frieß, M. Gil, J. Herman, C. Hermans, F. Hendrick, H. Irie, A. Merlaud, M. Navarro Comas, E. Peters, A.J.M. Piters, O. Puentedura, A. Richter, A. Schönhardt, R. Shaiganfar, E. Spinei, K. Strong, H. Takashima, M. Vrekoussis, T. Wagner, F. Wittrock, and S. Yilmaz. MAX-DOAS formaldehyde slant column measurements during CINDI: intercomparison and analysis improvement. Atmos. Meas. Tech. **6**(1), 167–185 (2013). https://doi.org/10.5194/amt-6-167-2013

G. Pinardi, M. van Roozendael, F. Hendrick, N. Theys, N. Abuhassan, A. Bais, F. Boersma, A. Cede, J. Chong, S. Donner, T. Drosoglou, A. Dzhola, H. Eskes, U. Frieß, J. Granville, J.R. Herman, R. Holla, J. Hovila, H. Irie, Y. Kanaya, D. Karagkiozidis, N. Kouremeti, J.-C. Lambert, J. Ma, E. Peters, A. Piters, O. Postylyakov, A. Richter, J. Remmers, H. Takashima, M. Tiefengraber, P. Valks, T. Vlemmix, T. Wagner, F. Wittrock, Validation of tropospheric NO_2 column measurements of GOME-2A and OMI using MAX-DOAS and direct sun network observations. Atmos. Meas. Tech. **13**(11), 6141–6174 (2020). https://doi.org/10.5194/amt-13-6141-2020

G. Pinardi, M. van Roozendael, F. Hendrick, A. Richter, P. Valks, R. Alwarda, K. Bognar, U. Frieß, J. Granville, M. Gu, P. Johnston, C. Prados-Roman, R. Querel, K. Strong, T. Wagner, F. Wittrock, M. Yela Gonzalez, Ground-based validation of the MetOp-A and MetOp-B GOME-2 OClO measurements. Atmos. Meas. Tech. **15**(11), 3439–3463 (2022). https://doi.org/10.5194/amt-15-3439-2022

J.P. Pinto, J. Dibb, B.H. Lee, B. Rappenglück, E.C. Wood, M. Levy, R.-Y. Zhang, B. Lefer, X.-R. Ren, J. Stutz, C. Tsai, L. Ackermann, J. Golovko, S.C. Herndon, M. Oakes, Q.-Y. Meng, J.W. Munger, M. Zahniser, J. Zheng, Intercomparison of field measurements of nitrous acid (HONO) during the SHARP campaign. J. Geophys. Res.: Atmos. **119**(9), 5583–5601 (2014). https://doi.org/10.1002/2013JD020287

J.N. Pitts, B.J. Finlayson, A.M. Winer, Optical systems unravel smog chemistry. Environ. Sci. Technol. **11**, 568–573 (1977)

J.N. Pitts Jr., Formation and fate of gaseous and particulate mutagens and carcinogens in real and simulated atmospheres. Environ. Health Perspect. **47**, 115–140 (1983)

J.N. Pitts Jr., H.W. Biermann, R. Atkinson, A.M. Winer, Atmospheric implications of simultaneous night-time measurements of NO_3 radicals and HONO. Geophys. Res. Lett. **11**, 557–560 (1984a)

J.N. Pitts, H.W. Biermann, A.M. Winer, E.C. Tuazon, Spectroscopic identification and measurement of gaseous nitrous acid in dilute auto exhaust. Atmos. Environ. **18**, 847–854 (1984b)

J.N. Pitts, E. Sanhueza, R. Atkinson, W.P.L. Carter, A.M. Winer, G.W. Harris, C.N. Plum, An investigation of the dark formation of nitrous acid in environmental chambers. Int. J. Chem. Kinet. **16**, 919–939 (1984)

J.N. Pitts, T.J. Wallington, H.W. Biermann, A.M. Winer, Identification and measurement of nitrous acid in an indoor environment. Atmos. Environ. **19**, 763–767 (1985)

J.M.C. Plane, C.-F. Nien, Study of night-time NO_3 chemistry by differential optical absorption spectroscopy, in *SPIE Proceedings* ed. by H.I. Schiff, vol. 1433, Measurement of Atmospheric Gases (The International Society for Optical Engineering) (Bellingham, 1991) pp. 8–20

J.M.C. Plane, C.F. Nien, Differential optical absorption spectrometer for measuring atmospheric trace gases. Rev. Sci. Instr. **63**, 1867–1876 (1992)

J.M.C. Plane, N. Smith, Atmospheric monitoring by differential optical absorption spectroscopy, in *Spectroscopy in Environmental Sciences*, ed. by R.J.H. Clark, R.E. Hester (Wiley Ltd., 1995), pp. 223–262

J.A. Plascyk, F.C. Gabriel, The Fraunhofer line discriminator MKII–an airborne instrument for precise and standardized ecological luminescence measurement. IEEE Trans. Instrum. Meas. **24**(4), 306–313 (1975). https://doi.org/10.1109/TIM.1975.4314448

U. Platt, Mikrometeorologische Bestimmung der SO_2-Abscheidung am Boden Dissertation, Uni Heidelberg (1977)

U. Platt, Dry deposition of SO_2. Atmos. Environ. **12**, 363–367 (1978)

U. Platt, D. Perner, H. Pätz, Simultaneous measurement of atmospheric CH_2O, O_3, and NO_2 by differential optical absorption. J. Geophys. Res. **84**, 6329–6335 (1979)

U. Platt, D. Perner, Direct measurements of atmospheric CH_2O, HNO_2, O_3, NO_2 and SO_2 by differential optical absorption in the near UV. J. Geophys. Res. **85**, 7453–7458 (1980)

U. Platt, D. Perner, G.W. Harris, A.M. Winer, J.N. Pitts, Detection of NO_3 in the polluted troposphere by differential optical absorption. Geophys. Res. Lett. **7**, 89–92 (1980a)

U. Platt, D. Perner, G.W. Harris, A.M. Winer, J.N. Pitts, Observations of nitrous acid in an urban atmosphere by differential optical absorption. Nature **285**, 312–314 (1980b)

U. Platt, D. Perner, J. Schröder, C. Kessler, A. Tönnissen, The diurnal variation of NO_3. J. Geophys. Res. **86**, 11965–11970 (1981)

U. Platt, D. Perner, C. Kessler, The importance of NO_3 for the atmospheric NOx cycle from experimental observations, in *Proceedings of the 2nd Symposium Composition of the Nonurban Troposphere, Williamsburg, May 82, 25–28* (1982)

U. Platt, D. Perner, Measurements of atmospheric trace gases by long path differential UV/visible absorption spectroscopy, in *Optical and Laser Remote Sensing, Springer Series in Optical Science*, vol. 39, ed. by D.K. Killinger, A. Mooradien (Springer, Berlin, 1983), pp.95–105

U. Platt, D. Perner, Ein Instrument zur spektroskopischen Spurenstoffmessung in der Atmosphäre. Fresenius Z. Anal. Chem. **317**, 309–313 (1984)

U. Platt, A.M. Winer, H.W. Biermann, R. Atkinson, J.N. Pitts, Measurement of nitrate radical concentrations in continental air. Environ. Sci. Techn. **18**, 365–369 (1984)

U. Platt, The origin of nitrous and nitric acid in the atmosphere, in *NATO ASI-Series*, ed. by W. Jäschke (Springer Verlag, Heidelberg), pp. 299–319 (1986)

U. Platt, M. Rateike, W. Junkermann, A. Hofzumahaus, D.H. Ehhalt, Detection of atmospheric OH radicals. Free. Rad. Res. Comms. **3**, 165–172 (1987)

U. Platt, M. Rateike, W. Junkermann, J. Rudolph, D.H. Ehhalt, New tropospheric OH measurements. J. Geophys. Res. **93**, 5159–5166 (1988)

U. Platt, D. Perner, S. Semke, Observation of nitrate radical concentrations and lifetimes in tropospheric air, in *Ozone in the Atmosphere. Proceedings of the Quadrennial Ozone Symposium 1988*, ed. by R.D. Bojkov, P. Fabian (A. Deepak Publishing, Hampton, 1989), pp. 512–515

U. Platt, G. LeBras, G. Poulet, J.P. Burrows, G. Moortgat, Peroxy radicals from night-time reaction of NO_3 with organic compounds. Nature **348**, 147–149 (1990)

U. Platt, Spectroscopic measurement of free radicals (OH, NO_3) in the atmosphere. Fresenius J. Anal. Chem. **340**, 633–637 (1991)

U. Platt, Differential optical absorption spectroscopy (DOAS), in *Air Monitoring by Spectroscopic Techniques*, ed. by M.W. Sigrist, Chemical Analysis Series, vol. 127, (Wiley Inc., New York, Chichester, Brisbane, Toronoto, Singapore, 1994), pp. 27–84

U. Platt, M. Hausmann, Spectroscopic measurement of the free radicals NO_3, BrO, IO, and OH in the troposphere. Res. Chem. Intermed. **20**, 557–578 (1994a)

U. Platt, F. Heintz, Nitrate radicals in tropospheric chemistry. Israel. J. Chem. **34**, 289–300 (1994b)

U. Platt, C. Janssen, Observation and role of the free radicals NO_3, ClO, BrO and IO in the troposphere. Faraday Discuss. **100**, 175–198 (1996a)

U. Platt, E. Lehrer, (eds.), Arctic tropospheric halogen chemistry. Final Report to EU (1996b)

U. Platt, E. Lehrer, Arctic tropospheric ozone chemistry, ARCTOC, final report of the EU-project EV5V-CT93-0318 (Heidelberg, 1997)

U. Platt, G. Le Bras, Influence of DMS on the NO_X - NO_Y partitioning and the NO_X distribution in the marine background atmosphere. Geophys. Res. Lett. **24**, 1935–1938 (1997)

U. Platt, L. Marquard, T. Wagner, D. Perner, Corrections for zenith scattered light DOAS. Geophys. Res. Lett. **24**, 1759–1762 (1997)

U. Platt, Modern methods of the measurement of atmospheric trace gases. J. Phys. Chem. Chem. Phys. 'PCCP' **1**, 5409–5415 (1999)

U. Platt, Air monitoring by differential optical absorption spectroscopy, in *Encyclopedia of Analytical Chemistry*. ed. by R.A. Meyers (Wiley, Chichester, 2000), pp.1936–1959

U. Platt, B. Alicke, R. Dubois, A. Geyer, A. Hofzumahaus, F. Holland, D. Mihelcic, T. Klüpfel, B. Lohrmann, W. Pätz, D. Perner, F. Rohrer, J. Schäfer, J. Stutz, Free radicals and fast photochemistry during BERLIOZ. J. Atmos. Chem. **42**, 359–394 (2002)

U. Platt, G. Hönninger, The role of halogen species in the troposphere. Chemosphere **52**(2), 325–338 (2003)

U. Platt, W. Allan, D. Lowe, Hemispheric average Cl atom concentration from $^{13}C/^{12}C$ ratios in atmospheric methane. Atmos. Chem. Phys. **4**, 2393–2399 (2004)

U. Platt, K. Pfeilsticker, M. Vollmer, Radiation and optics, in *the atmosphere, Chap. 19, in Springer Handbook of Lasers and Optics*. ed. by F. Träger (Springer, Heidelberg, 2007), pp.1165–1203

U. Platt, J. Meinen, D. Pöhler, T. Leisner, Broadband cavity enhanced differential optical absorption spectroscopy (CE-DOAS)–applicability and corrections. Atmos. Meas. Tech. **2**(2), 713–723 (2009). https://doi.org/10.5194/amt-2-713-2009

U. Platt, N. Bobrowski, *Quantification of Volcanic Reactive Halogen Emissions* (Cambridge University Press, 2015), pp. 115–132. https://doi.org/10.1017/CBO9781107415683.011

U. Platt, P. Lübcke, J. Kuhn, N. Bobrowski, F. Prata, M. Burton, C. Kern, Quantitative imaging of volcanic plumes—results, needs, and future trends. J. Volcanol. Geotherm. Res. **300**, 7–21 (2015). ISSN 0377-0273. https://doi.org/10.1016/j.jvolgeores.2014.10.006. https://www.sciencedirect.com/science/article/pii/S0377027314003047

U. Platt, N. Bobrowski, A. Butz, Ground-based remote sensing and imaging of volcanic gases and quantitative determination of multi-species emission fluxes. Geosciences **8**(2) (2018). https://doi.org/10.3390/geosciences8020044

U. Platt, J. Kuhn, Caution with spectroscopic NO_2 reference cells (cuvettes). Atmos. Meas. Tech. **12**(12), 6259–6272 (2019). https://doi.org/10.5194/amt-12-6259-2019

U. Platt, T. Wagner, J. Kuhn, T. Leisner, The "ideal" spectrograph for atmospheric observations. Atmos. Meas. Tech. **14**(10), 6867–6883 (2021). https://doi.org/10.5194/amt-14-6867-2021

C.N. Plum, E. Sanhueza, R. Atkinson, W.P.L. Carter, J.N. Pitts Jr., OH radical rate constants and photolysis rates of a-dicarbonyls. Environ. Sci. Technol. **17**, 479–484 (1983)

D. Pöhler, *Determination of Two Dimensional Trace Gas Distributions Using Tomographic LP-DOAS Measurements in the City of Heidelberg*. Ph.D. Thesis, University of Heidelberg (2010). http://www.ub.uni-heidelberg.de/archiv/10996

D. Pöhler, L. Vogel, U. Frieß, U. Platt, Halogen observations in the Amundsen Gulf, Arctic, during ozone depletion events using active longpath differential optical absorption spectroscopy. Proc. Natl. Acad. Sci. (PNAS) **107**(15), 6582–6587 (2010). https://doi.org/10.1073/pnas.0912231107

J.P. Pommereau, Observation of NO_2 diurnal variation in the stratosphere. Geophys. Res. Lett. **9**(8), 850–853 (1982)

J.P. Pommereau, F. Goutail, An advanced visible-UV spectrometer for atmospheric composition measurements. ESA **SP–270**, 197–200 (1987)

J.P. Pommereau, F. Goutail, O_3 and NO_2 ground-based measurements by visible spectrometry during the Arctic winter and spring 1988. Geophys. Res. Lett. **15**, 891–894 (1988a)

J.P. Pommereau, F. Goutail, Stratospheric O_3 and NO_2 Observations at the southern polar circle in summer and fall 1988. Geophys. Res. Lett. **15**, 895 (1988b)

J.-P. Pommereau, J. Piquard, Ozone and nitrogen dioxide vertical distributions by UV-visible solar occultation from balloons. Geophys. Res. Lett. **21**, 1227–1230 (1994a)

J.-P. Pommereau, J. Piquard, Observations of the vertical distribution of stratospheric OClO. Geophys. Res. Lett. **21**, 1231–1234 (1994b)

F.D. Pope, C.A. Smith, P.R. Davis, D.E. Shallcross, M.N.R. Ashfold, A.J. Orr-Ewing, Photochemistry of formaldehyde under tropospheric conditions. Faraday Discuss. **130**, 59–72 (2005a)

F.D. Pope, C.A. Smith, M.N.R. Ashfold, A.J. Orr-Ewing, High-resolution absorption cross sections of formaldehyde at wavelengths from 313 to 320 nm. Phys. Chem. Chem. Phys. **7**, 79–84 (2005b)

C. Popp, D. Brunner, A. Damm, M. van Roozendael, C. Fayt, B. Buchmann, High-resolution NO_2 remote sensing from the Airborne Prism EXperiment (APEX) imaging spectrometer. Atmos. Meas. Tech. **5**(9), 2211–2225 (2012). https://doi.org/10.5194/amt-5-2211-2012

D. Poppe, M. Wallasch, J. Zimmermann, The dependence of the concentration of OH on its precursors under moderately polluted conditions-a model study. J. Atmos. Chem. **16**, 61–78 (1993)

A. Porcar-Castell, E. Tyystjärvi, J. Atherton, C. van der Tol, J. Flexas, E.E. Pfündel, J. Moreno, C. Frankenberg, J.A. Berry, Linking chlorophyll a fluorescence to photosynthesis for remote sensing applications: mechanisms and challenges. J. Exp. Bot. **65**(15), 4065–4095 (2014). https://doi.org/10.1093/jxb/eru191

A. Porcar-Castell, Z. Malenovský, T. Magney, S. van Wittenberghe, B. Fernández-Marín, F. Maignan, Y. Zhang, K. Maseyk, J. Atherton, L.P. Albert, T.M. Robson, F. Zhao, J.-I. Garcia-Plazaola, I. Ensminger, P.A. Rajewicz, S. Grebe, M. Tikkanen, J.R. Kellner, J.A. Ihalainen, U. Rascher, B. Logan, Chlorophyll a fluorescence illuminates a path connecting plant molecular biology to Earth-system science. Nat. Plants **7**(8), 998–1009 (2021). https://doi.org/10.1038/s41477-021-00980-4

R.W. Portmann, S.S. Brown, T. Gierczak, R.K. Talukdar, J.B. Burkholder, A.R. Ravishankara, Role of nitrogen oxides in the stratosphere: a reevaluation based on laboratory studies. Geophys. Res. Lett. **26**(15), 2387–2390 (1999)

D.A. Potts, R. Timmis, E.J.S. Ferranti, J.D. Vande Hey, Identifying and accounting for the Coriolis effect in satellite NO_2 observations and emission estimates. Atmos. Chem. Phys. **23**(7), 4577–4593 (2023). https://doi.org/10.5194/acp-23-4577-2023

I.M. Povey, A.M. South, A. t'Kint de Roodenbeke, C. Hill, R.A. Freshwater, R.L. Jones, A broadband lidar for the measurement of tropospheric constituent profiles from the ground. J. Geophys. Res. **10**(D3), 3369–3380 (1998)

C. Prados-Roman, A. Butz, T. Deutschmann, M. Dorf, L. Kritten, A. Minikin, U. Platt, H. Schlager, H. Sihler, N. Theys, M. Van Roozendael, T. Wagner, K. Pfeilsticker, Airborne DOAS limb measurements of tropospheric trace gas profiles: case studies on the profile retrieval of O_4 and BrO. Atmos. Meas. Tech. **4**(6), 1241–1260 (2011). https://doi.org/10.5194/amt-4-1241-2011

C. Prados-Roman, C.A. Cuevas, T. Hay, R.P. Fernandez, A.S. Mahajan, S.-J. Royer, M. Galí, R. Simó, J. Dachs, K. Großmann, D.E. Kinnison, J.-F. Lamarque, A. Saiz-Lopez, Iodine oxide in the global marine boundary layer. Atmos. Chem. Phys. **15**(2), 583–593 (2015). https://doi.org/10.5194/acp-15-583-2015

W.H. Press, B.P. Flannery, S.A. Teukolsky, W.T. Vettering, *Numerical Recipes in C* (Cambridge University Press, Cambridge, 1986)

K.E. Preston, R.L. Jones, H.K. Roscoe, Retrieval of NO_2 vertical profiles from ground-based UV-visible measurements: method and validation. J. Geophys. Res. **102**(D15), 19089–19097 (1997)

J.K. Pribram, C.M. Penchina, Stray light in Czerny-Turner and Ebert spectrometers. Appl. Opt. **7**, 2005–2014 (1968)

P.N. Price, Pollutant tomography using integrated concentration data from non-intersecting optical paths. Atmos. Environ. **33**, 275–280 (1999)

R. Prinn, D. Cunnold, R. Rasmussen, P. Simmonds, F. Alyea, A. Crawford, P. Fraser, R. Rosen, Atmospheric trends in methylchloroform and the global average for the hydroxyl radical. Science **238**, 945–950 (1987)

A.A.P. Pszenny, W.C. Keene, D.J. Jacob, S. Fran, J.R. Maben, M.P. Zetwo, M. Sringer-Young, J.N. Galloway, Evidence of inorganic chlorine gases other than hydrogen chloride in marine surface air. Geophys. Res. Lett. **20**(8), 699–702 (1993)

J. Puķīte, S. Kühl, T. Deutschmann, U. Platt, T. Wagner, Extending differential optical absorption spectroscopy for limb measurements in the UV. Atmos. Meas. Tech. **3**(3), 631–653 (2010). https://doi.org/10.5194/amt-3-631-2010

J. Puķīte, T. Wagner, Quantification and parametrization of non-linearity effects by higher-order sensitivity terms in scattered light differential optical absorption spectroscopy. Atmos. Meas. Tech. **9**(5), 2147–2177 (2016). https://doi.org/10.5194/amt-9-2147-2016

J. Puķīte, C. Borger, S. Dörner, M. Gu, T. Wagner, OClO as observed by TROPOMI: a comparison with meteorological parameters and polar stratospheric cloud observations. Atmos. Chem. Phys. **22**(1), 245–272 (2022). https://doi.org/10.5194/acp-22-245-2022

I. Pundt, M. van Roozendael, T. Wagner, A. Richter, M.P. Chipperfield, J.P. Burrows, C. Fayt, F. Hendrick, K. Pfeilsticker, U. Platt, J.P. Pommereau, Simultaneous UV-vis measurements of BrO from balloon, satellite and ground: implications for tropospheric BrO, in Air Pollution Research Report 73, Proceedings 5th European Symposium on Polar Stratospheric Ozone, European Commission, Directorate General for Research Unit D.I. - Environment and Sustainable Development Programme, EUR 1934, ed. by N.R.P. Harris, M. Guirlet, G.T. Amanatis **2000**, 316–319 (1999)

I. Pundt, J.P. Pommereau, F. Lefevre, Investigation of stratospheric bromine and iodine oxides using the SAOZ balloon sonde, in *Proceedings of XVIII Quadrennial Ozone Symposium L'Aquila, International Ozone Commission Italy, 12–21 September*, ed. by R.D. Bojkov, D. Visconti (1996), pp. 575–578

I. Pundt, J.P. Pommereau, C. Phillips, E. Lateltin, Upper limit of iodine oxide in the lower stratosphere. J. Atmos. Chem. **30**, 173–185 (1998)

I. Pundt, J.-P. Pommereau, M.P. Chipperfield, M. van Roozendael, F. Goutail, Climatology of the stratospheric BrO vertical distribution by balloon-borne UV-visible spectrometry. J. Geophys. Res. **107**(D24), 4806 (2002). https://doi.org/10.1029/2002JD002230

P. Rairoux, H. Schillinger, S. Niedermeier, M. Rodriguez, F. Ronneberger, R. Sauerbrey, B. Stein, D. Waite, C. Wedekind, H. Wille, L. Wöste, C. Ziener, Remote sensing of the atmosphere using ultrashort laser pulses. App. Phys. B (Lasers and Optics) **71**, 573–580 (2000). https://doi.org/10.1007/s003400000375

B. Ramacher, J. Rudolph, R. Koppmann, Hydrocarbon measurements in the spring Arctic troposphere during the ARCTOC 95 campaign. Tellus **49B**, 466–485 (1997)

B. Ramacher, J. Rudolph, R. Koppmann, Hydrocarbon measurements during tropospheric ozone depletion events: evidence for halogen atom chemistry. J. Geophys. Res. **104**, 3633–3653 (1999)

U. Rascher, G. Agati, L. Alonso, G. Cecchi, S. Champagne, R. Colombo, A. Damm, F. Daumard, E. de Miguel, G. Fernandez, B. Franch, J. Franke, C. Gerbig, B. Gioli, J.A. Gómez, Y. Goulas, L. Guanter, O. Gutiérrez-de-la Cámara, K. Hamdi, P. Hostert, M. Jiménez, M. Kosvancova, D. Lognoli, M. Meroni, F. Miglietta, A. Moersch, J. Moreno, I. Moya, B. Neininger, A. Okujeni, A. Ounis, L. Palombi, V. Raimondi, A. Schickling, J.A. Sobrino, M. Stellmes, G. Toci, P. Toscano, T. Udelhoven, S. van der Linden, A. Zaldei. CEFLES2: the remote sensing component to quantify photosynthetic efficiency from the leaf to the region by measuring sun-induced fluorescence in the oxygen absorption bands. Biogeosciences **6**(7), 1181–1198 (2009). https://doi.org/10.5194/bg-6-1181-2009

U. Rascher, L. Alonso, A. Burkart, C. Cilia, S. Cogliati, R. Colombo, A. Damm, M. Drusch, L. Guanter, J. Hanus, T. Hyvärinen, T. Julitta, J. Jussila, K. Kataja, P. Kokkalis, S. Kraft, T. Kraska, M. Matveeva, J. Moreno, O. Muller, C. Panigada, M. Pikl, F. Pinto, L. Prey, R. Pude, M. Rossini, A. Schickling, U. Schurr, D. Schüttemeyer, J. Verrelst, F. Zemek, Sun-induced fluorescence–a new probe of photosynthesis: first maps from the imaging spectrometer hyplant. Global Change Biol. **21**(12), 4673–4684 (2015). https://doi.org/10.1111/gcb.13017

A.R.W. Raso, K.D. Custard, N.W. May, D. Tanner, M.K. Newburn, L. Walker, R.J. Moore, L.G. Huey, L. Alexander, P.B. Shepson, K.A. Pratt, Active molecular iodine photochemistry in the Arctic. Proc. Natl. Acad. Sci. **114**(38), 201702803 (2017). https://doi.org/10.1073/pnas.1702803114

L. Rayleigh, On the transmission of light through an atmosphere containing many small particles in suspension, and on the origin of the blue of the sky. Phil. Mag. **41**, 447–454. Also: in the scientific papers of Lord Rayleigh, vol. 4 (Dover, New York, 1964, 1899)

K.A. Read, A.S. Mahajan, L.J. Carpenter, M.J. Evans, B.V.E. Faria, D.E. Heard, J.R. Hopkins, J.D. Lee, S.J. Moller, A.C. Lewis, L. Mendes, J.B. McQuaid, H. Oetjen, A. Saiz-Lopez, M.J. Pilling, J.M.C. Plane, Extensive halogen-mediated ozone destruction over the tropical Atlantic Ocean. Nature **453**(7199), 1232–1235 (2008). https://doi.org/10.1038/nature07035

J. Reader, C.J. Sansonetti, J.M. Bridges, Irradiances of spectral lines in mercury pencil lamps. Appl. Opt. **35**, 78–83 (1996)

H.E. Redemann Fischer, F. Fergg, D. Rabus, Measurements of stratospheric NO_2 profiles using a gas correlation radiometer in the solar occultation mode. J. Atmos. Chem. **3**, 203–231 (1985)

A.R. Reisinger, G.J. Fraser, P.V. Johnston, R.L. McKenzie, W.A. Matthews, Slow-scanning DOAS system for urban air pollution monitoring, in *Atmospheric ozone, Proceedings of the XVIII Quadrennial Ozone Symposium, L'Aquila, Italy, September 1996*, ed. by R.D. Bojkov, G. Visconti, vol. 2 (Parco Scientifico e Tecnologico d'Abruzzo, L'Aquila, 1998), pp. 959–962

B. Ren, P. Xie, J. Xu, A. Li, X. Tian, Z. Hu, Y. Huang, X. Li, Q. Zhang, H. Ren, H. Ji, Use of the PSCF method to analyze the variations of potential sources and transports of NO_2, SO_2, and HCHO observed by MAX-DOAS in Nanjing, China during 2009. Sci. Total Environ. **782**, 146865 (2021). https://doi.org/10.1016/j.scitotenv.2021.146865

X. Ren, H. Harder, M. Martinez, R.L. Lesher, A. Oliger, J.B. Simpas, W.H. Brune, J.J. Schwab, K.L. Demerjian, Y. He, X. Zhou, H. Gao, OH and HO_2 chemistry in urban atmosphere of New York city. Atmos. Environ. **37**, 3639–3651 (2003)

X. Ren, W.H. Brune, J. Mao, M.J. Mitchell, R.L. Lesher, J.B. Simpas, A.R. Metcalf, J.J. Schwab, C. Cai, Y. Li, Behavior of OH and HO_2 in the winter atmosphere in New York City. Atmos. Environ. **40**, 252–263 (2006). https://doi.org/10.1016/j.atmosenv.2005.11.073

X. Ren, D. van Duin, M. Cazorla, S. Chen, J. Mao, L. Zhang, W.H. Brune, J.H. Flynn, N. Grossberg, B.L. Lefer, B. Rappenglück, K.W. Wong, C. Tsai, J. Stutz, J.E. Dibb, B.T. Jobson, W.T. Luke,

P. Kelley, Atmospheric oxidation chemistry and ozone production: results from SHARP 2009 in Houston, Texas. J. Geophys. Res.: Atmos. **118**(11), 5770–5780 (2013). https://doi.org/10.1002/jgrd.50342

J.-B. Renard, M. Pirre, C. Robert, G. Moreau, D. Huguenin, J.M. Russell III., Nocturnal vertical distribution of stratospheric O_3, NO_2 and NO_3 from balloon measurements. J. Geophys. Res. **101**, 28793–28804 (1996)

J.-B. Renard, M. Pirre, C. Robert, F. Lefèvre, E. Lateltin, B. Nozière, D. Huguenin, Vertical distribution of nighttime stratospheric NO_2 from balloon measurements: comparison with models. Geophys. Res. Lett. **24**, 73–76 (1997a)

J.B. Renard, F. Lefèvre, M. Pirre, D. Huguenin, Vertical profile of night-time stratospheric OClO. J. Atmos. Chem. **26**, 65–76 (1997b)

J.B. Renard, M. Pirre, C. Robert, D. Huguenin, The possible detection of OBrO in the stratosphere. J. Geophys. Res. **103**, 25383–25395 (1998)

J.-B. Renard, M. Chartier, C. Robert, G. Chalumeau, G. Berthet, M. Pirre, J.-P. Pommerreau, F. Goutail, SALOMON: a new, light balloonborne UV-visible spectrometer for nighttime observations of stratospheric trace-gas species. Appl. Opt. **39**(3), 386–392 (2000a)

J.-B. Renard, F.G. Taupin, E.D. Rivière, M. Pirre, N. Huret, G. Berthet, C. Robert, M. Chartier, Measurements and simulation of stratospheric NO_3 at Mid- and High-latitudes in the Northern Hemisphere. J. Geophys. Res. **106**, 32387–32399 (2000b)

H. Rhode, R. Charlson, E. Crawford, Svante Arrhenius and the greenhouse effect. Ambio **26**, 2–5 (1997)

A. Richter, Absorptionsspektroskopische Messungen atmosphärischer Spurengase über Bremen, 53°N, Doctoral Thesis, University of Bremen (1997)

A. Richter, F. Wittrock, M. Eisinger, J.P. Burrows, GOME observation of tropospheric BrO in northern hemispheric spring and summer 1997. Geophys. Res. Lett. **25**, 2683–2686 (1998)

A. Richter, M. Eisinger, A. Ladstätter-Weißenmayer, F. Wittrock, J.P. Burrows, DOAS zenith-sky observations: seasonal variations of BrO over Bremen (53°N) 1994–1995. J. Atmos. Chem. **32**, 83–99 (1999)

A. Richter, J.P. Burrows, Retrieval of tropospheric NO_2 from GOME measurements. Adv. Space Res. **29**(11), 1673–1683 (2002)

A. Richter, F. Wittrock, A. Ladstätter-Weißenmayer, J.P. Burrows, GOME measurements of stratospheric and tropospheric BrO. Adv. Space Res. **29**(11), 1667–1672 (2002)

A. Richter, J.P. Burrows, H. Nüß, C. Granier, U. Niemeier, Increase in tropospheric nitrogen dioxide over China observed from space. Nature **437**, 129–132 (2005a). https://doi.org/10.1038/nature04092

A. Richter, F. Wittrock, M. Weber, S. Beirle, S. Kühl, U. Platt, T. Wagner, W. Wilms-Grabe, J.P. Burrows, GOME observations of stratospheric trace gas distributions during the splitting vortex event in the Antarctic winter 2002 Part I: measurements. J. Atmos. Sci. **62**(3), 778–785 (2005b)

T.P. Riedel, N.L. Wagner, W.P. Dubé, A.M. Middlebrook, C.J. Young, F. Öztürk, R. Bahreini, T.C. VandenBoer, D.E. Wolfe, E.J. Williams, J.M. Roberts, S.S. Brown, J.A. Thornton, Chlorine activation within urban or power plant plumes: vertically resolved $ClNO_2$ and Cl_2 measurements from a tall tower in a polluted continental setting. J Geophys Res: Atmos **118**(15), 8702–8715 (2013). https://doi.org/10.1002/jgrd.50637

T.C.V.W. Riess, K.F. Boersma, J. van Vliet, W. Peters, M. Sneep, H. Eskes, J. van Geffen, Improved monitoring of shipping NO_2 with TROPOMI: decreasing NO_x emissions in European seas during the COVID-19 pandemic. Atmos. Meas. Tech. **15**(5), 1415–1438 (2022). https://doi.org/10.5194/amt-15-1415-2022

P. Rigaud, B. Leroy, G. Le Bras, G. Poulet, J.L. Jourdain, J. Combourieu, About the identification of some UV atmospheric absorptions laboratory study of ClO. Chem. Phys. Lett. **46**, 161 (1977)

P. Rigaud, J.-P. Naudet, D. Huguenin, Simultaneous measurements of vertical distribution of stratospheric NO_3 and O_3 at different periods of the night. J. Geophys. Res. **88**, 1463–1467 (1983)

C.P. Rinsland, E. Mahieu, R. Zander, N.B. Jones, M.P. Chipperfield, A. Goldman, J. Anderson, J.M. Russell III., P. Demoulin, J. Notholt, G.C. Toon, J.-F. Blavier, B. Sen, R. Sussmann, S.W. Wood, A. Meier, D.W.T. Griffith, L.S. Chiou, F.J. Murcray, T.M. Stephen, F. Hase, S. Mikuteit,

A. Schulz, T. Blumenstock, Long-term trends of inorganic chlorine from ground-based infrared solar spectra: past increases and evidence for stabilization. J. Geophys. Res. **108**(D8), 4252 (2003). https://doi.org/10.1029/2002JD003001

N.E. Rityn, Optics of corner cube reflectors. Sov. J. Opt. Tech. **34**, 198–201 (1967)

D. Ritz, M. Hausmann, U. Platt, An improved open-path multireflection cell for the measurement of NO_2 and NO_3, in *Proceedings Europto Series, Optical Methods in Atmospheric Chemistry, SPIE*, ed. by H.I. Schiff, U. Platt, vol. 1715 (1992), pp. 200–211

C. Robert, Simple, stable, and compact multiple-reflection optical cell for very long optical paths. Appl. Opt. **46**(22), 5408–5418 (2007). https://doi.org/10.1364/AO.46.005408

T. Röckmann, C.A.M. Brenninkmeijer, P.J. Crutzen, U. Platt, Short term variations in the $^{13}C/^{12}C$ ratio of CO as a measure of Cl activation during tropospheric ozone depletion events in the arctic. J. Geophys. Res. **104**, 16911697 (1999)

C.D. Rodgers, Characterisation and error analysis of profiles retrieved from remote sounding measurements. J. Geophys. Res. **95**(D5), 5587–5595 (1990)

C.D. Rodgers, *Inverse Methods for Atmospheric Sounding Theory and Practice* (World Scientific Publishing, Singapore, 2000)

F. Rohrer, B. Bohn, T. Brauers, D. Brüning, F.-J. Johnen, A. Wahner, J. Kleffmann, Characterisation of the photolytic HONO-source in the atmosphere simulation chamber SAPHIR. Atmos. Chem. Phys. Discuss. **4**, 7881–7915 (2004)

F. Rohrer, H. Berresheim, Strong correlation between levels of tropospheric hydroxyl radicals and solar ultraviolet radiation. Nature **442**(7099), 184–187 (2006). https://doi.org/10.1038/nature04924

H.K. Roscoe, J.A. Pyle, Measurements of solar occultation: the error in a native retrieval if the constituent's concentration changes. J. Atmos. Chem. **5**, 323–341 (1987)

H.K. Roscoe, D.J. Fish, R.L. Jones, Interpolation errors in UV-visible spectroscopy for stratospheric sensing: implications for sensitivity, spectral resolution, and spectral range. Appl. Opt. **35**, 427–432 (1996)

H.K. Roscoe, K.C. Clemitshaw, Measurement techniques in gas-phase tropospheric chemistry: a selective view of the past, present, and future. Science **276**, 1065–1072 (1997)

H.K. Roscoe, P.V. Johnston, M. Van Roozendael, A. Richter, A. Sarkissian, J. Roscoe, K.E. Preston, J.-C. Lambert, C. Hermans et al., Slant column measurements of O_3 and NO_2 during the NDSC intercomparison of zenith-sky UV-visible spectrometers in June 1996. J. Atmos. Chem. **32**, 281–314 (1999)

H.K. Roscoe, A.J. Charlton, D.J. Fish, J.G.T. Hill, Improvements to the accuracy of measurements of NO_2 by zenith-sky visible spectrometers II: errors in offset using a more complete chemical model. J. Quant. Spectrosc. Radiat. Trans. **68**, 337–349 (2001)

H.K. Roscoe, J.G.T. Hill, A.E. Jones, A. Sarkissian, Improvements to the accuracy of zenith-sky measurements of total ozone by visible spectrometers II: use of daily air-mass factors. J. Quant. Spectrosc. Radiat. Trans. **68**, 327–336 (2001)

M. Rossini, L. Nedbal, L. Guanter, A. Ač, L. Alonso, A. Burkart, S. Cogliati, R. Colombo, A. Damm, M. Drusch, J. Hanus, R. Janoutova, T. Julitta, P. Kokkalis, J. Moreno, J. Novotny, C. Panigada, F. Pinto, A. Schickling, D. Schüttemeyer, F. Zemek, U. Rascher, Red and far red sun-induced chlorophyll fluorescence as a measure of plant photosynthesis. Geophys. Res. Lett. **42**(6), 1632–1639 (2015). https://doi.org/10.1002/2014GL062943

E.-P. Röth, *Ozonloch, Ozonsmog: Grundlagen der Ozonchemie* (BI-Taschenbuchverlag, Mannheim, Mannheim, Germany, 1994)

K.W. Rothe, U. Brinkmann, H. Walther, Applications of tunable dye lasers to air pollution detection: measurements of atmospheric NO_2 concentrations by differential absorption. Appl. Phys. **3**, 115–119 (1974)

L.S. Rothman, R.R. Gamache, R.H. Tipping, C.P. Rinsland, M.A.H. Smith, D.C. Benner, V. Malathy Devi, J.-M. Flaud, C. Camy-Peyret, A. Perrin, A. Goldman, S.T. Massie, L.R. Brown, R.A. Toth, The HITRAN molecular database: editions of 1991 and 1992. J. Quant. Spectrosc. Radiat. Transf. **48**(5/6), 469–508 (1992)

L.S. Rothman, C.P. Rinsland, A. Goldman, S.T. Massie, D.P. Edwards, J.-M. Flaud, A. Perrin, C. Camy-Peyret, V. Dana, J.-Y. Mandin, J. Schroeder, A. McCann, R.R. Gamache, R.B. Wattson, K. Yoshino, K.V. Chance, K.W. Jucks, L.R. Brown, V. Nemtchinov, P. Varanasi, The HITRAN molecular spectroscopic database and HAWKS (HITRAN Atmospheric Workstation): 1996 edition. J. Quant. Spectrosc. Radiat. Transf. **60**(5), 665–710 (1998)

L.S. Rothman, A. Barbe, D. Chris Benner, L.R. Brown, C. Camy-Peyret, M.R. Carleer, K. Chance, C. Clerbaux, V. Dana, V.M. Devi, A. Fayt, J.-M. Flaud, R.R. Gamache, A. Goldman, D. Jacquemart, K.W. Jucks, W.J. Lafferty, J.-Y. Mandin, S.T. Massie, V. Nemtchinov, D.A. Newnham, A. Perrin, C.P. Rinsland, J. Schroeder, K.M. Smith, M.A.H. Smith, K. Tang, R.A. Toth, J. Vander Auwera, P. Varanasi, K. Yoshino, The HITRAN molecular spectroscopic database: edition of 2000 including updates through 2001. J. Quant. Spectros. Rad. Transf. **82**, 5–44 (2003)

L.S. Rothman, D. Jacquemart, A. Barbe, D. Chris Benner, M. Birk, L.R. Brown, M.R. Carleer, C. Chackerian, K. Chance, V. Dana, V.M. Devi, J.-M. Flaud, R.R. Gamache, A. Goldman, J.-M. Hartmann, K.W. Jucks, A.G. Maki, J.-Y. Mandin, S.T. Massie, J. Orphal, A. Perrin, C.P. Rinsland, M.A.H. Smith, J. Tennyson, R.N. Tolchenov, R.A. Toth, J. Vander Auwera, P. Varanasi, G. Wagner, The HITRAN 2004 molecular spectroscopic database. J. Quant. Spectrosc. Radiat. Transf. **96**, 139–204 (2005)

L.D. Rotstayn, U. Lohmann, Simulation of the tropospheric sulfur cycle in a global model with a physically based cloud scheme. J. Geophys. Res. **107**(D21), 4592 (2002). https://doi.org/10.1029/2002JD002128

V. Rozanov, D. Diebel, R. Spurr, J. Burrows, GOMETRAN: a radiative transfer model for the satellite project GOME-the plane-parallel version. J. Geophys. Res. **102**(D14), 16683–16695 (1997)

A. Rozanov, V.-V. Rozanov, J.-P. Burrows, Combined differential-integral approach for the radiation field computation in a spherical shell atmosphere: nonlimb geometry. J. Geophys. Res. **105**(D18), 22937–22942 (2000)

A. Rozanov, V.-V. Rozanov, J.-P. Burrows, A numerical radiative transfer model for a spherical planetary atmosphere: combined differential-integral approach involving the Picard iterative approximation. J. Quant. Spectrosc. Radiat. Transf. **69**(4), 491–512 (2001)

A. Rozanov, V. Rozanov, M. Buchwitz, A. Kokhanovsky, J. Burrows, SCIATRAN 2.0—a new radiative transfer model for geophysical applications in the 175–2400 nm spectral region. Adv. Space Res. **36**(5), 1015–1019 (2005). https://doi.org/10.1016/j.asr.2005.03.012

A. Rozanov, S. Kühl, A. Doicu, C. McLinden, J. Puķīte, H. Bovensmann, J.P. Burrows, T. Deutschmann, M. Dorf, F. Goutail, K. Grunow, F. Hendrick, M. von Hobe, S. Hrechanyy, G. Lichtenberg, K. Pfeilsticker, J.P. Pommereau, M. Van Roozendael, F. Stroh, T. Wagner, BrO vertical distributions from SCIAMACHY limb measurements: comparison of algorithms and retrieval results. Atmos. Meas. Tech. **4**(7), 1319–1359 (2011). https://doi.org/10.5194/amt-4-1319-2011

V.V. Rozanov, A.V. Rozanov, A.A. Kokhanovsky, J.P. Burrows, Radiative transfer through terrestrial atmosphere and ocean: software package SCIATRAN. J. Quant. Spectrosc. Radiat. Transf. **133**, 13–71 (2014). https://doi.org/10.1016/j.jqsrt.2013.07.004

V.V. Rozanov, T. Dinter, A.V. Rozanov, A. Wolanin, A. Bracher, J.P. Burrows, Radiative transfer modeling through terrestrial atmosphere and ocean accounting for inelastic processes: software package SCIATRAN. J. Quant. Spectrosc. Radiat. Transf. **194**, 65–85 (2017). https://doi.org/10.1016/j.jqsrt.2017.03.009

Y. Rudich, R. Talukdar, A.R. Ravishankara, Reactive uptake of NO_3 on pure water and ionic solutions. J. Geophys. Res. **D101**, 21023–21031 (1996)

J. Rudolph, B.R. Fu, T.A.K. Anlauf, J. Bottenheim, Halogen atom concentrations in the Arctic troposphere derived from hydrocarbon measurements: Impact on the budget of formaldehyde. Geophys. Res. Lett. **26**(19), 2941–2944 (1999)

A.G. Russell, G.R. Cass, J.H. Seinfeld, On some aspects of nighttime atmospheric chemistry. Environ. Sci. Technol. **20**, 1167–1172 (1986)

G. Russwurm, Differential optical absorption spectroscopy (DOAS)—A Status Report on the Instrumentation (1999)

R.G. Ryan, S. Rhodes, M. Tully, S. Wilson, N. Jones, U. Frieß, R. Schofield, Daytime HONO, NO_2 and aerosol distributions from MAX-DOAS observations in Melbourne. Atmos. Chem. Phys. **18**(19), 13969–13985 (2018). https://doi.org/10.5194/acp-18-13969-2018

A. Saiz-Lopez, J.M.C. Plane, J.A. Shillito, Novel iodine chemistry in the marine boundary layer. Geophys. Res. Lett. **31**, L04112 (2004a). https://doi.org/10.1029/2003GL019215

A. Saiz-Lopez, J.M.C. Plane, J.A. Shillito, Bromine oxide in the mid-latitude marine boundary layer. Geophys. Res. Lett. **31**, L03111.1–L03111.4 (2004b). https://doi.org/10.1029/2003GL018956

A. Saiz-Lopez, R.W. Saunders, D.M. Joseph, S.H. Ashworth, J.M.C. Plane, Absolute absorption cross-section and photolysis rate of I_2. Atmos. Chem. Phys. **4**, 1443–1450 (2004c)

A. Saiz-Lopez, A. Shillito, H. Coe, J.M.C. Plane, Measurements and modelling of I_2, IO, OIO, BrO and NO_3 in the mid-latitude marine boundary layer. Atmos. Chem. Phys. Discuss. **5**, 9731–9767 (2005). SRef-ID: 1680-7375/acpd/2005-5-9731

A. Saiz-Lopez, A.S. Mahajan, R.A. Salmon, S.J.-B. Bauguitte, A.E. Jones, H.K. Roscoe, J.M.C. Plane, Boundary layer halogens in coastal Antarctica. Science **317**(5836), 348–351 (2007). https://doi.org/10.1126/science.1141408

A. Saiz-Lopez, J.M.C. Plane, A.R. Baker, L.J. Carpenter, R. von Glasow, J.C.G. Martín, G. McFiggans, R.W. Saunders, Atmospheric chemistry of iodine. Chem. Rev. **112**(3), 1773–1804 (2012). https://doi.org/10.1021/cr200029u

A. Saiz-Lopez, R. von Glasow, Reactive halogen chemistry in the troposphere. Chem. Soc. Rev. **41**(19), 6448–6472 (2012). https://doi.org/10.1039/c2cs35208g

A. Saiz-Lopez, R.P. Fernandez, C. Ordóñez, D.E. Kinnison, J.C. Gómez Martín, J.-F. Lamarque, S. Tilmes, Iodine chemistry in the troposphere and its effect on ozone. Atmos. Chem. Phys. **14**(23), 13119–13143 (2014). https://doi.org/10.5194/acp-14-13119-2014

A. Saiz-Lopez, S. Baidar, C.A. Cuevas, T.K. Koenig, R.P. Fernandez, B. Dix, D.E. Kinnison, J.F. Lamarque, X. Rodriguez-Lloveras, T.L. Campos, R. Volkamer, Injection of iodine to the stratosphere. Geophys. Res. Lett. **42**(16), 6852–6859 (2015). https://doi.org/10.1002/2015GL064796

A. Saiz-Lopez, R.P. Fernandez, Q. Li, C.A. Cuevas, X. Fu, D.E. Kinnison, S. Tilmes, A.S. Mahajan, J.C. Gómez Martín, F. Iglesias-Suarez, R. Hossaini, J.M.C. Plane, G. Myhre, J.-F. Lamarque, Natural short-lived halogens exert an indirect cooling effect on climate. Nature **618**(7967), 967–973 (2023). https://doi.org/10.1038/s41586-023-06119-z

F. Sakamaki, S. Hatakeyama, H. Akimoto, Formation of nitrous acid and nitric oxide in the heterogeneous dark reaction of nitrogen dioxide and water vapor in a smog chamber. Int. J. Chem. Kinet. **XV**, 1013–1029 (1983)

R.J. Salawitch, T. Canty, T. Kurosu, K. Chance, Q. Liang, A. da Silva, S. Pawson, J.E. Nielsen, J.M. Rodriguez, P. K. Bhartia, X. Liu, L.G. Huey, J. Liao, R.E. Stickel, D.J. Tanner, J.E. Dibb, W.R. Simpson, D. Donohoue, A. Weinheimer, F. Flocke, D. Knapp, D. Montzka, J.A. Neuman, J.B. Nowak, T.B. Ryerson, S. Oltmans, D.R. Blake, E.L. Atlas, D.E. Kinnison, S. Tilmes, L.L. Pan, F. Hendrick, M. Van Roozendael, K. Kreher, P.V. Johnston, R.S. Gao, B. Johnson, T.P. Bui, G. Chen, R.B. Pierce, J.H. Crawford, D.J. Jacob, A new interpretation of total column BrO during Arctic spring. Geophys. Res. Lett. **37**(21) (2010). https://doi.org/10.1029/2010GL043798

R.J. Salawitch (Lead Author), L.A. McBride, C.R. Thompson, E.L. Fleming, R.L. McKenzie, K.H. Rosenlof, S.J. Doherty, D.W. Fahey, Twenty questions and answers about the ozone layer: 2022 update, scientific assessment of ozone depletion: 2022 (World Meteorological Organization, Geneva, Switzerland, 2023). 75 p

G. Salerno, M. Burton, C. Oppenheimer, T. Caltabiano, D. Randazzo, N. Bruno, V. Longo, Three-years of SO_2 flux measurements of Mt. Etna using an automated UV scanner array: comparison with conventional traverses and uncertainties in flux retrieval. J. Volcanol. Geotherm. Res. **183**(1), 76–83 (2009). https://doi.org/10.1016/j.jvolgeores.2009.02.013

B.E. Saltzman, Colorimetric microdetermination of nitrogen dioxide in the atmosphere. Anal. Chem. **26**, 1948–1955 (1954)

A. Sampaolo, P. Patimisco, M. Giglio, A. Zifarelli, H. Wu, L. Dong, V. Spagnolo, Quartz-enhanced photoacoustic spectroscopy for multi-gas detection: a review. Anal. Chim. Acta **1202**, 338894 (2022). https://doi.org/10.1016/j.aca.2021.338894

N.P. Sanchez, A. Saffari, S. Barczyk, B.K. Coleman, Z. Naufal, C. Rabideau, A.P. Pacsi, Results of three years of ambient air monitoring near a petroleum refinery in Richmond, California, USA. Atmosphere **10**(7), 385 (2019). https://doi.org/10.3390/atmos10070385

R. Sander, W.C. Keene, A.A.P. Pszenny, R. Arimoto, G.P. Ayers, E. Baboukas, J.M. Cainey, P.J. Crutzen, R.A. Duce, G. Hönninger, B.J. Huebert, W. Maenhaut, N. Mihalopoulos, V.C. Turekian, R. Van Dingenen, Inorganic bromine in the marine boundary layer: a critical review. Atmos. Chem. Phys. **3**, 1301–1336 (2003)

S.P. Sander, R.R. Friedl, Kinetics and product studies of the reaction ClO + BrO using flash photolysis-ultraviolet absorption. J. Phys. Chem. **93**, 4764–4771 (1989)

R. Sanders, S. Solomon, G. Mount, M.W. Bates, A. Schmeltekopf, Visible spectroscopy at McMurdo station, Antarctica: 3. Observation of NO_3. J. Geophys. Res. **92**, 8339–8342 (1987)

R.W. Sanders, S. Solomon, M.A. Carroll, A.L. Schmeltekopf, Visible and near-ultraviolet spectroscopy at McMurdo station, Antarctica 4. Overview and daily measurements of NO_2, O_3, and OClO during 1987. J. Geophys. Res. **94**(D9), 11381–11391 (1989)

R.W. Sanders, S. Solomon, J.P. Smith, L. Perliski, H.L. Miller, G.H. Mount, J.G. Keys, A.L. Schmeltekopf, Visible and near-UV spectroscopy at McMurdo station, Antarctica, 9. Observations of OClO from April to October 1991. J. Geophys. Res. **98**(D4), 7219–7228 (1993)

R.W. Sanders, Improved analysis of atmospheric absorption spectra by including the temperature dependence of NO_2. J. Geophys. Res. **101**, 20945–20952 (1996)

R.W. Sanders, S. Solomon, K. Kreher, P.V. Johnston, An intercomparison of NO_2 and OClO measurements at arrival heights, Antarctica during austral spring 1996. J. Atmos. Chem. **33**, 283–298 (1999)

S. Sanghavi, A.B. Davis, A. Eldering, Vsmartmom: a vector matrix operator method-based radiative transfer model linearized with respect to aerosol properties. J. Quant. Spectrosc. Radiat. Transf. **133**, 412–433 (2014). https://doi.org/10.1016/j.jqsrt.2013.09.004

S. Sanghavi, Raman scattering in the earth's atmosphere, part i: optical properties. J. Quant. Spectrosc. Radiat. Transf. **291**, 108328 (2022). https://doi.org/10.1016/j.jqsrt.2022.108328

S. Sanghavi, An efficient Mie theory implementation to investigate the influence of aerosols on radiative transfer, Diploma Thesis, University of Heidelberg (2003)

S. Sanghavi, C. Frankenberg, Raman scattering in the earth's atmosphere, part ii: radiative transfer modeling for remote sensing applications. J. Quant. Spectrosc. Radiat. Transf. **311**, 108791 (2023). https://doi.org/10.1016/j.jqsrt.2023.108791

C.J. Sansonetti, M.L. Salit, J. Reader, Wavelengths of spectral lines in mercury pencil lamps. Appl. Opt. **35**, 74–77 (1996)

A. Sarkissan, J.P. Pommereau, F. Goutail, Identification of polar stratospheric clouds from the ground by visible spectrometry. Geophys. Res. Lett. **18**, 779–782 (1991)

A. Sarkissian, D. Fish, M. Van Roozendael, M. Gil, H.B. Chen, P. Wang, J.P. Pommereau, J. Lenoble, Ozone and NO_2 air-mass factors for zenith sky spectrometers: intercomparison of calculations with different radiative transfer models. Geophys. Res. Lett. **21**, 1113–1116 (1995)

A. Sarkissian, G. Vaughan, H.K. Roscoe, L.M. Bartlett, F.M. O'Connor, D.G. Drew, P.A. Hughes, D. Moore, Accuracy of measurements of total ozone by a SAOZ ground-based zenith-sky visible spectrometer. J. Geophys. Res. **102**, 1379–1390 (1997)

C.V. Savigny, O. Funk, U. Platt, K. Pfeilsticker, Radiative smoothing in zenith-scattered skylight transmitted to the ground. Geophys. Res. Lett. **26**, 2949–2952 (1999)

A. Savitzky, M.J.E. Golay, Smoothing and differentiation of data by simplified least squares procedures. Anal. Chem. **36**, 1627–1639 (1964)

C.B. Schaaf, F. Gao, A.H. Strahler, W. Lucht, X. Li, T. Tsang, N.C. Strugnell, X. Zhang, Y. Jin, J.-P. Muller, P. Lewis, M. Barnsley, P. Hobson, M. Disney, G. Roberts, M. Dunderdale, C. Doll, R.P. d'Entremont, B. Hu, S. Liang, J.L. Privette, D. Roy, First operational BRDF, albedo nadir reflectance products from MODIS. Remote Sens. Environ. **83**(1), 135–148 (2002). https://doi.org/10.1016/S0034-4257(02)00091-3

C.B. Schaaf, J. Liu, F. Gao, A.H. Strahler, Aqua and Terra MODIS Albedo and Reflectance Anisotropy Products, in *Land Remote Sensing and Global Environmental Change: NASA's Earth Observing System and the Science of ASTER and MODIS, Remote Sensing and Digital*

Image Processing ed, by B. Ramachandran, C.O. Justice, M.J. Abrams. Springer, New York, NY (2011), pp. 549–561. https://doi.org/10.1007/978-1-4419-6749-7_24

C. Schall, K.G. Heumann, GC determination of volatile organoiodine and organobromine compounds in seawater and air samples. Fresenius J. Anal. Chem. **346**, 717–722 (1993)

S.M. Schauffler, E.L. Atlas, F. Flocke, R.A. Lueb, V. Stroud, W. Travnicek, Measurement of bromine-containing organic compounds at the tropical tropopause. Geophys. Res. Lett. **25**, 317–320 (1998)

H. Scheffler, H. Elsässer, *Physik der Sterne und der Sonne* (BI Wissenschaftsverlag, Mannheim/Wien/Zürich, 1974), pp.978–3411141722

V. M. E. Schenkeveld, G. Jaross, S. Marchenko, D. Haffner, Q.L. Kleipool, N.C. Rozemeijer, J.P. Veefkind, P.F. Levelt, In-flight performance of the Ozone Monitoring Instrument. Atmos. Meas. Tech. **10**(5), 1957–1986 (2017). https://amt.copernicus.org/articles/10/1957/2017/

R. Schermaul, J.W. Brault, A.A.D. Canas, R.C.M. Learner, O.L. Polyansky, N.F. Zobov, D. Belmiloud, J. Tennyson, Weak line water vapour spectrum in the regions 13200–15000 cm^{-1}. J. Mol. Spectrosc. **211**, 169–178 (2002)

H.I. Schiff, D.R. Karecki, G.W. Harris, D.R. Hastie, G.I. Mackay, A tunable diode laser system for aircraft measurements of trace gases. J. Geophys. Res. **95**, 10147–10154

S. Schill, R.S. McEwan, R. Moffet, J. Marrero, C. MacDonald, E. Winegar, Real-world application of open-path UV-DOAS, TDL, and FT-IR spectroscopy for air quality monitoring at industrial facilities. Spectrosc. Online **37**(S11), 18–22 (2022). https://www.spectroscopyonline.com/view/real-world-application-of-open-path-uv-doas-tdl-and-ft-ir-spectroscopy-for-air-quality-monitoring-at-industrial-facilities

C. Schiller, A. Wahner, U. Platt, H.-P. Dorn, J. Callies, D.H. Ehhalt, Near UV atmospheric absorption measurements of column abundances during airborne Arctic stratospheric expedition, January-February 1989: 2. OClO observations. Geophys. Res. Lett. **17**, 501–504 (1990)

E. Schlosser, B. Bohn, T. Brauers, H.-P. Dorn, H. Fuchs, R. Häseler, A. Hofzumahaus, F. Holland, F. Rohrer, L.O. Rupp, M. Siese, R. Tillmann, A. Wahner Intercomparison of two hydroxyl radical measurement techniques at the atmosphere simulation chamber SAPHIR. J. Atmos. Chem. **56**(2), 187–205 (2007).https://doi.org/10.1007/s10874-006-9049-3

M. Schmidt, *Von Christian Friedrich Schönbein bis zum Ozonloch, ein Abriß der Geschichte der Ozonforschung* (Max-Planck-Insitut für Aeronomie, Katlenburg-Lindau bei Göttingen, Germany, 1988)

J.A. Schmidt, D.J. Jacob, H.M. Horowitz, L. Hu, T. Sherwen, M.J. Evans, Q. Liang, R.M. Suleiman, D.E. Oram, M. Le Breton, C.J. Percival, S. Wang, B. Dix, R. Volkamer, Modeling the observed tropospheric BrO background: importance of multiphase chemistry and implications for ozone, OH, and mercury. J. Geophys. Res.: Atmos. **121**(19), 11,819–11,835 (2016). https://doi.org/10.1002/2015JD024229

J. Schmölling, Uebersicht über regulatorische Massnahmen zur Luftreinhaltung und deren Auswirkungen. AGF-Tagung "Luftreinhaltung–Luftverschmutzung", Bonn (1983), pp. 3–4

W. Schneider, G.K. Moortgat, G.S. Tyndall, J.P. Burrows, Absorption cross-sections of NO_2 in the UV and visible region (200–700 nm) at 298 K.J. Photochem. Photobiol. **40**, 195–217 (1987)

O. Schneising, P. Bergamaschi, H. Bovensmann, M. Buchwitz, J.P. Burrows, N.M. Deutscher, D.W.T. Griffith, J. Heymann, R. Macatangay, J. Messerschmidt, J. Notholt, M. Rettinger, M. Reuter, R. Sussmann, V.A. Velazco, T. Warneke, P.O. Wennberg, D. Wunch, Atmospheric greenhouse gases retrieved from SCIAMACHY: comparison to ground-based FTS measurements and model results. Atmos. Chem. Phys. **12**(3), 1527–1540 (2012). https://doi.org/10.5194/acp-12-1527-2012

O. Schneising, M. Buchwitz, J. Hachmeister, S. Vanselow, M. Reuter, M. Buschmann, H. Bovensmann, J.P. Burrows, Advances in retrieving XCH_4 and XCO from Sentinel-5 Precursor: improvements in the scientific TROPOMI/WFMD algorithm. Atmos. Meas. Tech. **16**(3), 669–694 (2023). https://doi.org/10.5194/amt-16-669-2023

R. Schofield, B.J. Connor, K. Kreher, P.V. Johnston, C.D. Rodgers, The retrieval of profile and chemical information from ground-based UV-visible spectroscopic measurements. J. Quant. Spectr. Rad. Transf. **86**, 115–131 (2004)

R. Schofield, P.V. Johnston, A. Thomas, K. Kreher, B.J. Connor, S. Wood, D. Shooter, M.P. Chipperfield, A. Richter, R. von Glasow, C.D. Rodgers, Tropospheric and stratospheric BrO columns over arrival heights, Antarctica, 2002. J. Geophys. Res. **111**, D22310 (2006). https://doi.org/10.1029/2005JD007022

T. Scholl, Photon path length distributions for cloudy skies—their first and second-order moments inferred from high resolution oxygen A-Band spectroscopy. Ph.D. Thesis, University of Heidelberg, Germany (2006)

C.F. Schönbein, Beobachtungen über den bei der Elektrolysation des Wassers and dem Ausströmen der gewöhnlichen Electrizität aus Spitzen sich entwickelnden Geruch. Ann. Phys. Chem. **50**, 616 (1840)

M. Schöne. Tropospheric BrO plumes in Arctic spring - A comparison of TROPOMI satellite observations and model results. Ph.D. Thesis, University of Heidelberg (2023). http://www.ub.uni-heidelberg.de/archiv/32838

Schott, Schott Optical Filter Datasheet Collection (2022). https://www.schott.com/en-gb/products/optical-filter-glass-P1000266/downloads. Accessed 18 Feb 2024

W.H. Schroeder, K.G. Anlauf, L.A. Barrie, J.Y. Lu, A. Steffen, D.R. Schneeberger, T. Berg, Arctic springtime depletion of mercury. Nature **394**, 331–332 (1998)

H.W. Schrötter, H.W. Klöckner, Raman scattering cross-sections in gases, in *Topics in Current Physics: Raman Spectroscopy of Gases and Liquids*. ed. by A. Weber (Springer, Verlag, 1979)

M. Schultz, M. Heitlinger, D. Mihelcic, A. Volz-Thomas, Calibration source for peroxy radicals with built-in actinometry using H_2O and O_2 photolysis at 185 nm. J. Geophys. Res. **100**, 18811–18816 (1995)

E.O. Schulz-DuBois, Generation of square lattice of focal points by a modified white cell. Appl. Opt. **12**, 1391–1393 (1973)

S.E. Schwartz, Are global cloud albedo and climate crontrolled by marine phytoplankton. Nature **336**, 441–445 (1988)

F. Schweitzer, P. Mirabel, C. George, Heterogeneous chemistry of nitryl halides in relation to tropospheric halogen activation. J. Atmos. Chem. **34**, 101–117 (1999)

J.H. Seinfeld, S.N. Pandis, *Atmospheric chemistry and physics: from air pollution to climate change*, 3rd edn. (Wiley, Hoboken, New Jersey, Hoboken, New Jersey, 2016)

S. Seisel, F. Caloz, F.F. Fenter, H. Van den Bergh, M.J. Rossi, The heterogeneous reaction of NO_3 with NaCl and KBr: a nonphotolytic source of halogen atoms. Geophys. Res. Lett. **24**, 2757–2760 (1997)

K. Seitz, J. Buxmann, D. Pöhler, T. Sommer, J. Tschritter, T. Neary, C. O'Dowd, U. Platt, The spatial distribution of the reactive iodine species IO from simultaneous active and passive DOAS observations. Atmos. Chem. Phys. **10**(5), 2117–2128 (2010). https://doi.org/10.5194/acp-10-2117-2010

T. Senne, J. Stutz, U. Platt, Measurement of the latitudinal distribution of NO_2 column density and layer height in Oct./Nov. 1993. Geophys. Res. Lett. **23**, 805–808 (1996)

J. Senzig, Troposphärische DOAS-Messungen stickstoffhaltiger und aromatischer Verbindungen in Heidelberg, (in German), Diploma Thesis, University of Heidelberg (1995)

S. Seo, A. Richter, A.-M. Blechschmidt, I. Bougoudis, J.P. Burrows, First high-resolution BrO column retrievals from TROPOMI. Atmos. Meas. Tech. **12**(5), 2913–2932 (2019). https://doi.org/10.5194/amt-12-2913-2019

S. Seo, A. Richter, A.-M. Blechschmidt, I. Bougoudis, J.P. Burrows, Spatial distribution of enhanced BrO and its relation to meteorological parameters in arctic and Antarctic sea ice regions. Atmos. Chem. Phys. **20**(20), 12285–12312 (2020). https://doi.org/10.5194/acp-20-12285-2020

R. Shaiganfar, S. Beirle, H. Denier van der Gon, S. Jonkers, J. Kuenen, H. Petetin, Q. Zhang, M. Beekmann, T. Wagner, Estimation of the Paris NO_x emissions from mobile MAX-DOAS observations and Chimere model simulations during the megapoli campaign using the closed integral method. Atmos. Chem. Phys. **17**(12), 7853–7890 (2017). https://doi.org/10.5194/acp-17-7853-2017

D. Shapley, Nitrosamines: scientists on the trail of prime suspect in. urban cancer. Science **191**, 268–270 (1976)

T.M. Sherwen, M.J. Evans, D.V. Spracklen, L.J. Carpenter, R. Chance, A.R. Baker, J.A. Schmidt, T.J. Breider, Global modelling of tropospheric iodine aerosol. Geophys. Res. Lett. **43** (2016). https://doi.org/10.1002/2016GL070062

B. Shirinzadeh, C.C. Wang, D.Q. Deng, Pressure dependence of ozone interference in the laser fluorescence measurements of OH in the atmosphere. Appl. Opt. **26**, 2102–2105 (1987)

Y. Shuai, J.G. Masek, F. Gao, C.B. Schaaf, An algorithm for the retrieval of 30-m snow-free albedo from Landsat surface reflectance and MODIS BRDF. Remote Sens. Environ. **115**(9), 2204–2216 (2011). https://doi.org/10.1016/j.rse.2011.04.019

M.W. Sigrist (ed.), *Air Monitoring by Spectroscopic Techniques, Chemical Analysis Series*, vol. 127 (Wiley, New York, 1994a)

M.W. Sigrist, Air monitoring by laser photoacoustic spectroscopy, in *Air Monitoring by Spectroscopic Techniques, Chemical Analysis Series*, vol. 127, ed. by M.W. Sigrist (Wiley, New York, 1994b), pp.163–238

H. Sihler, C. Kern, D. Pöhler, U. Platt, Applying light-emitting diodes with narrowband emission features in differential spectroscopy. Opt. Lett. **34**(23), 3716–3718 (2009). https://doi.org/10.1364/OL.34.003716

H. Sihler, U. Platt, S. Beirle, T. Marbach, S. Kühl, S. Dörner, J. Verschaeve, U. Frieß, D. Pöhler, L. Vogel, R. Sander, T. Wagner, Tropospheric BrO column densities in the Arctic derived from satellite: retrieval and comparison to ground-based measurements. Atmos. Meas. Tech. **5**(11), 2779–2807 (2012). https://doi.org/10.5194/amt-5-2779-2012

S. Sillman, J.A. Logan, S.C. Wofsy, The sensitivity of ozone to nitrogen oxides and hydrocarbons in regional ozone episodes. J. Geophys. Res. **95**, 1837–1852 (1990)

F.G. Simon, W. Schneider, G.K. Moortgat, J.P. Burrows, A study of the ClO absorption cross-section between 240 and 310 nm and the kinetics of the self-reaction at 300 K. J. Photochem. Photobiol., A: Chem. **55**, 1–23 (1990)

W.R. Simpson, Continuous wave cavity ring-down spectroscopy applied to in situ detection of dinitrogen pentoxide (N_2O_5). Rev. Sci. Instrum. **74**(7) (2003). https://doi.org/10.1063/1.1578705

W.R. Simpson, S.S. Brown, A. Saiz-Lopez, J.A. Thornton, R. von Glasow, Tropospheric halogen chemistry: sources, cycling, and impacts. Chem. Rev. **115**(10), 4035–4062 (2015). https://doi.org/10.1021/cr5006638

W.R. Simpson, P.K. Peterson, U. Frieß, H. Sihler, J. Lampel, U. Platt, C. Moore, K. Pratt, P. Shepson, J. Halfacre, S.V. Nghiem, Horizontal and vertical structure of reactive bromine events probed by bromine monoxide MAX-DOAS. Atmos. Chem. Phys. **17**(15), 9291–9309 (2017). https://doi.org/10.5194/acp-17-9291-2017

W.R. Simpson, U. Frieß, J.L. Thomas, J. Lampel, U. Platt, Polar nighttime chemistry produces intense reactive bromine events. Geophys. Res. Lett. **45**(18), 9987–9994 (2018). https://doi.org/10.1029/2018GL079444

H.B. Singh, G.L. Gregory, B. Anderson, E. Browell, G.W. Sachse, D.D. Davis, J. Crawford, J.D. Bradshaw, R. Talbot, D.R. Blake, D. Thornton, R. Newell, J. Merrill, Low ozone in the marine boundary layer of the tropical Pacific Ocean: photochemical loss, chlorine atoms, and entrainment. J. Geophys. Res. **101**, 1907–1917 (1996)

B.-M. Sinnhuber, D.W. Arlander, H. Bovensmann, J.P. Burrows, M.P. Chipperfield, C.-F. Ennel, U. Frieß, F. Hendrick, P.V. Johnston, R.L. Jones, K. Kreher, N. Mohamed-Tahrin, R. Müller, K. Pfeilsticker, U. Platt, J.P. Pommereau, I. Pundt, A. Richter, A. South, K.K. Toernkvist, M. Van Roozendael, T. Wagner, F. Wittrock, Comparison of measurements and model calculations of stratospheric bromine monoxide. J. Geophys. Res. **107**(D19), 4398 (2002). https://doi.org/10.1029/2001JD000940

R. Sinreich, U. Frieß, T. Wagner, U. Platt, Multi axis differential optical absorption spectroscopy (MAX-DOAS) of gas and aerosol distributions. Faraday Discuss. **130**, 153–164 (2005). https://doi.org/10.1039/B419274P

R. Sinreich, R. Volkamer, F. Filsinger, U. Frieß, C. Kern, U. Platt, O. Sebastián, T. Wagner, MAX-DOAS detection of glyoxal during ICARTT 2004. Atmos. Chem. Phys. **7**(5), 1293–1303 (2007). https://doi.org/10.5194/acp-7-1293-2007

R. Sinreich, S. Coburn, B. Dix, R. Volkamer, Ship-based detection of glyoxal over the remote tropical Pacific Ocean. Atmos. Chem. Phys. **10**(23), 11359–11371 (2010). https://doi.org/10.5194/acp-10-11359-2010

J. Sintermann, K. Dietrich, C. Häni, M. Bell, M. Jocher, A. Neftel, A miniDOAS instrument optimised for ammonia field measurements. Atmos. Meas. Tech. **9**, 2721–2734 (2016). https://doi.org/10.5194/amt-9-2721-2016

C.E. Sioris, W.F.J. Evans, Filling in of Fraunhofer and gas-absorption lines in sky spectra as caused by rotational Raman scattering. Appl. Opt. **38**(12), 2706–2713 (1999)

C.E. Sioris, C.S. Haley, C.A. McLinden, C. von Savigny, I.C. McDade, J.C. McConnell, W.F.J. Evans, N.D. Lloyd, E.J. Llewellyn, K.V. Chance, T.P. Kurosu, D. Murtagh, U. Frisk, K. Pfeilsticker, H. Bösch, F. Weidner, K. Strong, J. Stegman, G. Mégie, Stratospheric profiles of nitrogen dioxide observed by optical spectrograph and infrared imager system on the Odin satellite. J. Geophys. Res.: Atmos. **108**(D7) (2003). https://doi.org/10.1029/2002JD002672

A. Sjödin, Studies of the diurnal variation of nitrous acid in urban air. Environ. Sci. Technol. **22**, 1086–1089 (1988)

J. Slusser, K. Hammond, A. Kylling, K. Stamnes, L. Perliski, A. Dahlback, D. Anderson, R. DeMajistre, Comparison of air mass computations. J. Geophys. Res. **101**, 9315–9321 (1996)

J.R. Slusser, D.J. Fish, E.K. Strong, R.L. Jones, H.K. Roscoe, A. Sarkissian, Five years of NO_2 vertical column measurements at Faraday (65°S): evidence for the hydrolysis of $BrONO_2$ on Pinatubo aerosols. J. Geophys. Res. **102**, 12987–12993 (1997)

J. Smith, S. Solomon, Atmospheric NO_3: 3. Sunrise disappearance and the stratospheric profile. J. Geophys. Res. **95**, 13819–13827 (1990)

J. Smith, S. Solomon, R. Sanders, H. Miller, L. Perliski, J. Keys, A. Schmeltekopf, Atmospheric NO_3: 4. Vertical profiles at middle and polar latitudes at sunrise. J. Geophys. Res. **98**, 8983–8989 (1993)

P.L. Smith, C. Heise, J.R. Esmond, R.L. Kurucz, Atomic spectral line database from CD-ROM 23 (1995 Atomic Line Data (R.L. Kurucz, B. Bell) Kurucz CD-ROM No. 23. Cambridge, Mass.: Smithsonian Astrophysical Observatory) (1995)

N. Smith, J.M.C. Plane, C. Nien, O.A. Solomon, Nighttime radical chemistry in the San Joaquin valley. Atmos. Environ. **29**, 2887–2897 (1995)

K.M. Smith, D.A. Newnham, Near-infrared absorption spectroscopy of oxygen and nitrogen gas mixtures. Chem. Phys. Lett. **308**(1–2), 1–6 (1999)

F.G. Smith, T.A. King, *Optics and Photonics* (An Introduction. Wiley, New York, 2000). 0-471-48924-7

K.M. Smith, D.A. Newnham, Near-infrared absorption cross sections and integrated absorption intensities of molecular oxygen (O_2, O_2-O_2, and O_2-N_2). J. Geophys. Res. **105**(D6), 7383–7396 (2000)

K.M. Smith, D.A. Newnham, R.G. Williams, Collision-induced absorption of solar radiation in the atmosphere by molecular oxygen at 1.27 μm: field observations and model calculations. J. Geophys. Res. **106**(D7), 7541–7552 (2001)

C.A. Smith, F.D. Pope, B. Cronin, C.B. Parkes, A.J. Orr-Ewing, Absorption cross sections of formaldehyde at wavelengths from 300 to 340 nm at 294 and 245 K. J. Phys. Chem. A **110**(41), 11645–11653 (2006). https://doi.org/10.1021/jp063713y

M.T. Smith, Advances in understanding benzene health effects and susceptibility. Annu. Rev. Public Health **31**(1), 133–148 (2010). https://doi.org/10.1146/annurev.publhealth.012809.103646

N. Sobanski, M.J. Tang, J. Thieser, G. Schuster, D. Pöhler, H. Fischer, W. Song, C. Sauvage, J. Williams, J. Fachinger, F. Berkes, P. Hoor, U. Platt, J. Lelieveld, J.N. Crowley, Chemical and meteorological influences on the lifetime of NO_3 at a semi-rural mountain site during PARADE. Atmos. Chem. Phys. **16**(8), 4867–4883 (2016). https://doi.org/10.5194/acp-16-4867-2016

S. Solberg, N. Schmidtbauer, A. Semb, F. Stordal, Boundary-layer ozone depletion as seen in the Norwegian Arctic in spring. J. Atmos. Chem. **23**, 301–332 (1996)

S. Solomon, A.L. Schmeltekopf, R.W. Sanders, On the interpretation of zenith sky absorption measurements. J. Geophys. Res.: Atmos. **92**(D7), 8311–8319 (1987a)

S. Solomon, G. Mount, R.W. Sanders, A. Schmeltekopf, Visible spectroscopy at McMurdo station, Antarctica: 2, observation of OClO. J. Geophys. Res. **92**, 8329–8338 (1987b)

S. Solomon, G.H. Mount, R.W. Sanders, R.O. Jakoubek, A.L. Schmeltekopf, Observations of the nighttime abundance of OClO in the winter stratosphere above Thule. Greenland. Science **242**, 550–555 (1988)

S. Solomon, H.L. Miller, J.P. Smith, R.W. Sanders, G.H. Mount, A.L. Schmeltekopf, J.F. Noxon, Atmospheric NO_3, 1. Measurement technique and the annual cycle at 40°N. J. Geophys. Res. **94**, 11041–11048 (1989a)

S. Solomon, R.W. Sanders, M.A. Carroll, A.L. Schmeltekopf, Visible and near-ultraviolet spectroscopy at McMurdo Station, Antarctica, 5. Observations of the diurnal variations of BrO and OClO. J. Geophys. Res. **94**, 11393–11403 (1989b)

S. Solomon, R.W. Sanders, G.H. Mount, M.A. Carroll, R.O. Jakoubek, A.L. Schmeltekopf, Atmospheric NO_3, 2. Observations in polar regions. J. Geophys. Res. **94**(D13), 16423–16427 (1989c)

S. Solomon, R.W. Sanders, H.L. Jr. Miller, Visible and near-ultraviolet spectroscopy at McMurdo Station, Antarctica, 7. OClO diurnal photochemistry and implications for ozone destruction. J. Geophys. Res. **95**, 13807 (1990)

S. Solomon, J.P. Smith, R.W. Sanders, L. Perliski, H.L. Miller, G.H. Mount, J.G. Keys, A.L. Schmeltekopf, Visible and near-ultraviolet spectroscopy at McMurdo Station, Antarctica, 8. Observations of nighttime NO_2 and NO_3 from April to October 1991. J. Geophys. Res. **98**, 993–1000 (1993a)

S. Solomon, R.W. Sanders, R.R. Garcia, J.G. Keys, Increased chlorine dioxide over Antarctica caused by volcanic aerosols from Mount-Pinatubo. Nature **363**, 245–248 (1993b)

S. Solomon, R.W. Sanders, R.O. Jakoubek, K. Arpag, S.L. Stephens, J.G. Keys, R.R. Garcia, Visible and near-ultraviolet spectroscopy at McMurdo Station, Antarctica, 10. Reductions of NO_2 due to Pinatubo aerosols. J. Geophys. Res. **99**, 3509–3516 (1994a)

S. Solomon, R.R. Garcia, A.R. Ravishankara, On the role of iodine in ozone depletion. J. Geophys. Res. **99**, 20491–20499 (1994b)

S. Solomon, R.W. Portmann, R.W. Sanders, J.S. Daniel, Absorption of solar radiation by water vapor, oxygen, and related collision pairs in the Earth's atmosphere. J. Geophys. Res.: Atmos. **103**(D4), 3847–3858 (1998). https://doi.org/10.1029/97JD03285

S. Solomon, Stratospheric ozone depletion: a review of concepts and history. Rev. Geophys. **37**, 275–316 (1999)

R. Sommariva, W.J. Bloss, N. Brough, N. Carslaw, M. Flynn, A.-L. Haggerstone, D.E. Heard, J.R. Hopkins, J.D. Lee, A.C. Lewis, G. McFiggans, P.S. Monks, S.A. Penkett, M.J. Pilling, J.M.C. Plane, K.A. Read, A. Saiz-Lopez, A.R. Rickard, P.I. Williams, OH and HO_2 chemistry during NAMBLEX: roles of oxygenates, halogen oxides and heterogeneous uptake. Atmos. Chem. Phys. **6**(4), 1135–1153 (2006). https://doi.org/10.5194/acp-6-1135-2006

South Coast AQMD, Rule 1180: Refinery Fenceline and Community Air Monitoring (2017). http://www.aqmd.gov/home/rules-compliance/rules/support-documents/rule-1180-refinery-fenceline-monitoring-plans#

A.M. South, I.M. Povey, R.L. Jones, Broadband lidar measurements of tropospheric water vapour profiles. J. Geophys. Res. **103**, 31191–31202 (1998)

C.W. Spicer, E.G. Chapman, B.J. Finlayson-Pitts, R.A. Plastridge, J.M. Hubbe, J.D. Fast, C.M. Berkowitz, Unexpectedly high concentrations of molecular chlorine in coastal air. Nature **394**, 353–356 (1998)

C.W. Spicer, R.A. Plastridge, K.L. Foster, B.J. Finlayson-Pitts, J.W. Bottenheim, A.M. Grannas, P.B. Shepson, Molecular halogens before and during ozone depletion events in the Arctic at polar sunrise: concentrations and sources. Atmos. Environ. **36**, 2721–2731 (2002)

N. Spichtinger, M. Wenig, P. James, T. Wagner, U. Platt, A. Stohl, Satellite detection of a continental-scale plume of nitrogen oxides from boreal forest fires. Geophys. Res. Lett. **28**, 4579–4582 (2001)

P. Spietz, J. Gómez Martín, J.P. Burrows, Effects of column density on I_2 spectroscopy and a determination of I_2 absorption cross section at 500 nm. Atmos. Chem. Phys. **6**, 2177–2191 (2006)

E. Spinei, A. Whitehill, A. Fried, M. Tiefengraber, T.N. Knepp, S. Herndon, J.R. Herman, M. Müller, N. Abuhassan, A. Cede, D. Richter, J. Walega, J. Crawford, J. Szykman, L. Valin, D.J. Williams, R. Long, R.J. Swap, Y. Lee, N. Nowak, B. Poche, The first evaluation of formaldehyde column observations by improved Pandora spectrometers during the KORUS-AQ field study. Atmos. Meas. Tech. **11**(9), 4943–4961 (2018). https://doi.org/10.5194/amt-11-4943-2018

R.J.D. Spurr, Linearized radiative transfer theory: a general discrete ordinate approach to the calculation of radiances and analytic weighting functions, with application to atmospheric remote sensing. Ph.D. Thesis, Technical University of Eindhoven (2001)

R.J.D. Spurr, T.P. Kurosu, K.V. Chance, A linearized discrete ordinate radiative transfer model for atmospheric remote-sensing retrieval. J. Quant. Spectrosc. Radiat. Transf. **68**(6), 689–735 (2001). https://doi.org/10.1016/S0022-4073(00)00055-8

R.J.D. Spurr, LIDORT V2PLUS: a comprehensive radiative transfer package for UV/VIS/NIR nadir remote sensing, in *Remote Sensing of Clouds and the Atmosphere VIII*, vol. 5235 (SPIE, 2004), pp. 89–100. https://doi.org/10.1117/12.511103

R.J.D. Spurr, VLIDORT: a linearized pseudo-spherical vector discrete ordinate radiative transfer code for forward model and retrieval studies in multilayer multiple scattering media. J. Quant. Spectrosc. Radiat. Transf. **102**(2), 316–342 (2006). https://doi.org/10.1016/j.jqsrt.2006.05.005

R. Spurr, M. Christi, The LIDORT and VLIDORT linearized scalar and vector discrete ordinate radiative transfer models: updates in the last 10 years, in *Springer Series in Light Scattering: Volume 3: Radiative Transfer and Light Scattering*, ed. by A. Kokhanovsky. Springer Series in Light Scattering (Springer International Publishing, Cham, 2019), pp. 1–62. https://doi.org/10.1007/978-3-030-03445-0_1

R. Spurr, V. Natraj, S.F. Colosimo, J. Stutz, M. Christi, S. Korkin, VLIDORT-QS: a quasi-spherical vector radiative transfer model. J. Quant. Spectrosc. Radiat. Transf. **291**, 108341 (2022). https://doi.org/10.1016/j.jqsrt.2022.108341

J. Staehelin, J. Thudium, R. Bühler, A. Volz-Thomas, W. Graber, Trends in surface ozone concentrations at Arosa (Switzerland). Atmos. Environ. **28**, 75–87 (1994)

K. Stamnes, S.-C. Tsay, W. Wiscombe, K. Jayaweera, Numerically stable algorithm for discrete-ordinate-method radiative transfer in multiple scattering and emitting layered media. Appl. Opt. **27**(12), 2502–2509 (1988). https://doi.org/10.1364/AO.27.002502

K. Stamnes, S.C. Tsay, K. Jayaweera, W. Wiscombe, S. Stamnes, Z. Jin, Z. Lin, DISORT: discrete ordinate radiative transfer. Astrophys. Source Code Libr. ascl:1708.006 (2017). https://ui.adsabs.harvard.edu/abs/2017ascl.soft08006S

T. Stavrakou, J.-F. Müller, I. De Smedt, M. Van Roozendael, M. Kanakidou, M. Vrekoussis, F. Wittrock, A. Richter, J.P. Burrows, The continental source of glyoxal estimated by the synergistic use of spaceborne measurements and inverse modelling. Atmos. Chem. Phys. **9**(21), 8431–8446 (2009). https://doi.org/10.5194/acp-9-8431-2009

T. Stavrakou, J.-F. Müller, M. Bauwens, K.F. Boersma, J. van Geffen, Satellite evidence for changes in the NO_2 weekly cycle over large cities. Sci. Rep. **10**(1), 10066 (2020). https://doi.org/10.1038/s41598-020-66891-0

D. Stern, Global sulphur emissions from 1850 to 2000. Chemosphere **58**, 163–175 (2005) (update from November 2005, http://www.rpi.edu/~sternd/datasite.html)

R.K. Stevens, T.L. Vossler, DOAS (differential optical absorption spectroscopy) urban pollution, in *Measurements of Atmospheric Gases*, vol. 1433, ed. by H.I. Schiff (SPIE-The International Society of Optical Engineering, Bellingham, 1991), pp. 25–35

R.K. Stevens, R.J. Drago, Y. Mamane, A long path differential optical absorption spectrometer and EPA-approved fixed-point methods intercomparison. Atmos. Environ. **27B**, 231–236 (1993)

D.W. Steyert, J.M. Sirota, M.E. Mickelson, D.C. Reuter, Two new long-pass cells for infrared and visible spectroscopy. Rev. Sci. Instrum. **72**(12), 4337–4343 (2001). https://doi.org/10.1063/1.1409565

R.W. Stockwell, J.G. Calvert, The near ultraviolet absorption spectrum of gaseous HONO and N_2O_3. J. Photochem. **8**, 193–203 (1978)

W.R. Stockwell, J.G. Calvert, The mechanism of the OH-SO_2 reaction. Atmos. Environ. **17**, 2231–2235 (1983)

R.E. Stoiber, A. Jepsen, Sulfur-dioxide contributions to atmosphere by volcanos. Science **182**(4112), 577–578 (1973)

D. Stone, T. Sherwen, M.J. Evans, S. Vaughan, T. Ingham, L.K. Whalley, P.M. Edwards, K.A. Read, J.D. Lee, S.J. Moller, L.J. Carpenter, A.C. Lewis, D.E. Heard, Impacts of bromine and iodine chemistry on tropospheric OH and HO_2: comparing observations with box and global model perspectives. Atmos. Chem. Phys. **18**(5), 3541–3561 (2018). https://doi.org/10.5194/acp-18-3541-2018

R.J. Strutt, Ultra-violet transparency of the lower atmosphere, and its relative poverty in ozone. Proc. R. Soc. A **94**, 260–269 (1918)

F. Stuhl, H. Niki, Flash photochemical study of the reaction OH + NO + M using resonance fluorescent detection of OH. J. Chem. Phys. **57**, 3677–3679 (1972)

J. Stutz, U. Platt, Problems in using diode arrays for open path DOAS measurements of atmospheric species, in *Proceedings of the Europto Series. Optical Methods in Atmospheric Chemistry*, ed. by H.I. Schiff, U. Platt, vol. 1715 (1992), pp. 329–340

J. Stutz, Messung der Konzentration troposphärischer Spurenstoffe mittels Differentieller-Optischer Absorptionsspektroskopie: Eine neue Generation von Geräten und Algorithmen. Ph.D. Thesis, University of Heidelberg (1996)

J. Stutz, U. Platt, Numerical analysis and error estimation of differential optical absorption spectroscopy measurements with least squares methods. Appl. Opt. **35**, 6041–6053 (1996)

J. Stutz, U. Platt, A new generation of DOAS instruments, in *EUROTRAC Final Report*, ed. by J. Bösenberg, D.J. Brassington, P.C. Simon, vol. 8. Instrument Development for Atmospheric Research and Monitoring (1997a), pp. 370–378

J. Stutz, U. Platt, Improving long-path differential optical absorption spectroscopy (DOAS) with a quartz-fiber mode-mixer. Appl. Opt. **36**, 1105–1115 (1997b)

J. Stutz, E.S. Kim, U. Platt, P. Bruno, C. Perrino, A. Febo, UV-visible absorption cross-sections of nitrous acid. J. Geophys. Res. **105**, 14585–14592 (2000)

J. Stutz, R. Ackermann, J.D. Fast, L. Barrie, Atmospheric reactive chlorine and bromine at the Great Salt Lake, Utah. Geophys. Res. Lett. **29**, (2002a). https://doi.org/10.1029/2002GL014812

J. Stutz, B. Alicke, A. Neftel, Nitrous acid formation in the urban atmosphere: gradient measurements of NO_2 and HONO over grass in Milan. Italy. J. Geophys. Res. **107**(D22), 8192 (2002b). https://doi.org/10.1029/2001JD000390

J. Stutz, B. Alicke, R. Ackermann, A. Geyer, A. White, E. Williams, Vertical profiles of NO_3, N_2O_5, O_3, and NO_X in the nocturnal boundary layer: 1. Observations during the Texas Air Quality Study 2000. J. Geophys. Res. **109**, D12306 (2004a). https://doi.org/10.1029/2003JD004209

J. Stutz, B. Alicke, R. Ackermann, A. Geyer, S. Wang, A.B. White, E.J. Williams, C.W. Spicer, J.D. Fast, Relative humidity dependence of HONO chemistry in urban areas. J. Geophys. Res. **109**, D03307 (2004b). https://doi.org/10.1029/2003JD004135

J. Stutz, J. Thomas, S. Hurlock, M. Schneider, R. Von Glasow, M. Piot, K. Gorham, J. Burkhart, L. Ziemba, J. Dibb, B. Lefer, Longpath DOAS observations of surface BrO at Summit. Greenland. Atmos. Chem. Phys. **11**(18), 9899–9910 (2011). https://doi.org/10.5194/acp-11-9899-2011

J. Stutz, S.C. Hurlock, S.F. Colosimo, C. Tsai, R. Cheung, J. Festa, O. Pikelnaya, S. Alvarez, J.H. Flynn, M.H. Erickson, E.P. Olaguer, A novel dual-LED based long-path DOAS instrument for the measurement of aromatic hydrocarbons. Atmos. Environ. **147**(2), 121–132 (2016). https://doi.org/10.1016/j.atmosenv.2016.09.054

J. Stutz, B. Werner, M. Spolaor, L. Scalone, J. Festa, C. Tsai, R. Cheung, S.F. Colosimo, U. Tricoli, R. Raecke, R. Hossaini, M.P. Chipperfield, W. Feng, R.-S. Gao, E.J. Hintsa, J.W. Elkins, F.L. Moore, B. Daube, J. Pittman, S. Wofsy, K. Pfeilsticker, A new differential optical absorption spectroscopy instrument to study atmospheric chemistry from a high-altitude unmanned aircraft. Atmos. Meas. Tech. **10**(3), 1017–1042 (2017). https://doi.org/10.5194/amt-10-1017-2017

H. Su, Y. Cheng, R. Oswald, T. Behrendt, I. Trebs, F.X. Meixner, M.O. Andreae, P. Cheng, Y. Zhang, U. Pöschl, Soil nitrite as a source of atmospheric HONO and OH radicals. Science **333**(6049), 1616–1618 (2011). https://doi.org/10.1126/science.1207687

N. Sugimoto, A. Minato, Retroreflector with acute dihedral angles. Opt. Lett. **19**, 1660–1662 (1994)

R.M. Suleiman, K. Chance, X. Liu, G. González Abad, T.P. Kurosu, F. Hendrick, N. Theys, OMI total bromine monoxide (OMBRO) data product: algorithm, retrieval and measurement comparisons. Atmos. Meas. Tech. **12**(4), 2067–2084 (2019). https://doi.org/10.5194/amt-12-2067-2019

Y. Sun, L. Gu, J. Wen, C. van der Tol, A. Porcar-Castell, J. Joiner, C.Y. Chang, T. Magney, L. Wang, L. Hu, U. Rascher, P. Zarco-Tejada, C.B. Barrett, J. Lai, J. Han, Z. Luo, From remotely sensed solar-induced chlorophyll fluorescence to ecosystem structure, function, and service: Part I-Harnessing theory. Glob. Chang. Biol. **29**(11), 2926–2952 (2023). https://doi.org/10.1111/gcb.16634

Y. Sun, J. Wen, L. Gu, J. Joiner, C.Y. Chang, C. van der Tol, A. Porcar-Castell, T. Magney, L. Wang, L. Hu, U. Rascher, P. Zarco-Tejada, C.B. Barrett, J. Lai, J. Han, Z. Luo, From remotely-sensed solar-induced chlorophyll fluorescence to ecosystem structure, function, and service: Part II-Harnessing data. Glob. Chang. Biol. **29**(11), 2893–2925 (2023). https://doi.org/10.1111/gcb.16646

S. Svanberg, *Atomic and Molecular Spectroscopy, Springer Series on Atoms and Plasmas*, 2nd edn. (Springer, Berlin, Heidelberg, 1992)

R. Svensson, E. Ljungström, O. Lindqvist, Kinetics of the reaction between nitrogen dioxide and water vapour. Atmos. Environ. **21**, 1529–1539 (1987)

M.Q. Syed, A.W. Harrison, Ground based observations of stratospheric nitrogen dioxide. Can. J. Phys. **58**, 788–802 (1980)

F. Tack, A. Merlaud, M.-D. Iordache, T. Danckaert, H. Yu, C. Fayt, K. Meuleman, F. Deutsch, F. Fierens, M. Van Roozendael, High-resolution mapping of the NO_2 spatial distribution over Belgian urban areas based on airborne APEX remote sensing. Atmos. Meas. Tech. **10**(5), 1665–1688 (2017). https://doi.org/10.5194/amt-10-1665-2017

F. Tack, A. Merlaud, A.C. Meier, T. Vlemmix, T. Ruhtz, M.-D. Iordache, X. Ge, L. van der Wal, D. Schuettemeyer, M. Ardelean, A. Calcan, D. Constantin, A. Schönhardt, K. Meuleman, A. Richter, M. Van Roozendael, Intercomparison of four airborne imaging DOAS systems for tropospheric NO_2 mapping–the AROMAPEX campaign. Atmos. Meas. Tech. **12**(1), 211–236 (2019). https://doi.org/10.5194/amt-12-211-2019

T. Tajime, T. Saheki, K. Ito, Absorption characteristics of the γ-0 band of nitric oxide. Appl. Opt. **17**, 1290–1294 (1978)

H. Takashima, Y. Kanaya, S. Kato, M.M. Friedrich, M. Van Roozendael, F. Taketani, T. Miyakawa, Y. Komazaki, C.A. Cuevas, A. Saiz-Lopez, T. Sekiya, Full latitudinal marine atmospheric measurements of iodine monoxide. Atmos. Chem. Phys. **22**(6), 4005–4018 (2022). ISSN 1680-7316. https://doi.org/10.5194/acp-22-4005-2022

Y. Talmi, R.W. Simpson, Self-scanned photodiode array: a multichannel spectrometric detector. Appl. Opt. **19**, 1401–1414 (1980)

R.K. Talukdar, C.A. Longfellow, M.K. Gilles, A.R. Ravishankara, Quantum yields of $O(^1D)$ in the photolysis of ozone between 289 and 329 nm as a function of temperature. Geophys. Res. Lett. **25**, 143–146 (1998)

W. Tan, C. Liu, S. Wang, C. Xing, W. Su, C. Zhang, C. Xia, H. Liu, Z. Cai, J. Liu, Tropospheric NO_2, SO_2, and HCHO over the east china sea, using ship-based MAX-DOAS observations and comparison with OMI and OMPS satellite data. Atmos. Chem. Phys. **18**(20), 15387–15402 (2018). https://doi.org/10.5194/acp-18-15387-2018

Z. Tan, A. Hofzumahaus, K. Lu, S.S. Brown, F. Holland, L.G. Huey, A. Kiendler-Scharr, X. Li, X. Liu, N. Ma, K.-E. Min, F. Rohrer, M. Shao, A. Wahner, Y. Wang, A. Wiedensohler, Y. Wu, Z. Wu, L. Zeng, Y. Zhang, H. Fuchs, No evidence for a significant impact of heterogeneous chemistry on radical concentrations in the North China Plain in summer 2014. Environ. Sci. Technol. **54**(10), 5973–5979 (2020). https://doi.org/10.1021/acs.est.0c00525

T. Tang, J.C. McConnel, Autocatalytic release of bromine from arctic snow pack during polar sunrise. Geophys. Res. Lett. **23**, 2633–2636 (1996)

K. Tang, M. Qin, J. Duan, W. Fang, F. Meng, S. Liang, P. Xie, J. Liu, W. Liu, C. Xue, Y. Mu, A dual dynamic chamber system based on IBBCEAS for measuring fluxes of nitrous acid in agricultural fields in the north china plain. Atmos. Environ. **196**, 10–19 (2019). https://doi.org/10.1016/j.atmosenv.2018.09.059

K. Tang, M. Qin, W. Fang, J. Duan, F. Meng, K. Ye, H. Zhang, P. Xie, Y. He, W. Xu, J. Liu, W. Liu, Simultaneous detection of atmospheric HONO and NO_2 utilising an IBBCEAS system based on an iterative algorithm. Atmos. Meas. Tech. **13**(12), 6487–6499 (2020). https://doi.org/10.5194/amt-13-6487-2020

D.W. Tarasick, I.E. Galbally, O.R. Cooper, M.G. Schultz, G. Ancellet, T. Leblanc, T.J. Wallington, J.R. Ziemke, X. Liu, M. Steinbacher, J. Staehelin, C. Vigouroux, J.W. Hannigan, O.E. García, G. Foret, P. Zanis, E. Weatherhead, I. Petropavlovskikh, H.M. Worden, M.K. Osman, J. Liu, K. Chang, A. Gaudel, M. Lin, M.J. Granados-Muñoz, A.M. Thompson, S.J. Oltmans, J. Cuesta, G. Dufour, V. Thouret, B. Hassler, T. Trickl, J.L. Neu, Tropospheric ozone assessment report: tropospheric ozone from 1877 to 2016, observed levels, trends and uncertainties. Elem.: Sci. Anthropocene (2019). https://doi.org/10.1525/elementa.376

J. Tellinghuisen, Resolution of the visible-infrared absorption spectrum of I_2 into three contributing transitions. J. Chem. Phys. **58**, 2821–2834 (1973)

C.W. Tessum, D.A. Paolella, S.E. Chambliss, J.S. Apte, J.D. Hill, J.D. Marshall, PM $_{2.5}$ polluters disproportionately and systemically affect people of color in the United States. Sci. Adv. **7**(18), eabf4491 (2021). https://doi.org/10.1126/sciadv.abf4491

R. Thalman, R. Volkamer, Temperature dependent absorption cross-sections of O_2-O_2 collision pairs between 340 and 630 nm and at atmospherically relevant pressure. Phys. Chem. Chem. Phys. **15**(37), 15371–15381 (2013)

N. Theys, M. Van Roozendael, B. Dils, F. Hendrick, N. Hao, M. De Mazière, First satellite detection of volcanic bromine monoxide emission after the Kasatochi eruption. Geophys. Res. Lett. **36**(3) (2009). https://doi.org/10.1029/2008GL036552

N. Theys, I. De Smedt, J. van Gent, T. Danckaert, T. Wang, F. Hendrick, T. Stavrakou, S. Bauduin, L. Clarisse, C. Li, N. Krotkov, H. Yu, H. Brenot, M. Van Roozendael, Sulfur dioxide vertical column DOAS retrievals from the Ozone Monitoring Instrument: global observations and comparison to ground-based and satellite data. J. Geophys. Res.: Atmos. **120**(6), 2470–2491 (2015). https://doi.org/10.1002/2014JD022657

N. Theys, I. De Smedt, H. Yu, T. Danckaert, J. van Gent, C. Hörmann, T. Wagner, P. Hedelt, H. Bauer, F. Romahn, M. Pedergnana, D. Loyola, M. Van Roozendael, Sulfur dioxide retrievals from TROPOMI onboard Sentinel-5 precursor: algorithm theoretical basis. Atmos. Meas. Tech. **10**(1), 119–153 (2017). https://doi.org/10.5194/amt-10-119-2017

N. Theys, P. Hedelt, I. De Smedt, C. Lerot, H. Yu, J. Vlietinck, M. Pedergnana, S. Arellano, B. Galle, D. Fernandez, C. Carlito, C. Barrington, B. Taisne, H. Delgado-Granados, D. Loyola, M. Van Roozendael, Global monitoring of volcanic SO_2 degassing with unprecedented resolution from TROPOMI onboard Sentinel-5 Precursor. Sci. Rep. **9**, 2643 (2019). https://doi.org/10.1038/s41598-019-39279-y

N. Theys, R. Volkamer, J.-F. Müller, K.J. Zarzana, N. Kille, L. Clarisse, I. de Smedt, C. Lerot, H. Finkenzeller, F. Hendrick, T.K. Koenig, C.F. Lee, C. Knote, H. Yu, M. Van Roozendael, Global nitrous acid emissions and levels of regional oxidants enhanced by wildfires. Nat. Geosci. **13**(10), 681–686 (2020). https://doi.org/10.1038/s41561-020-0637-7

K. Thomas, A. Volz-Thomas, D. Kley, Zur Wechselwirkung von NO_3-Radikalen mit wässrigen Lösungen: Bestimmung des Henry- und des Massenakkommodationskoeffizienten. Berichte des Forschungszentrums Jülich (Dissertation K. Thomas University Wuppertal D468) 2755 (1993)

O. Thomsen, *Messung des Absorptionswirkungsquerschnitts von Schwefeldioxid im Wellenlängenbereich von 265 bis 298 nm* (Physik University Hamburg/GKSS-Geesthacht, Diplomarbeit, Fachber, 1990)

J.A. Thornton, P.J. Wooldridge, R.C. Cohen, M. Martinez, H. Harder, W.H. Brune, E.J. Williams, J.M. Roberts, F.C. Fehsenfeld, S.R. Hall, R.E. Shetter, B.P. Wert, A. Fried, Ozone production rates as a function of NO_X abundances and HO_X production rates in the Nashville urban plume. J. Geophys. Res. **107**(D12), 4146 (2002). https://doi.org/10.1029/2001JD000932

J.A. Thornton, P.J. Wooldridge, R.C. Cohen, E.J. Williams, D. Hereid, F.C. Fehsenfeld, J. Stutz, B. Alicke, Comparisons of in situ and long path measurements of NO_2 in urban plumes. J. Geophys. Res. **108**(D16), 4496 (2003). https://doi.org/10.1029/2003JD003559

J.A. Thornton, J.P. Kercher, T.P. Riedel, N.L. Wagner, J. Cozic, J.S. Holloway, W.P. Dubé, G.M. Wolfe, P.K. Quinn, A.M. Middlebrook, B. Alexander, S.S. Brown, A large atomic chlorine source inferred from mid-continental reactive nitrogen chemistry. Nature **464**(7286), 271–274 (2010). https://doi.org/10.1038/nature08905

H. Tian, R. Xu, J.G. Canadell, R.L. Thompson, W. Winiwarter, P. Suntharalingam, E.A. Davidson, P. Ciais, R.B. Jackson, G. Janssens-Maenhout, M.J. Prather, P. Regnier, N. Pan, S. Pan, G.P. Peters, H. Shi, F.N. Tubiello, S. Zaehle, F. Zhou, A. Arneth, G. Battaglia, S. Berthet, L. Bopp, A.F. Bouwman, E.T. Buitenhuis, J. Chang, M.P. Chipperfield, S.R.S. Dangal, E. Dlugokencky, J.W. Elkins, B.D. Eyre, B. Fu, B. Hall, A. Ito, F. Joos, P.B. Krummel, A. Landolfi, G.G. Laruelle, R. Lauerwald, W. Li, S. Lienert, T. Maavara, M. MacLeod, D.B. Millet, S. Olin, P.K. Patra, R.G. Prinn, P.A. Raymond, D.J. Ruiz, G.R. van der Werf, N. Vuichard, J. Wang, R.F. Weiss, K.C. Wells, C. Wilson, J. Yang, Y. Yao, A comprehensive quantification of global nitrous oxide sources and sinks. Nature **586**(7828), 248–256 (2020). https://doi.org/10.1038/s41586-020-2780-0

J.-L. Tirpitz, U. Frieß, F. Hendrick, C. Alberti, M. Allaart, A. Apituley, A. Bais, S. Beirle, S. Berkhout, K. Bognar, T. Bösch, I. Bruchkouski, A. Cede, K.L. Chan, M. den Hoed, S. Donner, T. Drosoglou, C. Fayt, M.M. Friedrich, A. Frumau, L. Gast, C. Gielen, L. Gomez-Martín, N. Hao, A. Hensen, B. Henzing, C. Hermans, J. Jin, K. Kreher, J. Kuhn, J. Lampel, A. Li, C. Liu, H. Liu, J. Ma, A. Merlaud, E. Peters, G. Pinardi, A. Piters, U. Platt, O. Puentedura, A. Richter, S. Schmitt, E. Spinei, D. Stein Zweers, K. Strong, D. Swart, F. Tack, M. Tiefengraber, R. van der Hoff, M. van Roozendael, T. Vlemmix, J. Vonk, T. Wagner, Y. Wang, Z. Wang, M. Wenig, M. Wiegner, F. Wittrock, P. Xie, C. Xing, J. Xu, M. Yela, C. Zhang, X. Zhao, Intercomparison of MAX-DOAS vertical profile retrieval algorithms: studies on field data from the CINDI-2 campaign. Atmos. Meas. Tech. **14**(1), 1–35 (2021). https://doi.org/10.5194/amt-14-1-2021

J.-L. Tirpitz. *Enhancing MAX-DOAS Atmospheric Remote Sensing by Multispectral Polarimetry*. Ph.D. Thesis, University of Heidelberg (2021). https://katalog.ub.uni-heidelberg.de/cgi-bin/titel.cgi?katkey=68771002

J.-L. Tirpitz, U. Frieß, R. Spurr, U. Platt, Enhancing MAX-DOAS atmospheric state retrievals by multispectral polarimetry–studies using synthetic data. Atmos. Meas. Tech. **15**(7), 2077–2098 (2022). https://doi.org/10.5194/amt-15-2077-2022

R.N. Tolchenov, O. Naumenko, N.F. Zobov, S.V. Shirin, O.L. Polyansky, J. Tennyson, M. Carleer, P.-F. Coheur, S. Fally, A. Jenouvrier, A.C. Vandaele, Water vapour line assignments in the 9250-26000 cm^{-1} frequency range. J. Mol. Spectrosc. **233**, 68–76 (2005)

K.K. Tørnkvist, D.W. Arlander, B.-M. Sinnhuber, Ground-based UV measurements of BrO and OClO over Ny-Ålesund during Winter 1996 and 1997 and Andøya during Winter 1998/1999. J. Atmos. Chem. **43**, 75–106 (2002)

R. Toumi, S. Bekki, The importance of the reactions between OH and ClO for stratospheric ozone. Geophys. Res. Lett. **20**, 2447–2450 (1993)

R. Toumi, BrO as a sink for dimethylsulphide in the marine atmosphere. Geophys. Res. Lett. **21**, 117–120 (1994)

H. Trad, P. Higelin, N. Djebaıli Chaumeix, C. Mounaim-Rousselle, Experimental study and calculations of nitric oxide absorption in the γ (0, 0) and γ (1, 0) bands for strong temperature conditions. J. Quant. Spectrosc. Radiat. Transf. **90**(3–4), 275–289 (2005)

H.G. Tremmel, W. Junkermann, F. Slemr, U. Platt, On the distribution of hydrogen peroxide in the lower troposphere over the Northeast United States during late summer 1988. J. Geophys. Res. **98**, 1083–1099 (1993)

S. Trick, The formation of nitrous acid on urban surfaces—a physical-chemical perspective. Ph.D. Thesis, University of Heidelberg (2004)

M. Trolier, R.L. Mauldin III., A.R. Ravishankara, Rate coefficients for the termolecular channel of the self-reaction of ClO (263 K). J. Phys. Chem. **94**, 4896–4907 (1990)

B. Trost, J. Stutz, U. Platt, UV-absorption cross sections of a series of monocyclic aromatic compounds. Atmos. Environ. **31**(23), 3999–4008 (1997). ISSN 1352-2310. https://doi.org/10.1016/S1352-2310(97)00214-8

C. Tsai, C. Wong, S. Hurlock, O. Pikelnaya, L.H. Mielke, H.D. Osthoff, J.H. Flynn, C. Haman, B. Lefer, J. Gilman, J. de Gouw, J. Stutz, Nocturnal loss of NO_X during the 2010 CalNex-LA study in the Los Angeles Basin. J. Geophys. Res.: Atmos. **119**(22), 13,004–13,025 (2014). https://doi.org/10.1002/2014JD022171

C. Tsai, M. Spolaor, S.F. Colosimo, O. Pikelnaya, R. Cheung, E. Williams, J.B. Gilman, B.M. Lerner, R.J. Zamora, C. Warneke, J.M. Roberts, R. Ahmadov, J. de Gouw, T. Bates, P.K. Quinn, J. Stutz, Nitrous acid formation in a snow-free wintertime polluted rural area. Atmos. Chem. Phys. **18**(3), 1977–1996 (2018). https://doi.org/10.5194/acp-18-1977-2018. https://acp.copernicus.org/articles/18/1977/2018/

E.C. Tuazon, A.M. Winer, R.A. Graham, J.N. Pitts, Atmospheric measurements of trace pollutants by kilometer-pathlength FT-IR spectroscopy. Environ. Sci. Technol. **10**, 259–299 (1980)

E.C. Tuazon, A.M. Winer, J.N. Pitts, Trace pollutant concentrations in a multiday smog episode in the California South Coast Air Basin by long path length Fourier transform IR spectroscopy. Environ. Sci. Technol. **15**, 1232–1237 (1981)

M.E. Tucceri, D. Hölscher, A. Rodriguez, T.J. Dillon, J.N. Crowley, Absorption cross section and photolysis of OIO. Phys. Chem. Chem. Phys. **8**, 834–846 (2006)

M. Tuckermann, Troposphärische DOAS-Messungen zum halogenkatalysierten Ozonabbau im arktischen Frühjahr (Ny-Ålesund, Svalbard). Diploma Thesis in physics, University of Heidelberg (1996)

M. Tuckermann, R. Ackermann, C. Gölz, H. Lorenzen-Schmidt, T. Senne, J. Stutz, B. Trost, W. Unold, U. Platt, DOAS-observation of halogen radical-catalyzed Arctic boundary layer ozone destruction during the ARCTOC-campaigns 1995 and 1996 in Ny-Ålesund. Spitsbergen. Tellus B **49**, 533–555 (1997)

K. Tuite, N. Brockway, S.F. Colosimo, K. Grossmann, C. Tsai, J. Flynn, S. Alvarez, M. Erickson, G. Yarwood, U. Nopmongcol, J. Stutz, Iodine catalyzed ozone destruction at the Texas coast and Gulf of Mexico. Geophys. Res. Lett. 7800–7807 (2018). https://doi.org/10.1029/2018GL078267

K. Tuite, J.L. Thomas, P.R. Veres, J.M. Roberts, P.S. Stevens, S.M. Griffith, S. Dusanter, J.H. Flynn, S. Ahmed, L. Emmons, S. Kim, R. Washenfelder, C. Young, C. Tsai, O. Pikelnaya, J. Stutz, Quantifying nitrous acid formation mechanisms using measured vertical profiles during the CalNex 2010 campaign and 1D column modeling. J. Geophys. Res.: Atmos. **126**(13) (2021). https://doi.org/10.1029/2021JD034689

G.S. Tyndall, J.J. Orlando, J.G. Calvert, Upper limit for the rate coefficient for the reaction $HO_2 + NO_2 \rightarrow HONO + O_2$. Environ. Sci. Technol. **29**, 202–206 (1995)

M. Tzortziou, J.R. Herman, A. Cede, N. Abuhassan, High precision, absolute total column ozone measurements from the Pandora spectrometer system: comparisons with data from a Brewer double monochromator and Aura OMI. J. Geophys. Res.: Atmos. **117**(D16) (2012). https://doi.org/10.1029/2012JD017814

V.S. Ulshöfer, Photochemische Produktion von Carbonylsulfid im Oberflächenwasser der Ozeane und Gasaustausch mit der Atmosphäre. Ph.D. Thesis, University of Heidelberg (1995)

V.S. Ulshöfer, M.O. Andreae, Carbonyl sulfide (COS) in the surface ocean and the atmospheric COS budget. Aquat. Geochem. **3**, 283–303 (1998)

W. Unold, Bodennahe Messungen von Halogenoxiden in der Arktis. Diploma Thesis, University of Heidelberg (1995)

F.P.J. Valero, A. Marshak, P. Minnis, Lagrange point missions: the key to next generation integrated earth observations: DSCOVR innovation. Front. Remote Sens. **2** (2021). https://doi.org/10.3389/frsen.2021.745938

A. Van Amerongen, H. Visser, R. Vink, T. Coppens, R. Hoogeveen, Development of immersed diffraction grating for the Tropomi-SWIR spectrometer, in *Sensors, systems, and next-generation satellites XIV*, vol. 7826 (SPIE, 2010), pp. 345–352

H.C. van de Hulst, *Multiple Light Scattering, Tables, Formulas and Applications*, vol. 1 (ISBN 0-12-710701-0) and vol. 2 (ISBN 0-12-710702-9) (Academic, New York, 1980)

R.J. van der A, D.H.M.U. Peters, H. Eskes, K.F. Boersma, M. van Roozendael, I. De Smedt, H.M. Kelder, Detection of the trend and seasonal variation in tropospheric NO_2 over China. J. Geophys. Res. **111**, D12317 (2006). https://doi.org/10.1029/2005JD006594

I.R. van der Velde, G.R. van der Werf, S. Houweling, H.J. Eskes, J.P. Veefkind, T. Borsdorff, I. Aben, Biomass burning combustion efficiency observed from space using measurements of CO and NO_2 by the TROPOspheric Monitoring Instrument (TROPOMI). Atmos. Chem. Phys. **21**(2), 597–616 (2021). ISSN 1680-7316. https://doi.org/10.5194/acp-21-597-2021

J.M. Van Doren, L.R. Watson, P. Davidovits, D.R. Worsnop, M.S. Zahniser, C.E. Kolb, Temperature dependence of the uptake coefficients of HNO_3, HCl and N_2O_5 by water droplets. J. Phys. Chem. **94**, 3265–3269 (1990)

M. Van Roozendael, C. Hermans, M. DeMaziere, P.C. Simon, Stratospheric NO_2 observations at the Jungfraujoch Station between June 1990 and May 1992. Geophys. Res. Lett. **21**, 1383–1386 (1994a)

M. Van Roozendael, M. de Maziere, S. Simon, Ground-based visible measurements at the Jungfraujoch station since 1990. J. Quant. Spectrosc. Radiat. Transf. **52**, 231–240 (1994b)

M. Van Roozendael, C. Fayt, D. Bolsee, P.C. Simon, M. Gil, M. Yela, J. Cacho, Ground-based stratospheric NO_2 monitoring at Keflavik (Iceland) during EASOE. Geophys. Res. Lett. **21**, 1379–1382 (1994c)

M. Van Roozendael, M. De Mazière, C. Hermans, P.C. Simon, J.P. Pommereau, F. Goutail, X.X. Tie, G.P. Brasseur, C. Granier, Ground-based observations of stratospheric NO_2 at high and mid-latitudes in Europe after the Mount Pinatubo eruption. J. Geophys. Res. **102**, 19171–19176 (1997)

M. Van Roozendael, T. Wagner, A. Richter, I. Pundt, D.W. Arlander, J.P. Burrows, M. Chipperfield, C. Fayt, P.V. Johnston, J.-C. Lambert, K. Kreher, K. Pfeilsticker, U. Platt, J.-P. Pommereau, B.-M. Sinnhuber, K.K. Tørnkvist, F. Wittrock, Intercomparison of BrO measurements from ERS-2 GOME, ground-based and balloon platforms. Adv. Space Res. **29**, 1661–1666 (2002)

M. Van Roozendael, C. Fayt, P. Post, C. Hermans, J.-C. Lambert, Retrieval of BrO and NO_2 from UV-visible observations, in *Sounding the Troposphere from Space: A New Era for Atmospheric Chemistry*, ed. by P. Borell, et al. (Springer, 2003)

A.C. Vandaele, M. Carleer, R. Colin, P.C. Simon, Detection of urban O_3, NO_2, H_2CO and SO_2 using Fourier transform spectroscopy, in *Proceedings of International Symposium on Environmental Sensing, European Optical Society, Berlin, 22–26 June, SPIE, Bellingham, USA* (1992), pp. 288–292. ISBN 0–8194–0880–8

A.C. Vandaele, P.C. Simon, J.M. Guilmot, M. Carleer, R. Colin, SO_2 absorption cross section measurement in the UV using a Fourier transform spectrometer. J. Geophys. Res. **99**, 25599–25605 (1994)

A. Vandaele, C. Hermans, P. Simon, M. Van Roozendael, J. Guilmot, M. Carleer, R. Colin, Fourier transform measurements of NO_2 absorption cross-section in the visible range at room temperature. J. Atmos. Chem. **25**, 289–305 (1996)

A.C. Vandaele, C. Hermans, P.C. Simon, M. Carleer, R. Colin, S. Fally, M.F. Merienne, A. Jenouvrier, B. Coquart, Measurements of the NO_2 absorption cross section from $42,000\,cm^{-1}$ to $10,000\,cm^{-1}$ (238–1,000 nm) at 220 K and 294 K. J. Quant. Spectrosc. Radiat. Transf. **59**, 171–184 (1998)

A.C. Vandaele, C. Hermans, S. Fally, M. Carleer, M.-F. Mérienne, A. Jenouvrier, B. Coquart, R. Colin, Absorption cross-sections of NO_2: simulation of temperature and pressure effects. J. Quant. Spectrosc. Radiat. Transf. **76**, 373–391 (2003). https://doi.org/10.1016/S0022-4073(02)00064-X

A. Vandaele, C. Hermans, S. Fally, Fourier transform measurements of SO_2 absorption cross sections: Ii.: temperature dependence in the 29000–44000 cm^{-1} (227–345 nm) region. J. Quant. Spectrosc. Radiat. Transf. **110**(18), 2115–2126 (2009). https://doi.org/10.1016/j.jqsrt.2009.05.006

J. Vaquero-Martinez, M. Anton, K.L. Chan, D. Loyola, Evaluation of water vapor product from TROPOMI and GOME-2 satellites against ground-based GNSS data over Europe. Atmosphere **13**(7) (2022). ISSN 2073-4433. https://doi.org/10.3390/atmos13071079

E. Vargas-Rodríguez, H. Rutt, Design of CO, CO_2 and CH_4 gas sensors based on correlation spectroscopy using a Fabry-Perot interferometer. Sens. Actuators, B Chem. **137**(2), 410–419 (2009). https://doi.org/10.1016/j.snb.2009.01.013

A. Vasilkov, W. Qin, N. Krotkov, L. Lamsal, R. Spurr, D. Haffner, J. Joiner, E.-S. Yang, S. Marchenko, Accounting for the effects of surface BRDF on satellite cloud and trace-gas retrievals: a new approach based on geometry-dependent Lambertian equivalent reflectivity applied to OMI algorithms. Atmos. Meas. Tech. **10**(1), 333–349 (2017). https://doi.org/10.5194/amt-10-333-2017

J. Vaughan, *The Fabry-Perot Interferometer* (Adam Hilger, Bristol and Philadelphia, 1989)

G. Vaughan, H.K. Roscoe, L.M. Bartlett, F.M. O'Connor, A. Sarkissian, M. Van Roozendael, J.C. Lambert, P.C. Simon, K. Karlsen, B.A. Kastad Hoiskar, D.J. Fish, R.L. Jones, R.A. Freshwater, J.P. Pommereau, F. Goutail, S.B. Andersen, D.G. Drew, P.A. Huges, D. Moore, J. Mellquist, E. Hegels, T. Klüpfel, F. Erle, K. Pfeilsticker, U. Platt, An intercomparison of ground-based UV-visible sensors of ozone and NO_2. J. Geophys. Res. **102**, 1411–1422 (1997)

VDI, Fernmessverfahren—Messungen in der bodennahen Atmosphäre nach dem DOAS-Prinzip—Messen gasförmiger Emissionen und Immissionen—Grundlagen, VDI Richtlinie VDI 4212, Blatt 1. (Remote sensing—atmospheric measurements near ground with DOAS—measurements of emissions and ambient air—fundamentals, VDI Guideline VDI 4212, 1) (2005)

J. Veefkind, I. Aben, K. McMullan, H. Förster, J. de Vries, G. Otter, J. Claas, H. Eskes, J. de Haan, Q. Kleipool, M. van Weele, O. Hasekamp, R. Hoogeveen, J. Landgraf, R. Snel, P. Tol, P. Ingmann, R. Voors, B. Kruizinga, R. Vink, H. Visser, P. Levelt, TROPOMI on the ESA Sentinel-5 Precursor: A GMES mission for global observations of the atmospheric composition for climate, air quality and ozone layer applications. Remote Sens. Environ. **120**, 70–83 (2012). https://doi.org/10.1016/j.rse.2011.09.027

J.P. Veefkind, J.F. de Haan, M. Sneep, P.F. Levelt, Improvements to the OMI O_2-O_2 operational cloud algorithm and comparisons with ground-based radar-lidar observations. Atmos. Meas. Tech. **9**(12), 6035–6049 (2016). https://doi.org/10.5194/amt-9-6035-2016

V. Veitel, O. Funk, C. Kurz, U. Platt, Pfeilsticker, K: Geometrical path length probability density function of the skylight transmitted by mid-latitude cloudy skies; some case studies. Geophys. Res. Lett. **25**, 3355–3358 (1998)

H. Veitel, Vertical profiles of NO_2 and HONO in the boundary layer. Ph.D. Thesis, University of Heidelberg (2002)

H. Veitel, B. Kromer, M. Mößner, U. Platt, New techniques for measurements of atmospheric vertical trace gas profiles using DOAS. Environ. Sci. Pollut. Res. **4**, 17–26 (2002)

T. Verhoelst, S. Compernolle, G. Pinardi, J.-C. Lambert, H.J. Eskes, K.-U. Eichmann, A.M. Fjæraa, J. Granville, S. Niemeijer, A. Cede, M. Tiefengraber, F. Hendrick, A. Pazmiño, A. Bais, A. Bazureau, K.F. Boersma, K. Bognar, A. Dehn, S. Donner, A. Elokhov, M. Gebetsberger, F. Goutail, M. Grutter de la Mora, A. Gruzdev, M. Gratsea, G.H. Hansen, H. Irie, N. Jepsen, Y. Kanaya, D. Karagkiozidis, R. Kivi, K. Kreher, P.F. Levelt, C. Liu, M. Müller, M. Navarro Comas, A.J.M. Piters, J.-P. Pommereau, T. Portafaix, C. Prados-Roman, O. Puentedura, R. Querel, J. Remmers, A. Richter, J. Rimmer, C. Rivera Cárdenas, L. Saavedra de Miguel, V.P. Sinyakov, W. Stremme, K. Strong, M. van Roozendael, J.P. Veefkind, T. Wagner, F. Wittrock, M. Yela González, C. Zehner, Ground-based validation of the Copernicus Sentinel-5P TROPOMI NO_2 measurements with the NDACC ZSL-DOAS, MAX-DOAS and Pandonia global networks. Atmos. Meas. Tech. **14**(1), 481–510 (2021). https://doi.org/10.5194/amt-14-481-2021

F. Vita, C. Kern, S. Inguaggiato, Development of a portable active long-path differential optical absorption spectroscopy system for volcanic gas measurements. J. Sens. Sens. Syst. **3**(2), 355–367 (2014). https://doi.org/10.5194/jsss-3-355-2014

A.L. Vitushkin, L.F. Vitushkin, Design of a multipass optical cell based on the use of a shift corner cubes and rightangle prisms. Appl. Opt. **37**, 162–165 (1998)

L. Vogel, H. Sihler, J. Lampel, T. Wagner, U. Platt, Retrieval interval mapping: a tool to visualize the impact of the spectral retrieval range on differential optical absorption spectroscopy evaluations. Atmos. Meas. Tech. **6**(2), 275–299 (2013). https://doi.org/10.5194/amt-6-275-2013

S.S. Vogt, R.G. Tull, P. Kelton, Self-scanned photodiode array: high performance operation in high dispersion astronomical spectrophotometry. Appl. Opt. **17**, 574–592 (1978)
R. Vogt, B.J. Finlayson-Pitts, Tropospheric HONO and reactions of oxides of nitrogen with NaCl. Geophys. Res. Lett. **21**(21), 2291–2294 (1994)
R. Vogt, P.J. Crutzen, R. Sander, A mechanism for halogen release from sea-salt aerosol in the remote marine boundary layer. Nature **383**, 327–330 (1996)
S. Voigt, J. Orphal, K. Bogumil, J.P. Burrows, The temperature dependence (203–293 K) of the absorption cross sections of O_3 in the 230–850 nm region measured by Fourier-transform spectroscopy. J. Photochem. Photobiol., A: Chem. **143**, 1–9 (2001)
S. Voigt, J. Orphal, J.P. Burrows, The temperature- and pressure-dependence of the absorption cross-section of NO_2 in the 250–800 nm region measured by Fourier-transform spectroscopy. J. Photochem. Photobiol., A: Chem. **149**, 1–7 (2002)
R. Volkamer, T. Etzkorn, A. Geyer, U. Platt, Correction of the oxygen interference with UV spectroscopic (DOAS) measurements of monocyclic aromatic hydrocarbons in the atmosphere. Atmos. Environ. **32**(21), 3731–3747 (1998). https://doi.org/10.1016/S1352-2310(98)00095-8
R. Volkamer, A DOAS study on the oxidation mechanism of aromatic hydrocarbons under simulated atmospheric conditions. Ph.D. Thesis, University of Heidelberg (2001)
R. Volkamer, U. Platt, K. Wirtz, Primary and secondary glyoxal formation from aromatics: experimental evidence for the bicycloalkyl-radical pathway from BTX. J. Phys. Chem. A **105**, 7865–7874 (2001)
R. Volkamer, B. Klotz, I. Barnes, T. Imamura, K. Wirtz, N. Washida, K.-H. Becker, U. Platt, OH-initiated oxidation of benzene: I. Phenol formation under atmospheric conditions. Phys. Chem. Chem. Phys. **4**, 1598–1610 (2002)
R. Volkamer, P. Spietz, J.P. Burrows, U. Platt, High-resolution absorption cross-section of glyoxal in the UV/vis and IR spectral ranges. J. Photochem. Photobiol., A: Chem. **172**, 35–46 (2005a). https://doi.org/10.1016/j.jphotochem.2004.11.011
R. Volkamer, L.T. Molina, M.J. Molina, T. Shirley, W.H. Brune, DOAS measurement of glyoxal as an indicator for fast VOC chemistry in urban air. Geophys. Res. Lett. **32**, L08806 (2005b). https://doi.org/10.1029/2005GL022616
R. Volkamer, S. Baidar, T.L. Campos, S. Coburn, J.P. DiGangi, B. Dix, E.W. Eloranta, T.K. Koenig, B. Morley, I. Ortega, B.R. Pierce, M. Reeves, R. Sinreich, S. Wang, M.A. Zondlo, P.A. Romashkin, Aircraft measurements of BrO, IO, glyoxal, NO_2, H_2O, O_2-O_2 and aerosol extinction profiles in the tropics: comparison with aircraft-/ship-based in situ and lidar measurements. Atmos. Meas. Tech. **8**(5), 2121–2148 (2015). https://doi.org/10.5194/amt-8-2121-2015
H. Volten, J.B. Bergwerff, M. Haaima, D.E. Lolkema, A.J.C. Berkhout, G.R. van der Hoff, C.J.M. Potma, R.J. Wichink Kruit, W.A.J. van Pul, D.P.J. Swart, Two instruments based on differential optical absorption spectroscopy (DOAS) to measure accurate ammonia concentrations in the atmosphere. Atmos. Meas. Tech. **5**(2), 413–427 (2012). https://doi.org/10.5194/amt-5-413-2012
A. Volz, D.H. Ehhalt, R.G. Derwent, Seasonal and latitudinal variation of ^{14}CO and the tropospheric concentration of OH radicals. J. Geophys. Res. **86**, 5163–5171 (1981)
A. Volz, D. Kley, Ozone measurements in the 19th century: an evaluation of the Montsouris series. Nature **332**, 240–242 (1988)
C. von Friedeburg, Derivation of trace gas information combining differential optical absorption spectroscopy with radiative transfer modelling. Ph.D. thesis, University of Heidelberg (2003)
C. von Friedeburg, I. Pundt, K.U. Mettendorf, T. Wagner, U. Platt, Multi-axis-DOAS measurements of NO_2 during the BAB II motorway emission campaign. Atmos. Env. **39**(5), 977–985 (2005)
C. von Friedeburg, T. Wagner, A. Geyer, N. Kaiser, B. Vogel, H. Vogel, U. Platt, Derivation of troposphere NO_3 profiles using Off-axis-DOAS measurements during sunrise and comparison with simulations. J. Geophys. Res. **107**, D13 (2002). https://doi.org/10.1029/2001JD000481
R. von Glasow, R. Sander, A. Bott, P.J. Crutzen, Modeling halogen chemistry in the marine boundary layer. 1. Cloud-free MBL. J. Geophys. Res. **107D**, 4341 (2002). https://doi.org/10.1029/2001JD000942

R. von Glasow, P.J. Crutzen, Tropospheric halogen chemistry, in *The Atmosphere: Treatise on Geochemistry Holland*, vol. 4, ed. by R.F. Keeling, K.K. Turekian (Elsevier-Pergamon, Oxford, 2003)

R. von Glasow, P.J. Crutzen, Model study of multiphase DMS oxidation with a focus on halogens. Atmos. Chem. Phys. **4**, 589–608 (2004)

R. von Glasow, R. von Kuhlmann, M.G. Lawrence, U. Platt, P.J. Crutzen, Impact of reactive bromine chemistry in the troposphere. Atmos. Chem. Phys. **4**, 2481–2497 (2004)

R. von Glasow, P.J. Crutzen, Tropospheric halogen chemistry, in *The Atmosphere*, ed. by R.F. Keeling, vol. 4 (Elsevier, 2007), pp. 21–64. https://doi.org/10.1016/B0-08-043751-6/04141-4

C. von Savigny, C.S. Haley, C.E. Sioris, I.C. McDade, E.J. Llewellyn, D. Degenstein, W.F.J. Evans, R.L. Gattinger, E. Griffioen, E. Kyrölä, N.D. Lloyd, J.C. McConnell, C.A. McLinden, G. Mégie, D.P. Murtagh, B. Solheim, K. Strong, Stratospheric ozone profiles retrieved from limb scattered sunlight radiance spectra measured by the OSIRIS instrument on the Odin satellite. Geophys. Res. Lett. **30**(14) (2003). https://doi.org/10.1029/2002GL016401

M. Vountas, V.V. Rozanov, J.P. Burrows, Impact of Raman scattering on radiative transfer in earth's atmosphere. J. Quant. Spectrosc. Radiat. Transf. **60**(6), 943–961 (1998)

M. Vountas, A. Richter, F. Wittrock, J.P. Burrows, Inelastic scattering in ocean water and its impact on trace gas retrievals from satellite data. Atmos. Chem. Phys. **3**, 1365–1375 (2003)

M. Vrekoussis, F. Wittrock, A. Richter, J.P. Burrows, Temporal and spatial variability of glyoxal as observed from space. Atmos. Chem. Phys. **9**(13), 4485–4504 (2009). https://doi.org/10.5194/acp-9-4485-2009

M. Vrekoussis, F. Wittrock, A. Richter, J.P. Burrows, GOME-2 observations of oxygenated VOCs: what can we learn from the ratio glyoxal to formaldehyde on a global scale? Atmos. Chem. Phys. **10**(21), 10145–10160 (2010). https://doi.org/10.5194/acp-10-10145-2010

H.E. Wagner, *Bestimmung der Spurengaszusammensetzung in der Troposphäre mit Langpfad-DOAS (Differenzielle Optische Absorptionsspektroskopie), Diploa thesis in physics* (Inst. f. Physikalische und Theoretische Chemie, TU München, 1990)

T. Wagner, T. Senne, F. Erle, C. Otten, J. Stutz, K. Pfeilsticker, U. Platt, Determination of cloud properties and cloud type with DOAS measurements, in *Poster, Third European Symposium on Polar Strat. Ozone, Schliersee, 18–22 September* (1995)

T. Wagner, U. Platt, Observation of tropospheric BrO from the GOME satellite. Nature **395**, 486–490 (1998)

T. Wagner, F. Erle, L. Marquard, C. Otten, K. Pfeilsticker, T. Senne, J. Stutz, U. Platt, Cloudy sky photon path lengths as derived from DOAS observations. J. Geophys. Res. **103**, 25307–25321 (1998)

T. Wagner, C. Otten, U. Platt, Observation of atmospheric NO_3 in the Arctic winter. Geophys. Res. Lett. **27**(21), 3441–3444 (2000)

T. Wagner, C. Leue, K. Pfeilsticker, U. Platt, Monitoring of the stratospheric chlorine activation by global ozone monitoring experiment (GOME) OClO measurements in the austral and boreal winters 1995 through 1999. J. Geophys. Res. **106**, 4971–4986 (2001a)

T. Wagner, C. Leue, M. Wenig, K. Pfeilsticker, U. Platt, Spatial and temporal distribution of enhanced boundary layer BrO concentrations measured by the GOME instrument aboard ERS-2. J. Geophys. Res. **106**, 24225–24235 (2001b)

T. Wagner, K. Chance, U. Frieß, M. Gil, F. Goutail, G. Hönninger, P.V. Johnston, K. Karlsen-Tørnkvist, I. Kostadinov, H. Leser, A. Petritoli, A. Richter, M. Van Roozendael, U. Platt, Correction of the Ring effect and I_o-effect for DOAS observations of scattered sunlight, ESA Technical Report (2002a)

T. Wagner, F. Wittrock, A. Richter, M. Wenig, J.P. Burrows, U. Platt, Continuous monitoring of the high and persistent chlorine activation during the Arctic winter 1999/2000 by the GOME instrument on ERS-2. J. Geophys. Res. **107**(D20), 8267 (2002b). https://doi.org/10.1029/2001JD000466

T. Wagner, V.C. Friedeburg, M. Wenig, C. Otten, U. Platt, UV/vis observation of atmospheric O_4 absorptions using direct moon light and zenith scattered sunlight under clear and cloudy sky conditions. J. Geophys. Res. **107**(D20), 4424 (2002c). https://doi.org/10.1029/2001JD001026

T. Wagner, J. Heland, M. Zöger, U. Platt, A fast H_2O total column density product from GOME-validation with in-situ aircraft measurements. Atmos. Chem. Phys. **3**, 651–663 (2003)

T. Wagner, B. Dix, C. von Friedeburg, U. Frieß, S. Sanghavi, R. Sinreich, U. Platt, MAX-DOAS O_4 measurements: a new technique to derive information on atmospheric aerosols-Principles and information content. J. Geophys. Res.: Atmos. **109**(D22) (2004). https://doi.org/10.1029/2004JD004904

T. Wagner, S. Beirle, M. Grzegorski, T. Deutschmann, U. Platt, Global distribution of atmospheric photon path lengths derived from GOME and SCIAMACHY (2007a). 'Envisat Symposium 2007', Montreux, Switzerland 23-27 April 2007 (ESA SP-636, July 2007)

T. Wagner, J.P. Burrows, T. Deutschmann, B. Dix, C. von Friedeburg, U. Frieß, F. Hendrick, K.-P. Heue, H. Irie, H. Iwabuchi, Y. Kanaya, J. Keller, C.A. McLinden, H. Oetjen, E. Palazzi, A. Petritoli, U. Platt, O. Postylyakov, J. Pukite, A. Richter, M. van Roozendael, A. Rozanov, V. Rozanov, R. Sinreich, S. Sanghavi, F. Wittrock, Comparison of box-air-mass-factors and radiances for multiple-axis differential optical absorption spectroscopy (MAX-DOAS) geometries calculated from different UV/visible radiative transfer models. Atmos. Chem. Phys. **7**(7), 1809–1833 (2007b). https://doi.org/10.5194/acp-7-1809-2007

T. Wagner, O. Ibrahim, R. Sinreich, U. Frieß, R. von Glasow, U. Platt, Enhanced tropospheric BrO over Antarctic sea ice in mid winter observed by MAX-DOAS on board the research vessel Polarstern. Atmos. Chem. Phys. **7**(12), 3129–3142 (2007). https://doi.org/10.5194/acp-7-3129-2007

T. Wagner, S. Beirle, T. Deutschmann, E. Eigemeier, C. Frankenberg, M. Grzegorski, C. Liu, T. Marbach, U. Platt, M.P. de Vries, Monitoring of atmospheric trace gases, clouds, aerosols and surface properties from UV/vis/NIR satellite instruments. J. Opt. A: Pure Appl. Opt. **10**(10), 104019 (2008). https://doi.org/10.1088/1464-4258/10/10/104019

T. Wagner, S. Beirle, T. Brauers, T. Deutschmann, U. Frieß, C. Hak, J.D. Halla, K.P. Heue, W. Junkermann, X. Li, U. Platt, I. Pundt-Gruber, Inversion of tropospheric profiles of aerosol extinction and HCHO and NO_2 mixing ratios from MAX-DOAS observations in Milano during the summer of 2003 and comparison with independent data sets. Atmos. Meas. Tech. **4**(12), 2685–2715 (2011). https://doi.org/10.5194/amt-4-2685-2011

T. Wagner, M.O. Andreae, S. Beirle, S. Dörner, K. Mies, R. Shaiganfar, MAX-DOAS observations of the total atmospheric water vapour column and comparison with independent observations. Atmos. Meas. Tech. **6**(1), 131–149 (2013a). https://doi.org/10.5194/amt-6-131-2013

T. Wagner, S. Beirle, H. Sihler, K. Mies, A feasibility study for the retrieval of the total column precipitable water vapour from satellite observations in the blue spectral range. Atmos. Meas. Tech. **6**(10), 2593–2605 (2013b). https://doi.org/10.5194/amt-6-2593-2013

T. Wagner, S. Beirle, N. Benavent, T. Bösch, K.L. Chan, S. Donner, S. Dörner, C. Fayt, U. Frieß, D. García-Nieto, C. Gielen, D. González-Bartolome, L. Gomez, F. Hendrick, B. Henzing, J.L. Jin, J. Lampel, J. Ma, K. Mies, M. Navarro, E. Peters, G. Pinardi, O. Puentedura, J. Puķīte, J. Remmers, A. Richter, A. Saiz-Lopez, R. Shaiganfar, H. Sihler, M. Van Roozendael, Y. Wang, and M. Yela. Is a scaling factor required to obtain closure between measured and modelled atmospheric O_4 absorptions? An assessment of uncertainties of measurements and radiative transfer simulations for 2 selected days during the MAD-CAT campaign. Atmos. Meas. Tech. **12**(5), 2745–2817 (2019). https://doi.org/10.5194/amt-12-2745-2019

T. Wagner, S. Dörner, S. Beirle, S. Donner, S. Kinne, Quantitative comparison of measured and simulated O_4 absorptions for one day with extremely low aerosol load over the tropical Atlantic. Atmos. Meas. Tech. **14**(5), 3871–3893 (2021a). https://doi.org/10.5194/amt-14-3871-2021

T. Wagner, S. Beirle, S. Dörner, C. Borger, R. Van Malderen, Identification of atmospheric and oceanic teleconnection patterns in a 20-year global data set of the atmospheric water vapour column measured from satellites in the visible spectral range. Atmos. Chem. Phys. **21**(7), 5315–5353 (2021b). https://doi.org/10.5194/acp-21-5315-2021

A. Wahner, R.O. Jakoubek, G.H. Mount, A.R. Ravishankara, A.L. Schmeltekopf, Remote sensing observations of nighttime OClO column during the airborne antarctic ozone experiment, September 8, 1987. J. Geophys. Res. **94**, 11405–11411 (1989a)

A. Wahner, R.O. Jakoubek, G.H. Mount, A.R. Ravishankara, A.L. Schmeltekopf, Remote sensing observations of daytime column NO_2 during the airborne Antarctic ozone experiment, August 22 to October 2, 1987. J. Geophys. Res. **94**, 16619–16632 (1989b)

A. Wahner, J. Callies, H.-P. Dorn, U. Platt, C. Schiller, Near UV atmospheric absorption measurements of column abundances during airborne Arctic stratospheric expedition, January–February 1989: 1. Technique and NO_2 observations. Geophys. Res. Lett. **17**, 497–500 (1990a)

A. Wahner, J. Callies, H.-P. Dorn, U. Platt, C. Schiller, Near UV atmospheric absorption measurements of column abundances during airborne Arctic stratospheric expedition, January-February 1989: 3. BrO observations. Geophys. Res. Lett. **17**, 517–520 (1990b)

A. Wahner, C. Schiller, Twilight variation of vertical column abundances of OClO and BrO in the North polar region. J. Geophys. Res. **97**(D8), 8047–8055 (1992)

A. Wahner, T.F. Mentel, M. Sohn, Gas-phase reaction of N_2O_5 with water vapor: importance of heterogeneous hydrolysis of N_2O_5 and surface desorption of HNO_3 in a large teflon chamber. Geophys. Res. Lett. **25**, 2169–2172 (1998a)

A. Wahner, T.F. Mentel, M. Sohn, J. Stier, Heterogeneous reaction of N_2O_5 on sodium nitrate aerosol. J. Geophys. Res. **103**, 31103–31112 (1998b)

H. Walker, D. Stone, T. Ingham, S. Hackenberg, D. Cryer, S. Punjabi, K. Read, J. Lee, L. Whalley, D.V. Spracklen, L.J. Carpenter, S.R. Arnold, D.E. Heard, Observations and modelling of glyoxal in the tropical Atlantic marine boundary layer. Atmos. Chem. Phys. **22**(8), 5535–5557 (2022). https://doi.org/10.5194/acp-22-5535-2022

P.R. Wamsley, J.W. Elkins, D.W. Fahey, G.S. Dutton, C.M. Volk, R.C. Myers, S.A. Montzka, J.H. Butler, A.D. Clarke, P.J. Fraser, L.P. Steele, M.P. Lucarelli, E.L. Atlas, S.M. Schauffler, D.R. Blake, F.S. Rowland, W.T. Sturges, J.M. Lee, S.A. Penkett, A. Engel, R.M. Stimpfle, K.R. Chan, D.K. Weisenstein, M.K.W. Ko, R.J. Salawitch, Distribution of halon-1211 in the upper troposphere and lower stratosphere and the 1994 total bromine budget. J. Geophys. Res. **103**(D1), 1513–1526 (1998). https://doi.org/10.1029/97JD02466

P. Wang, M. Bruns, A. Richter, J.P. Burrows, K.-P. Heue, I. Pundt, T. Wagner, U. Platt, Validation of SCIAMACHY with AMAXDOAS measurements from the DLR Falcon. Geophys. Res. **5**, 09341 (2003)

P. Wang, A. Richter, M. Bruns, V. Rozanov, J. Burrows, K.-P. Heue, T. Wagner, I. Pundt, U. Platt, Measurements of tropospheric NO_2 with an airborne multi-axis DOAS instrument. Atmos. Chem. Phys. **5**, 337–343 (2005)

P. Wang, A. Richter, M. Bruns, J.P. Burrows, R. Scheele, W. Junkermann, K.-P. Heue, T. Wagner, U. Platt, I. Pundt, Airborne multi-axis DOAS measurements of tropospheric SO_2 plumes in the Po-valley. Italy. Atmos. Chem. Phys. **6**(2), 329–338 (2006). https://doi.org/10.5194/acp-6-329-2006

P. Wang, P. Stammes, R. van der A, G. Pinardi, M. van Roozendael, FRESCO+: an improved O_2 A-band cloud retrieval algorithm for tropospheric trace gas retrievals. Atmos. Chem. Phys. **8**(21), 6565–6576 (2008). https://doi.org/10.5194/acp-8-6565-2008

S. Wang, J.A. Schmidt, S. Baidar, S. Coburn, B. Dix, T.K. Koenig, E. Apel, D. Bowdalo, T.L. Campos, E. Eloranta, M.J. Evans, J.P. DiGangi, M.A. Zondlo, R.-S. Gao, J.A. Haggerty, S.R. Hall, R.S. Hornbrook, D. Jacob, B. Morley, B. Pierce, M. Reeves, P. Romashkin, A. ter Schure, R. Volkamer, Active and widespread halogen chemistry in the tropical and subtropical free troposphere. Proc. Natl. Acad. Sci. **112**(30), 9281–9286 (2015). https://doi.org/10.1073/pnas.1505142112

S. Wang, K.A. Pratt, Molecular halogens above the Arctic snowpack: emissions, diurnal variations, and recycling mechanisms. J. Geophys. Res.: Atmos. **122**(21), 11,991–12,007 (2017). https://doi.org/10.1002/2017JD027175

H. Wang, A.H. Souri, G. González Abad, X. Liu, K. Chance, Ozone monitoring instrument (OMI) total column water vapor version 4 validation and applications. Atmos. Meas. Tech. **12**(9), 5183–5199 (2019a). https://doi.org/10.5194/amt-12-5183-2019

Y. Wang, S. Dörner, S. Donner, S. Böhnke, I. De Smedt, R.R. Dickerson, Z. Dong, H. He, Z. Li, Z. Li, D. Li, D. Liu, X. Ren, N. Theys, Y. Wang, Y. Wang, Z. Wang, H. Xu, J. Xu, T. Wagner,

Vertical profiles of NO_2, SO_2, HONO, HCHO, CHOCHO and aerosols derived from MAX-DOAS measurements at a rural site in the central western north china plain and their relation to emission sources and effects of regional transport. Atmos. Chem. Phys. **19**(8), 5417–5449 (2019). https://doi.org/10.5194/acp-19-5417-2019

X. Wang, D.J. Jacob, S.D. Eastham, M.P. Sulprizio, L. Zhu, Q. Chen, B. Alexander, T. Sherwen, M.J. Evans, B.H. Lee, J.D. Haskins, F.D. Lopez-Hilfiker, J.A. Thornton, G.L. Huey, H. Liao, The role of chlorine in global tropospheric chemistry. Atmos. Chem. Phys. **19**(6), 3981–4003 (2019b). https://doi.org/10.5194/acp-19-3981-2019

Y. Wang, A. Apituley, A. Bais, S. Beirle, N. Benavent, A. Borovski, I. Bruchkouski, K.L. Chan, S. Donner, T. Drosoglou, H. Finkenzeller, M.M. Friedrich, U. Frieß, D. Garcia-Nieto, L. Gómez-Martín, F. Hendrick, A. Hilboll, J. Jin, P. Johnston, T.K. Koenig, K. Kreher, V. Kumar, A. Kyuberis, J. Lampel, C. Liu, H. Liu, J. Ma, O.L. Polyansky, O. Postylyakov, R. Querel, A. Saiz-Lopez, S. Schmitt, X. Tian, J.-L. Tirpitz, M. Van Roozendael, R. Volkamer, Z. Wang, P. Xie, C. Xing, J. Xu, M. Yela, C. Zhang, T. Wagner, Inter-comparison of MAX-DOAS measurements of tropospheric HONO slant column densities and vertical profiles during the CINIDI-2 campaign. Atmos. Meas. Tech. **13**(9), 5087–5116 (2020). https://doi.org/10.5194/amt-13-5087-2020

S. Wang, H. Li, NO_3-initiated gas-phase formation of nitrated phenolic compounds in polluted atmosphere. Environ. Sci. Technol. **55**(5), 2899–2907 (2021). https://doi.org/10.1021/acs.est.0c08041

I. Wängberg, T. Etzkorn, I. Barnes, U. Platt, K.H. Becker, Absolute determination of the temperature behaviour of the $NO_2 + NO_3 + (M) \Leftrightarrow N_2O_5 + (M)$ equilibrium. J. Phys. Chem. A **101**(50), 9694–9698 (1997)

W. Wanner, A.H. Strahler, B. Hu, P. Lewis, J.-P. Muller, X. Li, C.L.B. Schaaf, M.J. Barnsley, Global retrieval of bidirectional reflectance and albedo over land from EOS MODIS and MISR data: theory and algorithm. J. Geophys. Res.: Atmos. **102**(D14), 17143–17161 (1997). https://doi.org/10.1029/96JD03295

S. Warnach, N. Bobrowski, S. Hidalgo, S. Arellano, H. Sihler, F. Dinger, P. Lübcke, J. Battaglia, A. Steele, B. Galle, U. Platt, T. Wagner, Variation of the BrO/SO_2 molar ratio in the plume of tungurahua volcano between 2007 and 2017 and its relationship to volcanic activity. Front. Earth Sci. **7** (2019). https://doi.org/10.3389/feart.2019.00132

S. Warnach, Bromine monoxide in volcanic plumes—a global survey of volcanic plume composition and chemistry derived from Sentinel-5 Precursor/TROPOMI data. Ph.D. Thesis, University of Heidelberg (2022). http://www.ub.uni-heidelberg.de/archiv/31910

S. Warnach, H. Sihler, C. Borger, N. Bobrowski, S. Beirle, U. Platt, T. Wagner, A new accurate retrieval algorithm of bromine monoxide columns inside minor volcanic plumes from Sentinel-5 Precursor/TROPOMI observations. Atmos. Meas. Tech. **16**(22), 5537–5573 (2023). https://doi.org/10.5194/amt-16-5537-2023

S. Watts, The mass budgets of carbonyl sulfide, dimethyl sulfide, carbon disulfide and hydrogen sulfide. Atmos. Environ. **34**, 761–779 (2000)

R.P. Wayne, I. Barnes, P. Biggs, J.P. Burrows, C. Canosa-Mas, J. Hjorth, G. Le Bras, G. Moortgat, D. Perner, G. Poulet, G. Restelli, H. Sidebottom, The nitrate radical: physics, chemistry and the atmosphere. Atmos. Environ. **25A**, 1–250 (1991)

R.P. Wayne, G. Poulet, P. Biggs, J.P. Burrows, R.A. Cox, P.J. Crutzen, G.D. Haymann, M.E. Jenkin, G. LeBras, G.K. Moortgat, U. Platt, R.N. Schindler, Halogen oxides: radicals, sources and reservoirs in the laboratory and in the atmosphere. Atmos. Environ. **29**, 2675–2884 (1995)

A. Weaver, S. Solomon, R.W. Sanders, K. Arpag, H.L. Miller, Atmospheric NO_3 5. Off-axis measurements at sunrise: estimates of tropospheric NO_3 at 40 degrees N. J. Geophys. Res. **101**(D13), 18605–18612 (1996)

A. Weber, Applications, in *The Raman Effect*, ed. by A. Anderson, vols. 1 and 2 (Dekker Inc., New York, 1973)

M. Weber, S. Dhomse, F. Wittrock, A. Richter, B.-M. Sinnhuber, J.P. Burrows, Dynamical control of NH and SH winter/spring total ozone from GOME observations in 1995–2002. Geophys. Res. Lett. **30**, 37.1–37.4 (2003). https://doi.org/10.1029/2002GL016799

P. Weibring, S. EdnerH, S. Svanberg, G. Cecchi, L. Pantani, R. Ferrara, T. Caltabiano, Monitoring of volcanic sulphur dioxide emissions using differential absorption lidar (DIAL), differential optical absorption spectroscopy (DOAS) and correlation spectroscopy (COSPEC). Appl. Phys. B **67**(4), 419–426 (1998)

F. Weidner, H. Bösch, H. Bovensmann, J.P. Burrows, A. Butz, C. Camy-Peyret, M. Dorf, K. Gerilowski, W. Gurlit, U. Platt, C. von Friedeburg, T. Wagner, K. Pfeilsticker, Balloon-borne limb profiling of UV/vis skylight radiances, O_3, NO_2, and BrO: technical set-up and validation of the method. Atmos. Chem. Phys. **5**, 1409–1422 (2005)

B. Weinstock, Carbon monoxide: residence time in the atmosphere. Science **166**, 224–225 (1969)

W. Weng, S. Li, M. Aldén, Z. Li, Ultraviolet absorption cross-sections of ammonia at elevated temperatures for nonintrusive quantitative detection in combustion environments. Appl. Spectrosc. **75**(9), 1168–1177 (2021). https://doi.org/10.1177/0003702821990445

M. Wenig, Satellite measurement of long-term global tropospheric trace distributions and source strength—algorithm development and data analysis. Ph.D. Thesis, University of Heidelberg (2001)

M. Wenig, N. Spichtinger, A. Stohl, G. Held, S. Beirle, T. Wagner, B. Jähne, U. Platt, Intercontinental transport of nitrogen oxide pollution plumes. Atmos. Chem. Phys. **3**, 387–393 (2003)

M. Wenig, S. Kühl, S. Beirle, E. Bucsela, B. Jähne, U. Platt, J. Gleason, T. Wagner, Retrieval and analysis of stratospheric NO_2 from the global ozone monitoring experiment. J. Geophys. Res. **109**, D04315 (2004). https://doi.org/10.1029/2003JD003652

M. Wenig, B. Jähne, U. Platt, Operator representation as a new differential optical absorption spectroscopy formalism. Appl. Opt. **44**(16), 3246–3253 (2005)

P.O. Wennberg, T.F. Hanisco, L. Jaeglé, D.J. Jacob, E.J. Hintsa, E.J. Lanzendorf, J.G. Anderson, R.-S. Gao, E.R. Keim, S.G. Donnelly, L.A. Del Negro, D.W. Fahey, S.A. McKeen, R.J. Salawitch, C.R. Webster, R.D. May, R.L. Herman, M.H. Proffitt, J.J. Margitan, E.L. Atlas, S.M. Schauffler, F. Flocke, C.T. McElroy, T.P. Bui, Hydrogen radicals, nitrogen radicals, and the production of O_3 in the upper troposphere. Science **279**, 49–53 (1998)

P.O. Wennberg, Bromine explosion. Nature **397**, 299–301 (1999)

P.W. Werle, K. Josek, F. Slemr, Application of FM spectroscopy in atmospheric trace gas monitoring: a study of some factors influencing the instrument design, in *Measurement of Atmospheric Gases*, vol. 1433, ed. by H.I. Schiff (SPIE, Bellingham, 1991), pp.128–135

P.W. West, G.C. Gaeke, Fixation of sulfur dioxide as disulfitomercurate(II) and subsequent colorimetric estimation. Anal. Chem. **28**, 1816–1819 (1956)

L.K. Whalley, K.L. Furneaux, T. Gravestock, H.M. Atkinson, C.S.E. Bale, T. Ingham, W.J. Bloss, D.E. Heard, Detection of iodine monoxide radicals in the marine boundary layer using laser induced fluorescence spectroscopy. J. Atmos. Chem. **58**(1), 19–39 (2007). https://doi.org/10.1007/s10874-007-9075-9

J.U. White, Long optical paths of large aperture. J. Opt. Soc. Am. **32**, 285–288 (1942)

J.U. White, Very long optical paths in air. J. Opt. Soc. Am. **66**, 411–416 (1976)

WHO (World Health Organization), Monitoring ambient air quality for health impact assessment. WHO Regional Publications, European Series, No. 85. WHO Regional Office for Europe, Copenhagen (1999)

T.C. Wilkes, A.J.S. McGonigle, J.R. Willmott, T.D. Pering, J.M. Cook, Low-cost 3D printed 1 nm resolution smartphone sensor-based spectrometer: instrument design and application in ultraviolet spectroscopy. Opt. Lett. **42**(21), 4323–4326 (2017). https://doi.org/10.1364/OL.42.004323

J.E. Williams, K.F. Boersma, P. Le Sager, W.W. Verstraeten, The high-resolution version of TM5-MP for optimized satellite retrievals: description and validation. Geosci. Model. Dev. **10**(2), 721–750 (2017). https://doi.org/10.5194/gmd-10-721-2017

S.B. Williams, Y. Shan, U. Jazzar, P.S. Kerr, I. Okereke, V.S. Klimberg, D.S. Tyler, N. Putluri, D.S. Lopez, J.D. Prochaska, C. Elferink, J.G. Baillargeon, Y.-F. Kuo, H.B. Mehta, Proximity to oil refineries and risk of cancer: a population-based analysis. JNCI Cancer Spectr. **4**(6), pkaa088 (2020). https://doi.org/10.1093/jncics/pkaa088

D.M. Wilmouth, T.F. Hanisco, N.M. Donahue, J.G. Anderson, Fourier transform ultraviolet spectroscopy of the $A(^2\Pi_{3/2}) \leftarrow X(^2\Pi_{3/2})$ transition of BrO. J. Phys. Chem. **103**, 8935–8944 (1999)

P.H. Wine, A.R. Ravishankara, D.L. Philen, D.D. Davis, R.T. Watson, High resolution absorption cross sections for the $A^2\Pi - X^2\Pi$ system of ClO. Chem. Phys. Lett. **50**, 101–106 (1977)

A.M. Winer, R. Atkinson, J.N. Pitts, Gaseous nitrate radical: possible night-time atmospheric sink for biogenic organic compounds. Science **224**, 156–159 (1984)

A.M. Winer, H.W. Biermann, Measurements of nitrous acid, nitrate radicals, formaldehyde, and nitrogen dioxide for the Southern California air quality study by differential optical absorption spectroscopy, in *Measurement of Atmospheric Gases*, ed. by H. Schiff, vol. 1433, (SPIE, Bellingham, 1991), pp. 44–57

A.M. Winer, H.W. Biermann, Long pathlength differential optical absorption spectroscopy (DOAS) measurements of gaseous HONO, NO_2 and HCHO in the California south coast air basin. Res. Chem. Intermed. **20**(3–5), 423–455 (1994)

O.W. Wingenter, M.K. Kubo, N.J. Blake, T.W. Smith, D.R. Blake, F.S. Rowland, Hydrocarbon and halocarbon measurements as photochemical and dynamical indicators of atmospheric hydroxyl, atomic chlorine, and vertical mixing obtained during Lagrangian flights. J. Geophys. Res. **101**, 4331–4340 (1996)

T. Winterrath, T. Kurosu, A. Richter, J.P. Burrows, O_3 and NO_2 in thunderstorm clouds: convection or production? Geophys. Res. Lett. **26**, 1291–1294 (1999)

W.J. Wiscombe, Improved Mie scattering algorithms. Appl. Opt. **19**, 1505–1509 (1980)

F. Wittrock, R. Müller, A. Richter, H. Bovensmann, J.P. Burrows, Measurement of iodine oxide (IO) above Spitsbergen. Geophys. Res. Lett. **27**(10), 1471–1474 (2000a)

F. Wittrock, A. Richter, A. Ladstätter-Weißenmayer, J.P. Burrows, Global observations of formaldehyde, in *Proceedings of ERS-ENVISAT Symposium, Gothenburg, October 2000*. ESA publication SP-461 (2000b)

F. Wittrock, H. Oetjen, A. Richter, S. Fietkau, T. Medeke, A. Rozanov, J.P. Burrows, MAX-DOAS measurements of atmospheric trace gases in Ny-Ålesund. Atmos. Chem. Phys. **4**, 955–966 (2004)

F. Wittrock, A. Richter, H. Oetjen, J.P. Burrows, M. Kanakidou, S. Myriokefalitakis, R. Volkamer, S. Beirle, U. Platt, T. Wagner, Simultaneous global observations of glyoxal and formaldehyde from space. Geophys. Res. Lett. **33**(16) (2006). https://doi.org/10.1029/2006GL026310

WMO, WDCGG Data Summary, WDCGG No. 29, GAW DATA, vol. IV-Greenhouse Gases and Other Atmospheric Gases. Japan Meteorological Agency in Co-Operation with World Meteorological Organisation (2005)

A. Wolanin, V.V. Rozanov, T. Dinter, S. Noel, M. Vountas, J.P. Burrows, A. Bracher, Global retrieval of marine and terrestrial chlorophyll fluorescence at its red peak using hyperspectral top of atmosphere radiance measurements: feasibility study and first results. Remote Sens. Environ. **166**, 243–261 (2015). https://doi.org/10.1016/j.rse.2015.05.018

K.W. Wong, H.-J. Oh, B.L. Lefer, B. Rappenglück, J. Stutz, Vertical profiles of nitrous acid in the nocturnal urban atmosphere of Houston. TX. Atmos. Chem. Phys. **11**(8), 3595–3609 (2011). https://doi.org/10.5194/acp-11-3595-2011

K.W. Wong, C. Tsai, B. Lefer, C. Haman, N. Grossberg, W.H. Brune, X. Ren, W. Luke, J. Stutz, Daytime HONO vertical gradients during SHARP 2009 in Houston. TX. Atmos. Chem. Phys. **12**(2), 635–652 (2012). https://doi.org/10.5194/acp-12-635-2012

E.C. Wood, P.J. Wooldridge, J.H. Freese, T. Albrecht, R.C. Cohen, Prototype for in situ detection of atmospheric NO_3 and N_2O_5 via laser-induced fluorescence. Environ. Sci. Technol. **37**(24), 5732–5738 (2003)

E.C. Wood, B.L. Deming, S. Kundu, Ethane-based chemical amplification measurement technique for atmospheric peroxy radicals. Environ. Sci. Technol. Lett. **4**(1), 15–19 (2017). https://doi.org/10.1021/acs.estlett.6b00438

C.Y.R. Wu, D.L. Judge, SO_2 and CS_2 cross section data in the ultraviolet region. Geophys. Res. Lett. **8**, 769–771 (1981)

T. Wu, Q. Zha, W. Chen, Z. Xu, T. Wang, X. He, Development and deployment of a cavity enhanced UV-LED spectrometer for measurements of atmospheric HONO and NO_2 in Hong Kong. Atmos. Environ. **95**, 544–551 (2014). https://doi.org/10.1016/j.atmosenv.2014.07.016

X. Yang, J. Tang, J.F. Mustard, J.-E. Lee, M. Rossini, J. Joiner, J.W. Munger, A. Kornfeld, A.D. Richardson, Solar-induced chlorophyll fluorescence that correlates with canopy photosynthesis on diurnal and seasonal scales in a temperate deciduous forest. Geophys. Res. Lett. **42**(8), 2977–2987 (2015). https://doi.org/10.1002/2015GL063201

H. Yang, X. Yang, Y. Zhang, M.A. Heskel, X. Lu, J.W. Munger, S. Sun, J. Tang, Chlorophyll fluorescence tracks seasonal variations of photosynthesis from leaf to canopy in a temperate forest. Glob. Chang. Biol. **23**(7), 2874–2886 (2017). https://doi.org/10.1111/gcb.13590

T. Yang, F. Si, Y. Luo, K. Zhan, P. Wang, H. Zhou, M. Zhao, W. Liu, Source contribution analysis of tropospheric NO_2 based on two-dimensional MAX-DOAS measurements. Atmos. Environ. **210**, 186–197 (2019). https://doi.org/10.1016/j.atmosenv.2019.04.058

Z. Yang, T. Albrow-Owen, W. Cai, T. Hasan, Miniaturization of optical spectrometers. Science **371**(6528), eabe0722 (2021). https://doi.org/10.1126/science.abe0722

D.A. Yates, T. Kuwana, Evaluation of a self-contained linear diode array detector for rapid scanning spectrophotometry. Anal. Chem. **48**, 510–514 (1976)

C. Ye, H. Gao, N. Zhang, X. Zhou, Photolysis of nitric acid and nitrate on natural and artificial surfaces. Environ. Sci. Technol. **50**(7), 3530–3536 (2016). https://doi.org/10.1021/acs.est.5b05032

Y. Yoshii, H. Kuze, N. Takeuchi, Long-path measurement of atmospheric NO_2 with an obstruction flashlight and a charge-coupled-device spectrometer. Appl. Opt. **42**(21), 4362–4368 (2003)

A.T. Young, On the Rayleigh-scattering optical depth of the atmosphere. J. Appl. Meteorol. **20**, 328–330 (1981)

C.J. Young, R.A. Washenfelder, P.M. Edwards, D.D. Parrish, J.B. Gilman, W.C. Kuster, L.H. Mielke, H.D. Osthoff, C. Tsai, O. Pikelnaya, J. Stutz, P.R. Veres, J.M. Roberts, S. Griffith, S. Dusanter, P.S. Stevens, J. Flynn, N. Grossberg, B. Lefer, J.S. Holloway, J. Peischl, T.B. Ryerson, E.L. Atlas, D.R. Blake, S.S. Brown, Evaluating evidence for Cl sources and oxidation chemistry in a coastal, urban environment. Atmos. Chem. Phys. Discuss. **13**(5), 13685–13720 (2013). https://doi.org/10.5194/acpd-13-13685-2013

C.J. Young, R.A. Washenfelder, P.M. Edwards, D.D. Parrish, J.B. Gilman, W.C. Kuster, L.H. Mielke, H.D. Osthoff, C. Tsai, O. Pikelnaya, J. Stutz, P.R. Veres, J.M. Roberts, S. Griffith, S. Dusanter, P.S. Stevens, J. Flynn, N. Grossberg, B. Lefer, J.S. Holloway, J. Peischl, T.B. Ryerson, E.L. Atlas, D.R. Blake, S.S. Brown, Chlorine as a primary radical: evaluation of methods to understand its role in initiation of oxidative cycles. Atmos. Chem. Phys. **14**(7), 3427–3440 (2014). https://doi.org/10.5194/acp-14-3427-2014

Y. Yu, A. Geyer, P. Xie, B. Galle, C. Limin, U. Platt, Observations of carbon disulfide by differential optical absorption spectroscopy in Shanghai. Geophys. Res. Lett. **31**, L11107 (2004). https://doi.org/10.1029/2004GL019543

S.A. Yvon, J.H. Butler, An improved estimate of the oceanic lifetime of atmospheric CH_3Br. Geophys. Res. Lett. **23**, 53–56 (1996)

S. Zabarnick, Kinetics of the reaction OH + NO + M $\rightarrow$ HONO + M as a function of temperature and pressure in the presence of argon, SF_6, and N_2 bath gas. Chem. Phys. **171**(1–2), 265–273 (1993)

R. Zellner, J. Hägele, A double-beam UV-laser differential method for monitoring tropospheric trace gases. Opt. Laser Technol. **4**, 79–82 (1985)

C.S. Zender, Global climatology of abundance and solar absorption of oxygen collision complexes. J. Geophys. Res. **104**, 24471–24484 (1999)

Z. Zhang, H. Jiang, J. Liu, X. Zhang, C. Huang, X. Lu, J. Jin, G. Zhou, An analysis of the global spatial variability of column-averaged CO_2 from SCIAMACHY and its implications for CO_2 sources and sinks. Int. J. Remote Sens. **35**(6), 2047–2066 (2014). https://doi.org/10.1080/01431161.2014.885151

J.J. Zhang, Y. Wei, Z. Fang, Ozone pollution: a major health hazard worldwide. Front. Immunol. **10** (2019). https://www.frontiersin.org/articles/10.3389/fimmu.2019.02518

C. Zhang, C. Liu, K.L. Chan, Q. Hu, H. Liu, B. Li, C. Xing, W. Tan, H. Zhou, F. Si, J. Liu, First observation of tropospheric nitrogen dioxide from the environmental trace gases monitoring instrument onboard the GaoFen-5 satellite. Light.: Sci. Appl. **9**(1), 66 (2020). https://doi.org/10.1038/s41377-020-0306-z

S. Zhang, S. Wang, R. Zhang, Y. Guo, Y. Yan, Z. Ding, B. Zhou, Investigating the sources of formaldehyde and corresponding photochemical indications at a suburb site in Shanghai from MAX-DOAS measurements. J. Geophys. Res.: Atmos. **126**(6), e2020JD033351 (2021). https://doi.org/10.1029/2020JD033351

W. Zhang, C. Zuo, T. Liao, G. Zhou, D. Zhao, Research on improving the performance of motor vehicle exhaust gas detection system based on DOAS technology, in *Ninth Symposium on Novel Photoelectronic Detection Technology and Applications*, vol. 12617 (SPIE, 2023), pp. 373–378. https://doi.org/10.1117/12.2664361

X. Zhang, L. Xi, H. Zhou, W. Wang, Z. Chang, F. Si, Y. Wang, High-resolution nitrogen dioxide measurements from an airborne fiber imaging spectrometer over Tangshan, China. Remote Sens. **16**(6) (2024). https://doi.org/10.3390/rs16061042

K. Zheng, C. Zheng, Y. Zhang, Y. Wang, F.K. Tittel, Review of incoherent broadband cavity-enhanced absorption spectroscopy (IBBCEAS) for gas sensing. Sensors **18**(11), 3646 (2018). https://doi.org/10.3390/s18113646

X. Zhou, H.J. Beine, R.E. Honrath, J.D. Fuentes, W. Simpson, P.B. Shepson, J.W. Bottenheim, Snowpack photochemical production of HONO: a major source of OH in the Arctic boundary layer in springtime. Geophys. Res. Lett. **28**, 4087–4090 (2001)

X. Zhou, Y. He, G. Huang, T. Thornberry, M. Carroll, S. Bertman, Photochemical production of nitrous acid on glass sample manifold surface. Geophys. Res. Lett. **29**(14), 1681 (2002a). https://doi.org/10.1029/2002GL015080

X. Zhou, K. Civerolo, H. Dai, G. Huang, J. Schwab, K. Demerjian, Summertime nitrous acid chemistry in the atmospheric boundary layer at a rural site in New York State. J. Geophys. Res. **107**, 4590 (2002b). https://doi.org/10.1029/2001JD001539

X. Zhou, H. Gao, Y. He, G. Huang, S. Bertman, K. Civerolo, J. Schwab, Nitric acid photolysis on surfaces in low-NO_X environments: significant atmospheric implications. Geophys. Res. Lett. **30**, 2217 (2003). https://doi.org/10.1029/2003GL018620

L. Zhu, D. Kellis, C.F. Ding, Photolysis of glyoxal at 193, 248, 308 and 351 nm. Chem. Phys. Lett. **257**(5–6), 487–491 (1996)

J. Zhu, S. Wang, S. Zhang, R. Xue, C. Gu, B. Zhou, Changes in NO_3 radical and its nocturnal chemistry in Shanghai from 2014 to 2021 revealed by long-term observation and a stacking model: impact of China's clean air action plan. J. Geophys. Res.: Atmos. **127**(24):e2022JD037438 (2022). https://doi.org/10.1029/2022JD037438

R. Zimmermann, Entwicklung eines miniaturisierten ballongetragenen Diodenlaser-Spektrometers zur Messung von stratosphärischen Methan- und Wasserdampfkonzentrationen. Ph.D. Thesis, University of Heidelberg (2003)

J. Zingler, U. Platt, Iodine oxide in the Dead Sea valley: evidence for inorganic sources of boundary layer IO. J. Geophys. Res. **110**, D07307 (2005). https://doi.org/10.1029/2004JD004993

P. Zoogman, X. Liu, R. Suleiman, W. Pennington, D. Flittner, J. Al-Saadi, B. Hilton, D. Nicks, M. Newchurch, J. Carr, S. Janz, M. Andraschko, A. Arola, B. Baker, B. Canova, C. Chan Miller, R. Cohen, J. Davis, M. Dussault, D. Edwards, J. Fishman, A. Ghulam, G. González Abad, M. Grutter, J. Herman, J. Houck, D. Jacob, J. Joiner, B. Kerridge, J. Kim, N. Krotkov, L. Lamsal, C. Li, A. Lindfors, R. Martin, C. McElroy, C. McLinden, V. Natraj, D. Neil, C. Nowlan, E. O'Sullivan, P. Palmer, R. Pierce, M. Pippin, A. Saiz-Lopez, R. Spurr, J. Szykman, O. Torres, J. Veefkind, B. Veihelmann, H. Wang, J. Wang, K. Chance, Tropospheric emissions: monitoring of pollution (TEMPO). J. Quant. Spectrosc. Radiat. Transf. **186**, 17–39 (2017). https://doi.org/10.1016/j.jqsrt.2016.05.008

Index

U. Platt and J. Stutz, *Differential Optical Absorption Spectroscopy*, Geophysics and Environmental Physics, https://doi.org/10.1007/978-3-662-68639-3

GPSR Compliance

The European Union's (EU) General Product Safety Regulation (GPSR) is a set of rules that requires consumer products to be safe and our obligations to ensure this.

If you have any concerns about our products, you can contact us on ProductSafety@springernature.com

In case Publisher is established outside the EU, the EU authorized representative is:

Springer Nature Customer Service Center GmbH
Europaplatz 3
69115 Heidelberg, Germany

Batch number: 10099960

Printed by Printforce, the Netherlands